Lecture Notes in Mathematics

Edited by A. Dold and B. Eckmann

1136

Quantum Probability and Applications II

Proceedings of a Workshop held in Heidelberg, West Germany, October 1–5, 1984

Edited by L. Accardi and W. von Waldenfels

Springer-Verlag
Berlin Heidelberg New York Tokyo

Editors

Luigi Accardi
Dipartimento di Matematica, Università di Roma II
Via Orazio Raimondo, 00173 Roma, Italy

Wilhelm von Waldenfels
Institut für Angewandte Mathematik, Universität Heidelberg
Im Neuenheimer Feld 294
6900 Heidelberg, Federal Republic of Germany

Mathematics Subject Classification (1980): 46L50, 46L55, 46L60, 47D05, 47D07, 47D45, 60Gxx, 60Hxx, 60Jxx, 81Bxx, 81C20, 82A05, 82A15

ISBN 3-540-15661-5 Springer-Verlag Berlin Heidelberg New York Tokyo
ISBN 0-387-15661-5 Springer-Verlag New York Heidelberg Berlin Tokyo

Printing and binding: Beltz Offsetdruck, Hemsbach/Bergstr.
2146/3140-543210

INTRODUCTION

The Second Workshop on Quantum Probability and Applications was held in Heidelberg, October 1-5, 1984. It was organized by the Sonderforschungsbereich 123 (Stochastische Mathematische Modelle) of the University of Heidelberg with a contribution from the University of Rome II.

Since the first Workshop on Quantum Probability, which was held in Villa Mondragone in 1982, many important achievements have been obtained in this branch of probability theory, concerning not only the inner mathematical development of the discipline, but also its applications to problems of quantum physics such as the quantum theory of irreversible processes, quantum optics, quantum field theory and the quantum theory of measurement. The Heidelberg meeting was an attempt to provide an overview of these results, as well as of open problems with particular emphasis on those arising from quantum physics.

It is a pleasure to thank the Deutsche Forschungsgemeinschaft and the University of Rome II for financial support; the lecturers and the participants for their contributions to the success of the conference. We would like to address a particular thank to Leo van Hemmen for his generous help in the organization of the workshop.

Luigi Accardi
Wilhelm von Waldenfels

TABLE OF CONTENTS

L. Accardi, S. Olla
On the polaron asymptotics at finite coupling constant 1

L. Accardi, K.R. Parthasarathy
Stochastic calculus on local algebras 9

S. Albeverio, Ph. Blanchard, Ph. Combe, R. Rodriguez, M. Sirugue, M. Sirugue-Collin
Trapping in stochastic mechanics and applications to covers of clouds and radiation belts 24

S. Albeverio, R. Høegh-Krohn
A remark on dynamical semigroups in terms of diffusion processes 40

D. Applebaum
Quasi-free stochastic evolutions 46

A. Barchielli, G. Lupieri
Dilations of operation valued stochastic processes 57

C. Barnett, I.F. Wilde
The Doob-Meyer decomposition for the square of Itô-Clifford L^2-martingales 67

J. Bertrand, B. Gaveau, G. Rideau
Poisson processes and quantum field theory: a model 74

O. Besson
The entropy of quantum Markov states 81

I. Bialynicki-Birula
Entropic uncertainty relations in quantum mechanics 90

Ph. Blanchard, Ph. Combe, M. Sirugue, M.Sirugue-Collin
Estimates of quantum deviations from classical mechanics using large deviation results 104

F. Casagrande, L.A. Lugiato, G. Strini
Adiabatic elimination technique for quantum dissipative systems 112

G. Casati
Limitations for chaotic motion in quantum mechanics 126

C. Cecchini
Non commutative L^p spaces and K.M.S. functions 136

C. D'Antoni
Normal product states and nuclearity: new aspects of algebraic quantum field theory 143

R. Dümcke
The low density limit for n-level systems 151

D.E. Evans
The C*-algebras of the two-dimensional Ising model 162

M. Fannes, J. Quaegebeur
Infinite divisibility and central limit theorems for completely positive mappings 177

G.W. Ford
Temperature-dependent Lamb shift of a quantum oscillator 202

A. Frigerio
Construction of stationary quantum Markov processes through quantum stochastic calculus 207

G.C. Ghirardi, A. Rimini, T. Weber
A model for a unified quantum description of macroscopic and microscopic systems 223

S. Goldstein
Conditional Expectations in L^p-spaces over von Neumann Algebras 234

V . Gorini, A. Frigerio, M. Verri
Quantum Gibbs states and the zeroth law of thermodynamics 240

H. Grabert
Dissipative quantum tunneling 248

F. Guerra
Carlen processes: a new class of diffusions with singular drifts 259

F. Haake, M. Lewenstein, R. Reibold
Adiabatic drag and initial slips for random processes with slow and fast variables 268

R.L. Hudson, J.M. Lindsay
Uses of non-Fock quantum Brownian motion and a quantum martingale representation theorem 276

A. Klein
Supersymmetry and a two-dimensional reduction in random phenomena 306

B. Kümmerer
On the structure of Markov dilations on W*-algebras 318

B. Kümmerer, W. Schröder
A new construction of unitary dilations: singular coupling to white noise 332

G. Lindblad
A new approach to quantum ergodicity and chaos 348

H. Maassen
Quantum Markov processes on Fock space described by integral kernels 361

H. Nakazawa
Quantization of Brownian motion processes in potential fields 375

W. Ochs
Gleason measures and quantum comparative probability 388

M. Ohya
State change and entropies in quantum dynamical systems 397

K.R. Parthasarathy
Some remarks on the integration of Schrödinger equation using the quantum stochastic calculus 409

A. Paszkiewicz
Convergence almost everywhere in W*-algebras 420

D. Petz
Properties of quantum entropy 428

G.A. Raggio, H.S. Zivi
Semiclassical description of n-level systems interacting with radiation fields 442

S. Scarlatti, M. Spera
The charge class of the vacuum state in a free massless Dirac field theory 453

G.L. Sewell
Derivation of classical hydrodynamics of a quantum Coulomb system 463

M. Schürmann
Positive and conditionally positive linear functionals on coalgebras 475

R.F. Streater
The Itô-Clifford integral, Part II 493

A. Verbeure
Detailed balance and equilibrium 504

W. von Waldenfels
Spontaneous light emission described by a quantum stochastic differential equation 516

ON THE POLARON ASYMPTOTICS AT FINITE COUPLING CONSTANT (*)

Luigi Accardi
Dipartimento di Matematica
Università di Roma II

Stefano Olla
Dipartimento di Matematica
Università di Roma II

1. Motivation and physical background

A Polaron is an electron in a ionic crystal coupled with the lattice vibration field produced by itself through polarization of the crystal. The problem is to compute the energy of the ground state of the Polaron under the assumptions (due to Fröhlich, cf. [8]) that the atomic structure of the crystal can be neglected; that it is possible to tract it as a continuum dielectric; and that the only phonon waves that interact with the electron have the same frequency. Under these assumptions the Fröhlich hamiltonian for the polaron (with all physical constants equal to 1) is:

$$H_F(\alpha) = \frac{1}{2} p^2 + \sum_{\bar{k}} a^+_{\bar{k}} a_{\bar{k}} + i(\sqrt{2}\alpha\pi)^{\frac{1}{2}} \sum_{k} \frac{1}{|\bar{k}|} (a^+_{\bar{k}} e^{-i\bar{k}\bar{x}} - a_{\bar{k}} e^{i\bar{k}\bar{x}}) \tag{1.1}$$

where $\bar{x}$ is the vector position of the electron, $\bar{p}$ its conjugate momentum, $a^+_{-\bar{k}}$ and $a_{\bar{k}}$ are the creation and annihilation operators of a phonon of momentum $\bar{k}$, and α is the coupling constant between the electron and the phonon field which depends on the dielectric constants of the ionic crystal (in NaCl α is about 5, and in general it runs from about 1 to 20).

The lowest eigenvalue $E_o(\alpha)$ of $H_F(\alpha)$ has been studied for many years by a lot of techniques. In the weak coupling limit (α small) perturbation treatment of (1.1) gives good results (cf. [7]).

For the strong coupling limit ($\alpha \to \infty$) a conjecture due to Pekar (cf. [18]) suggests that

$$E_o(\alpha) \underset{\alpha\to\infty}{\sim} -\gamma_p \alpha^2 \tag{1.2}$$

where

$$\gamma_p = - \inf_{\substack{\phi \in L^2(\mathbb{R}^3) \\ \|\phi\|_2 = 1}} \left\{ \frac{1}{2} \int_{\mathbb{R}^3} |\nabla\phi|^2 dx - \frac{1}{\sqrt{2}} \iint \frac{\phi^2(x)\phi^2(y)}{|x-y|} dxdy \right\} \tag{1.3}$$

The minimizing ϕ in (1.3) is the solution of the non-linear Shrödinger equation

$$-\frac{1}{2}\Delta\phi(x) - 2\sqrt{2}\left(\int \frac{|\phi(y)|^2}{|x-y|} dy\right)\phi(x) = e\phi(x) \tag{1.4}$$

which describes an electron trapped in its own hole. In [15] Lieb has shown that the minimizing ϕ in (1.3) exists and is unique up to translations (the problem was non trivial because of the non-convexity of the functional in (1.3)). In particular Lieb proved that the minimizing solution is infinitely differentiable and goes to zero at infinity. By numerical computations (cf. [17]) $\gamma_p \gtrsim 0.108513$, very close to $\frac{1}{3\pi}$ that is the value obtained in (1.3) when one uses gaussian functions as trial functions.

For intermediate coupling the most important techniques are those developed by Gross (cf. [10], who starts from the classical Riesz variational method) and of Feynman (cf. [6] and [7]), who reduced the problem to the asymptotic evaluation of a path integral. Using variational method (cf. [7]) he obtains numerical results for an upper bound for $E_0(\alpha)$, which are still the best known bounds up to some improvements due to Luttinger and Lu (cf. [16]) who take into account the second order correction in Feynmen's method.

Interesting direct numerical computations of the path integral were also performed using Monte Carlo techniques (cf. [9]).

The problem of the existence of a phase transition is first mentioned by Gross [10] who observes that the wave function of the polaron must be extended in the weak coupling region and localized for strong couplings. This corresponds to the intuitive idea that for α tending to zero the polaron dynamics tends to that of a free particle, and for $\alpha \to \infty$ a self localization effect arises.

The conjecture of Gross is that the transition between the localized and the extended function is abrupt, and this gives a discontinuity on the first derivative of $E_0(\alpha)$. The existence of such phase transition is an open problem at the moment.

In [16] Luttinger and Lu observe that in Feynman's harmonic approimation (an upper bound), a phase transition at $\alpha = 5.8$ exists (Feynman knew about the discontinuity at 5.8, cf. [6]; but he considered this rather a disadvantage of the approximation than a result).

The same result is also obtained by Gross in [10]. But it is not clear if this dynamical instability is just a property of $H_F(\alpha)$ or if it comes from the approximations introduced to obtain the estimate. Another conjecture, given by Lepine and Metz (cf. [13], 1979), is that there might be two phase transition points. One from an extended translational invariant wave function to a two dimensional localized wave function (with cylindrical symmetry), and another from this semilocalized wave function to a symmetric three dimensional localized wave function. This conjecture is suggested by Fock approximation mean field theory applied to $H_F(\alpha)$.

All these results concern only upper bounds for $E_0(\alpha)$, while the physical literature about lower bounds is very poor (cf. [11] and [14]) and limited to a short range of α.

2. Mathematical formulation of the problem

Using their powerful theory of large deviations for Markov processes [4], Donsker and Varadhan [5], found for $E_0(\alpha)$ the variational formula:

$$- E_0(\alpha) = g(\alpha) = \sup_{Q \in M_S(\Omega)} \{2\alpha Q(\int_0^\infty \frac{e^{-\sigma} d\sigma}{|w(\sigma)-w(o)|}) - H(Q|E^W_{0]}) \tag{2.1}$$

where $\Omega = D(\mathbb{R}, \mathbb{R}^3)$ is the space of all right-continuous functions $\mathbb{R} \to \mathbb{R}^3$ with only first kind discontinuities and left limits; $M_S(\Omega)$ denotes the space of all the stationary measures on Ω; $w(t)$ denotes the map:

$$w(t) : \omega \in \Omega \to w(t)(\omega) = \omega(t) \in \mathbb{R}^3 \tag{2.2}$$

$E^W_{0]}$ is the conditional expectation of the Wiener process onto the σ-algebra at time zero; and $H(Q|E^W)$ is the entropy of Q with respect to E^W - defined in [4] (cf. also [1]).

Using formula (2.1), Donsker and Varadhan were able to give a rigorous proof of Pekar's conjecture 5, i.e. the validity of (1.2) and (1.3).

In the present paper we want to exploit the identity (2.1) to study the behaviour of $E_0(\alpha)$ at finite α and the problem of the existence of a phase transition. At the moment we have not a full proof of the existence of a phase transition (cf. the remark at the end of section (3.)), however we obtain the following results:

i) We obtain a rigorous derivation of the results obtained in the physical literature. In particular we show that the expression obtained by Feynman [6], [1], Gross [10], Luttinger - Lu [16], in the harmonic approximation for $E_0(\alpha)$ is obtained from (2.1) by restriction of the sup to the Ornstein-Uhlenbeck processes (cf. section (5.)).

ii) We improve these estimates, deriving an explicit finite dimensional variational formula for the restriction of the sup in (2.1) to all distorted brownian motions (cf. section (3.)) and to all stationary gaussian processes (cf. section (4.).

iii) We obtain a general lower estimate for $E_0(\alpha)$ (upper estimate for $g(\alpha)$) which seems not to be present in the physical literature, and which allows to obtain both theoretical and numerical informations on the validity of the upper estimate (cf. the end of section (3.)).

Notational remark: following Donsker and Varadhan's notations we deal with $g(\alpha)$ rather than $E_0(\alpha)$ $(= -g(\alpha))$. So to compare our results with the ones mentioned in section (1.) one should keep in mind that the upper estimates in these ones correspond to our lower estimate, and conversely. Moreover, in the identity (2.1) we have rescaled α by a factor $\sqrt{8}$. Thus, to derive the numerical results in the physical literature from our ones, one should multiply α by the factor $\sqrt{8}$.

3. The Markovian approximation

Let us first recall the statement of Donsker-Varadhan's "contraction principle" (cf. [4]) namely: for any probability measure m on $\mathbb{R}^3$, one has:

$$\inf_{\substack{Q\in M_s(\Omega)\\ Q_0=m}} H(Q|E^W) = I_W(m) \tag{3.1}$$

where Q_0 is the marginal distribution of $Q\in M_s(\Omega)$, and:

$$I_W(m) = \begin{cases} \int_{\mathbb{R}^3} \frac{[\nabla f]^2}{f}(x)\,dx & \text{if } m(dx) \ll dx \text{ and } f = \frac{dm}{dx}\in C'(\mathbb{R}^3) \\ +\infty & \text{otherwise} \end{cases} \tag{3.2}$$

Denote

$$H = \{\phi\in L^2\cap C^2(\mathbb{R}^3)\ ,\ \|\phi\|_{L^2} = 1,\ \|\nabla\phi\|_{L^2} < +\infty,\} \tag{3.2/a}$$

where ∇ denotes the gradient.

In Appendix (A 1) it is shown that, denoting for each $\phi\in H$, $Q^\phi\in M_s(\Omega)$ the stationary Markovian measure with transition semi-group generated by

$$L_\phi = +\frac{\Delta}{2} + \frac{\nabla\phi}{\phi}\cdot\nabla \tag{3.3}$$

and with initial distribution $\phi^2(x)dx$ (the associated Markov process is called in the literature a "distorted Brownian Motion" (cf. for example [3])), then:

$$H(Q^\phi|E^W) = \int_{\mathbb{R}^3} |\nabla\phi|^2\,dx \tag{3.4}$$

namely the inf in (3.1) is realized by Q^ϕ for $m(dx) = \phi^2(x)dx$.

Therefore, restricting the sup in (2.1) to the class $\{Q^\phi, \phi\in H\}\subseteq M_s(\Omega)$ we obtain the following lower estimate:

Th (3.1) $\forall\ \alpha>0$

$$g(\alpha) \geq \sup_{\phi\in H} \{\alpha 2\int_{\mathbb{R}^3\times\mathbb{R}^3} \left(1-\frac{\Delta}{2}+\frac{\Delta\phi}{2\phi}(x)\right)^{-1}(x,y)\ \frac{\phi(x)\,\phi(y)}{|x-y|}\,dxdy - \frac{1}{2}\int_{\mathbb{R}^3} |\nabla\phi|^2 dx\ \} \tag{3.5}$$

Remark. It is possible to give a simple direct proof of the lower estimate (3.5) which does not make use of the full technical apparatus developed by Donsker and Varadhan. We will not include this here for lack of space.

Proof. From (2.1) and (3.5) it follows that:

$$\begin{aligned} g(\alpha) &= \sup_{Q\in M_s(\Omega)} \{2\alpha Q\Big(\int_0^\infty \frac{e^{-\sigma}\,d\sigma}{|w(\sigma)-w(0)|}\Big) - H(Q|E^W)\} \geq \\ &\geq \sup_{\phi\in H} \{2\alpha Q^\phi\Big(\int_0^\infty \frac{e^{-\sigma}}{|w(\sigma)-w(0)|}\Big) - \frac{1}{2}\int_{\mathbb{R}^3} |\nabla\phi|^2 dx\} \end{aligned} \tag{3.6}$$

The density of the marginal distribution of Q^ϕ is just $\phi^2(x)$, then by Fubini Tonelli's theorem:

$$\begin{aligned} Q^\phi\Big(\int_0^\infty \frac{e^{-\sigma}\,dt}{|w(\sigma)-w(0)|}\Big) &= \\ &= \int_{\mathbb{R}^3}\phi^2(x)dx \int_0^\infty e^{-\sigma} e^{+\sigma L_\phi}\Big(\frac{1}{|\cdot - x|}\Big)d\sigma = \\ &= \int_{\mathbb{R}^3} dx\phi^2(x) \int_{\mathbb{R}^3} dy\phi^2(y)\,(1-L_\phi)^{-1}(x,y)\ \frac{1}{|x-y|} \end{aligned} \tag{3.7}$$

where $(1-L_\phi)^{-1}(x,y)$ is the kernel of the resolvent $(1-L_\phi)^{-1}$ in $L^2(\mathbb{R}^3, \phi^2(x)dx)$.

Then by (3.7)

$$\begin{aligned} Q^\phi\Big(\int_0^\infty \frac{e^{-\sigma}\,d\sigma}{|w(\sigma)-w(0)|}\Big) &= \\ &= \int_{\mathbb{R}^3\times\mathbb{R}^3}\phi(x)\,(1-L_\phi)^{-1}(x,y)\,\phi(y)\frac{\phi(x)\,\phi(y)}{|x-y|}\,dxdy = \\ &= \int_{\mathbb{R}^3\times\mathbb{R}^3}[M_\phi(1-L_\phi)^{-1}M_\phi^{-1}](x,y)\ \frac{\phi(x)\,\phi(y)}{|x-y|}\,dxdy = \\ &= \int_{\mathbb{R}^3\times\mathbb{R}^3}[1-M_\phi L_\phi M_\phi^{-1}]^{-1}(x,y)\ \frac{\phi(x)\phi(y)}{|x-y|}\,dxdy \end{aligned} \tag{3.8}$$

where $M_\phi : f\in L^2(\mathbb{R}^3,\phi^2(x)dx) \to \phi f \in L^2(\mathbb{R}^3,dx)$ denotes the operator of multiplication by ϕ, and $[1-M_\phi L_\phi M_\phi^{-1}]^{-1}(x,y)$ is the kernel of $[1-M_\phi L_\phi M_\phi^{-1}]^{-1}$ in $L^2(\mathbb{R}^3,dx)$.
But

$$\begin{aligned} M_\phi L_\phi M_\phi^{-1} &= \phi\,[+\frac{1}{2}\nabla + \frac{\nabla\phi}{\phi}\cdot\nabla]\phi^{-1} = \\ &= +\frac{1}{2}\Delta - \frac{\Delta\phi}{2\phi} \end{aligned} \tag{3.9}$$

and (3.5) follows from (3.6) (3.8) (3.9).

To have an idea of how good the lower estimate (3.5) is, let us compare it with the upper estimate obtained as follows: start from Donsker and Varadhan's variational expression for $g(\alpha)$, i.e.:

$$\begin{aligned} g(\alpha) &= \sup_{Q\in M_s(\Omega)} \{2\alpha Q\Big(\int_0^\infty \frac{e^{-\sigma}\,d\sigma}{|w(\sigma)-w(0)|}\Big) - H(Q|E^W)\} \\ &= \sup_{Q\in M_{erg}(\Omega),\,H(Q|E^W)<\infty} \{2\alpha Q\Big(\int_0^\infty \frac{e^{-\sigma}\,d\sigma}{|w(\sigma)-w(0)|}\Big) - H(Q|E^W)\} \end{aligned}$$

$$= \sup_{\phi \in H} \sup'_{p(\cdot|\cdot)} \{2\alpha \int_0^\infty e^{-\sigma} d\sigma \int_{\mathbb{R}^3} \phi^2(x)dx \int_{\mathbb{R}^3} \frac{p(dy,\sigma|x,0)}{|y-x|} - \inf_{[\phi,p]} H(Q|E^W)\}$$

where we have used the following notations: for a stationary measure Q with marginal $\phi^2(x)dx$ and locally absolutely continuous with respect to the Wiener measure, we denote $p(dy,\sigma|x,0)$ the conditional probability of Q $F_0 \vee F_\sigma$ with respect to F_0; thus:

$$Q\left(\frac{1}{|w(\sigma)-w(0)|}\right) = \int_{\mathbb{R}^3} \phi^2(x)dx \int_{\mathbb{R}^3} \frac{p(dy,\sigma|x,0)}{|y-x|}$$

Moreover, $\sup'_{p(\cdot|\cdot)}$ denotes the sup taken over all the transition probabilities obtained as above; and $\inf_{[\phi,p]}$ denotes the inf taken over all the stationary (ergodic) measures Q whose marginal is $\phi^2(x)dx$ and whose associated transition probability density is p. With these notations we have the upper estimate:

$$g(\alpha) \le \sup_{\phi \in H} \{2\alpha \sup'_p \int_0^\infty e^{-\sigma} d\sigma \int_{\mathbb{R}^3} \phi^2(x)dx \int_{\mathbb{R}^3} \frac{p(dy,\sigma|x,0)}{|y-x|} - \inf_{Q_0=\phi^2(x)dx} H(Q|E^W)\}$$

$$= \sup_{\phi \in H} \{2\alpha(\sup'_p \int_0^\infty e^{-\sigma} d \int_{\mathbb{R}^3} \phi^2(x)dx \int_{\mathbb{R}^3} \frac{p(dy,\sigma|x,0)}{|y-x|}) - \int_{\mathbb{R}^3} |\nabla\phi|^2 dx\}$$

Thus the exactness of the lower estimate (3.5) would be implied by the fact that the sup of the integral

$$\int_0^\infty e^{-\sigma} d\sigma \int_{\mathbb{R}^3} \phi^2(x)dx \int_{\mathbb{R}^3} \frac{p(dy,\sigma|x,0)}{|y-x|}$$

over all stationary ergodic transition probabilities $p(dy,\sigma|x,0)$ is reached on those of the type $\exp \sigma L_\phi(x,y)dy$, i.e. on the kernels of the semi-group of the distorted brownian motion associated to the functions $\phi \in H$. We conjecture that this "conditional contraction principle" is true, but at the moment we have not a complete mathematical proof of it.

4. The Gaussian approximation

More explicit computations are possible if we choose as class of trial processes the stationary gaussian processes. Also here the problem is reduced to a finite dimensional one, because we can take the supremum over all covariance functions.

Let be $\rho_{ij}(t)$ the covariance matrix function of the stationary gaussian measure Q^ρ, corresponding to a stationary gaussian process $w(t) = (w_1(t), w_2(t), w_3(t))$: $: \Omega \to \mathbb{R}$ characterized by the condition:

$$Q^\rho(w_j(t+s)w_i(s)) = \rho_{ij}(t); \; s \in \mathbb{R}_+; \; j,i = 1,2,3 \tag{4.1}$$

The process $y_i(t) = |w_i(t) - w_i(0)|$ $(i = 1,2,3)$ is then gaussian with covariance matrix

$$\sigma_{ij}(t) = Q^\rho(y_i(t)y_j(t)) = 2(\rho_{ij}(0)-\rho_{ij}(t)) \tag{4.2}$$

Denoting $\{\lambda_j(t),\ j=1,2,3\}$ the eigenvalues of the matrix $(\sigma_{ij}(t))$, after some elementary computations one finds:

$$Q^\rho\left(\frac{1}{|w(t)-w(0)|}\right) = \frac{1}{(2\pi)^{3/2}} \frac{1}{(\prod_{j=1}^3 \lambda_j(t))^{1/2}} \int_{\mathbb{R}^3} \frac{e^{-\frac{1}{2}\sum_{i=1}^3 \lambda_j^{-1}(t)x_j^2}}{|x|} dx \tag{4.3}$$

In particular, if $\lambda_j(t) = \lambda(t)$, $j = 1,2,3$, then:

$$Q^{\rho}\left(\frac{1}{|w(t)-w(0)|}\right) = \frac{1}{(2\pi\lambda(t))^{3/2}} \int_{\mathbb{R}^3} \frac{e^{-\frac{1}{2}\lambda^{-1}(t)|x|^2}}{|x|} dx = \tag{4.4}$$
$$= \left(\frac{2}{\pi\lambda(t)}\right)^{1/2}$$

Using the well known formulae (cf. e.g. [12]) for the Radon-Nikodym derivatives of gaussian measures one can also give an explicit expression for the relative entropy of Q^{ρ} with respect to the Wiener measure. For lack of space we do not discuss here the general case (cf. [19]) and limit ourselves to an important particular case, to be discussed in the next section.

5. The Harmonic Approximation

In the intersection of the two classes of processes considered respectively in section (3) and (4) one finds the Ornstein-Uhlenbeck processes, i.e. the stationary gaussian markovian process with covariance matrix function

$$\rho_{ij}(t) = \frac{1}{2\beta} e^{-\beta t} \delta_{ij} \quad ; \; i,j = 1,2,3 \tag{5.1}$$

where $\beta > 0$ is a parameter characteristic of the process.

The density of the invariant measure is

$$\psi^2(x) = \left(\frac{\pi}{\beta}\right)^{-3/2} e^{-\beta x^2} \tag{5.2}$$

So β^{-1} is also the variance of the marginal distribution of the process.

The infinitesimal generator of the corresponding markovian semigroup is

$$L_{\beta} = + \frac{1}{2} \Delta - \beta \quad \cdot \nabla \tag{5.3}$$

so it was the form (3.3) and therefore the Ornstein-Uhlenbeck processes belong to the class of markovian processes considered in § 3, (3.4) and (5.2)

$$H(Q_{\beta}^{OU} \,|\, E^{W}) = \int_{\mathbb{R}^3} |\nabla\psi(x)|^2 dx = \frac{3}{4}\beta \tag{5.4}$$

In this case the matrix $\sigma_{ij}(t)$ defined by (4.2) is diagonal and we can put

$$\lambda(t) = \frac{1}{\beta}(1 - e^{-\beta t}) \tag{5.5}$$

(4.4)

$$Q_{\beta}^{OU}\left(\frac{1}{|w(t)-w(0)|}\right) = \sqrt{\frac{2}{\pi}} \frac{1}{(1-e^{-\beta t})^{1/2}} \tag{5.5}$$

Then it is possible to compute explicity also the action term in (21)

$$Q_{\beta}^{OU}\left(\int_0^{\infty} \frac{e^{-\sigma} d\sigma}{|w(t)-w(0)|}\right) = \sqrt{\frac{2}{\pi}} \int_0^{\infty} \frac{e^{-t} dt}{(1-e^{-\beta t})} = \tag{5.7}$$
$$= \delta 2\, \beta^{1/2} \frac{\Gamma(1+\frac{1}{\beta})}{\Gamma(\frac{1}{2}+\frac{1}{\beta})}$$

where Γ is the gamma function.

Then using O.U. processes as trialprocesses in (2.1) one obtains the following lower estimate

$$g(\alpha) \geq \sup_{\beta} \{\alpha\beta^{1/2}\frac{\Gamma(1+\frac{1}{\beta})}{\Gamma(\frac{1}{2}+\frac{1}{\beta})} - \frac{3}{4}\beta\} \tag{5.8}$$

The lower bound (5.8) is just that obtained by Feynmen in [6], Gross in [10], Luttingen-Lu [16], with completely different approaches.

APPENDIX (A1) Relative entropy of distorted Brownian Motion

Let $\phi \in C^2(\mathbb{R}^3) \cap L^2(dx)$ such that $\|\phi\|_{L^2} = 1$ and $\|\nabla\phi\|_{L^2} < +\infty$. Let $Q^{\phi} \in M_s(\Omega)$ the Markov process generated by

$$L_{\phi} = \frac{1}{2}\Delta + \frac{\nabla\phi}{\phi}\cdot\nabla$$

Let $E_o^{Q^{\phi}}$ the conditional expectation of Q^{ϕ} on the σ-algebra at time zero. Then by Cameron-Martin's formula and by Ito's formula:

$$\frac{dE_o^{Q^{\phi}}|_{F[0,t]}}{dE_o^{W}|_{F[0,t]}} = \frac{\phi(w_t)}{\phi(w_o)} \exp(-\frac{1}{2}\int_0^t \frac{\Delta\phi}{\phi}(w_s)ds) \tag{A.1}$$

Then by [4] and by the stationarity of Q^{ϕ}:

$$H(Q^{\phi}|E_o^{W}) = Q^{\phi}(\lg\frac{dE_o^{Q^{\phi}}|_{F[0,1]}}{dE_o^{W}|_{F[0,1]}}) = \tag{A.2}$$

$$= Q^{\phi}(\lg\phi(w_t) - \lg\phi(w_o) - \frac{1}{2}\int_0^1\frac{\Delta\phi}{\phi}(w_s)ds) =$$

$$= -\frac{1}{2}\int_0^1 Q^{\phi}(\frac{\Delta\phi}{\phi}(w_s))\, ds =$$

$$= -\frac{1}{2}\, Q^{\phi}(\frac{\Delta\phi}{\phi}(w_o)) = -\frac{1}{2}\int_{\mathbb{R}^3}\frac{\Delta\phi}{\phi}(x)\phi^2(x)\, dx =$$

$$= -\frac{1}{2}\int_{\mathbb{R}^3}\phi\Delta\phi\, dx = \frac{1}{2}\int_{\mathbb{R}^3}|\nabla\phi|^2\, dx$$

So we have the explicit formula:

$$H(Q^{\phi}|E^{W}) = \frac{1}{2}\int_{\mathbb{R}^3}|\nabla\phi|^2\, dx \tag{A.3}$$

References

1. L. Accardi, S. Olla - Donsker and Varadhan theory for stationary processes. Preprint (1984).

2. J. Adamowski, B. Goerlach, H. Leschke - Strong coupling limit of polaron energy, revisited - Physics Letters 79A, number 2,3, Sept. 1980 (249-251)

3. S. Albeverio, R. Hoegh-Krohn, L. Streit - Energy forms, Hamiltonians, and distorted Browian path - J. Math. Phys. 18, 5, 1977 (907-917)

4. M.D. Donsker, S.R.S. Varadhan - Asymptotic evaluation of certain Markov process expectations for large time, IV, - Comm. Pure Appl. Math. 36, 1983, (182-212)

5. M.D. Donsker, S.R.S. Varadhan - Asymptotics for the Polaron - Comm. Pure Appl. Math. 36, 1983, (505-528).

6. R.P. Feynman - Slow Electrons in a Polar Crystal - Phys. Rev. 97, 3, 1955 (660-665).

7. R.P. Feynman - Statistical Mechanics - W.A. Benjamin, Reading, MA, 1972

8. H. Fröhlich - Electrons in Lattice Fields
Advan. Phys. 3, 1954 (325- 361)

9. I.M. Gel'fand, N.M. Chentsov - The Numerical Calculation of Path Integrals - J.E.T.P. 31, 1956, (1106-107)

10. E.P. Gross - Analytical Methods in the Theory of Electron Lattice Interactions - Ann. Phys. 8, 1959, (78-99)

11. D.M. Larsen - Upper and Lower Bounds for the Intermediate - Coupling Polaron Grand-State Energy - Phys. Rev. 172, 1968 (967-971)

12. I. Guikhman, A. Skorokhod - The Theory of Stochastic Processes - Springer-Verlag

13. Y. Lepine, D. Metz - Mean Field Theory of a Single Fröhlich Polaron (Possible Existence of Phase Transitions) - Phys. Stat. Sol. (b) 96, 1979 (797-806)

14. E.H. Lieb, K. Yamazaki - Ground State Energy and Effective Mass of the Polaron - Phys. Rev. 111, 3; (1958) (728-733)

15. E.H. Lieb - Existence and Uniqueness of the Minimizing Solution of Choquard's Nonlinear Equation - Studies Appl. Math. 57, 1977 (93-105)

16. J.M. Luttinger, Chih-Yuan Lu - Generalized Path-Integral Formalism of the Polaron Problem - Phys. Rev. B 21, 10; 1980 (4251-4263)

17. S.J. Miyake - Strong-Coupling Limit of the Polaron Ground State - J. Phys. Soc. Japan 38, 1; 1975 (181-182)

18. S.I. Pekar - Theory of Polarons - Zh. Experim. i Tear. Fiz. 19, 1949 (796).

19. L.Accardi, S.Olla -Phase Transitions in the Gaussian Approximation for the Polaron -to appear

---

(*) Extended version of a talk given at the II-d workshop on Quantum Probability and Applications, Heidelberg 1-5 October, 1984 .

STOCHASTIC CALCULUS ON LOCAL ALGEBRAS (*)

Luigi Accardi
Dipartimento di Matematica
Università di Roma II
Roma, Italia

K.R. Parthasarathy
Indian Statistical Institute
Delhi Center
New Delhi, India

Introduction

We show that any "sufficiently regular" (cf. condition (2.1)) quantum process can be written as the sum of its initial value plus a bounded variation process plus a martingale. We then develop a stochastic calculus for "sufficiently regular" (cf. condition (2.10)) quantum martingales; we prove Ito's formula and give conditions for existence uniqueness and unitarity of certain linear equations. This machinery is then employed to show that, if the local algebras associated to disjoint intervals commute, then such a martingale defines in a canonical way a representation of the CCR over a pre-Hilbert space defined by the covariance and the corresponding state is necessarily a quasi-free state. This is a quantum version of Levy's martingale characterization of Brownian motion. Contrarily to the classical case, in which, up to random change of time and degeneracy, there is only one canonical form for a ("regular") martingale, here we find that, up to a 2-parameter random change of time, there are three canonical forms: one corresponds to the "Fock stochastic calculus over $L^2(\mathbb{R}_+)$", introduced in [6], another one to the "universal invariant stochastic calculus over $L^2(\mathbb{R}_+)$" discussed in [6], and the third one to the "quasi-free stochastic calculus" introduced in [1] (cf. D. Applebaum's paper in these proceedings for a presentation).

The present one is a structure theory for a general class of stochastic processes; in it both the quantum mechanical commutation relations and the quasi-free nature of the states are deduced and not postulated ab initio.

Starting from section (3) we extend the theory by considering: (i) semi-martingales instead of martingales; (ii) many "integrator processes" instead of one (and its adjoint). Using the basic estimate of section (3) we prove existence, uniqueness and regularity results for linear quantum stochastic differential equations. We define the "brackets" of two general quantum processes (section (6) and deduce Ito's formula as a necessary and sufficient condition for a space of stochastic differentials to be an algebra (§ 6). As an application we deduce the conditions for the unitarity of the solution of a quantum stochastic differential equation.

Simple stochastic integrals

In the following we shall deal with the structure defined by a triple $\{A,(A_{(s,t)}),(E_{t]})$ where:

- A is a $*$-algebra (algebra here will mean, unless otherwise stated, complex associative algebra with unit).
- $(A_{(s,t)})$ is a localization in A, i.e.

$$I \subseteq J \subseteq \mathbb{R}_+ = [0,\infty) \Rightarrow A_I \subseteq A_J$$

In particular $A_{t]} = A_{[0,t]}$ (resp. $A_{[t} = A_{[t,\infty)}$) are filtrations in A, i.e.

(*) Extended version of a talk given at the II-d workshop on Quantum Probability and Applications, Heidelberg 1-5 October, 1984.

$s \leq t \Rightarrow A_{s]} \subseteq A_{t]} \subseteq A$ (resp. $A \supseteq A_{[s} \supseteq A_{[t}$)

- $E_{t]} : A \to A_{t]}$ is a conditional expectation onto $A_{t]}$, i.e.

$$E_{t]}(a_{t]} \cdot a) = a_{t]} \cdot E_{t]}(a) ; \ \forall a_{t]} \in A_{t]} ; \ \forall a \in A \tag{1.1}$$

$$E_{t]}(1) = 1 \tag{1.2}$$

An additional element which plays an important role in the development of the theory is the shift, i.e. a 1-parameter semigroup (u_t) of endomorphisms of A compatible with the local structure on A, i.e.

$$u_t A_I = A_{I+t} \tag{1.3}$$

$$u_s E_{t]} = E_{s+t]} \cdot u_s \tag{1.4}$$

We also assume that each $u_t : A \to A_{[t}$ has a left inverse, denoted $u_t^* : A_{[t} \to A$

Many examples of such structures arise naturally in classical probability and in qunatum physics in connection with (stationary) stochastic processes or with representations of the CCR or of the CAR. Throughout this paper the index set will be chosen to be $\mathbb{R}_+ = [0,\infty)$. It will be clear from the context that most of the results hold for any interval $I \subseteq \mathbb{R}$.

We denote T the half-line $\mathbb{R}_+ = [0,\infty)$ and call adapted (or $(A_{t]})$-adapted if confusion is possible) a function $F : T \to A$, if for each $t \in T$, $F(t) \in A_{t]}$; the family of all adapted functions $T \to A$, will be denoted $F_<(T,A)$.

For $d \in \mathbb{N}$, $A^d = A \times A \times \ldots \times A$ (d-times) and $S_<(T,A^d)$ denotes the set of all functions $F : T \to A^d$ which can be written in the form:

$$F(t) = \sum_{k=0}^{\infty} F_{t_k]} \cdot \chi_{[t_k, t_{k+1})}(t) \tag{1.5}$$

where:

i) (t_k) is a strictly increasing sequence such that $\bigsqcup_k [t_k t_{k+1}) = T$

ii) $F_{t_k]} = (F_1(t_k), \ldots, F_d(t_k))$, with $F_\alpha(t_j) \in A_{t_j]}$.

iii) $F_{t_k]} \neq 0$, only for a finite number of t_k's.

and where, here and in the following, for any set I, χ_I will denote the characteristic function of I ($\chi_I(t) = 1$, if $t \in I$; $= 0$ if $t \notin I$).

The elements of the complex vector space $S_<(T,A^d)$ are adapted functions $T \to A^d$, they will be called simple adapted functions.

Definition (1.1) An A-valued stochastic differential (also called an additive functional) (with respect to the localization $(A_{(s,t)})$ is a finitely additive, localized, A-valued measure on $T = \mathbb{R}_+$, i.e. a map $(s,t) \subseteq T \to M_{(s,t)} = M(s,t) \in A$, such that:

$$M(s,t) \in A_{(s,t)} \tag{1.6}$$

$$r < s < t \Rightarrow M(r,t) = M(r,s) + M(s,t) \tag{1.7}$$

$$M(t,t) = 0 \tag{1.8}$$

Remark. To include the classical case in the discussion which follows, one should substitute for condition (1.6) a condition of the type:

$$M(s,t) \in A(s,t) \vee (\text{centre of } A_{s]}) \tag{1.9}$$

where V means generated by (algebraically or topologically according to the category to which A belongs). The stochastic calculus we are going to develop is still valid under the assumption (1.1) (weaker than (1.6)) if "centre" is understood in terms of the "ρ-commutant" (cf. further in this section).

To any set $M_1,\dots,M_d$ of A-valued stochastic differentials we can associate two linear maps:

$$L_M, R_M : S_<(T,A^d) \to F_<(T,A)$$

defined respectively by:

$$L_M(F)(t) = \sum_{k\geq 0} \sum_{\alpha=1}^{d} M_\alpha(t\wedge t_k, t\wedge t_{k+1}) F_\alpha(t_k) \tag{1.10}$$

$$R_M(F)(t) = \sum_{k\geq 0} \sum_{\alpha=1}^{d} F_\alpha(t_k) M_\alpha(t\wedge t_k, t\wedge t_{k+1}) \tag{1.11}$$

where

$$F = (F_1,\dots,F_d) = \sum_{k>0} F_{t_k]} \cdot \chi_{[t_k,t_{k+1})} \in S_<(T,A^d)$$

The elements in the range of L_M (resp. R_M) are called <u>simple left-stochastic</u> (resp. right-). We speak of M-stochastic integrals, if confusion might arise.

The strategy to define stochastic integrals for adapted functions $T \to A^d$, more general than the simple ones is to introduce topologies on $S_<(T,A^d)$ and $F_<(T,A)$ to prove the continuity if the maps L_M, R_M in these topologies; to extend the maps (1.10), (1.11) by continuity. In the following we realize this programme under some assumptions on the stochastic differentials M_α ($\alpha=1,\dots,d$). Namely we assume that, for each $\alpha=1,\dots,d$ there exists an automorphism $\rho_\alpha \in \mathrm{Aut}(A)$ such that:

$$\rho_\alpha(A_{t]}) \subseteq A_{t]} \quad ; \forall\alpha = 1,\dots,d \ ; \ \forall t \geq 0 \tag{1.12}$$

$$\rho_\alpha^2 = \mathrm{id} \tag{1.13}$$

$$a_{t]} \cdot M_\alpha(t,u) = M_\alpha(t,u)\rho_\alpha(a_{t]}); \ \forall a_{t]} \in \bar{A}_{t]} \ ; \ \forall\alpha = 1,\dots,d; \ \forall u > t \geq 0 \tag{1.14}$$

In this case we say that the stochastic differentials $M_1,\dots,M_d$ <u>satisfy a ρ-commutation relation</u>. Under this assumption, it is easy to see that the sets of right and left simple stochastic integral coincide.

If $X : T \to A$ is any map, we will use the notation:

$$dX(t) = X(dt) = X(t+dt) - X(t) \tag{1.15}$$

and $F = (F_1,\dots,F_d) \in S_<(T,A^d)$, then we use the notations:

$$dL_M(F)(t) = dM_\alpha(t) \cdot F_\alpha(t); \ dR_M(F)(t) = F_\alpha(t) dM_\alpha(t) \tag{1.16}$$

where, here and in the following, summation over repeated greek indices is understood.

From the assumptions above one easily deduces that for each $t \geq 0$ and $\alpha = 1,\dots,d$:

$$E_{t]}(dM_\alpha^{\varepsilon_1} dM_\alpha^{\varepsilon_2}) \in \text{centre } (A_{t]}) \tag{1.17}$$

where $\varepsilon_1, \varepsilon_2 = (\pm 1)$ and by convenction:

$$X^{(+1)} = X; \ X^{(-1)} = X^+ \tag{1.18}$$

In the following, for $F \in S_<(T,A^d)$, we will use the notations

$$\int_0^t dM \cdot F \quad ; \quad \int_0^t F \cdot dM \tag{1.19}$$

to denote the right hand sides of the identities (1.10), (1.11) respectively.

<u>Remark</u>. It is clear from (1.10) (resp. (1.11)) that the left (resp. right) simple stochastic integrals can be defined when A is a left (resp. right) H-module (H some space) and F takes values in H (cf. [4]).

2. A Levy representation theorem for operator valued processes

In the notations of section (1.), let T be a fixed locally convex topology on A. Let $F : \mathbb{R}_+ \to A$ be a map. Disregarding the relation of stochastic equivalence, such a map might be called a process. Following E.J. Nelson [10] we introduce regularity conditions on F:

R1) The map $t \to F(t)$ is T-continuous and the limit

$$\lim_{\substack{\Delta t \to 0 \\ \Delta t > 0}} E_{t]} \left(\frac{F(t+\Delta t) - F(t)}{\Delta t} \right) = D_+F(t) \tag{2.1}$$

exists in the T- topology for each $t \geq 0$. Moreover, the map $t \to D_+F(t)$ is T-continuous.

We will also assume that the conditional expectations $E_{t]}$ $(t \geq 0)$ are continuous in the T-topology.

In all the known examples, A can be concretely realized as an algebra of operators on a Hilbert space H, and the topology T is the topology of strong convergence on a dense sub-space of T.

The following Lemma is an easy generalization of Theorem(11.1) of [10].

Lemma (2.1) Let $F : \mathbb{R}_+ \to A$ be a (not necessarily adapted) function satisfying the regularity condition (R1). Then for any bounded interval $[a,b] \subseteq \mathbb{R}_+$:

$$E_{a]}(F(b) - F(a)) = E_{a]} \{\int_a^b D_+F(s)ds\} \tag{2.2}$$

A corollary of Lemma (2.1) is that the quantity:

$$M(a,b) = F(b) - F(a) - \int_a^b D_+F(s)ds \quad ; \ a \leq b \tag{2.3}$$

satisfies the following conditions:

DM1) $M(a,a) = 0$

DM2) $r < s < t \Rightarrow M(r,t) = M(r,s) + M(s,t)$

DM3) $E_{a]}(M(a,b)) = 0$

If F is adapted we have also:

$$M(a,b) \in A_{b]} \tag{2.4}$$

If F is localized in a strong sense, i.e. the $*$-algebra generated by the set $\{F(s), D_+F(s) : s \in [a,b]\}$ is contained in $A_{[a,b]}$, then we have also:

DM4) $M(a,b) \in A_{[a,b]}$ (DM4)

A map $(a,b) \subset \mathbb{R}_+ \to M(a,b) \in A$, satisfying the conditions (DM1), (DM2), (DM3), will be called a difference martingale. As remarked in [10], if the index set has a finite left end point t_o (in our case the index set is $\mathbb{R}_+$ and $t_o = 0$) then to every difference martingale $(M(a,b))$ we can associate a martingale

$$M_t = M(0,t) \tag{2.5}$$

such that

$$M_t \in A_{t]} = A_{[0,t]} \tag{2.6}$$

$$M(s,t) = M_t - M_s \tag{2.7}$$

and, if (2.4) rather than (DM4) is used to define a difference martingale, then the correspondence, given by (2.7), between martingales and difference martingales is one-to-one. If the index set is unbounded on the left, then one might have problems with the measurability condition (2.6).

In the remaining of this section we introduce the following two additional assumptions on the localization (A_I): for any two subintervals in $\mathbb{R}_+$, I, J, denoting $\overset{\circ}{I}$ the interior of I:

$$\overset{\circ}{I} \cap \overset{\circ}{J} = \emptyset \Rightarrow A_I \text{ commutes with } A_J \tag{2.8}$$

$$\forall t \geq 0; \text{ centre } (A_{t]}) = \mathbb{C}\, 1 \tag{2.9}$$

(the results of this section also hold in the anticommuting case, but the techniques if proof are quite different - cf.[2]).

A difference martingale (M(a,b)) will be called <u>regular</u> if there exist locally integrable functions (necessarily scalar, because of (1.17), (2.9)) $\sigma_{ij} : \mathbb{R}_+ \to \mathbb{C}$ satisfying:

$$E_{t]}\begin{pmatrix} dM.dM^+ & dM.dM \\ dM^+.dM^+ & dM^+.dM \end{pmatrix} = \begin{pmatrix} \sigma_{11}(t) & \sigma_{12}(t) \\ \sigma_{21}(t) & \sigma_{22}(t) \end{pmatrix} dt + o(dt) \tag{2.10}$$

where, here and in the following, the symbol o(dt) denotes a function $\varepsilon(t,dt)$ such that for any locally bounded (in the sense of the locally convex topology) function $G : \mathbb{R}_+ \to A$ and for any bounded interval $[a,b] \subseteq \mathbb{R}_+$, one has:

$$\lim_{\mathrm{Max}|t_{j+1}-t_j| \to 0} \sum G(t_j)\, \varepsilon(t_j, t_{j+1}-t_j) = 0 \tag{2.11}$$

where the limit is meant in the T-topology and (t_j) defines a partition of $[a,b]$.

In case of a regular difference martingale, the classical definition of $(L^2$-) stochastic integrals can be easily generalized. Let us outline the main steps of the construction: let in the notations of section (1)

$$F = \begin{pmatrix} F_1 \\ F_2 \end{pmatrix} = \sum F(t_j)\, \chi_{[t_i, t_{j+1})} = \sum \begin{pmatrix} F_1(t_j) \\ F_2(t_j) \end{pmatrix} \chi_{[t_j, t_{j+1})} \tag{2.12}$$

be a 2-component (i.e. A^2-valued) adapted step function. To it we associate the simple stochastic integral

$$\int (dM^+, dM) \begin{pmatrix} F_1 \\ F_2 \end{pmatrix} = \sum_j [M^+(t_j, t_{j+1}) F_1(t_j) + M(t_j, t_{j+1}) F_2(t_j)] \tag{2.13}$$

If ω is a state compatible with the family of conditional expectations $(E_{t]})$, i.e.

$$\omega\, E_{t]} = \omega \quad ; \forall t \geq 0 \tag{2.14}$$

then, if the topology on A defined by the semi-norm:

$$\|x\|^2_\omega = \omega(x^*x) \quad ; x \in A \tag{2.15}$$

is weaker than the T-topology, an easy calculation yields:

$$\left\| \int (dM^+, dM) \begin{pmatrix} F_1 \\ F_2 \end{pmatrix} \right\|^2_\omega = \int_0^\infty \omega\left[(F_1^+, F_2^+) \begin{pmatrix} \sigma_{11} & \sigma_{12} \\ \sigma_{21} & \sigma_{22} \end{pmatrix} \begin{pmatrix} F_1 \\ F_2 \end{pmatrix} \right] dt \tag{2.16}$$

So, denoting $L^2_<(\mathbb{R}_+, A^2; \omega\cdot\sigma\cdot dt)$ the space of adapted functions $F : \mathbb{R}_+ \to A^2$ $(F = \begin{pmatrix} F_1 \\ F_2 \end{pmatrix})$ such that:

$$\int_0^\infty \omega\left[(F_1^+, F_2^+) \begin{pmatrix} \sigma_{11} & \sigma_{12} \\ \sigma_{21} & \sigma_{22} \end{pmatrix} \begin{pmatrix} F_1 \\ F_2 \end{pmatrix} \right] dt < +\infty \tag{2.17}$$

and with $L^2(A,\omega)$ the GNS space of the pair (A,ω) (i.e. the completion of A - or of an appropriate quotient thereof - for the scalar product $(a,b) \to \omega(a^+ b)$) we obtain a unitary embedding:

$$M : L^2_<(\mathbb{R}_+, A^2; \omega\cdot\sigma\cdot dt) \to L^2(A,\omega) \tag{2.18}$$

characterized by the property that its restriction on

$$S_<(\mathbb{R}_+,A^2)\cap L^2_<(\mathbb{R}_+,A^2;\omega\cdot\sigma\cdot dt)$$

is given by (2.13).

If ω is faithful there is no difficulty in verifying that to each $F\in L^2_<(\mathbb{R}_+, A^2, \omega\cdot\sigma\cdot dt)$ one can associate a densely defined pre-closed operator affiliated to $\pi_\omega(A)''$ (π_ω - the GNS representation of (A,ω), denoted $\int(dM^+,dM)\cdot F$ and characterized by the property:

$$\{ \int(dM^+,dM)\cdot F \}\, a'\ \Omega \qquad = a'\cdot M(F)\Omega \ ;\quad \forall a'\in\pi_\omega(A)' \tag{2.19}$$

Moreover $\int(dM^+,dM)\cdot F$ is adapted in the sense that, if supp $F\subseteq I\subseteq\mathbb{R}_+$, then $\int(dM^+,dM)\cdot F$ is affiliated to $\pi_\omega(A_I)''$. If ω is not faithful, then one can use the BT theorem ([11]) to define $\int(dM^+,dM)\ F$ as a pre-closed operator affiliated to $\pi_\omega(A)''$ but more care is needed to guarantee adaptness. We won't discuss the non-faithful case here (cf [1]).

Now, let us denote

$$\sigma = \begin{pmatrix}\sigma_{11} & \sigma_{12}\\ \sigma_{21} & \sigma_{22}\end{pmatrix} \tag{2.20}$$

$$L^2_{loc}(\mathbb{R}_+,\sigma\cdot dt) = \{f:\mathbb{R}_+\to\mathbb{C} :\forall t\geq 0;\ \int_0^t (f^+,f)\cdot\sigma\cdot\binom{f^+}{f}dt<+\infty\ \} \tag{2.21}$$

For each $0<t<+\infty$ we can define the real (not complex!) pre-scalar product on $L^2_{loc}(\mathbb{R}_+,\sigma\cdot dt)$ by:

$$\langle e_{t]}f,\ e_{t]}g\rangle_{\sigma} = \int_0^t (f^+,f)\cdot\sigma\cdot\binom{g^+}{g}\, ds \tag{2.22}$$

where $(e_{t]}f)(s) = \chi_{[0,t]}(s)\cdot f(s)$, and $\chi_I(s)$ denotes, as usual, the characteristic function of I ($=0$ for $s\notin I$; $=1$ for $s\in I$).

We consider on $L^2_{loc}(\mathbb{R}_+,\sigma\cdot dt)$ the topology induced by this family of semi-norms. It can be shown (cf. [1]) that there exists a dense sub-space D of $L^2_{loc}(\mathbb{R}_+,\sigma\cdot dt)$ such that for each $f\in D$ and for each bounded interval $[0,T]$, there exists a unique unitary solution $U_f(t)$ ($t\in[0,T]$) of the equation

$$\begin{aligned} dU_f(t) &= \{i\ f^+dM + i\ f\ dM^+ - \tfrac{1}{2}(f^+,f)\cdot\sigma\cdot\binom{f^+}{f}dt\}U_f(t)\\ U_f(0) &= 1\end{aligned} \tag{2.23}$$

Now, for $f\in L^2_{loc}(\mathbb{R}_+,\sigma\cdot dt)$, let us introduce the notations:

$$f_1 = f^+\ ;\quad f_2 = f\quad ;\quad |f|^2_\sigma = f_j\sigma_{jk}f^+_k \tag{2.24}$$

(summation over repeated indices is understood). With these notations, one easily checks that for $f,g\in D$:

$$\begin{aligned} &dU_f(t)\ U_g(t) = (dU_f)\ U_g + U_f\ dU_g + dU_f\ dU_g = \\ &= [i(f+g)^+dM + i(f+g)dM^+ - \tfrac{1}{2}(f_j\sigma_{jk}f^+_k + g_j\sigma_{jk}g^+_k + 2\ f_j\sigma_{jk}g^+_k)dt]U_f\ U_g\end{aligned}$$

or equivalently

$$dU_f\ U_g = [i(f+g)^+dM + i(f+g)dM^+ - \tfrac{1}{2}|f+g|^2_\sigma dt + i\mathcal{I}m(f_j\sigma_{jk}g_k)dt]\ U_fU_g$$

and it is immediately verified that this equation, with the same initial condition, is also satisfied by

$$\begin{aligned} &U_{f+g}(t)\ \exp\{-\ i\mathcal{I}m\int_0^t f_j(s)\,\sigma_{jk}(s)g_k(s)ds\} = \\ &\quad = U_{f+g}(t)\ \exp\{-i\mathcal{I}m\ \langle e_{t]}f,\ e_{t]}g\rangle_\sigma\}\end{aligned} \tag{2.25}$$

From the uniqueness theorem we then conclude that, for each fixed $t\geq 0$, and for each $f,g\in D$, one has:

$$U_f(t)\, U_g(t) = U_{f+g}(t) \exp\{-i\, Im < e_{t]} f,\, e_{t]} g >\} \tag{2.26}$$

This means that, for each $t \in \mathbb{R}_+$, the map

$$f \in D \to U_f(t) \in A$$

realizes the representation of the CCR over the real linear space D with symplectic form

$$\sigma_t(f,g) = \sigma(e_{t]} f,\, e_{t]} g) = Im < e_{t]} f,\, e_{t]} g >_\sigma \tag{2.27}$$

From this it is not difficult (cf. [1] for detailed proofs) to deduce the identifications:

$$\sqrt{2}\, f^+ dM = a(e_{dt} f) \quad ; \quad \sqrt{2}\, f\, dM^+ = a^+(e_{dt} f) \quad ; \quad f \in D \tag{2.28}$$

where e_{dt} means $e_{[t,t+dt]}$ and $a(.)$, $a^+(.)$ denote the annihilation and creation operators associated to the above mentioned representation of the CCR over D (more precisely, over the sub-space D_o of D consisting of functions f with $\| f \|^2_\sigma < +\infty$). Finally, taking $E_{t]}$- expectations of both sides of (2.23) we obtain

$$E_{t]}(d\, U_f(t)) = -\frac{1}{2} |e_{dt} f|^2_\sigma\, E_{t]}(U_f(t)) \tag{2.29}$$

whence, taking ω-expectations of both sides of (2.29) for <u>any</u> state ω on the T-completion of A compatible with the family $(E_{t]})$ (i.e. satisfying (2.14)):

$$\frac{d}{dt}\omega(U_f(t)) = (-\frac{1}{2} f_j \sigma_{jk} f_k)\quad \omega(U_f(t))$$

or, since $\omega(U_f(0)) = 1$:

$$\omega(U_f(t)) = e^{-\frac{1}{2}\| e_{t]} f \|^2_\sigma} \tag{2.30}$$

hence ω is a quasi-free state.

<u>Remark</u> The preceeding discussion contains an heuristic argument, since we have freely identified $dM_i dM_j$ with $\sigma_{ij}(t)dt$ ($M_1 = M^+$, $M_2 = M$). Of course, as in the classical case, this identification requires some estimates. Starting from the following section we prove the necessary estimates in a more general framework. While we refer to [1]for the applications of these estimates to the deduction of Ito's formula.

3. <u>Operators valued semi-martingales: the basic estimate</u>

In this section no ρ-commutation relation will be assumed. Let A, $(A_{t]})$, $(A_{[s,t]})$, $(E_{t]})$ be as in section (1.), and let ω be a state on A compatible with $(E_{t]})$ (i.e. $\omega E_{t]} = $; $\forall t \geq 0$).

A <u>regular A^d-valued</u> ($d \in \mathbb{N}$) <u>semi-martingale</u> (with respect to the above structure) is an A^d-valued stochastic differential $M = (M_1, \ldots, M_d)$ (in the sense of Definition 1.1)) satisfying the additional conditions:

$$E_{t]}(M_\gamma(dt)) = m_\gamma(t)\quad d\nu(t) \tag{3.1}$$

$$E_{t]}(M^+_\alpha(dt)\, M_\beta(dt)) = E_{t]}(\sigma_{\alpha\beta}(t))d\nu(t) + o(dt) \tag{3.2}$$

for $\alpha, \beta, \gamma = 1, \ldots, d$ and for some measure ν which can always assumed to be positive and which will be assumed to take finite values on bounded intervals, and functions m_γ, $\sigma_{\alpha,\beta}$: $\mathbb{R}_+ \to A$ (m_γ is necessarily adapted).

If $F_j = (F_{j1}, \ldots, F_{jd}) : \mathbb{R}_+ \quad A^d$ ($j = 1,2$) are adapted functions, denoting

$$dN_j = dM_\alpha(t)\, F_{j\alpha}(t) \quad ; \quad j = 1,2 \tag{3.3}$$

and taking ω-expectations of both sides of the identity:

$$d(N^+_1 N_2) = dN^+_1 N_2 + N^+_1 dN_2 + dN^+_1 dN_2$$

one obtains

$$\omega(N_1^+(t)\, N_2(t)) - \omega(N_1^+(s)\, N_2(s)) =$$

$$= \int_s^t \omega(F_{1\alpha}^+ \cdot m_\alpha \cdot N_2)\, d\nu + \int_s^t \omega(N_1^+ \cdot m_\alpha \cdot F_{2\alpha})\, d\nu + \int_s^t \omega(F_{1\alpha}^+ \cdot \sigma_{\alpha\beta} \cdot F_{2\beta})\, d\nu$$

Therefore, choosing $N_1 = N_2 = N$ and denoting $F_{1\alpha} = F_{2\alpha} = F$ and:

$$\| N(t) \|_\omega^2 = \omega(|N(t)|^2) \tag{3.4}$$

one obtains:

$$\| N(t) \|_\omega^2 - \| N(s) \|_\omega^2 = \tag{3.5}$$

$$= 2\, Re \int_s^t \omega(N^+ \cdot m_\alpha \cdot F_\alpha) d\nu + \int_s^t \omega(F_\alpha^+ \cdot \sigma_{\alpha\beta} \cdot F_\beta)\, d\nu \leq$$

$$\leq \int_s^t 2\, \| N \|_\omega \cdot \| m_\alpha F_\alpha \| d\nu + \int_s^t (F_\alpha^+ \cdot \sigma_{\alpha\beta} \cdot F_\beta)\, d\nu \leq$$

$$\leq \int_s^t \| N(r) \|_\omega^2\, d\nu\,(r) + \int_s^t \| m_\alpha F_\alpha \|^2\, d\nu + \int_s^t \omega(F_\alpha^+ \sigma_{\alpha\beta}\, F_\beta)\, d\nu$$

$$= \int_s^t \| N(r) \|_\omega^2\, d\nu\,(r) + \int_s^t \omega(F_\alpha^+ \cdot \Delta_{\alpha\beta} \cdot F_\beta)\, d\nu$$

where we have put:

$$\Delta_{\alpha\beta}(r) = \sigma_{\alpha\beta}(r) + m_\alpha(r)^+ \cdot m_\beta(r) \tag{3.6}$$

One can prove the following generalization of Gronwall's lemma (cf. [1] for the proof):

<u>Lem (3.1)</u> Let f,c be non negative functions on $\mathbb{R}_+$ and let ν be a positive non atomic measure on $[0,\infty)$ such that:

$$f(t) - f(s) \leq \int_s^t f(r)\, d\nu\,(r) + \int_s^t c(r)\, d\nu\,(r) \tag{3.7}$$

Then

$$f(t) \leq e^{\nu(s,t)} f(s) + \int_s^t e^{\nu(r,t)}\, c(r)\, d\nu\,(r) \tag{3.8}$$

Applying Gronwall's inequality to (3.5), with $f(t) = \| N(t) \|_\omega^2$, one obtains:

$$\| N(t) \|_\omega^2 \leq e^{\nu(s,t)} \cdot \| N(s) \|_\omega^2 + \int_s^t e^{\nu(r,t)}\, c_\omega(r)\, d\nu(r) \tag{3.9}$$

with

$$c_\omega(r) = \sum_{\alpha\beta=1}^{d} \omega(F^+(r) \cdot \Delta_{\alpha\beta}(r) \cdot F_\beta(r)) \tag{3.10}$$

So, for any stochastic integral $N(t) = \int_0^t dM \cdot F$, one has:

$$\| N(t) \|_\omega^2 \leq 2\, \| N(0) \|_\omega^2 + 2 \int_0^t e^{\nu(r,t)}\, c_\omega(r)\, d\nu\,(r) \tag{3.11}$$

while, if $N(0) = 0$, we have the better estimate:

$$\omega(| \int_s^t dM \cdot F|^2) \leq \int_s^t e^{\nu(r,t)}\, \omega(F_\alpha^+(r) \cdot \Delta_{\alpha\beta}(r) \cdot F_\beta(r))\, d\nu\,(r) \tag{3.12}$$

Denoting $\{H_\omega, \pi_\omega, \Omega\}$ the GNS representation of $\{A, \omega\}$, the left hand side of (3.12) can be written as

$$\| \pi_\omega (\int_s^t dM \cdot F) \Omega \|^2$$

The inequality (3.12) allows to complete the space of simple left stochastic integrals defining a contraction

$$L_M : L^2_{<,loc}(\mathbb{R}_+, A^d, \omega \cdot \Delta \cdot e^\nu \cdot d\nu) \to L^2(A, \omega) \tag{3.13}$$

from the space of adapted functions $F : \mathbb{R}_+ \to A^d$ such that for any $0 \leq t < \infty$:

$$\int_0^t e^{\nu(r,t)}\, \omega(F_\alpha^+(r) \cdot \Delta_{\alpha\beta}(r) \cdot F_\beta(r))\, d\nu\,(r) < +\infty \tag{3.14}$$

to the GNS space of (A,ω). Again, as in section (2), if ω is faithful, this contraction defines a map from $L^2_{loc<}(\mathbb{R}_+, A^d, \omega\cdot\Delta\cdot e^{\nu}\cdot d\nu)$ to a space of densely defined, localized, pre-closed operators afiiliated to $\pi_\omega(A)''$. Again we refer to [1] for a discussion of the non-faithful case. For $F \in F^d_<(\mathbb{R}_+, A^d)$, satisfiying (3.14) (or the analogue of (3.14) in the interval (s,t)), we will use the notation:

$$\int_s^t dM\, F = L_M(\chi_{[s,t]}\cdot F) \in \pi_\omega(A_{[s,t]})'' = A^\omega_{[s,t]} \tag{3.15}$$

where $X \in BCB(H_\omega)$ means that X is a pre-closed operator affiliated to B.

Remark. One could "renormalize" a regular semi-martingale (dM_α) obtaining the martingale:

$$dN_\alpha(t) = dM_\alpha(t) - m_\alpha(t)\, d\nu(t) \tag{3.16}$$

In some cases (e.g. if $(E_{t]})$ is markovian or if $m_\alpha(t)$ is a scalar), each dN_α is a difference martingale in the sense of section (2.) (i.e. instead of (2.4) one has the stronger property (DM4.)). In such cases one obtains an estimate much better than (3.12). The relations between the stochastic integrals associated to (dM_α) and (dN_α) will be discussed in [1]

A map F, from $\mathbb{R}_+$ to the pre-closed operators affiliated to $A^\omega = \pi_\omega(A)''$ will be called adapted if, for each $t \geq 0$, F t) is affiliated to $A^\omega_{t]} = \pi_\omega(A_{t]})''$.

We sum up the results of our discussion in the following:

Theorem (3.2) In the above notations, let ω be a faithful state on A compatible with the family $(E_{t]})$ of conditional expectations. Then for each $F \in L^2_{loc<}(\mathbb{R}_+, A^d, \omega\cdot\Delta\cdot e^{\nu}\cdot d\nu)$ there exists a pre-closed operator $\int_{\mathbb{R}_+} dM\, F$ defined on the dense subspace $\pi_\omega(A)'\cdot\Omega \subseteq H_\omega$. If supp $F \subseteq [s,t]$, then: $\int_{\mathbb{R}_+} dM^+\cdot F$ is affiliated to $\pi_\omega(A_{[s,t]})''$ Moreover for each $0 \leq s \leq t < +\infty$, and for every $a' \in \pi_\omega(A)'$:

$$\langle a'\Omega, \left|\int_s^t dM\cdot F\right|^2 a'\Omega\rangle \leq \| a' \|^2_\infty \cdot \int_s^t e^{\nu(r,t)}\, \omega(F^+_\alpha(r)\Delta_{\alpha\beta}(r)\, F_\beta(r))\, d\nu(r) \tag{3.17}$$

4. Existence of solutions of linear stochastic differential equations

In this section no ρ-commutation relations will be assumed, however we assume the existence of a faithful state ω on A compatible with the family of conditional expectations $(E_{t]})$ (i.e. $\omega\, E_{t]} = \omega$, for each $t \geq 0$).

Under this assumption, keeping the notations of section (3.) we introduce the symbolic notation:

$$dU(t) = dM(t)\, F(t)\, U(t) \;;\; t \in [s,T] \;;\; U(s) = U_s \tag{4.1}$$

to mean that for each $t \in [s,T]$

$$U(t) - U(s) = \int_s^t dM(t)\, F(r)\, U(r) \tag{4.2}$$

where the stochastic integral on the right hand side of (4.2) is defined as in section (3.), and therefore is a densely defined preclosed operator on the GNS space H_ω of (A,ω), and the equality (4.2) is assumed to take place on a dense subspace of H_ω, independent on s, and t.

Thus to solve the stochastic differential equation (4.1) means to find and adapted function $t \to U(t)$ (cf. the end of section (3.)) from $\mathbb{R}_+$ to the pre-closed operators affiliated to A^ω such that (4.2) holds.

Theorem (4.1) Let $F = (F_1,\dots,F_d)$ be an adapted function $\mathbb{R}_+ \to A$ such that, for each $0 < T < +\infty$:

$$\sup_{s\in[0,T]} \| F^+_\alpha(s)\cdot\Delta_{\alpha\beta}(s)\, F_\beta(s) \|_\infty = \lambda_T = \lambda_T(F,M,\omega) < +\infty \tag{4.3}$$

Then the stochastic differential equation (4.1) has a solution U(t). Moreover, for each $a' \in \pi_\omega(A)'$, the map $t \to \| U(t)\, a'\Omega \|$ is Lipschitz and:

$$\|U(t)\|_{\omega} = \omega(|U(t)|^2)^{1/2} \le \|U_0\| + \{\int_0^t e^{\nu(s,t)} \cdot \omega(F^+_{\alpha}(s)\cdot\Delta_{\alpha\beta}(s)\cdot F_{\beta}(s))\cdot d\nu(s)\}^{1/2} \qquad (4.4$$

Proof We assume, without loss of generality, that s = 0 in (4.1). Define by induction:

$$U_n(t) = U_0 + \int_0^t dM\cdot F\cdot U_{n-1} \quad ; \; U_0(t) = U_0 \; ; \; t\in [0,T] \qquad (4.5$$

The integral in the right hand side of (4.4) is well defined, in fact due to (4.3) one has, for each $T \geq 0$:

$$\int_0^T e^{\nu(r,t)} \cdot \omega(F^+_{\alpha}(r)\cdot\Delta_{\alpha\beta}(r)\cdot F_{\beta}(r))\, d\nu(r) \le \lambda_T \cdot e^{\nu(0,T)} \cdot \nu(0,T) \qquad (4.6$$

Moreover, due to the basic estimate, for each $n \geq 1$:

$$\int_0^T e^{\nu(r,t)} \cdot \omega(U^+_n(r)\; F^+_{\alpha}(r)\Delta_{\alpha\beta}(r)\cdot F_{\beta}(r)\cdot U_n(r))\cdot d\nu(t) \le$$
$$\le 2\lambda^2_T e^{\nu(0,T)} \cdot \nu(0,T)\cdot \int_0^T \omega(U^+_{n-1}\cdot F^+_{\alpha}\Delta_{\alpha\beta}\cdot F_{\beta} U_{n-1})d\nu + 2\lambda_T \cdot e^{\nu(0,T)} \cdot \nu(0,T)\cdot \|U_0\|^2_{\omega} \qquad (4.7$$

therefore, by induction, for each $U_0 \in A_{0]}$, $(F_1 U_n, \dots, F_d U_n) = FU_n$ belongs to $L^2_{<,loc}(\mathbb{R}_+, A^d, \omega\cdot\Delta\cdot e^{\nu}\cdot d\nu)$ and therefore the (left) stochastic integral in (4.5) is well defined for each $t < +\infty$. Define now, for $n = 1,2,\dots$

$$D_n(t) = U_n(t) - U_{n-1}(t) \quad ; \quad t\in [0,T] \qquad (4.8$$

From the basic estimate and from (4.3), we obtain:

$$\omega(|D_n(t)|^2) \le \int_0^t \omega(D^+_{n-1}(s)\cdot F^+_{\alpha}(s)\cdot\Delta_{\alpha\beta}(s)\cdot F_{\beta}(s)\cdot D_{n-1}(s))\; e^{\nu(s,t)}\, d\nu(s)$$
$$\le \lambda_T \cdot \int_0^t e^{\nu(s,t)} \cdot \omega(|D_{n-1}(s)|^2)\; d\nu\,(s) \qquad (4.9$$

Thus iterating, for each $t \in [0,T]$:

$$\omega(|D_n(t)|^2) \le e^{\nu(0,T)} \cdot \lambda^n_T \frac{\nu(0,T)^n}{n!}\, \omega(|U_0|^2) \qquad (4.10$$

Therefore, for each $N,k \in \mathbb{N}$:

$$\|\pi_{\omega}(U_N(t))\cdot\Omega - \pi_{\omega}(U_{N+k}(t))\cdot\Omega\| \le \qquad (4.11$$
$$\le \sum_{n=N}^{N+k} \|\pi_{\omega}(D_n(t))\cdot\Omega\| = \sum_{n=N}^{N+k} \omega(|D_n(t)|^2)^{1/2}$$
$$\le e^{\nu(0,T)/2} \cdot \sum_{n=N}^{\infty} \frac{(\lambda_T\, \nu(0,T))^{n/2}}{(n!)^{1/2}}\, \omega(|U_0|^2)^{1/2}$$

Thus, for each $t \geq 0$, the sequence $(\pi_{\omega}(U_N(t))$ converges strongly on the dense set $\pi_{\omega}(A)'\cdot\Omega$ and defines a pre-closed operator U(t) affiliated to $A^{\omega}_{t]}$. It is easy to verify that U(t) satisfies (4.2) on the dense sub-space $\pi_{\omega}(A)'\cdot\Omega$.

5. Dependence of the solution on the coefficients

In this section, as in the preceeding one, no ρ-commutation relation is assumed, but ω is assumed to be faithful. We will use vector notations, i.e. for $F = (F_1,\dots,F_d) : \mathbb{R}_+ \to A^d$, we write

$$F^+\cdot\Delta\cdot F = F^+_{\alpha}\,\Delta_{\alpha\beta}\,F_{\beta} : \mathbb{R}_+ \to A \qquad (5.1$$

being given as usual by (3.6).

Consider two equations with the same initial data:

$$dU = dM\; FU \quad ; \quad U(0) = U_0 \qquad (5.2$$

$$dV = dM\; GV \quad ; \quad V(0) = U_0 \qquad (5.3$$

For each $0 < T < +\infty$, and for each $X : \mathbb{R}_+ \to A^d$ we will use the notation:

$$\|X\|_{\infty,T} = \sup_{0 \le s \le T} \max_{\alpha=1,\dots,d} \|X_\alpha(s)\|_\infty \tag{5.4}$$

Theorem (5.1) In the notations above, assume that for each $0 < T < +\infty$, one has:

$$\|F\|_{\infty,T} \; ; \; \|G\|_{\infty,T} \; ; \; \|\Delta\|_{\infty,T} < +\infty \tag{5.5}$$

Then, if $U(t)$, $V(t)$ are the solutions of the equations (5.2), (5.3) respectively, for each $0 < T < +\infty$ and for each $0 \le t \le T$, one has

$$\omega(|U(t) - V(t)|^2) \le a_T(F,G) \cdot \int_0^t \|F(s) - G(s)\|_\infty^2 d\nu(s) \tag{5.6}$$

with $a_T(F,G) = 2 \exp(b + \log b)$ and:

$$b = e^{\nu(0,T)} \cdot \nu(0,T)^{\frac{1}{2}} \|F\|_{\infty,T} \|G\|_{\infty,T} \|\Delta\|_{\infty,T} \tag{5.7}$$

Proof. cf. [1]

6. The algebraic closure of the stochastic differentials: Ito's table

In this section we show that the regularity condition (3.2) - i.e. Ito's multiplication table - arises as a necessary (and sufficient) condition for the vector space of stochastic integrals, associated to a given set of stochastic differentials, to be an algebra. In the notations of section (1), let (M_α) $(1 \le \alpha \le d)$ be a family of stochastic differentials and let us assume that a ρ - commutation relation is fulfilled (cf. (1.14)). We assume that the set (M_α) is self-adjoint, i.e. that for each index $\alpha = 1,\dots,d$ there exists a (unique) index - denoted α^+ - such that:

$$M_\alpha^+ = M_{\alpha^+} \tag{6.1}$$

From (1.14) one easily sees that, in this case:

$$\rho_{\alpha^+} = \rho_\alpha \quad ; \quad \alpha = 1,\dots,d \tag{6.2}$$

The space of simple left stochastic integrals will be denoted:

$$S_<(dM_\alpha) = S_<(\mathbb{R}_+, A, (dM_\alpha)) \tag{6.3}$$

It is a sub-space of A which is closed under involution (due to (6.1)) and which, due to the ρ-commutation relations, coincides with the vector space of all right simple stochastic integrals. We will investigate under which conditions the closureof $S_<(dM_\alpha)$ in a suitable topology is an algebra. To this goal, let us denote T a topology on A under which the stochastic differentials (M_α) integrate themselves on the right and on the left. More precisely: we assume that the closure $\bar{A}$ of A in the T- topology is still an algebra, and that for each $0 \le s \le t < +\infty$, and $\alpha,\beta = 1,\dots,d$, the following limits exist in the T- topoloogy and define elements of $\bar{A}$:

$$\lim_{|P_{s,t}| \to 0} \sum_k M_\alpha(0,t_k) M_\beta(t_k,t_{k+1}) = \int_s^t M_\alpha(0,r)\, dM_\beta(r) \tag{6.4}$$

$$\lim_{|P_{s,t}| \to 0} \sum_k M_\alpha(t_k,t_{k+1}) M_\beta(0,t_k) = \int_s^t dM_\alpha(r)\, M_\beta(0,r) \tag{6.5}$$

where $P_{s,t} = \{ < t_1 < \dots < t_k < \dots \}$ is a partition of (s,t) and $|P_{s,t}| = \max_j (t_{j+1} - t_j)$. From this assumption, using the identity:

$$d(M_\alpha M_\beta) = dM_\alpha M_\beta + M_\alpha dM_\beta + dM_\alpha dM_\beta$$

one deduces that the limit:

$$\lim_{|P_{s,t}|} \sum_k M_\alpha(t_k,t_{k+1})\, M_\beta(t_k,t_{k+1}) \tag{6.6}$$

exists in the T-topology and is equal to:

$$M_\alpha(0,t)\cdot M_\beta(0,t) - M_\alpha(0,s)\cdot M_\beta(0,s) - \int_s^t M_\alpha(0,r)\cdot dM_\beta(r) - \int_s^t dM_\alpha(r)\cdot M_\beta(0,r) \tag{6.7}$$

The limit (6.6) will be denoted, in the following, with the symbol $\langle M_\alpha,M_\beta\rangle(s,t)$. Now let us denote $\bar{S}_<(dM_\alpha)$ the pointwise closure of the space of simple stochastic integrals in the T-topology. From (6.6) and (6.7) it is clear that

$$M_\alpha(0,t)\cdot M_\beta(0,t) - M_\alpha(0,s)\cdot M_\beta(0,s) \in \bar{S}_<(dM_\alpha)$$

if and only if the limit (6.6) is itself a (dM_α)-stochastic integral, i.e. if and only if there exist adapted functions $c^\gamma_{\alpha\beta} : \mathbb{R}_+ \to \bar{A}$ such that

$$\lim_{|P_{s,t}|\to 0} \sum_k M_\alpha(t_k,t_{k+1})\cdot M_\beta(t_k,t_{k+1}) = \langle M_\alpha,M_\beta\rangle(s,t) = \int_s^t c^\gamma_{\alpha\beta}(r)\, dM_\gamma(r) \tag{6.8}$$

we will use the symbolic notation:

$$dM_\alpha\cdot dM_\beta = c^\gamma_{\alpha\beta}\, dM_\gamma + o(dt) = d\langle M_\alpha,M_\beta\rangle \tag{6.9}$$

to mean that the relation (6.8) holds. The relation (6.9) will be called the <u>Ito multiplication table</u> for the stochastic differentials $dM_1,\dots,dM_d$, and the maps $c^\gamma_{\alpha\beta}(t)$ will be called the <u>structure coefficients</u> associated to the stochastic differentials (dM_α).

Now assume that (6.4), (6.5) hold and let F_1, $F_2 \in S_<(\mathbb{R}_+, A^d)$ be simple adapted functions. Define:

$$dN_j = F_{j\alpha}\, dM_\alpha, \text{ i.e. } N_j(t) - N_j(s) = \int_s^t F_{j\alpha}\, dM_\alpha \; ; \; j = 1,2 \tag{6.10}$$

A simple computation shows that, under these assumptions:

$$\begin{aligned} N_1(t)\, N_2(t) - N_1(s)\, N_2(s) &= \int_s^t F_{1\alpha}(r)\cdot\rho_\alpha(N_2(r))\cdot dM_\alpha(r) + \\ &+ \int_s^t N_1(r), F_{2\beta}(r)\cdot dM_\beta(r) + \int_s^t F_{1\alpha}(r)\cdot\rho_\alpha(F_{2\beta}(r))\cdot d\langle M_\alpha,M_\beta\rangle(r) \end{aligned} \tag{6.11}$$

To give a precise meaning to the identity (6.9), let us introduce, in analogy with [ ] (chapter III), the following notations: for X, $Y \in \bar{S}_<(dM_\alpha)$ we introduce the equivalence relation:

$$X \sim Y \Leftrightarrow X(t) - X(s) = Y(t) - Y(s) \; ; \; \forall 0 \le s \le t < +\infty \tag{6.12}$$

The equivalence class containing X is called, as in [8], dX and called the stochastic differential of X. By definition $\int_s^t dX = X(t) - X(s)$, therefore our definition of dX is coherent with the notation introduced in (1.15). Comparing (6.11) with the algebraic identity

$$d(N_1 N_2) = dN_1\cdot N_2 + N_1\cdot dN_2 + dN_1\cdot dN_2 \tag{6.13}$$

(where dN means simply $N(t+dt)-N(t)$), we are naturally led to introduce in the space of simple stochastic differentials, denoted hereinafter $dS_<(dM_\alpha)$, the algebraic operations:

$$dN_1 + dN_2 = d(N_1 + N_2) \tag{6.14}$$

$$dN_1\cdot dN_2 = d\langle N_1,N_2\rangle = F_{1\alpha}\cdot\rho_\alpha(F_{2\beta})\cdot d\langle M_\alpha,M_\beta\rangle \tag{6.15}$$

(N_1,N_2 given by (6.10)), where the summation in α,β, in (6.15), is to be understood in the sense of (6.14). Remark that, with these notations:

$$dS_<(dM_\alpha) + dS_<(dM_\alpha) \subseteq dS_<(dM_\alpha) \tag{6.16}$$

but in general it is not even true that:

$$dS_<(dM) \cdot dS_<(dM) \subseteq d\bar{S}_<(dM) \qquad (6.17$$

In fact (6.11) implies that the above inclusion takes place if and only if there exist adapted functions $c^\gamma_{\alpha\beta} : \mathbb{R}_+ \to \bar{A}$ such that

$$< M_\alpha, M_\beta > (0,t) - < M_\alpha, M_\beta > (0,s) = \int_s^t c^\gamma_{\alpha\beta}(r)\, dM_\gamma(r) \qquad (6.18$$

or equivalently, in the notations introduced above:

$$d < M_\alpha, M_\beta > = c^\gamma_{\alpha\beta}\, dM_\gamma \qquad (6.19$$

If this is the case, we say that (6.19) is the Ito multiplication table for the family of stochastic differentials $dM_1, \dots, dM_d$, and the adapted functions $c^\gamma_{\alpha\beta}$ are called the structure coefficients of the multiplication table. The constraints on the $c^\gamma_{\alpha\beta}$'s (coming, e.g. from associativity) as well as an analysis of their possible structure, will be discussed elsewhere.

The space $dS_<(dM_\alpha)$ of simple stochastic differentials is also an $S_<(dM_\alpha)$-module for the ultiplication defined by:

$$X\, dN = d(X\, N) \quad ; \quad X \in S_<(dM_\alpha) \quad ; \quad dN \in dS_<(dM_\alpha) \qquad (6.20$$

Under some regularity conditions which will not be discussed here (e.g. continuity of the left and right multiplication in $\bar{A}$) the condition (6.19) implies not only the validity of (6.17), but also of

$$d\bar{S}_<(dM_\alpha) \cdot d\bar{S}_<(dM_\alpha) \subseteq d\bar{S}_<(dM_\alpha) \qquad (6.21$$

Under these conditions, the space $d\bar{S}_<(dM_\alpha)$ of stochastic differentials is a $*$-algebra.

Summing up:

Theorem (6.1) Let (dM_α) A, $\bar{A}$, and the topology T be as at the beginning of this section. Let $(dM_\alpha)_{\alpha=1,\dots,d}$ be a set of A-valued stochastic differentials such that the limits (6.4), (6.5) exist for each $0 \le s \le t$ and $\alpha, \beta = 1, \dots, d$. Assume that left and right multiplications in $\bar{A}$ are continuous in the T-topology. Then for any pair $dN_j = F_{j\alpha}\, dM_\alpha$ $(j = 1,2)$ of stochastic differentials in $d\bar{S}_<(dM_\alpha)$ there is a unique stochastic differential $d < N_1, N_2 >$, defined by (6.15). The space $d\bar{S}_<(dM_\alpha)$ of stochastic differentials is a $*$-algebra if and only if

$$d < M_\alpha, M_\beta > = c^\gamma_{\alpha\beta}\, dM_\gamma \qquad (6.22$$

for some adapted functions $c^\gamma_{\alpha\beta} : \mathbb{R}_+ \to \bar{A}$.

Proof. Clear from the discussion above.

A set of stochastic differentials will be called closed if it satisfies the Ito table (6.22) for some structure constants $c^\gamma_{\alpha\beta}$, and it is such that $f_\alpha \cdot dM_\alpha = 0$, for some adapted functions $f_\alpha : \mathbb{R}_+ \to \bar{A}$ $(\alpha = 1, \dots, d)$, for which the stochastic differentials $f_\alpha \cdot dM_\alpha$ are well defined, if and only if each $f_\alpha = 0$.

7. Unitarity of the solution

Consider the equation

$$dU = dM_\alpha \cdot F_\alpha U = dM\, FU \quad ; \quad U(0) = U_o \qquad (7.1$$

and write

$$f_\alpha(t, U(t)) = F_\alpha(t) \cdot U(t) \quad ; \quad t \ge 0 \qquad (7.2$$

In these notations equation (7.1) becomes:

$$dU = dM_\alpha \cdot f_\alpha(U) = dM \cdot f(U) \qquad (7.3$$

And we want to discuss under which conditions on the f_α's, the solution U(t) of (7.1) is unitary if the initial data U_0 is such. We use the notation (7.2) to suggest a natural generalization of the theory developed in sections (4.) and (5.). In fact most results in this sections (including the existence theorem) can be extended to the case in which the $f_\alpha(t,U(t))$ are sufficiently regular (say - LIpschitz in the ω-norm) non linear functions of U(t). We will assume that the set of stochastic differentials (dM_α) is self-adjoint in the sense explained at the beginning of section (6.). Assuming ρ-commutation relations, and that the set $\{dM_\alpha : \alpha = 1,\dots,d\}$ is a closed set of stochastic differentials in the sense of section (6.), the necessary condition for unitarity:

$$0 = d[U^+U] = dU^+\cdot U + U^+\cdot dU + dU^+\cdot dU$$

can be written, using Ito's table:

$$f_{\gamma^+}(U)^+\cdot\rho_\gamma(U) + U^+\cdot\rho_\gamma(f_\gamma(U)) + f_\alpha(U)^+\cdot c^\gamma_{\alpha^+\beta}\cdot\rho_\gamma(f_\beta(U)) = 0 \qquad (7.4$$

so, using (7.2), we find for the linear equation (7.1) the necessary condition for unitarity:

$$F^+_{\gamma^+} + \rho_\gamma(F_\gamma) + F^+_\alpha\cdot c^\gamma_{\alpha^+\beta}\cdot\rho_\gamma(F_\beta) = 0 \qquad (7.5$$

In several cases, in which the (dM_α) are explicitely given, one can show that condition (7.5) is also sufficient for the unitarity of the solution of (7.1). For example, using the explicit form (7.1), one easily finds

$$d(U^+U) = U^+\{[F^+_{\gamma^+} + \rho_\gamma(F_\gamma) + F^+_\alpha c^\gamma_{\alpha^+\beta}\rho_\gamma(F_\beta)]dM_\gamma\}\cdot U$$

which is zero if (7.5) holds, and unitarity follows from the uniquèness theorem.

References

1.) ACCARDI L., PARTHASARATHY K.R., Quantum stochastic calculus. To appear.

2.) ACCARDI L., APPLEBAUM D., QUAEGEBEUR J., Some representation theorems in quantum stochastic calculus. To appear.

3.) APPLEBAUM D., Quasi free stochastic evolutions. These proceedings.

4.) BARNETT C., STREATER R., WILDE I.F., The Ito Clifford Integral IV. J. Op. Theory 11 (1984) 255-271.

5.) DELLACHERIE C., On survoi de la theorie de l'integrale stochastique. Measure Theory Obervolfach 1979, Springer LN 794, 365-395.

6.) HUDSON R., PARTHASARATHY K.R., Quantum Ito's Formula and Stochastic evolutions. Comm. Math. Phys. 93, 301-323 (1984).

7.) HUDSON R., PARTHASARATHY K.R., Stochastic dilations of uniformly continuous completely positive semi-groups. Acta Math. Applicandae (to appear).

8.) IKEDA N., WATANABE S., Stochastic differential equations and diffusion processes. North-Holland 1981.

9.) MEYER P.A., Un cours sur les integrales stochastiques. Sem. Prob. X, Springer LN 511 (1979), 620-623.

10.) NELSON E., Dynamical theories of Brownian motion. Princeton University Press 1972.

11.) SAKAI S., C*-Algebras and W*-Algebras Springer Verlag 1971.

TRAPPING IN STOCHASTIC MECHANICS AND APPLICATIONS TO COVERS OF CLOUDS AND RADIATION BELTS

S. ALBEVERIO*, Ph. BLANCHARD**, Ph. COMBE***, R. RODRIGUEZ***, M. SIRUGUE, M. SIRUGUE-COLLIN****

Centre de Physique Théorique
CNRS - Luminy - Case 907
F-13288 MARSEILLE CEDEX 9 (France)

I. INTRODUCTION

There exists a lot of physical situations where a great number of particles are travelling in a medium which exerts rapidly varying forces on them (in space and time). First observations on particles in static fluids were made in the last century by R. Brown, leading to the discovery of Brownian motion [1]. One can also think of "particles" in a turbulent flow, e.g. the dispersion of smoke emitted by a stack in the lower atmosphere, or a cover of clouds. A different example is given by charged particles which are trapped in the magnetic field of planets. This magnetic field changes very rapidly on small scale. Other applications have been considered in different fields : Apes in a territory around some food source [2], colonies of Esterichia Coli in a Petri box [3]. Astronomical situations have been also treated in this spirit as for instance the formation of jet streams in the protosolar nebula [4] [5] [6] [7] and the morphology of Galaxies [8].

Statistical models are quite natural in such a situation although it is very hard to justify them from physical principle (the basic principles of fluid dynamics), for instance in the case of clouds in the atmosphere (see [9] and references therein).

In the situation previously considered, the forces acting on the particles have a deterministic smooth component e.g. gravitation or dipole like component of magnetic field around the earth.

This suggests, following [2-8] to use a Newtonian stochastic model which originally was initiated by E. Nelson to give an alternative description of quantum mechanics [10] [11] [12]. According to Nelson, it is possible to assign a stochastic acceleration to conservative stochastic diffusion processes. As a basic assumption, this stochastic acceleration is set equal to the deterministic smooth component of the external force acting on the particle, whereas the influences of the remainder is modelled by a diffusion coefficient. In some cases, it is possible to reduce the problem of solving the Fokker-Planck equation to a Schrödinger-like problem. Furthermore, we are interested in stationary situations which correspond to stationary solutions of the Schrödinger-like equation.

---

*Universität Bochum, R.F.A., **Universität Bielefeld, R.F.A.,
***Université d'Aix-Marseille II, France, ****Université de Provence, Marseille, France

These stationary solutions, in general, have nodal surfaces. It has been shown that these nodal surfaces correspond to impenetrable barriers for the diffusion process [6] [7] [13] [14] [15] [16] [17]. One of the basic physical assumption is that this barrier can be in some situation observed. In this paper, we shall make no attempt to justify the model on a deeper ground, but try to see whether it can account for the observation in two cases : the cover of clouds of planets and the radiation belts in the planetary magnetic field.

The paper is organized as follows.

In Section 2, we describe the basic properties of Newtonian Diffusion Stochastic Processes and indicate their connection with Schrödinger-like equations. Furthermore we give a heuristic interpretation of the nodal surfaces as impenetrable barriers for Newtonian Stochastic Diffusion Processes.

Section 3 concerns the possible applications to the observed average cloud covering in the planetary atmosphere, whereas in Section 4 we discuss the radiation belts (Van Allen Belts) along the previous ideas.

2. NEWTONIAN STOCHASTIC DIFFUSION PROCESSES

As we remarked in the introduction, we want to describe an assembly of "particles", which feel both an external deterministic field of forces and perturbations on a much lower scale (a more detailed description will follow). The net result of these influences is that the trajectory of an individual "particle" cannot be predicted in a precise way. Since we are not interested in the precise behaviour of a particle but only in the mean properties, it is tempting to use a probabilistic model : typical trajectories are trajectories of a stochastic process, i.e. one assigns a probability to the set of trajectories. This probability allows to compute all the statistical properties of the assembly of particles.

Equivalently, one can restrict the type of stochastic processes which are considered. The first restriction is that the probability is supported by continuous trajectories, which is a rather natural assumption. Furthermore one assumes that the stochastic process is a Markov process. This amounts to saying that the future of the system does not depend on its past but only on the present. The usual interpretation of these results is that on a short time scale a "particle" experiences a lot of perturbations and looses the memory of its past history. As a consequence of these two assumptions the stochastic process is a diffusion stochastic process X_t (see e.g. [18]).

For each t, X_t is a random variable viz. it depends on an event $\omega \in \Omega$ such that $X_t(\omega) = \omega(t)$ is the trajectory. The process X_t satisfies a stochastic differential equation of diffusion type

$$dX_t^i = \beta_+^i(X_t,t)\, dt + \sigma dW_t^i \tag{2.1}$$

where β_+ is called the forward drift, σ the diffusion constant and W_t is the standard Brownian motion in three dimensions.

The intuitive meaning of this equation is clear. If for instance $\sigma \equiv 0$ then we have to do with a purely deterministic equation, whereas if $\beta_+ \equiv 0$ and $\sigma = 1$, X_t is the standard Brownian motion.

This is not the most general stochastic differential equation of diffusion type however for the sake of simplicity we shall only consider this case. The general case can be treated along the same line (see e.g. [16]).

Under mild assumptions on β_+ the previous stochastic differential equation has a unique solution (see e.g. [18]). It suffices to say that a Lipschitz condition as in the classical theory of differential equations is sufficient to ensure both the existence and uniqueness of the solution of the stochastic differential equation at least for sufficiently small t (given β_+ and the initial condition $X_{t_0} = x_0$ a.s).

As in the case of the Wiener process, the trajectories of stochastic diffusion processes are continuous but nowhere differentiable with probability one. This makes it difficult to write a dynamical equation to constrain the forward drift as in classical mechanics. However it is possible to define substitutes for the total time derivative.

Let $E[\,.\,|X_t = x]$ be the conditional expectation given by $X_t = x$. Let furthermore F be a smooth function then

$$D_{\pm}F(x,t) = \lim_{t\downarrow 0} \pm (\Delta t)^{-1} E\left[F(X_{t\pm\Delta t}, t\pm\Delta t) - F(X_t,t)\,|\, X_t = x\right] \qquad (2.2)$$

defines respectively the forward or backward substitute for total time derivatives respectively and in general they differ. Indeed one has

$$D_{\pm}F(x,t) = \frac{\partial F}{\partial t}(x,t) + (\beta_{\pm}.\nabla)F(x,t) \pm \frac{1}{2}\sigma^2 \Delta F(x,t) \qquad (2.3)$$

If $F(x,t) \equiv x$

$$D_{\pm}X_t = \beta_{\pm}(X_t,t) \qquad \left(D_{\pm}\, x = \beta_{\pm}(x,t) \right) \qquad (2.4)$$

$\beta_+(x,t)$ (resp. $\beta_-(x,t)$) is interpreted as the mean velocity of outgoing (resp.ingoing) "particles" at point x at time t.

Forward and backward velocity $\beta_{\pm}$ are not unrelated. Indeed if we assume that the process X_t has a smooth density $\rho_t(x)$ in the sense that the expectation of a smooth function $F(X_t)$ is given at each time t by

$$E\left[F(X_t)\right] = \int F(x)\,\rho_t(x)\,dx \qquad (2.5)$$

then it is well known that $\rho_t(x)$ satisfies Fokker-Planck equation viz.

$$\frac{\partial}{\partial t}\rho_t = -\nabla.(\beta_\pm \rho_t) \pm \frac{\sigma^2}{2}\Delta\,\rho_t \tag{2.6}$$

which is in the classical case ($\sigma \equiv 0$) a continuity equation.
Comparison of these two equations shows that one can define the current and osmotic velocity v and u by

$$v = \frac{1}{2}(\beta_+ + \beta_-) \tag{2.7}$$

$$u = \frac{1}{2}(\beta_+ - \beta_-) \tag{2.8}$$

so the Fokker-Planck equations can be rewritten

$$\frac{\partial\rho}{\partial t} = -\nabla.(\rho v) \qquad \text{(continuity equation)} \tag{2.9}$$

$$\frac{\sigma^2}{2}\Delta\rho = \nabla.(\rho u) \qquad \text{(osmotic equation)} \tag{2.10}$$

Using the following integration by parts formula for functions f and g with compact support in time

$$\int E\left[D_+ f.g\right] dt = -\int E\left[F.D_- g\right] dt \tag{2.11}$$

and the definition of D_+, D_- and ρ it can be shown that the osmotic equation can always be integrated

$$u = \frac{\sigma^2}{2}\,\nabla\,\mathrm{Log}\,\rho = \frac{\sigma^2}{2}\,\frac{\nabla\rho}{\rho} \tag{2.12}$$

At this stage there is still no dynamical constraint on the drift. Following E. Nelson [10] [11] we introduce the mean stochastic acceleration

$$a = \frac{1}{2}(D_+D_- + D_-D_+)\,X_t \tag{2.13}$$

and we assume as a basic dynamical law, Newton's law in mean i.e.

$$\mu a = F \tag{2.14}$$

where F is an external force and μ the mass of the "particle".
A stochastic diffusion process which satisfies such an equation is called a Newtonian stochastic diffusion process.

It is easy to rewrite the Newton's law in the mean in terms of the velocities u and v

$$\frac{\partial v}{\partial t} + (V.\nabla)v - (u.\nabla)u - \frac{\sigma^2}{2}\,\Delta u = \frac{F}{\mu} \tag{2.15}$$

In the following we are interested in the case where the external force is of the form

$$F = -\nabla U + q(E + v \times B) \tag{2.16}$$

that is we allow for a force of Lorentz type, q being the charge of the particle,

U is a scalar potential, E and B the external electromagnetic fields. We denote by ϕ and A respectively the scalar and vector electromagnetic potentials, so that

$$E = -\nabla\phi - \frac{\partial}{\partial t} A \tag{2.17}$$

$$B = \nabla \times A \tag{2.18}$$

We assume furthermore that there exists a function S(x,t) such that

$$\mu \; v(u,t) = S(x,t) - q \; A(x,t) \tag{2.19}$$

Under these hypotheses and for the Coulomb gauge

$$\nabla . A = 0$$

the stochastic acceleration can be rewritten

$$\mu \; a = \nabla\left[\frac{\partial S}{\partial t} - \frac{\sigma^4\mu}{8} \nabla \operatorname{Log}\rho . \nabla \operatorname{Log}\rho + (\nabla S - qA)(\nabla S - qA)\frac{1}{2\mu} - \frac{\sigma^4\mu}{4} \Delta \operatorname{Log}\rho\right] + q\left(-\frac{\partial A}{\partial t} + v \times B\right) \tag{2.20}$$

We have at our disposal a couple of non linear coupled equations, Newton's law and the continuity equation

$$\frac{\partial S}{\partial t} - \frac{\sigma^4\mu}{2}\Delta(\operatorname{Log}\rho^{1/2}) + \frac{1}{2\mu}(\nabla S - qA)^2 - \frac{\sigma^4\mu}{2}(\nabla \operatorname{Log}\rho^{1/2})^2 + U(x) + q\phi(x) = 0 \tag{2.21}$$

$$\frac{\partial \rho^{1/2}}{\partial t} + \nabla\rho^{1/2} . \nabla S + \rho^{1/2}\Delta S + qA . \nabla\rho^{1/2} = 0 \tag{2.22}$$

In order to solve these equations we set [10]

$$\psi(x,t) = \rho^{1/2}(x,t) \; e^{\frac{i}{\mu\sigma^2} S(x,t)} \tag{2.23}$$

and it can be shown that ψ solves the linear equation

$$i\mu\sigma^2 \frac{\partial\psi(x,t)}{\partial t} = \frac{1}{2\mu}(-i\mu\sigma^2\nabla - qA)^2\psi(x,t) + (U(x,t) + q\phi(x,t))\,\psi(x,t) \tag{2.24}$$

which is of the Schrödinger type.

Conversely, given a square integrable solution of the previous equation, one can recover solution of the coupled equations(2.21), (2.22) and ultimately a stochastic diffusion process (see e.g. [19] [20]), the associated probability density being

$$\rho(x,t) = |\psi(x,t)|^2 \tag{2.25}$$

the current and osmotic velocities being given by

$$v(x,t) = \sigma^2 \; (\nabla \operatorname{Im} \operatorname{Log}(\psi))(x,t) \tag{2.26}$$

$$u(x,t) = \frac{\sigma^2}{2} \vec{\nabla} \{ \mathrm{Log} |\psi|^2(x,t) \} \tag{2.27}$$

Now let us consider the case where the electromagnetic field varies in time. Among the solutions of the Schrödinger-like equations, we are interested in the stationary ones, namely those which satisfy :

$$\psi_E(x,t) = \psi_E(x)\, e^{-i\frac{E}{\mu\sigma^2}t} \tag{2.28}$$

for some constant E ; $\psi_E(x)$ is solution of the stationary Schrödinger equation

$$\frac{1}{2\mu}(-i\mu\sigma^2\nabla - qA)^2\psi(x) + (U(x)+q\phi(x))\psi(x) = E\psi(x) \tag{2.29}$$

The corresponding density ρ and velocities u and v are time independent, i.e. the associated processes are stationary.

These solutions, except the one of lowest energy, have nodes, i.e. zeros. For a given ψ_E solution of (2.29), we consider the nodal surfaces N_ρ of the corresponding density

$$N_\rho = \{ x \in R^3 / |\psi_E(x)| = 0 \}$$

In general, nodal surfaces divide the space into disjoint subsets and it can be shown [13] [14] [15] [16] [17] that the stochastic process never crosses these nodal surfaces.

Intuitively, one can think of such phenomena as follows. In the neighbourhood of the nodal surface, the osmotic velocity u increases to infinity and is directed away from the surface, see Fig. 2.

3. APPLICATION TO CLOUD COVERING OF THE PLANETS [17]

In this section we make an attempt to apply the previous model to the large scale features of the covers of clouds of the planets. Indeed the now available pictures of planets with a substantial atmosphere exhibit regular structures, namely zonal bands on a large scale. This was already known by the end of the Seventeenth Century for Jupiter [22] when the first telescopic observations were made. Now direct exploration of the solar system is possible by means of automatic space crafts. (Pioneer and Voyagers encounters with Jupiter and Saturn [23] [24] [25] [26]).

We propose a mechanism as general as possible, which does not depend too much on different parameters of the planetary atmospheres but accounts for the general features of these large scale structures.

Think of clouds as being composed of "particles" either droplets or icy flakes (with typical size 1 μm). Apart from the gravitational forces, these "particles" feel very complicated forces from the surrounding turbulent atmosphere. We do not intend to take into account the detail of these influences but assume that it can be replaced by a diffusion mechanism. Turbulent diffusion is known to be a more efficient mechanism of diffusion than molecular diffusion [9].

Furthermore we shall make no precise statement about the overall force only assuming it is spherical symmetric and derives from a potential U(r). Under these assumptions, the associated stationary Schrödinger-like equation :

$$-\mu\frac{\sigma^2}{2}(\Delta\psi_E)(x) + U(r)\psi_E(x) = E\,\psi_E(x) \quad ; \quad r = |x| \,. \tag{3.1}$$

Furthermore we assume that the potential u(r) is "sufficiently strong" to ensure the existence of bound states.

These eigenstates, expressed in terms of spherical coordinates (r, θ, φ), have the form

$$\psi_{E_{nl}}(r,\theta,\varphi) = R_{nl}(r)\, P_l^m(\cos\theta)\, \exp\{\, im\varphi\} \tag{3.2}$$

for a discrete set of values of $E = E_{nl}$ labelled by a pair of positive integers n and l. P_l^m, $(|m| \leq \ell)$ is the associated Legendre function, which is real and such that $P_\ell^m = P_l^{-m}$.

Nodal surfaces are either spheres around the origin corresponding to the zeros $r_1, r_r, \ldots$ of the radial part R_{nl} or cones defined by the zeros $\theta_1, \theta_2 \ldots$ of P_ℓ^m . Possible zones of confinement are the annuli which are depicted in Fig. 1.

In tables 1, 2 and 3, there is an attempt to fit the observed zones of covers of clouds with the previous model.

The agreement of the model with observations is very good if one keeps in mind the following facts.

i) there are few free parameters involved, namely the integers n, l, m and it is nice that one can make a fit with relatively low numbers.

ii) the physical parameters in planetary atmosphere vary on a large range as far as the composition, temperature and pressure are concerned,

that means that one cannot hope to get a more precise fit to the observationsin this model.

Actually the model is too crude in the sense that it does not incorporate the basic mechanisms which are responsible for the physics of the atmosphere, e.g. temperature gradient. But it is a good indication that Newtonian diffusion in a sense is compatible with the geometry of the problem.

4. APPLICATION TO RADIATION BELTS [21]

In this section we apply our model to the confinement of charged particles in the terrestrial magnetic field. By the end of the fifties it was discovered that at very high altitude, typically between two and five earth radii there are electrically charged particles trapped in the earth magnetic field [27]. Their density is rather high and they display a toroïdal shape with a typical crescent shape section. It is obvious that the magnetic field is responsible for the trapping of these charged particles and the theory for the movement of charged particles in the dipole like earth magnetic field has been developed a long time ago [28]. However it seems difficult to ignore the rapid variation of the magnetic field which certainly perturbs the previous classical pictures [29]. This is the reason why it is tempting to apply the previous stochastic model to explain the main features of this confinement zone.

We assume that the magnetic field around the earth is dipole like, namely of the form

$$\begin{aligned} B_x(\underline{x}) &= -x \frac{\partial \mathcal{A}}{\partial z}(r,z) \\ B_y(\underline{x}) &= -y \frac{\partial \mathcal{A}}{\partial z}(r,z) \\ B_z(\underline{x}) &= 2\mathcal{A}(r,z) + r \frac{\partial \mathcal{A}}{\partial r}(r,z) \end{aligned} \tag{4.1}$$

where $r = (x^2+y^2)^{1/2}$.

This "Ansatz" corresponds to the most general magnetic field which has cylindrical symmetry and whose lines of forces are contained in meridian plane. $\mathcal{A}$ is a priori an arbitrary function. In Coulomb gauge, the vector potential has the following form

$$\begin{aligned} A_x(\underline{x}) &= -y\,\mathcal{A}(r,z) \\ A_y(\underline{x}) &= x\,\mathcal{A}(r,z) \qquad . \quad A_z(\underline{x}) = 0 \end{aligned}$$

It is important to notice that the lines of forces are defined by the equations

$$\alpha(r,z) = r^2 \mathcal{A}(r,z) = c^{te}$$

and

$$y + cx = 0$$

where c is an arbitrary constant.

As discussed previously, we look for stationary solutions of the Schrö-dinger-like equations

$$\frac{1}{2\mu}(-i\mu\sigma^2\nabla - qA)^2 \psi_E(\underline{x}) = E\,\psi_E(\underline{x})$$

where μ is the mass of particles, q the charge and σ a diffusion constant which models the random character of the environment.

In cylindear coordinates, the Schrödinger-like equation assumes the form

$$-\frac{\mu\sigma^4}{2}\left(\frac{\partial^2}{\partial z^2} + \frac{\partial^2}{\partial r^2} + \frac{\partial^2}{\partial \varphi^2} + \frac{1}{r}\frac{\partial}{\partial r}\right)\phi_i(r,z,\varphi)$$
$$+ i\left\{q\sigma^2 \mathcal{A}(r,z)\frac{\partial}{\partial\varphi} + \frac{q^2 r^2}{2\mu}\mathcal{A}^2(r,z) - E_i\right\}\phi_i(r,z,\varphi) = 0$$

where

$$\phi_i(r,z,\varphi) = \psi_{E_i}(\underline{x})$$

the solutions of this equation are linear combinations of solutions $\phi_l^{\pm}(r,z,\varphi)$ of the form

$$\phi_l^{\pm}(r,z,\varphi) = r^{-1/2}\,P_l(r,z)\,\exp\{\pm i\,\ell\varphi\}$$

the P_l satisfying the following equation

$$-\frac{\mu\sigma^4}{2}\left\{\frac{\partial^2}{\partial r^2} + \frac{\partial^2}{\partial z^2}\right\}P_l(r,z)$$
$$+ \frac{1}{2\mu r^2}\left(q\alpha(r,z) - \mu\sigma^2(\ell + \tfrac{1}{2})\right)\left(q\alpha(r,z) - \mu\sigma^2(\ell - \tfrac{1}{2})\right)P_l(r,z) = E_i P_l(r,z)$$

which is a two dimensional Schrödinger equation with an equivalent potential U_{eff} given by

$$U_{eff} = \frac{1}{2\mu r^2}\left(q\alpha(r,z) - \mu\sigma^2(\ell + \tfrac{1}{2})\right)\left(q\alpha(r,z) - \mu\sigma^2(\ell - 1/2)\right)$$

The existence of a bound state solution for this equation depends on the shape of the equivalent potential. If ql is positive the potential is purely repulsive and there is no solution. If ql is negative, then the existence of the solution depends on more precise assumptions on the behaviour of the magnetic field at infinity. We shall assume that these assumptions are satisfied. Consequently, the possible confinement zones have toroïdal shape defined by the equation

$$P_\ell(r,z) = 0$$

It is rather hard to make a precise statement in this general framework on the shape of these curves. However, the following semi-classical argument shows that for the low-lying eigenstate, the confinement zone has the expected shape. Indeed the effective potential U_{eff} is zero precisely on the line of forces (see Fig. 3)

$$\alpha(r,z) = (\ell \pm 1/2)\mu\sigma^2/q$$

and for small σ the density has a strong tendency to concentrate in between these lines.

As a final remark, the form of ψ_{E_i} indicates that the particle drift is either eastward or westward according to the charges. This is indeed the case in the Van Allen belts.

REFERENCES

[1] R. BROWN, A Brief Account of Microscopical Observations made in the Months of June, July, and August 1827, on the Particles Contained in the Pollen of Plants, and on the General Existence of Active Molecule in Organic and Inorganic Bodies, Philosophical Magazine N.S. 4, 161-173 (1828).

[2] M. NAGASAWA, Segregation of a Population in an Environment, J. Math. Biology 9, 213-235 (1980).

[3] M. NAGASAWA, An Application of the Segregation Model for Septation of Esterichia Coli, J. Theor. Biol. 90, 445-455 (1981).

[4] S. ALBEVERIO, Ph. BLANCHARD, R. HOEGH-KROHN, A Stochastic Model for the Orbits of Planets and Satellites. An Interpretation of Titius-Bode Law, Exp. Math. 4, 365-373 (1983).

[5] S. ALBEVERIO, Ph. BLANCHARD, R. HOEGH-KROHN, Processus de Diffusion, Confinement et Formation de "Jet-Streams" dans la Nébuleuse Protosolaire, Preprint TH 3536-CERN, Febr. 1983.

[6] S. ALBEVERIO, Ph. BLANCHARD, R. HOEGH-KROHN, Newtonian Diffusions and Planets, with a Remark on Non-Standard Dirichlet Forms and Polymers, to be published in Lecture Notes in Mathematics, Springer Verlag, Proceedings of the London Math. Soc. Symposium, Stochastic Analysis, Swansea, 1983.

[7] S. ALBEVERIO, Ph. BLANCHARD, R. HOEGH-KROHN, Reduction of Non-Linear Problems to Schrödinger or Heat Equations : Formation of Kepler Orbits, Singular Solutions for Hydrodynamical Equations, to be published in Lecture Notes in Mathematics, Springer Verlag, Heidelberg, Proceedings CIRM, April 1983.

[8] S. ALBEVERIO, Ph. BLANCHARD, R. HOEGH-KROHN, L. FERREIRA, L. STREIT, Work in preparation.

[9] A.S. MONIN, A.M. YAGLOM, Satistical Fluid Mechanics, Mechanics of Turbulence, Volume 1, English Translation, 1977, MIT Press.

[10] E. NELSON, Derivation of the Schrödinger Equation from Newtonian Mechanics, Phys. Rev. 150, 1079-1089 (1966).

[11] E. NELSON, Dynamical Theories of Brownian Motion, Princeton University Press, (1967).

[12] E. NELSON, Quantum Fluctuation, an Introduction, Physica 124A, 509-520 (1984).

[13] S. ALBEVERIO, R. HOEGH-KROHN, A Remark on the Connection between Stochastic Mechanics and the Heat Equation, J.M.P. 15, 1745-1747 (1974).

[14] S. ALBEVERIO, M. FUKUSHIMA, W. KARWOSKI, L. STREIT, Capacity and Quantum Mechanical Tunneling, Commun.Math.Phys. 81, 501-513 (1981).

[15] L. STREIT, Energy Forms, Schrödinger Theory, Processes, Physics Reports 77, 363-375 (1981).

[16] S. ALBEVERIO, Ph. BLANCHARD, R. HOEGH-KROHN, Diffusion sur une variété riemannienne : Barrières infranchissables et Applications, Astérisque, Société Mathématique de France (1985).

[16a] Ph. BLANCHARD, Trapping for Newtonian Diffusion Processes, in Stochastic Methods and Computer Techniques in Quantum Dynamics, H. Miller, L. Pittner Eds., Springer Verlag, 1984.

[17] S.ALBEVERIO, Ph. BLANCHARD, Ph. COMBE, R. HØEGH-KROHN, RODRIGUEZ, M. SIRUGUE, M. SIRUGUE-COLLIN, Zonal Wind Structure on the Planetary Atmospheres. A. Unified Stochastic Approach, preprint ZiF Bielefeld (1984)

[18] I.I. GIHMAN, A.V. SKOHOROD, The Theory of Stochastic Processes I, II and III, Springer Verlag, 1974.

[19] E. CARLEN, Conservative Diffusions, preprint Princeton University, Physics Department, January 1984.

[20] W. ZHENG, Semi-martingales dans les variétés et mécanique stochastique Nelson, Thèse, Juin 1984, IRMA Strasbourg 235/TE-25.

[21] S. ALBEVERIO, Ph. BLANCHARD, Ph. COMBE, R. HOEGH-KROHN, R. RODRIGUEZ, M. SIRUGUE, M. SIRUGUE-COLLIN, Magnetic Bottles in a Dirty Environment. A Stochastic Model for Radiation Belts, preprint ZIF Bielefeld 54 (1984).

[22] J. CASSINI, Ancient Memoir of "Académie des Sciences", tome 2, p. 104.

[23] Voyager I, Encounter with Jupiter, Science 204, n° 4396 (1979).

[24] Voyager II, Encounter with Jupiter, Science 206, n° 4421 (1979).

[25] B.A. SMITH et al., Encounter with Saturn, Voyager I, Imaging Science Results, Science 212, 163-190 (1981).

[26] B.A. SMITH et al., A New Look at the Saturn System, The Voyager II Images. Sciences 215, 504-536 (1982).

[27] J.A. VAN ALLEN, G.H. LUDWIG, E.C. RAY, C.E. MacILWAIN, Observation of High Intensity Radiation by Satellites, Jet Propulsion 28, 588-592 (1958).

[28] C. STORMER, The Polar Aurora, Oxford University Press (1955).

[29] M. WALT, T.A. FARLEY, in Handbook of Astronomy and Geophysics, vol. 1, The Earth Upper Atmosphere, Ionosphere and Magnetosphere (1978).

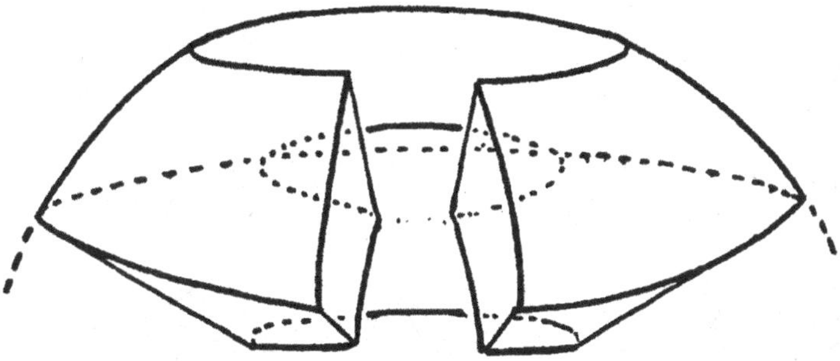

Figure 1

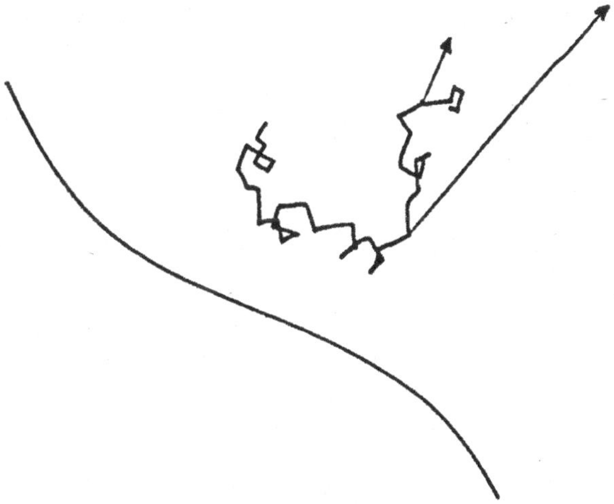

Figure 2

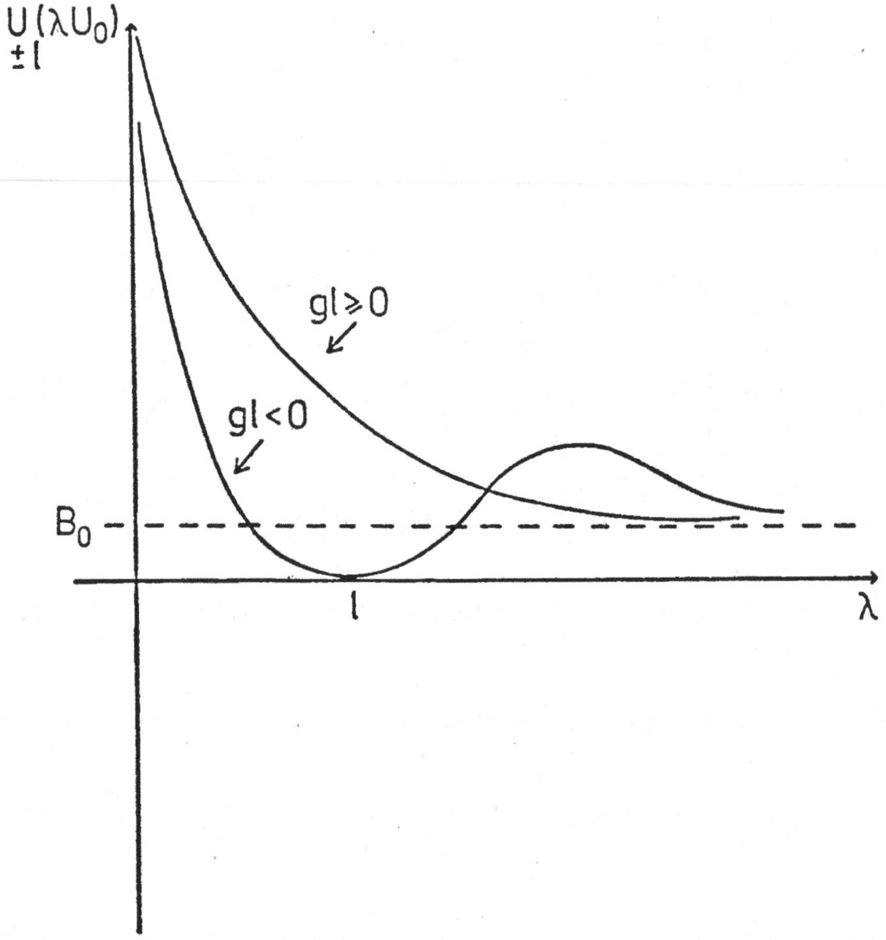

Fig. 3 The shapes of the equivalent potentials $U_{\pm \ell}$

Table 1:

Zonal Structure of the Atmospheres of the Telluric Planets

(Venus, the Earth, Mars) North Hemisphere

Planet Mass/ð Mass R_0 km $(R_1 - R_0)$ km	Period of Revolution	Observed Latitude of the Boundaries between Zones	Tentative ℓ	Tentative m	Position of the zeros of P_m^ℓ	General Direction of the Winds in the Zone (east or west wind)
VENUS .8 6050 65	243 d	$\Theta_1 = 0$ $\Theta_2 = 90$			$\Theta_1' = 0$ $\Theta_2' = 90$	east
THE EARTH 1 6378 15	24 H	$\Theta_1 = 0$ $\Theta_2 = 30$ $\Theta_3 = 60$ $\Theta_4 = 90$	6	1	$\Theta_1' = 0$ $\Theta_2' = 28$ $\Theta_3' = 56$ $\Theta_4' = 90$	east west east
MARS .1 3900 ~ 10	24 H	$\Theta_1 = 15$ $\Theta_2 = 90$	8	6	$\Theta_1' = 15$ $\Theta_2' = 90$	east west
MARS (inner structure) $\lesssim 5$		$\Theta_1 = 15$ $\Theta_2 = 60$ $\Theta_3 = 90$	5	1	$\Theta_1' = 16$ $\Theta_2' = 51$ $\Theta_3' = 90$	east west east

Table II:

Zonal Structure of Jupiter   North Hemisphere

Planet Mass/♁ Mass R_o km $(R_1 - R_o)$km	Period of Rotation	Observed Latitude of the Boundary between the Zones	entative ℓ	m	Position of the Zeros of P_m^ℓ	General Direction of the Winds in the Zone
JUPITER	10 H		31	1		
318					$\Theta_1' = 2.9$	east
72000					$\Theta_2' = 8.7$	
~8500 (estimated)		$\Theta_3 = 16$			$\Theta_3' = 14.5$	west
		$\Theta_4 = 20$			$\Theta_4' = 20.5$	east
		$\Theta_5 = 30$			$\Theta_5' = 26$	west
		$\Theta_6 = 34$			$\Theta_6' = 32$	east
		$\Theta_7 = 38$			$\Theta_7' = 37.6$	-id-
		$\Theta_8 = 42$			$\Theta_8' = 43.6$	
		$\Theta_9 = 50$			$\Theta_9' = 49.5$	
		$\Theta_{10} = 55$			$\Theta_{10}' = 55$	
		$\Theta_{11} = 59$			$\Theta_{11}' = 60.5$	
					$\Theta_{12}' = 67$	
					$\Theta_{13}' = 73.7$	
					$\Theta_{14}' = 78.5$	

Table III:

Zonal Structure of Saturn

North Hemisphere

Planet Mass/$\overset{+}{\delta}$ M ss R_o km $(R_1 - R_o)$km	Period of Rotation	Observed Latitude of the Boundary between the Zones	Tentative ℓ	Tentative m	Position of the Zeros of P_m^ℓ	General Direction of the Winds in the Zone
SATURN 95 60 000 ~ 30 000 (estimated)	11 H	$\Theta_3 = 38$ $\Theta_4 = 57$ $\Theta_5 = 70$	11	1	$\Theta_1' = 8$ $\Theta_2' = 23.6$ $\Theta_3' = 40$ $\Theta_4' = 55$ $\Theta_5' = 70$ $\Theta_6' = 90$	east - id -

A remark on dynamical semigroups in terms of diffusion processes

by

Sergio Albeverio and Raphael Høegh-Krohn

Mathematisches Institut
Ruhr-Universität Bochum
and Bielefeld-Bochum
Research Center Stochastics
of the Volkswagenstiftung

Matematisk Institutt
Universitetet i Oslo
and Université de Provence
and Centre de Physique Theorique,
CNRS, Marseille

ABSTRACT

We give a short exposition of results relating dynamical semigroups on a matrix algebra M_n to diffusion processes on SU(n) and to Markov processes on the state of M_n, running on pure states.

The question we want to discuss in this lecture is related to the basic one, in which sense can one speak of processes running on points in quantum probability? We are far from having an answer in general but we would like to point out that in the case of matrix algebras , at least, something like that can indeed be found. We basically report here about our original work with G. Olsen [1].

To explain what we mean by the above, let us first formulate the basic objects of the classical theory of "commutative Markov processes" in the following "algebraic" way.

We have a basic algebra, $\mathcal{A} = C(Z)$, Z being, say, a locally compact space with countable base for the topology. Moreover we have a submarkov semigroup $(P_t, t \in \mathbb{R}_+)$ of transformations of $\mathcal{A}$ into itself (such that $0 \leq f \leq 1 \rightarrow 0 \leq P_t f \leq 1$).

P_t is a Markov semigroup if $P_t 1 = 1$. In this case P_t is called conservative. P_t is symmetric (with respect to a reference Radon measure μ on Z if $P_t^* = P_t$ as operators in $L^2(Z,\mu)$. A Markov process with state space Z is a family $(\Omega, \mathcal{A}, (P^x)_{x \in \mathbb{R}_+}, (X_t)_{t \in \mathbb{R}_+})$, such that $(\Omega, \mathcal{A}, P^x)$ is a probability space, X_t is a stochastic process with state space Z (and underlying probability space $(\Omega, \mathcal{A}, P^x)$) and having the Markov property (see e.g. [2] and references therein). The relation between Markov processes and submarkov semigroups is just

$$(P_t f)(x) = E^x(f(X_t)) \tag{2.1}$$

where E^x means expectation with respect to P^x. Given P_t one can construct by this a Markov process as above, and viceversa from above process one gets a submarkov semigroup P_t.

The symmetry of P_t (resp. μ) is equivalent with the "detailed balance" condition (time reversal invariance) of (X_t, P^x) i.e. there exists a Radon measure μ s.t. $P_t^* = P_t$, with P_t defined through (X_t, P^x) by (2.1). In terms of the transition probabilities $P^x(X_t \in B) \equiv P_t(x,B)$ this condition reads $P_t(x,dy)\mu(dx) = P_t(y,dx)\mu(dy)$.

We assume (X_t, P^x) to be such that $P^x(X_o \in B) = \delta_x(B)$, i.e. (X_t, P^x) is the process started at time 0 in x (the initial distribution of the process is δ_x).

We write X_t^x for this process.

Let us assume that the map $f \in C(Z) \to f(X_t^x(\omega))$ for P^x a.e. ω, in its dependence on x, through the underlying probability P^x to X_t, is an endomorphism of $C(Z)$.

We shall use the notation $f(X_t^x(\omega)) \equiv \xi_t^x(f)(\omega)$. Whereas it is difficult to find a non commutative analogue of the object $X_t^x(\omega)$ we will show below that one can find non commutative analogues of $\xi_t^x(f)(\omega)$ and P_t.

$(\xi_t, t \in \mathbb{R}_+)$ is a semigroup of random endomorphisms: For P^x a.e. ω, $t \in \mathbb{R}_+$ ξ_t associates to the element $f \in C(Z)$, in a linear way, the element $x \to \xi_t^x(f)(\omega)$ of $C(Z)$. If $\xi_t^x(f)$ depends, for fixed f, continuously on (t,x) we speak of continuous semigroup of random endomorphisms.

Remark: The assumptions are all satisfied in the case where X_t^x is a solution of a Ito stochastic differential equation on R^d with bounded measurable Lifshitz coefficients, see e.g. [3]. In fact one can show in general situations that X_t^x is a homeomorphism of $\mathbb{R}^d$ (one-to-one onto, continuous in both directions, in x). (See e.g. [4], [5] , also for extensions to a "diffeomorphism theorem", on manifolds). In particular in such cases $\xi_t^x(f)$ is even a continuous semigroup of random automorphisms of $C(Z)$.

We shall now look for non commutative analogues of P_t and ξ_t^x.

As far as $(\mathcal{A}, P_t)$ is concerned, the corresponding non commutative object is well known. Let us take the simplest case of the C^*-algebra with unit consisting of all complex n x n matrices, denoted by $\mathcal{A} \equiv M_n$.

The analogue of P_t is a semigroup $(\varphi_t, t \in \mathbb{R}_+)$ of completely positive linear maps on $\mathcal{A}$ i.e. a so called "dynamical semigroup" (see e.g. [7], [8], [9]).

φ_t is called symmetric iff $\langle a, \varphi_t(b)\rangle = \langle(\varphi_t(a), b\rangle$, for all $a,b \in \mathcal{A}$, with $\langle a,b\rangle \equiv$

$\mathrm{tr}\,(a^*b)$ (this corresponds clearly to the symmetry condition $P_t^* = P_t$). Physically φ_t symmetry is again an expression of "detailed balance". φ_t is called conservative iff $\varphi_t(1) = 1$ (which corresponds to $P_t^1 = 1$).

The infinitesimal generator L of φ_t is by definition the infinitesimal generator of φ_t as a semigroup of linear maps of $\mathcal{A}$ as a Banach space. As usually one writes $\varphi_t = e^{tL}$. The structure of conservative dynamical semigroups is characterized by the following Theorem.

<u>Theorem 1</u> $\varphi_t = e^{tL}$ is a conservative symmetric dynamical semigroup on $\mathcal{A}$ if and only if $L = \sum_{i=1}^{k} (\mathrm{ad}\, \beta_i)^2$ for some $k \in \mathbb{N}$ and some $\beta_i \in su(n)$, where ad is the adjoint operation and su(n) denotes the Lie algebra of the group SU(n) of unitary n x n matrices of determinant 1.

<u>Proof:</u> This is basically the Gorini-Kossakowski-Sudarshan-Lindblad characterization of dynamical semigroups, coupled with Stinespring's theorem and an exploitation of the symmetry and conservation properties.
In fact the mentioned characterization yields $L(a) = \psi(a) + k^*a+ak$, $\forall a \in \mathcal{A}$, with ψ a completely positive map, for some $k \in \mathcal{A}$.
The symmetry of φ_t yields $k = k^*$.
By Stinespring's theorem and the symmetry $\psi(a) = \sum_{i=1}^{\ell} v_i^* a v_i + v_i a v_i^*$, $\forall a \in \mathcal{A}$, for some $\ell \in \mathbb{N}$, $v_i \in \mathcal{A}$.

A rewriting of ψ, introducing

$m_i \equiv \frac{1}{2}(v_i + v_i^*)$, $i = 1,\dots,\ell$; $m_i \equiv \frac{1}{2}(v_i - v_i^*)$, $i = \ell+1,\dots,2\ell$,

yields $\psi(a) = 2\sum_{i=1}^{2\ell} m_i\, a\, m_i^* = -\sum_{i=1}^{2\ell} (\mathrm{ad}\, m_i)^2(a) + ha + ah^*$,

with $h \equiv \sum_{i=1}^{2\ell} m_i^2 + k$, $m_i^* = m_i$.

The fact that φ_t is conservative yields $h = 0$.
Setting $\beta_j \equiv i m_j - i\,\mathrm{tr}\, m_j$ we have then $L(a) = \sum_i (\mathrm{ad}\beta_j)^2(a)$, with $\mathrm{tr}\beta_j = 0, \beta_j^* = -\beta_j$, hence $\beta_j \in su(n)$. □

There is an isomorphism of SU(n) resp. su(n) with the Lie group Aut $\mathcal{A}$ of * automorphismsof $\mathcal{A}$ resp. its Lie algebra aut $\mathcal{A}$. By this isomorphism to the element $\beta_i \in su(n)$ there corresponds uniquely the element ad $\beta_i \in$ aut $\mathcal{A}$. $\mathrm{ad}\beta_i$ can be looked upon as a left invariant vector field X_i on Aut $\mathcal{A}$.

We recall that any operator of the form $\tilde{L} = \sum_{i=1}^{K} X_i^2$ can be looked upon as the infinitesimal generator of a symmetric left invariant diffusion process on Aut $\mathcal{A}$, and viceversa any symmetrizable left invariant diffusion on Aut $\mathcal{A}$ has an operator of the form $\tilde{L}$ as infinitesimal generator.

Exploiting the above fact that X_i can be looked upon as $ad\beta_i$ with $\beta_i \in su(n)$ we get:

Theorem 2. The symmetric left invariant diffusions on Aut $\mathcal{A}$ have precisely infinitesimal generators of the form $L = \sum_{i=1}^{k} X_i^2$ and hence are in natural 1 - 1 correspondence with the generators $L = \sum_{i=1}^{k} (ad\beta_i)^2$ of symmetric conservative dynamical semigroups on $\mathcal{A}$.

Let us call X_t^e a symmetric left invariant diffusion on Aut $\mathcal{A}$, started at time zero from e. What is the relation coming from Theorem 2 between X_t^e and the dynamical semigroup $\varphi_t = e^{tL}$? It is not difficult to prove, using the fact that automorphisms are completely positive maps, that $a \to E(X_t^e(a))$ defines a dynamical semigroup on $\mathcal{A}$ (E means expectation). Viceversa, given a dynamical semigroup φ_t with infinitesimal generator L, by Theorem 2 we can associate to it a process X_t^e on Aut $\mathcal{A}$, such that $E(X_t^e(a)) = \varphi_t(a)$, see [1] for details. One has thus the:

Theorem 3. The symmetric left-invariant diffusions X_t^e on Aut $\mathcal{A}$, started at e, are in 1 - 1 correspondence with conservative symmetric dynamical semigroups φ_t on $\mathcal{A}$ by $\varphi_t(a) = E(X_t^e(a))$.

Remark For the left invariant Brownian motion X_t^e on Aut $\mathcal{A}$ = SU(n) one knows the heat semigroup $e^{t\tilde{L}}$, hence also φ_t, explicitely, see e.g. [10]. For connections with questions about the relation between quantum mechanics and classical mechanics see e.g. [11].

The above theorems give a connection between dynamical semigroups on $\mathcal{A}$ and diffusions on Aut $\mathcal{A}$. Can one also get processes on the states of $\mathcal{A}$, still connected with φ_t, and "running on points", i.e. on pure states?

The answer is yes. Let $S(\mathcal{A})$ be the state space of $\mathcal{A}$ and let φ_t be a dynamical semigroup on $\mathcal{A}$. For $\sigma \in S(\mathcal{A})$, $\sigma \to \sigma \circ \varphi_t$ is an affine map from $S(\mathcal{A})$ into itself, called a dynamical semigroup on $S(\mathcal{A})$. Let X_t^e be as in Theorem 3. Let $\partial_e S(\mathcal{A})$ be the extreme boundary of $S(\mathcal{A})$ i.e. the set of pure states.

Theorem 4 $\eta_t^\sigma \equiv \sigma \circ X_t^e$ is a Markov process on $S(\mathcal{A})$, which carries $\partial_e S(\mathcal{A})$ into $\partial_e S(\mathcal{A})$. $\sigma \to \eta_t^\sigma$ is an affine map on $S(\mathcal{A})$. One has $\sigma \circ \varphi_t = E(\eta_t^\sigma)$.

Remark Thus the process η_t^σ in the state space $S(\mathcal{A})$ of $\mathcal{A}$ "runs on points" inasmuch as if it is started at a point $\sigma \in \partial_e S(\mathcal{A})$ it will remain in $\partial_e S(\mathcal{A})$ (recall that the elements of $\partial_e S(\mathcal{A})$ can be looked upon as the "points" in $S(\mathcal{A})$).

Finally part of the above consideration can be extended to non symmetric situations. One can show in particular that $L' = \sum_{i=1}^{k} \partial_i^2 + \partial_o$, with ∂_i, ∂_o bounded *-derivations on $\mathcal{A}$, is the infinitesimal generator of a conservative dynamical semigroup. (This is proven using techniques similar to the above ones for each exp $(t\partial_i^2)$, together with Trotter's theorem.)

Define then $(\hat{\partial}_i f)(\sigma) \equiv \frac{d}{dt} f(\sigma \circ e^{t\partial_i})|_{t=0}$, $i = 0,\ldots, k$, $\sigma \in S(\mathcal{A})$ for f an arbitrary element in $P = UP_n$, P_n being the polynomials of degree n built with elements of the space Aff(S) of affine W^*-continuous functions on $S(\mathcal{A})$. Then P is a dense set of analytic vectors for $\hat{L} \equiv \sum_{i=1}^{k} \hat{\partial}_i^2 + \hat{\partial}_o$, P_n is an invariant subspace for $e^{t\hat{L}}$, and $e^{t\hat{L}}$ is a strongly continuous conservative Markov semigroup on $C(S(\mathcal{A}))$. To it there is associated a Markov process η_t^σ on $S(\mathcal{A})$ and one has

$$(e^{t\hat{L}} f)(\sigma) = E(f(\eta_t^\sigma))$$

for all $f \in C(S(\mathcal{A}))$. Moreover

$$e^{t\hat{L}} \restriction \mathrm{Aff}(S(\mathcal{A})) = \cdot \circ \varphi_t$$

and thus, by Theorem 4

$$e^{t\hat{L}}(\sigma) = \sigma \circ \varphi_t = E(\eta_t^\sigma)$$

with $\psi(\sigma) \equiv \sigma \circ \varphi_t$, so that $\psi \in \mathrm{Aff}(S(\mathcal{A}))$.

It would be very interesting to extend all of the above results to the case of more general C^*-algebras. For the study of spectral properties of the dynamical semigroups φ_t see e.g. [12], [13], [16].

For the connection of dynamical semigroups with Dirichlet forms on C^*-algebras see [14], [15].

Acknowledgements: The first named author is very grateful to the organizers for a kind invitation to a most interesting and stimulating meeting. We thank Gunnar Olsen for the joy of collaboration and Mrs. Mischke and Richter for the skilful typing.

References

[1] S. Albeverio, R. Høegh-Krohn, G. Olsen, Dynamical semigroups and Markov processes on C^*-algebras, J. Reine u. Angew. Math. 319, 25-37 (1980)

[2a] M. Fukushima, Dirichlet forms and diffusion processes, North Holland/ Kodansha, Amsterdam (1980)

[2b] D. Williams, Diffusions, Markov processes, and martingales, J. Wiley, New York 1979

[3] D.W. Stroock, S.R.S. Varadhan, Multidimensional diffusion processes, Springer, Berlin (1979)

[4] N. Ikeda, S. Watanabe, Stochastic differential equations and diffusion processes, North Holland/Kodansha, Amsterdam (1981)

[5] K.D. Elworthy, Stochastic differential equations on manifolds, Cambridge Univ. Press, Cambridge (1982)

[6] E.B. Davies, Quantum theory of open systems, Academic Press, London (1976)

[7] G. Lindblad, Non-equilibrium entropy and irreversibility, D. Reidel (1983)

[8] L. Accardi, A. Frigerio, V. Gorini, Quantum probability and applications to the quantum theory of inversible processes, Lect. Notes in Maths. 1055, Springer, Berlin (1984)

[9] W.A. Majewski, Dynamical semigroups in the algebraic formulation of statistical mechanics, Fortschr. d. Phys. 32, 89-133 (1984)

[10] T. Arede, Gêometrie du noyau de la chaleur sur les variêtés, Thèse de III cycle, Marseille 1983, and papers in preparation

[11] S. Albeverio, T. Arede, The relation between quantum mechanics and classical mechanics: a survey of some mathematical aspects, to appear in Proc. Como Conf. "Quantum Chaos", 1983, Ed. G. Casati, Plenum Press (1985)

[12] S. Albeverio, R. Høegh-Krohn, Frobenius theory for positive maps of von Neumann algebras, Commun. Math. Phys. 64, 83-94 (1978)

[13] M. Enomoto, Y. Watatani, A Perron-Frobenius type theorem for positive linear maps on C^*-algebras, Math. Jap. 24, 53-63 (1979)

[14] S. Albeverio, R. Høegh-Krohn, Dirichlet forms and Markov semigroups on C^*-algebras, Commun. Math. Phys. 56, 173-187 (1977)

[15] S. Albeverio, R. Høegh-Krohn, The method of Dirichlet forms, pp. 250-258 in "Stochastic behavior an classical and quantum Hamiltonian systems", Como 1977, Ed. G. Casati and J. Ford, Lect. Notes Phys. 93, Springer (1979).

[16] A. Frigerio, M. Verri, Long-time asymptotic properties of dynamical semigroups on W^*-algebras, Math. Z. 180, 275-286 (1982)

QUASI-FREE STOCHASTIC EVOLUTIONS

David Applebaum *
Dipartimento di Matematica
II Università degli Studi di Roma
Via Orazio Raimondo
00173 Roma

Abstract :

Examples of the theory of stochastic calculus on local algebras ([Ac Pa 1], [Ac Pa 2]) are constructed using a wide class of quasi-free states on the CCR and CAR algebras (see also [Ba St Wi] for a complementary analysis). Having established, in each case, the appropriate rule for multiplication of stochastic differentials, we investigate a class of stochastic differential equations whose solution is a family of unitary operators on an appropriate Hilbert space. We thus obtain a technique for constructing unitary dilations of a certain class of completely positive evolutions which generalises the unitary dilation scheme for quantum dynamical semigroups which have been constructed using the Fock state ([Hu Pa],[Ap Hu]) and extremal universally invariant (e.u.i.) quasi-free states [Hu Li].

We employ the following notation :

$$\mathbb{R}^+ = [0,\infty)$$

If N is a *-algebra and $A \in N$ then

A^+ denotes the adjoint of A.

For $A,B \in N$, we denote their commutator and anticommutator respectively

$$[A,B] = AB - BA$$
$$\{A,B\} = AB + BA$$

(1) The CCR Case

(1.1) The CCR Algebra and Conditional Expectation

Let h be a complex pre-Hilbert space. We denote by C(h) the C.C.R. polynomial *-algebra over h generated by the identity I and

$\{a(f)\ ,\ f \in h\}$ with each a(f) antilinear in f and

$$[a(f),\ a(g)] = 0\ ,\ [a(f),\ a^+(g)] = \langle f,g\rangle I \text{ for each } f,g \in h \qquad \ldots(1.1.1)$$

We will be interested in that class of gauge invariant quasi-free states ω on C(h) for which there exists a positive self-adjoint operator T on $\bar{h}$ with

$$\omega(a^+(f)\ a(g)) = \langle g,Tf\rangle \quad \text{for each } f,g \in h \qquad \ldots(1.1.2)$$

For simplicity we will take T to be bounded and assume that the state has no Fock part i.e. $T > 0$.

---

* Work begun when the author was supported by a CNR Visiting Professorship and completed when supported by an SERC European Fellowship.

Let $\Gamma_s(h)$ denote symmetric Fock space over h. For each $f,g \in h$, $A(f)$ and $A^+(g)$ will denote (respectively) annihilation and creation operators on $\Gamma_s(h)$ and $\Omega^s=(1,0,0,\dots)$ the vacuum vector in $\Gamma_s(h)$.

We write $\quad H^s = \Gamma_s(h) \otimes \Gamma_s(h)$

and $\quad \psi^s = \Omega^s \otimes \Omega^s$

It is well known (see e.g. [BraRo]) that C(h) may be realized as a *-algebra of (unbounded) operators on H^s via the prescription

$$a(f) = A(\sqrt{I+T}\, f) \otimes I + I \otimes A^+(J\sqrt{T}\, f) \qquad \text{for } f \in h$$

where J is an antilinear involution on h satisfying $\langle Jf, Jg\rangle = \langle g,f\rangle$ and ω takes the form $\omega(X) = \langle \psi^s, X\psi^s\rangle$ for $X \in C(h)$.

Let $\quad I_n = \{1,\dots,n\}$ and $I_m = \{1,\dots,m\}$.
For each $J = \{j_1,\dots,j_p\} \subseteq I_n$ and $K = \{k_1,\dots,k_q\} \subseteq I_m$, let J' and K' denote their respective complements in I_n and I_m.
ϕ will, as usual, denote the empty set.

For any $P \in B(\bar{h})$, we define the Wick monomials $X^P_{J,K} \in C(h)$ by the prescription

$$\left.\begin{aligned} X^P_{J,K} &= a^+(Pf_{j_p})\dots a^+(Pf_{j_1})a(Pg_{k_q})\dots a(Pg_{k_1}) \\ X^P_{\phi,K} &= a(Pg_{k_q})\dots a(Pg_{k_1}) \quad , \qquad X^P_{\phi,\phi} = I \end{aligned}\right\} \quad \dots(1.1.3)$$

where $f_{j_p},\dots,f_{j_1},g_{k_q},\dots,g_{k_1} \in h$.
Let E be a projection in h with range D and let $E^\perp$ denote the projection I-E with range $D^\perp$. We assume that $[E,T] = 0$.

C(h) is spanned by $\{X^I_{J,K}\ ;\ J \subseteq I_n,\ K \subseteq I_m\ ;\ n,m \in \mathbb{N}\}$ and we define the <u>conditional expectation</u> $\mathbb{E}^\omega_D$ from C(h) to C(D) by linear extension of

$$\mathbb{E}^\omega_D\,(X^I_{J,K}) = \sum_{J\subseteq I_n}\sum_{K\subseteq I_m} \omega(X^{E^\perp}_{J',K'})X^E_{J,K} \qquad \dots(1.1.4)$$

<u>Remark 1</u>. To understand how (1.1.4) works, observe that there is a *-isomorphism between C(h) and $C(D) \otimes C(D^\perp)$ for which each a(f) is mapped to $a(Ef) \otimes I + I \otimes a(E^\perp f)$.

<u>Remark 2</u>. Let $L^s = \Gamma_s(D) \otimes \Gamma_s(D)$ and $L^s_\perp = \Gamma_s(D^\perp) \otimes \Gamma_s(D^\perp)$. Let $\Omega^s_\perp$ be the vacuum vector in $B(\Gamma_s(D^\perp)$ and write $\psi^s_\perp = \Omega^s_\perp \otimes \Omega^s_\perp$
Following [HuLi] we extend (1.1.4) to a conditional expectation from $B(H^s)$ to $B(L^s)$ by continuous linear extension of the following : let $A \in B(L^s)$ and $B \in B(L^s_\perp)$ then

$$\mathbb{E}^\omega_D\,(A \otimes B) = \langle \psi^s_\perp,\ B\,\psi^s_\perp\rangle A \otimes I \qquad \dots(1.1.5)$$

<u>Remark 3</u>. It is easily verified that the map $\mathbb{E}^\omega_D$ satisfies all the usual properties of a conditional expectation, in particular

$$\left.\begin{aligned} &M \text{ a subspace of } D \Rightarrow \mathbb{E}^\omega_M \circ \mathbb{E}^\omega_D = \mathbb{E}^\omega_M \\ &X,Z \in C(D) \text{ (or } B(L^s)) \text{ and } Y \in C(h) \text{ (or } B(H^s)) \Rightarrow \mathbb{E}^\omega_D(XYZ) = X\,\mathbb{E}^\omega_D(Y)Z \\ &\omega \circ \mathbb{E}^\omega_D = \omega \end{aligned}\right\} \dots(1.1.6)$$

(1.2) Itô Calculus in C(h)

Let $\mathcal{I}$ be an interval in $\mathbb{R}$ and let $\{\ell_{t]},\ t\in\mathcal{I}\}$ be a strongly continuous family of projections in h, such that the prescription $\ell_{[s,t]} = \ell_{t]}-\ell_{s]}$ $(s\leq t,\ s,t\in\mathcal{I})$ generates a spectral measure on h.

We assume that $[\ell_{t]},\ T] = 0$ ...(1.2.1)

for all $t\in\mathcal{I}$ and denote by $\mathbb{E}^{\omega}_{t]}$ the conditional expectation from C(h) to $C(\ell_{t]}h)$ as given by (1.1.4).

Let $f\in h$ and write $\ell_{dt]}f = \ell_{[t,t+dt]}f$

We define
$$\begin{aligned} a_f(dt) &= a(\ell_{dt]}f) \\ a_f^+(dt) &= a^+(\ell_{dt]}f) \\ \lambda_f(dt) &= \langle\ell_{dt]}f,f\rangle \end{aligned} \qquad ...(1.2.2)$$

It is easily verified that these are stochastic differentials in the sense of [Ac Pa 1] and [Ac Pa 2] (indeed λ_f is a Borel measure on $\mathcal{I}$).

By (1.1.4) we have $\mathbb{E}^{\omega}_{t]}(a_f(dt)) = \omega(a_f(dt)) = 0$ and $\mathbb{E}^{\omega}_{t]}(a_f^+(dt)) = 0$

We similarly find that conditional expectations of all products of pairs of stochastic differentials vanish with the sole exception of the following

$$\begin{aligned} \mathbb{E}^{\omega}_{t]}(a_f^+(dt)\ a_f(dt)) &= \omega(a_f^+(dt)\ a_f(dt)) && \text{by (1.1.4)} \\ &= \langle\ell_{dt]}\ f,\ T\ \ell_{dt]}f\rangle && \text{by (1.1.2)} \\ &= \langle\ell_{dt]}\ f,Tf\rangle && \text{by (1.2.1)} \end{aligned}$$

By (1.1.1) and linearity of $\mathbb{E}^{\omega}_{t]}$,

$$\mathbb{E}^{\omega}_{t]}(a_f(dt)a_f^+(dt)) = \langle\ell_{dt]}f,\ \ell_{dt]}f\rangle+\omega(a_f^+(dt)a_f(dt)) = \langle\ell_{dt]}\ f,\ (I+T)f\rangle$$

Define
$$\left.\begin{aligned} \lambda_{sf}(dt) &= \langle\ell_{dt]}f,\ Tf\rangle \\ \lambda_{cf}(dt) &= \langle\ell_{dt]}f,\ (I+T)f\rangle \end{aligned}\right\} \qquad ...(1.2.3)$$

λ_{sf} and λ_{cf} are Borel measures on $\mathcal{I}$ satisfying the identity

$$\lambda_{cf}(dt) - \lambda_{sf}(dt) = \lambda_f(dt) \qquad ...(1.2.4)$$

Remark : The notation in (1.2.3) is stimulated by the following observation. Let $C=\sqrt{I+T}$ and $S=\sqrt{T}$, then $C^2 - S^2 = I$ and

$$\begin{aligned} \lambda_{sf}(dt) &= \langle\ell_{dt]}Sf\ ,\ Sf\rangle \\ \lambda_{cf}(dt) &= \langle\ell_{dt]}Cf\ ,\ Cf\rangle \end{aligned}$$

Both λ_{sf} and λ_{cf} are absolutely continuous with respect to λ_f for e.g.

$$\begin{aligned} \lambda_{sf}(dt) &= \langle\ell_{dt]}f,\ Tf\rangle \\ &= \langle\ell_{dt]}f,\ T\ \ell_{dt]}f\rangle && \text{by (1.2.1)} \\ &\leq \|\ell_{dt]}\ f\|\ \|T\ \ell_{dt]}f\| && \text{by the Schwarz inequality} \\ &\leq \|T\|\ \|\ell_{dt]}\ f\|^2 \\ &= \|T\|\ \lambda_f(dt). \end{aligned}$$

Thus we may write

$$\mathbb{E}_{t]}(a_f^+(dt)a_f(dt)) = \frac{d\lambda_{sf}}{d\lambda_f}(t)\ \lambda_f(dt)$$

$$\mathbb{E}_{t]}(a_f(dt)a_f^+(dt)) = \frac{d\lambda_{cf}}{d\lambda_f}(t)\ \lambda_f(dt) \qquad \ldots(1.2.5)$$

The Radon-Nikodym derivatives $\frac{d\lambda_{sf}}{d\lambda_f}(t)$ and $\frac{d\lambda_{cf}}{d\lambda_f}(t)$ are the "generalized structure constansts" of [Ac Pa 2]. From (1.2.4) we see that they satisfy the relation

$$\frac{d\lambda_{cf}}{d\lambda_f}(t) - \frac{d\lambda_{sf}}{d\lambda_f}(t) = 1 \qquad \ldots(1.2.6)$$

We summarize in the following table, the rules for products of stochastic differentials

	$a_f(dt)$	$a_f^+(dt)$	$\lambda_f(dt)$
$a_f^+(dt)$	$\lambda_{sf}(dt)$	0	0
$a_f(dt)$	0	$\lambda_{cf}(dt)$	0
$\lambda_f(dt)$	0	0	0

...(1.2.7)

Example 1.1 (HuLi) Let $h = L^2_{loc}(\mathbb{R}^+)$, $\mathscr{I} = \mathbb{R}^+$ and $f = 1$.
Choose $T = \sinh^2\gamma\, I(\gamma \in \mathbb{R}^+, \gamma \neq 0)$ so that ω is e.u.i.
We take $\ell_{t]}$ to be multiplication by the indicator function $\chi_{[o,t]}$ so that the operator valued stochastic differentials arise from quantum Brownian motion $\{a(\chi_{[o,t]}), a^+(\chi_{[o,t]})\ ;\ t\in\mathbb{R}^+\}$ of variance $\cosh 2\gamma$ [Co Hu] and λ_f is Lebesgue measure.
For the "structure constants", we obtain

$$\frac{d\lambda_{cf}}{d\lambda_f}(t) = \cosh^2\gamma \qquad \frac{d\lambda_{sf}}{d\lambda_f}(t) = \sinh^2\gamma$$

(1.3) Unitary Stochastic Differential Equations

Let h_o be a complex separable Hilbert space and choose $\mathscr{I} = \mathbb{R}^+$. We introduce a local structure on the von Neumann algebra $N = B(h_o \otimes H^S)$ by defining $N_{s,t} = B(h_o \otimes H^S_{s,t})$ where $H^S_{s,t} = \Gamma_s(\ell_{[s,t]}h) \otimes \Gamma_s(\ell_{[s,t]}h)$. We write $N_{t]} = N_{o,t}$ and $H^S_{t]} = H^S_{o,t}$.
Let $\{j_t\ ;\ t \in \mathbb{R}^+\}$ be that family of injections of $B(h_o)$ into $N_{t]}$ given by

$$j_t(X) = X \otimes I_{t]} \quad \text{for } X \in B(h_o)$$

where $I_{t]}$ is the identity in $B(H^S_{t]})$.

For $\alpha = 1,2,3$, let $F_\alpha(t)$ be weakly measurable maps from $\mathbb{R}^+$ to $B(h_o)$ and consider the stochastic differential equation in N given by

$$dU = \sum_{\alpha=1}^3 U_t\ j_t(F_\alpha(t))\ dX_\alpha \qquad \text{with } U_o = I \qquad \ldots(1.3.1)$$

where $dX_1 = a_f(dt)$, $dX_2 = a_f^+(dt)$, $dX_3 = \lambda_f(dt)$.

We assume that the $F_\alpha(t)$'s satisfy the conditions given in [Ac Pa 1] which ensure existence and uniqueness of the solution to (1.3.1), hence each $U_t \in N_{t]}$. We establish conditions for each U_t to be a unitary operator in N (see also [Ac Pa 1])

$$U_t \text{ unitary} \Rightarrow U_t^+ U_t = I$$

$$\Rightarrow d(U_t^+ U_t) = 0$$

$$\Rightarrow dU_t^+ U_t + U_t^+ dU_t + dU_t^+ dU_t = 0$$

$$\Rightarrow a_f^+(dt)\, j_t(F_1^+(t)) + a_f(dt)\, j_t(F_2^+(t)) + j_t(F_3^+(t))\, \lambda_f(dt)$$

$$+ j_t(F_1(t))\, a_f(dt) + j_t(F_2(t))\, a_f^+(dt) + j_t(F_3(t))\, \lambda_f(dt)$$

$$+ j_t(F_1^+(t)\, F_1(t))\, \lambda_{sf}(dt) + j_t(F_2^+(t)\, F_2(t))\, \lambda_{cf}(dt) = 0$$

$$\left.\begin{aligned}&\Rightarrow F_1^+(t) + F_2(t) = 0\\ &F_3(t) + F_3^+(t) + \frac{d\lambda_{sf}}{d\lambda_f}(t)\, F_1^+(t)\, F_1(t) + \frac{d\lambda_{cf}}{d\lambda_f}(t)\, F_2^+(t)\, F_2(t) = 0\end{aligned}\right\} \quad \ldots(1,3,2)$$

where we have repeatedly used commutativity of $a_f(dt)$ and $a_f^+(dt)$ with elements of $N_{t]}$, (1.2.7) to evaluate the "Itô correction term" $dU_t^+\ dU_t$ and finally independence of the stochastic differentials to equate coefficients.

We write $F_1(t) = F_t$ and let $\mathscr{H}_t$ be a weakly measurable map from $\mathbb{R}^+$ to $B(h_o)$ satisfying

$$\mathscr{H}_t = \mathscr{H}_t^+ \quad \text{for all } t \in \mathbb{R}^+.$$

The general solution of (1.3.2) is given by

$$\begin{aligned}F_2(t) &= -F_t^+\\ F_3(t) &= i\mathscr{H}_t - \frac{1}{2}\frac{d\lambda_{sf}}{d\lambda_f}(t)\, F_t^+F_t - \frac{1}{2}\frac{d\lambda_{cf}}{d\lambda_f}(t)\, F_tF_t^+\end{aligned} \quad \ldots(1.3.3)$$

We obtain an evolution on N by defining for $s,t \in \mathbb{R}^+$, $s \leq t$,

$$U_{s,t} = U_s^+ U_t \in N_{s,t}$$

so that for $r \leq s \leq t$

$$U_{r,s}\, U_{s,t} = U_{r,t} \qquad \text{and } U_{s,s} = I \quad \ldots(1.3.4)$$

By (1.3.4) we see that the $U_{s,t}$'s are <u>multiplicative functionals</u> in the sense of [Nel] which furthermore satisfy the stochastic differential equation :

$$dU_{s,t} = U_{s,t}(j_t(F_t)a_f(dt) - j_t(F_t^+)a_f^+(dt) + (i\, j_t(\mathscr{H}_t)$$

$$- \frac{1}{2}\frac{d\lambda_{sf}}{d\lambda_f}(t)\, j_t(F_t^+F_t) - \frac{1}{2}\frac{d\lambda_{cf}}{d\lambda_f}(t)\, j_t(F_tF_t^+))\, \lambda_f(dt)) \quad \ldots(1.3.5)$$

(1.4) <u>Unitary Dilations of Completely Positive Evolutions</u>

We extend $\mathbb{E}_{t]}^{\omega}$, as defined by (1.1.5), to a conditional expectation from N into $N_{t]}$ by the requirement that it leaves $B(h_o)$ invariant.
The main result of this section is the following

<u>Theorem 1</u>. The prescription

$$P_{s,t}(X) = j_s^{-1}\, \mathbb{E}_{s]}^{\omega}\, (U_{s,t}\, j_t(X)\, U_{s,t}^+) \quad \ldots(1.4.1)$$

where $X \in B(h_o)$ yields a completely positive evolution on $B(h_o)$ of the form

$$P_{s,t}(X) = \mathsf{T}\exp\{\int_s^t \mathscr{L}_\tau(X)\, \lambda_f(d_\tau)\} \quad \ldots(1.4.2)$$

where for each $t \in \mathbb{R}^+$

$$\begin{aligned}\mathscr{L}_t(X) = i[\mathscr{H}_t, X] &+ \frac{d\lambda_{cf}}{d\lambda_f}(t)\quad (F_tXF_t^+ - \frac{1}{2}\{F_tF_t^+, X\}) +\\ &+ \frac{d\lambda_{sf}}{d\lambda_f}(t)\quad (F_t^+XF_t - \frac{1}{2}\{F_t^+F_t, X\})\end{aligned} \quad \ldots(1.4.3)$$

<u>Remark</u>. The right hand side of (1.4.2) is a time-ordered exponential i.e.

$$\mathsf{T}\exp\{\int_s^t \mathscr{L}_\tau(X)\lambda_f(d_\tau)\} = I + \int_s^t \mathscr{L}_\tau(X)\lambda_f(d_\tau) + \int_s^t(\int_s^t \mathscr{L}_\tau(\mathscr{L}_\sigma(X))\lambda_f(d_\sigma))\lambda_f(d_\tau) + \ldots$$

If the above series converges, it uniquely determines the solution of the equation

$$dP_{s,t}(X) = P_{s,t}(\mathscr{L}_t(X))\lambda_f(dt) \qquad \text{with } P_{s,s}(X) = X\ .$$

Proof. $P_{s,t}$ is clearly completely positive since conditional expectation and unitary conjugation are both completely positive.

To show that $P_{s,t}$ is an evolution, let $r \leq s \leq t$, then $P_{r,s}(P_{s,t}(X))$

$$= j_r^{-1}\,\mathbb{E}^{\omega}_{r]}[U_{r,s}j_s(j_s^{-1}\,\mathbb{E}^{\omega}_{s]}(U_{s,t}j_t(X)U^+_{s,t}))U^+_{r,s}]$$

$$= j_r^{-1}\,\mathbb{E}^{\omega}_{r]}[U_{r,s}\,\mathbb{E}^{\omega}_{s]}(U_{s,t}j_t(X)U^+_{s,t})U^+_{r,s}] = j_r^{-1}\,\mathbb{E}^{\omega}_{r]}\,\mathbb{E}^{\omega}_{s]}(U_{r,s}U_{s,t}j_t(X)U^+_{s,t}U^+_{r,s})$$

$$= j_r^{-1}\,\mathbb{E}^{\omega}_{r]}(U_{r,t}j_t(X)U^+_{r,t}) = P_{r,t}(X)$$

where we have used (1.1.6), (1.3.4) and adaptedness of $U_{r,s}$ to $N_{s]}$.

It is easily seen from (1.4.1) that $P_{s,s}(X) = X$.

We compute $d(U_{s,t}j_t(X)U^+_{s,t})$

$$= dU_{s,t}j_t(X)U^+_{s,t} + U_{s,t}j_t(X)dU^+_{s,t} + dU_{s,t}j_t(X)dU^+_{s,t}$$

$$\begin{aligned} &= U_{s,t}\{[j_t(F_t)a_f(dt) - j_t(F_t^+)a_f^+(dt) + (ij_t(\mathscr{H}_t) - \tfrac{1}{2}\tfrac{d\lambda_{sf}}{d\lambda_f}(t)j_t(F_t^+F_t) \\ &- \tfrac{1}{2}\tfrac{d\lambda_{cf}}{d\lambda_f}(t)j_t(F_tF_t^+))\lambda_f(dt)]j_t(X) + j_t(X)[j_t(F_t^+)a_f^+(dt) - j_t(F_t)a_f(dt) + (-ij_t(\mathscr{H}_t) \\ &- \tfrac{1}{2}\tfrac{d\lambda_{sf}}{d\lambda_f}(t)j_t(F_t^+F_t) - \tfrac{1}{2}\tfrac{d\lambda_{cf}}{d\lambda_f}(t)j_t(F_tF_t^+))\lambda_f(dt)] + (\tfrac{d\lambda_{cf}}{d\lambda_f}(t)j_t(F_t^+XF_t) \\ &+ \tfrac{d\lambda_{sf}}{d\lambda_f}(t)j_t(F_tXF_t^+))\lambda_f(dt)\}U^+_{s,t} \end{aligned} \qquad \ldots(1.4.4)$$

where the last two terms arise from the evaluation of $dU_{s,t}j_t(X)dU^+_{s,t}$ using (1.2.7).

Take $\mathbb{E}^{\omega}_{s]}$ conditional expectations of both sides of (1.4.4) noting that e.g.

$$\begin{aligned} \mathbb{E}^{\omega}_{s]}(U_{s,t}j_t(F_t)a_f(dt)j_t(X)U^+_{s,t}) &= \mathbb{E}^{\omega}_{s]}\,\mathbb{E}^{\omega}_{t]}(U_{s,t}j_t(F_t)a_f(dt)j_t(X)U^+_{s,t}) \\ &= \mathbb{E}^{\omega}_{s]}(U_{s,t}j_t(F_t)\,\mathbb{E}^{\omega}_{t]}(a_f(dt))j_t(X)U^+_{s,t}) \quad \text{by (1.1.6)} \\ &= 0 \quad \text{since } \mathbb{E}^{\omega}_{t]}(a_f(dt)) = 0\ . \end{aligned}$$

So we obtain from (1.4.4)

$$d\,\mathbb{E}^{\omega}_{s]}(U_{s,t}j_t(X)U^+_{s,t}) = \mathbb{E}^{\omega}_{s]}(U_{s,t}j_t(\mathscr{L}_t(X))U^+_{s,t})\lambda_f(dt)$$

whence $dP_{s,t}(X) = P_{s,t}(\mathscr{L}_t(X))\lambda_f(dt)$

and the result follows by iteration (Convergence of the series of iterates is guaranteed by the given conditions on F_t and $\mathscr{H}_t$). □

The result of theorem 1 enables us to construct a <u>unitary dilation</u> in N of completely positive evolutions of the form (1.4.2) acting in $B(h_o)$, in the sense that the following diagram commutes for all $s,t \in \mathbb{R}^+$ with $s \leq t$.

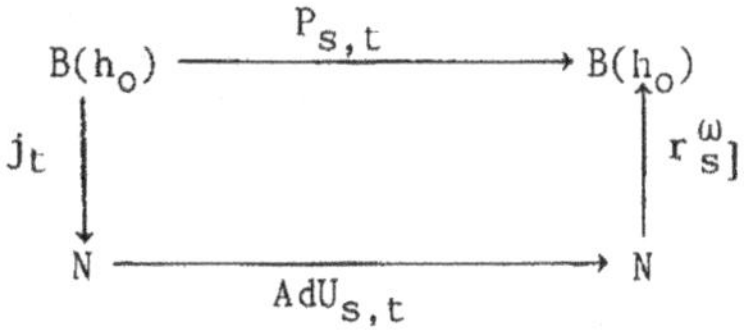

where $r^{\omega}_{s]}: N \to B(h_o)$, $s \in \mathbb{R}$, are a family of morphisms satisfying $j_s \circ r^{\omega}_{s]} = \mathbb{E}^{\omega}_{s]}$.

It is an open problem to find the most general class of completely positive evolutions on a type I factor for which such a unitary dilation exists. In this regard, it is interesting to compare (1.4.3) with the form found by Lindblad [Lind.] for the generator of any quantum dynamical semigroup on a von Neumann algebra.

Example 1.2 We suppose that the spectrum of T contains a discrete part and choose f to be an eigenvector of T. Now (1.2.3) yields

$$\lambda_{sf}(dt) = \sinh^2\gamma\ \lambda_f(dt) \qquad \lambda_{cf}(dt) = \cosh^2\gamma\ \lambda_f(dt)$$

for $\gamma \in \mathbb{R}^+ (\gamma \neq 0)$.
We may now obtain solutions to (1.3.1) with $F_\alpha(t)$ ($\alpha = 1,2,3$) all constant. Furthemore, (1.4.2) takes the form

$$P_{s,t}(X) = e^{\lambda_f[s,t]\mathscr{L}}(X)$$

where $\mathscr{L}(X) = i[\mathscr{H},X] + \cosh^2\gamma\,(FXF^+ - \frac{1}{2}\{F,F^+,X\}) + \sinh^2\gamma\,(F^+XF - \frac{1}{2}\{F^+F,X\})$
with $F, \mathscr{H} \in B(h_o)$ ...(1.4.5)

An important special case of example 1.2 is the following.

Example 1.3 [Hu Li] We adopt the formalism of example 1.1 and obtain in (1.2.3)

$$\lambda_{sf}(dt) = \sinh^2\gamma\ dt \qquad \lambda_{cf}(dt) = \cosh^2\gamma\ dt$$

whence $P_{o,t} = e^{t\mathscr{L}}(X)$ is a quantum dynamical semigroup with $\mathscr{L}$ given by (1.4.5).

(2) The C.A.R. Case

The analysis of the CAR case is similar to that of the CCR and we will shorten our exposition by concentrating on those areas where they differ. In particular, we will omit details of arguments and proofs where these are identical to the CCR case and retain the notation of § 1 whenever feasible.

(2.1) The CAR Algebra and Conditional Expectation

Let h be a complex pre-Hilbert space and let U(h) denote the C.A.R. C*-algebra over h generated by I together with $\{b(f),\ f \in h\}$ for which the map $f \to b(f)$ is antilinear and

$$\{b(f), b(g)\} = 0,\ \{b(f), b^+(g)\} = \langle f,g\rangle I \quad \text{for each } f,g \in h \qquad ...(2.1.1)$$

We denote by ρ, the *-automorphism of U(h) whose action on generators is given by

$$\rho(b(f)) = -b(f) \qquad ...(2.1.2)$$

Let ω be a gauge invariant, quasi-free state on U(h) whence [Bra Ro] there exists a positive, self adjoint operator R on $\bar{h}$ with $0 \leq R \leq 1$ and

$$\omega(b^+(f)b(g)) = \langle g, Rf\rangle \qquad \text{for all } f,g \in h \qquad ...(2.1.3)$$

We will assume that $S > 0$ and $S < 1$ so that ω has no Fock or anti-Fock parts.

Let $\Gamma_A(h)$ denote anti-symmetric Fock space over h, let B(f) and $B^+(g)$ be, respectively, annihilation and creation operators in $\Gamma_A(h)$ and $\Omega^A = (1,0,0,\ldots)$ be the Fock vacuum. Write $H^A = \Gamma_A(h) \otimes \Gamma_A(h)$ and $\psi^A = \Omega^A \otimes \Omega^A$.

There exists [Bra Ro] a representation Π_ω of $(U(h),\omega)$ on (H^A,ψ^A) for which

$$\Pi_\omega(b(f)) = B(\sqrt{I-T}\,f) \otimes I + \Phi \otimes B^+(J\sqrt{T}\,f)$$

where ϕ is the parity operator on $\Gamma_A(h)$ and J is an antilinear involution on h satisfying $\langle Jf,Jg\rangle = \langle g,f\rangle$.

Let E be a projector in h for which $[E,R] = 0$ and let D , $D^\perp$, $E^\perp$ be as in § 2.1. We write $L^A = \Gamma_A(D)\otimes\Gamma_A(D)$, $L_\perp^A = \Gamma_A(D^\perp)\otimes\Gamma_A(D^\perp)$, and $\psi_\perp^A = \Omega_\perp^A\otimes\Omega_\perp^A$ where $\Omega_\perp^A$ is the vacuum vector in $\Gamma_A(D^\perp)$.

The ω-compatible conditional expectation $\mathbb{E}_D^\omega$ from U(h) to U(D) has been constructed by Evans ([Eva] see also [Ba St Wi]) as follows: let Π denote the injective *-morphism from U(h) into $U(D)\otimes B(L_\perp^A)$ whose action on generators is given by

$$\Pi(b(f)) = b(Ef)\otimes\phi' + I\otimes\Pi_\omega(b^+(E^\perp f))$$

where ϕ' is the parity operator in $\Gamma_A(D^\perp)$.

Let $\xi : U(D)\times B(L_\perp^A)\to U(D)$ be the completely positive map given by continuous, linear extension of the prescription

$$\xi(A\otimes B) = \langle\psi_\perp^A , B\psi_\perp^A\rangle A \qquad \text{where } A\in U(D) \text{ and } B\in B(L_\perp^A), \text{ then}$$

$$\mathbb{E}_D^\omega = \xi\circ\Pi \qquad \ldots(2.1.4)$$

Since Π_ω is irreducible in H^A, we may extend $\mathbb{E}_D^\omega$ by weak continuity to a conditional expectation from $B(H^A)$ into $B(L^A)$.

(2.2) Itô Calculus in U(h)

As in (1.2) we require that $\{\ell_{t]}, t\in\mathscr{I}\}$ satisfy $[\ell_{t]},R] = 0$ for each $t\in\mathscr{I}$

Let $f\in h$ be fixed and define the stochastic differentials

$$b_f(dt) = b(\ell_{dt]}f) \qquad b_f^+(dt) = b^+(\ell_{dt]}f) \qquad \nu_f(dt) = \langle\ell_{dt]}f,f\rangle \quad \ldots(2.2.1)$$

We compute conditional expectations using (2.1.4) and find

$$\mathbb{E}_{t]}^\omega(b_f(dt)) = 0 = \mathbb{E}_{t]}^\omega(b_f(dt))$$

and $\mathbb{E}_{t]}^\omega$ vanishes on all products of pairs of stochastic differentials except

$$\mathbb{E}_{t]}^\omega(b_f^+(dt)b_f(dt)) = \omega(b_f^+(dt)b_f(dt)) = \langle\ell_{dt]}f,R\ell_{dt]}f\rangle \text{ by (2.1.3) and}$$

$$\mathbb{E}_{t]}^\omega(b_f(dt)b_f^+(dt)) = \langle\ell_{dt]}f,(I-R)\ell_{dt]}f\rangle \text{ by (2.1.1)}$$

Let $\quad \nu_{sf}(dt) = \langle\ell_{dt]}f,Rf\rangle \qquad \nu_{cf}(dt) = \langle\ell_{dt]}f,(I-R)f\rangle \qquad \ldots(2.2.2)$

Clearly $\quad \nu_{sf}(dt) + \nu_{cf}(dt) = \nu_f(dt) \qquad \ldots(2.2.3)$

The Borel measures ν_{sf} and ν_{cf} are each absolutely continuous with respect to ν_f and their Radon Nikodym derivatives satisfy

$$\frac{d\nu_{sf}}{d\nu_f}(t) + \frac{d\nu_{cf}}{d\nu_f}(t) = 1 \qquad \ldots(2.2.4)$$

So the rules for computing products of stochastic differentials are given by

	$b_f(dt)$	$b_f^+(dt)$	$\nu_f(dt)$
$b_f^+(dt)$	$\nu_{sf}(dt)$	0	0
$b_f(dt)$	0	$\nu_{cf}(dt)$	0
$\nu_f(dt)$	0	0	0

...(2.2.5)

The C.A.R. analogue of example 1.1 is the following

<u>Example</u> 2.1 . Let h, $\mathscr{I}$, f and $\ell_{t]}$ be as in example 1.1 and choose

$$T = \sin^2\beta\, I \qquad (\beta\in(0,\pi/2))$$

so that ω is e.u.i.

$b_f(dt)$ and $b_f^+(dt)$ are differentials of fermion Brownian motion of variance $\sigma^2 = \cos 2\beta$ [App], ν_f is Lebesgue measure and

$$\frac{d\nu_{cf}}{d\nu_f} = \cos^2\beta\ , \qquad \frac{d\nu_{sf}}{d\nu_f} = \sin^2\beta\ .$$

(2.3) Unitary Stochastic Differential Equations and Dilations

Let h_o be a complex, separable, $\mathbb{Z}_2$-graded Hilbert space i.e. $h_o = h_{o,+} \oplus h_{o,-}$ where $h_{o,+}$, $h_{o,-}$ are called the even and odd subspaces respectively. It follows that $B(h_o)$ is a $\mathbb{Z}_2$ graded algebra in the sense of [Chev] by the prescription that $T \in B(h_o)$ is even if $Th_{o,\pm} \subseteq h_{o,\pm}$ and odd if $Th_{o,\pm} \subseteq h_{o,\mp}$. Hence we may write $B(h_o) = B(h_o)_+ \oplus B(h_o)_-$ (*)
Let ρ_o be the automorphisms of $B(h_o)$ defined by linear extension of

$$\rho_o(X) = X \quad \text{when } X \in B(h_o)_+ \qquad \rho_o(X) = -X \quad \text{when } X \in B(h_o)_-$$

ρ_o is unitarily implementable on $B(h_o)$ by the self-adjoint unitary operator θ which acts as I on $h_{o,+}$ and $-I$ on $h_{o,-}$. Clearly $\theta^2 = I$.

We take $\mathcal{I} = \mathbb{R}^+$, define $M = B(h_o \otimes H)$ and associate a local structure to M by the prescription $M_{s,t} = B(h_o \otimes H^A_{s,t})$ where $H^A_{s,t} = \Gamma_A(\ell_{[s,t]}h) \otimes \Gamma_A(\ell_{[s,t]}h)$.

We define injections $j_t : B(h_o) \to M_{t]}$ by the prescription $j_t(X) = X \hat{\otimes} I_{t]}$, $X \in B(h_o)$ where $\hat{\otimes}$ denotes the "anticommuting tensor product" of Chevalley ([Chev], see also [Ap Hu] and [App]).

The map $\rho' = \rho_o \otimes \rho$ is an automorphism of period 2. We find that for arbitrary $Y_t \in M_{t]}$

$$b_f^{\#}(dt)Y_t = \rho'(Y_t)b_f^{\#}(dt) \qquad \ldots(2.3.1)$$

where $b_f^{\#}(dt)$ denotes $b_f(dt)$ or $b_f^+(dt)$.

We consider the stochastic differential equation in N given by

$$dU = \sum_{\alpha=1}^{3} U_t j_t(F_\alpha(t))dX_\alpha \ , \quad U_o = I \qquad \ldots(2.3.2)$$

where $dX_1 = b_f(dt)$, $dX_2 = b_f^+(dt)$, $dX_3 = \nu_f(dt)$.

We require conditions for each U_t to be unitary whence $d(U_t^+U_t) = 0$ and so

$$b_f^+(dt)j_t(F_1^+(t)) + b_f(dt)j_t(F_2^+(t)) + j_t(F_3^+(t))\nu_f(dt) + j_t(F_1(t))b_f(dt) + j_t(F_2(t))b_f^+(dt)$$
$$+ j_t(F_3(t))\nu_f(dt) + b_f^+(dt)j_t(F_1^+(t)F_1(t))b_f(dt) + b_f(dt)j_t(F_2^+(t)F_2(t))b_f^+(dt) = 0$$

So, by (2.2.5), (2.3.1) and the definition of j_t and ρ' we obtain the conditions

$$\left.\begin{aligned} &F_2^+(t) + \rho_o(F_1(t)) = 0 \\ &F_3^+(t) + F_3(t) + \frac{d\nu_{sf}}{d\nu_f}(t)\rho_o(F_1^+(t)F_1(t)) + \frac{d\nu_{cf}}{d\nu_f}(t)\rho_o(F_2^+(t)F_2(t)) = 0 \end{aligned}\right\} \quad \ldots(2.3.3)$$

We write $F_1(t) = F_t$ and let $\mathcal{H}_t = \mathcal{H}_t^+ \in B(h_o)$ for each $t \in \mathbb{R}^+$. The general solution of (2.3.3) is given by

$$\begin{aligned} F_2(t) &= -\rho_o(F_t^+) \\ F_3(t) &= i\mathcal{H}_t - \frac{1}{2}\frac{d\nu_{sf}}{d\nu_f}(t)\rho_o(F_t^+F_t) - \frac{1}{2}\frac{d\nu_{cf}}{d\nu_f}(t)F_tF_t^+ \end{aligned} \qquad \ldots(2.3.4)$$

The formula $U_{s,t} = U_s^+U_t$ $(s \leq t)$ again yields on evolution adapted to $M_{s,t}$ with

(*) N.B. $B(h_o)_+ \neq B(h_{o,+})$

$$dU_{s,t} = U_{s,t}(j_t(F_t)b_f(dt) - j_t(\rho_o(F_t^+))b_f^+(dt) + (ij_t(\mathscr{H}_t) - \frac{1}{2}\frac{d\nu_{sf}}{d\nu_f}(t)j_t(\rho_o(F_t^+F_t) - \frac{1}{2}\frac{d\nu_{cf}}{d\nu_f}(t)j_t(F_tF_t^+))\nu_f(dt)) \qquad \ldots(2.3.5)$$

Theorem 2. $P_{s,t}(X) = j_s^{-1}\mathbf{E}_{s]}^{\omega}(U_{s,t}j_t(X)U_{s,t}^+)$, for $X \in B(h_o)$, yields a completely positive evolution on $B(h_o)$ of the form

$$P_{s,t}(X) = \mathsf{T}\exp(\int_s^t \mathscr{L}_\tau(X)\nu_f(d\tau))$$

with, for $t \in \mathbb{R}^+$.

$$\mathscr{L}_t(X) = i[\mathscr{H}_t, X] + \frac{d\nu_{cf}}{d\nu_f}(t)(F_t\theta X\theta F_t^+ - \frac{1}{2}\{F_tF_t^+, X\}) + \frac{d\nu_{sf}}{d\nu_f}(t)(\theta F_t^+XF_t\theta - \frac{1}{2}\{\theta F_t^+F_t\theta, X\}) \qquad \ldots(2.3.5)$$

Proof A similar argument to that of theorem 1 shows that $P_{s,t}$ is a completely positive evolution on $B(h_o)$. By stochastic differentiation, we obtain:

$$d(U_{s,t}XU_{s,t}^+)$$
$$= U_{s,t}\{[j_t(F_t)b_f(dt) - j_t(\rho_o(F_t^+))b_f^+(dt) + (ij_t(\mathscr{H}_t) - \frac{1}{2}\frac{d\nu_{sf}}{d\nu_f}(t)j_t(\rho_o(F_t^+F_t))$$
$$- \frac{1}{2}\frac{d\nu_{cf}}{d\nu_f}(t)j_t(F_tF_t^+))\nu_f(dt)]j_t(X) + j_t(X)[b_f^+(dt)j_t(F_t^+) - b_f(dt)j_t(\rho_o(F_t)) + (-ij_t(\mathscr{H}_t)$$
$$- \frac{1}{2}\frac{d\nu_{sf}}{d\nu_f}(t)j_t(\rho_o(F_t^+F_t) - \frac{1}{2}\frac{d\nu_{cf}}{d\nu_f}(t)j_t(F_tF_t^+))\nu_f(dt)] + j_t(F_t)b_f(dt)j_t(X)b_f^+(dt)j_t(F_t^+) +$$
$$j_t(\rho_o(F_t^+)b_f^+(dt)j_t(X)b_f(dt)j_t(\rho_o(F_t))\}U_{s,t}^+ \qquad \ldots(2.3.6)$$

We observe that

$$\begin{aligned} j_t(F_t)b_f(dt)j_t(X)b_f^+(dt)j_t(F_t^+) &= j_t(F_t)\rho'(j_t(X))b_f(dt)b_f^+(dt)j_t(F_t^+) && \text{by (2.3.1)} \\ &= \frac{d\nu_{cf}}{d\nu_f}(t)j_t(F_t\rho_o(X)F_t^+)\nu_f(dt) && \text{by (2.2.5)} \\ &= \frac{d\nu_{cf}}{d\nu_f}(t)j_t(F_t\theta X\theta F_t^+)\nu_f(dt) && \ldots(2.3.7) \end{aligned}$$

and $\quad j_t(\rho_o(F_t^+)b_f^+(dt)j_t(X)b_f(dt)j_t(\rho_o(F_t))$

$$\begin{aligned} &= j_t(\rho_o(F_t^+))\rho'(j_t(X))b_f^+(dt)b_f(dt)j_t(\rho_o(F_t)) && \text{by (2.3.1)} \\ &= \frac{d\nu_{sf}}{d\nu_f}(t)j_t(\rho_o(F_t^+XF_t))\nu_f(dt) && \text{by (2.2.5)} \\ &= \frac{d\nu_{sf}}{d\nu_f}(t)j_t(\theta F_t^+XF_t\theta)\nu_f(dt) && \ldots(2.3.8) \end{aligned}$$

We substitute (2.3.7) and (2.3.8) into (2.3.6). The result now follows by a similar argument to that of theorem 1. □

C.A.R. analogues of examples 1.2 and 1.3 are easily established via the correspondence

$$\sin\gamma \to \sinh\beta \quad \cosh\gamma \to \cosh\beta \qquad F \to F\theta \quad F^+ \to \theta F^+$$

As in the CCR case, we see that $U_{s,t}$ yields a unitary dilation of the evolution $P_{s,t}$. We take a trivial $\mathbb{Z}_2$-grading on h_o i.e. $h_{o,+} = h_o$, $h_{o,-} = \{0\}$ and choose F_t and $\mathscr{H}_t$ to be the same in (1.4.3) and (2.3.5). It follows that the evolution $P_{s,t}$ possesses two distinct unitary dilations into the von Neumann algebras M and N, respectively. The existence of other unitary dilations remains an open problem.

Acknowledgement: I am grateful to Luigi Accardi for many helpful and stimulating discussions and to Robin Hudson for valuable comments.

References

[Ac Pa 1] L.Accardi and K.R.Parthasarathy, Stochastic Calculus on Local Algebras (these proceedings).

[Ac Pa 2] L.Accardi and K.R.Parthasarathy, Stochastic Calculus on Local Algebras (preprint).

[App] D.Applebaum, Fermion Stochastic Calculus, Univ. of Nottingham Ph.D. thesis (1984).

[Ap Hu] D.Applebaum and R.L.Hudson, Fermion Itô's Formula and Stochastic Evolutions (to appear in Commun. Math. Phys.)

[Ba St Wi] C.Barnett, R.F.Streater and I.F.Wilde, Quasi-Free Quantum Stochastic Integrals for the CAR and CCR, J. Func. Anal., 52, 19, (1983).

[Bra Ro] O.Bratteli and D.W.Robinson, Operator Algebras and Quantum Statistical Mechanics II, Springer-Verlag (New York) (1979).

[Chev] C.Chevalley, The Construction and Study of Certain Important Algebras, Publ. Math. Soc. Japan I (1955)

[Co Hu] A.Cockcroft and R.L.Hudson, Quantum Mechanical Wiener Processes, J. Multivariate Anal. 7, 107 (1978).

[Eva] D.E.Evans, Completely Positive Quasi-Free Maps on the CAR Algebra , Commun. Math. Phys., 70, 50 (1979).

[Hu Li] R.L.Hudson and J.M.Lindsay, The Classical Limit of Reduced Quantum Stochastic Evolutions, (preprint).

[Hu Pa] R.L.Hudson and K.R.Parthasarathy, Quantum Itô's Formula and Stochastic Evolutions, Commun. Math. Phys., 93, 301, (1984).

[Lind] G.Lindblad, On the Generators of Quantum Dynamical Semigroups, Commun. Math. Phys., 48, 119, (1976).

[Nel] E.Nelson, Quantum Fields and Markoff Fields, Amer. Math. Soc. Summer Institute on Partial Differential Equations held in Berkeley, 1971.

DILATIONS OF OPERATION VALUED STOCHASTIC PROCESSES

A. BARCHIELLI and G. LUPIERI

Dipartimento di Fisica dell'Università di Milano
Istituto Nazionale di Fisica Nucleare, Sezione di Milano
Via Celoria ,16 - 20133 Milan - Italy

1. Operation valued stochastic processes.

In the last twenty years a very flexible formulation of quantum mechanics has been developed, based on the notions of effect and operation (and of effect and operation valued measures) which generalize the concept of observable and the Von Neumann reduction postulate (see the contribution by J.T. Lewis in this volume and ref.[1]). In this framework continual measurements can be consistently introduced in quantum mechanics. The first works on continual measurements are due to Davies, who treated the theory of counting processes under the name of "quantum stochastic processes" (a complete list of references is given in [1-3]). Later, a more general formulation of the problem of continual measurements was obtained by introducing the notion of operation valued stochastic process (OVSP) [4-6].

The most general setup for treating continually measured quantities is that of generalized stochastic processes [7]. Let $\mathcal{D}$ be the nuclear space of the n-component, real C^∞-functions $\vec{h}(t)$ on $\mathbb{R}$ with compact support. Let $\mathcal{D}'$ be the topological dual space of $\mathcal{D}$; for $x \in \mathcal{D}'$ and $h \in \mathcal{D}$, we denote by x_h the distribution x applied to the test function h. The subsets of $\mathcal{D}'$ of the form $\{x \in \mathcal{D}' : (x_{h^{(1)}}, \dots, x_{h^{(s)}}) \in B\}$, where B is a Borel subset of $\mathbb{R}^s$, are called cylinder sets. We equipe $\mathcal{D}'$ with the family of σ-algebras $\{\Sigma_{(t_1,t_2)}, t_1, t_2 \in \mathbb{R}, t_1 < t_2\}$, where $\Sigma_{(t_1,t_2)}$ is the σ-algebra generated by the cylinder sets defined by test functions with supports contained in the time interval (t_1,t_2). The space $\mathcal{D}'$ is interpreted as the set of the a priori possible trajectories for the continually measured quantities; $x \in N$ $(N \in \Sigma_{(t_1,t_2)})$ represents a possible outcome of a measurement in the time interval (t_1,t_2).

For any Hilbert space $\mathcal{h}$, let us denote by $T(\mathcal{h})$ the Banach space of trace-class operators on $\mathcal{h}$ and by $B(\mathcal{h})$ the Banach space of bounded operators on $\mathcal{h}$. An OVSP on $\mathcal{h}$ [4] (where $\mathcal{h}$ is the Hilbert space associated with the measured system) is defined to be a family $\{\mathcal{F}(t_2,t_1;N), t_2 > t_1, N \in \Sigma_{(t_1,t_2)}\}$ of linear maps from $T(\mathcal{h})$ into itself with the following properties:

i) $\mathcal{F}(t_2,t_1;N)$, $N \in \Sigma_{(t_1,t_2)}$ is completely positive;

ii) $\mathcal{F}(t_2,t_1;\cdot)$ is strongly σ-additive on $\Sigma_{(t_1,t_2)}$;

iii) $\mathcal{F}(t_2,t_1,\mathcal{D}')$ is trace preserving;
iv) the following Markov property holds

$$\mathcal{F}(t_3,t_2;M)\mathcal{F}(t_2,t_1;N)=\mathcal{F}(t_3,t_1;M\cap N)\ ,\ \forall N\in\Sigma_{(t_1,t_2)},\forall M\in\Sigma_{(t_2,t_3)},\ t_1<t_2<t_3 . \tag{1.1}$$

By properties i)-iii), $\mathcal{F}(t_2,t_1,\cdot)$ is an operation valued measure on $\Sigma_{(t_1,t_2)}$ for any time interval of measurement (t_1,t_2). If ρ is the statistical operator for the system at time t_1, then the quantity

$$P(N|\rho,t_1):=\mathrm{Tr}\,\{\mathcal{F}(t_2,t_1;N)(\rho)\} \tag{1.2}$$

is the probability of finding the result $x\in N$ in the measurement interval (t_1,t_2). Property iv) ensures the consistency of measurements referring to different time intervals. Roughly speaking, a measurement in the time interval (t_1,t_2) followed by a measurement in (t_2,t_3) gives the same result as a measurement in (t_1,t_3).

An OVSP is said to be time-translation invariant if

$$\mathcal{F}(t_2,t_1;N)=\mathcal{F}(t_2+\bar{t},t_1+\bar{t};N_{\bar{t}})\quad ,\forall\bar{t}\in\mathbb{R}\ ,\forall N\in\Sigma_{(t_1,t_2)}\ , \tag{1.3}$$

where $N_{\bar{t}}\in\Sigma_{(t_1+\bar{t},\,t_2+\bar{t})}$ is obtained from N by time-translation; time-translation in $\mathcal{D}'$ is naturally defined by duality starting from the time shifting in $\mathcal{D}$

$$\vec{\varphi}\xrightarrow{\bar{t}}\vec{\varphi}^{(\bar{t})}\ ,\quad \vec{\varphi}^{(\bar{t})}(t)=\varphi(t-\bar{t}) . \tag{1.4}$$

The analog of the characteristic functional of a generalized stochastic process [7] is the notion of <u>characteristic operator</u> $\mathcal{G}(t_2,t_1;[\vec{\varphi}])$ of an OVSP, defined as the mean value of $\exp(ix_\varphi)$, $\vec{\varphi}\in\mathcal{D}_{(t_1,t_2)}$, with respect to the operation valued measure $\mathcal{F}(t_2,t_1;\ \cdot\ )$; $\mathcal{D}_{(t_1,t_2)}$ is the subspace of $\mathcal{D}$ of the functions with supports contained in (t_1,t_2). An OVSP is uniquely determined by its characteristic operator as stated by the following result:

<u>Theor. 1.1.</u> The family of bounded linear maps $\{\mathcal{G}(t_2,t_1;[\vec{\varphi}]),\ t_1<t_2,\ \vec{\varphi}\in\mathcal{D}_{(t_1,t_2)}\}$ from $T(\mathcal{h})$ into itself is the characteristic operator of a time-translation invariant OVSP if and only if
a) it is normalized, i.e.

$$\mathrm{Tr}\,\{\mathcal{G}(t_2,t_1;[\vec{0}])(X)\}=\mathrm{Tr}\,\{X\}\quad ,\forall X\in T(\mathcal{h})\ ; \tag{1.5}$$

b) it is of completely positive type, i.e. the quantities

$$\sum_{i,j=1}^{n} \alpha_i^* \mathcal{G}(t_2,t_1;[\vec{\varphi}_i-\vec{\varphi}_j])\,\alpha_j$$

are completely positive for any choice of the integer n, of the complex numbers α_i and of the test functions $\vec{\varphi}_i(t)$;

c) $\mathcal{G}(t_2,t_1;[\vec{\varphi}])$ is strongly continuous in $\vec{\varphi}$ $(\vec{\varphi}\in\mathcal{D}_{(t_1,t_2)})$;

d) the following composition law holds: $\forall\vec{\varphi}_1\in\mathcal{D}_{(t_1,t_2)}$, $\forall\vec{\varphi}_2\in\mathcal{D}_{(t_2,t_3)}$,

$$\mathcal{G}(t_3,t_2;[\vec{\varphi}_2])\,\mathcal{G}(t_2,t_1;[\vec{\varphi}_1]=\mathcal{G}(t_3,t_1;[\vec{\varphi}_1+\vec{\varphi}_2]) \qquad ,t_1<t_2<t_3; \tag{1.6}$$

e) it is time-translation invariant, i.e.

$$\mathcal{G}(t_2+\bar{t},\, t_1+\bar{t};[\vec{\varphi}^{(\bar{t})}])=\mathcal{G}(t_2,t_1;[\vec{\varphi}]) \qquad ,\ \forall\vec{\varphi}\in\mathcal{D}_{(t_1,t_2)}\ , \tag{1.7}$$

where $\vec{\varphi}^{(\bar{t})}$ is given by eq. (1.4).

A demonstration of this result, though not completely rigorous, can be found in ref. [4] .

The quantity

$$\mathcal{G}(t_2-t_1):=\mathcal{F}(t_2,t_1;\mathcal{Y}')\equiv\mathcal{G}(t_2,t_1;[\vec{0}]) \tag{1.8}$$

turns out to be a quantum dynamical semigroup; as $\mathcal{G}(t_2-t_1)(\rho)$ represents the state of the system at time t_2, when no selection is made in the time interval (t_1,t_2), $\mathcal{G}(t)$ gives the dynamics of the system subject to continual measurement.

The recent results about unitary dilations of dynamical semigroups by means of the powerful techniques of quantum stochastic calculus [8,9] have suggested to us the study of analogous dilations of OVSPs. What one wants to obtain in this way is

1) to construct a class of OVSPs, more general of that ones of refs. [1-3] (counting processes, analogous to classical Poisson processes) and of ref. [4,5] (continual measurements with "Gaussian" instruments);

2) to find models for measured system + measuring apparatus + environment such that the dynamics for the global system be unitary;

3) to obtain a dilation of the whole operation valued measure $\mathcal{F}(t_2,t_1;N)$, which could be useful for a better understanding of the theory of OVSPs.

2. Construction of operation valued stochastic processes.

We first introduce a continuous tensor product structure. Let $\mathcal{L}(-\infty,+\infty)\equiv\Gamma(L^2_k(-\infty,+\infty))$ be the symmetric Fock space over the space of the square-

-integrable functions taking values in a complex separable Hilbert space k. We denote by $\psi(f)$, $f \in L^2_k(-\infty,+\infty)$, the exponential vectors in the Fock space

$$\psi(f) = (1, f, \dots, (n!)^{-\frac{1}{2}} f \otimes \dots \otimes f, \dots) . \tag{2.1}$$

Introducing similarly the spaces $\mathcal{L}(s,t) \equiv \Gamma(L^2_k(s,t))$ for $s < t$, one has the canonical identification

$$\mathcal{L}(-\infty,+\infty) = \mathcal{L}(-\infty,s) \otimes \mathcal{L}(s,t) \otimes \mathcal{L}(t,+\infty) \tag{2.2}$$

in which, for $f_1 \in L^2_k(-\infty,s)$, $f_2 \in L^2(s,t)$, $f_3 \in L^2(t,+\infty)$,

$$\psi(f_1 \oplus f_2 \oplus f_3) = \psi(f_1) \otimes \psi(f_2) \otimes \psi(f_3) . \tag{2.3}$$

Using the structure (2.2), we introduce the W*-algebras

$$\mathcal{B}(s,t) := \mathbb{1} \otimes B(\mathcal{L}(s,t)) \otimes \mathbb{1} . \tag{2.4}$$

The strongly continuous one-parameter group of unitary time shifting operators $\{S_\tau, \tau \in R\}$ [10],

$$S_\tau \psi(f) = \psi(f_\tau) \quad , \quad f_\tau(t) = f(t+\tau) , \tag{2.5}$$

leaves the vacuum invariant and translates the algebras $\mathcal{B}(s,t)$

$$S_\tau \psi(0) = \psi(0) , \tag{2.6}$$

$$S_\tau^+ \mathcal{B}(s,t) S_\tau = \mathcal{B}(s+\tau, t+\tau) . \tag{2.7}$$

Let now h be another Hilbert space and consider $h \otimes \mathcal{L}(-\infty,+\infty)$. Any operator M in h (resp. N in $\mathcal{L}(-\infty,+\infty)$) will be identified with $M \otimes \mathbb{1}_{\mathcal{L}}$ (resp. $\mathbb{1}_h \otimes N$) whenever convenient. The vacuum conditional expectation map E_o: $B(h \otimes \mathcal{L}(-\infty,+\infty)) \to B(h)$ is defined by

$$\langle u, E_o(J) v \rangle = \langle u \otimes \psi(0), J\, v \otimes \psi(0) \rangle \quad , u,v \in h \; , J \in B(h \otimes \mathcal{L}(-\infty,+\infty)) . \tag{2.8}$$

Identifying h with the Hilbert space of the measured system, dilations of OVSPs can be constructed in $h \otimes \mathcal{L}(-\infty,+\infty)$; the main ingredients are two suitable families of unitary operators $U(t,s)$ and $V[\vec{\varphi}]$. $\{U(t,s),\ s \le t,\ s,t \in R\}$ is a covariant adapted unitary evolution [10,11], i.e. a two-parameter family of unitary operators on $h \otimes \mathcal{L}(-\infty,+\infty)$ with the properties

$$S_\tau^+ U(t,s) S_\tau = U(t+\tau, s+\tau) , \tag{2.9a}$$

U(t,s) strongly continuous in s and t, (2.9b)

$$U(t,s) \in B(h) \otimes \mathcal{B}(s,t) , \tag{2.9c}$$

$$U(t,s) U(s,r) = U(t,r) \quad , \quad r \le s \le t . \tag{2.9d}$$

$\{V[\vec{\varphi}], \vec{\varphi} \in \mathcal{D}\}$ is a family of unitary operators on $\mathcal{L}(-\infty,+\infty)$ with the properties

$$V^+[\vec{\varphi}] = V[-\vec{\varphi}] , \tag{2.10a}$$

$$V[\vec{\varphi}_1 + \vec{\varphi}_2] = V[\vec{\varphi}_1] V[\vec{\varphi}_2] , \tag{2.10b}$$

$$V[\vec{\varphi}] \in \mathcal{B}(t_1,t_2) \quad if \quad supp(\vec{\varphi}) \subset (t_1,t_2) , \tag{2.10c}$$

$V[\vec{\varphi}]$ strongly continuous in $\vec{\varphi}$, (2.10d)

$$S_{\bar{t}} V[\vec{\varphi}^{(\bar{t})}] S_{\bar{t}}^+ = V[\vec{\varphi}] , \tag{2.10e}$$

where $supp(\vec{\varphi}) = \bigcup_{s=1}^{n} supp(\varphi_s)$. Remark that these properties guarantee that $L[\vec{\varphi}] := \langle \psi(0), V[\vec{\varphi}] \psi(0) \rangle$ is the characteristic functional of a (classical) generalized stochastic process with independent values at every point [7]; in particular, by eq. (2.10c) we have $L[\vec{\varphi}_1 + \vec{\varphi}_2] = L[\vec{\varphi}_1] L[\vec{\varphi}_2]$ for $supp(\vec{\varphi}_1) \cap supp(\vec{\varphi}_2) = \emptyset$.

Theor. 2.1 [6] . Let $V[\vec{\varphi}]$ and U(t,s) be two families of unitary operators with properties (2.9) and (2.10). We define the operator $\mathcal{G}'(t,t_0;[\vec{\varphi}])$ for $supp(\vec{\varphi}) \subset (t_0,t)$, from B(h) into itself, by

$$\mathcal{G}'(t,t_0;[\vec{\varphi}])(X) = E_0\left(U(t,t_0)^+ X V[\vec{\varphi}] U(t,t_0)\right) \quad , \forall X \in B(h). \tag{2.11}$$

Then, $\mathcal{G}'(t,t_0;[\vec{\varphi}])$ is the adjoint of a bounded operator $\mathcal{G}(t,t_0;[\vec{\varphi}])$ on T(h) satisfying conditions a)-e) of Theor. 1.1; therefore it defines a time-translation invariant OVSP.

Proof (sketch):

From the definition (2.11), it follows that $\mathcal{G}'(t,t_0;[\vec{\varphi}])$ is a linear operator on $B(\mathfrak{h})$, continuous in the ultraweak topology [1]. This guarantees that it is the adjoint of a bounded linear operator $\mathcal{J}(t,t_0;[\vec{\varphi}])$ from $T(\mathfrak{h})$ into itself.

Properties a), b) and e) follow by direct verification; for property a) the fact that $V[\vec{0}]=\mathbb{1}$ (which follows from eqs. (2.10) and unitarity) is used, for property b) eq. (2.10b) is used and for property e) eq. (2.10e).

Property c) can be easily proved by using the fact that $V[\vec{\varphi}_\alpha]Y \to V[\vec{\varphi}]Y$ in the trace norm, $\forall Y \in T(\mathfrak{h} \otimes \mathcal{L}(-\infty,+\infty))$, for every net $\{\vec{\varphi}_\alpha, \alpha \in I\}$ in $\mathcal{D}(t_1,t_2)$ converging to $\vec{\varphi}$ (see for instance ref.[1]).

Note that, for $C_1 \in \mathcal{B}(t_1,t_2)$, $C_2 \in \mathcal{B}(t_2,t_3)$, $t_1<t_2<t_3$, the equation

$$\langle \psi(0), C_1 C_2 \psi(0)\rangle = \langle \psi(0), C_1 \psi(0)\rangle \langle \psi(0), C_2 \psi(0)\rangle$$

holds; then using eqs. (2.9c), (2.9d), (2.10c), we have

$$\mathcal{J}'(t_2+\bar{t}, t_1+\bar{t}; [\vec{\varphi}^{(\bar{t})}])(X) = E_0\left(U(t_2+\bar{t}, t_1+\bar{t})^+ X\, V[\vec{\varphi}^{(\bar{t})}]\, U(t_2+\bar{t}, t_1+\bar{t})\right) =$$

$$= E_0\left(S_{\bar{t}}^+ U(t_2,t_1)^+ X\, S_{\bar{t}}\, V[\vec{\varphi}^{(\bar{t})}]\, S_{\bar{t}}^+\, U(t_2,t_1)\, S_{\bar{t}}\right) =$$

$$= E_0\left(U(t_2,t_1)^+ X\, V[\vec{\varphi}]\, U(t_2,t_1)\right) = \mathcal{J}'(t_2,t_1;[\vec{\varphi}])(X)\ ;$$

therefore, property d) holds. ■

A very powerful tool for constructing covariant adapted unitary evolutions [10] is <u>quantum stochastic calculus</u> of Hudson and Parthasarathy. Consider now for simplicity the Fock space over the space of square-integrable functions with values in $\mathbb{C}^N$, so that our global Hilbert space becomes $\mathfrak{h} \otimes \mathcal{L}(-\infty,+\infty) \equiv \mathfrak{h} \otimes \Gamma(\mathbb{C}^N \otimes L^2(\mathbb{R}))$. In this space <u>quantum stochastic differential equations</u> [9] can be introduced in terms of N independent quantum Brownian motions $A_t^{(j)}$, $A_t^{(j)+}$ and gauge processes $\Lambda_t^{(j)}$.

Consider the following two stochastic differential equations

$$dU(t,t_0) = \Big\{\sum_{j=1}^{N}[-R_j^+\, dA_t^{(j)} + R_j\, dA_t^{(j)+} - \tfrac{1}{2}R_j^+R_j\, dt] - iH\,dt\Big\}\, U(t,t_0), \quad (2.12)$$

$$dV(t,t_0;[\vec{\varphi}]) = \Big\{\sum_{j=1}^{M}(e^{i\vec{\alpha}^{(j)}\cdot\vec{\varphi}(t)}-1)\, d\Lambda_t^{(j)} + \sum_{j=M+1}^{N}[i\vec{\alpha}^{(j)}\cdot\vec{\varphi}(t)\,(dA_t^{(j)} + dA_t^{(j)+}) -$$

$$- \tfrac{1}{2}(\vec{\alpha}^{(j)}\cdot\vec{\varphi}(t))^2 dt] + i\vec{\varepsilon}\cdot\vec{\varphi}(t)\, dt\Big\}\, V(t,t_0;[\vec{\varphi}]), \quad (2.13)$$

where $M \le N$, $R_j, H \in B(\mathfrak{h})$, $H=H^+$, $\vec{\varphi} \in \mathcal{D}$, $\vec{\alpha}^{(j)} \in \mathbb{R}^n$, $\vec{\varepsilon} \in \mathbb{R}^n$.

The solution of these two equations with initial conditions

$$U(t_0,t_0)=\mathbb{1} \quad , \quad V(t_0,t_0;[\vec{\varphi}])=\mathbb{1} \tag{2.14}$$

s unique [9]; $U(t_2,t_1)$ is a covariant adapted unitary evolution [10] and it s easy to prove [6] that

$$V[\vec{\varphi}]=\lim_{\substack{t\to+\infty\\ t_0\to-\infty}} V(t,t_0;[\vec{\varphi}]) \tag{2.15}$$

satisfies all conditions (2.10); therefore, through eq. (2.11) we construct an OVSP.

Using the rules of quantum stochastic calculus one obtains for $\xi(\ldots)$ the following differential equation [6]

$$\frac{\partial}{\partial t}\,\xi(t,t_0;[\vec{\varphi}]) = \mathcal{K}(\vec{\varphi}(t))\,\xi(t,t_0;[\vec{\varphi}]) \ , \tag{2.16}$$

$$\mathcal{K}(\vec{\varphi})(\rho) = \mathcal{L}(\rho)+\sum_{j=1}^{M} R_j\rho R_j^+ \left(e^{i\vec{\alpha}^{(j)}\cdot\vec{\varphi}}-1\right)+ \\ +\sum_{j=M+1}^{N}\left[(R_j\rho+\rho R_j^+)\, i\vec{\alpha}^{(j)}\cdot\vec{\varphi}-\tfrac{1}{2}(\vec{\alpha}^{(j)}\cdot\vec{\varphi})^2\rho\right]+i\vec{z}\cdot\vec{\varphi}\,\rho \ , \tag{2.17}$$

$$\mathcal{L}(\rho)=\sum_{j=1}^{N}\left(R_j\rho R_j^+-\tfrac{1}{2}R_j^+R_j\rho-\tfrac{1}{2}\rho R_j^+R_j\right)-i[H,\rho] \ . \tag{2.18}$$

Here, $\mathcal{L}$ is the generator of the quantum dynamical semigroup (1.7) and $\mathcal{K}(\vec{\varphi})$ is the generator of the characteristic operator. In the OVSP defined by the characteristic operator constructed in this way one can identify a "Poisson contribution" (due to the terms with j=1,...,M in $\mathcal{K}(\vec{\varphi})$) and a "Gaussian contribution" (due to the terms with j=M+1,...,N in $\mathcal{K}(\vec{\varphi})$). OVSPs of pure Poisson type correspond [6] to the counting processes of Davies and Srinivas [1-3], whilst the class of OVSPs of pure Gaussian type was found in ref.[4]. Obviously from eq. (2.11) one can obtain more general examples of OVSPs; for instance, one can have a characteristic operator satisfying eq. (2.16) with a generator containing also time derivatives of $\vec{\varphi}(t)$.

3. The dilation.

In the previous section we have constructed a certain class of OVSPs on $\mathcal{h}$ by giving a dilation for them on $\mathcal{h}\otimes\mathcal{L}(-\infty,+\infty)$. Therefore, we can say that we have a global system composed by the measured system with Hilbert space $\mathcal{h}$ interacting with an auxiliary system with Hilbert space $\mathcal{L}(-\infty,+\infty)$. Now we want to show the main features of this dilation, which can be so summarized: 1) on the global system the continual measurement is represented by a projection valued measure, as in the standard formulation of quantum mechanics;

2) the measured observables are operators on $\mathcal{L}(-\infty,+\infty)$;

3) the dynamics of the global system is given by a one-parameter group of unitary operators.

By the last two points, the global system can be considered as isolated and the auxiliary system can be interpreted as a rough model of measuring apparatus plus environment.

Let us divide the interval of measurement (t_0,t_f) in subintervals of amplitude τ $(t_r:=t_0+r\tau,\ t_f=t_m\equiv t_0+m\tau)$ and consider the time averages $\bar{x}_i^{(r)}:= = \frac{1}{\tau}\int_{t_{r-1}}^{t_r} dt\, x_i(t)$ of the continually measured quantities $\vec{x}(t)$. First we have

$$P(\vec{\bar{x}}^{(1)}\in B_1,\dots,\vec{\bar{x}}^{(m)}\in B_m\,|\,\rho,t_0)=\omega\Big(U(t_f,t_0)^+\Big(\prod_{r=1}^{m}P_{(t_{r-1},t_r)}(B_r)\Big)U(t_f,t_0)\Big), \tag{3.1}$$

where B_r is a Borel set in $\mathbb{R}^n$, $\omega=\rho\circ E_0$, $\rho(\cdot)=Tr_h(\cdot\,\rho)$ and $P_{(s,t)}(B)$ is the projection valued measure

$$P_{(s,t)}(B):=\int_B d_n\vec{x}\,\frac{1}{(2\pi)^n}\int d_n\vec{k}\,e^{-i\vec{k}\cdot\vec{x}} \times$$

$$\times\exp\Big\{\frac{i}{t-s}\sum_{i=1}^{n}k_i\int_s^t\Big[\sum_{j=1}^{M}\alpha_i^{(j)}d\Lambda_{t'}^{(j)}+\sum_{j=M+1}^{N}\alpha_i^{(j)}(dA_{t'}^{(j)}+dA_{t'}^{(j)+})+c_i\,dt'\Big]\Big\}. \tag{3.2}$$

The formal justification of this result is as follows. We have

$$P(\vec{\bar{x}}^{(1)}\in B_1,\dots,\vec{\bar{x}}^{(m)}\in B_m|\rho,t_0)=\int_{B_1\times\dots\times B_m}d_n\vec{x}^{(1)}\dots d_n\vec{x}^{(m)}\frac{1}{(2\pi)^{nm}}\int d_n\vec{k}^{(1)}\dots d_n\vec{k}^{(m)}\,e^{-i\sum_{r=1}^{m}\vec{k}^{(r)}\cdot\vec{x}^{(r)}}$$

$$\times Tr_h\Big\{\mathcal{G}\big(t_f,t_0;\Big[\sum_{r=1}^{m}\frac{1}{\tau}\vec{k}^{(r)}\chi_{[t_{r-1},t_r]}\Big]\big)(\rho)\Big\}=\int_{B_1\times\dots\times B_m}d_n\vec{x}^{(1)}\dots d_n\vec{x}^{(m)} \tag{3.3}$$

$$\times\frac{1}{(2\pi)^{nm}}\int d_n\vec{k}^{(1)}\dots d_n\vec{k}^{(m)}\,e^{-i\sum_{r=1}^{m}\vec{k}^{(r)}\cdot\vec{x}^{(r)}}\,\omega\Big(U(t_f,t_0)^+\Big(\prod_{r=1}^{m}V\big[\tfrac{1}{\tau}\vec{k}^{(r)}\chi_{[t_{r-1},t_r]}\big]\Big)U(t_f,t_0)\Big).$$

The first step is a consequence of the definition of characteristic operator, whilst in the second step eqs. (2.11) and (2.10b) have been used. Equations (3.1) and (3.2) follow by inserting into eq. (3.3) the formal solution of eq. (2.13), which is given by

$$V(t_f,t_0;[\vec{\varphi}])=\exp\Big\{i\int_{t_0}^{t_f}\sum_{i=1}^{n}\varphi_i(t)\Big[\sum_{j=1}^{M}\alpha_i^{(j)}d\Lambda_t^{(j)}+\sum_{j=M+1}^{N}\alpha_i^{(j)}(dA_t^{(j)}+dA_t^{(j)+})+c_i\,dt\Big]\Big\}. \tag{3.4}$$

By suitable limits and changements of variables, all probabilities of the kind (1.2) can be obtained from the probabilities for the time averages $\vec{\bar{x}}^{(r)}$. Correspondingly, suitable projection valued measures will appear in r.h.s. of eq. (3.1) (point 1).

Now, let us introduce the unitary operators on $h\otimes\mathcal{L}(-\infty,+\infty)$

$$T(t) := S_t\, U(t,0) \quad \text{for } t\geq 0, \quad T(t) := T(|t|)^+ \quad \text{for } t<0; \tag{3.5}$$

$\{T(t),\ t\in \mathbb{R}\}$ is a strongly continuous one-parameter group [7]; we have also

$$S_t\, U(t,t_0)\, S_{t_0}^+ = T(t-t_0). \tag{3.6}$$

Then, using eqs. (2.9c), (2.9d), (3.6) and the fact that $P_{(s,t)}(B) \in \mathbb{1} \otimes \mathcal{b}(s,t)$, we have

$$U(t_f,t_0)^+ P_{(t_{r-1},t_r)}(B_r)\, U(t_f,t_0) = U(t_r,t_0)^+ P_{(t_{r-1},t_r)}(B_r)\, U(t_r,t_0)$$
$$= S_{t_0}^+ T(t_r-t_0)^+ S_{t_r} P_{(t_{r-1},t_r)}(B_r)\, S_{t_r}^+ T(t_r-t_0)\, S_{t_0}. \tag{3.7}$$

Finally, by eq. (2.6) and the fact that (cf. eq. (2.7))

$$S_\tau\, d\Lambda_t^{(j)} S_\tau^+ = d\Lambda_{t-\tau}^{(j)}, \quad \text{etc.} \tag{3.8}$$

we can write

$$P(\vec{\bar{x}}^{(1)}\in B_1, \dots, \vec{\bar{x}}^{(m)}\in B_m \mid \rho, t_0) = \omega\Big(\prod_{r=1}^{s} \tilde{P}(B_r,t_r)\Big), \tag{3.9}$$

where

$$\tilde{P}(B_r,t_r) = T(t_r-t_0)^+ P_{(-\tau,0)}(B_r)\, T(t_r-t_0). \tag{3.10}$$

Therefore, we see that the dynamics of the global system is given by the unitary group $\{T(t), t\in\mathbb{R}\}$ (point 3). Note that Heisenberg picture is used; at the initial time t_0 the projectors $\tilde{P}(B,t)$ belong to $\mathbb{1} \otimes B(\mathcal{b}(-\infty,+\infty))$, as one sees from eqs. (3.10) and (3.2) (point 2). Moreover, eq. (3.2) shows that the continually measured quantities are associated with some linear combinations of $\Lambda_t^{(j)}$, $A_t^{(j)}$, $A_t^{(j)+}$.

Note that, because of the interaction between the two systems, from the results of the observation on the auxiliary system we can extract information on some observable of the measured system. What observables are actually measured by this procedure is apparent from the expression of mean values. Indeed, from eqs. (2.16), (2.17) and the definition of characteristic operator, we have for the mean values [4-6]

$$\langle x_i(t)\rangle = -i\frac{\delta}{\delta\varphi_i(t)} \mathrm{Tr}_{\mathcal{H}}\big(\mathcal{G}(t_f,t_0;[\vec{\varphi}])(\rho)\big)\Big|_{\vec{\varphi}=\vec{0}} = \mathrm{Tr}_{\mathcal{H}}(C_i\,\rho(t)), \tag{3.11}$$

where

$$\rho(t) = \exp\left[\mathcal{L}(t-t_0)\right](\rho) \; , \tag{3.12}$$

$$C_i = \sum_{j=1}^{M} \alpha_i^{(j)} R_j^+ R_j + \sum_{j=M+1}^{N} \alpha_i^{(j)} (R_j + R_j^+) + c_i \; . \tag{3.13}$$

Equation (3.11) is the standard quantum formula for mean values, so that the continually observed quantity $x_i(t)$ can be interpreted as associated with the selfadjoint operator C_i on $\mathcal{h}$.

References.

1. E.B. Daviès, Quantum Theory of Open Systems (Academic Press, London, 1976).

2. M.D. Srinivas and E.B. Davies, Optica Acta 28, 981-996 (1981).

3. M.D. Srinivas, in Quantum Probability and Applications to the Quantum Theory of Irreversible Processes, ed. by L. Accardi, A. Frigerio and V. Gorini, Lecture Notes in Mathematics 1055 (Springer, Berlin, 1984) pp. 356-364.

4. A. Barchielli, L. Lanz and G.M. Prosperi, Found. of Physics 13, 779-812 (1983).

5. G.M. Prosperi, in Quantum Probability and Applications to the Quantum Theory of Irreversible Processes, ed. by L. Accardi, A. Frigerio and V. Gorini, Lecture Notes in Mathematics 1055 (Springer, Berlin, 1984)pp. 301-326.

6. A. Barchielli and G. Lupieri, Quantum sochastic calculus, operation valued stochastic processes and continual measurements in quantum mechanics, preprint IFUM 306/FT (Milan, Oct. 1984).

7. I.M. Gel'fand and N.Ya. Vilenkin, Generalized Functions, vol. 4, Applications of Harmonic Analysis (Academic Press, New York and London, 1964).

8. R.L. Hudson and K.R. Parthasarathy, in Quantum Probability and Applications to the Quantum Theory of Irreversible Processes, ed. by L. Accardi, A. Frigerio and V. Gorini, Lecture Notes in Mathematics 1055 (Springer, Berlin, 1984) pp. 171-198.

9. R.L. Hudson and K.R. Parthasarathy, Commun. Math. Phys. 93, 301-323 (1984).

10. A. Frigerio, Covariant Markov dilations of quantum dynamical semipreprint IFUM 302/FT (Milan, May 1984) and the contribution in this volume.

11. R.L. Hudson, P.D.F. Ion and K.R. Parthasarathy, Commun. Math. Phys. 83, 261-280 (1982).

The Doob-Meyer decomposition for the square of Itô-Clifford L^2-martingales

C. Barnett,
Dept. of Mathematics,
Imperial College,
London SW7 2AZ.

I.F. Wilde,
Dept. of Mathematics,
King's College,
London WC2R 2LS.

§0. Introduction

If $(Y_t)_{t\in\mathbb{R}^+}$ is a sufficiently well-behaved submartingale over a filtered probability space, then it can be uniquely written as the sum of a martingale and an increasing (natural = predictable) process (null at t = 0). This is the Doob-Meyer decomposition of (Y_t). If (X_t) is an L^2-martingale, then $|X_t|^2$ is an L^2-submartingale and the Doob-Meyer decomposition can be used to characterize stochastic integrals with respect to (X_t) (see, for example, [8]).

Analogous results within the framework of non-commutative probability theory have been discussed in [2,3,5]. In [2], the existence of such a decomposition for the square of any Itô-Clifford stochastic integral was established, analogues of class D were defined for general non-commutative processes and such submartingales were shown to have a Doob-Meyer decomposition. Using this, a general theory of stochastic integration with respect to non-commutative L^2-martingales was set up. These results were extended and refined in [3].

The characterization of the Itô-Clifford stochastic integral X_t in terms of bracket processes was a result of [5]. However, the uniqueness of the decomposition of $|X_t|^2$ was not known, and so bracket processes were defined via the particular decomposition given in [2]. We shall show here that the Doob-Meyer decomposition of $|X_t|^2$ is, indeed, unique. In fact, we shall see that if A_t and B_t are any increasing L^1-processes null at t = 0 and such that $|X_t|^2-A_t$ and $|X_t|^2-B_t$ are both martingales, then $A_t = B_t$. It turns out that <u>any</u> increasing L^1-process null at 0 is "nearly natural"; so also is any process of locally bounded variation and null at 0. (See [6] for details of this last result.)

In addition to establishing uniqueness, we shall also give a simpler proof of the existence of the Doob-Meyer decomposition than that given in [2]. For further details and results relating to uniqueness and natural processes over a general probability gauge space, we refer to [6].

§1. Notation

Let $\Psi(u)$, for $u \in L^2_{\mathbb{R}}(\mathbb{R}^+)$, denote the Fermi-fields on the anti-symmetric Fock space over $L^2(\mathbb{R}^+)$. For $t \in \mathbb{R}^+$, $\mathcal{C}_t$ denotes the von Neumann algebra generated by the fields $\Psi(u)$, where u has support in $[0,t]$, and $\mathcal{C}$ is that generated by all the $\mathcal{C}_t$, $t \in \mathbb{R}^+$. Let Ω denote the Fock vacuum vector and write $m(\cdot) = (\Omega, \cdot\Omega)$. One can show that m is a central state on $\mathcal{C}$. For $1 \leq p < \infty$, $L^p(\mathcal{C})$ is the completion of $\mathcal{C}$ with respect to the norm $x \mapsto \|x\|_p = m(|x|^p)^{1/p}$. The elements of $L^p(\mathcal{C})$ can be identified with (possibly unbounded) operators on the Fock space and so it makes sense to speak of positive elements of $L^p(\mathcal{C})$. Similarly, one defines $L^p(\mathcal{C}_t)$ to be the appropriate completion of $\mathcal{C}_t$, $t \in \mathbb{R}^+$. Since $\mathcal{C}_s \subseteq \mathcal{C}_t$ for $0 \leq s \leq t$, we have $L^p(\mathcal{C}_s) \subseteq L^p(\mathcal{C}_t)$ for $0 \leq s \leq t$. Denote by M_t the conditional expectation map: $L^p(\mathcal{C}) \to L^p(\mathcal{C}_t)$. Then, for each $t \geq 0$, M_t is a contraction (onto $L^p(\mathcal{C}_t)$) and the map $t \mapsto M_t$ is strongly continuous on $L^2(\mathcal{C})$. We also have $M_s M_t = M_s$ for $0 \leq s \leq t$. The collection $(\mathcal{C} \equiv L^\infty(\mathcal{C}), m, \{\mathcal{C}_t\}_{t \in \mathbb{R}^+})$ gives a non-commutative analogue of a filtered probability space.

Definition. An L^p-process is a collection $(X_t)_{t \in \mathbb{R}^+}$ where $X_t \in L^p(\mathcal{C}_t)$ for each $t \in \mathbb{R}^+$; thus processes are by definition adapted. Evidently, one can also think of an L^p-process as a map from $\mathbb{R}^+$ into $L^p(\mathcal{C})$. If $M_s X_t = X_s$ for any $0 \leq s \leq t$, (X_t) is called an L^p-martingale. If $X_s \leq M_s(X_t)$ for $0 \leq s \leq t$, then (X_t) is an L^p-submartingale.

For $t \geq 0$, set $\Psi_t = \Psi(\chi_{[0,t]})$. Then one sees that (Ψ_t) is an L^∞-martingale [2]. We shall be concerned with L^1 and L^2 processes. For further details see [2,6] and references therein.

§2. The existence of the Doob-Meyer decomposition

Let (X_t) be an $L^2(\mathcal{C})$-martingale. From [2], we deduce that there is a (unique) element $\hat{X} \in L^2_{loc}(\mathbb{R}^+, dx; L^2(\mathcal{C}))$ such that $\hat{X}(s) \in L^2(\mathcal{C}_s)$ a.e. and

$$X_t = X_0 + \int_0^t d\Psi_s \hat{X}(s) \tag{2.1}$$

where the integral is the Itô-Clifford stochastic integral.

Note that eqn.(2.1) can be rewritten as

$$X_t = X_0 + \int_0^t \tilde{X}(s) d\Psi_s \tag{2.2}$$

with $\tilde{X}(s) = \beta(\hat{X}(s))$, where β is the parity operator (see [2,5]). Eqn. (2.1) is more convenient than eqn. (2.2) for our purposes here in that it avoids using β.

<u>Theorem</u>. Let (X_t) be an $L^2(\mathcal{C})$-martingale. Then there is an increasing L^1-process (A_t) such that $Z_t = X_t^*X_t - A_t$ is an L^1-martingale.

<u>Proof</u>. First note that since $X_t \in L^2(\mathcal{C}_t)$, $X_t^*X_t$ is an L^1-process. Without loss of generality, we may suppose that $X_0 = 0$. Hence we may write $X_t = \int_0^t d\Psi_s \hat{X}(s)$ for some process $\hat{X}$ in $L^2_{loc}(\mathbb{R}^+, dx; L^2(\mathcal{C}))$.

Let $t > 0$ and suppose first that $\hat{X}$ is simple on $[0,t)$: i.e. $\hat{X}(s) = \sum_{j=0}^n h_j \chi_{[t_j,t_{j+1})}(s)$ on $[0,t)$ for some $0 = t_0 < t_1 < \ldots < t_{n+1} = t$ and $h_j \in L^2(\mathcal{C}_{t_j})$. Then $\int_0^t d\Psi_s \hat{X}(s) = \sum_{j=0}^n \Delta\Psi_j h_j$ where $\Delta\Psi_j = \Psi_{t_{j+1}} - \Psi_{t_j} = \Psi(\chi_{(t_j,t_{j+1}]})$.

Thus
$$\begin{aligned} X_t^*X_t &= (\sum_j \Delta\Psi_j h_j)^*(\sum_k \Delta\Psi_k h_k) \\ &= \sum_{j<k} h_j^*\Delta\Psi_j\Delta\Psi_k h_k + \sum_{j>k} h_j^*\Delta\Psi_j\Delta\Psi_k h_k \\ &\quad + \sum_j h_j^* \Delta\Psi_j \Delta\Psi_j h_j \end{aligned} \tag{2.3}$$

Now, using the canonical anticommutation relations, we see that $(\Delta\Psi_j)^2 = \int_{t_j}^{t_{j+1}} ds = t_{j+1} - t_j \equiv \Delta t_j$ and so

$$\begin{aligned} \sum_j h_j^*\Delta\Psi_j\Delta\Psi_j h_j &= \sum_j h_j^* h_j \Delta t_j \\ &= \int_0^t |\hat{X}(s)|^2 ds \\ &\equiv A_t. \end{aligned}$$

Suppose $0 \leqslant s \leqslant t$. Without loss of generality, we may suppose that $s = t_m$, $0 \leqslant m \leqslant n+1$. If $k \geqslant m$, $M_{t_m}M_{t_k} = M_{t_m}$ and so, for $j < k$,

$$\begin{aligned} M_{t_m}(h_j^*\Delta\Psi_j\Delta\Psi_k h_k) &= M_{t_m}M_{t_k}(h_j^*\Delta\Psi_j\Delta\Psi_k h_k) \\ &= M_{t_m}(h_j^*\Delta\Psi_j(M_{t_k}(\Delta\Psi_k))h_k) \\ &= 0, \text{ since } (\Psi_t) \text{ is a martingale.} \end{aligned}$$

It follows that

$$M_s(\sum_{j<k} h_j^*\Delta\Psi_j\Delta\Psi_k h_k) = \sum_{\substack{j<k \\ k\leqslant m}} h_j^*\Delta\Psi_j\Delta\Psi_k h_k$$

and, similarly,

$$M_s\Big(\sum_{j>k} h_j^* \Delta\Psi_j \Delta\Psi_k h_k\Big) = \sum_{\substack{j>k \\ j\leq m}} h_j^* \Delta\Psi_j \Delta\Psi_k h_k .$$

The sum of these two expressions is therefore equal to $X_s^* X_s - A_s$: that is, we have

$$M_s(X_t^* X_t - A_t) = X_s^* X_s - A_s .$$

Now, simple $L^2(\mathcal{C})$-processes on $[0,t)$ are dense in the set of processes in $L^2([0,t],dx;L^2(\mathcal{C}))$[2], and so one sees that in the general case we have also that

$$M_s(X_t^* X_t - A_t) = X_s^* X_s - A_s \tag{2.4}$$

where $A_t = \int_0^t |\hat{X}(s)|^2 ds$ with $\hat{X}$ given by eqn.(2.1). (If $X_t^{(n)} = \int_0^t d\Psi_s \hat{X}^{(n)}(s)$ and $\hat{X}^{(n)} \to \hat{X}$ in $L^2([0,t],dx;L^2(\mathcal{C}))$, then $X_t^{(n)} \to X_t$ in $L^2(\mathcal{C})$ and $X_t^{(n)*} X_t^{(n)} \to X_t^* X_t$ in $L^1(\mathcal{C})$. Moreover, one checks that $|\hat{X}^{(n)}|^2 \to |\hat{X}|^2$ in $L^1([0,t],dx;L^1(\mathcal{C}))$ and so $A_t^{(n)} = \int_0^t |\hat{X}^{(n)}(s)|^2 ds \to A_t = \int_0^t |\hat{X}(s)|^2 ds$ in $L^1(\mathcal{C})$). Putting $Z_t = X_t^* X_t - A_t$, eqn.(2.4) implies that $M_s(Z_t) = Z_s$ for $0 \leq s \leq t$ and so (Z_t) is an L^1-martingale. It is clear that (A_t) is increasing.

QED.

<u>Remark</u>. Eqn.(2.3) is the Itô-formula for X^*X. Indeed, if $Y = \sum_j \Delta\Psi_j g_j$, then

$$Y_{t_{k+1}} X_{t_{k+1}} - Y_{t_k} X_{t_k} = \Big(\sum_{j<k} \Delta\Psi_j g_j\Big) \Delta\Psi_k h_k$$

$$+ \Delta\Psi_k g_k \Big(\sum_{j<k} \Delta\Psi_j h_j\Big) + \Delta\Psi_k g_k \Delta\Psi_k h_k$$

i.e. $d(YX) = YdX + (dY)X + dYdX$.

Integration with respect to dX has been discussed in [2,3,5]. This quantum Itô-formula has been proved for fermions by Appelbaum and Hudson [1] and for bosons by Hudson and Parthasarathy [7] (-provided the integrands are bounded). It was this formula which motivated the above simplified proof of the existence of the Doob-Meyer decomposition of $|X_t|^2$. Notice that the integrand $\hat{X}(s)$ need <u>not</u> be bounded. Moreover, as far as the Doob-Meyer decomposition is concerned, we need only consider the <u>sum</u> of the first and second terms on the r.h.s. of eqn. (2.3). This is just $Z_t^{(n)} = X_t^{(n)*} X_t^{(n)} - A_t^{(n)}$ which we have seen converges to Z_t in $L^1(\mathcal{C})$ (as the simple approximations to $\hat{X}$ converge to $\hat{X}$). To prove an Itô-formula, one would have to consider the terms <u>separately</u>.

This, of course, is a much harder problem.

§3. The uniqueness of the Doob-Meyer decomposition

The results of this section are taken from [6]. The idea of the proof is to use a discrete form of Meyer's notion of naturalness [9].

Lemma. Let $Y \in \mathcal{C}$ be such that $s \mapsto M_s(Y)$ is operator norm continuous and let (A_t) be any increasing L^1-process with $A_0 = 0$. Let $\sigma = \{0 = s_0 < s_1 < \ldots < s_{n+1} = t\}$ be a finite subdivision of $[0,t]$. Then

$$m\Big(\sum_{j=0}^{n} M_{s_j}(Y)(A_{s_{j+1}} - A_{s_j})\Big) \to m(YA_t) \qquad (3.1)$$

as mesh$\sigma \to 0$.

Proof. If ΔA_j denotes $A_{s_{j+1}} - A_{s_j}$, then

$$\begin{aligned}\sum_j m(M_{s_{j+1}}(Y)\Delta A_j) &= \sum_j m(M_{s_{j+1}}(Y\Delta A_j)) \\ &= \sum_j m(Y\Delta A_j) \\ &= m(YA_t) \quad \text{since} \quad A_0 = 0.\end{aligned}$$

Hence
$$\begin{aligned}&|m(\textstyle\sum M_{s_j}(Y)\Delta A_j) - m(YA_t)| \\ &= |m(\textstyle\sum (M_{s_j}(Y) - M_{s_{j+1}}(Y))\Delta A_j| \\ &\leq \textstyle\sum \|M_{s_j}(Y) - M_{s_{j+1}}(Y)\| m(\Delta A_j) \\ &< \varepsilon \textstyle\sum m(\Delta A_j)\end{aligned}$$

for sufficiently small meshσ, by uniform continuity of $s \to M_s(Y)$ on $[0,t]$

$$= \varepsilon\, m(A_t).$$

The result follows.

QED.

Remark. If Y is any Wick monomial in the fields $\Psi(u)$, $u \in L^2(\mathbb{R}^+)$, then $s \mapsto M_s(Y)$ is operator norm continuous. In fact, $M_s(:\Psi(u_1)\ldots\Psi(u_n):) = :\Psi_s(u_1)\ldots\Psi_s(u_n):$, where $\Psi_s(u) = \Psi(u\chi_{[0,s]})$. It follows that there is an ultraweakly dense set of Y's in $\mathcal{C}$ for which $s \mapsto M_sY$ is norm continuous, namely, the linear span of $\mathbb{1}$ and all Wick monomials (-this is equal to the algebra of polynomials in the fields). Thus

$(\mathcal{C}, m, \{\mathcal{C}_t\})$ has an abundance of norm continuous L^∞-martingales. It is not clear how special this is. For example, if $(\Omega, \mu, (\mathcal{F}_t))$ is the usual filtered space of Brownian motion, does there exist an element Y of $L^\infty(\Omega, \mu, \mathcal{F}_\infty)$ such that $s \mapsto \mathbb{E}_s Y$ is L^∞-continuous? This seems unlikely.

We are now in a position to establish the uniqueness.

Theorem. Let (X_t) be an $L^2(\mathcal{C})$-martingale. Then there is a unique L^1-martingale (Z_t) and a unique increasing L^1-process (A_t) with $A_0 = 0$ such that

$$X_t^* X_t = Z_t + A_t \quad \text{for} \quad t \geq 0.$$

Proof. We have shown the existence of such a decomposition, so suppose

$$X_t^* X_t = Z_t + A_t = Z_t' + A_t'.$$

Then $Z_t - Z_t' = A_t' - A_t$. But each of A_t and A_t' satisfies eqn.(3.1) and so, therefore, does their difference. Thus $Z_t - Z_t' \equiv \hat{Z}_t$ also satisfies eqn.(3.1). However, since $(\hat{Z}_t)$ is a martingale, $m(M_{s_j}(Y)(\hat{Z}_{s_{j+1}} - \hat{Z}_{s_j})) = 0$ and we deduce that $m(Y\hat{Z}_t) = 0$ for an ultraweakly dense set of Y's in $\mathcal{C}$. It follows that $\hat{Z}_t = 0$. That is $Z_t = Z_t'$ and so $A_t = A_t'$.

QED.

Remark 1. The lemma says that any increasing process null at 0 is "nearly natural", and so our uniqueness result is somewhat stronger than the classical one. Perhaps this is to be expected in view of the "smoothness" in the Itô-Clifford theory.

Remark 2. The bracket process $\langle X, X^* \rangle_t$ is the uniquely determined increasing $L^1(\mathcal{C})$-process (A_t). By polarization one defines $\langle X, Y \rangle_t$ for $L^2(\mathcal{C})$-martingales (X_t) and (Y_t): one finds that $\langle X, Y \rangle_t = \int_0^t \hat{Y}(s)^* \hat{X}(s)\, ds$. For suitable vector-valued integrands f, one can define the Bartle integral $\int_0^t f d\langle X, Y \rangle$ and the stochastic integral $\int_0^t f dX$. One sees that $\int_0^t f dX$ is the unique centred $L^2(\mathcal{C})$-martingale (Z_t), say, such that

$$m(Z_t Y_t) = m\left(\int_0^t f d\langle Y, X^* \rangle\right)$$

for all $t \geq 0$ and all martingales (Y_t). (For details see [5,6].) This characterizes $(\int_0^t f dX)$ as in the classical theory.

Remark 3. An analogous uniqueness-existence result holds for $Y_t^* Y_t$ where $Y_t = \int_0^t db f + db^* g$ is the fermion quantum stochastic integral of [4] with f and g both simple and bounded. The proofs are essentially the same as those above.

References

1. Appelbaum, D. and Hudson, R.: Fermion Itô's formula and stochastic evolutions, preprint 1984.
2. Barnett, C., Streater, R.F. and Wilde, I.F.: The Itô-Clifford integral, J. Funct. Anal. 48, 172-212 (1982).
3. Barnett, C., Streater, R.F. and Wilde, I.F.: Stochastic integrals in an arbitrary probability gage space, Math. Proc. Camb. Phil. Soc. 94, 541-551 (1983).
4. Barnett, C., Streater, R.F. and Wilde, I.F.: Quasi-free quantum stochastic integrals for the CAR and CCR, J. Funct. Anal. 52, 19-47 (1983).
5. Barnett, C., Streater, R.F. and Wilde, I.F.: The Itô-Clifford integral. IV: A Radon Nikodym theorem and bracket processes, J. Operator Theory 11, 255-271 (1984).
6. Barnett, C. and Wilde, I.F.: Natural processes and Doob-Meyer decompositions over a probability gage space, J. Funct. Anal. 58, 320-334 (1984).
7. Hudson, R. and Parthasarathy, K.R.: Quantum Itô's formula and stochastic evolutions, Commun. Math. Phys. 93, 301-323 (1984).
8. Kopp, P.E.: Martingales and stochastic integrals, Camb. Univ. Press, London, 1984.
9. Meyer, P.A.: Probability and potentials, Blaisdell, Waltham, Mass., 1966.

POISSON PROCESSES AND QUANTUM FIELD THEORY : A MODEL

J. BERTRAND(*), B. GAVEAU(**) and G. RIDEAU(*)

(*) Université Paris VII, Laboratoire de Physique théorique et mathématique, Tour 33-43, 1e étage, 2, place Jussieu - 75251 Paris Cedex 05 - FRANCE.
(**) Université Paris VI - Département de Mathématiques, Tour 45-46, 5e étage - 4, place Jussieu - 75230 Paris Cedex 05 - FRANCE.

1. Introduction

The constructive theory of quantum fields is based on brownian processes defined by the free fields. Beside the use of imaginary time, this approach has the defect of treating the interaction as a perturbation. But the perturbation being generally more singular than the free hamiltonian, the converse point of view should be more realistic. This has become possible for the Schrödinger equation itself through the use of Maslov's work [1] associating Poisson processes to certain classes of potentials. This approach has been further extended in many ways [2] and, in particular, to Quantum Field Theory [3] though essentially for bounded interactions.

In the present paper, we start with a similar point of view. However we show, on a simple model of field on a lattice, that changing p-representation into occupation number representation allows to treat an unbounded interaction with the same Poisson processes techniques. Now, the stochastic Poisson measures have a simple physical meaning : they count the jumps of occupation numbers for each field mode.

2. Notations

All work will be done in a box V of side L in $\mathbb{R}^n$.

The free boson field momenta are on the lattice $\mathcal{L}$ of vectors $k = \frac{2\pi}{L}(p_1,\dots,p_n)$ where p_i are integers. The state space $\mathfrak{F}$ is the Fock space generated as a Hilbert space by the orthonormal set of vectors :

$$| (n_k)_{k\in\mathcal{L}} \rangle = | n_{k_1}, n_{k_2}, \dots \rangle$$

Here n_k is a non negative finite integer which is equal to zero except for a finite set of k ; we shall denote $(\mathbb{N}^{\mathcal{L}})_0$ the set of all such sequences $(n_k)_{k\in\mathcal{L}}$ (with bounded support).

Creation and annihilation operators a^*_k, a_k are defined as usual by

$$(1)\qquad \begin{aligned} a^*_k |(n_\ell)_{\ell\in\mathcal{L}}\rangle &= \sqrt{n_k + 1}\; |(n_\ell + \delta_{\ell k})_{\ell\in\mathcal{L}}\rangle \\ a_k |(n_\ell)_{\ell\in\mathcal{L}}\rangle &= \sqrt{n_k}\; |(n_\ell - \delta_{\ell k})_{\ell\in\mathcal{L}}\rangle \end{aligned}$$

The free field hamiltonian is given by

$$(2)\qquad H_F = \sum_{k\in\mathcal{L}} \varphi(k)\, a^*_k\, a_k$$

where $\varphi(k)$ is some positive function on $\mathcal{L}$.

A quantum particle interacting with the above field is described in the state space $L^2(\mathcal{L}) \otimes \mathcal{F}$, where $L^2(\mathcal{L})$ is the space of square summable functions on the lattice $\mathcal{L}$ of momenta.

We consider the hamiltonian

$$(3)\qquad H = H_F + \frac{|P|^2}{2m} + W$$

where

$$(4)\qquad \begin{cases} W = g \sum_{k\in\mathcal{L}} \dfrac{1}{2\,\varphi(k)^{1/2}} \left(a^*_k\, \tau_{-k} + a_k\, \tau_k\right) \\ (\tau_k f)(p) = f(p-k) \end{cases}$$

We will assume throughout the cut-off condition

$$(5)\qquad \mu = \frac{1}{2} \sum_{k\in\mathcal{L}} [\varphi(k)]^{-1/2} < \infty$$

Denoting $F(p,(n_k))$ the wave function associated to a state $|\psi\rangle$, we can describe the system in the space of functions F on $\mathcal{L} \times (\mathbb{N}^{\mathcal{L}})_0$ with the obvious L^2-norm.

We want to solve the Cauchy problem for the Schrödinger equation

$$(6)\qquad \begin{cases} \dfrac{1}{i}\dfrac{\partial F}{\partial t} = \tilde{H}\, F_t \\ F_t\big|_{t=0} = F_0 \end{cases}$$

where $\tilde{H}$, obtained from (3) by duality and the use of (1), reads

$$(\tilde{H} F_t)\ (p^\circ, (n^\circ_\ell)_{\ell\in\mathcal{L}}) = \Big(\sum_{k\in\mathcal{L}} \varphi(k)\, n^\circ_k + \frac{|p^\circ|^2}{2m}\Big)\ F_t(p^\circ, (n^\circ_\ell))$$

$$+\ g \sum_{k\in\mathcal{L}} \sum_{\varepsilon=\pm 1} \frac{1}{2}\ [\varphi(k)]^{-1/2}\ \Big(n^\circ_k + \frac{1+\varepsilon}{2}\Big)^{1/2}\ F_t\ (p^\circ+\varepsilon k, n^\circ_\ell + \delta_{\ell k}\varepsilon)$$

3. Construction of Poisson processes and solution of Cauchy problem.

The interaction hamiltonian W consists of two types of terms which can be visualized as follows :

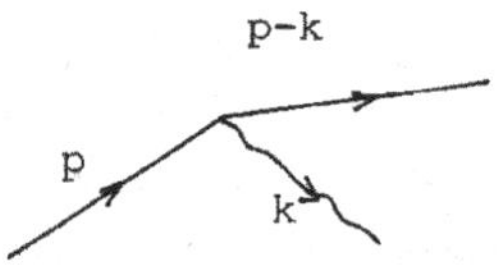

"event" $(k, \varepsilon = +1)$

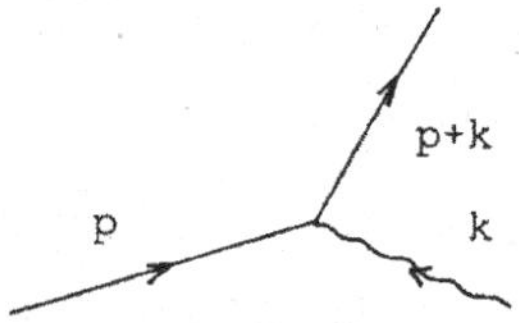

"event" $(k, \varepsilon = -1)$

where $\longrightarrow$ represents the quantum particle and $\rightsquigarrow$, the boson.

We shall construct a Poisson measure $\nu_\omega(A)$ on the borel sets of $S = \mathbb{R} \times \mathcal{L} \times \{+1,-1\}$ which counts the number of fundamental "events" (k,ε) occuring in time dt. More precisely, we define on S the measure

$$(7)\quad \Lambda(dt \times k \times \varepsilon) = dt \times \sum_{\ell\in\mathcal{L}} [\varphi(\ell)]^{-1/2}\ \delta(k-\ell) \times \frac{1}{2} \sum_{\eta=\pm 1} \delta_{\eta\varepsilon}$$

By condition (5), we have

$$(8)\quad \Lambda([o,t] \times \mathcal{L} \times \{+1,-1\}) = \mu\, t < \infty$$

$\nu_\omega(A)$ is then defined as the Poisson measure on S whose mathematical expectation is

$$(9)\quad E\ \nu_\omega(A) = \Lambda(A)$$

where ω is the sample in the probability space.

Now, we define two right continuous Markov processes with independent increments

$$(10)\qquad \begin{aligned} p_\omega(t) &= p_\omega(o) + \int_o^t \sum_{k\in\mathcal{L}} \sum_{\varepsilon=\pm 1} (-\varepsilon)k\ \nu_\omega(ds \times k \times \varepsilon) \\ n_{k\omega}(t) &= n_{k\omega}(o) + \int_o^t \sum_{k\in\mathcal{L}} \sum_{\varepsilon=\pm 1} \varepsilon\ \nu_\omega(ds \times k \times \varepsilon) \end{aligned}$$

$p_\omega(t)$ is in the momentum space of the particle.
$n_{k\omega}(t)$ can take on negative values even though $n_{k\omega}(o) \geqslant 0$. This is unimportant since no trajectory going to negative values of one of the n_k will ever contribute to the expectation considered below.

Let $\mathfrak{J}(t)$ denote the total number of jumps of Poisson measure up to time t :

$$(11)\qquad \mathfrak{J}(t) = \nu_\omega([o,t] \times \mathcal{L} \times \{+1, -1\})$$

$\mathfrak{J}(t)$ is finite almost surely since

$$(12)\qquad E\ \mathfrak{J}(t) = \mu t$$

Moreover, we have

$$(13)\qquad \text{Prob}\ (\mathfrak{J}(t) = n) = e^{-\mu t} \frac{(\mu t)^n}{n!}$$

The solution of the Cauchy problem for wave functions is then given by the following result.

<u>Theorem.</u> Let $\mathcal{B}$ denote the set of functions F such that there exists an integer $N \geqslant 0$, depending on F, with $F(p,(n_k)) = 0$ whenever $\sum_{k\in\mathcal{L}} n_k > N$. Then the solution of Cauchy problem (6) with an initial data in $\mathcal{B}$ is the wave function F_t given by

$$(14)\qquad F_t(p^\circ,(n_k^\circ)) = e^{\mu t}\ E\left\{ e_1(t)e_2(t)e_3(t)F_o(p_\omega(t),(n_{k\omega}(t)) \,\Big|\, p_\omega(o) = p^\circ,\ n_{k\omega}(o) = n_k^\circ \right\}$$

where

$$e_1(t) = \exp\left\{ i\int_o^t \sum_k \varphi(k) n_{k\omega}(s)\,ds + \frac{i}{2m}\int_o^t |p_\omega(s)|^2\,ds \right\}$$

$$e_2(t) = (ig)^{\mathfrak{J}(t)}$$

$$e_3(t) = \exp\left\{ \int_o^t \sum_{k\in\mathcal{L}} \sum_{\varepsilon=\pm 1} \log\left[(n_{k\omega}(s^-) + \tfrac{1+\varepsilon}{2})^{1/2} \right] \nu_\omega(ds \times k \times \varepsilon) \right\}$$

$$n_{k\omega}(s^-) = \lim_{u \nearrow s^-} n_{k\omega}(u)$$

Proof. We shall give here a few indications. More details can be found in [4] .

Step 1. All quantities inside the expectation sign are well defined.

Problems could arise with trajectories $n_k(t)$ going from 0 to (-1) at some time T. But such a jump is governed by $\nu_\omega(ds, k, -1)$, which leads to the value $e_3(T) = 0$. Thus these trajectories will not contribute to E.

Then, e_1, e_2 and e_3 are finite almost surely since $\mathcal{J}(t)$ is ; this also implies that $F_o(p(t), n_k(t))$ is well defined for $F_o \in \mathcal{B}$.

Step 2. The expectation is absolutely convergent.

We need only work on the quantity :

$$(15)\quad \Phi_t(\overset{\circ}{n}_k) = E\left\{ e_3(t) g^{\mathcal{J}(t)} F_o(p_\omega(t), n_{k\omega}(t)) \,\middle|\, p_\omega(o) = \overset{\circ}{p}, n_{k\omega}(o) = \overset{\circ}{n}_k \right\}$$

In fact, we shall prove a more general result that is needed in step 3.

First we notice that, due to property (12), $\Phi_t(\overset{\circ}{n}_k)$ is defined only for sequences $\{\overset{\circ}{n}_k\}$ of bounded support.

a) Let us consider $F_o \in \mathcal{B}$; F_o is different from zero for those trajectories $\{n_k(t)\}$ with $\sum_k n_k(t) = Q < Q_o$ where Q_o is given. We choose an integer $\overset{\circ}{P} > \overset{\circ}{Q}$ as large as necessary and look for a bound on Φ_t for all values of $P = \sum_k \overset{\circ}{n}_k$. For trajectories performing n jumps and ending up in $\{n_k(t)\}$ such that $\sum_k n_k(t) < Q_o$, we have the estimate

$$(16)\qquad e_3(t) \leqslant \left[(Q_o + 1)\ldots(Q_o + n)\right]^{1/2}$$

Using (13) and (16), we obtain the bounds

$$(17)\quad \sup_{P<P_o} |\Phi_t(\overset{\circ}{n}_k)| \leqslant \sum_{n\geqslant 0} e^{-\mu t} \frac{(g\mu t)^n}{\sqrt{n!}} Q_o^{n/2} \sup |F_o| < \infty$$

$$(18)\quad \sup_{P\geqslant P_o} |\Phi_t(\overset{\circ}{n}_k)| \leqslant \frac{Q_o^{1/2(P_o - Q_o)}}{\sqrt{(P_o - Q_o)!}} C(t, Q_o) \sup |F_o| .$$

where C is a function of t and Q_o only.

We will say that functions Φ_t with the properties just obtained belong to class $\mathcal{K}$.

b) Now, we consider a function F_o belonging to class $\mathcal{K}$. Using the estimate

(19) $e_3(t) \leqslant \left[(P+1)\ldots(P+n)\right]^{1/2}$

valid on trajectories such that $\sum_k n_k^\circ = P$ and $\mathcal{J}(t) = n$, we obtain :

(20) $$\sup_{\sum_k n_k^\circ = P} \left|\Phi_t(n_k^\circ)\right| < \sum_{n \geqslant 0} \sup |F_\circ| \frac{(g\mu t)^n}{\sqrt{n!}} P^{\frac{n}{2}} e^{-\mu t} < \infty$$

Step 3. Formula (14) gives a semi-group in t.

This can be verified using Markov property and results (17), (18) and (20).

We can then use the notation

$$F_t = e^{it\tilde{H}} F_\circ$$

Step 4. F_t tends to $F_{t_\circ}$ if $t \longrightarrow t_\circ^+$

Using the semi-group property, we have only to prove this for $t_\circ = 0$ and $F_\circ$ in class $\mathcal{K}$. Let T_1 be the first time of jump of the Poisson process ; T_1 is strictly positive since in any finite interval there is almost surely a finite number of jumps. If $t < T_1$:

$$e_3(t) e_2(t) F_\circ(p_\omega(t), n_{k\omega}(t)) = F_\circ(p_\circ, n_k^\circ)$$

and $e_1(t)$ tends to 1 when $t \longrightarrow 0^+$.

Thus the product inside the E sign tends to $F_\circ$ almost surely. Moreover it is dominated by $g^{\mathcal{J}(t)} e_3(t) F_\circ(p_\omega(t), n_{k\omega}(t))$ which is integrable by the above results. This concludes step 4 by Lebesgue theorem.

Step 5. F_t satisfies the evolution equation.

Again it suffices to check this when $t = 0$ and $F_\circ \in \mathcal{K}$. It is done through a tedious computation which uses Markov property and condition (5).

Thus we have proposed a new image of quantum field theory in a box, giving the occupation numbers of the fields an interpretation in terms of Poisson stochastic processes. The approach is simple enough and can be extended to interacting boson and fermion fields with discrete momenta. This way of presenting field theory allows to do renormalization.

References

[1] V.P. Maslov and A.P. Chebotarev, Viniti Itogui Nauki, Vol 15 (1978) 5.

[2] See for instance : Ph. Combe, R. Hoegh-Krohn, R. Rodriguez, M. Sirugue and M. Sirugue-Collin, Commun. Math. Phys. 77 (1980) 269 ; J. Math. Phys. 23 (1982) 405.

For extension to phase space, see also :

J. Bertrand, B. Gaveau, C.R. Acad. Sci. 295 (1982) 189
J. Funct. Anal. 50 (1983) 81

J. Bertrand, G. Rideau, Lett. Math. Phys. 7 (1983) 327

Ph. Combe, F. Guerra, R. Rodriguez, M. Sirugue, M. Sirugue-Collin, Physica 124A (1984) 567.

[3] Ph. Blanchard, Ph. Combe, R. Hoegh-Krohn, M. Sirugue, Commun. Math. Phys. (to appear)

[4] J. Bertrand, B. Gaveau, G. Rideau, Lett. Math. Phys. 9 (1985) 73.

The entropy of quantum Markov states.

O. Besson
Institut de Mathématiques, Univ. de Neuchâtel
Chantemerle 20, CH-2000 Neuchâtel

In this paper we compute the Connes - Stǿrmer entropy of a non-ommutative shift automorphism associated with a quantum Markov state efined by L. Accardi.

Our result is that under some hypothesis the entropy of this non-ommutative Markov shift is equal to the equilibrium quantum mechanical ntropy of the associated quantum Markov state.

This result classify some Heisenberg models with nearest neighbour nteraction in the sense that if the entropies of two such dynamical ystems differ they are not isomorphic.

Moreover this result is an attempt to study non-product states on -algebras and give new examples with computable entropy.

In the first section we recall the definition and the properties the Connes - Stǿrmer entropy given in [CS]. A detailed motivation this notion can be found in [B1].

In the second section we recall the definition of a stationary uantum Markov state and prove the result announced above. As an example give the computation for the one dimensional Ising model with nearest eighbour interaction.

I am grateful to the Swiss National Fund for Scientific Research or his financial support.

The entropy of an automorphism.

In all this paper the letter η designates the continuous function $\in [0,\infty) \to -t\log t \in \mathbb{R}$.

Let us first recall the definition of the entropy in the classical heory. Let $(X,\mathcal{B},\mu)$ be a probability space and $P = \{p_1,\dots,p_n\}$ be a nite mesurable partition of X. Then the entropy of the partition P is

$$h(P) = \sum_1^n \eta\mu(p_i).$$

$Q = \{q_1,\dots,q_m\}$ is another partition, then the relative entropy of

P given Q is

$$h(P|Q) = h(P \vee Q) - h(Q)$$

where $P \vee Q = \{p_i \cap q_j\}_{i,j}$ is the partition generated by P and Q.

Let T be a non-singular transformation of X preserving the mesure μ (i.e. a mesurable invertible mapping of X that preserves the null sets of μ) and put

$$h(P,T) = \lim_{n\to\infty} n^{-1}\, h(P \vee TP \vee \ldots \vee T^n P)$$

$$h(T) = \sup_P h(P,T)$$

h(T) is called the entropy of the transformation T.

These definitions extend naturally to the case of <u>abelian</u> von Neumann algebras. But since two finite dimensional non-abelian algebras can fail to generate a finite dimensional algebra, there is no immediate analogue of the entropy in the non-commutative frame. From the definition of h(P,T) we see that it is necessary to look for a replacement of the quantity $h(P_1 \vee P_2 \vee \ldots \vee P_n)$.

Let M be a finite von Neumann algebra with faithful, normal and normalized trace τ. For each $n \in \mathbb{N}$ let S_n be the set of all families $x = (x_{i_1,\ldots,i_n})_{i_j \in \mathbb{N}}$ of positive elements of M, zero except for a finite number of indices and satisfying

$$\sum_{i_1,\ldots,i_n} x_{i_1,\ldots,i_n} = 1.$$

For $x \in S_n$, $k \in \{1,\ldots,n\}$ and $i_k \in \mathbb{N}$ we put

$$x^k_{i_k} = \sum_{\substack{i_1,\ldots,i_{k-1} \\ i_{k+1},\ldots,i_n}} x_{i_1,\ldots,i_n}$$

If P is a von Neumann subalgebra of M, we denote by E_P the unique faithful normal conditional expectation of M onto P which preserves the trace τ.

The definitions proposed by Connes and Størmer in [CS] are

<u>Definition 1.</u> Let $P_1,\ldots,P_n$ be finite dimensional von Neumann subalgebras of M, then the <u>entropy of the family</u> $(P_1,\ldots,P_n)$ is

$$H(P_1,\dots,P_n) = \sup_{x\in S_n} \sum_{i_1,\dots,i_n} \eta\tau(x_{i_1,\dots,i_n}) - \sum_{k=1}^{n} \sum_{i_k} \tau\eta E_{P_k}(x^k_{i_k})$$

efinition 2. Let P and Q be finite dimensional von Neumann subalgebras f M, then the relative entropy of P given Q is

$$H(P|Q) = \sup_{x\in S_1} \sum_i \tau\eta E_Q(x_i) - \tau\eta E_P(x_i).$$

t is clear that $H(P_1,\dots,P_n)$ is symetric and positive and that $H(P|Q)$ s positive. Moreover Connes and Størmer have proved the following proerties analogous to the properties of the classical entropy.

A) $H(P_1,\dots,P_n) \leq H(Q_1,\dots,Q_n)$ if $P_j \subset Q_j$ $j = 1,\dots,n$

B) $H(P_1,\dots,P_n,P_{n+1},\dots,P_k) \leq H(P_1,\dots,P_n) + H(P_{n+1},\dots,P_k)$

C) $P_1,\dots,P_n \subset P \Longrightarrow H(P_1,\dots,P_n,P_{n+1},\dots,P_k) \leq H(P,P_{n+1},\dots,P_k)$

D) For any family (e_j) of minimal projections of P such that $\sum_j e_j = 1$ one has

$$H(P) = \sum_j \eta\tau(e_j)$$

E) If $(P_1 \cup \dots \cup P_n)''$ is generated by pairwise commuting von Neumann ubalgebras Q_j of P_j then

$$H(P_1,\dots,P_n) = H((P_1 \cup \dots \cup P_n)'')$$

F) $H(P_1,\dots,P_n) \leq H(Q_1,\dots,Q_n) + \sum_{j=1}^{n} H(P_j|Q_j)$

G) $H(P|Q) \leq H(P|N) + H(N|Q)$

H) $H(P|Q)$ is increasing in P and decreasing in Q

I) If P and Q commute then

$$H(P|Q) = H((P\cup Q)''|Q) = H((P\cup Q)'') - H(Q)$$

emark 3. a) From the properties B, D and H it follows that the entropy nd the relative entropy have finite values.
) Using properties D, E and I we see that when M is abelian the above ntropies coincide with the classical one.

The most difficult result proved by Connes and Størmer is the trong continuity of the relative entropy. More precisely if P and Q are on Neumann subalgebras of M we shall write $P \overset{\delta}{\subset} Q$ for $\delta > 0$ if for all $x \in P$, $\|x\| \leq 1$, there exists $y \in Q$, $\|y\| \leq 1$, $\|x-y\|_2 < \delta$ where $\|a\|_2 = \tau(a^*a)^{1/2}$.

Theorem 4. For each integer n and each $\varepsilon > 0$, there is a $\delta > 0$ such that for any pair of finite dimensional von Neumann subalgebras P, Q of M

$$(\dim P = n,\ P \overset{\delta}{\subset} Q) \Rightarrow H(P|Q) < \varepsilon.$$

Let θ be an automorphism of M preserving the trace τ. Similary to the classical case, it is now easy to define the entropy of θ.

Definition 5. Let P be a finite dimensional von Neumann subalgebra of M, we put

a) $H(P, \theta) = \lim_{n \to \infty} n^{-1} H(P, \theta(P), \ldots, \theta^n(P))$

b) $H(\theta) = \sup_P H(P, \theta)$

We call $H(\theta)$ the entropy of the automorphism θ.

Notice that the limit in a) exists because of property B. It is clear from the definitions that the entropy is an invariant of conjugacy for automorphism (i.e. if σ and θ are automorphisms such that $\theta = \alpha\sigma\alpha^{-1}$ for some automorphism α of M then $H(\theta) = H(\sigma)$).

As in the classical theory one of the most useful tool for the computation of the entropy of an automorphism is the Kolmogorov-Sinai theorem which can be stated only when M is hyperfinite.

Theorem 6. [CS] Assume that M is hyperfinite and let (N_k) be an increasing sequence of finite dimensional von Neumann subalgebras of M with $\bigcup_k N_k$ weakly dense in M. Then

$$H(\theta) = \lim_{k \to \infty} H(N_k, \theta).$$

The proof of this theorem is an easy consequence of theorem 4.

Remark 7. a) Another possible candidate for the entropy of an automorphism is the abelian entropy

$$H_a(\theta) = \sup h(\theta|A)$$

where the supremum is taken over all abelian von Neumann subalgebras A of M with $\theta(A) = A$ and $h(\theta|A)$ is the entropy of $\theta|A$ defined in the classical theory. But S. Popa [P] has proved that there exist automorhisms leaving no abelian von Neumann subalgebras invariant.

b) In [PP] M. Pimsner and S. Popa have found an interesting relaion between the index, defined by V. Jones [J], of a subfactor of a inite factor and the relative entropy.

. Non-commutative stationnary Markov states.

Let $M_o = M_d(\mathbb{C})$, $d > 1$, and C be the C*-algebra $C = \bigotimes_{\mathbb{Z}} M_o$. Let J_n e the canonical injection of M_o into the n-th factor of C. For $I \subset \mathbb{Z}$ e put $M_I = \bigvee_{n\in I} J_n(M_o)$ where $A \vee B$ is the C*-algebra generated by the *-algebras A and B.

efinition 1. Let ϕ_o be a state on M_o and $E: M_o \otimes M_o \to M_o$ be a completely ositive linear map. Then (ϕ_o, E) is called a Markov pair if

1) $E(1 \otimes 1) = 1$

2) $\phi_o(E(a \otimes 1)) = \phi_o(a)$ for all $a \in M_o$

3) $\phi_o(E(1 \otimes E(E(a \otimes b) \otimes c))) = \phi_o(E(a \otimes E(b \otimes c)))$
for all $a, b, c \in M_o$

4) $\phi_o(E(a_o \otimes E(a_1 \otimes \ldots \otimes E(a_n \otimes 1)))) =$
$$\phi_o(E(a_o \otimes E(a_1 \otimes \ldots \otimes E(a_{n-1} \otimes a_n))))$$
for all $a_o, a_1, \ldots, a_n \in M_o$.

sing properties 2), 3) and 4) it is easy to see that

$$\phi_o(E(1 \otimes E(a \otimes b))) = \phi_o(E(a \otimes b))$$
or all $a, b \in M_o$.

If (ϕ_o, E) is a Markov pair, for $p, q \in \mathbb{Z}$, $p < q$, we define a state $\phi_{[p,q]}$ on $M_{[p,q]}$ by

$$\phi_{[p,q]}(J_p(a_p)\ldots J_q(a_q)) = \phi_o(E(a_p \otimes E(a_{p+1} \otimes \ldots \otimes E(a_{q-1} \otimes a_q))))$$
or all $a_p, \ldots, a_q \in M_o$.

efinition 2. [A] The state ϕ on C defined by the sequence $\{\phi_{[-n,n]}\}_{n\geq 1}$ s called a stationary quantum Markov state.

Let α be the shift on C (i.e. $\alpha(J_n(a)) = J_{n+1}(a)$) then because of he properties of the Markov pair (ϕ_o, E), the above quantum Markov tate ϕ is α-invariant.

From now we assume that ϕ is faithful. Let M be the von Neumann algebra obtained from the GNS representation of the state ϕ on C and let N be the centralizer of ϕ in M (i.e. $N = \{x \in M;\ \phi(xy) = \phi(yx)$ for all $y \in M\}$). Then N is a finite von Neumann algebra and the restriction of ϕ to N is a faithful, normal and normalized trace.

<u>Definition 3.</u> Since the Markov state ϕ on M is α-invariant, the automorphism θ of N obtained from the restriction of α to N is well defined and is preserving the trace ϕ on N. We call it the <u>non-commutative Markov shift</u> associated with the stationnary quantum Markov state ϕ.

If σ^{ϕ} is the modular group for ϕ in M and for $p,\ q \in \mathbb{Z}$, $p < q$, $\sigma^{(p,q)}$ is the modular group for $\phi_{[p,q]}$ in $M_{[p,q]}$, we define

$$N_{[p,q]} = \{x \in M_{[p,q]};\ \sigma_t^{(p-1,q+1)}(x) = x \text{ for all } t \in \mathbb{R}\}$$

$$Z_{[p,q]} = \{x \in M_{[p,q]};\ \sigma_t^{(p,q)}(x) = x \text{ for all } t \in \mathbb{R}\}$$

<u>Lemma 4.</u> a) For all $p,\ q \in \mathbb{Z}$, $p < q$, we have

$$\sigma_t^{(p-k,q+j)}(x) = \sigma_t^{(p,q)}(x)$$

for all $x \in M_{[p+1,q-1]}$, all $k,\ j \geq 0$ and all $t \in \mathbb{R}$. Therefore

$$N_{[p,q]} \subset Z_{[p,q]}.$$

b) The sequence $N_{[-n,n]}$ of finite dimensional von Neumann algebras is increasing and generates N.

<u>Proof.</u> a) The first assertion is a consequence of Theorem 4.2 of [AF]. Indeed this theorem says that

$$\sigma_t^{(p,q+1)}(x) = \sigma_t^{(p,q)}(x)$$

for all $x \in M_{[p,q-1]}$ and all $t \in \mathbb{R}$. In the same way we get

$$\sigma_t^{(p-1,q)}(x) = \sigma_t^{(p,q)}(x)$$

for all $x \in M_{[p+1,q]}$ and all $t \in \mathbb{R}$. Hence

$$\sigma_t^{(p-k,q+j)}(x) = \sigma_t^{(p,q)}(x)$$

for all $x \in M_{[p+1,q-1]}$, all $k,\ j \geq 0$ and all $t \in \mathbb{R}$.
Notice that property 4 (resp. property 3) of the Markov pair (ϕ_o, E) is used to prove the first (resp. second) equality.

ıerefore if $n > \max\{|p|, |q|\}$ then

$$\sigma_t^{(-n,n)}(x) = \sigma_t^{(p-1,q+1)}(x)$$

ɔr all $x \in M_{[p,q]}$ and all $t \in \mathbb{R}$. Since $\sigma_t^{(-n,n)}$ converges strongly to σ_t^{ϕ} ıen $n \to \infty$ we get

$$\sigma_t^{\phi}(x) = \sigma_t^{(p-1,q+1)}(x)$$

ɔr all $x \in M_{[p,q]}$ and all $t \in \mathbb{R}$, so $N_{[p,q]} \subset Z_{[p,q]}$.

b) Follows from a).

q.e.d.

Let $W_{[p,q]}$ be the density matrix of $\phi_{[p,q]}$ in $M_{[p,q]}$ and write

$$W_{[p,q]} = \sum_i w_i(p,q) e_i(p,q)$$

ıere $e_i(p,q)$ are minimal projections in $M_{[p,q]}$ with sum 1. Then $e_i(p,q)$ $Z_{[p,q]}$ and the abelian von Neumann algebra $A_{[p,q]}$ generated by the (p,q) is maximal abelian in $Z_{[p,q]}$.

Recall that for $x \in M_n(\mathbb{C})$, $x \geq 0$, the quantum mechanical entropy x is defined by

$$S(x) = \mathrm{Tr}(\eta(x)) = - \mathrm{Tr}(x \log x)$$

ıere Tr is the usual (non-normalized) trace on $M_n(\mathbb{C})$.

Now we can state our result.

ıeorem 5. Assume that $A_{[p,q]} \subset N_{[p,q]}$ for all $p,q \in \mathbb{Z}$, $p < q$. Then the ɔnnes - Størmer entropy of the non-commutative Markov shift θ associa-ed with the stationnary quantum Markov state ϕ is

$$H(\theta) = \lim_{k\to\infty} k^{-1} S(W_{[0,k]}).$$

roof. By the Kolmogorov - Sinai theorem and lemma 4b) we have

$$H(\theta) = \lim_{n\to\infty} H(N_{[-n,n]}, \theta)$$

ıd for a fixed n

$$H(N_{[-n,n]}, \theta) = \lim_{q\to\infty} q^{-1} H(N_{[-n,n]}, \theta(N_{[-n,n]}), \ldots, \theta^q(N_{[-n,n]}))$$

$$= \lim_{q\to\infty} q^{-1} H(N_{[-n,n]}, N_{[-n+1,n+1]}, \dots, N_{[-n+q,n+q]}).$$

Using lemma 4a) we see that

$$N_{[-n+j,n+j]} \subset N_{[-n,n+q]}$$

for all $j = 0, \dots, q$. Since $A_{[-n,n+q]} \subset N_{[-n,n+q]}$, by the properties C and D given in section 1 we get

$$H(N_{[-n,n]}, N_{[-n+1,n+1]}, \dots, N_{[-n+q,n+q]}) \leq H(N_{[-n,n+q]})$$
$$= H(A_{[-n,n+q]}) = H(A_{[0,n+2q]}).$$

Moreover for all $k > 0$ we have

$$H(A_{[0,k]}) = -\sum_i \phi_{[0,k]}(e_i(0,k)) \log \phi_{[0,k]}(e_i(0,k))$$
$$= -\sum_i Tr(W_{[0,k]} e_i(0,k)) \log Tr(W_{[0,k]} e_i(0,k))$$
$$= -\sum_i w_i(0,k) \log w_i(0,k)$$

because the $e_i(0,k)$ are minimal projections in $M_{[0,k]}$. So

$$H(A_{[0,k]}) = -Tr(W_{[0,k]} \log W_{[0,k]}) = S(W_{[0,k]})$$

and we get

$$H(N_{[-n,n]}, \theta(N_{[-n,n]}), \dots, \theta^q(N_{[-n,n]})) \leq S(W_{[0,2n+q]}).$$

Therefore

$$H(N_{[-n,n]}, \theta) \leq \lim_{q\to\infty} q^{-1} S(W_{[0,2n+q]}) = \lim_{k\to\infty} k^{-1} S(W_{[0,k]})$$

and then

$$H(\theta) \leq \lim_{k\to\infty} k^{-1} S(W_{[0,k]}).$$

For the converse inequality we have by definition

$$H(\theta) \geq H(A_{[-n,n]}, \theta)$$

for all $n \geq 0$ and by the above computation

$$H(A_{[-n,n]}, \theta) = \lim_{k\to\infty} k^{-1} S(W_{[0,k]}).$$

q.e.d.

Remark that without assuming $A_{[p,q]} \subset N_{[p,q]}$ we easily have $H(\theta) \leq \lim_{k\to\infty} k^{-1} S(W_{[0,k]})$ since $N_{[p,q]} \subset Z_{[p,q]}$.

xample 6. [B2] The one dimensional Ising model

Let $K = e^{k \otimes k} \in M_o \otimes M_o$ with $k = k^* \in M_o$ and with $\overline{Tr_1}(K^2) = 1$ here $\overline{Tr_1} : M_o \otimes M_o \to M_o$ is the conditional expectation defined by $\overline{Tr_1}(x \otimes y) = xTr(y)$.

Put $k = \sum_i k_i e_i$ where the e_i are minimal projections in M_o with sum ne. If $P = K^2 = \sum_{i,j} p_{ij} e_i \otimes e_j$, then the matrix $p = (p_{ij}) \in M_o$ is a tochastic matrix. let $\Lambda = (\lambda_1, \ldots, \lambda_d)$, $\lambda_i \geq 0$, $\sum_i \lambda_i = 1$, be the left igenvector of p for the eigenvalue 1 and consider the state ϕ_o on M_o efined by the density matrix $W_o = \sum_i \lambda_i e_i$. Let $E : M_o \otimes M_o \to M_o$ be the ompletely positive linear map given by

$$E(x) = \overline{Tr_1}(KxK)$$

or all $x \in M_o \otimes M_o$. Then it is easy to see that (ϕ_o, E) is a Markov air and that the assumption in Theorem 5 is satisfied.

If ϕ is the Markov state given by this pair and θ is the induced arkov shift, one can show that

$$H(\theta) = - \sum_{i,j} \lambda_i p_{ij} \log p_{ij}.$$

oreover using a result of W. Krieger [K] one can see that the von Neumann lgebra N is the hyperfinite II_1-factor.

eferences.

A] L. Accardi; Non commutative Markov chains; Proc. Int. School, Univ. Camerino (1974) 268-295.

AF] L. Accardi, A. Frigerio; Markovian cocycle; Proc. R. Ir. Acad. 83 A (1983) 251-263.

B1] O. Besson; Sur l'entropie des automorphismes des algèbres de von Neumann finies; Thèse Univ. de Neuchâtel (1982)

B2] O. Besson; A note on non commutative Markov states; preprint Univ. of Oslo (1984).

CS] A. Connes, E. Størmer; Entropy for automorphism of II_1-von Neumann algebras; Acta Math. 134 (1975) 289-306.

J] V.F.R. Jones; Index for subfactors; Invent. Math. 72 (1983) 1-25.

K] W. Krieger; On the finitary isomorphism of Markov shifts that have finite expected coding time; Z. Wahrscheinlichkeitstheorie verw. Geb. 65 (1983) 323-328.

PP] M. Pimsner, S. Popa; Entropy and the index for subfactors; preprint INCREST (1983).

P] S. Popa; Maximal injective subalgebras in factors associated with free groups; Adv. in Math. 50 (1983) 27-48.

Entropic Uncertainty Relations in Quantum Mechanics

Iwo Bialynicki-Birula[1]

Certum est quia impossibile est

Tertullian

New uncertainty relations based on the information entropy are reviewed and contrasted with the traditional uncertainty relations, which were based on the dispersions of the the physical variables. Improved lower bounds are given for the position-momentum and the angle-angular momentum pairs. Novel uncertainty relation for the angular distribution and angular momentum in the three-dimensional space is introduced.

KEY WORDS: Uncertainty relations, information, entropy, quantum measurement.

1. INTRODUCTION

Every notable physical theory has its trademark - an eye catching symbol or equation - that can be put on a book-cover or a T-shirt. In Newtonian mechanics it is $\vec{F} = m\vec{a}$, in relativity theory it is $E = mc^2$, in the theory of elementary particles it is a Feynman diagram, and in quantum mechanics it is, of course, the Heisenberg relation

$$\Delta x \cdot \Delta p \sim h \quad , \qquad (1)$$

expressing the uncertainty principle.

[1] Institute for Theoretical Physics, Lotnikow 32/46, 02-668 Warsaw, Poland.

This relation was discovered by Heisenberg in 1927, already after the foundations of matrix mechanics and wave mechanics were laid down. The first formulation of the uncertainty principle appeared in the paper(1) "On the intuitive content of the quantum-theoretic kinematics and mechanics", where Heisenberg explained the uncertainty principle (1) in the following way: "The more accurately the position is determined, the less accurately the momentum is known and conversely." In his uncertainty principle Heisenberg saw a "direct intuitive interpretation" of the commutation relations between the position and the momentum operators that were in those early days of quantum mechanics still full of mystery.

The derivation of the uncertainty principle from the mathematical formalism of wave mechanics was given by Kennard(2). In his formula

$$\Delta x \cdot \Delta p \geqslant \hbar/2 \tag{2}$$

the quantities Δx and Δp are already precisely defined as standard deviations or variances in mathematical terminology. Shortly afterwards Robertson(3) proved the general uncertainty principle, valid for an arbitrary pair of the physical quantities A and B described by two operators $\hat{A}$ and $\hat{B}$. The Robertson inequality reads:

$$\Delta A \cdot \Delta B \geqslant \frac{\hbar}{2} |\langle \hat{C} \rangle| , \tag{3}$$

where the operator $\hat{C}$ is defined as

$$\hat{C} = (\hat{A}\hat{B} - \hat{B}\hat{A})/i\hbar , \tag{4}$$

the brackets <> denote the average values and the standard deviations are defined, as usual, by the formula

$$(\Delta A)^2 = \langle (A - \langle A \rangle)^2 \rangle \quad . \tag{5}$$

An improved version of the Robertson inequality is due to Schrödinger[4]. The Schrödinger inequality is stronger than the Robertson inequality, because it contains an additional term on the right hand side,

$$(\Delta A)^2 \cdot (\Delta B)^2 \geqslant (\langle \hat{A}\hat{B} + \hat{B}\hat{A} \rangle / 2 - \langle \hat{A} \rangle \langle \hat{B} \rangle)^2 + \frac{\hbar^2}{4} |\langle \hat{C} \rangle|^2 \ . \tag{6}$$

In all inequalities listed above the dispersion or its square root - the mean standard deviation - are used as a measure of the spread (uncertainty) of the physical quantities around their mean values. It is true that standard deviations play a very important role in statistics, but they are not the only available measure of uncertainty. For example, one uses also the mean deviation. In principle, one may use all kinds of moments and expressions related to various moments[5] to describe the spreading of values around the mean value, although the dispersion (defined in terms of second moments) is the easiest to use; all proofs are very simple.

Still, the characteristics of the distribution of the values based on the calculation of moments look somewhat arbitrary and seem to be lacking a more fundamental significance. Standard uncertainty relations have also been criticized for not being adequate for some interesting physical situations[5].

2. ENTROPIC UNCERTAINTY RELATIONS

There exists a measure of spreading that is clearly of fundamental importance. It is the information entropy used by Shannon[6] to build the modern theory of communication.

The information entropy H is defined by the formula

$$H = - \sum_i p_i \ln p_i \ , \qquad (7)$$

where p_i is the probability to find the i-th result of a measurement or the a priori probability of the i-th message. In information theory one uses the logarithms to the base 2 (H is then measured in bits), but a change of the base results only in a change of the scale; H gets multiplied by a constant factor.

The information entropy may serve as a very accurate measure of uncertainty and it has even been called by that name[7]. The information entropy serves at the same time as a measure of uncertainty and as a measure of information. One may choose one of these two interpretations depending on whether one is dealing with the situation before the experiment had been performed, when one wants to determine the uncertainty about its outcome, or whether one wants to evaluate the amount of information that has been gained in the experiment. In an ideal experiment (an analog of a reversible process) these two measures are equal; information is transferred without loss and the information gain cancels the uncertainty.

There is strong evidence that the information entropy is a much better measure of uncertainty or spreading than the dispersion. First of all, one should mention in this respect the noiseless coding theorem (cf., for example, Ref.7): Average number of elementary questions (i.e. the questions to which the answer is only yes or no) needed to discover the "truth" hidden in one of the N boxes with the probability distribution p_i is bounded from below by H and by a proper choice of the strategy one may approach H arbitrarily close. Another confirmation that H is the correct measure of uncertainty comes from experimental psychology. For example, Hyman[8] measured the time needed to process the information (reaction time) transmitted by light signals. Sets of lamps were being lighted according to certain patterns that were governed by probabilistic distributions. It turned out that the reaction time of the recipients of these light signals varied as a linear function of the uncertainty in the stimulus measured by H. Finally, a beautiful confirmation that the entropy (7) gives the correct measure of information comes from general relativity. As has been shown by Beckenstein and Hawking (cf. Ref.9 for a thorough nontechnical review of these problems), the information that disappears into a black hole must be added to other forms of the entropy in the Universe in order to guarantee the validity of the second law of thermodynamics. I may also add that the expression (7) can be derived from a set of fairly natural axioms (cf., for example, Ref.7).

Information entropy seems to be the best measure of information and uncertainty in the analysis of various phenomena: from communication lines to human reaction times. Information entropy is a perfect example of a crucial importance in science of a good definition. The whole new branch of science - information theory - has sprung up from this one definition.

3. ENTROPIC UNCERTAINTY RELATION FOR POSITION AND MOMENTUM

I shall derive now the uncertainty relation for the position and momentum in quantum mechanics with the information entropy as the measure of uncertainty. Let me begin with a few words on the history of the subject.

It all started almost 30 years ago, when a physicist Everett[10] and a mathematician Hirschman[11], independently and almost at the same time, conjectured that the wave function $\psi(x)$ and its Fourier transform $\tilde{\psi}(p)$,

$$\tilde{\psi}(p) = (2\pi\hbar)^{-\frac{1}{2}} \int dx\ e^{-ipx/\hbar}\ \psi(x) \quad , \tag{8}$$

satisfy the following inequality

$$-\int dx\ |\psi|^2 \ln|\psi|^2 - \int dp\ |\tilde{\psi}|^2 \ln\left(|\tilde{\psi}|^2 \hbar\right) \geq 1 + \ln\pi \quad . \tag{9}$$

Their conjecture was supported by the observation that this inequality is saturated by all Gaussian functions and that the variation of the left hand side around the Gaussian function vanishes. For those who are interested in the problems of interpretation of quantum theory, I would like to add a side remark. Everett derived the inequality (9) in the expanded version[10] (published some 15 years after it has been written) of his Ph.D. thesis. Everett's thesis[12] contained the, now famous, many-worlds interpretation of quantum mechanics.

The first proof of the inequality (9) was given by Mycielski and myself[13] and independently by Beckner[14] almost 20 years after its discovery. This inequality represents by itself an important and fundamental mathematical relation, but it can not be treated as the entropic uncertainty relation, because the integrals appearing in it do not have a direct physical interpretation as measures of the uncertainty. This inequality is, however, instrumental in the derivation of the proper entropic uncertainty principle, expressed in terms of the uncertainty measures H for the position and momentum.

To set a general framework for entropic uncertainty relations, let me consider a physical quantity A described in the formalism of quantum mechanics by the self-adjoint operator $\hat{A}$. For each such operator there exists a spectral family of projection operators, say P_i^A. These operators project on the subspaces of the Hilbert space, characterized by a certain partition of the spectrum of $\hat{A}$ into cells or bins (in the terminology of Partovi[16]). In the simplest case, when the operator $\hat{A}$ has a pure point spectrum, the projection operators P_i may be chosen as projectors on the eigenspaces of $\hat{A}$. In the general case, in the i-th bin one may have values from the point spectrum and from the continuous spectrum. In turn, to each projection operator P_i^A one may assign the probability p_i^A that for a given state of the system the measurement of A will yield a value from the i-th bin $\Delta\alpha_i$. For a pure state , described by a state vector Ψ, the formula for the probability p_i^A reads:

$$p_i^A = (\Psi | P_i^A \Psi) \quad . \tag{10}$$

The generalization to mixed states, each described by a density matrix ρ, is given by the formula:

$$p_i^A = \mathrm{Tr}\{ P_i^A \rho \} \quad . \tag{11}$$

From the set of all p_i^A's defined for a given state by either (10) or (11), one may construct the measure of uncertainty H^A that characterizes the measurements of the physical quantity A in a given state and based on the chosen partition of the spectrum of A into the bins $\Delta\alpha_i$. Let us notice that the minimal value of the uncertainty - the certainty - is attained only when all the values α that result from measurements on the chosen state of the system belong to just one bin, i.e. when the state vector or the density operator ρ of the chosen state lie in the subspace defined by the projection operator P_i^A.

The formulation of the uncertainty principle in terms of the uncertainty measure (7) was introduced recently by Deutsch(15), for the purely point spectrum, and by Partovi(16), for the general case.

The starting point of Deutsch's analysis was a critique of the Robertson-Schrödinger inequalities, on the grounds that their right hand side depends in general on the state of the system through the expectation values of the operators. Deutsch was seeking a measure of uncertainty that would be independent of the state of the system and found it in H. The inequality that he derived for a pair of physical quantities A and B, described by the operators $\hat{A}$ and $\hat{B}$ with pure point spectrum, reads:

$$H^A + H^B \geq 2\ \ln\frac{2}{1+\sup\{|\langle a|b\rangle|\}} \quad , \tag{12}$$

where $|a\rangle$ and $|b\rangle$ are the eigenvectors of the operators $\hat{A}$ and $\hat{B}$ and the supremum is taken with respect to all pairs of the eigenvectors. The inequality (12) is nontrivial (i.e. the right hand side is greater than zero) only when the operators $\hat{A}$ and $\hat{B}$ do not have a common eigenvector.

Partovi extended this approach to cover the general case and obtained certain bounds on the sums of the two uncertainties for two important pairs of physical quantities: position-momentum and angle-angular momentum. I shall not describe his results here, because I will be able to improve upon them significantly with the help of our inequalities that were proven in Ref.13.

Let me begin with the analysis of the position-momentum pair in one dimension. For the position x, the probabilities p_i^x for a pure state are given by the expressions

$$p_i^x = \int_{x_i}^{x_{i+1}} dx\ |\psi|^2 \quad . \tag{13}$$

In this case the projection operators cut out from the wave function that part which has its support in the interval (x_i, x_{i+1}). The finer the partition, the greater the uncertainty; H^x tends to infinity when the size of the largest bin tends to zero. In the opposite limiting case, when one bin covers the whole real axis, the measured value of the coordinate will for sure be found in this bin and the uncertainty H^x will be equal to zero for every wave function.

In an analogous manner one may define the probability p_i^p to find the value of the momentum in a given interval,

$$p_i^p = \int_{p_i}^{p_{i+1}} dp\ |\tilde{\psi}|^2 \quad . \tag{14}$$

Both measures of uncertainty, H^x and H^p can be separately made equal to zero by choosing the wave function or its Fourier transform in such a way that the support of it lies in only one bin. One can not achieve this, however, <u>simultaneously</u> for both quantitites when the bins are of finite size. The entropic uncertainty relation is a precise, quantitative expression of this fact. I shall restrict myself first to the simplest case, which happens to be also the most interesting one, when all the bins for the measurements of the coordinate x are equal to Δx and those for the measurements of momentum are equal to Δp. From now on, Δx and Δp will denote the bin sizes, not the standard deviations. As has been emphasized by Partovi[(16)], Δx and Δp describe the resolutions of the measuring devices.

On the basis of dimensional analysis one may infer that the sum of the two uncertainties H^x and H^p is bounded from below by a certain function of the dimensionless parameter γ,

$$\gamma = \Delta x \cdot \Delta p / h \quad . \tag{15}$$

I shall determine this function in the most interesting case, when the value of γ is much smaller then 1. To this end, I shall make use of the following basic inequality for convex functions (cf., for example, Ref.17)

$$\langle f(x) \rangle \geqslant f(\langle x \rangle) \quad , \tag{16}$$

which simply states that the average value of the convex function may never exceed the value of this function at the average value of the argument. This inequality is most easily proven by averaging the Taylor expansion with the remainder in the Lagrange form,

$$f(x) = f(\langle x \rangle) + (x - \langle x \rangle) f'(\langle x \rangle) + \tfrac{1}{2}(x - \langle x \rangle)^2 f''(c) \quad , \tag{17}$$

where c lies between x and <x>, and then using the fact that the second derivative of the convex function is nonnegative. Since $x\ln x$ is a convex function, one obtains from (16):

$$\frac{1}{\Delta x}\int_{\Delta x} dx\ |\psi|^2 \ln|\psi|^2 \geqslant \frac{1}{\Delta x}\int_{\Delta x} dx\ |\psi|^2 \ln\Big(\frac{1}{\Delta x}\int_{\Delta x} dx\ |\psi|^2\Big) \quad , \tag{18}$$

or

$$-\int_{\Delta x} dx\ |\psi|^2 \ln\Big(\int_{\Delta x} dx\ |\psi|^2\Big) \geqslant -\int_{\Delta x} dx\ |\psi|^2 \ln|\psi|^2 - \ln(\Delta x)\int_{\Delta x} dx |\psi|^2 \quad . \tag{19}$$

Let us notice that the difference between both sides of the inequality (19) goes to zero faster that Δx, when Δx tends to zero.

I shall now add up the inequalities (19) for the coordinate wave function and the analogous inequalities for the momentum wave functions. Assuming that the wave function is normalized to one, I obtain

$$H^x + H^p \geqslant -\int dx\ |\psi|^2 \ln|\psi|^2 - \int dp\ |\tilde{\psi}|^2 \ln\Big(|\tilde{\psi}|^2\hbar\Big) - \ln(2\pi\gamma) \quad . \tag{20}$$

Finally, I shall use our result obtained with Mycielski to obtain the entropic uncertainty relation for the position-momentum pair,[19]

$$H^x + H^p \geqslant 1 - \ln 2 - \ln(\Delta x \Delta p/h) \quad . \tag{21}$$

It follows from the remark made right after the formula (19) that the inequality (21) becomes exact in the limit when γ tends to zero. For example, for $\gamma = .05$, the relative difference between the left and the right hand side of (21) for the Gaussian wave function is only 3%.

Partovi has obtained an estimate of the form

$$H^x + H^p \geqslant -2\ln\Big(\frac{1+\mu}{2}\Big) \quad , \tag{22}$$

where μ varies from $\sqrt{\gamma}$ to 1, when γ changes over its whole range from zero to infinity. In the most interesting case of small γ his inequality is much weaker than mine, whereas for large γ it is stronger.

4. ENTROPIC UNCERTAINTY RELATION FOR ANGLE AND ANGULAR MOMENTUM

In this Section, I shall describe the entropic uncertainty relation for the angle-angular momentum pair. This is a very interesting case, because one can not handle it with the help of the standard methods that employ the dispersions of the relevant variables. The point is that there is no self-adjoint operator to represent the angle. As a result, the angle ϕ and the z-component of the angular momentum L_z do not form a canonically conjugate pair. This is, of course, directly related to the fact that the angular variable returns to its original value after the change by 2π, which leads to the quantization of the angular momentum. The traditional uncertainty relations may be written down only for periodic functions of the angle, but not for the angle itself . Modified uncertainty relations that overcome the problem of the absence of the angle operator have also been proposed[18], but they were based on a rather arbitrary definition of the uncertainty measure for the angle.

As it turns out this case is even better suited to the description in terms of the entropic uncertainty measures, because in this case one may easily obtain the optimal lower bound on the right hand side of the uncertainty relation. Since the reasoning that leads to the final inequality is very similar to that for the position-momentum pair, I shall give only a brief sketch of the proof.

The analog of the inequality (9), applicable to the angle-angular momentum pair, has also been derived in our paper with Mycielski[13]. It has the form

$$- \int_0^{2\pi} d\phi \; |\psi|^2 \; \ln|\psi|^2 \; - \sum_m |c_m|^2 \; \ln|c_m|^2 \geq \ln(2\pi) \quad , \tag{23}$$

where ψ is the wave function depending on the angular variable ϕ and c_m's are its expansion coefficients into the set of eigenfunctions of L_z,

$$\psi(\phi) = (2\pi)^{-\frac{1}{2}} \sum_{-\infty}^{\infty} e^{im\phi} \; c_m \quad . \tag{24}$$

With the use of our inequality one obtains the following entropic uncertainty relation for the angle and the angular momentum[19]

$$H^{\phi} + H^{L_z} \geq -\ln\frac{\Delta\phi}{2\pi} \quad , \tag{25}$$

where $\Delta\phi$ is the size of the bin in the angular variable. The measures of uncertainty H^{ϕ} and H^{L_z} are constructed according to the general prescription (7) from the probabilities p_i^{ϕ},

$$p_i^{\phi} = \int_{\Delta\phi_i} d\phi \; |\psi|^2 \quad , \tag{26}$$

and from the probabilities $p_m^{L_z}$,

$$p_m^{L_z} = |c_m|^2 \quad . \tag{27}$$

The inequality (25) is optimal; it is saturated by all eigenfunctions of the angular momentum operator L_z.

There exists also the entropic uncertainty relation characterizing the measurements of two angular variables on the sphere (angular distributions in in physical space) and the measurements of the z-component of the angular momentum together with the total angular momentum. This entropic uncertainty relation reads [20]

$$H^{(\phi,\theta)} + H^{(L_z,L^2)} \geqslant -\ln\frac{\Delta\Phi}{2\pi} \quad , \tag{28}$$

where the relevant probabilities are defined as follows

$$\psi(\phi,\theta) = \sum_{lm} c_{lm} Y_l^m(\phi,\theta) \quad , \tag{29}$$

$$p_i^{(\phi,\theta)} = \int_{\Delta\Omega_i} d\phi \sin\theta \, d\theta \, |\psi|^2 \quad , \tag{30}$$

$$p_i^{(L_z,L)} = |c_{lm}|^2 \quad , \tag{31}$$

and $\Delta\Phi$ is the largest angular size of the bins in the (ϕ,θ) variables, as measured along great circles.

5. CONCLUSIONS

In the presentation of the entropic uncertainty relations for (x,p) and (ϕ,L_z) pairs, I have restricted myself to the simplest case of equal bins. Generalization of these relations to the case of unequal bin sizes is, however, straightforward. The entropic uncertainty relations for the position-momentum and the angle-angular momentum retain their forms (21) and (25), respectively, but in the general case Δx, Δp, and $\Delta\phi$ denote the sizes of <u>the largest bins</u>. I am sure that other generalizations and refinements of the entropic uncertainty relations will be discovered in the future.

Entropic uncertainty relations require, in general, more advanced mathematical methods in their derivations, but can compete with the standard relations as far as the depth of the notions used and the elegance of the final results are concerned. I am certain that these relations will in the future find their way into the texbooks of quantum mechanics.

REFERENCES

1. W. Heisenberg, Z. Phys. 43, 172 (1927).
2. E.H. Kennard, Z. Phys. 44, 326 (1927); Phys. Rev. 31, 344 (1928).
3. H.P. Robertson, Phys. Rev. 34, 163 (1929).
4. E. Schrödinger, Sitzungber. Preuss. Akad. Wiss. 296 (1930).
5. J.B.M. Uffink and J. Hilgevoord, Phys. Lett. 105A, 176 (1984).
6. C.E. Shannon, Bell System Tech. J. 27, 379, 623 (1948).
7. R.B. Ash, *Information Theory* (Interscience, New York, 1965).
8. R. Hyman, J. Exper. Psych. 45, 188 (1955).
9. J.D. Beckenstein, Physics Today 33, 24 (1980).
10. H. Everett, in *The Many-Worlds Interpretation of Quantum Mechanics*, B.S. DeWitt and N. Graham, eds. (Princeton U. Press, Princeton, 1973).
11. I.I. Hirshman, Amer. J. Math. 79, 152 (1957).
12. H. Everett, Rev. Mod. Phys. 29, 454 (1957).
13. I. Bialynicki-Birula and J. Mycielski, Comm. Math. Phys. 44, 129 (1975).
14. W. Beckner, Ann. Math. 102, 159 (1975).
15. D. Deutsch, Phys. Rev. Lett. 50, 631 (1983).
16. M.H. Partovi, Phys. Rev. Lett. 50, 1883 (1983).
17. E.F. Beckenbach and R. Bellman, *Inequalities* (Springer, Berlin, 1961).
18. D. Judge, Phys. Lett. 5, 189 (1963).
    D. Judge and J.T. Lewis, Phys. Lett. 5, 190 (1963).
    D. Judge, N.Cim. 31, 332 (1964).
    P. Carruthers and M.M. Nieto, Rev. Mod. Phys. 40, 411 (1968).
19. I. Bialynicki-Birula, Phys. Lett. 103A, 253 (1984).
20. I. Bialynicki-Birula and J. Madajczyk (to be published).

ESTIMATES OF QUANTUM DEVIATIONS FROM CLASSICAL MECHANICS USING LARGE DEVIATION RESULTS

Ph. BLANCHARD*, Ph. COMBE**, M. SIRUGUE, M. SIRUGUE-COLLIN***

Centre de Physique Théorique
CNRS - Luminy - Case 907
F-13288 MARSEILLE CEDEX 9 (France)

I. INTRODUCTION

Classical and quantum motions of particles are very different. Indeed one cannot assign a trajectory to a quantum particle. However one has the physical intuition that quantum particles wander around the corresponding classical path in phase space by an amount of h. Deviations come from the uncertainty principle.

Underlying this picture, there are two ideas :

i) there is some kind of probability associated with possible paths ;

ii) this probability concentrates around the classical path in a region whose magnitude is related to h.

The first attempt to incorporate these ideas in an operative schema was made by R.P. Feynman in the late forties [1],[2]. He wrote the transition probability amplitude as an integral over all possible paths γ damped by a factor $\exp\left\{\frac{i}{\hbar} S(\gamma)\right\}$, $S(\gamma)$ being the classical action. This factor strongly enhances the classical path as much as h is small.

As it is well known, it is difficult to understand this representation on a rigorous mathematical basis. For years it has been only a useful heuristic tool. Two decades ago it got a renewed interest and a rigorous definition in the Euclidean region and it was at the origin of the strong development of probabilistic ideas and techniques in the field, see e.g. [3] and references therein.

Some years ago Maslov and Chebotarev remarked that it was possible to interpret Feynman's representation in a bona fide sense even in the real time region [4] [5]. Subsequently these ideas were extended to many domains and especially to field theory [6] [7] [8].

---

* and Universität Bielefeld, RFA

** and Université d'Aix-Marseille II, Luminy, Marseille, France

*** and Université de Provence, Marseille, France

In what follows we want to show that there exists in phase space a probabilistic schema which accounts in a much more transparent way for the intuitive ideas alluded above.

To achieve this program it is convenient to deal with the description of quantum states not in term of wave function but in term of Wigner functions. It allows a more symmetrical treatment of phase-space variables. Also it is expected to be a less singular object than the wave function in the classical limit.

Second section is devoted to describe the elementary results concerning Wigner function. It also describes the time development equation of these functions.

In the third section we describe the quantum flow which represents the quantum dynamics in phase space.

In the last section, we study the asymptotics of this flow showing that it tends in a suitable sense to the classical flow. The most useful tool for this study is the Ventsel's theory of large deviations [9] [10] [11]. This theory already proved to be very efficient in the study of classical limit, see e.g. [12] [13] [14] [15] [16]. However it has not been used for jump processes up to now.

2. WIGNER FUNCTION AND THE PHASE SPACE DESCRIPTION OF QUANTUM MECHANICS

As it is well known the state of a quantum non relativistic particle is completely described by a wave function $\psi(x,t)$, $x \in R^N$, $t \in R$. It is not directly observed but it is used to compute the expected value of observables in this state. The Weyl quantization rule associates with an observable whose classical analogue is a function f on the phase space its quantum expectation value $\langle f \rangle_Q$ in the state ψ,

$$\langle f \rangle_Q = (\hbar\pi)^{-N} \int_{R^{2N}} dq dp \, f(q,p) \int_{R^N} d\xi \, \bar{\psi}(q+\xi) \, e^{\frac{2ip.\xi}{\hbar}} \psi(\xi - q) \tag{2.1}$$

provided this expression makes sense (e.g. if $f \in L_1$).

The Wigner's function W_ψ :

$$W_\psi(q,p) = (\hbar\pi)^{-N} \int_{R^N} d\xi \, \bar{\psi}(q+\xi) \, e^{2i\frac{p.\xi}{\hbar}} \psi(\xi - q) \tag{2.2}$$

contains all the information about the state. It appears as a kind of density on phase space. However, if it is real, it is not positive and consequently has not quite the interpretation of a statistical state of classical mechanics. Only in the limit $\hbar \searrow 0$ can one prove that in some case it approaches a classical state. The most general Wigner function is a convex combination of the previous ones, which correspond to pure states.

This state evolves with time according to the quantum dynamics. In what follows we shall consider dynamics given by a hamiltonian H such that :

$$H = \sum_{i=1}^{N} \left\{ \frac{P_i^2}{2m} + \frac{m\omega^2}{2} Q_i^2 \right\} + V \tag{2.3}$$

where P_i and Q_i are the usual momentum and position operators of quantum mechanics whereas V is of the form

$$V = \int d\mu(q,p) \exp\left\{ iqP - ipQ \right\} \tag{2.4}$$

μ being a bounded measure on phase space $\mathcal{S}$ .t.

$$d\mu(q,p) = d|\mu|(q,p) \exp\left\{ i\Phi(q,p) \right\} \tag{2.5}$$

$|\mu|$ is a bounded symmetric measure and Φ a smooth antisymmetric function.

According to the Weyl quantization prescription, it corresponds to the classical Hamiltonian function $\mathcal{H}$

$$\mathcal{H}(q,p) = \sum_{i=1}^{N} \left\{ \frac{p_i^2}{2m} + m \frac{\omega^2}{2} q_i^2 \right\} + V(q,p) \tag{2.6}$$

where

$$V(q,p) = \int_{R^{2N}} d|\mu|(q'.p') \cos(q'p - qp' + \Phi(q',p')) \tag{2.7}$$

It allows for a velocity dependent Hamiltonian.

It is a matter of simple computation to derive the time development equation for the Wigner functions, viz. one has the

<u>Proposition</u> : let W be a Wigner function which is once differentiable with bounded derivative then it satisfies the equation [17]:

$$\frac{\partial W}{\partial t}(q,p) - \sum_{i=1}^{N} \frac{p_i}{m} \frac{\partial}{\partial q_i} W(q,p) - m\omega^2 q_i \frac{\partial W}{\partial p_i}(q,p)$$
$$- \frac{i}{\hbar} \int_{R^{2N}} d|\mu|(q',p') \sin(q'p - qp' + \Phi(q',p')) W(q + \frac{\hbar}{2} q', p + \frac{\hbar}{2} p') \tag{2.8}$$

This equation is a version of the Moyal equation [18]. On a formal way, it approaches the classical Liouville equation when $\hbar$ goes to zero

(2.9) $$\frac{\partial W}{\partial t} + \{\mathcal{H}, W\} = 0 \tag{2.9}$$

where $\mathcal{H}$ has been defined in (2.6). It is tempting to prove that the solution of equation (2.8) approaches the solution of the classical equation which is given by

$$W_t(q,p) = W_o(q_t,p_t) \tag{2.10}$$

q_t, p_t being the solution of the classical equations (Hamilton's equations) with initial values q,p .
Equation (2.9) defines a flow for the functions on phase space, the classical flow.

The Schrödinger equation (2.8) defines as well a stochastic flow not on phase space but better on an extended phase space. Indeed let us remark that the integral operation in equation (2.8) looks formally like a Markov generator except for the positivity. However it is the trace of a true Markov generator in a phase space with one more dimension.

Let F be a function from $R^{2N} \times [0,T]$ to $\mathbb{C}$. Let it correspond to the function $\overline{F}$:

$$\overline{F}(q,p,s,t) = \exp\left\{4\,\frac{(T-t)}{\hbar}\,\|\mu\| + i\,\frac{s}{\hbar}\right\} F(q,p,t) \tag{2.11}$$

$\|\mu\|$ being the total mass of the measure $|\mu|$.

Schrödinger equation for $\overline{F}$ rewrites as follows

$$\frac{\partial \overline{F}}{\partial t}(q,p,s,t) + \sum_{i=1}^{N}\left\{\frac{p_i}{m}\frac{\partial \overline{F}}{\partial q_i}(q,p,s,t) - m\,\omega^2\, q_i \frac{\partial \overline{F}}{\partial p_i}(q,p,s,t)\right\}$$
$$+\int_{R^{2N+1}} d\nu_{q,p,s}(q',p',s')\left(\overline{F}(q+q',\ p+p',\ s+s',t) - \overline{F}(q,p,s,t)\right) = 0 \tag{2.12}$$

where $\nu_{q,p,s}$ is a positive bounded measure on R^{2N+1} defined as follows :

$$d\,\nu_{q,p,s}(q',p',s') = \frac{1}{\hbar}\, d|\mu|\left(\frac{2q'}{\hbar}, \frac{2p'}{\hbar}\right)\Big(1$$
$$- \sin\left(\frac{2}{\hbar}(q'p - qp') + \Phi\left(\frac{2}{\hbar}q', \frac{2}{\hbar}p'\right)\right)\Big)\left(\delta_{\hbar}(s') + \delta_{-\hbar}(s')\right) \tag{2.13}$$
$$+ \frac{1}{\hbar}\, d|\mu|\left(\frac{2}{\hbar}q', \frac{2}{\hbar}p'\right)\left(\delta_{\hbar/2}(s') + \delta_{-\hbar/2}(s')\right)$$

Equation (2.14) has the form of a backward Kolmogorov equation [19] whose solution has a probabilistic representation :

<u>Proposition</u> [17] [20] : there exists an infinitely divisible Markov process $(Q_T(t), P_T(t), S_T(t))$ $t \leq T$ in R^{2N+1} such that

i) $(Q_T(T), P_T(T), S_T(T)) = (q,p,s)$ a.s.

ii) its generator has a modification A_t such that

$$(A_t f)(q,p,s) = \sum_{i=1}^{N} \frac{p_i}{m}\frac{\partial}{\partial q_i} f(q,p,s) - m\omega^2 q_i \frac{\partial}{\partial p_i} f(q,p,s)$$
$$+ \int_{R^{2N+1}} d\nu_{q,p,s}(q',p',s') \left\{ f(q+q', p+p', s+s') - f(q,p,s) \right\}$$

iii) the solution of equation (2.12) with final condition

$$\overline{F}(q,p,s,T) = \overline{F}_o(q,p,s)$$

is given by

$$\overline{F}(q,p,s,t) = \mathbb{E}\left[F_o(Q_T(t), P_T(t), S_T(t))\right] \tag{2.14}$$

Taking into account formula (2.11) for Wigner functions, we have :

Proposition [17], [20] : the Wigner function at time t has the following representation :

$$W_t(q,p) = e^{4\frac{t}{\hbar}\|\mu\|}\, \mathbb{E}\left[e^{\frac{i}{\hbar}S_t(o)} W_o(Q_t(o), P_t(o))\right]$$

where $(Q_T(t), P_T(t), S_T(t))$ is the stochastic process described in the previous proposition.

If one has in mind that expectation $\mathbb{E}$ represents an integration over paths one realizes that the previous formula is really a Feynman Path Integral formula.

3. ASYMPTOTIC OF THE QUANTUM FLOW

As previously mentioned the Schrödinger equation in the limit $\hbar \searrow 0$ approaches the classical Liouville equation. It is tempting to prove that in probability the paths of the stochastic process defining the quantum evolution approaches the classical path.

Let us consider the following naïve example : the solution of the equation :

$$\partial_t f(x,t) = \frac{1}{h}\left(f(x+h,t) - f(x,t)\right) \tag{3.1}$$

is given for bounded $f(x,o) = f_o(x)$ by

$$f(x,t) = \mathbb{E}\left(f(x+N_t^h)\right) \tag{3.2}$$

where N_t^h is the standard Poisson process such that

$$\mathbb{E}(N_t^h) = \frac{t}{h} \tag{3.3}$$

starting at zero. When h goes to zero equation (3.1) approaches the equation

$$\partial_t f(x,t) = \partial_x f(x,t) \tag{3.4}$$

However one can prove that the probability for the trajectories of N_t^h to be far from $x_t = x+t$ is exponentially small. Hence depending on the continuity module of $f_o(x)$, $f(x,t)$ approaches the solution of the limiting equation (3.4). This equation is the prototype of the Schrödinger-Moyal equation and similar results can be obtained using Ventsel's theory of large deviations [9].

An important tool for this is an action functional which is defined as follows : $G_\hbar(q,p,s,z)$, $z = (z_1, z_2, z_3) \in R^{2N+1}$ is the exponential moment

$$G_\hbar(q,p,s,z) = \sum_{i=1}^{N} z_{1i}\frac{p_i}{m} - m\omega^2 z_{2i}q_i + \int_{R^{2N+1}} d\nu_{q,p,s}(q',p',s') \exp\{q'.z_1+p'.z_2+s'.z_3\} \tag{3.5}$$

We assume it exists at least for small z. Equivalently we assume that the classical potential can be extended to imaginary arguments. Obviously $G_\hbar$ is a convex function.

<u>Theorem</u>: the Legendre transform $G_\hbar^*$ of $G_\hbar$:

$$G_\hbar^*(q,p,s,\xi) = \operatorname*{Sup}_{z\in R^{2N+1}} (z.\xi - G_\hbar(q,p,s,z)) \tag{3.6}$$

is also convex and reaches its minimum for :

$$z_1 = \frac{\partial \mathcal{H}}{\partial p}(q,p) \qquad z_2 = -\frac{\partial \mathcal{H}}{\partial q}(q,p) \qquad z_3 = 0 \tag{3.7}$$

independently of $\hbar$.

The proof is almost obvious.

As a result of Ventsel's theory, $G_\hbar^*$ also controls the behaviour of the trajectories of the stochastic process for small $\hbar$. Namely let us observe that $G_1(q,p,s,z) = \hbar G_\hbar(q,p,s,\frac{1}{\hbar}z)$ is independent of $\hbar$ and is again a convex function. G_1^* its Legendre transform is convex and its minimum defines the classical trajectory. Furthermore let φ be an application from $[0,T]$ to R^{2N+1}. If we define [9]

$$I(\varphi) = \int_0^t d\tau\, G_1^*(\varphi(\tau),\dot\varphi(\tau)) \tag{3.8}$$

the associated variational principle defines the classical trajectory.

The properties of I allows to define a tublet $\Phi_{qps}(\epsilon)$ around the classical path

$$\Phi_{qps}(\varepsilon) = \{\varphi;\ \varphi(o) = (q,p,s)\ ;\ I(\varphi) \leq \varepsilon\} \tag{3.9}$$

Ventsel's theory of large deviations allows a strong control of trajectories viz. with respect to the uniform distance :

$$\rho_T(\varphi, \psi) = \sup_{t \in [0,T]} \left(\sum_{i=1}^{2N+1} |\varphi_i(t) - \psi_i(t)|^2 \right)^{1/2} \tag{3.10}$$

Using these definitions it is possible to prove the following :

Proposition [20] : For any positive $\delta, \gamma, \mathcal{E}_o$, for sufficiently small $\hbar$, for all $(q,p,s) \in R^{2N+1}$ and for $\epsilon \leq \mathcal{E}_o$, one has the following estimate on the probability $P^{\hbar}_{qps}$ for the trajectories to have excursions outside a tublet centred around the classical path.

$$P^{\hbar}_{qps} \left\{ \rho_T((Q_{\hbar}, P_{\hbar}, S_{\hbar}), \Phi_{qps}(\epsilon)) \geq \delta \right\} \tag{3.11}$$

$$\leq \quad \exp\left\{ -\frac{1}{\hbar}(\epsilon - \gamma) \right\}$$

One can as well get an estimate of the same kind from below.

Corollary. The quantum flow defined by (2.14) tends to the classical flow viz. for $F \in C_b^1(R^{2N+1})$

$$\lim_{\hbar \downarrow 0} \quad F(Q_t(o), P_t(o), S_t(o)) \tag{3.12}$$

$$= F(q_t, p_t, o)$$

where (q_t, p_t) are the solutions of the classical equations of motion.

The proof of the previous proposition requires both the use of Ventsel's result and the construction of an interaction schema to avoid the growth of the drift.

Previous results can be used to study the classical limit of the Wigner functions [20].

REFERENCES

[1] R.P. FEYNMAN, Space Time Approach to Non Relativistic Quantum Mechanics, Rev. Mod. Phys. 20 (1948), 367-387.

[2] R.P. FEYNMAN and A.R. HIBBS, Quantum Mechanics and Path Integral, (Mac Graw-Hill, New York 1965).

[3] A. JAFFE and J. GLIMM, Quantum Physics. A Functional Integral Point of View, (Springer-Verlag, New York 1981).

[4] A.M. CHEBOTAREV and V.P. MASLOV, Processus à sauts et leurs applications dans la mécanique quantique, in Feynman Path Integral, Proceedings Marseille Conference 1978, Lecture Notes in Physics 106 (Springer-Verlag Berlin, New York, 1979).

[5] V.P. MASLOV and A.M. CHEBOTAREV, Jump Type Processes and their Application to Quantum Mechanics, Journal of Soviet Mathematics 13 (1980), 315-357.

[6] Ph. COMBE, R. HOEGH-KROHN, R. RODRIGUEZ, M. SIRUGUE, M. SIRUGUE-COLLIN, Poisson Processes on Group and Feynman Path Integral, Commun.Math.Phys. 77 (1980), 269-288.

[7] Ph. COMBE, R. RODRIGUEZ, M. SIRUGUE, M. SIRUGUE-COLLIN, High Temperature Behaviour of Thermal Functionals, Publications of the RIMS (Kyoto University), 19 (1983), 355-365.

[8] S. ALBEVERIO, Ph. BLANCHARD, R. HOEGH-KROHN, M. SIRUGUE, Local Relativistic Invariant Flows for Quantum Fields, Commun.Math.Phys. 90 (1983), 329-351.

[9] A.D. VENTSEL', Rough Limit Theorems on Large Deviations for Markov Stochastic Processes,
I. Theory Prob. Applications 21 (1976), 227-242.
II. " " " 21 (1976), 499-512.
III. " " " 24 (1979), 675-692.
IV. " " " 25 (1982), 215-234.

[10] M.I. FRIEDLIN, A.D. VENTSEL', Random Perturbations of Dynamical Systems, Springer-Verlag, New York, 1984.

[11] R. AZENCOTT, Grandes Déviations et Applications, Cours de Probabilité de Saint-Flour, Lecture Notes in Mathematics 744, Springer-Verlag (1978).

[12] G. JONA-LASINIO, F. MARTINELLI, E. SCOPPOLA, New Approach to the Semi-Classical Limit of Quantum Mechanics I. Multiple Tunneling in One Dimension, Commun. Math.Phys. 80 (1981), 223-254.

[13] G. JONA-LASINIO, F. MARTINELLI, E. SCOPPOLA, The Semi-Classical Limit of Quantum Mechanics : A Qualitative Theory via Stochastic Mechanics, Physics Reports 77 (1981), 313-327.

[14] W.G. FARIS and G. JONA-LASINIO, Large Fluctuations for a Non-Linear Heat Equation with Noise, J. Phys. A 15 (1982), 3025-3055.

[15] B. SIMON, Instantons, Double Wells and Large Deviations, Bull. AMS, March 1983.

[16] R. AZENCOTT, H. DOSS, L'équation de Schrödinger quand h $\searrow$ 0. Une approche probabiliste, IIème Rencontre Franco-Allemande entre Physiciens et Mathématiciens, CIRM, Mars 1983, to appear in Lecture Notes in Mathematics, Springer.

[17] Ph. COMBE, F. GUERRA, R. RODRIGUEZ, M. SIRUGUE, M. SIRUGUE-COLLIN, Quantum Dynamical Time Evolutions as Stochastic Flows in Phase Space, Physica 124 A (1984), 561-574.

[18] J.E. MOYAL, Quantum Mechanics as a Statistical Theory, Proc. Cambridge Phil. Soc. (1949), 99-124.

[19] I.I. GIHMAN and A.V. SKOROHOD, The Theory of Stochastic Processes I, II and III, Springer-Verlag, New York 1974.

[20] Ph. BLANCHARD, M. SIRUGUE, Large Deviations from Classical Paths. Hamiltonian Flows as Classical Limits of Quantum Flows, Preprint Z.I.F. (1983), to appear in Commun.Math.Phys.

Adiabatic elimination technique for quantum dissipative systems

F. Casagrande, L. A. Lugiato and G. Strini

Dipartimento di Fisica dell'Università

Via Celoria 16, 20133 Milano, Italy

A systematic method to perform the adiabatic elimination of fast variables in quantum dissipative systems is illustrated in the framework of the single-mode laser model with injected signal.

1. INTRODUCTION: THE PROBLEM OF THE ADIABATIC ELIMINATION OF FAST VARIABLES

Since we consider dissipative systems, our starting point will be a master equation which incorporates suitable damping terms. This master equation will be taken for granted, without discussing its derivation or its limits of validity. Furthermore, we shall put ourselves on a macroscopic level of description, i.e. the master equation describes the dynamics of the macroscopic variables of the system, but it describes also their fluctuations and correlations, which arise from the underlying microscopic structure.

To illustrate the problem of adiabatic elimination, let us consider first a purely deterministic description that neglects fluctuations and correlations, i.e. a semiclassical description in which all quantum effects are absent. At this level, the dynamics of the system will be described typically by a set of nonlinear differential equations for the n macroscopic variables of the system $x_1, x_2, \ldots, x_n$:

$$\dot{x}_i = f_i(x_1, x_2, \ldots, x_n), \quad i=1,2,\ldots,n \tag{1}$$

Most often, the set of eqs (1) is exceedingly complicated and contains such a huge number of parameters that it is impossible to explore it systematically, even by numerical methods. A crucial simplification can be introduced whenever one can subdivide the variables in two distinct groups, such that the variables in one group (<u>fast</u> variables) vary in time much more rapidly than the variables in the other group (<u>slow</u> variables). Namely, one identifies two well-separated time scales τ_{fast} and τ_{slow} which caracterize the two groups, so that system (1) splits accordingly in two subsystems for the slow variables x_S and the fast variables x_F:

$$\dot{x}_S = f_S(x_S, x_F) \tag{2a}$$

$$\dot{x}_F = f_F(x_S, x_F) \tag{2b}$$

In this situation, at times $t \gg \tau_{fast}$ usually the fast variables have relaxed to a state of instantaneous equilibrium with the slow variables, in which they follow adiabatically the evolution of the slow variables. This quasistationary state is obtained by dropping the time derivative with respect to time of the fast variables in the group of eqs. (2b). Thus, these equations reduce to a set of algebraic equations, which can be solved to find the expression of the fast variables as function of the slow ones:

$$x_F(t) = \Phi(x_S(t)) \tag{3}$$

On substituting the expression (3) for the fast variables in the group of eqs. (2a), one obtains a closed-form set of differential equations for the slow variables:

$$\dot{x}_S = f_S(x_S, \Phi(x_S)) \tag{4}$$

This procedure, which is called adiabatic elimination of the fast variables, produces a substantial reduction of the complexity of the dynamical problem which is considered. For this reason the adiabatic elimination principle has been put by Haken as one of the cornerstones in the foundations of his Synergetics [1].

The problem that we treat in this paper is that of the adiabatic elimination in quantum systems, which exhibit fluctuations created by the intrinsic quantum nature of the system, i.e. quantum noise. The problem of the adiabatic elimination is similar to that of deriving a closed form dynamical equation for a system in contact with a thermal reservoir from the Hamiltonian dynamics of the compound system. In fact, in that case one must eliminate the variables of the reservoir. There are various methods available to perform the adiabatic elimination in quantum systems. E.g., we mention Zwanzig's projection operator technique [2], the strategy devised by Drummond, Gardiner and Walls [3] and the method recently elaborated by Haake and Lewenstein [4]. Here we illustrate a technique that was introduced by one of us (L.A.L.) in 1975 [5] and recently improved in [6]. The properties of this technique are that i) it is systematic, ii) it is closely related to the standard adiabatic elimination procedure in the semiclassical equations, that has been sketched in this section; iii) in its first-order approximation it does not imply any weak coupling approximation; iv) it does not introduce elements which require some arbitrary choice, as e.g. the choice of the projection operator in the projection technique.

The paper is organized as follows. In Section 2 we introduce the model of the one-mode laser with injected signal. In Section 3 we perform the adiabatic elimination of the atomic variables at the semiclassical level by applying a precise adiabatic elimination limit. On the basis of these results we treat the fully quantum-mechanical problem in Section 4.

2. THE ONE-MODE LASER WITH INJECTED SIGNAL

In this section we discuss a model which exhibits quantum fluctuations, namely the one-mode laser model, as formulated by Haken's school [7] and especially by Weidlich and Haake [8], and finally extended by Bonifacio and Lugiato [9] to include the possibility of a coherent injected field. In a laser system we have a resonant cavity with mirrors, of which one is semitransparent, with transmission coefficient T. This cavity contains N atoms homogeneously distributed in a pencil-shaped sample of length L. The atoms are assumed to be two-level atoms with transition frequency ω. Furthermore we assume that there is a cavity mode whose frequency coincides exactly with the atomic frequency; this allows the neglect of all the other cavity modes. If we call W the statistical operator of the system atoms + resonant mode, the time evolution of W is governed by a master equation which presents three distinct groups of terms, which describe the dynamics of the atoms and of the single mode field, and the interaction between atoms and field:

$$dW/dt = (dW/dt)_A + (dW/dt)_F + (dW/dt)_{AF} \tag{5}$$

Let us consider the three groups in Eq. (5) separately, starting with the N two-level atoms. The i-th atom is associated with the raising and lowering operators r_i^+, r_i^- andwith the inversion operator $r_{3i} = (1/2)(r_i^+ r_i^- - r_i^- r_i^+)$. These operators obey the angular momentum commutation relations

$$\left[r_i^+ , r_j^-\right] = 2\, r_{3i}\,\delta_{ij} \qquad \left[r_{3i}, r_j^\pm\right] = \pm\, r_i^\pm\,\delta_{ij} \tag{6}$$

The collection of atoms is associated with three collective operators $R^\pm$, R_3, which describe the macroscopic atomic polarization and the total population inversion, and again obey angular momentum commutation relations:

$$R^{\pm} = \sum_{i=1}^{N} r_i^{\pm} \qquad R_3 = \sum_{i=1}^{N} r_{3i}$$
$$\left[R^{+}, R^{-}\right] = 2R_3 \qquad \left[R_3, R^{\pm}\right] = \pm R^{\pm} \tag{7}$$

As long as we do not consider the interaction with the cavity mode, the atoms evolve independently of one another. This time evolution arises from the free evolution, from the decay due to spontaneous emission, and from the pump action that we exert on the atoms. Hence we obtain the following atomic dynamics:

$$(dW/dt)_A = -i \mathcal{L}_A W + \Lambda_A W \tag{8a}$$

$$\mathcal{L}_A W = \omega \sum_{i=1}^{N} \left[r_{3i}, W\right] = \omega \left[R_3, W\right] \tag{8b}$$

$$\Lambda_A W = \sum_{i=1}^{N} \Big\{ (\gamma{\downarrow}/2)\left(\left[r_i^{-}, W r_i^{+}\right] + \left[r_i^{-} W, r_i^{+}\right] \right) + (\gamma{\uparrow}/2)\left(\left[r_i^{+}, W r_i^{-}\right] + \left[r_i^{+} W, r_i^{-}\right] \right) \Big\} \tag{8c}$$

In Eqs. (8a-c) $\mathcal{L}_A$ describes the free evolution of the atoms, while Λ_A is a dissipative term which describes the downward transitions due to spontaneous emission, with a rate $\gamma{\downarrow}$, and the upward transition due to the pump, with a rate $\gamma{\uparrow}$. Two parameters that are immediately connected with this Liouvillian Λ_A are the global transition rate γ and the population inversion per atom σ which arises from the balance of pump and decay:

$$\gamma = \gamma{\uparrow} + \gamma{\downarrow} \quad ; \qquad \sigma = (\gamma{\uparrow} - \gamma{\downarrow})/(\gamma{\uparrow} + \gamma{\downarrow}) \tag{9}$$

Let us now consider the dynamics of the cavity mode. Let A (A^{+}) be the annihilation (creation) operator of this mode, with the harmonic oscillator commutation relation

$$\left[A, A^{+}\right] = 1 \tag{10}$$

The time evolution of the cavity mode arises from the free time evolution, and from the fact that the photons escape from the cavity with a rate k equal to the inverse of the transit time of photons in the cavity, times the transmittivity coefficient of the mirrors

$$k = (c/L)\, T \tag{11}$$

Hence we obtain the following dynamics for the cavity mode:

$$(dW/dt)_F = -i \mathcal{L}_F W + \Lambda_F W \tag{12a}$$

$$\mathcal{L}_F W = \omega \left[A^+ A , W \right] \tag{12b}$$

$$\Lambda_F W = k \left(\left[A , W A^+ \right] + \left[A W , A^+ \right] \right) \tag{12c}$$

In Eqs. (12a-c) $\mathcal{L}_F$ is the free time evolution term and the damping term Λ_F describes the escape of photons.

Finally, the term which describes the interaction of the atoms with a cavity mode is a conservative term given simply by the commutator with the interaction Hamiltonian, taken in the dipole and rotating-wave approximations

$$(dW/dt)_{AF} = -i \mathcal{L}_{AF} W = -(i/\hbar) \left[H_{AF} , W \right] \tag{13a}$$

$$H_{AF} = i\hbar g \sum_{i=1}^{N} (A^+ r_i^- - A r_i^+) = i\hbar g (A^+ R^- - A R^+) \tag{13b}$$

with g being the atom-field coupling constant.

This model has been generalized [9] to include the possibility of an external coherent field, with the same frequency ω, injected into the cavity. In such a way one obtains the so-called laser with injected signal, in which the incident field can be utilized e.g. to control the phase and the polarization of the output field. The injected field is taken into account by simply adding to the r.h.s. of the field eq. (12a) a contribution $-i \mathcal{L}_{ext} W$, where

$$\mathcal{L}_{ext} W = i k \alpha \left(\left[A^+ , W \right] e^{-i\omega t} - \left[A , W \right] e^{i\omega t} \right), \tag{14}$$

α being the real, constant amplitude of the incident field. The inclusion of the injected signal has the additional advantage that by the same model we can treat not only the laser, but also the so-called optical bistability (OB). In the case of OB, the atomic system is not pumped, so that in the atomic term Λ_A (Eq.(8c)) we have $\gamma\uparrow = 0$, i.e. $\gamma = \gamma\downarrow$ and $\sigma = -1$ (Eq.(9)).

In this situation, we have a coherent field that is transmitted by the optical cavity, which is filled by a medium that, contrary to the laser case, does not amplify but absorbs the radiation. When the medium is dense enough, the steady state curve of transmitted versus incident field intensity exhibits a hysteresis cycle with two distinct states of transmission, that is just optical bistability.

By passing to the interaction picture, we get rid of the free time evolution terms $\mathcal{L}_A$ (eq (8b)) and $\mathcal{L}_F$ (eq (12b)), and the external signal term (14) becomes time independent (i.e., it loses the factors $\exp(\pm i\omega t)$).

3. SEMICLASSICAL ADIABATIC ELIMINATION

From the model described in the previous section, we can immediately derive the time evolution equations for the mean values of the macroscopic quantities $R^{\pm}, R_3$, A, A^+. In the semiclassical approximation in which one neglects all fluctuations and correlations, and therefore all the mean values of products factorize into products of mean values, we obtain the following closed-form set of nonlinear macroscopic equations

$$d\langle A\rangle/dt = g\langle R^-\rangle - k(\langle A\rangle - \alpha) \tag{15a}$$

$$d\langle R^-\rangle/dt = 2g\langle A\rangle\langle R_3\rangle - (\gamma/2)\langle R^-\rangle \tag{15b}$$

$$d\langle R_3\rangle/dt = -g(\langle A\rangle\langle R^+\rangle + \langle A^+\rangle\langle R^-\rangle) - \gamma(\langle R_3\rangle - \sigma N/2) \tag{15c}$$

The equations for $\langle A^+\rangle$, $\langle R^+\rangle$ are the complex conjugates of Eqs. (15a) and (15b), respectively.

In a large class of optical systems the photon damping constant is much smaller than the atomic decay rate,

$$k \ll \gamma \tag{16}$$

In the limit (16), one usually performs the adiabatic elimination of the atomic variables by dropping the time derivatives $d\langle R^-\rangle/dt$, $d\langle R_3\rangle/dt$. Clearly, this is a very rude way of doing because in the dynamical equations there are not only the damping terms but also the interaction terms and one must discuss their order of magnitude. A more rigorous and systematic discussion of the adiabatic elimination in these semiclassical models is given in [10], which gives a precise definition of the adiabatic elimination limit. As we shall see, the use of this definition is essential also in the fully quantum statistical case.

The crucial step is the introduction of normalized variables for the output field and for the atomic polarization and inversion

$$\langle x\rangle = 2\sqrt{2}\,(g/\gamma)\langle A\rangle \tag{17a}$$

$$\langle P^{\pm}\rangle = -(\sqrt{2}/N)\langle R^{\pm}\rangle \tag{17b}$$

$$\langle P_3\rangle = (2/N)\langle R_3\rangle \tag{17c}$$

In terms of the new variables (17), the set (15) becomes

$$d\langle x\rangle/dt = k\,(-2C\langle P^-\rangle - \langle x\rangle + y) \tag{18a}$$

$$d\langle P^-\rangle/dt = -(\gamma/2)\,(\langle x\rangle\langle P_3\rangle + \langle P^-\rangle) \tag{18b}$$

$$d\langle P_3\rangle/dt = \gamma\Big[(1/2)(\langle x\rangle\langle P^+\rangle + \langle x^+\rangle\langle P^-\rangle) - \langle P_3\rangle + \sigma\Big] \tag{18c}$$

where

$$y = 2\sqrt{2}\,(g/\gamma)\,\alpha \tag{19a}$$

$$C = g^2\,N/k\gamma \tag{19b}$$

The virtue of the normalization (17) is that now in each dynamical equation (18a-c) all terms scale in the same way, all of them being proportional to the same rate constant. Furthermore, by this normalization the number of effective parameters in play has been reduced to a minimum; besides the rate constants k, γ and the inversion parameter σ, we have only the normalized incident field y (19a) and the parameter C (19b), which is the pump parameter in the laser case and the instability parameter in the case of OB.

Now let us introduce the smallness parameter ε of the adiabatic problem and the time variable τ normalized to the slow rate constant k

$$\varepsilon = k/\gamma \tag{20a}$$

$$\tau = kt \tag{20b}$$

The dynamical eqs. (18) take the form

$$d\langle x\rangle/d\tau = 2C\langle P^-\rangle - \langle x\rangle + y \tag{21a}$$

$$\varepsilon\, d\langle P^-\rangle/d\tau = -\langle x\rangle\langle P_3\rangle - \langle P^-\rangle \tag{21b}$$

$$\varepsilon\, d\langle P_3\rangle/d\tau = (1/2)(\langle x\rangle\langle P^+\rangle + \langle x^+\rangle\langle P^-\rangle) - \langle P_3\rangle + \sigma \tag{21c}$$

Next, we define the <u>adiabatic elimination limit</u>:

$$\begin{aligned} &\varepsilon \to 0 \\ &\langle x\rangle,\ \langle P^\pm\rangle,\ \langle P_3\rangle,\ y,\ C = O(\varepsilon^0) \end{aligned} \tag{22}$$

Therefore at order zero in ε the time derivatives in the atomic eqs. (21b,c) drop and we obtain the adiabatic elimination in the usual way, arriving at the following closed-form dinamical equation for the field variable

$$d\langle x\rangle/d\tau = y - \langle x\rangle + 2C\sigma\langle x\rangle/(1 + |\langle x\rangle|^2) \tag{23}$$

Since the parameter C remains finite in the adiabatic limit (22), by re-writing it as $C = (N/\varepsilon)(g/\gamma)^2$, it follows that

$$(g/\gamma)^2 = O(\varepsilon) \tag{24}$$

This result, albeit obtained in the framework of the semiclassical theory, has an immediate consequence on the fluctuations of the system. In fact, as we shall see the fluctuations of the electric field scale as the inverse of the so-called saturation photon number N_s, where

$$N_s^{-1} = 8\,(g/\gamma)^2 = O(\varepsilon) \tag{25}$$

This show that <u>fluctuations are small in the adiabatic limit.</u>

One more comment is in order. Even if $(g/\gamma)^2 = O(\varepsilon)$, the adiabatic elimination does not correspond with a weak coupling approximation. This is clearly seen from the closed-form eq. (23), derived in the adiabatic limit. In fact, the variable x is proportional to the coupling constant g (eq. (17a)), and therefore the coupling constant appears in the dynamical eq. (23) at all orders.

4. QUANTUM - MECHANICAL ADIABATIC ELIMINATION

Let us consider the master equation (5). First of all, in order to exploit the results of the semiclassical analysis, we transform this operator equation into a classical - looking partial differential equation. We do that only with respect to the variables that we do not eliminate, i.e. the field variables, while we leave the atomic variables in operator form.

The transformation is performed by using the Wigner distribution, i.e. we introduce

$$\widetilde{W}(\beta,\beta^*,t) = \pi^{-1}\int d^2u\,\exp\left\{-iu\beta - iu^*\beta^*\right\}\mathrm{Tr}_F\left(\exp\left\{iuA + iu^*A^+\right\} W(t)\right) \tag{26}$$

where Tr_F means partial trace with respect to the Hilbert space of the field. Thus, $\widetilde{W}$ is an operator in the Hilbert space of the atomic system, and a function with respect to the C - number variables β,β^* that correspond with the operators A, A^+.

On the basis of the semiclassical analysis, we normalize the field variables and the atomic operators as follows (cfr. Eqs. (17)):

$$x = 2\sqrt{2}\,(g/\gamma)\,\beta \quad ;$$
$$P^- = -(\sqrt{2}/N)\,R^- \quad ; \qquad (27)$$
$$P_3 = (2/N)\,R_3$$

The generator L of the master equation (5) becomes now a differential operator with respect to the field variables, and on reexpressing it in terms of the normalized variables and operators (27) one finds that L can be subdivided into three different parts L_o, L_1, L_2 of order zero, one and two, respectively, with respect to the smallness parameter ε :

$$\begin{aligned} d\widetilde{W}/dt &= L\widetilde{W} = (L_o + L_1 + L_2)\widetilde{W} \\ L_o\widetilde{W} &= \Lambda_A\widetilde{W} + (N\gamma/4)\left\{\left[x\,P^+, \widetilde{W}\right] + h.c.\right\} \\ L_1\widetilde{W} &= (\gamma\varepsilon/2)\left\{(\partial/\partial x)(x - y)\widetilde{W} + C(P^-\widetilde{W} + \widetilde{W}P^-) + h.c.\right\} \\ L_2\widetilde{W} &= (\gamma\varepsilon/2N_s)\,\partial^2\widetilde{W}/\partial x^*\partial x \end{aligned} \qquad (28)$$

This type of subdivision of the Liouvillian in the framework of the adiabatic elimination problem was first introduced by Haake and Lewenstein [4].

Now, in order to derive a closed - form time evolution equation for the Wigner distribution of the cavity mode

$$P_W(x, x^*; t) = \mathrm{Tr}_A\,\widetilde{W}(x, x^*; t) \qquad (29)$$

where Tr_A is the trace over the atomic Hilbert space, the first step is to take Tr_A on both sides of the master equation (28), obtaining

$$\frac{\partial P_W}{\partial \tau} = \left\{\left(\frac{\partial}{\partial x}(x - y) + c.c.\right) + \frac{1}{N_s}\,\frac{\partial^2}{\partial x^*\partial x}\right\}P_W$$
$$+ 2C\left(\frac{\partial W^{(-)}}{\partial x} + \frac{\partial W^{(+)}}{\partial x^*}\right) \qquad (30a)$$

where

$$W^{(\pm)}(x, x^*; t) = \mathrm{Tr}\left\{P^{\pm}\widetilde{W}(x, x^*; t)\right\} \qquad (30b)$$

and the scaled time $\tau = kt$ has been used.

Eq. (30a) is not closed, due to the presence of the quantities $W^{(+)}$, $W^{(-)}$. The next step is to derive from the master equation (28) the time evolution equations for $W^{(+)}$ and $W^{(-)}$. Clearly, this procedure leads to a hierarchy of equations for the Wigner distribution P_W, the functions $W^{(\pm)}$, and a hierarchy of functions, whose lowest - order elements are

$$\begin{aligned}
W_3(x, x^*, t) &= \mathrm{Tr}\left\{ P_3 \widetilde{W}(x, x^*, t)\right\} \\
W^{(+-)}(x, x^*, t) &= \mathrm{Tr}\left\{ (1/2)(P^+P^- + P^-P^+) \widetilde{W}(x, x^*, t)\right\} \\
W^{(--)}(x, x^*, t) &= \mathrm{Tr}\left\{ P^-P^- \widetilde{W}(x, x^*, t)\right\} \\
W^{(-3)}(x, x^*, t) &= \mathrm{Tr}\left\{ (1/2)(P^-P_3 + P_3P^-) \widetilde{W}(x, x^*, t)\right\} \\
W^{(33)}(x, x^*, t) &= \mathrm{Tr}\left\{ P_3P_3 \widetilde{W}(x, x^*, t)\right\}
\end{aligned} \tag{31}$$

where for the sake of definiteness symmetric products of atomic operators are considered. Now we show that the adiabatic limit introduces, at each order in the smallness parameter ε, an exact truncation of the hierarchy of equations. To see this, let us consider first the time evolution equations for the quantities $W^{(-)}$, $W^{(3)}$

$$\begin{aligned}
\varepsilon \frac{d W^{(-)}}{d\tau} = &- (W^{(-)} + xW^{(3)}) + \varepsilon\left\{\left[\frac{\partial}{\partial x}(x-y) + c.c.\right] + \frac{1}{N_s}\frac{\partial^2}{\partial x^* \partial x}\right\} W^{(+)} \\
&+ 2C\varepsilon\left\{\frac{\partial W^{(+-)}}{\partial x^*} + \frac{\partial W^{(--)}}{\partial x}\right\}
\end{aligned} \tag{32}$$

$$\begin{aligned}
\varepsilon \frac{dW^{(3)}}{d\tau} = &\left\{\frac{1}{2}(x^*W^{(-)} + x\,W^{(+)}) + \sigma P_W - W^{(3)}\right\} \\
&+\varepsilon\left\{\left[\frac{\partial}{\partial x}(x-y) + c.c.\right] + \frac{1}{N_s}\frac{\partial^2}{\partial x^* \partial x}\right\} W^{(3)} + 2C\varepsilon\left\{\frac{\partial W^{(+3)}}{\partial x^*} + \frac{\partial W^{(-3)}}{\partial x}\right\}
\end{aligned} \tag{33}$$

The equation for $W^{(+)}$ is the complex conjugate of eq. (32). Note the immediate correspondence of eq. (30a), (32) and (33) with the semiclassical equations (18a - c). In fact, if we take into account that

$$\begin{aligned}
\langle x\rangle(t) &= \int d^2x\, x\, P_W(x, x^*, t) \\
\langle P^\pm\rangle(t) &= \int d^2x\, W^{(\pm)}(x, x^*, t) \\
\langle P_3\rangle(t) &= \int d^2x\, W^{(3)}(x, x^*, t)
\end{aligned} \tag{34}$$

from eqs. (30a), (32), (33) we derive a set of equations for the mean values (34) which coincide with the semiclassical eqs. (18a - c) if one factorizes the mean values of products into the products of the mean values and uses definitions (19a,b).

Now we exploit the existence of the smallness parameter ε by expanding all quantities $W^{(-)}$, $W^{(3)}$, ... in power of this parameter:

$$W^{(\dots)} = \sum_{m=0}^{\infty} W_m^{(\dots)} \quad , \quad W_m^{(\dots)} = O(\varepsilon^m) \tag{35}$$

In such a way, by eliminating the auxiliary quantities $W^{(\dots)}$, we derive a closed - form time evolution equation for the Wigner distribution, in which the generator Λ is expanded in powers of the smallness parameter:

$$\partial P_W/\partial\tau = \Lambda P_W = (\Lambda_0 + \varepsilon\Lambda_1 + \varepsilon^2\Lambda_2 + \dots)\, P_W \tag{36}$$

Let us consider first the zeroth order term of expansion (36). In this case we must only calculate $W^{(\pm)}$, $W^{(3)}$ at zero order in ε, obtaining the simple set of algebraic equations

$$\begin{aligned} 0 &= W_0^{(-)} + x\, W_0^{(3)} \\ 0 &= (1/2)(x^* W_0^{(-)} + x\, W_0^{(+)}) + \sigma P_W - W_0^{(3)} \end{aligned} \tag{37}$$

System (37) is immediately solved, so that at zero order we obtain the following equation for the Wigner function

$$\frac{\partial P_W}{\partial\tau} = \left\{ \frac{\partial}{\partial x}\left(x - y - \frac{2C\,x}{1+|x|^2}\right) + \text{c.c.} \right\} P_W \tag{38}$$

Eq. (38) can be solved by the method of the characteristic equation, which turns out to coincide with the semiclassical eq. (23) obtained by adiabatically eliminating the atomic variables. Hence at order zero we obtain an equation that does not include fluctuations, as it must be because fluctuations vanish in the limit $\varepsilon \to 0$ (see eq. (25)).

Therefore we go to first order in ε. The equations for $W^{(-)}$, $W^{(3)}$ at first order are:

$$W_1^{(-)} = -x\, W_1^{(3)} + \varepsilon \left\{ 2C \left(\frac{\partial W_1^{(+-)}}{\partial x^*} + \frac{\partial W_0^{(--)}}{\partial x} \right) - \left[\frac{\partial}{\partial \tau} - \frac{\partial}{\partial x}(x-y) - \frac{\partial}{\partial x^*}(x^* - y) \right] W_0^{(-)} \right\} \tag{39a}$$

$$W_1^{(3)} = \frac{1}{2}\left(x\, W_1^{(+)} + x^* W_1^{(-)} \right) + \frac{\varepsilon}{2} \left\{ 2C \left(\frac{\partial W_0^{(-3)}}{\partial x} + \frac{\partial W_0^{(+3)}}{\partial x^*} \right) - \left[\frac{\partial}{\partial \tau} - \frac{\partial}{\partial x}(x-y) - \frac{\partial}{\partial x}(x^* - y) \right] W_0^{(3)} \right\} \tag{39b}$$

From eqs. (39a,b) we see that we must go on in the hierarchy and calculate other quantities, namely $W^{(+-)}$, $W^{(--)}$, $W^{(-3)}$. However, at order zero these quantities are the solution of simple algebraic equations that do not imply any new quantity, and therefore <u>the hierarchy is truncated</u>. This truncation is due to the fact that the zeroth - order term L_0 in the Liouvillian (28) does not introduce any new element of the hierarchy. In such a way, one calculates $W^{(\pm)}$, $W^{(3)}$ at first order and arrives at the following Fokker - Planck equation for the Wigner function:

$$\frac{\partial P_W(x, x^*, \tau)}{\partial \tau} = \left\{ \frac{\partial}{\partial x}\left(x - y - \frac{2C\sigma x}{1+|x|^2} + \theta \right) + \text{c.c.} - \frac{C}{2N_s} \frac{\partial^2}{\partial x^2} x^2 f(|x|^2) + \text{c.c.} + \frac{1}{N_s} \frac{\partial^2}{\partial x^* \partial x} g(|x|^2) \right\} P_W(x, x^*, \tau) \tag{40a}$$

where

$$f = \left[(1+|x|^2)^2 + 3\sigma^2 \right] (1+|x|^2)^{-3}$$

$$g = 1 + C\left[(1+|x|^2)^2 (2+|x|^2) - 3\sigma^2 |x|^2 \right. (1+|x|^2)^{-3} \tag{40b}$$

$$\theta = O(\varepsilon)$$

and θ is a small correction to the right term that we do not write explicitly. In principle, we can go on with our procedure to arbitrary order in ε, but it is enough to consider eq. (40), that is the most complete Fokker - Planck equation that describes fluctuations in a single - mode laser with two - level atoms. In the threshold region, it reduces to the classic Risken equation [11], but it incorporates also the saturation effects that become important well above threshold. Eq. (40) was first derived in [12] in linearized form in the framework of OB, and was later generalized to this complete form in [13].

The adiabatic elimination technique described here has been illustrated in a specific example, but we believe it can be applied to a fairly general class of quantum dissipative systems. The main elements of this technique are: i) the precise definition of the adiabatic elimination limit and the systematic use of the related smallness parameter, ii) the translation of the generator of the original operator master equation into a differential operator with respect to the variables that are not eliminated. This translation can be accomplished by using the Wigner distribution, or the Glauber distribution, or a Kramers - Moyal expansion, and so on.

A problem of mathematical interest is to work out a rigorous formulation of the method described in this paper. Also, it would be interesting to establish the connections beetween this technique and the recent works of Frigerio, Lewis and Pulé [14,15].

The Fokker - Planck equations, derived from the single mode laser model and from some related models, have been used e.g. to study the quantum effect of squeezing in optical systems [16].

REFERENCES

1. H. Haken, "Synergetics - an introduction" (Springer, Berlin, 1977)
2. R. Zwanzig, Lect. Theor. Phys. (Boulder) 3, 10b (1960), J. Chem. Phys. 33, 1338 (1960)
3. P. D. Drummond and C. W. Gardiner, J. Appl. Phys. A 13, 2353 (1980); P. D. Drummond, C. W. Gardiner and D. F. Walls, Phys. Rev. A 24, 916 (1981)
4. F. Haake and M. Lewenstein, Z. Phys. B 48, 37 (1982); Phys. Rev. A 27, 1013 (1983)
5. L. A. Lugiato, Physica 81 A, 565 (1975); Physica 82 A, 1 (1976)
6. F. Casagrande, E. Eschenazi and L. A. Lugiato, Phys. Rev. A 29, 239 (1984)
7. H. Haken, "Handbuch der Physik", vol. XXV 12c (Springer, Berlin, 1970)
8. W. Weidlich and F. Haake, Z. Phys. 185, 30 (1965); 186, 203 (1965)
9. R. Bonifacio and L. A. Lugiato, Phys. Rev. A 18, 1129 (1978)

10. L. A. Lugiato, P. Mandel and L. Narducci, Phys. Rev. A 29, 1438 (1984)

11. H. Risken, Z. Phys. 186, 85 (1965)

12. L. A. Lugiato, Nuovo Cimento B 50, 89 (1979)

13. L. A. Lugiato, F. Casagrande and L. Pizzuto, Phys. Rev. A 26, 3438 (1982)

14. A. Frigerio, J. T. Lewis and J. V. Pulé, Adv. Appl. Math. 2, 456 (1981)

15. A. Frigerio, J. T. Lewis and J. V. Pulé, preprint

16. F. Casagrande, E. Eschenazi, L. A. Lugiato and G. Strini, in Coherence and Quantum Optics V, ed. by L. Mandel and E. Wolf. Plenum Press, New York 1984 p. 801

LIMITATIONS FOR CHAOTIC MOTION IN QUANTUM MECHANICS

Giulio Casati
Dipartimento di Fisica, Università di Milano
Via Celoria 16, 20133 Milano - ITALY

1. Introduction

As it was pointed out in several occasions the main difficulty in building a theory of stochastic motion for bounded quantum systems stems from the discreteness of the energy spectra which entails quasi-periodicity in time of any quantity related to the undisturbed evolution of the system. On the other hand, Fourier analysis of trajectories of bounded conservative classical systems often unveils a continuous spectrum, with the consequence that the motion, while perfectly deterministic, displays many truly stochastic features. In spite of this severe difficulty, we indicate two possible ways to search for disordered or random motion in Quantum Mechanics:

1) Focusing the attention on the structure of spectra rather than on time evolution - in other words, inquiring whether the distinction between quantum systems that in the classical limit are chaotic and systems that in the same limit become integrable can be described in terms of some peculiarities of their spectral sequences. Since one major practical problem with (classically) chaotic systems is how to get some semiclassical description of their quantum spectra, analogous to Einstein-Brillouin-Keller quantization for integrable systems, success in this inquiry would be of great practical relevance.

2) Subjecting systems with a discrete spectrum to external perturbations periodic in time. In this class of non conservative classical systems, the onset of self-generated stochasticity is a very common phenomenon with physically observable consequences.

Unlike the conservative case, for such quantum systems there is no a priori spectral prescription forbidding erratic behaviour. Moreover, this class of systems is closest to applications: for instance, laser chemistry of complex molecules and microwave ionization of highly excited atoms[1].

2. Fluctuations properties of spectra

Our present understanding of quantum dynamics is modelled after a very few exactly solvable cases, more complex cases being treated perturbatively. Therefore, the present state of quantum mechanics can be compared to the state of classical mechanics before the advent of qualitative (or topological) methods, which where able to unveil the amazing richness of classical motion lying beyond the scope of perturbative theories. As a major outcome of these new perspective in classical mechanics, we are now aware of the possibility and necessity of statistical methods even in 'simple' cases, involving only a small number of degrees of freedom.

One important task of quantum mechanics for bounded, conservative systems is finding their energy spectra. For classically integrable systems, this task has been succesfully carried out. In this case, the Schrödinger equation is often exactly solvable. Moreover semiclassical quantization rules of the Bohr-Sommerfeld type can be found, relying on the existence of invariant tori in the classical phase space. Such rules provide an efficient algorithm by which the spectrum (in the semiclassical region at least) can be directly computed, thus by-passing the difficulties inherent in the solution of partial differential equations.

Due to nonexistence - or, better to say, exceptionality - of invariant tori in the phase space of classically chaotic systems, it is very doubtful that analogous rules can ever be found for this class of systems. Thus we are led to suspect that energy spectra will look much less "regular" than in integrable cases, and also that an overall description of such spectra will be obtained only in statistical terms: much in the same way that is being used since a long time in the study of nuclear spectra[2,5].

Once this step of giving up a precise description of spectra, in favour of a statistical description is taken, the following questions arise:

1) Is there a real difference in "regularity" between integrable and non-integrable spectra?

2) Is it possible to devise statistical tests in order to discriminate the spectra of non - integrable systems from the spectra of integrable ones?

As to the second point, the statistical tests of practical use in the analysis of energy spectra are:

i) The "smoothed density" $\rho(E)$ which is defined as the local density of energy levels in a neighbourhood of the value E of the energy.

ii) The distribution of spacings $p(E,s)$, which is defined in such a way that $p(E,s)ds$ gives the probability - i.e., the relative frequency - that a spacing between two consecutive levels lying in a neighbourhood of the energy E lies between s and s+ds.

iii) The Δ_3- statistic of Dyson and Mehta,[3,5] that measures the so-called 'rigidity' of the spectrum:

$$\Delta_3(L;x) = \frac{1}{L} \min_{a,b} \int_x^{x+L} \left[N(E) - aE - b \right]^2 dE.$$

The function $\Delta_3(L;x)$ is a measure of the fluctuations of the staircase cumulative density $N(E)$ from a best-fitting straight-line under the assumption of constant level density (one can always reduce to this case via an "unfolding" procedure[16]). If we assume that the spectrum is traslational invariant, then the average $\Delta_3(L)$ over the spectral sequence will be independent of x. This function provides an indication of the degree of correlation between different levels. In the extreme case of a perfectly regular sequence, with equally spaced levels, one has $\Delta_3(L)$=const.; one says that the spectrum is 'rigid'. Instead, in the opposite case of a completely random sequence, with a Poisson distribution of spacings (i.e., $p(E,s) = e^{-s}$) one finds $\Delta_3(L) \sim L/15$. In all other cases the dependence of Δ_3 on L should be intermediate between these two extreme cases.

It is natural to compute the above statistical parameters for simple quantum systems whose classical behaviour is known to be integrable or stochastic. The most widely used class of model systems are the two-dimensional billiards - i.e., particles in two-dimensional potential wells. Indeed, depending on the shape of the boundary, billiards can be, on the classical side, integrable or chaotic.

The density of levels $\rho(E)$ of quantum billiards has a well - known asymptotics $\rho(E) \sim \frac{1}{4\pi}(A - \frac{L}{\sqrt{E}})$; A, L being the area of the billiard and its perimeter, regardless of the shape of the boundary. Therefore, any difference in statistics between integrable and non - integrable spectra must be sought in fluctuation properties, i.e., in the quantities $p(E,s)$ and $\Delta_3(L)$.

Concerning $p(E,s)$, a general analytical argument[6] as well as numerical experiments[6,15] suggest that $p(E,s) = e^{-s}$ for integrable billiards. This result has been often contrasted with Wigner's formula[3] $p(E,s) = \frac{\pi}{2} s\, e^{-\frac{\pi}{4}s^2}$ which fits well with available empirical data about levels of heavy nuclei[15,17]. Nuclei do not have a classical analog and so, at first sight, do not fit adequately into our picture. Nevertheless, it is usually assumed that nuclear spectra call for statistical analysis because nuclei are 'complex' systems. It is a natural conjecture that this 'quantum complexity' is essentially the same quality that distinguishes non-integrable from integrable systems, with much the same effects on the texture of spectra. In fact, $p(E,s)$ has been numerically computed[7,15,18] for some 'chaotic' billiards, and the result is in good qualitative agreement with Wigner's formula.

As to the Δ_3-statistic, when applied to chaotic billiards it exhibits a logarithmic increase of $\Delta_3(L)$ - again in accordance with data available from nuclear spectroscopy.[15]

It is important to recall here that the theoretical model for the statistics of nuclear spectra is Random Matrix Theory[2,3,16]. In this theory one assumes that the fluctuation properties of nuclear spectra are the same as for the spectra of matrices picked at random in suitable statistical ensembles. The latter fluctuation properties can be analitically predicted and give the experimentally observed p(E,s) and $\Delta_3(L)$.

In summary:

"Complicated" Quantum Systems - heavy nuclei, or Classically Chaotic Systems - obey Wigner's formula. In particular, they exhibits "level repulsion", i.e., the probability p(E,s) of small spacings s tends to 0 as $s \rightarrow 0$. Moreover they have "rigid" spectra - in other words, $\Delta_3(L)$ increases much more slowly than in a random sequence. This means that levels are significantly correlated.

In contrast, the distribution of spacings, for 'generic' integrable systems is Poisson, i.e. $p(E,s) = e^{-s}$. In particular, there is no level repulsion.

As to the question (1) we have taken a direct approach by making use of the notion of randomness formalized by Algorithmic Complexity Theory[19-21]. Roughly speaking, the complexity of a finite string of numbers is the minimum length of a binary sequence that must be fed into an ideal computing machine in order to get the given string at the output. Now, one should not call "random" a given infinite numerical sequence, unless it were found that the complexity of finite segments of the string is asymptotically of the same order as their length.

With this background, we have considered a simple integrable billiard with a rectangular boundary with incommensurate sides. Numerical estimates of p(E,s) give a Poisson distribution with a good accuracy. We have been able to show that the sequence of energy levels of this billiard is not random in the sense sketched above[11], and we have reasons to submit that an analogous conclusion holds for all integrable systems of practical interest. In contrast, we have analized the complexity of the spectra of random matrices in the so-called Orthogonal Circular Ensemble, and we have found that these spectra are random. Then, if we accept the widespread idea that the spectra of random matrices provide a paradhygm for the statistical properties of the spectra of chaotic systems, we must conclude that a distinction between integrable and non-integrable spectra is indeed possible, the former type of spectra being regular and the latter random - though not as random as a completely uncorrelated sequence.

3. Systems under external perturbations

Model systems that, while retaining the essential features of quantum systems of physical interest, are amenable to effective numerical and/or theoretical analysis, play here an important role.

An interesting model in this class is the δ-kicked rotator, namely a quantum rotator subject to periodic δ-like pulses[22].

The dynamics of this, periodically perturbed, quantum system is defined by the structure of its quasi-energy spectrum. Indeed, if the quasi-energy spectrum is discrete, then the system displays recurrent behaviour and no energy growth. On the other hand a continuous component in the q.e. spectrum, would cause the energy to grow.

We now know that there is no a priori reason to exclude such continuous components. On the contrary we have shown[23] that as the external perturbation is turned on, the quasi-energy spectrum undergoes a transition from pure point to partially continuous spectrum. This transition has many similarities with the Kolmogorov-Arnold-Moser theorem of classical mechanics. Moreover there are strong indications that the quasi-energy spectrum is singular continuous rather than absolutely continuous: this would imply that the motion is not fully chaotic but corresponds to some kind of mixing behaviour. This property has to be reflected in the average growth of the rotator energy.

As a matter of fact, the numerical results so far available on the kicked rotator indicate that the growth of the energy is strongly reduced in the quantum model, thus confirming the original conclusion[22] that there are limitations of quantum origin to the appearence of chaos in periodically perturbed systems.

This raises the question whether such limitations to quantum chaotic motion are peculiar to the kind of perturbations studied or are inherent to the structure of quantum mechanics: one might even wonder whether any type of stochastic motion is possible at all in quantum mechanics.

In order to investigate this problem let us consider a more realistic system namely an hydrogen atom under a microwave field. In recent experiments[1] the ionization rate of highly-excited hydrogen atoms by a microwave monochromatic field has been measured. Even though the microwave frequency ($\sim$10GHz) was about one half of the frequency necessary for the resonant transition from the initial level $n_o=66$ to the next level n=67, nevertheless a noticeable ionization was detected for a microwave field intensity much below the classical ionization threshold and also of the quantal tunneling threshold.

A quantum numerical study has been given in refs. (26-28) in which the hydrogen atom excitation was studied starting with initial states very extended along the field direction, i.e. states with parabolic quantum numbers $n_1 \gg n_2 \sim 1$ and m=o. This situation is conveniently described by the one-dimensional Hamiltonian

$$H = \frac{P_z^2}{2} - \frac{1}{|z|} + \varepsilon z \cos \omega t \qquad (1)$$

where ε, ω are the field strenght and frequency in atomic units. Hamiltonian (1) was used in refs. (29) to study surface state electrons bounded to the surface of liquid Helium by their image charge (in this case there is also an infinite repulsive barrier at the surface, which is due to the Pauli exclusion principle).

Due to the high quantum numbers, a classical description may seem appropriate and indeed in refs. (24,25) the relevance of chaotic motion in the diffusion mechanism leading to ionization has been stressed. As it is now well known the Kolmogorov-Arnold-Moser theorem ensures that as the strenght ε of the perturbation increases, the classical motion undergoes a transition from stable, near-integrable motion, to chaotic motion. In the latter situation the electron executes a random-like motion in action space leading to diffusion and ionization. The critical value ε_c of the perturbation may be found using the Chirokov's overlapping criterium. To this end, introducing the action-angle variables of the unperturbed Hamiltonian and expanding the perturbation in a Fourier series, we may rewrite Hamiltonian (1) in the form:

$$H = \frac{-Z^2}{2I^2} + \varepsilon \sum_{n=-\infty}^{\infty} z_n(I) \cos(n\theta - \omega t) \qquad (2)$$

where

$$z_n(I) = \frac{1}{2\pi} \int_0^{2\pi} z(\theta, I)\, e^{in\theta} d\theta = J_n'(n) \frac{I^2}{Zn} \qquad (3)$$

and

$$\dot{\vartheta} = \Omega(I) = Z^2/I^3 \tag{4}$$

is the frequency of the unperturbed motion.

The perturbation strongly affects the unperturbed motion at resonances

$$r\dot{\vartheta} - \omega = 0 \tag{5}$$

namely at values

$$I_r = \left(r Z^2/\omega\right)^{1/3} \tag{6}$$

In the vicinity of each resonance I_r , the motion is approximately described by a pendulum Hamiltonian. Indeed, expanding the Hamiltonian near I_r , and using the time-dependent generating function $F(\mathcal{J}, \vartheta, t) = (I_r + \mathcal{J})(r\vartheta - \omega t)/r$ we have

$$H \simeq -\frac{1}{2}\left(\frac{3Z^2}{I_r^4}\right)\mathcal{J}^2 + \varepsilon z_r(I_r)\cos(r\varphi) \tag{7}$$

where

$$\begin{aligned} \mathcal{J} &= I - I_r \\ \varphi &= (r\vartheta - \omega t)/r \end{aligned} \tag{8}$$

Each resonance has a width (libration region) given by the maximum distance of the two pendulum separatrices:

$$\Delta_r = 4\sqrt{\left(\frac{\varepsilon z_r(I_r)}{3}\right)} \cdot \frac{I_r^2}{Z} \tag{9}$$

On the other hand, the distance between two consecutive resonances is

$$\delta_r = I_{r+1} - I_r \sim I_r/3r \tag{10}$$

For sufficiently small ε, the distance δ_r between resonances is larger than the width Δ_r and the motion is trapped inside the resonance island.

As the strength of the perturbation increases, the resonance width grows larger and larger until a critical value ε_c is reached at which two consecutive resonances overlap. The value ε_c is determined by the relation

$$\delta_r = \frac{1}{2}\left(\Delta_r + \Delta_{r+1}\right) \tag{11}$$

From Eqs. (6), (9), (10) and using the asymptotic expression

$$z_r(I_r) = J_r'(r)\, I^2/Z r \sim 0.4\, I^2 \tag{12}$$

we obtain

$$\varepsilon_c \simeq Z^{11/3}/\left(20\,\omega^{1/3} I^5\right) \tag{13}$$

When the perturbation exceeds the critical value ε_c the resonance islands overlap and the electron orbits freely move in the action space through the islands chain in a diffusive - like manner.

Numerical studies confirm the above anlytical predictions within a factor ~ 2. Indeed the estimate do not take into account secondary effects such as higher order islands which increase the width of each resonance and therefore decrease the critical field value ε_c for the transition to stochastic motion.

We numerically studied the quantum motion by integrating the Schrödinger equation for Hamiltonian (1) and we found that also in this case quantum mechanics introduces severe limitations to classical chaotic motion. In particular the distribution fuction $f_n(t)$, namely the probability of occupation of the n-th level at time t, remains localized around the initial value $n=n_0$ (except for a multiphoton "plateau" with a definite resonant structure) even for intesities of the microwave field well above the critical classical value ε_c for transition to chaotic and diffusive motion. On the other hand we have also found a quantum delocalization border ε_q above which the quantum packet delocalizes and infinite diffusion and ionization takes place. For the one-dimensional hydrogen atom we have[28]

$$\varepsilon_q = \omega^{7/6}/2n \qquad (14)$$

Numerical experiments are now in progress in order to check the above estimate. In any event it is presently not clear whether for $\varepsilon > \varepsilon_q$ the quantum motion has any distinctive feature of randomness which will permit to term "quantum diffusion" the spreading of the wave packet.

References

1) J.E. Bayfield and P.M. Koch, Phys.Rev.Lett. 33, (1974) 258; J.E. Bayfield, L.D. Gardner and P.M. Koch, Phys.Rev.Lett. 39, (1977) 76; P.M. Koch, J. Phys. (Paris), Coolq. 43, C2-187 (1982); R.J. Damburg and V.V. Kolosov, J. Phys. B12, (1979) 2637; P.M. Koch and D.R. Mariani, Phys.Rev.Lett. 46, (1981) 1275.

2) Statistical Theory of spectra: Fluctuations ed. C. E. Porter (Academic Press, New York, 1965).

3) M.L. Mehta, Random Matrices, Academic, New York, 1967.

4) E.P. Wigner, Math.Ann. 53 (1951) 36; 62 (1955) 548; 65 (1957) 203; 67 (1958) 325.

5) F.J. Dyson, J.Math.Phys. 3 (1962) 140, 157, 166.

6) M.V. Berry, M. Tabor Proc.Roy.Soc. London, 356 (1977), 375.

7) S.W. McDonald, A.N. Kaufman Phys.Rev.Lett. 42 (1979), 1189.

8) V. Buch, R.B. Gerber, M.A. Rather J.Chem.Phys. 76 (1982), 5397.

9) H.S. Camarda, P.D. Georgopolus Phys.Rev.Lett., 50 (1983), 492.

10) E. Haller, H. Koppel, L.S. Cederbaum. Chem.Phys.Lett., 101 (1983), 215; Phys.Rev.Lett., 52 (1984), 1665.

11) H.Hirooka, Y. Yotsuga, Y. Kobayashi, N. Saito, "New Representation of Quantum Chaos", Phys. Lett. 101A (1984) 115.

12) T. Ishikawa, T. Yukawa, "Transition from Regular to Irregural Spectra in the Quantum Billiards". Preprint KEK-TH85 (1984).

13) G. Casati, B.V. Chirikov, I. Guarneri "Energy level statistics of integrable quantum systems. Preprint.

NON COMMUTATIVE L^p SPACES AND K.M.S. FUNCTIONS

by

Carlo CECCHINI

Istituto di Matematica dell'Università di Genova - Via L.B.Alberti, 4 - GENOVA

1. Introducation. In 2 a theory for non commutative $L(p;\mathcal{M},\omega)$ spaces $(1\leq p<\infty)$ with respect to a von Neumann algebra $\mathcal{M}$ acting on a Hilbert space $\mathcal{H}$, and on which a normal faithful state ω is defined has been developed. These $L(p;\mathcal{M},\omega)$ spaces are Banach spaces of complex linear combinations of positive forms (or complex forms) on $D(\mathcal{H},\omega)=\{\xi\in\mathcal{H}: \|a\xi\|^2<\alpha\,\omega(a^+a)$ for some $\alpha>0$ and all $a\in\mathcal{M}\}$, which is a linear dense subspaces of $\mathcal{H}$. They are representations of the L^p spaces defined as interpolation spaces between $\mathcal{M}$ and $\mathcal{M}_*$ by Terp [8] and by Zoletarev [9], and generalize the theory developed by Sherstnev (see, for a review paper, [6]) for L^1 spaces. If we introduce an auxiliary normal, faithful state ω' on the commutant $\mathcal{M}'$ of $\mathcal{M}$, they are closely connected to the spaces $L^p(\mathcal{M},\omega')$ defined and studied by Connes [4] and Hilsum [5]. However, if $p_1<p_2$, then $L(p_2;\mathcal{M},\omega)$ is contained and norm dense in $L(p_1;\mathcal{M},\omega)$.

By using the above mentioned connection with the spaces $L^p(\mathcal{M},\omega')$, it is possible to define $L(p_3;\mathcal{M},\omega)$ valued products between elements of $L(p_1;\mathcal{M},\omega)$ and $L(p_2;\mathcal{M},\omega)$, whenever $1\leq p_1,p_2,p_3\leq+\infty$ and $p_1^{-1}+p_2^{-1}=p_3^{-1}$. Those products have the classical continuity properties, can be used to obtain the usual explicit duality relation between $L(p;\mathcal{M},\omega)$ and $L(p';\mathcal{M},\omega)$ if $p^{-1}+p'^{-1}=1$, and are shown to be independent from the particular auxiliary ω' we are using. They however depend on p_1 and p_2, in the sense that if, for instance, we take $q_1,q_2\in L(\infty;\mathcal{M},\omega)$ (and so, of course, in all spaces $L(p;\mathcal{M},\omega)$), then all the products which we can consider by looking at q_1 as at an element of $L(p_1;\mathcal{M},\omega)$ and q_2 as an element of $L(p_2;\mathcal{M},\omega)$ with $(p_1,p_2)\in\{(x,y): 1\leq x,\ y\leq\infty,\ x^{-1}+y^{-1}\leq 1\}$ are in general different.

The purpose of this note is to show that it is possible to give an equivalent intrinsic (i.e. with no reference to ω') definition of those products by using K.M.S. functions, and in the process to clarify the reason and the way the products depend on p_1 and p_2.

2. Preliminaries.

Let $\mathcal{M}$ be a von Neumann algebra acting on a Hilbert space $\mathcal{H}$, $\mathcal{M}'$ its commutant and ω (ω') a normal faithful state defined on $\mathcal{M}$ ($\mathcal{M}'$) with modular automorphism roup σ^t_ω ($\sigma^t_{\omega'}$). The triple $(\pi_\omega, \mathcal{H}_\omega, \Omega)$ ($(\pi_{\omega'}, \mathcal{H}_{\omega'}, \Omega')$) is the result f G.N.S. construction with ω (ω').

We summarize now some results from [4], [5] and [3].

Set $D(\mathcal{H}, \omega) = \{\xi \in \mathcal{H} : \|a\xi\|^2 < c\,\omega(a^+a)$ for some $\alpha > 0$ and all $a \in \mathcal{M}\}$. $D(\mathcal{H}, \omega)$ is a dense linear subspaces of $\mathcal{H}$, and for each $\xi \in D(\mathcal{H}, \omega)$ there is a unique bounded linear operator $R_\omega(\xi) : \mathcal{H}_\omega \longrightarrow \mathcal{H}$ such that $R_\omega(\xi)\, \pi_\omega(a)\Omega = a\xi$. The correspondence $\xi \longrightarrow R_\omega(\xi)$ is linear, and, for all $\xi, \eta \in D(\mathcal{H}, \omega)$, the operator $R_\omega(\xi) R_\omega(\eta)^+$ is in $\mathcal{M}'$ and the operator $R_\omega(\xi)^+ R_\omega(\eta)$ is in the von Neumann algebra of the operators of the form $\pi'_\omega(a) = J_\omega \pi_\omega(a) J_\omega$ (J_ω is the isometrical involution associated to the triple $(\pi_\omega, \mathcal{H}_\omega, \Omega)$).

So, for each $\xi \in D(\mathcal{H}, \omega')$, there are two positive bounded linear operators:

$$H_\omega(\xi) = |R_\omega(\xi)^+|^2 \in \mathcal{M}' \qquad \text{and}$$

$$K_\omega(\xi) = \pi_\omega^{-1}\left(J_\omega |R_\omega(\xi)|^2 J_\omega \right) \quad .$$

Exchanging now the roles of $\mathcal{M}$ and $\mathcal{M}'$ (resp. ω and ω') for each $\varphi \in \mathcal{M}^+_*$ the equality

$$q_\varphi(\xi) = \varphi(H_{\omega'}(\xi)) \qquad (\xi \in D(\mathcal{H}, \omega))$$

defines a lower semicontinuous positive form on $D(\mathcal{H}, \omega)$, to which a positive self adjoint operator $\frac{d\omega}{d\omega'}$ is associated. We shall set $\frac{d\omega}{d\omega'} = d$; under our hypothesis $\frac{d\omega'}{d\omega} = d^{-1}$. It turns out that if $\xi \in D(\mathcal{H}, \omega)$, then $d^{-1/2}\xi \in D(\mathcal{H}, \omega')$ and that $K_\omega(\xi) = H_{\omega'}(d^{-1/2}\xi)$.

In [5] the spaces $L^p(\mathcal{M}, \omega')$ $(1 \leq p < \infty)$ are defined as the sets of all closed, densely defined operators on $\mathcal{H}$ with polar decomposition $T = u|T|$ such that $u \in \mathcal{M}$ and $|T|^p = \frac{d\varphi}{d\omega'}$ for some $\varphi \in \mathcal{M}^+_*$.

If $\psi \in \mathcal{M}_*$ has a polar decomposition $\psi = u|\psi|$, then we set $T_{\omega'}(\psi) = u\frac{d|\psi|}{d\omega'}$ and $\int T_{\omega'}(\psi)\, d\omega' = \psi(1)$.

The spaces $L^p(\mathcal{M}, \omega')$ $(1 \leq p < \infty)$ are proved to be Banach spaces ($L^2(\mathcal{M}, \omega')$ is a Hilbert space) under the norm $\|T\|_p = \left(\int |T|^p\, d\omega'\right)^{1/p}$ if by sum (and, later, by product) of unbouded operators we take the strong sum (product).

To define our spaces $L(p; \mathcal{M}, \omega)$ for $1 \leq p < \infty$ we first define $\mathcal{H}(p)$ as the Hilbert space completion of the domain of $d^{-1/2p}$ under the inner product

$$\langle \xi, \eta \rangle_p = \langle d^{-1/2p}\xi, d^{-1/2p}\eta \rangle_{\mathcal{H}}$$

and $\mathcal{H}(\infty) \equiv \mathcal{H}$. There is a unique unitary operator $V(p_2, p_1) : \mathcal{H}(p_1) \longrightarrow \mathcal{H}(p_2)$ such that $V(p_2, p_1) = d^{-(p_1^{-1} - p_2^{-1})/2}\xi$ for $\xi \in D(\mathcal{H}, \omega)$ and for $1 \leq p_1 < p_2 \leq \infty$.

Let $1 \leq p < \infty$. We set $L(p;\mathcal{H},\omega)$ for the set of all complex forms (i.e. complex linear combinations of positive forms) defined on $D(\mathcal{H},\omega)$ of the type:
$q(T)\ (\xi) = < |T|^{1/2}\ V(p,\infty)^{+}\ u^{+}\ V(p,\infty),\ |T|^{1/2}\xi > p$
for $\xi \in D(\mathcal{H},\omega)$. Here T is a closed densely defined operator on $\mathcal{H}$ (p) with polar decomposition $T = V(\infty,\ p)^{+}\ u\ V(\infty,\ p)\ |T|$ (u partial isometry in $\mathcal{H}$ and $V(\infty,\ p)\ T\ V\ (\infty,\ p)^{+} \in L^{p}\ (\mathcal{H},\ \omega'))$.
For $p = \infty$ we set $L(\infty\ ;\mathcal{H},\omega) = \left\{q(a)\ :\ a \in \mathcal{H}\right\}$, where $q(a)\ (\xi) = <\xi,\ a\xi>$.
The mapping $\lambda_{p}\ :\ L(p;\mathcal{H},\omega) \longrightarrow L^{p}\ (\mathcal{H},\omega'),\ \lambda_{p}\ :\ q(T) \longrightarrow V(\infty,\ p)T\,V(\infty,p)^{+}$ is an isomorphism for $1 \leq p \leq \infty$, and we define the norm on $L(p;\mathcal{H},\omega)$ by requiring it to be an isometry. Thus they become Banach spaces (in particular $L(2;\mathcal{H},\omega)$ is a Hilbert space) which are proved in [2] to be independent from the choice of ω' (which can even be a normal faithful semifinite weight).
The mapping $i\ :\ \mathcal{H}_{*} \longrightarrow L(1;\mathcal{H},\ \omega)$ defined as $\left[i\ (\psi)\right](\xi) = \psi(K_{\omega}(\xi)\ )$ establishes the canonical isometrical isomorphism between $\mathcal{H}_{*}$ and $L(1;\mathcal{H},\omega)$, and if $p_1 < p_2$ then $L(p_2;\mathcal{H},\omega)$ is contained and norm dense in $L(p_1;\mathcal{H},\omega)$. In particular if $q = q(a) \in L(p_1;\mathcal{H},\omega)$, then $\lambda_{p}(q) = d^{1/2p}\ a^{1/2p}.L(p;\mathcal{H},\omega)$ is an interpolation space between $L(1;\mathcal{H},\omega)$ and $L(\infty\ ;\mathcal{H},\omega)$. For $1 \leq p_1,\ p_2,\ p_3 \leq \infty$, $p_2^{-1} + p_2^{-1} = p_3^{-1}$ we define a mapping (or product) from $L(p_1;\mathcal{H},\omega) \times L(p_2;\mathcal{H},\omega)$ to $L(p_3;\mathcal{H},\omega)$ by mapping the couple (q_1,q_2) into $q_1(p_1)\ q_2(p_2)$ where $q_1(p_1)q_2(p_2)$ =

$$(\lambda_{p_3}(\lambda_{p_1}(q_1)\ \lambda_{p_2}(q_2))\ .$$

As remarked in the introduction, this product is continuous in both variables; it does not depend on the choice of ω'. If $q_1,q_2 \in L(\infty;\mathcal{H},\omega)$, however, in general $q_1(p_1)q_2(p_2) \neq q_1(p_3)q_2(p_4)$, even when both members make sense.

3. The main result.

In the following $S = \left\{z \in \mathbb{C}\ :\ 0 \leq \ \mathrm{Re}z \leq 1\right\}$, $\overset{\circ}{S}$ is the interior of S. For all $a_1,\ a_2$ we shall call $F_{(a_1,a_2)}$ the K.M.S. function on S for the couple (a_1,a_2), that is the complex function bounded and continuous on S and analytic on $\overset{\circ}{S}$ such that $F(it) = \omega(\ \sigma_{\omega}^{t}\ (a_1)\ a_2),\ \ F(1+it) = \omega\ (a_2\ \ \sigma_{\omega}^{t}\ (a_2))$.

3.1. Lemma Let $q_1,q_2 \in L(\infty;\mathcal{H},\omega)$, $q_1 = q(a_1)\ q_2 = q(a_2)\ (a_1,a_2 \in \mathcal{H})$ and $\xi \in \mathcal{H}$. Then

$$\left[q_1(1)\ q_2(\infty)\right]\ (\xi) = F_{(a_1,a_2K_{\omega}(\xi))}(1/2)$$

$$\left[q_1(\infty)\ \ q_2(1)\right]\ (\xi) = F_{(a_2,\ K_{\omega}(\xi)\ a_1)}(1/2)$$

Proof. It suffices to prove the first equality, since the proof of the second is the same. We have then:

$$[q_1(1)\, q_2(\infty)]\,(\xi) = [\lambda_1^{-1}(\lambda_1(q_1)\, \lambda_\infty(q_2))]\,(\xi) =$$

$$= [\lambda_1^{-1}(d^{1/2}\, a_1\, d^{1/2}\, a_2)]\,(\xi) = \int d^{1/2}\, a_1\, d^{1/2}\, K_\omega(\xi)\, d\omega' =$$

$$= \int a_2\, K_\omega(\xi)\, d^{1/2}\, a_1\, d^{1/2}\, d\omega' \text{ by [5], prop. 7, as } d \in L^1(\mathcal{M}, \omega').$$

Now corollary 6 in [5] again implies our statement.

3.2. Lemma The set of the vectors in $L^2(\mathcal{M}, \omega')$ of the form $T = d^{1/2} R_{\omega'}(\eta)\, R_{\omega'}(\xi)^+$ for $\xi, \eta \in D(\mathcal{H}, \omega')$ is dense in $\mathcal{H}$.

Proof. Let S $L^2(\mathcal{M}, \omega')$. Then

$$< d^{1/2} R_{\omega'}(\eta)\, R_{\omega'}(\xi)^+, S >_{L^2(\mathcal{M}, \omega')} =$$

$$= \int R_{\omega'}(\xi)\, R_{\omega'}(\eta)^+\, d^{1/2}\, S\, d\omega' =$$

$$= < d^{1/2} S, R_{\omega'}(\xi)\, R_{\omega'}(\eta) >_{L^1(\mathcal{M}, \omega') \times L^\infty(\mathcal{M}, \omega')} = < d^{1/2}\xi, S\eta >_{\mathcal{H}}$$

by [5], prop. 5. So, if $< d^{1/2} R_{\omega'}(\eta)\, R_{\omega'}(\xi)^+ d^{1/2}, S >_{L^2(\mathcal{M}, \omega')} = 0$ for all $\xi, \eta \in D(\mathcal{H}, \omega)$, this implies by the density of the set $\{ d^{1/2}\xi : \xi \in D(\mathcal{H}, \omega)\}$ that $S\eta = 0$ for all $\eta \in D(\mathcal{H}, \omega)$ and so $S = 0$.

3.3. Lemma The mapping

$$z \longrightarrow \varphi(z) = \int T\, d^{z/2}\, a\, d^{(1-z)/2}\, d\omega'$$

is bounded and continuous on S and analytic on $\mathring{S}$ for each $T \in L^2(\mathcal{M}, \omega')$ and $a \in \mathcal{M}$ with polar decomposition $a = u|a|$.

Proof. As $d \in L^1(\mathcal{M}, \omega)$, applying the results of [5], $d^{z/2}\, a\, d^{(1-z)/2} \in L^2(\mathcal{M}, \omega')$ for $z \in S$. So $\varphi(z) = < T^+, d^{z/2}\, a\, d^{(1-z)/2} >_{L^2(\mathcal{M}, \omega')}$ and

$$|\varphi(z)| \le \|T\|_{L^2(\mathcal{M}, \omega')} \|d^{z/2}\, a\, d^{(1-z)/2}\|_{L^2(, ')} \le$$

$$\le \|T\|_{L^2(\mathcal{M}, \omega')} \|d^{\mathrm{Re} z/2}\|_{L^{2/\mathrm{Re} z}(\mathcal{M}, \omega')} \|a\|_{\mathcal{M}} \|d^{(1-\mathrm{Re} z)/2}\|_{L^{2/1-\mathrm{Re} z}(\mathcal{M}, \omega')} =$$

$= \|T\|_{L^2(\mathcal{H},\omega')} \|a\|_{\mathcal{H}}$ for all $z \in S$.

Note now that for $\xi, \eta \in D(\mathcal{H}, \omega')$ the mapping

$z \longrightarrow < \xi, d^{z/2} a\, d^{(1-z)/2} \eta > = < d^{z/2} \xi, a\, d^{(1-z)/2} \eta >$ is continuous on S and analytic on $\mathring{S}$. Since $D(\mathcal{H}, \omega')$ is dense in $\mathcal{H}$, then for all $\xi \in \mathcal{H}$ we can choose a sequence $\xi_n \longrightarrow \xi$ in $\mathcal{H}$, and so the sequence $< \xi_n, d^{z/2} a\, d^{(1-z)/2} \eta >$ con- vergerces uniformly on S to $< \xi, d^{z/2} a\, d^{(1-z)/2} \eta >$ which implies the continuity on S and analiticity on $\mathring{S}$ of this last function for all $\xi \in \mathcal{H}$ Let now $T = d^{1/2} R_{\omega'}(\xi) R_{\omega'}(\eta)^+$, with $\xi, \eta \in D(\mathcal{H}, \omega')$. Then $T \in L^2(\mathcal{H}, \omega)$ and

$$< d^{1/2} R_{\omega'}(\xi) R_{\omega'}(\eta)^+, d^{z/2} a\, d^{(1-z)/2} >_{L^2(\mathcal{H},\omega')} =$$

$$= \int R_{\omega'}(\eta) R_{\omega'}(\xi)^+ d^{1/2} d^{z/2} a\, d^{(1-z)/2} d\omega' =$$

$$= \int d^{(1+z)/2} a\, d^{(1-z)/2} R_{\omega'}(\eta) R_{\omega'}(\xi)^+ d\omega' =$$

$$= < d^{1/2} \xi, d^{z/2} a\, d^{(1-z)/2} \eta >_{\mathcal{H}'}$$

which has just been proved to be continuous on S and analytic on $\mathring{S}$.

We have proved therefore the statement for the particular case in which T is of the type $d^{1/2} R_{\omega'}(\xi), R_{\omega'}(\eta)^+$, with $\xi, \eta \in D(\mathcal{H}, \omega')$.

An application of lemma 3.2 allows us to take a sequence $d^{1/2} R_{\omega'}(\xi_n) R_{\omega'}(\eta_n)^+$ converging to a general T in $L^2(\mathcal{H}, \omega')$, with $\xi_n, \eta_n \in D(\mathcal{H}, \omega')$. Then the se- quence $< d^{1/2} R_{\omega'}(\xi_n) R_{\omega'}(\eta_n)^+, d^{z/2} a\, d^{(1-z)/2} >_{L^2(\mathcal{H},\omega)}$ converges unifor ly to $\varphi(z)$ on S, and we get our statement.

__3.4 Theorem__ Let $q_1, q_2 \in L(\alpha; \mathcal{H}, \omega)$ with $q_1 = q(a_1)$, $q_2 = q(a_2)$ $(a_1, a_2 \in \mathcal{H})$ and $\xi \in D(\mathcal{H}, \omega)$. There is then a mapping $G\colon S \times S \longrightarrow \mathbb{C}$, which is bounded and continuous on S and analytic on $\mathring{S}$ in each of the variables whenever the other one is fixed, and such that, for $1 \leq p_1, p_2 \leq +\infty$ and $t_1, t_2 \in \mathbb{R}$

$$G(1/p_1, 1/p_2) = [q_1(p_1)\; q_2(p_2)](\xi),$$

$$G(it_1, it_2) = F^{(1/2)}_{(K_\omega(\xi),\, \sigma_\omega^{-t_2/2}(a_1)\, \sigma_\omega^{t_1/2}(a_2))}$$

$$G(\,it_1;\ 1 + it_2) = F^{(1/2)}_{(\quad t_1/2(a_2),\ K\ (\ )\quad -t_2/2(a_1))}$$

$$G(1 + it_1,\ it_2) = F^{(1/2)}_{(\sigma_\omega^{-t_2/2}(a_1),\ \sigma_\omega^{t_1/2}(a_2)\ K_\omega(\xi))}$$

$$G(1 + it_1,\ 1 + it_2)\ F^{(1/2)}_{(I,\ \sigma_\omega^{t_1/2}(a_2)\ K_\omega(\xi)\ \sigma_\omega^{-t_2/2}(a_1))}$$

roof. The function

$$(z_1, z_2) = \int d^{(1-z_2/2}\, a_1\, d^{(z_1 + z_2)/2}\, a_2\, d^{(1-z_1)/2}\, K_\omega(\xi)\, d\omega'$$

as the above claimed properties.

ndeed, by [5] $d^{1-z_2/2}\, a_1\, d^{z_2/2}$ and $d^{z_1/2}\, a_2\, d^{1-z\ /2}$ are in $L^2(\mathcal{H}, \omega')$, and so lemma 3.3) above gives us the above stated regularity properties.

e have:

$$(1/p_1,\ 1/p_2) = \int d^{(1-1/p_2)/2}\, a_1\, d^{(1/p_1+1/p_2)/2}\, a_2\, d^{(1-1/p_1)/2}\, K_\omega(\xi)\, d\omega' =$$

$$\int d^{(1-1/p_1\ -\ 1/p_2)/2}\, \lambda_{p_1}(q_1)\, \lambda_{p_2}(q_2)\, d^{(1-1/p_1\ )/2}\, K_\omega(\xi)\, d\omega' =$$

$$[q_1\,(p_1)\ q_2(p_2)]\,(\xi)\ ;$$

$$(it_1,\ it_2) = \int d^{1/2}\, \sigma_\omega^{-t_2/2}(a_1)\ \sigma_\omega^{t_1/2}(a_2)\, d^{1/2}\, K_\omega(\xi)\, d\omega'$$

nd the claim follows by an application of [5], cor. 6, as in the proof of lemma 3.1. he expressions involving $G(it_1,\ 1 + it_2)$ and $G(1 + it_1,\ it_2)$ follow again by substi ution and application of lemma 3.1 as above and

$$(1 + it_1\ ,\ 1 + it_2) = \int d^{\,-it_2/2}\, a_1\, d^{1+i(t_1+t_2)/2}\, a_2\, d^{-it_1/2}\, K_\omega(\xi)\, d\omega' =$$

$$\int \sigma_\omega^{-t_2/2}(a_1)\, d\ \sigma_\omega^{t_1/2}(a_2)\, K_\omega(\xi)\ d\omega' =$$

$\int d^{1/2}\, I\, d^{1/2}\ \sigma_\omega^{t_1/2}(a_2)\, K_\omega(\xi)\ \sigma_\omega^{-t_2/2}\, d\omega'$, which gives our claim, after, nother application of [5] , cor. 6

.5 Remark Note that theorem 3.4 can be used to define explicitely the products $q_1(p)\ q_2\,(q)]$ for $q_1,\ q_2 \in L(\infty; \mathcal{H}, \omega)$, using the K.M.S. functions and avoiding the ntroductions of any auxiliary state ω' on $\mathcal{H}'$.

he continuity of the products in each of the variables allows us then to define hem for the general case.

Let now $\mathcal{M}_1$ be von Neumann subalgebra of $\mathcal{M}$ and $\omega_1 = \omega|_{\mathcal{M}_1}$. Then (cfr.[3]) we have $D(\mathcal{H}, \omega_1) \supseteq D(\mathcal{H},\omega)$, and the mapping $K : q \longrightarrow k(q) = q|_{D(\mathcal{H},\omega)}$ from $L(p;\mathcal{M}_1,\omega_1)$ to $L(p;\mathcal{M},\omega)$ is a linear contraction. Theorem 3.4 implies immediately the following

<u>3.6 Corollary</u> If $\sigma_{\omega}^{t}(a) = \sigma_{\omega_1}^{t}(a)$ for all $a \in \mathcal{M}_1$ (or, equivalently, cfr.[1] and [7], if there is a conditional expectation preserving ω from $\mathcal{M}$ to $\mathcal{M}_1$), then $\left[[k(q_1)](p_1)\,[k(q_2)](p_2)\right] = k(\left[q_1(p_1), q_2(p_2)\right])$ for each $q_1 \in L(p_1;\mathcal{M}_1,\omega_1)$, $q_2 \in L(p_2;\mathcal{M}_1,\omega_1)$ with $1 \leq p_1, p_2, p_1^{-1} + p_2^{-1} \leq \infty$.

Bibliography

1. L.Accardi, C.Cecchini "Conditional expectations in von Neumann algebras and a theorem of Takesaki "J. Funct. anal. 45 (1982), 215-273.

2. C.Cecchini, "Non commutative integration for states on von Neumann algebras" - preprint, to be published on J. Op. Th.

3. C.Cecchini, D.Petz "Norm convergence of generalized martingales in L^p spaces over von Neumann algebras" - preprint, to be published on Acta Sc.Math.

4. A.Connes "On the spatial theory of von Neumann algebras" J;Fucnt. Anal. 35 (1980) 153-164.

5. M.Hilsum "Les espaces L^p d'une algèbre de von Neumann définies par la derivée spatiale". J.Funct. Anal. 40 (1981), 151-169.

6. A.N. Shertsnev "A general theory of measure and integration in von Neumann algebras" (in Russian) Matematika 8 (1982), 20-35.

7. M.Takesaki "Conditional expectations in von Neumann algebras" J.Funct.Anal. 9 (1972), 306-321.

8. M.Terp "Interpolation spaces between a von Neumann algebras and its dual" J. Op. Theory (1982), 327-360.

9. A.A.Zoletarev "L^p spaces on von Neumann algebras and interpolation" (in Russian) Matematika, 8 (1982), 36-43.

NORMAL PRODUCT STATES AND NUCLEARITY: NEW ASPECTS OF ALGEBRAIC QUANTUM FIELD THEORY

C. D'Antoni
Istituto Matematico
Università dell'Aquila
L'Aquila - Italy

Introduction

Let us recall the framework of algebraic Quantum Field Theory [10]

(1) a net of von Neumann algebras on a Hilbert space $\mathcal{H}$, $\mathcal{A}(O)$ indexed by bounded regions of Minkowski space

(2) a unitary continuous representation $\mathcal{U}$ of a compact group G, the gauge group, on $\mathcal{H}$ inducing automorphisms α_g on each $\mathcal{R}(O)$.

(3) a unitary continuous representation of $\mathbb{R}^4$, satisfying the spectrum condition.

(4) a distinguished vector $\Omega \in \mathcal{H}$: $\mathcal{U}_g \Omega = \Omega \ \forall_g \in G$ and $\mathcal{U}_x \Omega = \Omega \ \forall x \in \mathbb{R}^4$.

We denote by $\mathcal{A}$ the C*-algebra generated by $\bigvee_\theta \mathcal{A}(O)$.

This scheme has been successful in deriving properties of Q FT that are consequences of general principles: existence of scattering states, structure of superselection sectors,... . It has been however not yet equally useful for discussion of a particular property of a given theory. Let us mention among the others the difficulty of constructing the net explicitely, the absence of a notion of similarity for different nets, the absence of a perturbative treatment of the algebraic structure.

We will study here a very significative example. The Superselection sectors are obtained as equivalence classes of representations of $\mathcal{A}$ describing situations of interest in physics. Various choices of the relevant representations have been proposed by Borchers [1], Doplicher, Haag and Roberts [2] and by Buchholz and Fredenhagen [3].

The quantum numbers labelling the Superselection Sectors can be considered as eigenvalues of charge operators. These are global operators and do not belong to $\mathcal{A}$. In Lagrangean Field Theory one has observable (Wightman) fields which are supposed to measure approximately the charge. Can we derive a similar structure from first principles? Can we construct observables which are the analogues of space integrals of regularized densities? In other terms: can we construct a local implementation of the gauge automorphism group?

Local implementation problem: given an action of a compact group G by automorphisms α_g such that $\alpha_g(\mathcal{A}(O)) = \mathcal{A}(O)$ $\forall O$, is there a region $\hat{O} \supset O$ and a continuous unitary representation of G V_G, such that $V_g \in \mathcal{A}(\hat{O})$; $V_g A V_g^* = \alpha_g(A)$ if $A \in \mathcal{A}(O)$?

A new imput is needed to solve this problem. A condition of local character has been proposed, not in connection with this question by Borchers long ago which is now called split condition [6].

Split condition: a net $\{\mathcal{A}(O)\}$ satisfies the split condition if when $O_1 \subset\subset O_2$ there is a type I factor $\mathcal{N}$ such that $\mathcal{A}(O_1) \subset \mathcal{N} \subset \mathcal{A}(O_2)$. Moreover there is a vector Ω cyclic and separating for $\mathcal{A}(O_1), \mathcal{A}(O_2), \mathcal{A}(O_1)' \wedge \mathcal{A}(O_2)$. We say, equivalently, that $(\mathcal{A}(O\ ), \mathcal{A}(O_2), \Omega)$ is a standard split inclusion.

This condition has been checked by D. Buchholz [4] in the free fields case but had no developments till [5] and [6,7]. In the first paper the split condition has been formalized as a property of pairs of von Neumann algebras and related to other interesting mathematical concepts as the existence of normal product states, the implementability of the flip and the statistical independence. In the other papers the problem of local implementation is solved and a rich mathematical structure discovered. In [8] the existence of charge operator associated to spacelike cones is discussed and the split property for pairs of spacelike cones.

We would like to complement the discussion giving some theorems that guarantees the non existence of interpolating type I factors. These are variations of a theorem of Driessler [9], [10].

Theorem 1. Let $\mathcal{A} \subset \mathcal{B}$ be type III von Neumann algebras, α_n a sequence of endomorphisms of $\mathcal{B}$ such that $\alpha_n(\mathcal{A}) \subset \mathcal{A}$ and α_n is asymptotically abelian on $\mathcal{B}$, then there is not type I factor $\mathcal{N}$ s.t. $\mathcal{A} \subset \mathcal{N} \subset \mathcal{B}$.

Proof. Let $\mathcal{N}$ be a type I factor s.t. $\mathcal{A} \subset \mathcal{N} \subset \mathcal{B}$ and $E \in \mathcal{N}$ a projection s.t. $E\mathcal{N}E$ is finite. Let τ be a trace on $E\mathcal{N}E$.

If $A \in \mathcal{A}$ define $\omega(A) = \text{LIM}\ \tau(E\alpha_n(A)E)$ (LIM denotes a limit taken along a ultrafilter).

If $A,B \in \mathcal{A}$ $\omega(AB) = \text{LIM}\ \tau(E\alpha_n(A)\alpha_n(B)E) = \text{LIM}\ \tau(E\alpha_n(A)E\alpha_n(B)E) +$
$\text{LIM}\ \tau(E\alpha_n(A)[\alpha_n(B),E]E) = \text{LIM}\ \tau(E\alpha_n(B)E\alpha_n(A)E) =$
$\text{LIM}\ \tau(E\alpha_n(B)\alpha_n(A)E) = \omega(BA)$.

That is ω is a trace. Absurd.

In the case of Quantum Field Theory, we have an invariant state ω.

Theorem 2. Let $\mathcal{A} \subset \mathcal{B}$ be type III von Neumann algebras, Ω a cyclic and separating vector for both algebras, $\omega(A) = (\Omega\ A\Omega)$, α_n a sequence of endomorphisms of $\mathcal{B}$ s.t. $\alpha_n(\mathcal{A}) = \mathcal{A}$ α_n asymptotically abelian on $\mathcal{B}$, $\omega \cdot \alpha_n = \omega$, then there is no type I factor $\mathcal{N}$ s.t. $\mathcal{A} \subset \mathcal{N} \subset \mathcal{B}$

Proof. Let $\mathcal{N}$ be a type I factor s.t. $\mathcal{A} \subset \mathcal{N} \subset \mathcal{B}$ then there is a minimal projection $P \in \mathcal{B}$ s.t. $\mathcal{N} = \mathcal{A}\ \bar{V}\ \{P\}$ and $PAP = \omega(A)P$ for $A \in \mathcal{A}$. For any unit vector $\Psi \in P\mathcal{H}$ and for $A,B \in \mathcal{A}$

$$(\Psi, \alpha_n(AB)\Psi) = (\Psi, P\alpha_n(AB)P\ \Psi) = \omega(\alpha_n(AB)) = \omega(AB) \qquad \forall_n \geq 0$$

on the other hand

$$\lim_{n\to\infty} (\Psi, \alpha_n(AB)\Psi) = \lim_{n\to\infty}(\Psi, P\alpha_n(A)\alpha_n(B)P\Psi) = \lim_{n\to\infty}(\Psi, P\alpha_n(A)P\alpha_n(B)P\Psi) =$$

$$= \lim_n \omega(\alpha_n(A))\ \omega(\alpha_n(B)) = \omega(A)\omega\ (B)$$

that is ω is a trace on $\mathcal{A}$

It is instructive to check the asymptotic abelianess required in Theorem 2 in the case of dilation invariance.

Let α_n be dilations, S_1 a spacelike cone and $S = S_1 + a$ for some $a \in \mathbb{R}^4$ $S \subset S_1$. Then $\alpha_n(S_1) = S_1 \forall_n$, $\alpha_n(S) \subset S$ $n \geq 0$.

If $B, C \in \mathcal{R}(O)$, $O \subset S' \; S_1$ and $B \in \mathcal{R}(S)$

$$\text{w-}\lim_n \alpha_n^{-1}(A) = \omega(A). \quad [19].$$

$$\text{w-}\lim_n \alpha_n(A) = \lim_n (B\Omega, \alpha_n(A)C\Omega) = \lim_n (\Omega, B^* C \alpha_n(A)\Omega) =$$

$$\lim_n (\Omega, \alpha^{-1}(B^*C)A\Omega) = (\Omega, A\Omega)(\Omega, B^*C\Omega)$$

From this we can derive that if $A \in \mathcal{R}(S)$, $B \in \mathcal{R}(S_1)$, $B_\varepsilon \in \mathcal{R}(O)$, O as before, such that $\|(B_\varepsilon - B)\Omega\| \leq \varepsilon$ then

$$\lim_n \| [\alpha_n(A), B]\Omega \| \leq \lim_n \| (\alpha_n(A)B_\varepsilon - B\alpha_n(A))\Omega \| + \varepsilon \| A \| =$$

$$= \lim_n \| (B_\varepsilon - B)\alpha_n(A)\Omega \| + \varepsilon \| A \| = \lim_n \| \alpha_n^{-1}(B_\varepsilon - B)A\Omega \| + \varepsilon \| A \| =$$

$$\| (B_\varepsilon - B)\Omega \| \cdot \| A\Omega \| + \varepsilon \| A \| \leq 2\varepsilon \| A \|.$$

Since $[\alpha_n(A); B]$ is uniformly bounded by $\| A \| \cdot \| B \|$ and Ω is separating for $\mathcal{R}(S_1)$ we have the asymptotic abelianess need.

Comparison of charge operators

Let $\mathcal{A}$ be a net with the split property and G a compact group acting on $\mathcal{A}$. We will show how the local implementation problem is solved. The interpolating type I factors can be chosen in a functorial way: if $(\mathcal{A}, \mathcal{B}, \Omega)$ and $(\mathcal{A}_1, \mathcal{B}_1, \Omega_1)$ are standard split inclusions and Ψ is an isomorphism of $\mathcal{A}$ on $\mathcal{A}_1$ such that $\Psi(\mathcal{B}) = \mathcal{B}_1$ and $(\Omega_1, \Psi(A)\Omega_1) = (\Omega, A\Omega)$ $A \in \mathcal{A}$ then $\Psi(\mathcal{N}) = \mathcal{N}_1$.

Applying this result to the standard split inclusion $(\mathcal{A}(O_1), \mathcal{A}(O_2)$ $\Omega)$ we have that $\alpha_g(\mathcal{N}) = \mathcal{N}$ so, being up to isomorphism, an automorphism of some $\mathcal{B}(\mathcal{H})$, α_g is inner (in $\mathcal{N}$) $\forall g \in \mathcal{G}$. Cohomological problems are solved by means of Araki's cones techniques [11] and we have a continuous unitary representation $\mathcal{U}_g$ with $\mathcal{U}_g \in \mathcal{A}(O_2)$

If G is a Lie group, generators of $\mathcal{U}_g$ can be defined as selfadjoint operators affiliated to $\mathcal{A}(O_2)$. As expected on physical grounds they have discrete spectrum; we call these operators "canonical harges".

The split property solves automatically spectrum problems, coho-ological problems and affiliation problems. It is interesting to compare, when they both exist, these canonical charges with the charges perators that, following Noether's theorem, are built from the basic ields.

We will make a brief digression on local properties of selfadjoint extensions to give an idea of the difficulties connected with he affiliation problem.

Let n_+, n_-, P_+, P_-, be the deficiency indices and the projections n the deficiency subspaces as a closed operator.

emma 1. Let A be a closed operator affiliated to a von Neumann algebra $\mathcal{R}(A \eta \mathcal{R})$. If $n_+ = n_-$ and P_+ is equivalent to P_- in $\mathcal{R}$ then there s a selfadjoint extension of A affiliated to $\mathcal{R}$.

rof. $A \eta \mathcal{R}$ implies $A^* \eta \mathcal{R}$ hence $(A \pm i)(A \pm i)^* \eta \mathcal{R}$ and selfadjoint. Deine $P_\pm$ as the projection on the kernel of $(A \mp i)(A^* \pm i)$ then $P_\pm \in \mathcal{R}$.

If $P_+ \sim P_-$ in $\mathcal{R}$, there is a partial isometry $V \in \mathcal{R}$. $VV^* = P_+$ $V^*V = P_-$ hen $(A+i)(A-i)^{-1}(1-P_+) + V^* = W$ is a unitary in $\mathcal{R}$. $\hat{A} = -i(1-W)^{-1}(1+W)$ is the desired extension of A.

emma 2. Let $\mathcal{A}(O)$ be a local algebra in a theory with the Reeh-chlieder property. Let A be a closed operator, $A \eta \mathcal{R}(O)$ then if $n_+ = n_-$ here is a selfadjoint extension of A affiliated to $\mathcal{R}(O_1)$, $O_1 \supset \bar{O}$.

roof. By a classical result of Borchers [12] any projection $E \in \mathcal{R}(O)$ s equivalent to 1 in $\mathcal{R}(O_1)$ $O_1 \supset \bar{O}$.

We consider now as in [13] quadratic forms A on $C^\infty(H) = \bigcap_K (H+I)^{-K} \mathcal{H}$, where H us the Hamiltonian operator, such that

) $\| (H+I)^{-K} A (H+I)^{-K} \| < \infty$, some $K > 0$

) $\forall \mathcal{O}$ neighbourhood of the origin $\forall C \in \mathcal{R}(O)'$

$C C^\infty(H) \subset C^\infty(H)$ and $C^* C^\infty(H) \subset C^\infty(H)$

$CA = AC$ as quadratic forms.

Conditions 1) and 2) imply that the minimal closed extensions of $A(f) = \int f(x) \mathcal{U}_x A \mathcal{U}_x^{-1} dx, f \in \mathcal{D}$ are affiliated to the local von Neumann algebras.

Lemma 3. Let A be a quadratic form satisfying conditions 1) and 2) and $f \in S(\mathbb{R}^4)$ with supp $f \subset O$, some bounded open region O, then $A(f)^{-}\eta\ R(O)$. If it has a selfadjoint extension then it is also affiliated to $R(O)$.

Proof. Lemma 1 plus Lemma 2.2 of [13].

Nuclearity property

In view of the richness of the structure that can be derived from it we would like to relate the split property to some physical property of general validity. In fact D. Buchholz and E. Wichmann [14] isolated a property of local nature that should hold in a great variety of physically plausible theories: nuclearity. R. Haag and J. Swieca]19] proposed a compactness criterium to be satisfied by a theory with a particle interpretation. Its heuristic foundation is the old argument that the number of quantum states in a finite phase space of volume Ω is finite [16] (namely $\frac{\Omega}{h^3}$). Their precise statement is: $P_E\ A(O_1)\ \Omega$ is a compact set in H; where P_E is the projection on the states of energy less than E.

D. Buchholz and E. Wichmann propose a refinement of Haag's and Swieca's condition. They take as heuristic example the behaviour of the partition function of a system of particles in a box of volume V

$$Z(V,s) \equiv Tr(e^{-sH}) \precsim e^{c\,Vs^{-3}} \qquad s \searrow 0$$

and abstract the following criterium:

Nuclearity: a net satisfies nuclearity if

1) $e^{-sH}\ A(O)_1\ \Omega \equiv X(s)$ is a nuclear set in the following sense: [17]:

There is a set of functionals $\{\ell_n\} \subset X(s)^*$ and a set of unit vectors $\{e_n\} \subset H$ s.t. for any $x \in X(s)$ $x = \sum_n (x,\ell_n)e_n$

moreover if $\lambda_n = \sup_{x \in X(s)} |(x,\ell_n)|$, $N(X(s)) = \sum_n \lambda_n < \infty$

this last number is called the nuclear dimension of X(s).

2) $N(X(s)) \leq a\, e^{bs^{-n}}$ some a,b,n >0 as $s \searrow 0$

Among the results that have been obtained for nuclear sets [18] let us mention:

1. If $T \in B(H)$, ker T = {0} then $N(TX) \leq \|T\|\ N(X)$

2. In $A \otimes B$, $N(e^{-s(H\otimes 1+1\otimes H)}(A \otimes B)_1, \Omega \otimes \Omega) \leq N(e^{-sH}A_1 \Omega)\, N(e^{-sH}B_1 \Omega)$

3. In $\overset{\infty}{\underset{n=1}{\otimes}} A_n$ $N(e^{-sHn}(An)_1 \Omega) \to 0$ fast enough guarantees the analogous result.

Nuclearity implies split property [14] and permits to solve the following problem: given a set with the split property and a local implementation of an automorphic action, what can be said about

$$\lim_{O_1, O_2 \nearrow R^4} U_g ?$$

Nuclearity implies that the local implementation converges to the global when the distance between O_1 and O_2 also grows sufficiently fast [18].

REFERENCES

1 Borchers,H.J.: Local rings and the connection of spin with statistics. Commun. Math. Phys. 1 281 (1965)

2 Doplicher,S., Haag,R., Roberts,J.E.: Local observables and particles statistics I. Commun, Math. Phys. 23, 198 (1971): II. Commun. Math. Phys. 35, 49 (1974)

3 Buchholz,D., Fredenhagen,K.: Locality and the structure of particle states Comm. Math. Phys. 84, 1 (1982)

4 Buchholz,D.: Product states for local algebras. Commun. Math. Phys. 36 287 (1974)

5 D'Antoni,C., Longo,R.: Interpolation by type I factors and the flip automorphism. J. Funct. Anal. 51, 361 (1983)

6 Doplicher,S.: Local aspects of superselection rules. Commun. Math. Phys. 85 73 (1982).
Doplicher, S., Longo, R.: Local aspects of superselection rules II Commun. Math. Phys 88 399 (1983).

7 Doplicher,S.,Longo,R.: Standard and split inclusions of von Neumann algebras. Invent. Math. 75, 493 (1984).

8 D'Antoni,C., Fredenhagen, K.: Charge in spacelike cones Commun. Math. Phys. 94 537 (1984)

9 Driessler,W.: Type of local algebras. Commun. Math. Phys. 53 295 (1977)

10 Longo, R.: Algebraic and modular structure of von Neumann algebras of Physics. Proceedings of Symposia in PUrė Math. 38 551 (1982) part 2

11 Araki,H.: Positive cones,...Proceedings of the Int. School E. Fermi Course LX, Kastler D. (ed) Soc. Italiana di Fisica. North

Holland 1976

12 Borchers,H.J.: A remark on a theorem of B. Misra. Commun. Math. Phys. 4 315 (1967)

13 Fredenhagen,K., Hertel, J.: Local algebras of observables and pointlike localized fields. Commun. Math. Phys. 80 555 (1981)

14 Buchholz,D.,Wichmann,E.: Causal independence and Energy-level density of localized states in Quantum Field Theory (preprint)

15 Haag,R., Swieca J.A.: When does a Quantum Field Theory describe particle? Commun. Math. Phys. 1, 308 (1965)

16 Weyl,H.:Das asymptotische Verteilungsgesetz der Eigenwerte linearer partieller Differential gleichungen Math. Ann. 71, 441 (1911)

17 Grothendieck, A.: Produits tensoriels topologiques et espaces nucléaires Memoirs A.M.S. 16 (1955)

18 D'Antoni,C., Fredenhagen,K.,Doplicher,S.,LOngo,R.: Convergence of local charges and continuity properties for standard W*-inclusions (preprint)

19 Roberts, J.E. Some applications of dilation invariance to structural questions in the theory of local observables. Comm. Math. Phys. 37 273 (1974).

THE LOW DENSITY LIMIT FOR N-LEVEL SYSTEMS

R. Dümcke
Fachbereich Physik
Universität München
Federal Republic of Germany

. Introduction

In this talk a quantum system coupled to a free quantum gas t low density is considered. It is shown that the reduced dynamics f the system converges to a quantum dynamical semigroup when the eservoir density tends to zero. The limiting semigroup may be regarded s a linear quantum Boltzmann equation.

For classical systems there has been considerable progress on he rigorous derivation of the Boltzmann equation from microscopic ynamics in the last ten years. Lanford /1/ and King /2/ - using ideas f Grad /3/ - proved the convergence of the one particle distribution unction for a system of particles interacting via short range forces o the solution of the Boltzmann equation in the low density limit.

The corresponding quantum problem is still open. In the physical iterature there is some agreement that the quantum Boltzmann equation s obtained if one replaces the classical cross section by the proper uantum mechanical cross section. The distribution function appearing n the equation is interpreted as the Wigner function reduced to the ne particle space.

A derivation of the quantum Boltzmann equation along the lines f Grad and Lanford seems to difficult, at present. Therefore, the uch simpler problem of a test particle in a quantum gas is considered. or technical reasons the test particle is further simplified to an -level system.

2. The Model

The formal Hamiltonian for the problem is of the form

$$H = H_S \otimes \mathbb{1} + \mathbb{1} \otimes H_B + H_I \quad , \tag{1}$$

where H_S is self adjoint on the finite dimensional system Hilbert space $\mathcal{H}_S$. The bath Hamiltonian $H_B = \int d^3k\ k^2/2\ a^+(k)a(k)$ is defined on the bath Hilbert space $\mathcal{H}_B = \mathcal{F}_{\pm}(\mathcal{H}_e)$, where $\mathcal{F}_{\pm}$ denotes the Fermion or Boson Fock space over the one paticle space $\mathcal{H}_e = L^2(R^3)$. The interacion Hamiltonian is of the form $H_I = Q \otimes \sum_i \alpha_i\ a^+(f_i)a(f_i)$, where Q is self adjoint on $\mathcal{H}_S$ and $A := \sum_i \alpha_i |f_i\rangle\langle f_i|$ is in the trace class $\mathcal{T}(\mathcal{H}_e)$. For a Fermionic reservoir H_I is bounded. The interaction describes elastic scattering, no particles are created or annihilated. The initial state is of the form $\omega = \omega_\rho \otimes \omega_{n,\beta}$, where $\omega_\rho(A) = \mathrm{tr}\ A\rho$ and $\omega_{n,\beta}$ is the thermal equilibrium state of the bath at particle density n and inverse temperature β.

For a Fermionic reservoir the dynamics of the system is obtained as a perturbation series with respect to H_I. Let $\alpha_o(t)$ be the automorphism group describing the free dynamics and $L_I \cdot = [H_I, \cdot]$. Then the full dynamics for all $t > 0$ is given by

$$\alpha(t) = \sum_{n=0}^{\infty} i^n \int\limits_{0 \le t_1 \le \ldots \le t_n \le t} dt_1 \ldots dt_n\ \alpha_o(t-t_n) L_I \ \ldots\ L_I \alpha_o(t_1) \quad . \tag{2}$$

In the following the reduced dynamics of the system is considered, defined by

$$\mathrm{tr}\ X\ T(t)\rho = \omega(\alpha(t)\ X \otimes 1) \quad , \qquad X \in \mathcal{B}(\mathcal{H}_S) \ . \tag{3}$$

For Bosons H_I is unbounded and therefore one has to face domain problems in the series (2). However, (2) makes still sense when only the expectation values needed for (3) are considered.

3. The Limit Theorem

The scaling for the low density limit is as follows. The density of the reservoir is scaled as $n_\varepsilon=\varepsilon n$. The mean free path $(\sigma n_\varepsilon)^{-1}$ is of the order ε^{-1}. To obtain a fixed collision rate one scales $t_\varepsilon=\varepsilon^{-1}t$. With this scaling one obtains the reduced dynamics

$$\operatorname{tr} X\, T_\varepsilon(t)\rho \;=\; \omega_\rho \otimes \omega_{n_\varepsilon,\beta}(\alpha(\varepsilon^{-1}t)\; X\otimes\mathbb{1}) \;, \quad X\in \mathcal{B}(\mathcal{H}_S). \tag{4}$$

To formulate the limit theorem one needs two further assumptions:

A) $H_1=H_S\otimes\mathbb{1}+\mathbb{1}\otimes H_e+Q\otimes A$ has only finitely many bound states,

B) $\int_{-\infty}^{\infty}\|A\exp(-iH_e t)\phi\|\,dt<\infty$ for all $\phi\in\mathcal{D}$, $\mathcal{D}$ dense in $\mathcal{H}_e$.

If the system would interact with the reservoir particles by a potential, then condition (A) states that the interaction is short range, namely that the potential decreases faster then r^{-2} for $r\to\infty$. Assumption (A) may be regarded as a substitute of a short range condition for non potential interactions. For long range potentials one cannot expect a Boltzmann equation in the low density limit, therefore a condition of this kind is essential. Assumtion (B) states that freely moving reservoir particles leave the interaction region sufficiently fast.

Theorem: If (A) and (B) hold, then there is some finite time $T>0$, such that for all $t\in[0,T)$ and all $\rho\in\mathcal{T}(\mathcal{H}_S)$

$$\lim_{\varepsilon\downarrow 0}\|T_\varepsilon(t)\rho - T^{\#}_\varepsilon(t)\rho\|_1 = 0\;, \tag{5}$$

where

$$T^{\#}_\varepsilon(t) = \exp\{(-i\varepsilon^{-1}L_S + K^{\#})t\}\;. \tag{6}$$

In order to define the generator $K^{\#}$, some further notation has to be introduced. Let $T:=Q\otimes A\,\Omega$ denote the T operator of the scattering process of one bath particle with the system, where $\Omega=\lim_{t\to\infty}\exp(-iH_1t)\cdot\exp(iH_ot)$, $H_o=H_S\otimes 1+1\otimes H_e$, and let $T_{nn'}(\underline{k},\underline{k}')=\langle n\underline{k}|T|n'\underline{k}'\rangle$ denote the T-matrix element, where n is the n-th eigenstate of H_S and $\underline{k}$ the momentum of the bath particle. For $\omega\in sp(L_S)$ define $T_\omega(\underline{k},\underline{k}'):=$

$:= \sum_{\omega_m - \omega_n = \omega} T_{mn}(\underline{k},\underline{k}') \; |m\rangle\langle n|$. $R^o(\underline{k}) := n(2\pi/\beta)^{3/2} \exp(-\beta k^2/2)$ denotes the Maxwell distribution. The generator is of the form

$$K^{\#} \rho = -i\Big[\sum_{\substack{n,n' \\ \omega_n = \omega_{n'}}} \int d^3\underline{k} \; R^o(\underline{k}) \; (T_{nn'}(\underline{k},\underline{k}) + \overline{T_{nn'}(\underline{k},\underline{k})}) \; |n\rangle\langle n'| \; , \; \rho \Big]$$

$$+ \; 2\pi \sum_{\omega \in sp(L_S)} \int d^3\underline{k} \int d^3\underline{k}' \;\; \delta(k'^2/2 - k^2/2 + \omega) \; R^o(\underline{k})$$

$$\{ T_\omega(\underline{k}',\underline{k}) \; \rho \; T^*_\omega(\underline{k}',\underline{k}) \tag{7}$$

$$- 1/2 \left[T^*_\omega(\underline{k}',\underline{k}) T_\omega(\underline{k}',\underline{k}) \rho \; + \; \rho \, T^*_\omega(\underline{k}',\underline{k}) T_\omega(\underline{k}',\underline{k}) \right] \} \; .$$

The first term is a Hamiltonian term leading to a shift of the systems energy levels. The second term describes the elastic scattering of bath particles at the system. The scattering is to all final states $\underline{k}'$ which satisfy energy conservation.

Remarks:

1. The generator contains the full quantum scattering cross section, in contrast to the weak coupling limit, where only the Born approximation enters.
2. The statistics of the reservoir does not enter into the generator, which is expected physically for a system at low density. In the weak coupling limit the two point function of the reservoir contains information on the statistics.
3. $K^{\#}$ is a convex superposition of generators of Lindblad type, and therefore generates a quantum dynamical semigroup.

4. The BBGKY-Hierarchy

The perturbation series (2) defining the dynamics is not appropriate for performing the low density limit. One uses a perturbation series of the BBGKY-hierarchy.

The reduced n-particle density matrix R_n on $\mathcal{H}_n = \mathcal{H}_S \otimes (\overset{n}{\underset{1}{\otimes}} \mathcal{H}_e)$ is implicitely defined by

$$(|f_o\rangle\langle g_o| \otimes a^+(f_n) \ldots a^+(f_1)a(g_1) \ldots a(g_n)) :=$$

$$:= (g_o \otimes g_1 \otimes \ldots \otimes g_n, R_n\, f_o \otimes f_1 \otimes \ldots \otimes f_n) \tag{8}$$

$$n = 0,1,2, \ldots$$

he reduced density matrices corresponding to the initial state $=\omega_\rho \otimes \omega_{n,\beta}$ are $R_n = \rho \otimes \sum_\pi \begin{Bmatrix} \mathrm{sgn}\,\pi \\ 1 \end{Bmatrix} U_\pi R \otimes \ldots \otimes R$ for $\begin{Bmatrix} \text{Fermions} \\ \text{Bosons} \end{Bmatrix}$. π is permutation of n elements and U_π the corresponding permutation perator on $\overset{n}{\underset{1}{\otimes}} \mathcal{H}_e$. A straightforward formal calculation leads to a hierrchical system of equations for the reduced density matrices, the uantum BBGKY-hierarchy:

$$\frac{d}{dt} R_n(t) = -i\left[H_n, R_n(t)\right] + \underline{C}_{n\,n+1} R_{n+1}(t) , \tag{9}$$

here

$$H_n := H_S + \sum_{j=1}^{n} (H_{Bj} + H_{Ij}) \tag{10}$$

nd

$$\underline{C}_{n\,n+1} R_{n+1} := -i\, tr_{n+1}\left[H_{In+1}, R_{n+1}\right] . \tag{11}$$

efining the scaled n-particle dynamics as $\underline{U}_n(t)\cdot = \exp(-i\varepsilon^{-1}H_n t)\cdot$ xp$(i\varepsilon^{-1}H_n t)$, the scaled reduced density matrix $R_n(t)$ may be repreented by the perturbation series

$$R^\varepsilon_n(t) = \sum_{m=0}^{\infty} \int_{0\le t_m \le \ldots \le t_1 \le t} dt_1 \ldots dt_m \;\; \underline{U}^\varepsilon_n(t-t_1)\,\underline{C}_{n\,n+1} \ldots \underline{C}_{n+m-1\;n+m}\,\underline{U}^\varepsilon_{n+m}(t_m) \; \varepsilon^{-m} R^\varepsilon_{n+m}(0). \tag{12}$$

In /4/ it is rigorously shown that the dynamics defined by (12) is quivalent to the time evolution given by (2). One shows first that the equivalence holds for reservoirs in a finite volume, where one can work n Fock space. Then one takes the thermodynamic limit.

The perturbation series (12) is the starting point of the proof f the limit theorem, which runs briefly as follows:

. One easily finds a majorant of the series for $R^\varepsilon_o(t) = T_\varepsilon(t)\rho$, which is uniform in ε on the finite time interval $[0,T)$. Therefore the problem is reduced to discussing the limits of the individual terms of the series, for which tools from multi particle scattering theory are used.

. One proves that $U^\varepsilon_n(s)$ may be replaced by $\underline{\Omega}_n \underline{U}^\varepsilon_{no}(s)$, where $\underline{\Omega}_n$ is the wave operator for the scattering process of the n-th bath particle with the system, and $\underline{U}^{\varepsilon}_{no}(s)$ denotes the free dynamics of n particles.

3. One shows that terms in $R^{\varepsilon}_{n}(0) = \rho \otimes \sum_{\pi} \left\{ {sgn\ \pi \atop 1} \right\} U_{\pi} R^{\varepsilon} \otimes \ldots \otimes R^{\varepsilon}$ with non-trivial π lead to rapidly oscillating terms, which vanish in the limit. This proves that there is no contribution of the statistics in the limit. At this stage the perturbation series for $R^{\varepsilon}_{o}(t)$ is reduced to the form

$$\sum_{m=0}^{\infty} \int_{0 \leq t_1 \leq \ldots \leq t_m \leq t} dt_1 \ldots dt_m \quad \underline{U}^{\varepsilon}_{oo}(t-t_1)\, \underline{C}_{01}\, \underline{\Omega}_1\, \underline{U}^{\varepsilon}_{o1}(t_1-t_2) \ldots \ldots \underline{C}_{m-1\,m}\, \underline{\Omega}_m\, \underline{U}^{\varepsilon}_{om}(t_m)\, \rho \otimes R^{o} \otimes \ldots \otimes R^{o} \; . \tag{13}$$

One notes that (13) is equivalent to

$$\sum_{m=0}^{\infty} \int_{0 \leq t_1 \leq \ldots \leq t_m \leq t} dt_1 \ldots dt_m \quad \underline{U}_S(t-t_1)\, K \ldots K\, \underline{U}_S(t_m)\rho \; , \tag{14}$$

where

$$K\rho := -i\ tr_1 \left[Q \otimes A,\ \Omega\, \rho \otimes R^{o}\, \Omega^{*} \right] . \tag{15}$$

Clearly, (14) is the perturbation series of the semigroup $\exp\{(-i\varepsilon^{-1}L_S + K)t\}$.

4. To obtain a completely positive semigroup one averages the generator K with respect to the free dynamics U_S. The averaging procedure is well known from the weak coupling limit /5/.

5. A Sketch of the Proof

Steps 2. and 3. of the proof, which contain all the essential difficulties, are considered in more detail.

As $\lim_{\varepsilon \downarrow 0} \| \varepsilon^{-1} R^{\varepsilon} - R^{o} \| = 0$, it is clear that $\varepsilon^{-1} R^{\varepsilon}_{n}(0)$ may be replaced by $R^{o}_{n} := \rho \otimes \sum_{\pi} \left\{ {sgn\ \pi \atop 1} \right\} U_{\pi} R^{o} \otimes \ldots \otimes R^{o}$. From $\dim \mathcal{H}_S < \infty$ follows that it is sufficient to consider the limit of expectation values with arbitrary system operators for the terms of the perturbation series of $R^{\varepsilon}_{o}(t)$. The limit is performed step by step proceeding from right to left. For $k=1,\ldots,m$ and arbitrary $X \in \mathcal{B}(\mathcal{H}_S)$ one has to prove:

$$\int_{\leq t_m \leq \ldots \leq t_1 \leq t} dt_1 \ldots dt_m \;\; tr_S\, X\, \underline{U}^{\varepsilon}_{oo}(t-t_1)\, \underline{C}_{01}\, \underline{\Omega}_1\, \underline{U}^{\varepsilon}_{o1}(t_1-t_2) \ldots$$

$$\ldots\, \underline{C}_{k-1\,k} \underbrace{\left[\underline{U}^{\varepsilon}_{k}(t_k-t_{k+1}) - \underline{\Omega}_k\, \underline{U}^{\varepsilon}_{ok}(t_k-t_{k+1})\right]} \qquad (16)$$

$$\underline{C}_{k\,k+1}\, \underline{U}^{\varepsilon}_{k+1}(t_{k+1}-t_{k+2}) \ldots \underline{C}_{m-1\,m}\, \underline{U}^{\varepsilon}_{m}(t_m)\, \tilde{R}^{o}_{m}$$

$$\xrightarrow{\varepsilon \to 0} 0 \quad .$$

t is sufficient to consider the limit of the integrand. The transition rom $\underline{U}^{\varepsilon}_{k}(t_k-t_{k+1})$ to $\underline{\Omega}_k \underline{U}^{\varepsilon}_{ok}(t_k-t_{k+1})$ in the underlined term is done in everal steps shown in the following diagram:

$$\underline{U}^{\varepsilon}_{k}(t_k-t_{k+1}) \qquad (17)$$

$$\downarrow \text{ step 1}$$

$$\underline{U}^{\varepsilon}_{1k}(t_k-t_{k+1}) \qquad (18)$$

$$= \underbrace{\underline{U}^{\varepsilon}_{1k}(t_k-t_{k+1})\, \underline{U}^{\varepsilon}_{ok}(t_{k+1}-t_k)}\, \underline{U}^{\varepsilon}_{ok}(t_k-t_{k+1})$$

$$\downarrow \text{ step 2}$$

$$\underline{P}_k \underline{U}^{\varepsilon}_{1k}(t_k-t_{k+1})\, \underline{U}^{\varepsilon}_{ok}(t_{k+1}-t_k)\, \underline{U}^{\varepsilon}_{ok}(t_k-t_{k+1}) \qquad (19)$$

$$\downarrow \text{ step 3}$$

$$\underline{\Omega}_k\, \underline{U}^{\varepsilon}_{ok}(t_k-t_{k+1}) \quad . \qquad (20)$$

The various quantities in the diagram are defined as follows: ok $:= H_S + \sum_{j=1}^{k} H_{Bj}$ and $H_{1k} := H_S + \sum_{j=1}^{k} H_{Bj} + H_{Ik}$ denote the free Hamiltonian n $\mathcal{H}_n$ and the Hamiltonian where only the k-th particle interacts with he system, respectively. The corresponding unitary groups are $U^{\varepsilon}_{ok}(t)$ = $\exp(-i\varepsilon^{-1}H_{ok}t)$ and $U^{\varepsilon}_{1k}(t) := \exp(-i\varepsilon^{-1}H_{1k}t)$. Now one puts $\underline{U}^{\varepsilon}_{ok}(t)\,\cdot$ = $U^{\varepsilon}_{ok}(t) \cdot U^{\varepsilon}_{ok}(-t)$ and $\underline{U}^{\varepsilon}_{1k}(t)\cdot := U^{\varepsilon}_{1k}(t) \cdot U^{\varepsilon}_{1k}(-t)$. The wave operators re defined by $\Omega_k := \text{s-}\lim_{\varepsilon \downarrow 0} U^{\varepsilon}_{1k}(t)\, U^{\varepsilon}_{ok}(-t)$, $t > 0$, and $\underline{\Omega}_k \cdot := \Omega_k \cdot \Omega_k^*$. enote by P_k the projector onto the absolutely continuous subspace of 1k and put $\underline{P}_k \cdot := P_k \cdot P_k$.

In step 1 the multi particle scattering problem is reduced to a wo particle scattering problem of the system and the k-th bath paricle. The term underlined in (18) should converge to the wave operator

$\underline{\Omega}_k$. However, wave operators converge in the strong sense. As $\underline{U}^{\varepsilon}_{1k}(t_k-t_{k+1})\,\underline{U}^{\varepsilon}_{ok}(t_{k+1}-t_k)$ acts on a bounded operator, one cannot hope to have a limit, in general. Therefore, by cyclic permutation under the trace one passes to the Heisenberg picture. Then the operators act on the object on their left, and this may be shown to be a trace class operator. Using the strong convergence on Hilbert space one proves norm convergence in the trace class. In the Heisenberg picture one has to consider the convergence of $U^{\varepsilon}_{ok}(t)\,U^{\varepsilon}_{1k}(-t)$, which converges strongly only on the absolutely continuous subspace of H_{1k}. Therefore step 2 has to be introduced before one passes finally to the wave operator in step 3.

To perform step 1 one first notes the identity

$$\underbrace{\mathrm{tr}_S\, X\, \underline{U}^{\varepsilon}_{oo}(t-t_1)\,\underline{C}_{01}\,\underline{\Omega}_1\,\underline{U}^{\varepsilon}_{o1}(t_1-t_2)\cdots \cdots \underline{C}_{k-1\,k}}\,\underline{U}^{\varepsilon}_{k}(t_k-t_{k+1})\,\underbrace{\underline{C}_{k\,k+1}\cdots \underline{C}_{m-1\,m}\,\underline{U}^{\varepsilon}_{m}(t_m)\,\tilde{R}^{o}_{m}}_{=:\,Y^{\varepsilon}_k} \tag{21}$$

$$= \underbrace{\mathrm{tr}_{\mathcal{H}_k}\, X^{\varepsilon}_k\, \underline{U}^{\varepsilon}_{B1}(t_1-t_2)\underline{U}^{\varepsilon}_{B2}(t_2-t_3)\cdots \underline{U}^{\varepsilon}_{Bk-1}(t_{k-1}-t_k)}\;\underline{U}^{\varepsilon}_{k}(t_k-t_{k+1})\;Y^{\varepsilon}_k\,,$$

where $\underline{U}^{\varepsilon}_{Bj}(t)\cdot := U^{\varepsilon}_{Bj}(t)\cdot U^{\varepsilon}_{Bj}(-t)$ and $U^{\varepsilon}_{Bj}(t) := \exp\{-i\varepsilon^{-1}(\sum_{l=1}^{j} H_{Bl})t\}$. In the first underlined expression one splits U^{ε}_{oj} in the free dynamics of the system and the free dynamics of the bath. U^{ε}_{Bj} is commuted to the right, and in the remaining expression one passes to the Heisenberg picture, which gives X^{ε}_k. Thus one obtains the second underlined expression. For $\varepsilon \in (0,1]$ the operators X^{ε}_k vary in a compact set in $\mathcal{T}(\mathcal{H}_k)$. Therefore results on strong Hilbert space convergence can be lifted to $\mathcal{T}(\mathcal{H}_k)$. To prove step 1 one uses a version of the cluster theorem /6/ for multi particle scattering:

$$\lim_{\varepsilon\downarrow 0} \| [U^{\varepsilon}_{k}(\underbrace{t_{k+1}-t_k}_{<0}) - U^{\varepsilon}_{1k}(t_{k+1}-t_k)]\, U^{\varepsilon}_{Bk-1}(\underbrace{t_k-t_{k-1}}_{<0}) \cdots \cdots U^{\varepsilon}_{B1}(\underbrace{t_2-t_1}_{<0})\phi \| = 0\,. \tag{22}$$

In the proof (/4/,Theorem 5.1) condition (B) is used. The meaning of the theorem is that particles evolved freely (backward in time) for a sufficiently long time will leave the interaction region. For outgoing particles far from the interaction region free and interacting otion almost coincide.

In step 2 only the second term of the series is considered. The asic idea for the other terms is the same, but the proof is technicaly more involved. The expression

$$\int_0^t dt_1 \; tr_S \; X \; \underline{U}^{\varepsilon}_{oo}(t-t_1) \, \underline{C}_{01} \, \underline{U}^{\varepsilon}_{11}(t_1) \; (1-\underline{P}_1) \, \tilde{R}^o_1 \tag{23}$$

ay be equivalently written as

$$-i \int_0^t dt_1 \; tr_{H_1} \; X \otimes 1 \; \underline{U}^{\varepsilon}_{oo}(t-t_1) \, \underline{L}_{I1} \, \underline{U}^{\varepsilon}_{11}(t_1) \, (1-\underline{P}_1) \, \tilde{R}^o_1 \tag{24}$$

$$= \quad tr_{H_1} \; X \otimes 1 \; (\underline{U}^{\varepsilon}_{11}(t) - \underline{U}^{\varepsilon}_{oo}(t)) \; \underbrace{(1-\underline{P}_1) \, \tilde{R}^o_1}.$$

ere the perturbation formula $\underline{U}^{\varepsilon}_{11}(t) = \underline{U}^{\varepsilon}_{oo}(t) - i\varepsilon^{-1} \int_0^t dt_1 \; \underline{U}^{\varepsilon}_{oo}(t-t_1) \cdot$ $\underline{L}_{I1} \, \underline{U}^{\varepsilon}_{11}(t_1)$ was used. The underlined expression is written as $(1-P_1)\tilde{R}^o_1 + P_1 \tilde{R}^o_1 (1-P_1)$. By condition (A) $1-P_1$ is in $\mathcal{T}(H_1)$ and therefore $(1-\underline{P}_1)\tilde{R}^o_1$ is trace class. The trace is bounded in ε and because of the refactor ε the whole expression converges to zero for $\varepsilon \downarrow 0$.

Step 3 is now straightforward.

In order to prove that the statistics does not contribute in the imit, one has to show that

$$\lim_{\varepsilon \downarrow 0} \int_{0 \le t_m \le \ldots \le t_1 \le t} dt_1 \ldots dt_m \quad tr_S \; X \; \underline{U}^{\varepsilon}_{oo}(t-t_1) \, \underline{C}_{01} \, \underline{\Omega}_1 \; \ldots \; \underline{C}_{m-1\,m} \, \underline{\Omega}_m \, \underline{U}^{\varepsilon}_{om}(t_m) \tag{25}$$

$$\otimes \; U_{\pi} R^o \otimes \ldots \otimes R^o \; = \; 0$$

or all non trivial permutations π. The integrand is written as

$$tr_{H_m} \; \underbrace{X^{\varepsilon}_m \; \underline{\Omega}_m \; U^{\varepsilon}_{oo}(t_m) \; \rho \otimes R^o \otimes \ldots \otimes R^o} \; V^{\varepsilon}_m(\underline{t}_m) \tag{26}$$

ith

$$V^{\varepsilon}_m(\underline{t}_m) = U^{\varepsilon}_{B1}(t_1-t_2) \; \ldots \; U^{\varepsilon}_{Bm}(t_m) \; U_{\pi} \; U^{\varepsilon}_{Bm}(-t_m) \; \ldots \; U^{\varepsilon}_{B1}(t_2-t_1)$$

$$= \exp\{-i\varepsilon^{-1} \; (\sum_{j=1}^{m} H_{Bj}(t_{\pi(j)}-t_j)\} \quad . \tag{27}$$

If $\sum_{j=1}^{m} H_{Bj}(t_{\pi(j)}-t_j) \neq 0$, then it has absolutely continuous spectrum, and for $\phi, \psi \in H_m$ holds $\lim_{\varepsilon \downarrow 0} (\phi, V^{\varepsilon}_m(\underline{t}_m)\psi) = 0$, which is easily deduced from the spectral calculus and ______ lemma. As in step 1 one also has $\lim_{\varepsilon \downarrow 0} tr_{H_m} \; X^{\varepsilon}_m \, V^{\varepsilon}_m(\underline{t}_m) = 0$ if $\overline{\{X^{\varepsilon}_m \mid \varepsilon \in (0,1]\}}$ is a compact set in $\mathcal{T}(H_m)$.

Written in the Heisenberg picture, the underlined operator is of that type, and therefore (26) converges to zero in the limit $\varepsilon\downarrow 0$.

6. Discussion

The ergodic properties of the limiting semigroup are given by the following

Theorem(/4/): If

(a) H_S has non degenerate spectrum,

(b) $T_{mn}(\underline{k},\underline{k}') = T_{nm}(-\underline{k}',-\underline{k})$ for $k^2/2+\omega_m=k'^2/2+\omega_n$ (microreversibil y),

(c) for each pair m,n there is a pair $\underline{k},\underline{k}'$ with $k^2/2+\omega_m=k'^2/2+\omega_n$ satisfying $T_{mn}(\underline{k},\underline{k}')\neq 0$ (efficient coupling to the reservoir),

then

(1) $\rho_{eq} = \exp(-\beta H_S)/\mathrm{tr}\,\exp(-\beta H_S)$ is stationary for $T_\varepsilon^{\#}$ and

(2) $\lim_{t\to\infty} T_\varepsilon^{\#}(t)\rho = \rho_{eq}$ for all $\rho\in\mathcal{T}(\mathcal{H}_S)$.

A first analysis of the low density limit for an open system was given by Palmer /7/. However, in this work some terms of order ε^2 were dropped, and therefore only the second Born approximation to the scattering cross section was obtained.

Some care has to be taken if the reservoir is Bosonic. For $T>0$ and sufficiently low density one has no condensate and the methods used here apply. For $T=0$, the system is condensed at all densities, and the proof of the limit theorem breaks down.

The low density limit was proved for a reservoir in three dimensional space. The proof still holds for space dimension $d\geq 3$. For $d<3$ condition (B) is violated, in general.

The finite radius of convergence of the series (12) is due to the many terms coming from the statistics. For Boltzmann statistics the radius of convergence is infinite. In the low density limit the statistics does not contribute. Therefore, if one could give better estimates for the contribution of the statistics, there is some hope to obtain an infinite radius of convergence.

One would also like to consider the limit of the higher reduced density matrices $R_n^\varepsilon(t)$, $n>0$. As $R_n^\varepsilon(t)$ contains the n-particle motion, which becomes very fast as $\varepsilon\downarrow 0$, one can at most expect $U_{on}^\varepsilon(-t)R_n^\varepsilon(t)$ to have a limit. However, there is a problem. Technically, the cluster theorem used in the proof does not apply in this case. This may indicate that the convergence holds only in a weaker sense. This situation is familiar in the classical case, where convergence holds only for those initial conditions, which do not lead to recollision events.

References

/1/ Lanford, O.E.: Time Evolution of Large Classical Systems. In: Dynamical Systems, Theory and Applications, edited by J. Moser. Berlin: Springer 1975
Lanford, O.E.: On a Derivation of the Boltzmann Equation. Astérisque 40, Soc. Math. de France(1976)
/2/ King, F.: Ph. D. Thesis, University of California, Berkeley(1975)
/3/ Grad, H.: Principles of the Kinetic Theory of Gases. In: Handbuch der Physik, Vol. 12, edited by S. Flügge. Berlin: Springer 1958
/4/ Dümcke, R.: The Low Density Limit for an N-Level System Interacting with a Free Bose or Fermi Gas (to appear in Commun. math. Phys.)
/5/ Davies, E.B.: Markovian Master Equations II. Math. Ann. 219, 147--158(1976)
/6/ Hunzicker, W.: Cluster Properties of Multiparticle Systems. J. Math. Phys. 6, 6-10(1965)
/7/ Palmer, P.F.: The Rigorous Theory of Infinite Quantum Mechanical Systems - Master Equations and the Dynamics of Open Systems. D. Phil. Thesis, Oxford University(1976)

THE C*-ALGEBRAS OF THE TWO-DIMENSIONAL ISING MODEL

David E. Evans
Mathematics Institute, University of Warwick, Coventry CV4 7AL

§1. INTRODUCTION

The two dimensional Ising model with nearest neighbour interactions is well known to possess a phase transition [30, 21, 28, 29, 18, 19, 27, 35, 34, 26, 1, 14]. More precisely, there is a finite non-zero inverse critical temperature β_c, such that the spontaneous magnetisation m* vanishes for $\beta < \beta_c$, whilst $m^* \neq 0$, for $\beta > \beta_c$. Moreover for $\beta < \beta_c$, there is an unique equilibrium state, whilst for each $\beta > \beta_c$ there are precisely two distinct extremal equilibrium states. This model is classical in the sense that it is formulated entirely within commutative C*-algebras. It concerns the behaviour of certain probability measures, or equilibrium states, on a certain compact Hausdorff space, the space $P = \{\pm 1\}^{\mathbf{Z}^2}$ of all configurations on the two dimensional lattice. The transfer matrix formalism [21] allows one to reduce the classical model set in the commutative C*-algebra of all continuous functions on configuration space, $C(P) = \otimes_{\mathbf{Z}^2}\mathbb{C}^2$, to a "quantum mechanical model" in one dimension, described with the aid of a non commutative C*-algebra $\otimes_{\mathbf{Z}} M_2$, which we shall refer to as the Pauli algebra A^P.(M_q will denote the algebra of $q \times q$ complex matrices). This reduction of dimension will take an equilibrium state $< \cdot >_\beta$ to a state ϕ_β on the non-commutative algebra A^P, and a local observable F on configuration space to a "quantum observable" F_β in A^P such that the classical expectation values can be computed using a knowledge of the state ϕ_β on A^F:

$$<F>_\beta = \phi_\beta(F_\beta) \tag{1.1}$$

Here we describe a C*-algebraic approach to phase transition in the two dimensional Ising model, through a study of the family of states ϕ_β on A^P, and see how the phase transition manifests itself in this picture. More precisely, (imposing free boundary conditions) the following was obtained in [4]:

Theorem 1 [4]. The states ϕ_β are pure for $0 \leq \beta \leq \beta_c$, and a non-trivial mixture of two non-equivalent pure states for $\beta > \beta_c$.

Here β_c is the same inverse critical temperature computed by Onsager [28]. The idea in [4] was to first understand the situation at zero and infinite temperatures with states ϕ_∞, ϕ_0 respectively. Then the structure of the states ϕ_β at non zero finite temperature was obtained with the aid of a mod 2 relative index between a projection E_β, depending on the inverse temperature, and its Hilbert transform. The mathematical mechanism behind the phase transition was that this index remained constant in each region $\beta_c < \beta \leq \infty$ and $0 \leq \beta < \beta_c$ but jumped at the critical temperature. This idea of pulling out the information at zero and infinite temperatures to finite non-zero temperatures was developed in [13] to show:

Theorem 2 [13]. There exists a family $\{\nu_\beta : \beta \neq \beta_c\}$ of automorphisms of $A^P)$ such that

$$\phi_\beta = \begin{cases} \phi_\infty \circ \nu_\beta & \beta > \beta_c \\ \phi_0 \circ \nu_\beta & \beta < \beta_c \end{cases}$$

In particular, one can recover Theorem 1, when $\beta \neq \beta_c$. In this survey, we discuss the C*-algebraic formulation of the two-dimensional Ising model, the mathematical tools behind the manifestation of the phase transition in the above theorems, and some connections with other recent work.

§2. THE C*-ALGEBRAIC FORMULATION

We consider the two dimensional Ising model with the Hamiltonian (using free boundary conditions):

$$H^{LM}(\xi) = -\left(\sum_{i=-L}^{L-1} \sum_{j=-M}^{M} J_1 \xi_{ij}\xi_{i+1,j} + \sum_{i=-L}^{L} \sum_{j=-M}^{M-1} J_2 \xi_{ij}\xi_{i,j+1}\right) \tag{2.1}$$

where $\xi_{ij} = \pm 1$ is the classical spin at the lattice site $(i,j) \in \mathbf{Z}^2$, and J_1 and J_2 are positive constants. Then the Gibbs ensemble average is given by

$$\begin{aligned} \langle F\rangle_{LM} &= Z_{LM}^{-1} \sum_\xi F(\xi) e^{-\beta H^{LM}(\xi)}, \\ Z_{LM} &= \sum_\xi e^{-\beta H^{LM}(\xi)}, \end{aligned} \tag{2.2}$$

where the sum is over all configurations $\xi_{ij} = \pm 1$, and $\beta \geq 0$, and the local observable F is a function of ξ_{ij} for $|i| \leq \ell$, $|j| \leq m$ and some $\ell \leq L$, $m \leq M$.

The transfer matrix treatment of this Hamiltonian [28, 18, 19] can be reformulated in terms of the Fermion algebra [35]. To do this we break up a configuration

$$\xi = \{\xi_{ij}\} \in \{\pm 1\}^{\Lambda_{LM}}$$

where Λ_{LM} denotes the finite lattice

$$\Lambda_{LM} = \{(i,j) : |i| \leq L,\ |j| \leq M\}$$

as

$$\xi = \begin{pmatrix} \xi^L \\ \vdots \\ \xi^{-L} \end{pmatrix}$$

if $\xi^i = (\xi_{i,-M},\ldots,\xi_{iM}) \in \{\pm 1\}^{2M+1}$ denotes the configuration of ξ along the ith row, for $|i| \le L$. We then have a decomposition

$$H^{LM}(\xi) = \sum_{i=-L}^{L} S(\xi^i) + \sum_{i=-L}^{L-1} I(\xi^{i+1},\xi^i) \tag{2.3}$$

in terms of the internal energies of the rows and the interaction energies between neighbouring rows if

$$S(\bar{\xi}) = -J_2 \sum_{j=-M}^{M-1} \bar{\xi}_j\bar{\xi}_{j+1} \tag{2.4}$$

$$I(\bar{\xi},\bar{\xi}') = -J_1 \sum_{j=-M}^{M} \bar{\xi}_j\bar{\xi}'_j \tag{2.5}$$

and $\bar{\xi}, \bar{\xi}' \in \{\pm 1\}^{2M+1}$ are row configurations. The transfer matrix $T = T_M$ is defined as the symmetric $2^{2M+1} \times 2^{2M+1}$ array

$$T(\bar{\xi},\bar{\xi}') = \exp - \beta\{[S(\xi) + S(\xi')]/2 + I(\xi,\xi')\} \tag{2.6}$$

if $\bar{\xi},\bar{\xi}' \in \{\pm 1\}^{2M+1}$. It allows one to pull down the two dimensional Λ_{LM}, or $C(\{\pm 1\}^{\Lambda_{LM}}) = \otimes_{\Lambda_{LM}} \mathbb{C}^2$ onto one row $\{j : |j| \le M\}$, but the price we must pay for that is we have to work with the non-commutative Pauli algebra $A_M^p = \otimes_{-M}^{M} M_2$, generated by the Pauli spin matrices

$$\sigma_x^{(j)} = \begin{pmatrix} 1 & 0 \\ 0 & -1 \end{pmatrix}, \quad \sigma_y^{(j)} = \begin{pmatrix} 0 & i \\ -i & 0 \end{pmatrix}, \quad \sigma_z^{(j)} = \begin{pmatrix} 0 & 1 \\ 1 & 0 \end{pmatrix} .$$

For example, the partition function Z_{LM} can be expressed as

$$Z_{LM} = \| (T_M)^L \Omega_M \|^2 \tag{2.7}$$

if $$\Omega_M(\bar{\xi}) = \exp[-\beta S(\bar{\xi})/2]. \tag{2.8}$$

We will identify the transfer matrix T_M as an element of the Pauli algebra A_M^p so that

$$T_M = (2 \sinh 2K_1)^{M+\frac{1}{2}} V^{\frac{1}{2}} W V^{\frac{1}{2}} \tag{2.9}$$

if

$$V = \exp [K_2 \sum_{j=-M}^{M-1} \sigma_x^{(j)} \sigma_x^{(j+1)}] \tag{2.10}$$

$$W = \exp\,[K_1^* \sum_{j=-M}^{M} \sigma_z^{(j)}] \tag{2.11}$$

and $K_j = \beta J_j, \quad j = 1,2$

$$K_1^* = \tfrac{1}{2} \log\,[\coth K_1]. \tag{2.12}$$

Now the vector state

$$\phi^{LM} = \langle T_M^L \Omega_M, \cdot T_M^L \Omega_M \rangle / Z_{LM} \text{ on } A_M^P \tag{2.13}$$

can be used to compute the expectation values in (2.2). More precisely, given a local observable F as in (2.2), then there is a "quantum observable" $F_{\beta M}$ in A_M^P such that

$$\langle F \rangle_{LM} = \phi_\beta^{LM}(F_{\beta M}) \tag{2.14}$$

(For example, if $F = \prod_{i=-\ell}^{\ell} F_i$ where each F_i is a function of the ith row alone, then

$$F_{\beta M} = T_M^{-\ell}\, \hat{F}_{-\ell} T_M \hat{F}_{-\ell+1} \cdots T_M \hat{F}_\ell\, T_M^{-\ell}, \text{ if } \hat{F}_i = F_i(\sigma_x^{(-m)}, \ldots, \sigma_x^{(m)}).)$$

We note from (2.6), that T_M has strictly positive entries, so that by the Perron Frobenius theorem, it possesses an unique unit vector $\Omega^M = \Omega^M(\xi)$, $\Omega^M(\xi) > 0$, belonging to the largest eigenvalue. (However note that if we regard K_2 and K_1^* as independent parameters in (2.9), then the largest eigenvalue of T_M is degenerate when $K_1^* = 0$, $K_2 < \infty$). By (2.8) we see $\langle \Omega^M, \Omega_M \rangle > 0$, and so as $L \to \infty$

$$\langle T_M^L \Omega_M, \cdot T_M^L \Omega_M \rangle Z_{LM}^{-1} \tag{2.15}$$

will pick out the largest eigenspace so that

$$\lim_{L\to\infty} \langle F \rangle_{LM} = \langle \Omega^M, F_{\beta M} \Omega^M \rangle. \tag{2.16}$$

Then if $A^P = \lim_{\to} A_M^P$ the Pauli algebra generated by all the spin matrices on $\mathbb{Z}$, we have

$$\langle F \rangle_\beta = \phi_\beta(F_{\beta M}) \tag{2.17}$$

where $F_\beta = \lim_{M\to\infty} F_{\beta M}$, and

$$\phi_\beta = \lim_{M\to\infty} \langle \Omega^M, \cdot \Omega^M \rangle \tag{2.18}$$

is a state on A^P.

In [4,13], K_2 and K_1^* are essentially regarded as independent parameters in (2.9). Then $K_2 = 0$, $K_1^* > 0$ corresponds to $\beta = 0$, and $K_1^* = 0$, $K_2 > 0$ to $\beta = \infty$. The corresponding states ϕ_0 and ϕ_∞ can be described as follows. First, if α is a unit vector in $\mathbb{C}^2$, let $\omega(\alpha) = \langle\alpha, \cdot\alpha\rangle$ denote the corresponding vector state on M_2, and then let $\alpha(\pm)$ denote unit eigenvectors of σ_α corresponding to ± 1 respectively. Then ϕ_0 is the pure state $\bigotimes_{-\infty}^{\infty} \omega_{z(+)}$ and $\phi_\infty = \frac{1}{2}\left(\bigotimes_{-\infty}^{\infty} \omega_{x(+)} + \bigotimes_{-\infty}^{\infty} \omega_{x(-)}\right)$, a mixture of inequivalent pure states on A^P. When $K_2 = 0$ or $K_1^* = 0$, the transfer matrix in (2.9) reduces to a scalar multiple of W or V respectively. The largest eigenspace of W is spanned by $\bigotimes_{-M}^{M} z(+)$, whilst largest eigenspace of V is degenerate, and spanned by $\bigotimes_{-M}^{M} x(+)$, $\bigotimes_{-M}^{M} x(-)$, (corresponding to all spins up and all spins down respectively). Thus the structure of ϕ_β at zero and infinite temperatures is clear, and in [4,13] it was shown that the same structure persists in the region $0 \le \beta \le \beta_c$ (corresponding to $0 \le K_2 < K_1^*$) and in $\beta_c < \beta \le \infty$ (corresponding to $0 \le K_1^* < K_2$) where W and V respectively dominate in (2.9).

In practice, when working with the state ϕ_β, when $0 < \beta < \infty$, one takes a Jordan Wigner transformation to a Fermion or Clifford algebra description. However, for a two sided lattice, infinitely extended in both directions, there are serious difficulties in getting the Jordan Wigner prescription to give an isomorphism between A^P and the Fermion algebra $A^F = C^*(c_i; i \in \mathbb{Z})$, generated by annihilation operators c_i, (satisfying the canonical anti-commutation relations $[c_i, c_j]_+ = 0$, $[c_i, c_j^*]_+ = \delta_{ij}$) which preservesat least some of the quasi-local structure. On a finite sublattice, the Jordan Wigner isomrophsim γ_M between A_M^P and $A_M^F = C^*(c_i; |i| \le M)$ is given by:

$$\sigma_z^{(j)} = 2c_j^* c_j - 1,$$

$$\sigma_x^{(j)} = TS_j(c_j + c_j^*), \quad \sigma_y^{(j)} = TS_j i(c_j - c_j^*),$$

$$S_j = \begin{cases} \prod_{k=1}^{j-1} \sigma_z^{(k)} & \text{if } j > 1, \\ 1 & \text{if } j = 1, \\ \prod_{k=0}^{j} \sigma_z^{(k)} & \text{if } j < 1. \end{cases} \tag{2.19}$$

$$\text{if} \qquad T = \prod_{k=-M}^{0} \sigma_z^{(k)} \tag{2.20}$$

However the isomorphisms γ_M, γ_{M+1} are not compatible under the inclusions of A_M^P in A_{M+1}^P and A_M^F in A_{M+1}^F. (Note that this is not the case for a one-sided lattice, where there is an isomorphism of

$$A^P[1,\infty) = C^*(\sigma_i^\alpha : i = 1,2,\dots)$$

with $A^F[1,\infty) = C^*(c_i : i = 1,2,\dots)$

which identifies $C^*(\sigma_i^\alpha : 1 \le i \le m)$ with $C^*(c_i : 1 \le i \le m)$ for all M). Also, when taking $M \to \infty$, the tail $\prod_{-\infty}^{o} \sigma_z^{(i)}$ does not converge in A^P. (Again, in the case of a one sided lattice, the problem would not arise).

When looking at the two-sided one-dimensional XY model, Araki [3] introduced a technique for getting around this problem, which has also proved useful in the Ising model [4,13]. Instead of considering $\prod_{-M}^{o} \sigma_z^{(j)}$, one looks instead at the automorphism Θ_- on A_M^P or A_M^F which this self adjoint unitary implements:

$$\Theta_-\sigma_z^{(j)} = \sigma_z^{(j)}, \quad \Theta_-\sigma_x^{(j)} = \begin{cases} \sigma_x^{(j)} & j \ge 1, \\ -\sigma_x^{(j)} & j < 1 \end{cases} \quad \Theta_-\sigma_y^{(j)} = \begin{cases} \sigma_y^{(j)} & j \ge 1 \\ -\sigma_y^{(j)} & j < 1 \end{cases} \tag{2.21}$$

$$\Theta_- c_j = \begin{cases} c_j & j \ge 1 \\ -c_j & j < 1 \end{cases} \tag{2.22}$$

These formulae (2.21) and (2.22) also define (outer) automorphisms Θ_- on A^P and A^F. One can then take any representation of A^F, where there is a self adjoint unitary T on the Hilbert space of the representation such that

$$\Theta_-(a) = TaT, \quad a \in A^F$$

This operator T can be substituted in formulae (2.19) (for all j) to define an embedding of A^P in the C*-algebra $\hat{A} = A^F + TA^F$, generated by A^F and T. (Since Θ_- is outer, the choice of T is not important, and $\hat{A}$ is in fact $A^F \times_{\Theta_-} \mathbb{Z}_2$, the crossed product of A^F by the period two automorphism Θ_-). We thus regard A^F and A^P as two distinct C*-algebras of A. (Note that $\hat{A}$ can also be identified with $A^P \times_{\Theta_-} \mathbb{Z}_2$, but in applications to the Ising model, it is preferable to express everything in terms of A^F, where things can be computed).

Now there is a grading Θ, or a period two automorphism, of $\hat{A}$ given by $\Theta T = T$, $\Theta c_i = -c_i$, which gives gradings of A^F and A^P. (Note, that $\Theta\sigma_z^{(j)} = \sigma_z^{(j)}$, $\Theta\sigma_y^{(j)} = -\sigma_y^{(j)}$, $\Theta\sigma_x^{(j)} = -\sigma_x^{(j)}$). The embedding of (2.19) identifies the even algebra A_+^P with A_+^F. Note also that the even algebras are generated by

$$\begin{aligned} \sigma_z^{(i)} &= 2c_i^* c_i - 1 \\ \sigma_x^{(i)}\sigma_x^{(i+1)} &= (c_i - c_i^*)(c_{i+1} + c_{i+1}^*), \end{aligned} \tag{2.23}$$

It is a consequence of [13] that this identification of A_+^P with A_+^F cannot be extended to an isomorphism between A^P and A^F.

There is an affine correspondence between even states $\phi = \phi_0 \Theta$ on A^P and even states $\tilde{\phi} = \tilde{\phi} \circ \Theta$ on A^F such that $\phi|_{A_+^P} = \tilde{\phi}|_{A_+^F}$. Now since the transfer matrix T_M is even, the state ϕ_β on A^P is even. The structure of ϕ_β on A^P will be observed by studying the associated even state $\tilde{\phi}_\beta$ on A^F. The quadratic form of the exponents in the transfer matrix (see (2.9-11) and (2.23)) will mean that $\omega_\beta = \tilde{\phi}_\beta$ are quasi-free states. In fact they are Fock states, and hence pure for all temperatures, and so the phase transition does not manifest itself in the Fermion algebra picture. (The situation with a half lattice is different. Here, as we have noted previously, the corresponding Pauli $A^P(1,\infty)$ and Fermion $A^F[1,\infty)$ are canonically isomorphic. Again, the corresponding states are quasi-free, but this time are non-Fock. They are in fact the restrictions of ω_β on A^F to $A^F[1,\infty)$. Using the criterion of [25, 33, 36] on the structure of quasi-free states, one can show that they are primary for $\beta < \beta_c$, and non primary for $\beta > \beta_c$ [24]). To give a precise description of the quasi-free states ω_β on A^F, it is convenient to adopt the self dual formalism of [2], so that A^F is generated by the range of a linear map B on $\ell^2 \oplus \ell^2$ given by

$$B(h) = \sum_{-\infty}^{\infty} (c_j^* f_j + c_j g_j) \qquad h = \binom{f}{g} \quad f = (f_j) \quad g = (g_j) \tag{2.24}$$

Here B satisfies

$$[B(h_1)^*, B(h_2)]_+ = \langle h_1, h_2 \rangle 1 \ , \ B(h_1)^* = B(\Gamma h)$$

where

$$\Gamma\binom{f}{g^*} = \binom{g}{f^*}.$$

A basis projection E is a projection E on $\ell_2 \oplus \ell_2$ such that $\Gamma E \Gamma = 1 - E$. Any basis projection E gives rise to an unqiue state ω on A^F such that $\omega[B(f)B(f)^*] = 0$, $f \in E(\ell_2 \oplus \ell_2)$. Then we write ω_E for ω, and it is a quasi-free Fock state, irreducible and satisfies

$$\omega_E[B(f)^*B(g)] = \langle f, Eg \rangle \quad f,g \in \ell_2 \oplus \ell_2.$$

The basis projections E_β of the quasi-free Fock states ω_β can be described as follows, after identifying ℓ_2 with $L^2(\mathbf{T})$ using Fourier series. First $\gamma(\theta) \geq 0$ is determined by

$$\cosh 2K_1^* \cosh 2K_2 - \sinh 2K_1^* \sinh 2K_2 \cos\theta = \cosh\gamma(\theta), \tag{2.25}$$

and $\delta(\theta) = \theta^*(\theta) - \theta$ is determined by

$$\cos \delta(\theta) = (\sinh\gamma(\theta))^{-1}(\cosh 2K_1^* \sinh 2K_2 - \sinh 2K_1^* \cosh 2K_2 \cos \theta) \tag{2.26}$$

$$\sin \delta(\theta) = (\sinh\gamma(\theta))^{-1} \sinh 2K_1^* \sin\theta . \tag{2.27}$$

Then E_β is the multiplication operator

$$E_\beta(\theta) = \frac{1}{2}\begin{pmatrix} 1 - \cos \vartheta(\theta) & i \sin \vartheta(\theta) \\ -i \sin \vartheta(\theta) & 1 + \cos \vartheta(\theta) \end{pmatrix} \tag{2.28}$$

In particular

$$E_0 = \begin{pmatrix} 1 & 0 \\ 0 & 0 \end{pmatrix} \qquad E_\infty = \frac{1}{2}\begin{pmatrix} 1 - \cos \theta & i \sin \theta \\ -i \sin \theta & 1 + \cos \theta \end{pmatrix} \tag{2.29}$$

In the self dual formalism, a Bogoluibov automorphism $\tau(u)$ is defined on A^F for unitaries u on $\ell_2 \oplus \ell_2$, which commute with Γ, and given by $\tau(u)B(f) = B(uf)$. In particular, note that $\Theta = \tau(-1)$, $\Theta_- = \tau(\theta_-)$, if

$$(\theta_- f)_j = \begin{cases} f_j & j \geq 1 \\ -f_j & j < 0 \end{cases} \tag{2.30}$$

For details on these matters see [32, 37, 23, 24, 22, 12, 4, 13].

§3. MANIFESTATION OF THE PHASE TRANSITION

The approach in [4] was to first derive an abstract criterion to decide when an even state ϕ of A^P is pure in terms of properties of the associated even state $\tilde{\phi}$ of A^F :

Theorem 3 [4] Assume that $\tilde{\phi}$ is an even pure state of A^F. Then ϕ is not pure if and only if:

(3.1) $\tilde{\phi}$ and $\tilde{\phi}\circ\Theta_-$ are equivalent.

(3.2) $\tilde{\phi}|_{A_+}$, and $\tilde{\phi}\circ\Theta_-|_{A_+}$ are not equivalent.

If ϕ is not pure, it is a mixture of two non-equivalent pure states.

When applying this to the Ising model, we have to decide when the Fock states ω_{E_β} and $\omega_{\theta_- E_\beta \theta_-}$ are equivalent on A^F, and their restrictions on A_+. Now ω_{E_β} and $\omega_{\theta_- E_\beta \theta_-}$ are equivalent if and only if $E_\beta - \theta_- E_\beta \theta_-$ is Hilbert Schmidt [33,2]. A computation in [4] shows that $E_\beta - \theta_- E_\beta \theta_-$ is Hilbert Schmidt if and only if $\beta \neq \beta_c$. We deduce using Theorem 3, that ϕ_{β_c} is pure, but we have to do more to decide what the situation is away from β_c. In fact we need a criterion to decide when the restriction of two Fock states to the even algebra are equivalent. To deal with this, we introduced in [4] the following (symmetric) $\mathbb{Z}_2$ index between two basis projections E_1, E_2 such that $E_1 - E_2$ is Hilbert Schmidt:

$$\sigma(E_1,E_2) = (-1)^{\dim E_1 \wedge (1-E_2)} \quad . \tag{3.3}$$

Theorem 4 [4] The restrictions of Fock states ω_{E_1} and ω_{E_2} of A^F to the even algebra A_+^F are equivalent if and only if

(3.4) $E_1 - E_2$ is in the Hilbert Schmidt class.

(3.5) $\sigma(E_1,E_2) = 1$.

Theorem 1 then follows from:

$$\sigma(E_\beta,\theta_- E_\beta \theta_-) = \begin{cases} 1 & \beta < \beta_c \\ -1 & \beta > \beta_c \end{cases} \tag{3.6}$$

An evaluation of this index can be explicitly achieved when $K_2 = 0$, $K_1^* > 0$, $(\beta = 0)$ and $K_1^* = 0$, $K_2 > 0$ $(\beta = \infty)$. The index σ is then shown to be continuous in the norm topology, and that E_β is norm continuous in K_1^* and K_2

(regarded as independent parameters) in the regions $K_1^* > K_2 \geq 0$ (or $\beta < \beta_c$) and $K_2 > K_1^* \geq 0$ (or $\beta > \beta_c$) . Thus an explicit computation at zero and infinite temperatures is enough.

Note that condition (3.4) is saying that ω_{E_1} and ω_{E_2} are equivalent on A^F [33,2]. The presence of the mod 2 index as a second condition (namely (3.5)) comes from the following fact. Suppose π is a Fock representation of A^F on a Fock space F , regarded as a space of antisymmetric tensors. Then F splits as $F_- \oplus F_-$ where F_+ and F_- are the subspaces of even and odd tensors respectively. Then $\pi|_{A_+}$ leaves $F_\pm$ invariant, and the mod 2 index enters because these two representations are not equivalent [4, Lemma 4.2]. Carey [11] has given an interpretation of the index map of (3.3) in terms of the K_1-group of a certain Banach algebra. If E is a real separable infinite dimensional Hilbert space, with fixed complex structure J , let $B_2(E)$ denote the Banach algebra of bounded real linear operators A on E which commute with J up to a Hilbert Schmidt operator (i.e. $AJ-JA$ is Hilbert Schmidt), equipped with the norm $|||A||| = ||A|| + ||AJ-JA||_{HS}$. Then

$$K_1(B_2(E)) = \mathbb{Z}_2 \quad .$$

In fact if $\mathcal{O}$ denotes the invertible orthogonal elements of $B_2(E)$, then $\pi_0(\mathcal{O}) = \mathbb{Z}_2$, via the map

$$j(A) = \dim \mathrm{Ker}_{\mathbb{C}}(AJ + JA) \bmod 2 \;, \qquad A \in \mathcal{O} \tag{3.7}$$

which is essentially the same as the index map σ in (3.3), [11]. Theorem 4, and in particular the $\mathbb{Z}_2$ index of (3.3) has also been successfully used in [5] to help find the number of pure ground states in the one-dimensional XY-model. We also refer the reader to [38,6-9] for other recent work on the equivalence of restrictions of states to fixed point algebras (e.g. the gauge invariant CAR algebra).

The idea of exploiting the explicit and simple information available at zero and infinite temperature was carried a stage further in [13]. The starting point of that investigation was the duality between high and low temperatures which Kramers and Wannier used to locate the critical temperature [21]. Mathematically, this duality is effected by the following automorphism κ on A_+ :

$$\text{(3.8)} \qquad \kappa(\sigma_z^{(j)}) = \sigma_x^{(j)} \, \sigma_x^{(j+1)}$$

$$\text{(3.9)} \qquad \kappa(\sigma_x^{(j)} \, \sigma_x^{(j+1)}) = \sigma_z^{(j+1)}$$

(See the remarks of Onsager on Kramers-Wannier duality [28, page 123]). Recall that $\sigma_z^{(j)}{}_{\infty}$ and $\sigma_x^{(j)}\sigma_x^{(j+1)}$ generate A_+. Then κ^2 is the restriction of the shift on $\otimes_{-\infty}^{\infty} M_2$ to A_+. In [13] it was shown that κ does not extend to A^P, but does extend to A^F. In fact we can extend κ from A_+ to A^F by taking κ to be the Bogoluibov automorphism $\tau(W)$ on A^F where

$$W = i/2\begin{pmatrix} 1 - U^* & 1 + U^* \\ -1-U^* & U^* - 1 \end{pmatrix} \tag{3.10}$$

and U is the shift $(Uf)_k = f_{k+1}$ on ℓ_2. Then $W^2 = U^* \oplus U^*$, and $W^* E_0 W = E_\infty$. In particular $\omega_0 o\kappa = \omega_\infty$ or κ takes the infinite temperature state to the zero temperature state, as we would expect from (2.9-11). To relate states at other temperatures, define

(3.11) $$U_\beta = e^{-i\boldsymbol{\theta}}$$

where $\boldsymbol{\theta}$ is as in (2.26-27), and

(3.12) $$W_\beta = i/2\begin{pmatrix} 1 - U_\beta^* & 1 + U_\beta^* \\ -1-U_\beta^* & U_\beta^* - 1 \end{pmatrix} .$$

Then $W_\beta^* E_0 W_\beta = E_\beta$, and so

(3.13) $$\omega_0 o\tau(W_\beta) = \omega_\beta$$

(3.14) $$\omega_\infty o\tau(W^* W_\beta) = \omega_\infty .$$

To obtain Theorem 2, one needs a criterion to let us decide when the restriction of a Bogoluibov automorphism from A^F to A_+ extends to a graded automorphism of A^P. This is provided by the following. (We say that an automorphism of a graded algebra is graded, if it leaves invariant the even and odd parts, and an inner automorphism of a graded algebra is even (respectively odd) if it can be implemented by an even (respectively odd) unitary).

<u>Proposition 5</u>. [13] Let ν be a graded automorphism of A^F such that $\Theta_-\nu\Theta_-\nu^{-1}$ is an even inner automorphism of A^F. Then $\nu|_{A_+}$ extends to a graded automorphism of A^P.

To apply this to the Bogoluibov automorphisms $\tau(W_\beta)$ and $\tau(W^* W_\beta)$ of the Ising model we need the following characterisation of [2] of inner Bogoluibov automorphisms:

Theorem 6 [2] A Bogoluibov automorphism $\tau(U)$ of A^F is inner if and only if one of the following conditions hold

(3.15) $1 - U$ is trace class and $\det(U) = 1$

(3.16) $1 + U$ is trace class and $\det(-U) = -1$.

In which case, if (3.15) (respectively (3.16)) holds, then $\tau(U)$ is even (respectively odd).

From Proposition 5, it is seen that (3.15) is the condition relevant for our purposes. In their computation of the spontaneous magnetisation, Montroll, Potts and Ward [27] evaluated the Fourier coefficients of θ and δ . With the aid of those computations, one can show that

(3.17) $1 - \theta_- W_\beta \theta_- W_\beta^*$

(3.18) $1 - \theta_- W^* W_\beta \theta_- W_\beta^* W$

are all trace class if $\beta \neq \beta_c$. (The computations of [4] allow one to deduce that they are Hilbert Schmidt if and only if $\beta \neq \beta_c$, but this is not enough to be able to apply Theorem 6). Next one shows that

(3.19) $$\det \theta_- W_\beta \theta_- W_\beta^* = \begin{cases} 1 & \beta < \beta_c \\ -1 & \beta > \beta_c \end{cases}$$

(3.20) $$\det \theta_- W^* W_\beta \theta_- W_\beta^* W = \begin{cases} 1 & \beta > \beta_c \\ -1 & \beta < \beta_c \end{cases}$$

As in the earlier work [4] described above, these determinants are computed by regarding K_1^* and K_2 as independant parameters, showing that (3.17) and (3.18) are continuous (in the trace class norm) in the regions $K_1^* > K_2 \geq 0$ and $K_2 > K_1^* \geq 0$, making an explicit computation at infinite and zero temperatures, and using continuity of the determinant. Theorem 2 then follows from Proposition 5, (3.15), (3.19) and (3.20).

§4. THE q-STATE POTTS MODEL

The Kramers Wannier automorphism κ on the even part of the Pauli algebra, and their analogues in the q-state Potts model have appeared recently in work of Jones [15-17] and Pimsner and Popa [31] on the index and entropy of subfactors, and braid groups. The q-state Potts model is a generalisation [10] of the Ising model, where the spin at any lattice site can point in any one of q-equally spaced directions. It has been shown to possess a first order phase transition for large q [20]. The local transfer matrix of this model in two dimensions can be regarded as an element of the UHF algebra $F_q = \bigotimes_{-\infty}^{\infty} M_q$, and can be expressed in terms of a sequence $\{e_i\}_{-\infty}^{\infty}$ of projections in F_q which are defined as follows. First, let $\{E_{ij} : i,j = 1,..,q\}$ be matrix units for M_q, and

$$f = \sum_{i,j=1}^{q} E_{ij}/q \quad , \qquad g = \sum_{i=1}^{q} E_{ii} \otimes E_{ii}$$

be projections in M_q and $M_q \otimes M_q$ respectively. Then define

(4.1) $$e_{2i-1} = \ldots .\, 1 \otimes 1 \otimes \underset{i\text{th position}}{f} \otimes 1 \otimes \ldots \ldots .$$

(4.2) $$e_{2i} = \ldots .\, 1 \otimes 1 \otimes \underset{i,i+1 \text{ positions}}{g} \otimes 1 \otimes \ldots .$$

In the Ising model, where $q = 2$,

(4.3) $$e_{2i-1} = (\sigma_z^{(i)} + 1)/2$$

(4.4) $$e_{2i} = (\sigma_x^{(i)} \sigma_x^{(i+1)} + 1)/2 \quad .$$

The family $\{e_i\}$ satisfy the relations:

(4.5) $$e_i e_j = e_j e_i \qquad |i-j| \geq 2$$

(4.6) $$e_i e_{i\pm 1} e_i = \frac{1}{q} e_i$$

(4.7) $$\mathrm{tr}\, x\, e_i = \frac{1}{q} \mathrm{tr}\, x \qquad \text{if } x \in C^*\text{-algebra generated by } \{e_j\}_{-\infty}^{i-1} \quad .$$

The local transfer matrix in the Potts model can be expressed, up to a scalar, as $X^{\frac{1}{2}}\, Y\, X^{\frac{1}{2}}$, where

(4.8) $$X = \exp 2K_2 \Sigma e_{2i}$$

(4.9) $$Y = \exp 2K_1^* \Sigma e_{2i-1}$$

and $K_j = \beta J_j$, $(e^{2K_1^*}-1)(e^{2K_1}-1) = q$, [39,10]. Families of projections satisfying relations (4.5) - (4.6) (even when q is not necessarily an integer) have played a prominent role in recent analysis [15-17,31] on the structure of subfactors. Such relations have also been used to get powerful new invariants and techniques for studying braids and links. The substitution

$$(4.10)\qquad \sigma_i = \sqrt{t}(te_i - (1 - e_i)) \quad , \quad t/(1 + t)^2 = 1/q$$

gives a representation of the braid group:

$$(4.11)\qquad \sigma_i\sigma_j = \sigma_j\sigma_i \qquad |i - j| \geq 2$$

$$(4.12)\qquad \sigma_i\sigma_{i+1}\sigma_i = \sigma_{i+1}\sigma_i\sigma_{i+1} \quad .$$

In particular, if b is a braid, $tr(b)$, when suitably normalised, becomes a polynomial in t and a link invariant [17]. The finite volume partition function of the Potts model, can then be interpreted as Jones' polynomial invariant for certain links [17]. In the Potts model the natural representation of the family $\{e_i\}$ in (4.5) - (4.6) is in F_q. However Temperley and Lieb [39] (or see [10]) gave another representation in $F_2 = A^P$, the Pauli algebra, which they used to obtain an equivalence with an ice type model. This representation (which is a $*$-representation only for $q \geq 4$) was independently rediscovered by Pimsner and Popa, [31] who used it to interpret the Kramers Wannier automorphism $\kappa_q : e_i \to e_{i+1}$ as a non-commutative Bernoulli shift in F_2, if $q \geq 4$.

References.

1. M. Aizenman. Commun. math. Phys. 73 (1980) 83-94.
2. H. Araki. Publ. RIMS Kyoto Univ. 6 (1970) 385-442.
3. H. Araki. Publ. RIMS Kyoto Univ. 20 (1984) 277-296.
4. H. Araki, D.E. Evans. Commun. math. Phys. 91 (1983) 489-503.
5. H. Araki, T. Matsui. Ground states of the XY-model. Preprint Kyoto 1984.
6. B.M. Baker. Trans. Amer. Math. Soc. 237 (1978) 35-61, 254 (1974) 133-155.
7. B.M. Baker. J. Funct. Anal. 35 (1980) 1-25.
8. B.M. Baker, R.T. Powers. J. Funct. Anal. 50 (1983) 229-266.
9. B.M. Baker, R.T. Powers. J. Operator Theory. 10 (1983) 365-393.
10. R.J. Baxter. Exactly solved models in Statistical Mechanics. Academic Press. London 1982.
11. A.L. Carey. Some infinite dimensional groups and bundles. Preprint ANU 1983.
12. D.E. Evans, J.T. Lewis. Commun. math. Phys. 92 (1984), 309-327.

13. D.E. Evans, J.T. Lewis. On a C^*-algebra approach to phase transition in the two dimensional Ising model II.

14. Y. Higuchi. On the absence of non-translationally invariant states for the two dimensional Ising model. Colloquia Societatis Janos Bolyai. 27, Esztergom, Hungary 1979.

15. V.F.R. Jones. Invent Math. 72 (1983) 1-25.

16. V.F.R. Jones. Braid groups, Hecke algebras and type II_1 factors. Proceedings Japan US Conference 1983 (to appear).

17. V.F.R. Jones. A polynomial invariant for Knots via von Neumann algebras. Preprint Berkeley 1984.

18. B. Kaufman. Phys. Rev. 76 (1949) 1232-1243.

19. B. Kaufman, L. Onsager. Phys. Rev. 76 (1949) 1244-1252.

20. R. Kotecky, S.B. Shlosman. Commun. math. Phys. 83 (1982) 493-515.

21. H.A. Kramers, G.H. Wannier. Phys. Rev. 60 (1941) 252-262.

22. R. Kuik. Doctoraals dissertation. Gröningen, 1981.

23. J.T. Lewis, PNM Sisson. Commun. math. Phys. 44 (1975) 279-292.

24. J.T. Lewis, M. Winnink. The Ising model phase transition and the index of states on the Clifford algebra. Colloquia Mathematica Societatis. Janos Bolyai 27, Random fields. Esztergom, Hungary 1979.

25. J. Manuceau, A. Verbeure. Commun. math. Phys. 18 (1970) 319-326.

26. A. Messager, S. Miracle-Sole. Commun. math. Phys. 40 (1975) 187-196.

27. E. Montroll, R.B. Potts, J.C. Ward, J. Math. Phys. 4 (1963) 308-322.

28. L. Onsager. Phys. Rev. 65 (1944) 117-149.

29. L. Onsager. Il Nuovo Cimento. Suppl 6 (1949) 261-262.

30. R. Peierls. Proc. Camb. Philos. Soc. 32 (1936) 477-481.

31. M. Pimsner, S. Popa. Entropy and index for subfactors. Preprint INCREST 1983.

32. S. Pirogov. Theor. Math. Phys. 11 (3) (1972) 614-617.

33. R.T. Powers, E. Størmer. Commun. math. Phys. 16 (1970) 1-33.

34. D. Ruelle. Ann. Phys. 69 (1972) 364-374.

35. T.D. Schultz, D.C. Mattis, E. Lieb, Rev. Mod. Phys. 36 (1964) 856-871.

36. M. Sirugue, M. Winnink. Commun. math. Phys. 19 (1970) 161-168.

37. P.N.M. Sisson. Ph.D. Thesis. Dublin University 1975.

38. S. Stratila, D. Voiculescu. Math. Ann. 235 (1978) 87-110.

39. H.N.V. Temperley, E.H. Lieb. Proc. Roy. Soc. (London) A 322 (1971) 251-280.

INFINITE DIVISIBILITY AND CENTRAL LIMIT THEOREMS FOR COMPLETELY POSITIVE MAPPINGS

M. FANNES* and J. QUAEGEBEUR**
Instituut voor Theoretische Fysica
Universiteit Leuven, B-3030 Leuven, Belgium

Introduction

In this contribution we present a generalization of the study of infinitely divisible positive definite functions $f : G \to \mathbb{C}$ on a group G. Instead of complex valued functions, we consider completely positive (C.P.) mappings $\Phi : G \to B(H)$, taking values in the bounded operators on some hilbertspace H. Hence, as $\mathbb{C}$ is replaced by $B(H)$, we treat the fully non-commutative case.

In the first chapter, some definitions and useful properties of C.P. mappings are listed.

In chapter II the notion of infinite divisibility is generalized to C.P. mappings.

In chapter III we introduce central limits of sequences of C.P. mappings and study the relation with infinite divisibility.

Finally in the last chapter, we want to characterize infinitely divisible C. P. mappings and investigate their structure. First we generalize a result for functions which is known as the "Araki-Woods embedding theorem" and which says, roughly speaking, that the GNS representation of an infinitely divisible positive definite function lives on a Fock space. We find that, if a C.P. mapping is infinitely divisible, then the mapping itself as well as its Stinespring representation (which is a generalized GNS representation) live on Fock spaces.

Another known property for functions is that a function is infinitely divisible if and only if it is the exponential of some conditionally positive definite function. Also this result can be extended to C.P. mappings. Starting from an infinitely divisible C.P. mapping, we construct a "logarithm" which is a conditionally completely positive mapping and which, after "exponentiating" yields the C.P. mapping we started from.

I. COMPLETELY POSITIVE MAPPINGS ON GROUPS

Definition I.1

(i) A mapping $\Phi : G \to B(H)$ from a group G into the bounded operators $B(H)$ on a hilbertspace H is called *completely positive* (C.P.) if

$$\sum_{g,g' \in G} \langle \xi_g | \Phi(g^{-1}g')\xi_{g'} \rangle \geq 0$$

for all choices of functions $g \in G \to \xi_g \in H$ vanishing everywhere but on a finite number of elements of G.

(ii) A C.P. mapping is *normalized* if $\Phi(e) = \mathbb{1}$ where e is the neutral element of G and $\mathbb{1}$ is the identity operator on H.

(iii) If G is a topological group, then Φ is called *continuous* if $g \in G \to \langle \xi | \Phi(g)\xi' \rangle \in \mathbb{C}$ is continuous for all $\xi, \xi' \in H$

---

* Bevoedgverklaard Navorser N.F.W.O., Belgium

** Onderzoeker I.I.K.W., Belgium

Examples

(i) If $H = \mathbb{C}$ then the definition of a C.P. mapping reduces to the definition of a positive definite function $\Phi : G \to \mathbb{C}$; e.g. take $G = \mathbb{R}$ and consider a $\mathbb{R}$-valued random variable X, then its characteristic function $\Phi(t) = \langle e^{itX} \rangle$ is a positive definite function on $\mathbb{R}$.

(ii) Every unitary representation of G is a C.P. mapping

(iii) If G is the unitary group of some unital C*-algebra A, then a C.P. mapping $\Phi : G \to B(H)$ extends uniquely to a linear mapping $\Phi : A \; B(H)$ which is C.P. in the usual sense [1]

Stinespring decomposition

Any normalized C.P. mapping can be decomposed into an isometry and a unitary representation in the following way:

Theorem I.2

(i) Let $\Phi : G \to B(H)$ be a normalized C.P. mapping, then there exists

. a hilbertspace K

. a unitary representation $\pi : G \to U(K)$

. an isometry $V : H \to K$

such that

$$\phi(g) = V^* \pi(g) V \quad , \quad g \in G$$

(ii) If the minimality condition

$$\overline{\{\pi(G)\}'' VH} = K$$

is satisfied, then the triplet (K, π, V) is unique (up to unitary equivalence).

Proof [2]

The minimal triplet (K, π, V) is called the *Stinespring triplet* of Φ. If Φ is a positive definite function on G, then the Stinespring decomposition reduces to the well known GNS representation theorem.

Remarks

Using the Stinespring triplet (K, π, V) of a normalized C.P. mapping $\Phi : G \to B(H)$ some useful properties of Φ can easily be derived.

(i) $\Phi(g)^* = (V^*\pi(g)V)^* = V^*\pi(g)^*V = V^*\pi(g^{-1})V = \Phi(g^{-1})$

(ii) For $g_1, g_2 \in G$ and $\xi \in H$:

$$\| (\Phi(g_1) - \Phi(g_2))\xi \|^2 = \| V^*(\pi(g_1) - \pi(g_2))V\xi \|^2$$

$$\leq \| (\pi(g_1) - \pi(g_2))V\xi \|^2$$

$$= 2 \langle \xi | (\mathbb{1} - \Phi(g_1^{-1} g_2))\xi \rangle \qquad (1.1)$$

hence weak continuity of $g \to \Phi(g)$ at $g = e$ implies strong continuity of $g \to \Phi(g)$ everywhere.

(iii) Let $g \in G \to \xi_g \in H$ be everywhere zero but at a finite number of group elements, then using $VV^* \leq \mathbb{1}$:

$$\sum_{g,g'} \langle \xi_g | \Phi(g^{-1}) \Phi(g') \xi_{g'} \rangle$$

$$= \sum_{g,g'} \langle \xi_g | V^*\pi(g^{-1}) VV^* \pi(g') V\xi_{g'} \rangle$$

$$= \langle \sum_g \pi(g) V\xi_g | VV^* \sum_g \pi(g) V\xi_g \rangle$$

$$\leq \sum_{g,g'} \langle \xi_g | V^* \pi(g^{-1}g') V \xi_{g'} \rangle$$

$$= \sum_{g,g'} \langle \xi_g | \Phi(g^{-1}g') \xi_{g'} \rangle \tag{1.2}$$

This inequality is known as the 2-positivity inequality. In particular, it implies

$$\| \Phi(g) \| \leq 1 \tag{1.3}$$

II. INFINITE DIVISIBILITY FOR C.P. MAPPINGS ON GROUPS

As we will need in the sequel a lot of cyclicity conditions we introduce the following notions:

Definition II.1

(i) We call (H,Φ,Ω) a *C.P. triplet* on a group G if

- . H is a hilbertspace
- . $\Phi : G \to B(H)$ is a normalized C.P. mapping
- . $\Omega \in H$ is a normalized vector which is cyclic for $\{\Phi(G)\}''$

(ii) A C.P. triplet (H, Φ, Ω) on a topological group is called *continuous* if Φ is continuous

(iii) Two C.P. triplets (H_i, Φ_i, Ω_i), i=1,2, on a group G are called *unitarily equivalent* (notation $(H_1, \Phi_1, \Omega_1) \cong (H_2, \Phi_2, \Omega_2)$) if there exists a unitary $U : H_1 \to H_2$ such that

$$\Phi_2(g) = U \, \Phi_1(g) U^*$$

$$\Omega_2 = U \, \Omega_1$$

Now we want to define the notion of infinite divisibility for C.P. triplets. Since these objects are the generalizations of characteristic functions of random variables, we will need the notion of product of C.P. triplets.

Proposition II.2

Let (H_i, Φ_i, Ω_i) $(i=1,\ldots,n)$ be C.P. triplets on a group G.
Put $H = \{\otimes_{i=1}^{n} (\Phi_i(g)) \,|\, g \in G\}'' \otimes_{i=1}^{n} \Omega_i$

$$\Phi(g) = \otimes_{i=1}^{n} (\Phi_i(g))|_H$$

$$\Omega = \otimes_{i=1}^{n} \Omega_i$$

then (H, Φ, Ω) is a C.P. triplet which is called the *product triplet*.
Notation: $(H, \Phi, \Omega) = \bigotimes_{i=1}^{n} (H_i, \Phi_i, \Omega_i)$

Proof

Let (H_i, π_i, V_i) be the Stinespring triplet for Φ_i. Then, $V = \otimes_{i=1}^{n} V_i$ is an isometry and $\pi = \otimes_{i=1}^{n} \pi_i$ is a unitary representation of G, and as $\Phi(g) = V^* \pi(g) V|_H$, the mapping Φ is C.P. By definition of H, Ω is cyclic for $\{\Phi(G)\}''$.

∎

Definition II.3

A C.P. triplet (H,Φ,Ω) on a group G is *infinitely divisible* if for all $n \in \mathbb{N}$, there exists a C.P. triplet $(H^{1/n}, \Phi^{1/n}, \Omega^{1/n})$ which is an nth root for (H,Φ,Ω) in the sense that

$$(H, \Phi, \Omega) \cong \otimes^n (H^{1/n}, \Phi^{1/n}, \Omega^{1/n})$$

Remark that for the special case of unitary representations or positive definite functions ($H = \mathbb{C}, \Omega = 1 \in \mathbb{C}$) this definition reduces to the usual one as given in [3], [4], [5]; e.g. a function $\Phi : G \to \mathbb{C}$ is infinitely divisible if for all $n \in \mathbb{N}$, there exists a positive definite function $\Phi^{1/n} : G \to \mathbb{C}$ such that $(\Phi^{1/n}(g))^n = \Phi(g)$ for all $g \in G$.

A non trivial example of an infinitely divisible C.P. mapping will be given in the next chapter.

III. CENTRAL LIMIT THEOREMS FOR C.P. MAPPINGS ON GROUPS

First we introduce a notation which will be very useful in the sequel.

Notation

Denote by $\tilde{G}$ the set of all n-tuples $(g_1,\dots,g_n)$ of group elements $g_i \in G$, $(i = 1,\dots,n;\ n \in \mathbb{N})$ where a o-tuple is the empty set $\emptyset$. A composition law in G can be defined by juxtaposition

$$(\Delta,\Delta') \in \tilde{G} \times \tilde{G} \to \Delta \times \Delta' \in \tilde{G}$$

where $\Delta = (g_1,\dots,g_n)$, $\Delta' = (g'_1,\dots,g'_{n'})$ and $\Delta \times \Delta' = (g_1,\dots,g_n, g'_1,\dots,g'_{n'})$. By this $\tilde{G}$ becomes a semigroup with neutral element $\emptyset$. Furthermore the group inversion in $\tilde{G}$ induces a natural involution in $\tilde{G}$:

$$* : \Delta = (g_1,\dots,g_n) \in \tilde{G} \to \Delta^* = (g_n^{-1},\dots,g_1^{-1}) \in \tilde{G}$$

If χ is a function on G with values in the (possibly unbounded) linear operators on some hilbertspace, we will use the notation

$$\chi(\Delta) \equiv \chi(g_1)\,\chi(g_2)\,\dots\,\chi(g_n) \equiv \prod_{g \in \Delta}^{\rightarrow} \chi(g) \quad , \ \Delta = (g_1,\dots,g_n)$$

if such a product makes sense. ($\overset{\rightarrow}{\prod}_{g\in\Delta}$ denotes the ordered product). Finally by convention $\chi(\emptyset) = 1$;

Definition III.1

A net $(H_\alpha,\Phi_\alpha,\Omega_\alpha)_{\alpha \in I}$ of C.P. triplets on a group G *converges weakly* to a C.P. triplet (H,Φ,Ω) on G (notation: $\text{w-lim}_\alpha (H_\alpha,\Phi_\alpha,\Omega_\alpha) = (H,\Phi,\Omega)$) if

$$\lim_\alpha \langle\Omega_\alpha|\Phi_\alpha(\Delta)\Omega_\alpha\rangle = \langle\Omega|\Phi(\Delta)\Omega\rangle$$

for all $\Delta \in \tilde{G}$.

(H,Φ,Ω) is called the *weak limit* of $(H_\alpha,\Phi_\alpha,\Omega_\alpha)_{\alpha \in I}$.

Clearly, if the weak limit of a net exists, then it is unique up to unitary equivalence.

Definition III.2

A sequence $(H_n,\Phi_n,\Omega_n)_{n \in \mathbb{N}}$ of C.P. triplets on a group G has a *central limit* (H,Φ,Ω) if

$$\text{w-lim}_n \otimes^n (H_n,\Phi_n,\Omega_n) = (H,\Phi,\Omega).$$

Examples

A. Very elementary case

Take $G = \mathbb{R}$, $H = \mathbb{C}$ and $\Omega = 1 \in \mathbb{C}$. Consider a $\mathbb{R}$-valued random variable X with mean $\langle X\rangle = 0$ and variance $\langle X^2\rangle = \sigma^2$. Take $\Phi(t) = \langle e^{itX}\rangle$ $(t \in \mathbb{R})$, and $\Phi_n(t) = \Phi(\frac{t}{\sqrt{n}})$.

Then the central limit of $(\mathbb{C},\Phi_n,1)_{n\in\mathbb{N}}$ is given by $(\mathbb{C},\Phi_\sigma,1)$ where Φ_σ is the characteristic function of a gaussian random variable with mean zero and variance σ^2, i.e. $\Phi_\sigma(t) = \exp -\sigma^2t^2/2$. Remark that by choosing a different n-dependence for Φ_n, one gets different limit theorems (e.g. $\Phi_n(t) = \Phi(\frac{t}{n})$ yields the law of large numbers).

Before we go on with a non-trivial example of a central limit theorem, we remind the reader to the definition of Fock space. Let $\mathcal{H}$ be any hilbertspace. Then the (symmetric) *Fock space* $S(\mathcal{H})$ over $\mathcal{H}$ is defined by

$$S(\mathcal{H}) = \bigoplus_{n\geq 0}(\otimes^n \mathcal{H})_s$$

where $(\otimes^n\mathcal{H})_s$ is the symmetric n-fold tensor product of H. The so called *exponential* or *coherent* vectors in $S(\mathcal{H})$ are given by

$$\mathrm{Exp}\ \xi = \bigoplus_{n\geq 0}\frac{\otimes^n\xi}{\sqrt{n!}} \quad ,\ \xi\in\mathcal{H}$$

and they satisfy

$$\langle \mathrm{Exp}\ \xi|\mathrm{Exp}\ \eta\rangle = \exp\langle\xi|\eta\rangle \quad , \quad \xi,\eta\in\mathcal{H} \tag{3.1}$$

Moreover it is well known that $\{\mathrm{Exp}\ \xi|\xi\in\mathcal{H}\}$ is a linearly independent and total set in $S(\mathcal{H})$ [5].

B. A central limit theorem for C.P. mappings on the Heisenberg group (CCR-case)

Here we take the group G to be the Heisenberg group which is constructed as follows.

Let $\mathcal{H}$ be a hilbertspace, put $G = \mathcal{H}\times\mathbb{R}$, then G is a group for the multiplication

$$(\xi,\theta)\ (\xi',\theta') = (\xi+\xi',\theta+\theta' - \mathrm{Im}\langle\xi|\xi'\rangle)$$

In G we can define scaling automorphisms α_λ by

$$\alpha_\lambda : G\to G \quad (\xi,\theta)\to(\lambda\xi,\lambda^2\theta) \qquad \lambda\in\mathbb{R}.$$

Consider now a C.P. mapping $\Phi : G\to B(H)$ such that

$$\Phi(\xi,\theta) = e^{ic\theta}T(\xi) \tag{3.2}$$

for some $c\in\mathbb{R}$. Furthermore let $\Omega\in H$ be a normalized vector which is cyclic for $\{T(\mathcal{H})\}''$

Our aim is to calculate the central limit of the sequence

$$(H_n = H, \Phi_n = \Phi\circ\alpha_{1/\sqrt{n}}, \Omega_n = \Omega)_{n\in\mathbb{N}_0} \tag{3.3}$$

Therefore, we will have to compute, among others, limits of the type

$$\lim_{n\to\infty}\langle\Omega_n|\Phi_n(\xi,\theta)\Omega_n\rangle^n = e^{ic\theta}\lim_{n\to\infty}\langle\Omega|\Phi(\frac{\xi}{\sqrt{n}},0)\Omega\rangle^n$$

Introducing the Stinespring triplet (K,π,V) of Φ this limit can be rewritten as

$$e^{ic\theta}\lim_{\lambda\to 0}\langle V\Omega|\pi(\lambda\xi,0)V\Omega\rangle^{1/\lambda^2}$$

In order to ensure the existence of this limit it will therefore be natural to assume the following regularity conditions on Φ.

R.1 The 1-parameter group $\lambda\in\mathbb{R}\to\pi(\lambda\xi,0)$ is strongly continuous for all $\xi\in H$ (or equivalently, the mapping $\lambda\in\mathbb{R}\to\Phi(\lambda\xi+\xi',0)$ is weakly continuous for all $\xi,\xi'\in\mathcal{H}$).

R.2 For all $\xi \in \mathcal{H}$, the generator $B(\xi)$ of $\lambda \in \mathbb{R} \to \pi(\lambda\xi, 0)$ has $V\Omega$ in its domain (this corresponds to the condition of finite variance).

R.3 For all $\xi \in \mathcal{H}$: $\langle V\Omega | B(\xi) V\Omega \rangle = 0$ (this corresponds to mean zero)

Althought it is not strictly necessary, we will here, for technical convenience, assume that the underlying hilbertspace $\mathcal{H}$ of the Heisenberg group G is finite dimensional. For the same reason we will also strengthen regularity condition R.3. We will namely assume that

$$V^* B(e^{i\phi} \xi) V\Omega = e^{i\phi} V^* B(\xi) V\Omega \quad , \phi \in \mathbb{R} \tag{3.4}$$

(this amount to gauge covariance)

For a more general and detailed treatment we refer to [6] where the central limit theorem is proved in a slightly different setting.

Notice that the generators $B(\xi)$, $\xi \in \mathcal{H}$ satisfy the following equation

$$B(\xi) V\Omega + B(\eta) V\Omega = B(\xi+\eta) V\Omega \tag{3.5}$$

Indeed, (3.5) follows from

$$\begin{aligned} &[B(\xi) + B(\eta) - B(\xi+\eta)] V\Omega \\ &= \lim_{\lambda \to 0} \frac{1}{i\lambda}[\pi(\lambda\xi,0)\,\pi(\lambda\eta,0)\,\pi(-\lambda(\xi+\eta),0) - \mathbb{1}] V\Omega \\ &= \lim_{\lambda \to 0} \frac{1}{i\lambda}[e^{-ic\lambda^2 \mathrm{Im}\langle\xi|\eta\rangle} - \mathbb{1}] V\Omega = 0 \end{aligned}$$

where we used that fact that (3.2) implies that $\pi(\xi,\theta) = e^{ic\theta}\pi(\xi,0)$.

Now we are able to compute the central limit of $(H_n, \Phi_n, \Omega_n)_{n \in \mathbb{N}}$. First we calculate the limit

$$\lim_{n \to \infty} \langle \Omega_n | \Phi_n(\Delta)\, \Omega_n \rangle^n \quad , \quad \Delta \in \tilde{G}$$

and then we will have to identify this limit with the corresponding matrix element $\langle \bar{\Omega}, \bar{\Phi}(\Delta)\bar{\Omega} \rangle$ of some C.P. triplet $(\bar{H}, \bar{\Phi}, \bar{\Omega})$

Take $\Delta = ((\xi_1, \theta_1), \ldots, (\xi_k, \theta_k))$ and put

$$g(\lambda) = \langle \Omega | V^* e^{i\lambda B(\xi_1)} V V^* e^{i\lambda B(\xi_2)} V \ldots V^* e^{i\lambda B(\xi_k)} V\Omega \rangle \ , \ \lambda \in \mathbb{R}$$

Then we have $\langle \Omega_n | \Phi_n(\Delta)\, \Omega_n \rangle^n = e^{ic(\theta_1 + \ldots + \theta_k)} g(1/\sqrt{n})^n$ and as $\lim_{n \to \infty} g(1/\sqrt{n}) = 1$ we have that

$$|g(1/\sqrt{n}) - 1 - \ell n\, g(1/\sqrt{n})| < |g(1/\sqrt{n}) - 1|^2$$

for n large enough (where the branch cut for the logarithmic function is chosen along the negative real half axis). Since $g(1/\sqrt{n})^n = \exp n\, \ell n\, g(1/\sqrt{n})$ it will therefore be sufficient to examine the limit $\lim_{\lambda \to 0} (g(\lambda) - 1)/\lambda^2$.

Clearly we have that

$$\begin{aligned} &\frac{1}{\lambda^2}(g(\lambda) - 1) \\ &= \sum_{\ell=1}^{k} \sum_{1 \le j_1 < j_2 < \ldots < j_\ell \le k} \frac{1}{\lambda^2} \langle \Omega | (V^* e^{i\lambda B(\xi_{j_1})} V - \mathbb{1}) \ldots (V^* e^{i\lambda B(\xi_{j_\ell})} V - \mathbb{1}) \Omega \rangle \end{aligned}$$

Consider first the limit of the terms on the right hand side with $\ell = 1$. It follows from the regularity conditions R.2 and R.3 that the functions $\lambda \to f_j(\lambda) = \langle\Omega|(V^* e^{i\lambda B(\xi_j)}V - \mathbb{1})\Omega\rangle$ $(j=1,\ldots,k)$ are twice differentiable in $\lambda = 0$ and $f'_j(0)=0$ and $f''_j(\lambda) = -\|B(\xi_j)V\Omega\|^2$. Hence

$$\lim_{\lambda \to 0} \frac{1}{\lambda^2} \langle\Omega|V^* e^{i\lambda B(\xi_j)} V\Omega\rangle = -\frac{1}{2} \|B(\xi_j)V\Omega\|^2 \tag{3.6}$$

If $\ell \geqslant 2$, then as $\lim_{\lambda \to 0} \frac{1}{\lambda} (V^* e^{i\lambda B(\xi_j)}V - \mathbb{1})\Omega = i\, V^*B(\xi_j)V\Omega$, we have that for any uniformly bounded strongly continuous one parameter family $\lambda \to x_\lambda$ of operators on H:

$$\lim_{\lambda \to 0} \frac{1}{\lambda^2} \langle\Omega|(V^* e^{i\lambda B(\xi_{j_1})} V - \mathbb{1}) x_\lambda (V^* e^{i\lambda B(\xi_{j_\ell})}V - \mathbb{1})\Omega\rangle$$
$$= -\langle V^* B(\xi_{j_1})V\Omega|x_o V^*B(\xi_{j_\ell})V\,\Omega\rangle$$

Taking $x_\lambda = \mathbb{1}$ in the case where $\ell=2$ and $x_\lambda = (V^*e^{i\lambda B(\xi_{j_2})}V-\mathbb{1})\ldots(V^*e^{i\lambda B(\xi_{j_{\ell-1}})}V-\mathbb{1})$ when $\ell>2$ we find that

$$\lim_{\lambda \to 0} \frac{1}{\lambda^2} \langle\Omega|(V^*e^{i\lambda B(\xi_{j_1})}V - \mathbb{1})(V^* e^{i\lambda B(\xi_{j_2})}V^* - \mathbb{1})\Omega\rangle$$
$$= -\langle V^*B(\xi_{j_1})V\Omega|V^*B(\xi_{j_2})V\Omega\rangle \tag{3.7}$$

and

$$\lim_{\lambda \to 0} \frac{1}{\lambda^2} \langle\Omega|V^* e^{i\lambda B(\xi_{j_1})}V - \mathbb{1})\ldots(V^* e^{i\lambda B(\xi_{j_\ell})}V - \mathbb{1})\Omega\rangle = 0 \text{ if } \ell > 2 \tag{3.8}$$

Summarizing (3.6), (3.7) and (3.8) we end up with

$$\lim_{n \to \infty} \langle\Omega_n|\Phi_n(\Delta)\Omega_n\rangle^n$$
$$= \exp\left[ic \sum_{j=1}^{k} \theta_j - \frac{1}{2}\sum_{j=1}^{k} \|B(\xi_j)V\Omega\|^2\right.$$
$$\left. - \sum_{1 \leqslant j_1 < j_2 \leqslant k} \langle V^*B(\xi_{j_1})V\Omega|V^*B(\xi_{j_2})V\Omega\rangle\right] = \langle\bar{\Omega}|\bar{\Phi}(\Delta)\bar{\Omega}\rangle \tag{3.9}$$

Notice that up to now, we only used conditions R.1, R.2 and R.3. Hence, these conditions are sufficient to ensure the existence of the central limit of (3.3). However, we will use the additional assumptions to identify this limit in a comfortable way. Indeed, by (3.4) and (3.5)

$$(\xi,\eta) \to \langle V^* B(\xi)V\Omega|V^*B(\eta)V\Omega\rangle$$

is a positive sesquilinear form on $\mathcal{H}$ and since we assume that $\dim \mathcal{H} < \infty$, we have

$$\langle V^*B(\xi)V\Omega|V^*B(\eta)V\Omega\rangle = \langle A\xi|A\eta\rangle \tag{3.10}$$

for some $A = A^* \in \mathcal{B}(\mathcal{H})$

Next by (3.5)

$$(\xi,\eta) \to \mathrm{Re}\,\langle B(\xi)V\Omega|B(\eta)V\Omega\rangle$$

is a positive $\mathbb{R}$-bilinear form on $\mathcal{H}$. Consider now $\mathcal{H}$ as a real hilbertspace with scalar product $Re\langle .|.\rangle$, then

$$Re\langle B(\xi)V\Omega|B(\eta)V\Omega\rangle = Re\langle\xi|D\eta\rangle \tag{3.11}$$

for some positive $D\in B_{\mathbb{R}}(\mathcal{H})$ (= $\mathbb{R}$-linear operators on $\mathcal{H}$). Clearly, as $VV^*\leq 1\!\!1$, we have $D\geq A^*A$. Put $D-A^*A = Q\in B_{\mathbb{R}}(\mathcal{H})$, then $Q\geq 0$ and using (3.10) and (3.11) we can rewrite (3.9) as

$$\langle\bar\Omega|\bar\Phi(\Delta)\bar\Omega\rangle = \exp\Big[ic\sum_{j=1}^{k}\theta_j - \frac{1}{2}\sum_{j=1}^{k}\langle\xi_j|(A^*A+Q)\xi_j\rangle - \sum_{1\leq j_1<j_2\leq k}\langle A\xi_{j_1}|A\xi_{j_2}\rangle\Big] \tag{3.12}$$

Since we get as a result exponentials of scalar products, keeping (3.2) in mind, we are tempted to try to identify $\bar\Phi$ as a mapping which lives in the Fockspace $S(\mathcal{H})$. Indeed, one can easily check that the following is consistent with (3.12)

- $\bar H = S(\mathcal{H}_A)$ where $\mathcal{H}_A$ = Range A^-
- $\bar\Phi = \Phi_{A,Q,c}$ where

$$\Phi_{A,Q,c}(\xi,\theta)\ \text{Exp}\ \eta = \exp\Big[ic\theta - \frac{1}{2}\langle\xi|(A^*A+Q)\xi\rangle - \langle A\xi|\eta\rangle\Big]\ \text{Exp}\ (\eta+A\xi)$$

- $\bar\Omega = \text{Exp}\ 0$

Hence $(S(\mathcal{H}_A), \Phi_{A,Q,c}, \text{Exp}\ 0)$ is the central limit of (3.3). Notice that the central limit is infinitely divisible, indeed

$$(S(\mathcal{H}_A), \Phi_{A,Q,c}, \text{Exp}\ 0) \overset{\sim}{=} \otimes^n (S(\mathcal{H}_A), \Phi_{A/\sqrt{n},\, Q/n,\, c/n}, \text{Exp}\ 0)$$

Remark also that the mapping $\Phi_{A,Q,c}$ we have found by the central limit construction is a so called quasi-free mapping. These mappings where introduced for CCR-algebra's in [7]. Due to the particular choice of the scaling in 3.3, we can thus say that the quasi-free mappings arise as non-commutative generalizations of the gaussian distributions.

In the preceding examples, the central limit turned out to be infinitely divisible. In fact this property holds in general.

Theorem III.3

i) Every infinitely divisible C.P. triplet is the central limit of at least one sequence

ii) Every central limit is infinitely divisible.

Comment. This theorem shows that a central limit theorem provides a tool to construct explicitly infinitely divisible C.P. triplet on a given group.

Notation

Given a positive kernel $(x,y)\in X\times X\to k(x,y)\in\mathbb{C}$ on a set X, one constructs in a standard way a hilbertspace. k extends to a positive sesquilinear form on the complex free vector space V(X) generated by X. Let $V_0 = \{u\in V(X)|k(u,u) = 0\}$. By positivity V_0 is a linear subspace. By hil(X,k) we denote the completion of $V(X)/V_0$ for the scalar product induced by k. By abuse of notation we will denote the elements of $V(X)/V_0$ by u instead of $u + V_0$. Also the scalar product in hil(X,k) will be denoted in the conventional way.

Proof of Theorem III.3

(i) Every infinitely divisible C.P. triplet is the central limit of the sequence of its nth roots.

(ii) Let $(H,\Phi,\Omega) = \text{w-}\lim_{n\to\infty} \otimes^n (H_n,\Phi_n,\Omega_n)$. For any $k\in \mathbb{N}_0$ we want to construct a kth-root for (H,Φ,Ω). A candidate for a kth root is obviously given by:

$$\text{w-}\lim_{n\to\infty} (\tilde{H}_n,\tilde{\Phi}_n,\tilde{\Omega}_n)$$

where $(\tilde{H}_n,\tilde{\Phi}_n,\tilde{\Omega}_n) = \otimes^n (H_{nk},\Phi_{nk},\Omega_{nk})$.

However, this limit doesn't exist in general, but, using a compactness argument, we can show the existence of a net $(\tilde{H}_\alpha,\tilde{\Phi}_\alpha,\tilde{\Omega}_\alpha)_{\alpha\in I}$ reaching infinitely many points of $(\tilde{H}_n,\tilde{\Phi}_n,\tilde{\Omega}_n)_{n\in\mathbb{N}}$, and a C.P. triplet $(H^{1/k},\Phi^{1/k},\Omega^{1/k})$ such that

$$\text{w-}\lim_\alpha (\tilde{H}_\alpha,\tilde{\Phi}_\alpha,\tilde{\Omega}_\alpha) = (H^{1/k},\Phi^{1/k},\Omega^{1/k})$$

Indeed, consider the sequence of functions $f_n : \tilde{G}\times\tilde{G}\to\mathbb{C}$ given by

$$f_n(\Delta,\Delta') = \langle\Omega_n\,|\,\Phi_n(\Delta^*\times\Delta')\Omega_n\rangle$$

By (1.3) the sequence $(f_n)_{n\in\mathbb{N}}$ lies in the set $F = \{f : \tilde{G}\times\tilde{G}\to\mathbb{C}\,|\,|f|\leq 1\}$ which is compact for the topology of pointwise convergence. Hence there exists a net $(f_\alpha)_{\alpha\in I}$ in F reaching infinitely many points of $(f_n)_{n\in\mathbb{N}}$, and a function $f\in F$ such that $(f_\alpha)_{\alpha\in I}$ converges pointwise to f. Moreover f is a positive kernel on G, hence we can define a hilbertspace $H^{1/k} = \text{hil}(G,f)$. Define $\Phi^{1/k} : G\to B(H^{1/k})$ by $\Phi^{1/k}(g)\,\Delta = (g)\times\Delta$. One can check that $\Phi^{1/k}$ is well defined and C.P.. Finally, put $\Omega^{1/k} = \emptyset\in H^{1/k}$. By construction $(H^{1/k},\Phi^{1/k},\Omega^{1/k})$ is a kth-root for (H,Φ,Ω). ∎

IV. CHARACTERIZATION OF CONTINUOUS INFINITELY DIVISIBLE C.P. MAPPINGS

A. Introduction

In this chapter we will tackle the following problem:

Let (H,Φ,Ω) be a continuous infinitely divisible C.P. triplet with continuous roots on an arcwise connected group G. What can then be said about the structure of Φ and of its Stinespring triplet?

For the special case of normalized continuous infinitely divisible functions $f : G\to\mathbb{C}$ with continuous roots on an arcwise connected group G, the following results are known:

1 The "Araki-Woods embedding theorem" [5], [3]

The GNS representation π of f lives on a Fock space $S(H)$; it is of type S (i.e. $\pi(g)$ maps a coherent vector into a multiple of a coherent vector) and the cyclic vector is given by the vacuum Exp 0 $\in S(H)$. Explicitly:

$$f(g) = \langle \text{Exp } 0\,|\,\pi(g)\text{ Exp } 0\rangle$$

where

$$\pi(g)\text{ Exp }\xi = c_g \exp\left[-\frac{1}{2}\|b_g\|^2 - \langle b_g|U_g\xi\rangle\right] \text{Exp }(U_g\xi + b_g)$$

with . $g\in G\to U_g\in U(H)$ a unitary representation

. $g\in G\to b_g\in H$ such that $b_{g_1g_2} = b_{g_1} + U_{g_1}b_{g_2}$

: $g\in G\to c_g\in\{z\in\mathbb{C}\,|\,|z|=1\}$ such that $c_{g_1g_2} = c_{g_1}c_{g_2}\,e^{i\,\text{Im}\langle U_{g_1}b_{g_2}|b_{g_1}\rangle}$

2 The existence of a conditionally positive definite logarithm for f [4] [5]

A continuous $f : G \to \mathbb{C}$ is infinitely divisible if and only if f can be written as $f = \exp v$ where $v : G \to \mathbb{C}$ satisfies:

(i) Conditional positive definiteness

$$\text{i.e.} \quad \sum_{g,g'} \bar{\lambda}_g \lambda_{g'} \, v(g^{-1}g') \geqslant 0 \quad \text{if} \quad \sum_g \lambda_g = 0 \;, \quad \lambda_g \in \mathbb{C} \tag{4.1}$$

(ii) $v(e) = 0$

(iii) v continuous

3 Lévy-Khinchine formulae

i.e. explicit forms for all continuous conditionally positive definite functions on a given group. These formulae are known e.g. for

- $G = \mathbb{R}^n$ [8]
- G compact [9]
- G locally compact [10]

Now the aim is to follow the same program for C.P. triplets. In subsection B we will prove a structural result about infinitely divisible triplets which generalizes the "Araki-Woods embedding theorem". In subsection C a conditionally completely positive logarithm of an infinitely divisible C.P. triplet will be constructed, generalizing point 2, and also the inverse, say "exponential", construction will be given. However, as far as Lévy-Khinchine formulae are concerned, we have up to now no results.

B. A structural theorem for continuous infinitely divisible C.P. triplets on an arcwise connected group

First we generalize the well known result of Parthasarathy that a normalized continuous infinitely divisible positive definite function on a connected group never vanishes [9].

Lemma IV.1

Let (H,Φ,Ω) be a continuous infinitely divisible C.P. triplet on a connected group G, then we have that

$$\langle \Omega \,|\, \Phi(\Delta)\Omega \rangle \neq 0$$

for all $\Delta \in \tilde{G}$

Proof

The proof uses an induction argument on the length $\#(\Delta)$ of the n-tuple $\Delta \in \tilde{G}$

For $\#(\Delta) = 1$, the situation reduces to that of infinitely divisible positive definite functions which was proved in [9].

Suppose that the result holds all $\Delta' \in \tilde{G}$ with $\#(\Delta') \leqslant k$. Now choose $\Delta \in \tilde{G}$ with $\#(\Delta) = k+1$. We show that the result is also valid for Δ. Write $\Delta = \{g\} \times \Delta'$, $g \in G$, $\Delta' \in \tilde{G}$ with $\#(\Delta') = k$. Let $(K^{1/n}, \pi^{1/n}, V^{1/n})$ be the Stinespring triplet for the nth-root $(H^{1/n}, \Phi^{1/n}, \Omega^{1/n})$ of (H,Φ,Ω). Choose $\lambda, \mu \in \mathbb{C}$ such that $|\lambda| = |\mu| = 1$, then by the triangle inequality

$$\begin{aligned} &\| \lambda V^{1/n}\Omega^{1/n} - \mu\pi^{1/n}(g)V^{1/n}\Phi^{1/n}(\Delta')\Omega^{1/n} \|^2 \\ &\leqslant 2\, \| \lambda V^{1/n}\Omega^{1/n} - \pi^{1/n}(g)V^{1/n}\Omega^{1/n} \|^2 \\ &\quad + 2\, \| \pi^{1/n}(g)V^{1/n}\Omega^{1/n} - \mu\pi^{1/n}(g)V^{1/n}\Phi^{1/n}(\Delta')\Omega^{1/n} \|^2 \end{aligned} \tag{4.2}$$

We now compute the different terms which appear in (4.2):

$$\| \lambda V^{1/n}\Omega^{1/n} - \mu\pi^{1/n}(g)V^{1/n}\Phi^{1/n}(\Delta')\Omega^{1/n} \|^2$$

$$= 1 + \| \Phi^{1/n}(\Delta')\Omega^{1/n} \|^2 - 2\, Re\, \lambda\mu <\Omega^{1/n}| \Phi^{1/n}(\Delta)\Omega^{1/n}> \tag{4.3}$$

$$\| \lambda V^{1/n}\Omega^{1/n} - \pi^{1/n}(g)V^{1/n}\Omega^{1/n} \|^2$$

$$= 2 - 2\, Re\, \bar{\lambda} <\Omega^{1/n}|\Phi^{1/n}(g)\Omega^{1/n}> \tag{4.4}$$

$$\| \pi^{1/n}(g)V^{1/n}\Omega^{1/n} - \mu\pi^{1/n}(g)V^{1/n}\Phi^{1/n}(\Delta')\Omega^{1/n} \|^2$$

$$= 1 + \| \Phi^{1/n}(\Delta')\Omega^{1/n} \|^2 - 2\, Re\, \mu <\Omega^{1/n}|\Phi^{1/n}(\Delta')\Omega^{1/n}> \tag{4.5}$$

By choosing appropriate phases for λ and μ and inserting (4.3), (4.4) and (4.5) into (4.2) we then obtain the following inequality

$$1 - |<\Omega^{1/n}|\Phi^{1/n}(\Delta)\Omega^{1/n}>|$$

$$\leq 2\,(1 - |<\Omega^{1/n}|\Phi^{1/n}(g)\Omega^{1/n}>|)$$

$$+ \; 2\,(1 - |<\Omega^{1/n}|\Phi^{1/n}(\Delta')\Omega^{1/n}>|) \tag{4.6}$$

As $(H^{1/n},\Phi^{1/n},\Omega^{1/n})$ is an $n^{\underline{th}}$ root of (H,Φ,Ω), we have for any $X \in \tilde{G}$

$$|<\Omega^{1/n}|\Phi^{1/n}(X)\Omega^{1/n}>| = |<\Omega|\Phi(X)\Omega>|^{1/n} \tag{4.7}$$

Multiplying (4.6) by n, using (4.7) and taking the limit $n \to \infty$ we then obtain

$$-\frac{1}{2}\,\ell n\,|<\Omega|\Phi(\Delta)\Omega>| \leq -\,\ell n\,|<\Omega|\Phi(g)\Omega>| - \ell n\,|<\Omega|\Phi(\Delta')\Omega>|$$
$$< \infty$$

where the last inequality follows from the induction hypothesis. Hence

$$|<\Omega|\Phi(\Delta)\Omega>| > 0$$

∎

<u>Lemma IV.2</u>

Let $f : X \to \mathbb{C}$ be a continuous function on a connected topological space X such that $f(x) \neq 0$ for all $x \in X$ and $f(x_o) = 1$ for some $x_o \in X$.

(i) If for some $n \in \mathbb{N}_o$, there exists a continuous function $f_1 : X \to \mathbb{C}$ such that

. $f_1^n(x) = f(x)$ for all $x \in X$

. $f_1(x_o) = 1$ (4.8)

then f_1 is the only continuous function that satisfies (4.8).

(ii) If X is arcwise connected and for all $n \in \mathbb{N}_o$ there exists a continuous function $f_n : X \to \mathbb{C}$ such that $(f_n)^n = f$ and $f_n(x_o) = 1$ then there exists a unique continuous function $v : X \to \mathbb{C}$ such that

$$f = \exp v \tag{4.9}$$
$$v(x_o) = 1$$

Moreover, for all $x \in X$ we have

$$v(x) = \lim_{n\to\infty} n(f_n(x) - 1) \tag{4.10}$$

<u>Proof</u>

(i) Let f_2 be another continuous function satisfying (4.8). Put $f_j(x) = |f_j(x)|e^{i\theta_j(x)}$, then as $|f_j(x)| > 0$, the function $x \to e^{i\theta_j(x)}$ is continuous on X. Since $f_1^n(x) = f_2^n(x)$, we have $|f_1(x)| = |f_2(x)|$ and $e^{in(\theta_1(x)-\theta_2(x))} = 1$. Hence $\theta_1(x)-\theta_2(x) = 2\pi k(x)/n$ with $k(x) \in \mathbb{Z}$. But the function $x \to e^{i(\theta_1(x)-\theta_2(x))} = e^{2\pi i k(x)/n}$ is continuous on the connected space X and takes the value 1 in x_o; therefore $k(x) = \ell(x)n$ with $\ell(x) \in \mathbb{Z}$ which implies $f_1(x) = f_2(x)$.

(ii) Fix $x \in X$ and let $\gamma_1 : [0,1] \to X$ be a path in X connecting x_o and x (i.e. $\gamma_1(0)=x_o$ and $\gamma_1(1)=x$). Then $t \to \tilde{\gamma}_1(t) = f(\gamma_1(t))/|f(\gamma_1(t))|$ is a path in the 1-dimensional torus with $\gamma_1(0)=1$, and hence, by the Covering Path Property, there exists a unique path $\phi_1 : [0,1] \to \mathbb{R}$ with $\phi_1(0) = 0$ such that $\tilde{\gamma} = e^{i\phi_1}$. Put now $\theta_1(x) = \phi_1(1)$. Then $f(x) = |f(x)|e^{i\theta_1(x)}$. We have to show that $\theta_1(x)$ is independent of the choice of γ_1. Consider therefore another path γ_2 in X connecting x_o and x and let ϕ_2 be the corresponding path in $\mathbb{R}$ and $\theta_2(x)$ the corresponding number. As $f(x)=|f(x)|e^{i\theta_1(x)} = |f(x)|e^{i\theta_2(x)}$ and $|f(x)| > 0$, there exists a $k \in \mathbb{Z}$ such that

$$\theta_2(x) = \theta_2(x) + 2\pi k \tag{4.11}$$

For j=1,2 the functions $t \to |f_{|k|+1}(\gamma_j(t))| \exp i\phi_j(t)/(|k|+1)$ and $t \to f_{|k|+1}(\gamma_j(t))$ are both continuous $(|k|+1)^{th}$ roots of $t \to f(\gamma_j(t))$ taking the value 1 in $t = 0$. Hence by (i) we have that $f_{|k|+1}(\gamma_j(t)) = |f_{|k|+1}(\gamma_j(t))| \exp i\phi_j(t)/(|k|+1)$. Taking $t = 1$ and using (4.11) we find

$$\exp i\theta_1(x)/(|k|+1) = \exp i\theta_2(x)/(|k|+1)$$
$$= \exp [i\theta_1(x)/(|k|+1) + 2\pi ik/(|k|+1)]$$

which implies $k = 0$ and therefore $\theta_1(x) = \theta_2(x)$.
Furthermore, it can easily be seen that $x \to \theta_2(x)$ is continuous.

Summarizing, we have now shown that there exists a unique continuous function $x \in X \to \theta(x) \in \mathbb{R}$ such that

$$\theta(x_o) = 0$$
$$f(x) = |f(x)|\, e^{i\theta(x)}$$
$$f_n(x) = |f_n(x)|e^{i\theta(x)/n}$$

Now it follows immediately that $x \to v(x) = \ln|f(x)| + i\theta(x)$ is the unique continuous logarithm for f with $v(x_o) = 0$ and clearly also (4.10) holds.

■

From lemmas IV.1 and IV.2 (i) we deduce:

Corollary IV.3

If (H,Φ,Ω) is a continuous infinitely divisible C.P. triplet with continuous roots on a connected group G, then the continuous roots are unique (up to unitary equivalence).

From lemmas IV.1 and IV.2 (ii) we get:

Corollary IV.4

Let (H,Φ,Ω) be a continuous infinitely divisible C.P. triplet with continuous roots $(H^{1/n},\Phi^{1/n},\Omega^{1/n})$ on an arcwise connected group G. Then there exists a unique continuous function d

$$d : \tilde{G} \to \mathbb{C} : \Delta \to d_\Delta$$

(where $\tilde{G}$ has the natural topological structure induced by G) such that

$$\text{(i)} \quad \langle\Omega\,|\Phi(\Delta)\Omega\rangle = e^{d_\Delta} \tag{4.12}$$

$$\text{(ii)} \quad d_{\Delta_e} = 0 \text{ where } \Delta_e = (e,\dots,e) \tag{4.13}$$

Furthermore

$$<\Omega^{1/n}|\Phi^{1/n}(\Delta)\,\Omega^{1/n}> = e^{d_\Delta/n} \qquad (4.14)$$

Using this function d we can now construct two positive kernels.

Lemma IV.5

Let (H,Φ,Ω) be an infinitely divisible C.P. triplet with continuous roots on an arcwise connected group G, and let $d : \tilde{G} \to \mathbb{C}$ be as above, then

(i) $$k_1((g,\Delta),(g',\Delta')) = d_{\Delta^* \times (g^{-1}g') \times \Delta'} - d_{\Delta^* \times (g^{-1})} - d_{(g')\times \Delta'} \qquad (4.15)$$

is a positive kernel on $G \times \tilde{G}$

(ii) $$k_2(\Delta,\Delta') = d_{\Delta^* \times \Delta'} - d_{\Delta^*} - d_{\Delta'} \qquad (4.16)$$

is a positive kernel on $\tilde{G}$

Proof

(i) Let $(H^{1/n},\Phi^{1/n},\Omega^{1/n})$ be the nth root of (H,Φ,Ω) and take

$$\chi_n = \sum_{g,\Delta} \lambda_{g,\Delta} (\Phi^{1/n}((g)\times\Delta)\Omega^{1/n} - \Omega^{1/n}) \in H^{1/n}$$

Then,

$$\begin{aligned} 0 &\leqslant n \, \| \chi_n \|^2 \\ &= n \sum_{\substack{g,\Delta \\ g',\Delta'}} \bar{\lambda}_{g,\Delta} \lambda_{g',\Delta'} [<\Omega^{1/n}|\Phi^{1/n}(\Delta^* \times (g^{-1}) \times (g) \times \Delta')\Omega^{1/n}> + 1 \\ &\qquad - <\Omega^{1/n}|\Phi^{1/n}(\Delta^* \times (g^{-1}))\Omega^{1/n}> - <\Omega^{1/n}|\Phi^{1/n}((g')\times\Delta')\Omega^{1/n}>] \end{aligned}$$

By the 2-positivity inequality (1.2) the first term can be majorized to get

$$\begin{aligned} 0 < n \sum_{\substack{g,\Delta \\ g',\Delta'}} \bar{\lambda}_{g,\Delta} \lambda_{g',\Delta'} [&(<\Omega^{1/n}|\Phi^{1/n}(\Delta^* \times (g^{-1}g')\times\Delta')\Omega^{1/n}> - 1) \\ &- (<\Omega^{1/n}|\Phi^{1/n}(\Delta^* \times (g^{-1}))\Omega^{1/n}> - 1) \\ &- (<\Omega^{1/n}|\Phi^{1/n}((g')\times\Delta')\Omega^{1/n}> - 1)] \end{aligned}$$

Taking now the limit $n \to \infty$ and using (4.14) we get

$$\sum_{\substack{g,\Delta \\ g',\Delta'}} \bar{\lambda}_{g,\Delta} \lambda_{g',\Delta'} (d_{\Delta^* \times (g^{-1}g')\times\Delta} - d_{\Delta^* \times (g^{-1})} - d_{(g')\times\Delta}) \geqslant 0$$

(ii) follows immediately from (i) by putting $g = g' = e$ and observing that $\Phi(e) = 1$. ∎

Now we are able to proof the following generalization of the "Araki-Woods embedding theorem" if (H,Φ,Ω) is infinitely divisible, then Φ and its Stinespring representation live on Fock spaces. More explicitly we have the following two theorems:

Theorem IV.6

Let (H,Φ,Ω) be a continuous infinitely divisible C.P. triplet with continuous roots on an arcwise connected group G and let $H_1 = \mathrm{hil}(G \times \tilde{G}, k_1)$ where k_1 is the positive kernel given by (4.15)

(i) $V_1 : H \to S(H_1) : \Phi(\Delta)\Omega \to \langle\Omega|\Phi(\Delta)\Omega\rangle \operatorname{Exp}(e,\Delta)$ (4.17)

extends to an isometry from H into $S(H_1)$

(ii) $U : G \to U(H_1) : h \to U_h$, with

$$U_h(g,\Delta) = (hg,\Delta) - (h,\phi) \quad , \ (g,\Delta) \in G \times \tilde{G} \tag{4.18}$$

is a continuous unitary representation of G on H_1.

(iii) $\pi : G \to U(S(H_1)) : h \to \pi_h$ with

$$\pi_h \operatorname{Exp} \eta = \langle\Omega|\Phi(h)\Omega\rangle \exp\langle(h^{-1},\emptyset)|\eta\rangle \operatorname{Exp}(U_h\eta + (h,\emptyset)), \quad \eta \in H_1 \tag{4.19}$$

defines a continuous unitary representation of G on $S(H_1)$.

(iv) $\Phi(g) = V_1^* \pi_g V_1$, $g \in G$ (4.20)

Hence $(S(H_1)_o, \pi_o, V_1)$, where $S(H_1)_o = \{\pi(G)\}'' \, V_1 H^- \subset S(H_1)$ and π_o is the restriction of π to $S(H_1)_o$, is the Stinespring triplet for (H, Φ, Ω).

Proof

(i) For $\Delta, \Delta' \in \tilde{G}$, using (4.12) and (4.15), we have

$$\langle \langle\Omega|\Phi(\Delta)\Omega\rangle \operatorname{Exp}(e,\Delta) \,|\, \langle\Omega|\Phi(\Delta')\Omega\rangle \operatorname{Exp}(e,\Delta')\rangle$$

$$= \langle\Omega|\Phi(\Delta^*)\Omega\rangle \langle\Omega|\Phi(\Delta')\Omega\rangle \exp \langle(e,\Delta)|(e,\Delta')\rangle$$

$$= \langle\Omega|\Phi(\Delta^*)\Omega\rangle \langle\Omega|\Phi(\Delta')\Omega\rangle \exp(d_{\Delta^*\times\Delta'} - d_{\Delta^*} - d_{\Delta'})$$

$$= \langle \Phi(\Delta)\Omega|\Phi(\Delta')\Omega\rangle$$

(ii) For $h,g,g' \in G$, and $\Delta, \Delta', \in \tilde{G}$, using (4.15) one checks that

$$\langle (hg,\Delta) + (h,\phi) \,|\, (hg',\Delta') - (h,\phi)\rangle = \langle (g,\Delta)|(g',\Delta')\rangle$$

Hence U_h is well defined by (4.18) and isometric. Also $U_h U_{h'} = U_{h h'}$ and as $U_e = 1$ U is a unitary representation. The continuity of U follows immediately from the continuity of d.

(iii) As

.U is a continuous unitary representation

.$(h h',\phi) = U_h(h',\phi) + (h,\phi)$ and $h \in G \to (h,\phi) \in H_1$ is continuous

.$c_{h h'} = c_h c_{h'} \exp i \operatorname{Im} \langle U_h(h',\phi)|(h,\phi)\rangle$ where $c_h = \exp i \operatorname{Im} d_{(h)}$, and

$h \in G \to c_h \in \mathbb{C}$ is continous,

it follows that π as defined in (4.19) is a continuous representation of type S. (For a study of type S representations, see [5])

(iv) By a straightforward computation one gets

$$V_1^* \operatorname{Exp}(h,\Delta) = \frac{1}{\langle\Omega|\Phi((h)\times\Delta)\Omega\rangle} \Phi((h)\times\Delta)\Omega \ , \quad h \in G, \quad \Delta \in \tilde{G}$$

Hence,

$$V_1^* \pi_h V_1 \ \Phi(\Delta)\Omega = \langle\Omega|\Phi(\Delta)\Omega\rangle V_1^* \pi_h \operatorname{Exp}(e,\Delta)$$

$$= \langle\Omega|\Phi((h)\times\Delta)\Omega\rangle V_1^* \operatorname{Exp}(h,\Delta)$$

$$= \Phi(h)\ \Phi(\Delta)\Omega$$

Theorem IV.7

Let (H, Φ, Ω) be a continuous infinitely divisible C.P. triplet with continuous roots on an arcwise connected group G, and let $H_2 = \text{hil}(G, k_2)$ where k_2 is given by (4.16)

(i) $V_2 : H \to S(H_2) : \Phi(\Delta)\Omega \to \langle\Omega|\Phi(\Delta)\rangle \ \text{Exp}\ \Delta$ extends to an isometry

(ii) $\tilde{\Phi} : G \to B(S(H_2)) : g \to \tilde{\Phi}(g)$ with $\tilde{\Phi}(g)$ given by

$$\tilde{\Phi}(h)\ \text{Exp}\,\zeta = \exp\,(d_{(h)} + \langle (h^{-1})|\zeta\rangle)\ \text{Exp}\,(A_n\,\zeta + (h))\ , \zeta \in H_2$$

where $A_h \in B(H_2)$ such that $A_h\ \Delta = (h)\times\Delta - (h)$, is a continuous normalized C.P. mapping

(iii) $\Phi(h) = V_2^*\ \tilde{\Phi}(h) V_2\ ,\quad h \in G$

Proof

(i) The proof is analogous to the one for theorem IV.6 (i)

(ii) There exists a unique isometry $W : H_2 \to H_1$ such that $W\Delta = (e, \Delta)$.

Then $W^* : H_1 \to H_2$ is given by $W^*(h, \Delta) = (h)\times\Delta$. Hence the mapping $V_3 : S(H_2) \to S(H_1)$ given by $V_3 \text{Exp}\,\eta = \text{Exp}\,W\eta, \eta \in H_2$ is an isometry and $V_3^*\ \text{Exp}\,\zeta = \text{Exp}\,W^*\zeta,\ \zeta \in H_1$.

Let now $\tilde{\Phi}(h) = V_3^*\,\pi_h\,V_3$ where π is the representation given by (4.19). Clearly $\tilde{\Phi}$ is C.P., normalized and continuous. Moreover

$$\begin{aligned} W^*U_h W\,\Delta &= W^*U_h\ (e, \Delta) \\ &= W^*((h, \Delta) - (h, \emptyset)) \\ &= (h)\times\Delta - (h) = A_h\,\Delta \end{aligned}$$

Hence, for all $\eta \in H_2$

$$\begin{aligned} \tilde{\Phi}(h)\text{Exp}\,\eta &= V_3^*\,\pi_h\,V_3\ \text{Exp}\,\eta \\ &= V_3^*\,\pi_h\ \text{Exp}\,W\eta \\ &= \langle\Omega|\,\Phi(h)\Omega\rangle\,\exp\langle(h^{-1}, \emptyset)\,|W\eta\rangle\quad V_3^*\ \text{Exp}\,(U_h W\eta + (h, \emptyset)) \\ &= \langle\Omega|\,\Phi(h)\,\Omega\rangle\,\exp\langle W^*(h^{-1}, \emptyset)\,|\,\eta\rangle\ \text{Exp}(W^*U_h W\eta + W^*(h, \emptyset)) \\ &= \exp\,(d_{(h)} + \langle(h^{-1})\,|\,\eta\rangle)\ \text{Exp}(A_h\,\eta + (h)) \end{aligned}$$

(iii) The proof follows from a straightforward computation. ∎

C. Logarithms of infinitely divisible C.P. triplets

Here we will be concerned with the generalization of the second point of the program proposed in subsection A, namely we tackle the following problem: can we characterize infinitely divisible C.P. triplets as those triplets which have, in some sense, a conditionally completely positive logarithm.

In order to find a hint how to construct a logarithm for an infinitely divisible C.P. triplet, we first consider the case of a continuous normalized infinitely divisible positive definite function $f : G \to \mathbb{C}$ with continuous roots f_n on an arcwise connected group G. We know from lemma IV.2 that the logarithm v of f is given by the following limit

$$v = \lim_{n\to\infty} n(f_n - 1) \tag{4.21}$$

From this expression it is clear that v is conditionally positive definite (i.e. (4.1) holds). This notion of conditional positive definiteness for functions generalizes in a natural way to the notion of conditional complete positivity for mappings:

Definition IV.8

A mapping $\Psi : G \to L_D(H)$ (= set of possibly unbounded linear operators on some hilbertspace H which have some common dense domain $D \subset H$) is said to be *conditionally completely positive* if

$$\sum_{g,g'} \langle \xi_g | \Psi(g^{-1}g') \xi_{g'} \rangle \geq 0$$

for all choices of $g \to \xi_g \in D$ such that $\sum_g \xi_g = 0$

Now we want to mimic (4.21) to construct a logarithm for a continuous infinitely divisible C.P. triplet (H,Φ,Ω) with continuous roots $(H^{1/n},\Phi^{1/n},\Omega^{1/n})$.

In a first naive attempt one could consider the following limit

$$\text{w-}\lim_{n\to\infty} n(\Phi^{1/n}(g) - \mathbb{1})$$

where this limit is meant in the weak sense (cfr. central limit). However, one immediately sees that this attempt fails. Indeed, consider for instance:

$$\langle \Omega^{1/n} | n(\Phi^{1/n}(g) - \mathbb{1}) n(\Phi^{1/n}(h) - \mathbb{1}) \Omega^{1/n} \rangle$$
$$= n^2 [\langle \Omega^{1/n} | \Phi^{1/n}(g) \Phi^{1/n}(h) \Omega^{1/n} \rangle - \langle \Omega^{1/n} | \Phi^{1/n}(g) \Omega^{1/n} \rangle$$
$$\langle \Omega^{1/n} | \Phi^{1/n}(h) \Omega^{1/n} \rangle + 1]$$

Using the function $d : \tilde{G} \to \mathbb{C}$ introduced in corollary IV.4 this can be written as

$$n^2 [\exp d_{(g,h)}/n - \exp d_{(g)}/n - \exp d_{(h)}/n + 1]$$

and clearly this has only a finite limit if $d_{(g,h)} = d_{(g)} + d_{(h)}$ which is generally not true as soon as Φ is not a complex valued function.

In the next lemma it will become clear that the correct generalization of (4.21) for C.P. mappings is given by

$$\text{w-}\lim_{n\to\infty} \Gamma_n(g) \tag{4.22}$$

where

$$\Gamma_n(g) = (\Phi^{1/n}(g) - \mathbb{1}) \otimes \mathbb{1} \otimes \ldots \otimes \mathbb{1} + \mathbb{1} \otimes (\Phi^{1/n}(g) - \mathbb{1}) \otimes \ldots \otimes \mathbb{1} + \ldots$$
$$\ldots + \mathbb{1} \otimes \ldots \otimes \mathbb{1} \otimes (\Phi^{1/n}(g) - \mathbb{1}) \tag{4.23}$$

Remark that in the special case of functions (i.e. $H = \mathbb{C}$) (4.22) is identical to (4.21)

Lemma IV.9

Let (H,Φ,Ω) be a continuous infinitely divisible C.P. triplet with continuous roots $(H^{1/n},\Phi^{1/n},\Omega^{1/n})$ on an arcwise connected group G. Then using the notation (4.23),

(i) $f(\Delta) \equiv \lim_{n\to\infty} \langle \otimes^n \Omega^{1/n} | \Gamma_n(\Delta) \otimes^n \Omega^{1/n} \rangle$ exists

$$\text{and } f(\Delta) = \sum_{p\in\mathcal{P}_\Delta} \sum_{\Lambda\in p} \prod_{X\subset\Lambda} (-1)^{\#(\Lambda\setminus X)} d_X \tag{4.24}$$

where $\mathcal{P}_\Delta$ is the set of ordered partition p of Δ into non empty sets Λ and where d_X is given by (4.12)

(ii) $\tilde{f}(\Delta,\Delta') = f(\Delta^* \times \Delta')$ is a positive kernel on $\tilde{G}$.

Remark: If Φ is a function (i.e. $H = \mathbb{C}$), then $f(\Delta) = \prod_{g \in \Delta} v(g)$, where $\Phi = e^v$

Proof

(i) Using the notation

$$\eta_n(g) = \Phi^{1/n}(g) - 1\!\!1 \qquad (4.25)$$

we can write

$$f(\Delta) = \lim_{n\to\infty} \langle \otimes^n \Omega^{1/n} | \prod_{g\in\Delta}^{\rightarrow} \sum_{j=1}^{n} (1\!\!1 \otimes \ldots \otimes \eta_n(g)_j \otimes \ldots 1\!\!1) \otimes^n \Omega^{1/n} \rangle$$

$$= \lim_{n\to\infty} \sum_{p\in\mathcal{P}_\Delta} \frac{n!}{(n-\#(p))!} \prod_{\Lambda\in p} \langle \Omega^{1/n} | \eta_n(\Lambda)\Omega^{1/n} \rangle$$

$$= \lim_{n\to\infty} \sum_{p\in\mathcal{P}_\Delta} \frac{n!}{(n-\#(p))!} \prod_{\Lambda\in p} \sum_{X\subset\Lambda} (-1)^{\#(\Lambda\setminus X)} \langle \Omega^{1/n} | \Phi^{1/n}(X)\Omega^{1/n} \rangle$$

Now, as $\Lambda \in p$ is a non-empty set, we have $\sum_{X\subset\Lambda} (-1)^{\#(\Lambda\setminus X)} = 0$ and we can, using (4.14), rewrite $f(\Delta)$ as follows

$$f(\Delta) = \lim_{n\to\infty} \sum_{p\in\mathcal{P}_\Delta} \frac{n!}{(n-\#(p))!} \prod_{\Lambda\in p} \sum_{X\subset\Lambda} (-1)^{\#(\Lambda\setminus X)} [\exp(d_X/n) - 1]$$

$$= \sum_{p\in\mathcal{P}_\Delta} \prod_{\Lambda\in p} \sum_{X\subset\Lambda} (-1)^{\#(\Lambda\setminus X)} d_X$$

which proves (i). ∎

(i) The proof of (ii) is straightforward.

As will be seen in the next theorem, the following notions arise naturally in the study of logarithms of infinitely divisible C.P. triplets.

Definition IV.10

(i) We call (H,Ψ,Ω°) a *conditionally completely positive (C.C.P.) triplet* on a group G if $\Psi : G \to L(K)$ is a mapping from G into the (possibly unbounded) linear operators on a hilbertspace K and $\Omega^\circ \in K$ is a normalized vector such that:

- $\Omega^\circ \in D(\Psi(g))$ and $\Psi(\Delta)\Omega^\circ \in D(\Psi(g))$ for all $g \in G$; $\Delta \in \tilde{G}$ ($D(\Psi(g))$ is the domain of $\Psi(g)$)
- $D = \text{span}\{\Psi(\Delta)\Omega^\circ | \Delta \in \tilde{G}\}$ is dense in K
- $\Psi^\cdot : G \to L_D(K)$ is conditionally completely positive in the sense of definition IV.8

(ii) A C.C.P. triplet (K,Ψ,Ω°) is *continuous* if $g \to \langle \xi | \Psi(g)\eta \rangle$ is continuous for all $\xi,\eta \in D$

(iii) A C.C.P. triplet (K,Ψ,Ω°) is called *hermitian* if $\Psi(g^{-1}) \subset \Psi(g)^*$

(iv) A C.C.P. triplet (K,Ψ,Ω°) on G is called *infinitely additive* if for all $n \in \mathbb{N}_\circ$, there exists a C.C.P. triplet $(K_n,\Psi_n,\Omega_n^\circ)$ on G and an isometry $U_n : K \to \otimes^n K_n$ such that

- $U_n D \subset \otimes^n D_n$ where $D_n^\circ = \text{span}\{\Psi_n(\Delta)\Omega_n^\circ | \Delta \in \tilde{G}\}$
- $U_n \Omega^\circ = \otimes^n \Omega_n^\circ$
- $\Psi(g) = U_n(\Psi_n(g) \otimes 1\!\!1 \otimes \ldots \otimes 1\!\!1 + \ldots + 1\!\!1 \otimes \ldots \otimes 1\!\!1 \otimes \psi_n(g))U_n$ (on D) (4.26)

We call (K_n,Ψ_n,Ω_n) an nth *part* of (K,Ψ,Ω°)

Remark

If $K = \mathbb{C}$ and $\Omega° = 1 \in \mathbb{C}$ we recover the usual definition of a conditionally positive definite function $\Psi : G \to \mathbb{C}$. Notice that in this case the notion of infinite additivity trivializes. Indeed, take $K_n = \mathbb{C}$, $\Omega°_n = 1 \in \mathbb{C}$, U_n the isomorphism between $\mathbb{C}$ and $\otimes^n \mathbb{C}$ and

$$\Psi_n(g) = \Psi(g)/n \tag{4.27}$$

the clearly (4.26) holds.

Before we can prove in the next theorem the existence of a logarithm for a infinitely divisible C.P. triplet and list all the properties of that logarithm, we still need another technical tool, namely a non-commutative version of the cumulants :

Definition IV.11

Given a mapping $\Psi : G \to L(K)$ and a normalized vector $\Omega° \in K$ such that $\Omega°$ and $\psi(\Delta)\Omega°$ belong to domain of $\Psi(g)$ for all $g \in G$, $\Delta \in \tilde{G}$, the *cumulants* P^{Ψ}_{Λ}, $\Lambda \in \tilde{G}$ of Ψ with respect to $\Omega°$ are uniquely determined by the following propositions

- $P^{\Psi}_{\emptyset} = 0$
- P^{Ψ}_{Λ} is a polynomial in $\langle\Omega°|\Psi(X)\Omega°$, $X \in \Lambda$, homogeneous of degree $\#(\Lambda)$ (where $\deg\langle\Omega°|\Psi(X)\Omega°\rangle = \#(X)$), with coefficients independent of Ψ and $\Omega°$.
- $$\langle\Omega°|\Psi(\Delta)\Omega°\rangle = \sum_{p \in \mathcal{P}_{\Delta}} \prod_{\Lambda \in p} P^{\Psi}_{\Lambda} \tag{4.28}$$

  where $\mathcal{P}_{\Delta}$ was defined in lemma IV.9

Remarks

1) This definition allows us to compute P^{Ψ}_{Λ} by induction on $\#(\Lambda)$ e.g.

$$P^{\Psi}_{(g)} = \langle\Omega°|\Psi(g)\Omega°\rangle$$

$$P^{\Psi}_{(g,h)} = \langle\Omega°|\Psi(g)\Psi(h)\Omega°\rangle - \langle\Omega°|\Psi(g)\Omega°\rangle\langle\Omega°|\Psi(h)\Omega°\rangle$$

2) P^{Ψ}_{Λ} generalizes the usual cumulants P_n which are defined by

$$\langle\Omega|e^{itA}\Omega\rangle = \exp \sum_{n \geq 1} \frac{(it)^n}{n!} P_n \quad , \quad A \in L(K)$$

Indeed, take $\Psi(g) = A$ for all $g \in G$, then $P_n = P^{\Psi}_{\Lambda}$ whenever $\#(\Lambda) = n$.

3) If Ψ is a complex valued function, then $P^{\Psi}_{\Lambda} = 0$ for all Λ with $\#(\Lambda) > 1$.

Lemma IV.12

If $(K, \Psi, \Omega°)$ is an infinitely additive C.C.P. triplet with n^{th} parts $(K_n, \Psi_n, \Omega°_n)$ on a group G, then the cumulants of Ψ w.r.t. $\Omega°$ satisfy

$$\text{(i)} \quad P^{\Psi}_{\Delta} = \lim_{n \to \infty} n \langle\Omega°_n|\Psi_n(\Delta)\Omega°_n\rangle , \qquad \Delta \in \tilde{G}\setminus\{\emptyset\} \tag{4.29}$$

$$\text{(ii)} \quad P^{\Psi_n}_{\Delta} = \frac{1}{n} P^{\Psi}_{\Delta}, \qquad \Delta \in \tilde{G} \tag{4.30}$$

Proof

First we prove the existence of the limit in (4.29) by induction on $\#(\Delta)$. For $\#(\Delta) = 1$, say $\Delta = (g)$, we have by infinite addivity of $(K, \Psi, \Omega°)$ that $n\langle\Omega°_n|\Psi_n(g)\Omega°_n\rangle = \langle\Omega°|\Psi(g)\Omega°\rangle$. Suppose now that the limit exists for all $\Delta \in \tilde{G}$ with $\#(\Delta) \leq m$. Take then a $\Delta \in \tilde{G}$ with $\#(\Delta) = m+1$. Now note that by infinite additivity we have for all $\Delta \in \tilde{G}$

$$\langle\Omega^\circ|\Psi(\Delta)\,\Omega^\circ\rangle = \langle\otimes^n\Omega^\circ_n|\ \prod_{g\in\Delta}^{\rightarrow}\ (\sum_{j=1}^{n} 1\!\!1\otimes\dots\otimes\psi_n(g)_j\otimes\dots\otimes 1\!\!1)\ \otimes^n\Omega^\circ_n\rangle$$

$$= \sum_{p\in\mathcal{P}_\Delta}\frac{n!}{(n-\#(p))!}\prod_{\Lambda\in p}\langle\Omega^\circ|\Psi^\circ_n(\Lambda)\Omega^\circ_n\rangle \tag{4.31}$$

Hence

$$n\,\langle\Omega^\circ_n|\Psi_n(\Delta)\Omega^\circ_n\rangle = \langle\Omega^\circ|\Psi(\Delta)\,\Omega^\circ\rangle - \sum_{\substack{p\in\mathcal{P}_\Delta\\ p\neq\{\Delta\}}}\frac{n!}{(n-\#(p))!}\prod_{\Lambda\in p}\langle\Omega^\circ_n|\psi_n(\Lambda)\Omega^\circ_n\rangle \tag{4.32}$$

The limit for $n\to\infty$ of the right hand side of (4.32) exists by the induction hypothesis since all Λ's that appear have $\#(\Lambda)\leqslant m$; so the limit of the left hand side of (4.32) must exist as well. Denoting this limit by P^Ψ_Λ and taking the limit $n\to\infty$ of (4.31) it is also clear that (4.28) holds.

Moreover, again by an induction argument on $\#(\Delta)$ it follows immediately from (4.32) that P^Ψ is homogeneous polynomial of degree $\#(\Delta)$ in $\langle\Omega^\circ|\Psi(X)\Omega^\circ\rangle$, $X\subset\Delta$. Hence, the $P^{\Psi,\Delta}_\Delta$'s given by (4.29) are actually the cumulants of Ψ w.r.t. Ω°.

(ii) It is straightforwardly checked that $(K_n, \Psi_n, \Omega^\circ_n)$ is infinitely additive and that its kth parts are given by $(K_{nk}, \Psi_{nk}, \Omega_{nk})$. So by (i):

$$P^{\Psi_n}_\Delta = \lim_{k\to\infty} k\langle\Omega^\circ_{nk}|\Psi_{nk}(\Delta)\Omega^\circ_{nk}\rangle = \frac{1}{n}\lim_{k\to\infty} nk\langle\Omega^\circ_{nk}|\Psi_{nk}(\Delta)\Omega^\circ_{nk}\rangle = \frac{1}{n}P^\Psi_\Delta$$

■

<u>Theorem IV.13</u>

Let (H,Φ,Ω) be a continuous infinitely divisible C.P. triplet with continuous roots on an arcwise connected group G. Using the notation of lemma IV.9, let $K = \mathrm{hil}(\tilde{G},\tilde{f})$.

(i) For all $g\in G$

$$\Psi(g)\ :\ \Delta\in K \to (g)\times\Delta\in K \tag{4.23}$$

defines a linear operator on the dense subspace $D = \mathrm{span}\{\Delta\,|\,\Delta\in\tilde{G}\}$ of K

(ii) Put $\Omega^\circ = \emptyset\in K$, then (K,Ψ,Ω°) is a continuous hermitian C.C.P. triplet for which $\Psi(e) = 0$

(iii) (K,Ψ,Ω°) is infinitely additive and has continuous hermitian n^{th} parts

(iv) (K,Ψ,Ω°) satisfies an additional positivity condition:

$$((g,\Delta),(g'\Delta')) \to \sum_{\Lambda\subset\Delta^*\times(g^{-1}g')\times\Delta'} P^\Psi_\Lambda \tag{4.34}$$

is a conditionally positive kernel on $G\times\tilde{G}$

<u>Proof</u>

(i) If $\sum_\Delta\lambda_\Delta\Delta = 0$ in K, then $\sum_\Delta\lambda_\Delta(g)\times\Delta = 0$ as well since

$$\|\sum_\Delta\lambda_\Delta(g)\times\Delta\|^2 = \sum_{\Delta,\Delta'}\bar{\lambda}_\Delta\lambda_{\Delta'}\,f(\Delta^*\times(g^{-1},g)\times\Delta')$$

$$= \sum_{\Delta,\Delta'}\bar{\lambda}_\Delta\lambda_{\Delta'}\ f(((g^{-1},g)\times\Delta)^*\times\Delta')$$

$$= \langle\sum_\Delta\lambda_\Delta(g^{-1},g)\times\Delta\ |\ \sum_{\Delta'}\lambda_{\Delta'}\Delta'\rangle = 0$$

Hence $\Psi(g)$ is well defined by (4.33)

(ii) By construction $\Omega^\circ = \emptyset\in D(\Psi(g))$ and $\Delta = \Psi(\Delta)\,\Omega^\circ\in D(\Psi(g))$. Also D =

$= \text{span } \{\Psi(\Delta)\Omega^\circ \mid \Delta \in \tilde{G}\}$ is dense in K.

Furthermore, for all $g \to \xi_g = \sum_\Delta \lambda_{g,\Delta} \Delta \in D$ with $\sum_g \xi_g = 0$, we have, using notation (4.23)

$$\sum_{g,g'} \langle \xi_g \mid \Psi(g^{-1}g')\xi_{g'} \rangle$$

$$= \sum_{\substack{g,g' \\ \Delta,\Delta'}} \bar{\lambda}_{g,\Delta} \lambda_{g',\Delta'} f(\Delta^* \times (g^{-1}g') \times \Delta')$$

$$= \lim_{k\to\infty} \sum_{\substack{g,g' \\ \Delta,\Delta'}} \bar{\lambda}_{g,\Delta} \lambda_{g',\Delta'} \langle \Gamma_k(\Delta) \otimes^k \Omega^{1/k} \mid \Gamma_k(g^{-1}g')\Gamma_k(\Delta') \otimes^k \Omega^{1/k} \rangle$$

$$= \lim_{k\to\infty} \sum_{\substack{g,g' \\ \Delta,\Delta'}} \bar{\lambda}_{g,\Delta} \lambda_{g',\Delta'} \langle \Gamma_k(\Delta) \otimes^k \Omega^{1/k} \mid \sum_{j=1}^{k} (1\otimes\ldots\otimes\Phi^{1/k}(g^{-1}g')_j \otimes\ldots\otimes 1)\Gamma_k(\Delta') \otimes^k \Omega^{1/k} \rangle$$

$$- \lim_{k\to\infty} \sum_{\substack{g,g' \\ \Delta,\Delta'}} \bar{\lambda}_{g,\Delta} \lambda_{g',\Delta'} \langle \Gamma_k(\Delta) \otimes^k \Omega^{1/k} \mid \Gamma_k(\Delta') \otimes^k \Omega^{1/k} \rangle$$

The first term is positive by complete positivity of $\Phi^{1/k}$, whereas the second term tends to $\sum_{\substack{g,g' \\ \Delta,\Delta'}} \bar{\lambda}_{g,\Delta} \lambda_{g',\Delta'} f(\Delta^* \times \Delta') = \| \sum_g \xi_g \|^2 = 0$

Hence (K, Ψ, Ω°) is a C.C.P. triplet.

From (4.24) and the continuity of $X \in \tilde{G} \to d_X \in \mathbb{C}$, it follows that

$$g \in G \to f(\Delta^* \times (g) \times \Delta') = \langle \Delta \mid \Psi(g)\Delta' \rangle$$

is continuous for all $\Delta, \Delta' \in \tilde{G}$. Hence (K, Ψ, Ω°) is continuous.

Moreover, as $\langle \Delta \mid \psi(g)\Delta' \rangle = f(\Delta^* \times (g) \times \Delta') = f(((g^{-1}) \times \Delta)^* \times \Delta') = \langle \Psi(g^{-1})\Delta \mid \Delta' \rangle$, (K, Ψ, Ω°) is hermitian.

Finally, since $f(\Delta) = 0$ as soon as $e \in \Delta$, it is clear that $\Psi(e) = 0$

(iii) Let $(H^{1/n}, \Phi^{1/n}, \Omega^{1/n})$ be the continuous n^{th} root of (H, Φ, Ω). Because $(H^{1/n}, \Phi^{1/n}, \Omega^{1/n})$ is also infinitely divisible and has continuous roots, we can by (i) and (ii) construct with it a continuous hermitian C.C.P. triplet which we denote by $(K_n, \Phi_n, \Omega^\circ_n)$. We prove that $(K_n, \Psi_n, \Omega^\circ_n)$ is the nth part of (K, Ψ, Ω°).

Note therefore that (use notation (4.25))

$$\langle \Psi(\Delta)\Omega^\circ \mid \Psi(\Delta')\Omega^\circ \rangle$$

$$= \lim_{k\to\infty} \langle \otimes^k \Omega^{1/k} \mid \Gamma_k(\Delta^* \times \Delta') \otimes^k \Omega^{1/k} \rangle$$

$$= \lim_{k\to\infty} \langle \otimes^{kn} \Omega^{1/kn} \mid \Gamma_{kn}(\Delta^* \times \Delta') \otimes^{kn} \Omega^{1/kn} \rangle$$

$$= \lim_{k\to\infty} \langle \otimes^{kn} \Omega^{1/kn} \mid \overset{\to}{\prod_{g \in \Delta^* \times \Delta'}} \sum_{\ell=0}^{n-1} \Big(\sum_{j=\ell k+1}^{(\ell+1)k} 1\otimes\ldots\otimes\eta_{kn}(g)_j \otimes\ldots\otimes 1 \Big) \otimes^{kn} \Omega^{1/kn} \rangle$$

$$= \lim_{k\to\infty} \sum_{p \in \mathcal{P}_{\Delta^* \times \Delta'}} \frac{n!}{(n-\#(p))!} \prod_{\Lambda \in p} \langle \otimes^k \Omega^{1/kn} \mid \overset{\to}{\prod_{g \in \Lambda}} \sum_{j=1}^{k} (1\otimes\ldots\otimes\eta_{kn}(g)_j \otimes\ldots\otimes 1) \otimes^k \Omega^{1/kn} \rangle$$

$$= \lim_{k\to\infty} \sum_{p \in \mathcal{P}_{\Delta^* \times \Delta'}} \frac{n!}{(n-\#(p))!} \prod_{\Lambda \in p} \langle \otimes^k \Omega^{1/kn} \mid \Gamma_{kn}(\Lambda) \otimes^k \Omega^{1/kn} \rangle$$

$$= \sum_{p \in \mathcal{P}_{\Delta^*\times\Delta'}} \frac{n!}{(n-\#(p))!} \prod_{\Lambda \in p} \langle \Omega_n^\circ | \Psi_n(\Lambda)\Omega_n^\circ \rangle$$

$$= \langle \otimes^n \Omega_n^\circ | \prod_{g\in\Delta^*\times\Delta'}^{\rightarrow} \left(\sum_{j=1}^{n} \mathbb{1}\otimes\ldots\otimes\Psi_n(g)_j\otimes\ldots\otimes\mathbb{1}\right) \otimes^n \Omega_n^\circ \rangle$$

$$= \langle \prod_{g\in\Delta}^{\rightarrow} \left(\sum_{j=1}^{n} \mathbb{1}\otimes\ldots\otimes\Psi_n(g)_j\otimes\ldots\otimes\mathbb{1}\right) \otimes^n \Omega_n^\circ | \prod_{g'\in\Delta'}^{\rightarrow} \left(\sum_{j=1}^{n} \mathbb{1}\otimes\ldots\otimes\Psi_n(g')\otimes\ldots\otimes\mathbb{1}\right)\otimes^n \Omega_n^\circ \rangle$$

This implies that the mapping

$$\Psi(\Delta)\Omega^\circ \in K \rightarrow \prod_{g\in\Delta} \left(\sum_{j=1}^{n} \mathbb{1}\otimes\ldots\otimes\Psi_n(g)_j\otimes\ldots\mathbb{1}\right)\otimes^n \Omega_n^\circ \in \otimes^n K_n$$

is well defined and can be extended to an isometry $U_n : K \rightarrow \otimes^n K_n$. It is now clear that (4.26) is satisfied. Hence (K, Ψ, Ω°) is infinitely additive and has continuous n^{th} parts.

(iv) By comparing (4.28) and (4.24), recalling that $f(\Delta) = \langle\Omega^\circ|\Psi(\Delta)\Omega^\circ\rangle$ and observing that $P^\Psi_\emptyset = d_\emptyset = 0$, we have for all $\Lambda \in \tilde{G}$

$$P^\Psi_\Lambda = \sum_{X\subset\Lambda} (-1)^{\#(\Lambda\setminus X)} d_X \tag{4.35}$$

So, summing (4.25) over $\Lambda \subset \Delta$ one gets

$$\sum_{\Lambda\subset\Delta} P^\Psi_\Lambda = \sum_{\Lambda\subset\Delta}\sum_{X\subset\Lambda} (-1)^{\#(\Lambda\setminus X)} d_X = \sum_{X\subset\Delta}\left(\sum_{X\subset\Lambda\subset\Delta} (-1)^{\#(\Lambda\setminus X)}\right) d_X$$

Since $\sum_{X_1\subset\Lambda\subset X_2} (-1)^{\#(\Lambda)} = 0$ if $X_1 \neq X_2$, we end up with

$$d_\Delta = \sum_{\Lambda\subset\Delta} P^\Psi_\Lambda \tag{4.36}$$

Hence in order to prove (iv), we have to show that $((g,\Delta),(g',\Delta')) \rightarrow d_{\Delta^*\times(g^{-1}g')\times\Delta'}$ is a conditionally positive kernel. But this follows immediately from the fact it is hermitian and that its exponential (i.e. $((g,\Delta),(g'\Delta') \rightarrow \langle\Omega\,|\,\Phi(\Delta^*\times(g^{-1}g')\times\Delta)\,\Omega\rangle$) is a positive kernel.

∎

<u>Remark</u>

In the case of conditionally positive definite functions $\Psi : G\rightarrow\mathbb{C}$ the propositions (iii) and (iv) of the preceding theorem are trivially satisfied, since then the notion of infinite additivity trivialises (see 4.27), and the additional positivity condition turns out to be equivalent with conditional positive definiteness of Ψ itself. However, for mappings ($\dim K \geqslant 1$) (iii) and (iv) are non-trivial properties.

<u>Definition IV.14</u>

Let (H,Φ,Ω) and (K,Ψ,Ω°) be as in theorem IV.13. We call (K,Ψ,Ω°) the *logarithm* of (H,Φ,Ω). Notation: $(K,\Psi,\Omega^\circ) = \ell n(H,\Phi,\Omega)$

Clearly in the special case of a continuous infinitely divisible complex valued function $f = e^v$, we recover the usual definition of the logarithm: $\ell n(\mathbb{C},f,1) = (\mathbb{C},v,1)$.

As we have now found an infinitely additive C.C.P. triplet as a logarithm for an infinitely divisible C.P. triplet, an obvious question arises: can we conversely "exponentiate" an infinitely additive C.C.P. triplet in some way, to end up with an infinitely divisible C.P. triplet. This will be the problem we will solve in the rest of this contribution.

Again, to find a way to construct an exponential of a hermitian infinitely additive C.C.P. triplet, we consider the special case of complex valued functions. If $v : G \to \mathbb{C}$ is a conditionally positive definite function with $v(e) = 0$ and $\overline{v(g)} = v(g^{-1})$, then

$$e^{v} = \lim_{n\to\infty} \left(1 + \frac{v}{n}\right)^n \tag{4.37}$$

is an infinitely divisible normalized positive definite function. If (K,Ψ,Ω°) is a hermitian infinitely additive C.C.P. triplet with n^{th} part $(K_n,\Psi_n,\Omega^\circ_n)$, it is clear from (4.27) that $\frac{v}{n}$ should be replaced by Ψ_n. Therefore we generalize (4.37) by

$$\text{w-}\lim_{n\to\infty} \otimes^n (\mathbb{1} + \Psi_n(g)).$$

<u>Lemma IV.15</u>

Let (K,Ψ,Ω°) be a hermitian infinitely additive C.C.P. triplet with n^{th} parts $(K_n,\Psi_n,\Omega^\circ_n)$ on a group G

(i) $F(\Delta) \equiv \lim_{n\to\infty} \langle \otimes^n \Omega^\circ_n | \overrightarrow{\prod_{g\in\Delta}} (\otimes^n(\mathbb{1} + \Psi_n(g)) \otimes^n \Omega^\circ_n \rangle$

exists and

$$F(\Delta) = \exp \sum_{\Lambda\subset\Delta} P^{\Psi}_{\Lambda} \tag{4.38}$$

(ii) $\tilde{F}(\Delta,\Delta') = F(\Delta^{*}\times\Delta')$ is a positive kernel on $\tilde{G}$

<u>Proof</u>

(i) We have

$$\langle \otimes^n \Omega^\circ_n | \overrightarrow{\prod_{g\in\Delta}} \otimes^n (\mathbb{1} + \Psi_n(g)) \otimes^n \Omega^\circ_n \rangle$$

$$= \langle \Omega^\circ_n | \overrightarrow{\prod_{g\in\Delta}} (\mathbb{1} + \Psi_n(g)) \Omega^\circ_n \rangle^n$$

$$= \Big[\sum_{\Lambda\subset\Delta} \langle \Omega^\circ_n | \Psi_n(\Lambda) \Omega^\circ_n \rangle \Big]^n$$

$$= \Big[1 + \frac{1}{n} \sum_{\substack{\Lambda\subset\Delta \\ \Lambda\neq\emptyset}} n \langle \Omega^\circ_n | \Psi_n(\Lambda)\, \Omega^\circ_n \rangle \Big]^n$$

Now use (4.29) and the fact that $P^{\Psi}_{\emptyset} = 0$ to get (4.38)

(ii) follows straightforwardly.

<u>Theorem IV.16</u>

Let (K,Ψ,Ω°) be a hermitian infinitely additive C.C.P. triplet on a group G such that $\Psi(e) = 0$ and the additional positivity condition (4.34) is satisfied. Using the notation of lemma IV.16, let $H = \text{hil}\,(\tilde{G},\tilde{F})$.

(i) For all $g \in G$, $\Phi(g) : \Delta \in H \to (g)\times\Delta \in H$ defines a bounded linear operator on H

(ii) Put $\Omega = \emptyset \in H$, then (H,Φ,Ω) is an infinitely divisible C.P. triplet on G

(iii) If (K,Ψ,Ω°) is continuous and has continuous n^{th} parts, then also (H,Φ,Ω) is continuous and it has continuous roots.

Proof

(i) If $\sum_\Delta \lambda_\Delta \Delta = 0$ in H, then $\sum_\Delta \lambda_\Delta((g)\times\Delta) = 0$ as well, because

$$\|\sum_\Delta \lambda_\Delta((g)\times\Delta)\|^2 = \sum_{\Delta,\Delta'} \bar\lambda_\Delta \lambda_{\Delta'} F(\Delta^*\times(g^{-1},g)\times\Delta')$$

$$= \langle \sum_\Delta \lambda_\Delta \Delta | \sum_{\Delta'} \lambda_{\Delta'} (g^{-1},g)\times\Delta'\rangle = 0$$

Hence $\Phi(g)$ is well defined as a linear operator on the dense subspace $D = \operatorname{span}\{\Delta | \Delta \in G\}$. Since $F(\Delta^*\times(g)\times\Delta') = F(((g^{-1})\times\Delta)^*\times\Delta')$, it is clear that $\Phi(g^{-1}) \subset \Phi(g)^*$.

Moreover, as $\Psi(e) = 0$ we have $\Psi_n(e) = 0$ and so

$$\langle\Delta|\Phi(e)\Delta'\rangle = F(\Delta^*\times(e)\times\Delta')$$

$$= \lim_{n\to\infty} \langle\Omega^\circ_n| \prod^{\rightarrow}_{g\in\Delta^*}(\mathbb{1}+\Psi_n(g))(\mathbb{1}+\Psi_n(e))\prod^{\rightarrow}_{g\in\Delta'}(\mathbb{1}+\Psi_n(g'))\Omega^\circ_n\rangle^n$$

$$= \lim_{n\to\infty} \langle\Omega^\circ_n| \prod^{\rightarrow}_{g\in\Delta^*\times\Delta'}(\mathbb{1}+\Psi_n(g))\Omega^\circ_n\rangle^n$$

$$= F(\Delta^*\times\Delta') = \langle\Delta|\Delta'\rangle$$

which means that $\Phi(e) = \mathbb{1}$.

To prove boundedness of $\Phi(g)$, we first show complete positivity of $g \to \Phi(g)$ on D. Let $\xi_g = \sum_\Delta \lambda_{g,\Delta}\Delta \in D$, then

$$\sum_{g,g'} \langle\xi_g|\Phi(g^{-1}g')\xi_{g'}\rangle$$

$$= \sum_{\substack{g,g'\\ \Delta,\Delta'}} \bar\lambda_{g,\Delta}\lambda_{g',\Delta'} F(\Delta^*\times(g^{-1}g')\times\Delta)$$

$$= \sum_{\substack{g,g'\\ \Delta,\Delta'}} \bar\lambda_{g,\Delta}\lambda_{g',\Delta}\exp \sum_{\Lambda\subset\Delta^*\times(g^{-1}g')\times\Delta'} P^\Psi_\Lambda \geqslant 0 \tag{4.39}$$

where the inequality follows from the additional positivity property of Ψ and the fact that the exponential of a hermitian conditionally positive kernel is positive.

Now (4.39) implies for all $\xi,\eta \in D$

$$\langle\xi|\Phi(g)\eta\rangle + \langle\Phi(g)\eta|\xi\rangle + \langle\xi|\xi\rangle + \langle\eta|\eta\rangle \geqslant 0$$

Take $\|\xi\| = \|\eta\| = 1$ and multiply ξ and η with an appropriate phase factor to get that $|\langle\xi|\Phi(g)\eta\rangle| \leqslant 1$ for all normalized $\xi,\eta \in D$ and since D is dense in H this implies $\|\Phi(g)\| \leqslant 1$.

(ii) In the proof of (i) we have already shown that $g \to \Phi(g)$ is C.P. on D and by continuity also on the whole of H. By construction, $\|\Omega\| = 1$ and Ω is cyclic for $\{\Phi(G)\}''$. Hence (H,Φ,Ω) is a C.P. triplet.

Now we show that it is infinitely divisible. The nth part $(K_n,\Psi_n,\Omega^\circ_n)$ of (K,Ψ,Ω°) is clearly infinitely additive as well and by (4.30) it also satisfies the additional positivity condition (4.34). Therefore we can construct with it a C.P. triplet $(H^{1/n},\Phi^{1/n},\Omega^{1/n})$ in the same way as (H,Φ,Ω) was made out of (K,Ψ,Ω°). It can easily be seen that $(H^{1/n},\Phi^{1/n},\Omega^{1/n})$ is an nth root for (H,Φ,Ω). Indeed, notice that

$$
\begin{aligned}
<\Omega|\,\Phi(\Delta)\,\Omega> &= F(\Delta)\\
&= \exp \sum_{\Lambda\subset\Delta} P_\Lambda^{\Psi}\\
&= \exp n \sum_{\Lambda\subset\Delta} P_\Lambda^{\Psi_n}\\
&= F_n(\Delta)^n\\
&= <\Omega^{1/n}|\,\Phi^{1/n}(\Delta)\,\Omega^{1/n}> = <\otimes^n\Omega^{1/n}|\otimes^n\Phi(\Delta)\,\otimes^n\Omega^{1/n}>
\end{aligned}
$$

Hence $(H,\Phi,\Omega) \cong \otimes^n (H^{1/n},\Phi^{1/n},\Omega^{1/n})$

(iii) The continuity of (K,Ψ,Ω°) and (K,Ψ_n,Ω_n) yields the continuity of $g \to F(\Delta^* \times (g) \times \Delta')$ and $g \to F_n(\Delta^* \times (g) \times \Delta')$ and this clearly implies the continuity of Φ and $\Phi^{1/n}$.

Definition IV.17

Let (K,Ψ,Ω°) and (H,Φ,Ω) be as in theorem IV.16. We call (H,Φ,Ω) the *exponential* of (K,Ψ,Ω°). Notation : $(H,\Phi,\Omega) = \exp(K,\Psi,\Omega^\circ)$.

Remark

In the special case of a conditionally positive definite function $v : G \to \mathbb{C}$ we have $\exp(\mathbb{C},v,1) = (\mathbb{C},e^{v},1)$.

The logarithmic construction of theorem IV.13 and the exponential of theorem IV.16 are mutually inverse. In fact we have:

Theorem IV. 18

(i) If (H,Φ,Ω) is a continuous infinitely divisible C.P. triplet with continuous roots on an arcwise connected group, we have

$$\exp(\ln(H,\Phi,\Omega)) = (H,\Phi,\Omega)$$

(up to unitary equivalence).

(ii) If (K,Ψ,Ω°) is a continuous hermitian infinitely additive C.C.P. triplet on an arcwise connected group, satisfying $\Psi(e) = 0$ and the additional positivity condition (4.34) and having continuous parts, then we have

$$\ln(\exp(K,\Psi,\Omega^\circ)) = (K,\Psi,\Omega^\circ)$$

(up to unitary equivalence).

Proof

(i) Let $(K,\Psi,\Omega^\circ) = \ln(H,\Phi,\Omega)$ and $(\tilde H,\tilde\Phi,\tilde\Omega) = \exp(K,\Psi,\Omega^\circ)$. Then using (4.38) and (4.36) we have

$$<\tilde\Omega|\,\tilde\Phi(\Delta)\,\tilde\Omega> = \exp \sum_{\Lambda\subset\Delta} P_\Lambda^{\Psi} = \exp d_\Delta = <\Omega\,|\,\Phi(\Delta)\,\Omega>$$

Hence $(\tilde H,\tilde\Phi,\tilde\Omega) \cong (H,\Phi,\Omega)$.

(ii) Let $(H,\Phi,\Omega) = \exp(K,\Psi,\Omega^\circ)$ and $(\tilde K,\tilde\Psi,\tilde\Omega^\circ) = \ln(H,\Phi,\Omega)$. Let $d:\tilde G \to \mathbb{C}$ be the function satisfying $<\Omega|\,\Phi(X)\,\Omega> = \exp d_X$. Then it follows from the construction of Φ and (4.38) that

$$d_X = \sum_{Y\subset X} P_Y^{\Psi}$$

Hence, using (4.24) and (4.28) one gets

$$<\tilde{\Omega}^\circ|\tilde{\Psi}(\Delta)\tilde{\Omega}^\circ> = \sum_{p\in\mathcal{P}_\Delta} \prod_{\Lambda\in p} \sum_{X\subset\Lambda} (-1)^{\#(\Lambda\setminus X)} d_X$$

$$= \sum_{p\in\mathcal{P}_\Delta} \prod_{\Lambda\in p} \sum_{X\subset\Lambda} (-1)^{\#(\Lambda\setminus X)} \sum_{Y\subset X} P_Y^\Psi$$

$$= \sum_{p\in\mathcal{P}_\Delta} \prod_{\Lambda\in p} \sum_{Y\subset\Lambda} \Big(\sum_{Y\subset X\subset\Lambda} (-1)^{\#(\Lambda\setminus X)} \Big) P_Y^\Psi$$

$$= \sum_{p\in\mathcal{P}_\Delta} \prod_{\Lambda\in p} P_\Lambda^\Psi$$

$$= <\Omega^\circ|\Psi(\Delta)\,\Omega^\circ>$$

■

whic means $(\tilde{K},\tilde{\Psi},\tilde{\Omega}^\circ) \cong (\tilde{K},\tilde{\Psi},\tilde{\Omega}^\circ)$.

References

[1] M. TAKESAKI; Theory of operator algebras I, Springer-Verlag Berlin/Heidelberg New York (1979)

[2] W.F. STINESPRING; Positive functions on C*-algebras, Proc. Amer. Math. Soc., 6 (1955), 211-216

[3] R.F. STREATER; Current commutation relations, continuous tensor products and infinitely divisible group representations; Rendiconti di Sc. Int. di Fisica E. Fermi, Vol. XI (1069) 247-263.

——;A continuum analogue of the lattice gas, Comm. Math. Phys. 12 (1969) 226-232

[4] K.R. PARTHASARATHY, K. SCHMIDT; Positive definite kernels, continuous tensor products, and central limit theorems of probability theory, Lecture Notes in Mathematics, no. 272. Springer-Verlag, Berlin/Heidelberg/New York (1972)

[5] A. GUICHARDET; Symmetric hilbert spaces and related topics, Lecture Notes in Mathematics no. 261. Springer-Verlag Berlin/Heidelberg/New York (1972)

[6] J. QUAEGEBEUR; A non commutative central limit theorem for CCR-algebras, J. Funct. Anal. 57 (1984), 1-20

[7] B. DEMOEN,P. VANHEUVERZWIJN, A. VERBEURE; Completely positive maps on the CCR algebra, Lett. Math. Phys. 2 (1977) 161-166

[8] B.V. GNEDENKO, N. KOLMOGOROV; Limit distributions for sums of independent random variables, Addison-Wesley, Reading, Mass. (1954)

[9] K.R. PARTHASARATHY; Infinitely divisible representantions and positive definite functions on a compact group, Comm. Math. Phys. 16 (1970) 148-156

[10] HEYER; Probability measures on locally compact groups, Ergebnisse der Mathematik and ihrer Grenzgebiete 94 (1977).

TEMPERATURE-DEPENDENT LAMB SHIFT OF A QUANTUM OSCILLATOR

G.W. Ford
Institut Laue-Langevin
156X, 38042 Grenoble Cedex, France

A quantum stochastic process in physics corresponds in general to a random operator variable belonging to a small system, this small system itself being coupled to a large (usually infinite) system called the heat bath. The essential point here is that both the small system and the heat bath must be dynamical systems whose motion is described by quantum mechanical equations of motion, and that this imposes restrictions upon the form of the description of the quantum stochastic process. Professor Lewis, in his opening talk[1], has spoken in a general way about these restrictions. Here I want to discuss these restrictions in a much more specific framework, that of the quantum Langevin equation, and to illustrate them with an explicit calculation of a recently observed phenomenon : the temperature dependent Stark shift of levels in Rydberg atoms.

Consider a particle, which we may picture as a Brownian particle and which for convenience I take to be moving in one dimension, which is coupled to a heat bath. Within the approximation that the coupling gives rise to a linear response, the quantum mechanical motion of the particle is described by the generalized quantum Langevin equation :

$$m\ddot{x} + \int_{-\infty}^{t} dt' \, \mu(t-t') \, \dot{x}(t') + V'(x) = F(t) \quad . \tag{1}$$

This is an equation for the time-dependent Heisenberg operator x(t), the displacement operator for the particle. In this equation V(x) is an external potential, V'(x) = dV/dx is the corresponding force. The coupling to the heat bath corresponds to two terms : the reaction force characterized by the memory function μ(t), and the fluctuating term characterized by the operator valued random force F(t). The properties of this random force, as in the classical case[2], are completely characterized by the memory function. Thus, the symmetric autocorrelation of the force is

$$\frac{1}{2}\langle F(t)F(t') + F(t')F(t)\rangle = \frac{\hbar}{\pi}\int_0^\infty d\omega \, \omega \coth\frac{\hbar\omega}{2kT} \, \mathrm{Re}\,\{\hat{\mu}(\omega + io^+)\} \cos\omega(t-t'), \tag{2}$$

and the commutator is

$$[F(t), F(t')] = \frac{2\hbar}{i\pi}\int_0^\infty d\omega \, \omega \, \mathrm{Re}\,\{\hat{\mu}(\omega + io^+)\} \sin\omega(t-t') \tag{3}$$

In these expressions,

$$\hat{\mu}(\omega) = \int_0^\infty dt \, e^{i\omega t} \mu(t) \quad , \quad \mathrm{Im}\,\omega > o, \tag{4}$$

the Fourier transform of the memory function. This characterization of the random rce is completed when we require the Gaussian property : symmetric correlations in- lving an odd number of factors of F vanish, those involving an even number of factors e equal to the sum over all pairings of the pair correlation (2).

ese properties of the random force can be straightforwardly derived from simple dels of the heat bath[3]. The properties (2) and (3) can also be derived in a model dependent way from the fluctuation-dissipation theorem, using a procedure entirely milar to that used by Kubo for the classical Langevin equation[2].

re, however, I want to emphasize the central role played by the memory function, or ther by its Fourier transforms $\hat{\mu}(\omega)$. It is clear from (4) that $\hat{\mu}(\omega)$ is analytic in e upper half ω-plane. In addition, energy conditions require that

$$\mathrm{Re}\,\{\hat{\mu}(\omega + io^{+})\} > o \quad , \tag{5}$$

ere ω is here on the real axis. This positivity condition is of fundamental physical portance; it is necessary if the power spectrum of the random force (2) is to be sitive, and its violation amounts to a violation of the second law of thermodynamics[4]. ese two properties, analyticity and positivity, characterize a class of what are rmed positive functions[4,5]. This is a very restrictive class with many special pro- rties : positive functions have positive real part in the upper half-plane, they ve neither zeros nor poles in the upper half-plane, on the real axis they have only mple zeros, the reciprocal of a positive function is a positive function, etc. The int here is that given $\hat{\mu}(\omega)$ is a positive function it then characterizes completely e quantum Langevin equation, i.e., not only the memory term in the equation itself t also the correlation and commutator of the random force.

an application of these ideas I consider a physical effect which has recently been served experimentally: the temperature dependence of the Lamb shift in Rydberg atoms[6]. Rydberg atom is an atom in which an outer electron has been excited to move in a rge circular orbit. The energy levels for such an atom are then given by the Rydberg rmula, $E_n = Ry/n^2$, with n large. In this case the levels are closely and nearly iformly spaced and are therefore well approximated by harmonic oscillator levels. e Lamb shift is a shift in the electron energy levels due to the coupling with the ectromagnetic field. The largest contribution to this shift arises from fluctuations the field[7]. The question here is : what is the temperature dependence of this fect ? We can answer this using quantum stochastic methods in which we use for the at bath the fluctuating electormagnetic field in a blackbody cavity. In this case e memory function can be calculated in a manner following the treatment of classical liation reaction found in standard textbooks of electrodynamics. For a simple model the electron form-factor[8], this takes the form

$$\hat{\mu}\,(\omega) = 2e^2\Omega^2\omega \;/\; 3c^3(\omega + i\Omega) \quad , \tag{6}$$

where Ω is a large cutoff frequency. Note that this is a positive function.

As a simplification, and in order to make a simple closed form calculation, I consider the case of the linear oscillator. As I remarked above, for the high Rydberg levels this should be a good approximation to the atomic systems. In this case the quantum Langevin equation takes the form

$$m\ddot{x} + \int_{-\infty}^{t} dt' \, \mu(t - t') \, \dot{x}(t') + kx = F(t) \quad , \tag{7}$$

where $\hat{\mu}(\omega)$ is given by (7) and k is the oscillator force constant. As I have emphasized above, the coupling with the heat bath (i.e., the radiation field) is characterized by $\hat{\mu}(\omega)$. For our purposes, however, it is convenient to introduce an equivalent quantity, the generalized admittance, which is formed by taking the Fourier transform of (7) and writing the result in the form

$$-i\omega \, \hat{x}(\omega) = Y(\omega)\hat{F}(\omega) \quad . \tag{8}$$

Here $Y(\omega)$ is the generalized admittance,

$$Y(\omega) = [-i\omega m + ik/\omega + \hat{\mu}(\omega)]^{-1} \quad . \tag{9}$$

It is not difficult to see from this expression that $Y(\omega)$ is a positive function.

The system of oscillator coupled to the radiation field has a well defined energy. The part of this energy ascribed to the oscillator, U_0, is the energy of this coupled system minus the energy of the radiation field in the absence of the oscillator. For this energy we have the remarkable formula :

$$U_0(T) = \frac{1}{\pi} \int_0^{\infty} d\omega \frac{\hbar\omega}{\exp(\hbar\omega(k_B T) - 1} \, \mathrm{Im} \, \{\frac{d \ln Y(\omega + io^+)}{d\omega}\} \quad . \tag{10}$$

I call this formula remarkable because it expresses the energy of the interacting oscillator in terms of this same function $\hat{\mu}(\omega)$ which characterizes the Langevin equation. It can be obtained using the following heuristic argument. Since $Y(\omega)$ is a positive function it can have only simple zeros and poles on the real axis. If the normal modes of the system are discrete these will be the only singularities of $Y(\omega)$, the poles being at the normal mode frequencies of the interacting system and the zeros being at the normal mode frequencies of the radiation field in the absence of the oscillator. This should be clear from the defining relation (8) : if $Y(\omega) = 0$ there can be a free motion of the radiation field with no $\hat{x}$, while if $Y(\omega)^{-1} = 0$ there can be a motion of $\hat{x}$ with no force. Therefore, one can write

$$Y(\omega) \propto \prod_i (\omega - \omega_i^o) \; / \; \prod_j (\omega - \omega_j) \tag{11}$$

where the numerator is the product over normal modes of the free radiation field and the denominator is the product over those of the interacting system. If now one recalls the formula : $\mathrm{Im}\,\{1/(x + io^+)\} = -\pi\,\delta(x)$, one sees that

$$\frac{1}{\pi}\,\mathrm{Im}\,\{\frac{d\,\ln Y\,(\omega + io^{+})}{d\omega}\} = \sum_j \delta(\omega - \omega_j) - \sum_i \delta(\omega - \omega_i^{o}) \quad . \tag{12}$$

th this we see that (10) can be written

$$U_0\,(T) = \sum_j u(\omega_j,\ T) - \sum_i u(\omega_i^{o},\ T) \quad , \tag{13}$$

ere $u(\omega,\ T) = \hbar\omega\ /\ [\exp(\hbar\omega/k_B T) - 1]$ is the Planck energy of a single (normal mode) cillator with frequency ω and at temperature T. This form shows that $U_0(T)$ is indeed e difference between the energy of the coupled systems and that of the free radiation eld.

e calculation is now straightforward. We evaluate (10) with $Y(\omega)$ given by (9) in ich $\hat{\mu}(\omega)$ is given by (6). That is,

$$Y(\omega) = \frac{\omega(\omega + i\Omega)}{-im\omega^3 + M\Omega\omega^2 + ik(\omega + i\Omega)} \quad , \tag{14}$$

ere M is the renormalized electron mass,

$$M = m + 2e^2\,\Omega\ /\ 3c^3 \quad . \tag{15}$$

is expression for $Y(\omega)$ does not show the structure of zeros and poles evoked in the gument of the previous paragraph. This is because the normal mode frequencies of e radiation field are continuously distributed. The real axis then becomes a "branch t" and the poles and zeros of (14) in the lower half plane are on the "unphysical eet" reached by analytically continuing through the cut. Nevertheless the formula 0) still holds, excepting only that the zero at $\omega = 0$ gives no contribution.

e denominator in (14) can be factored to write

$$Y(\omega) = \frac{\omega(\omega + i\Omega)}{im(\omega + i\Omega')\ (\omega_o^2 - \omega^2 - i\gamma\omega)} \quad , \tag{16}$$

ere

$$\frac{1}{\Omega} = \frac{\gamma}{\omega_o^2} + \frac{1}{\Omega'}\ ,\ \frac{k}{M} = \omega_o^2\,\frac{\Omega'}{\Omega' + \gamma}\ ,\ \frac{M - m}{M\Omega} = \frac{\gamma}{\omega_o^2} + \frac{\gamma}{\Omega'(\Omega' + \gamma)} \quad . \tag{17}$$

ese last relations can be viewed as expressions for the parameters Ω, k, M in terms new parameters Ω', ω_o, γ which when substituted in (14) give (16). When the form 6) is put in (10) the result can be written

$$U_0\,(T) + U_0^{o}(T) + \Delta U_0\,(T) \quad , \tag{18}$$

ere

$$U_0^{o}(T) = \frac{1}{\pi}\int_0^\infty d\omega\,\frac{\hbar\omega}{\exp(\hbar\omega/k_B T) - 1}\,\frac{\gamma(\omega_o^2 + \omega^2)}{(\omega_o^2 - \omega^2)^2 + \gamma^2\omega^2} \quad , \tag{19}$$

d

$$\Delta U_0(T) = \frac{1}{\pi}\int_0^\infty d\omega \frac{\hbar\omega}{\exp(\hbar\omega/k_BT) - 1} \left(\frac{\Omega'}{\omega^2 + \Omega'^2} - \frac{\Omega}{\omega^2 + \Omega^2}\right) . \quad (20)$$

The expression (19) is familiar. It is exactly what one obtains for the energy of a quantum oscillator with natural frequency ω_o and width γ, i.e., what one obtains if in (9) one puts $\hat{\mu} = m\gamma$ a constant and $k = m\omega_o^2$ [9]. Therefore the term $\Delta U_0(T)$ corresponds to a uniform temperature-dependent shift in the energy of each quantum level of the oscillator. Since the cutoff frequency is large, $\hbar\Omega >> kT$, in which case (20) can be evaluated to give

$$\Delta U_0(T) = -\left(\frac{1}{\Omega} - \frac{1}{\Omega'}\right) \frac{\pi(k_BT)^2}{6\hbar} . \quad (21)$$

Now, from (17) we see that in the limit of large cutoff (more strictly, in the limit $\Omega' \to \infty$),

$$\frac{1}{\Omega} - \frac{1}{\Omega'} \cong \frac{M - m}{M\Omega} = \frac{2e^2}{3MC^3} , \quad (22)$$

where for this last (15) has been used. Hence, one gets

$$\Delta U_0(T) = -\pi e^2(k_BT)^2/9\hbar MC^3 . \quad (23)$$

In order to compare with the experiments this result should be multiplied by a factor of three for the three dimensions of space.

The experimental results are consistent with this result except for the sign; the observed shift in the energy of a photon absorbed in the transition to the Rydberg state is positive. However, this apparent discrepency is resolved when one recalls that the work done in an isothermal transition is the change in free energy, not the energy. The relation between energy, U, and free energy, F, is

$$U = F - T\frac{\partial F}{\Delta T} . \quad (24)$$

From this it is clear that a term proportional to T^2 will have the same magnitude but opposite sign in F and U. Thus, although the shift in energy is negative that in free energy is positive and in accord with the observations.

References

1. J.T. Lewis, talk at conference.
2. R. Kubo, Rep. Progr. Theor. Phys. 29 (1966) 255
3. G.W. Ford, M. Kac and P. Mazur, J. Math. Phys. 6 (1965) 504
4. J. Meixner, "Linear Passive Systems", in "Statistical Mechanics of Equilibrium and Non-Equilibrium", ed. J. Meixner (North-Holland, Amsterdam 1965)
5. E.A. Guillemin, "Synthesis of Passive Networks" (Wiley, New York 1957)
6. L. Hollberg and J.L. Hall, Phys. Rev. Lett. 53 (1984) 230
7. T.A. Welton, Phys. Rev. 74 (1948) 1157
8. P. Ullersma, Physica 32 (1966) 27
9. G.W. Ford, M. Kac and P. Mazur, J. Math. Phys. 6 (1965) 504

CONSTRUCTION OF STATIONARY QUANTUM MARKOV PROCESSES THROUGH QUANTUM STOCHASTIC CALCULUS

Alberto Frigerio

Dipartimento di Fisica, Sezione Fisica Teorica, Università di Milano,

Via Celoria 16, I - 20133 Milano, Italy;

and INFN, Sezione di Milano.

1. Dilations as singular perturbations.

The theory of unitary dilations of quantum dynamical semigroups [1—5] has received a great impulse from the development of quantum stochastic calculus [6—10]. For an arbitrary norm continuous dynamical semigroup (one-parameter semigroup of completely positive identity preserving normal linear maps) $T_t = \exp[L\,t]$ on the algebra $\mathcal{B}(\mathcal{H})$ of all bounded linear operators on a separable Hilbert space $\mathcal{H}$, it has been shown [8—10] that there exists an auxiliary Hilbert space $\mathcal{F}$, a group $\{\alpha_t : t \in \mathbb{R}\}$ of *-automorphisms of $\mathcal{B}(\mathcal{H} \otimes \mathcal{F})$ and a conditional expectation E_0 of $\mathcal{B}(\mathcal{H} \otimes \mathcal{F})$ onto $\mathcal{B}(\mathcal{H}) \otimes 1\!\!1_{\mathcal{F}}$ such that

$$T_t(X) \otimes 1\!\!1_{\mathcal{F}} = E_0(\alpha_t(X \otimes 1\!\!1_{\mathcal{F}})) \quad : X \in \mathcal{B}(\mathcal{H}) , \; t \in \mathbb{R}^+ . \qquad (1.1)$$

The evolution α_t is a "singular perturbation" of the "free evolution" α_t^0 on $\mathcal{B}(\mathcal{F})$, of the form

$$\alpha_t(.) = U(t)\,\alpha_t^0(.)\,U(t)^* , \qquad (1.2)$$

where $\{U(t) : t \in \mathbb{R}^+\}$ satisfies the cocycle condition

$$U(t)\,\alpha_t^0(U(s)) = U(s+t) \quad : \quad t , s \in \mathbb{R}^+, \qquad (1.3)$$

and is the solution of a noncommutative stochastic differential equation [6—10]. Here we give a brief illustration of this result.

The general form of the (bounded) generator L of T_t is [11]

$$L(X) = K^* X + X K + \sum_{j=1}^{\infty} V_j^* X V_j \quad : \quad X \in \mathcal{B}(\mathcal{H}) , \qquad (1.4)$$

where K , V_j are operators in $\mathcal{B}(\mathcal{H})$ such that

$\sum_{j=1}^{\infty} V_j^* V_j$ converges ultraweakly to $-K-K^*$. For the sake of notational simplicity, we shall begin with the simplest case:

$$L(X) = i\left[H, X\right] - \frac{1}{2}\left[V^*V, X\right]_+ + V^* X V \quad : \quad X \in \mathcal{B}(\mathcal{H}), \tag{1.5}$$

where H is a self-adjoint element of $\mathcal{B}(\mathcal{H})$.

Following Hudson and Parthasarathy [6], we shall take the auxiliary Hilbert space $\mathcal{F}$ to be the symmetric Fock space over $L^2(\mathbb{R})$, generated by the exponential vectors

$$\Psi(f) = (1, f, \ldots, (n!)^{-1/2} f \otimes \ldots \otimes f, \ldots) \quad : \quad f \in L^2(\mathbb{R}), \tag{1.6}$$

and define the <u>annihilation process</u> $\{A(t) : t \in \mathbb{R}^+\}$, the <u>creation process</u> $\{A^*(t) : t \in \mathbb{R}^+\}$, and the <u>gauge process</u> $\{\Lambda(t) : t \in \mathbb{R}^+\}$ by

$$A(t)\,\Psi(f) = \left(\int_0^t f(\tau)\, d\tau\right) \Psi(f), \tag{1.7}$$

$$A^*(t)\,\Psi(f) = \frac{d}{d\varepsilon} \Psi(f + \varepsilon\, \chi_{[0,t]})\Big|_{\varepsilon = 0}, \tag{1.8}$$

$$\Lambda(t)\,\Psi(f) = \frac{d}{d\varepsilon} \Psi(\exp\{\varepsilon\, \chi_{[0,t]}\} f)\Big|_{\varepsilon = 0}. \tag{1.9}$$

We identify any operator X in $\mathcal{B}(\mathcal{H})$ with the corresponding operator $X \otimes \mathbf{1}_{\mathcal{F}}$ in $\mathcal{B}(\mathcal{H} \otimes \mathcal{F})$, and any operator Y with domain $\mathcal{D} \subseteq \mathcal{F}$ with the algebraic tensor product $\mathbf{1}_{\mathcal{H}} \underline{\otimes} Y$ with domain $\mathcal{H} \underline{\otimes} \mathcal{D}$; and we consider the noncommutative stochastic differential equation [6]

$$dU(t) = U(t)\left[(W - \mathbf{1}_{\mathcal{H}})\, d\Lambda(t) + i\, V^*\, dA(t) + i\, W\, V\, dA^*(t) + (i\, H - \frac{1}{2} V^*V)\, dt\right], \tag{1.10}$$

with initial condition $U(0) = \mathbf{1}$, where H, V are as in (1.5), and where W is an arbitrary unitary operator on $\mathcal{H}$. By [6, Theorem 7.1], Equation (1.10) has a unique solution which is a continuous adapted process consisting of unitary operators. Upon defining the conditional expectation E_0 of $\mathcal{B}(\mathcal{H} \otimes \mathcal{F})$ onto $\mathcal{B}(\mathcal{H}) \otimes \mathbf{1}_{\mathcal{F}}$ by

$$E_0(.) = E(.) \otimes \mathbf{1}_{\mathcal{F}},$$

$$E(X \otimes Y) = \langle \Psi(0) | Y\, \Psi(0) \rangle\, X \quad : \quad X \in \mathcal{B}(\mathcal{H}),\ Y \in \mathcal{B}(\mathcal{F}), \tag{1.11}$$

we have also, by [6 , Theorem 8.1],

$$E_0(U(t)\, X \otimes \mathbb{1}_{\mathfrak{F}}\, U(t)^*) = \exp[L\, t](X) \otimes \mathbb{1}_{\mathfrak{F}} : t \in \mathbb{R}^+ , X \in \mathcal{B}(\mathcal{H}) , \qquad (1.12)$$

where L is given by (1.5), independently of the choice of the unitary operator W. Note that, in order for (1.12) to hold, it is important that $d\{U(t)^*[\phi \otimes \psi(0)]\}$ should <u>not</u> be a multiple of dt in general: indeed, if this were the case, a term like $V^* X V$ in $L(X)$ could never arise. This is the reason why we have spoken of "singular" perturbations.

For a physicist, Equation (1.10) in the special case $W = \mathbb{1}_{\mathcal{H}}$ has a transparent meaning: it is the differential equation for the time evolution operator of a quantum system with a singular coupling to a boson reservoir, in the interaction picture with respect to the free evolution of the reservoir; the origin of the "Ito correction" $-(1/2)\, V^*V\, dt$ can be traced to the operation of Wick ordering [12]. In order to obtain a group dilation, it is therefore necessary to introduce the "free evolution" on the auxiliary Hilbert space $\mathfrak{F}$, cf. [4 , 9 , 10].

Let $\{S_t : t \in \mathbb{R}\}$ be the strongly continuous one-parameter unitary group on $L^2(\mathbb{R})$ defined by

$$(S_t f)(x) = f(x - t) \quad : \quad f \in L^2(\mathbb{R}) , \; x , t \in \mathbb{R} ; \qquad (1.13)$$

let $\{\hat{S}_t : t \in \mathbb{R}\}$ be its second quantization on $\mathcal{H} \otimes \mathfrak{F}$, defined by

$$\hat{S}_t[\phi \otimes \psi(f)] = \phi \otimes \psi(S_t f) : \quad \phi \in \mathcal{H} , f \in L^2(\mathbb{R}) , t \in \mathbb{R} , \qquad (1.14)$$

and consider the group $\{\alpha^0_t : t \in \mathbb{R}\}$ of *-automorphisms of $\mathcal{B}(\mathcal{H} \otimes \mathfrak{F})$ defined by

$$\alpha^0_t(.) = \hat{S}_t(.)\, \hat{S}_{-t} \quad : \quad t \in \mathbb{R} . \qquad (1.15)$$

Then it can be shown, as in [8 , Theorem 7.1], that for all s in $\mathbb{R}^+$ $\{\alpha^0_{-s}[U(s)^*\, U(s + t)] : t \in \mathbb{R}^+\}$ is a continuous adapted process satisfying the same stochastic differential equation (1.10) as $\{U(t) : t \in \mathbb{R}^+\}$, with the same initial condition; hence both processes coincide and the cocycle condition (1.3) holds. It follows that the family $\{\alpha_t : t \in \mathbb{R}\}$ of *-automorphisms of $\mathcal{B}(\mathcal{H} \otimes \mathfrak{F})$ defined by

$$\alpha_t(.) = U(t)\alpha^0_t(.)\, U(t)^* \; ; \; \alpha_{-t} = (\alpha_t)^{-1} \quad : \quad t \in \mathbb{R}^+ , \qquad (1.16)$$

is a (weakly* continuous) group, so that a physicist may interpret it as the reversible time evolution of an isolated system made up of the original quantum system and a boson reservoir (cf. [9 , 10]).

The generalization to the case of a generator L containing finitely many V_j's is straightforward, involving just finitely many independent copies of the annihilation, creation and gauge processes [6]. Also the general case (1.4) can be handled with the technique of [8]. Similar results are obtained by means of fermion stochastic differential equations [7]. See also the contributions by Accardi, Applebaum, Hudson, Lindsay, Maassen in this volume.

2. Covariant Markov processes.

The structure constructed in the preceding Section determines a W*-stochastic process [13] $(\mathcal{A}, \{j_t : t \in \mathbb{R}\}, \varphi)$ over $\mathcal{B}(\mathcal{H})$ as follows: j_t is a faithful normal *-representation of $\mathcal{B}(\mathcal{H})$ into $\mathcal{B}(\mathcal{H} \otimes \mathcal{F})$ defined by

$$j_t(X) = \alpha_t(X \otimes \mathbb{1}_{\mathcal{F}}) \qquad : X \in \mathcal{B}(\mathcal{H}),\ t \in \mathbb{R}$$

$$\left(\begin{array}{ll} = U(t)\, X \otimes \mathbb{1}_{\mathcal{F}}\, U(t)^* & \text{for } t \text{ in } \mathbb{R}^+ , \\ = \alpha_t^0(U(|t|)^*)\, X \otimes \mathbb{1}_{\mathcal{F}}\, \alpha_t^0(U(|t|)) & \text{for } t \text{ in } \mathbb{R}^- \end{array} \right), \qquad \begin{array}{r} (2.1a) \\ (2.1b) \end{array}$$

a state φ may be defined on $\mathcal{B}(\mathcal{H} \otimes \mathcal{F})$ by

$$\varphi(.) = \mathrm{Tr}_{\mathcal{H}}[\rho E(.)] \qquad (E \text{ defined in } (1.11)) , \tag{2.2}$$

where ρ is an arbitrary density operator on $\mathcal{H}$, and then restricted to a C*-subalgebra $\mathcal{A}$ of $\mathcal{B}(\mathcal{H} \otimes \mathcal{F})$, which is most conveniently chosen to be

$$\mathcal{A} = \Big[\bigcup_{I \text{ bounded interval in } \mathbb{R}} \{ j_t(\mathcal{B}) : t \in I \}'' \Big]^- , \tag{2.3}$$

where $\{\ldots\}''$ denotes double commutant and $[\ldots]^-$ denotes norm closure. A family $\{\mathcal{A}_J : J \text{ interval in } \mathbb{R}\}$ of local algebras in $\mathcal{A}$ is defined by

$$\mathcal{A}_J = \Big[\bigcup_{I \text{ bounded subinterval of } J} \{ j_t(\mathcal{B}) : t \in I \}'' \Big]^- . \tag{2.4}$$

As emphasized by Kümmerer [2—4] (see also [13]) , it is important that a unitary dilation $(\mathcal{A}, \{\alpha_t : t \in \mathbb{R}\}, E_0)$ of a quantum dynamical semigroup T_t should define a Markov process, in some sense. We show that this is the case for the dilation constructed in the preceding Section (cf. [9 , 10]). We need some definitions:

Definition 2.1. Let $\{\mathcal{A}_J : J \subseteq \mathbb{R}\}$ be an isotonic family of C*-algebras of operators on the same Hilbert space $\tilde{\mathcal{H}}$, indexed by the intervals in the real line, such that $\mathcal{A}_I = \mathcal{A}_I''$ when I is bounded and

$$\mathcal{A}_J = \Big[\bigcup_{I \text{ bounded subinterval of } J} \mathcal{A}_I\Big]^- .$$

Suppose that there exists a group $\{\alpha_t : t \in \mathbb{R}\}$ of *-automorphisms of $\mathcal{A} \equiv \mathcal{A}_{\mathbb{R}}$ such that $\alpha_t(\mathcal{A}_J) = \mathcal{A}_{J+t}$ for all $t \in \mathbb{R}$, $J \subseteq \mathbb{R}$, and that there exists a family $\{E_{s]} : s \in \mathbb{R}^+\}$ of conditional expectations in $\mathcal{A}$ such that

$$E_{s]}(\mathcal{A}) = \mathcal{A}_{(-\infty,s]} \quad : \quad s \in \mathbb{R}^+ ; \tag{2.5}$$

$$E_{r]}E_{s]} = E_{r\wedge s]} \quad : \quad r, s \in \mathbb{R}^+ ; \tag{2.6}$$

$$\alpha_t E_{s]} \alpha_{-t} = E_{s+t]} \quad : \quad s, s+t \in \mathbb{R}^+ ; \tag{2.7}$$

and satisfying the Markov property

$$E_{s]}(\mathcal{A}_{[s,+\infty)}) = \mathcal{A}_{\{s\}} \quad : \quad s \in \mathbb{R}^+ . \tag{2.8}$$

Suppose also that the maps α_t and $E_{s]}$ are locally normal and that the functions $t \longmapsto \alpha_t(A) : A \in \mathcal{A}$ are continuous in the ultraweak topology of $\mathcal{B}(\tilde{\mathcal{H}})$. Then $(\mathcal{A}_J, \alpha_t, E_{s]})$ is said to be a <u>covariant Markov structure</u> on $\tilde{\mathcal{H}}$. If the $\mathcal{A}_J$ come from a W*-stochastic process $(\mathcal{A}, \{j_t : t \in \mathbb{R}\}, \varphi)$ as in (2.4) and if the conditional expectations $E_{s]}$ satisfying (2.5)—(2.8) are compatible with φ in the sense that

$$\varphi = (\varphi \restriction \mathcal{A}_{(-\infty,s]}) \circ E_{s]} \quad : \quad s \in \mathbb{R}^+ , \tag{2.9}$$

then we shall say that $(\mathcal{A}, \{j_t : t \in \mathbb{R}\}, \varphi)$ is a <u>covariant Markov process</u> (a <u>stationary Markov process</u> if in addition φ is stationary under $\{\alpha_t : t \in \mathbb{R}\}$).

If $\mathcal{F}$ is the boson Fock space over $L^2(\mathbb{R})$, it is easily seen that a covariant Markov structure on $\tilde{\mathcal{H}} = \mathcal{H} \otimes \mathcal{F}$ is given by $(\mathcal{A}_J^0, \alpha_t^0, E_{s]}^0)$, where

$$\mathcal{A}_I^0 = \mathcal{B}(\mathcal{H}) \otimes \{A(t) - A(s), A^*(t) - A^*(s) : [s,t] \subseteq I\}'' \tag{2.10}$$

for bounded intervals $I \subseteq \mathbb{R}^+$, α_t^0 is defined by (1.15), and where

$$E_{s]}^0(X \otimes \exp[i\int\{\bar{f}\, dA + f\, dA^*\}])$$

$$= \exp[-\tfrac{1}{2}\int_s^\infty |f(\tau)|^2 d\tau]\; X \otimes \exp[i\int \chi_{(-\infty,s]}\{\bar{f}\, dA + f\, dA^*\}] \tag{2.11}$$

for all X in $\mathcal{B}(\mathcal{H})$ and f in $L^2(\mathbb{R})$ [1, Section 10]; note that $E_{s]}^0$ is compatible with the state φ defined by (2.2). Then the proof that the W*-stochastic process defined by (2.1)——(2.3) is covariant Markov is based on the following Theorem, which is an extension (cf. [4, 9]) of an idea originally due to Accardi [14]:

Theorem 2.2. Let $(\mathcal{A}_J^0, \alpha_t^0, E_{s]}^0)$ be a covariant Markov structure on $\tilde{\mathcal{H}}$. Suppose that $\{U(t) : t \in \mathbb{R}^+\}$ is a strongly continuous family of unitary operators on $\tilde{\mathcal{H}}$, with $U(t) \in \mathcal{A}_{[0,t]}^0 : t \in \mathbb{R}^+$, satisfying the cocycle condition (1.13). Then:

(i) $\{E_{s]}^0 : s \in \mathbb{R}^+\}$ satisfies the covariance condition (2.7) also with respect to the automorphism group $\{\alpha_t : t \in \mathbb{R}\}$ defined by (1.16);

(ii) for all t_1, t_2 in $\mathbb{R}^+$, we have

$$\mathcal{A}_{[-t_1,t_2]} \equiv \{\alpha_t(\mathcal{A}_{\{0\}}^0) : -t_1 \leq t \leq t_2\}'' \subseteq \mathcal{A}_{[-t_1,t_2]}^0 \quad ;$$

(iii) the Markov property holds for the new local algebras, i. e.

$$E_{s]}^0(\mathcal{A}_{[s,t]}) = \mathcal{A}_{\{s\}} \quad : s \leq t \in \mathbb{R}^+ .$$

Proof (Sketch). For A in $\mathcal{A}_{\mathbb{R}}^0$, we have

$$\alpha_t E_{s]}^0(A) = U(t)\alpha_t^0(E_{s]}^0(A))U(t)^* = U(t)\, E_{s+t]}^0(\alpha_t^0(A))U(t)^*$$

by (2.7). For $s \geq 0$, we have $U(t) \in \mathcal{A}_{[0,t]}^0 \subseteq \mathcal{A}_{(-\infty,s+t]}^0$, hence $U(t)$ and $U(t)^*$ can be moved inside $E_{s+t]}^0$, and (i) follows. (ii) is obvious from (2.1a,b), since, for all t in $\mathbb{R}^+$, we have $U(t) \in \mathcal{A}_{[0,t]}^0$, $\alpha_{-t}^0(U(t)) \in \mathcal{A}_{[-t,0]}^0$. By (i), it suffices to prove (iii) for $0 = s \leq t$; and this is an immediate consequence of (ii) and (2.8). □

Theorem 2.3. Let $\{U(t) : t \in \mathbb{R}^+\}$ be the solution of the stochastic differential equation (1.10), and define a W*-stochastic process $(\mathcal{A}, \{j_t : t \in \mathbb{R}\}, \varphi)$

as in (2.1)—(2.3). Then $(\mathcal{A}, \{j_t : t \in \mathbb{R}\}, \varphi)$ is a covariant Markov process, with conditional expectations $\{E_{s]} = E^0_{s]} \restriction \mathcal{A} : s \in \mathbb{R}^+\}$, and we have

$$E_{0]}\alpha_t(X \otimes 1_{\mathfrak{Z}}) = E_0\, \alpha_t(X \otimes 1_{\mathfrak{Z}}) = T_t(X) \otimes 1_{\mathfrak{Z}} : \; X \in \mathcal{B}(\mathcal{H}) \; , \; t \in \mathbb{R}^+ \; , \tag{2.12}$$

where $T_t = \exp[L\, t]$, with L given by (1.5).

Proof (Sketch). It follows from [6 , 8] that $\{U(t) : t \in \mathbb{R}^+\}$ satisfies the assumptions of Theorem 2.2, and it can be shown [9] that $E^0_{s]}$ maps $\mathcal{A}$ onto $\mathcal{A}_{(-\infty,s]}$ for all s in $\mathbb{R}^+$. Taking into account (1.12), the result follows.□

Theorem 2.4. Also $(\mathcal{A}, \{j_{-t} : t \in \mathbb{R}\}, \varphi)$ is a covariant Markov process, with group of automorphisms $\{\alpha_{-t} : t \in \mathbb{R}\}$ and with reversed conditional expectations $\{E_{[-s} = E^0_{[-s} \restriction \mathcal{A} : s \in \mathbb{R}^+\}$ (with obvious notations), and we have

$$E_{[0}\alpha_{-t}(X \otimes 1_{\mathfrak{Z}}) = E_0\alpha_{-t}(X \otimes 1_{\mathfrak{Z}}) = T_{-t}(X) \otimes 1_{\mathfrak{Z}} : \; X \in \mathcal{B}(\mathcal{H}) \; , \; t \in \mathbb{R}^+ , \tag{2.13}$$

where $T_{-t} = \exp[L_- t]$, with L_- given by

$$L_-(X) = -i[H , X] - \tfrac{1}{2}[V^*V , X]_+ + V^* W^* X W V \; : \; X \in \mathcal{B}(\mathcal{H}) \; . \tag{2.14}$$

Proof (Sketch). Also $(\mathcal{A}^0_J, \alpha^0_{-t}, E^0_{[-s})$ is a covariant Markov structure, and $\{U(-t) \equiv \alpha^0_{-t}(U(t)^*) : t \in \mathbb{R}^+\}$ satisfies the assumptions of Theorem 2.2 relative to it. Then $(\mathcal{A}, \{j_{-t} : t \in \mathbb{R}\}, \varphi)$ is a covariant Markov process, hence (2.13) defines a dynamical semigroup $\{T_{-t} : t \in \mathbb{R}^+\}$ on $\mathcal{B}(\mathcal{H})$. The explicit form (2.14) of its generator is obtained by differentiating (2.13) at $t = 0$ and using the quantum Ito's formula of [6].□

The conclusions of Theorems 2.3 , 3.4 hold also for the generalizations of (1.10) mentioned at the end of Section 1 ; hence every norm continuous dynamical semigroup on $\mathcal{B}(\mathcal{H})$ has a unitary dilation which defines a covariant Markov process through (2.1)—(2.3) (a covariant Markov dilation in the following); for details, see [9]. We refer to Kümmerer's contribution in the present volume for a discussion of the "converse" problem of the general structure of Markov dilations.

3. Stationary Markov dilations: detailed balance.

When the dynamical semigroup T_t on $\mathcal{B}(\mathcal{H})$ has a stationary state ρ, it is an interesting problem whether the state φ defined by (2.2) is stationary under $\{\alpha_t : t \in \mathbb{R}\}$, thus making $(\mathcal{A}, \{j_t : t \in \mathbb{R}\}, \varphi)$ a stationary Markov dilation of T_t, rather than just a covariant Markov dilation. As we shall see, this is not always the case (cf. [2]), but it will always be possible to find a state φ_∞ on $\mathcal{A}$ such that $(\mathcal{A}, \{j_t : t \in \mathbb{R}\}, \varphi_\infty)$ is a stationary Markov process ; φ_∞ will not be of the form (2.2), in general (see the next Section). We begin with the discussion of some examples.

Example 3.1 (relaxation to the ground state). Let L be of the form (1.5), and let Φ_0 be a unit vector in $\mathcal{H}$, such that

$$H \Phi_0 = 0 , \quad V \Phi_0 = 0 . \tag{3.1}$$

Then the pure state $\langle \Phi_0 | (.) \Phi_0 \rangle$ on $\mathcal{B}(\mathcal{H})$ is stationary under the dynamical semigroup $T_t = \exp[L\, t]$, and the pure state $\langle \Phi_0 \otimes \Psi(0) | (.) \Phi_0 \otimes \Psi(0) \rangle$ on $\mathcal{B}(\mathcal{H} \otimes \mathcal{F})$ is stationary under the automorphism group α_t.

Proof. It suffices to show that the functions

$$t \longmapsto \langle \Phi_0 \otimes \Psi(0) | \alpha_t(Y)\ \Phi_0 \otimes \Psi(0) \rangle \quad : \quad Y \in \mathcal{B}(\mathcal{H} \otimes \mathcal{F})$$

are constant on the negative half-axis. We have, for t in $\mathbb{R}^+$,

$$\langle \Phi_0 \otimes \Psi(0) | \alpha_{-t}(Y)\ \Phi_0 \otimes \Psi(0) \rangle = \langle \Phi_0 \otimes \Psi(0) | \hat{S}_{-t}\ U(t)^*\ Y\ U(t)\ \hat{S}_t\ \Phi_0 \otimes \Psi(0) \rangle$$
$$= \langle U(t)[\Phi_0 \otimes \Psi(0)] \,|\, Y\ U(t)[\Phi_0 \otimes \Psi(0)] \rangle .$$

By (1.10), we have

$$d\{U(t)[\Phi_0 \otimes \Psi(0)]\}$$
$$= U(t) \Big\{ (W - \mathbf{1}_{\mathcal{H}})\Phi_0 \otimes d\Lambda(t)\ \Psi(0) + i\ V^* \Phi_0 \otimes dA(t)\ \Psi(0)$$
$$- i\ W\ V\Phi_0 \otimes dA^*(t)\ \Psi(0) + (i\ H\ \Phi_0 - \tfrac{1}{2}\ V^*V\ \Phi_0) \otimes \Psi(0)\ dt \Big\} . \tag{3.2}$$

The first two terms in braces vanish by (1.9) and (1.7), and the remaining ones vanish by (3.1) Since $dA^*(t)\ \Psi(0)$ and $\Psi(0)\ dt$ are linearly independent, (3.1) is also a necessary condition for the vanishing of (3.2). □

More generally, we have

Theorem 3.2. Let Φ_0 be a unit vector in $\mathcal{H}$, and let $T_t = \exp[L\ t]$

be a norm continuous dynamical semigroup on $\mathcal{B}(\mathcal{H})$. Then the following are equivalent:

(i) $\langle\phi_0|T_t(X)\phi_0\rangle = \langle\phi_0|X\phi_0\rangle$ for all X in $\mathcal{B}(\mathcal{H})$ and t in $\mathbb{R}^+$;

(ii) L can be written in the form (1.4) with $K\phi_0 = 0$ (then also $V_j\phi = 0$ for all j and $K^*\phi_0 = 0$);

(iii) $\langle\phi_0\otimes\psi(0)|\alpha_t(Y)\,\phi_0\otimes\psi(0)\rangle = \langle\phi_0\otimes\psi(0)|Y\,\phi_0\otimes\psi(0)\rangle$ for all Y in $\mathcal{B}(\mathcal{H}\otimes\mathcal{F})$ and t in $\mathbb{R}$.

<u>Proof</u> (Sketch). The equivalence of (ii) with (iii) can be shown by an extension of the above reasoning. Obviously (iii) implies (i). The proof of the implication (i) $\Rightarrow$ (ii) is as follows:

L remains unchanged if K is replaced by $K + i a\mathbb{1} + \sum_{j=1}^{\infty} \bar{b}_j V_j$ and each V_j is replaced by $V_j - b_j\mathbb{1}$ at the same time $(a\in\mathbb{R}, \{b_j\}\in\ell^2)$. Then we may assume that $\langle\phi_0|V_j\phi_0\rangle = 0$ for all j and $\langle\phi_0|K\phi_0\rangle$ is real, and we have

$$0 = \langle\phi_0|L(|\phi_0\rangle\langle\phi_0|)\phi_0\rangle = 2\,\mathrm{Re}\langle\phi_0|K\phi_0\rangle ,$$

hence $\langle\phi_0|K\phi_0\rangle = 0$. We have also

$$\sum_{j=1}^{\infty} \|V_j\phi_0\|^2 = -2\,\mathrm{Re}\langle\phi_0|K\phi_0\rangle = 0 , \tag{3.3}$$

hence $V_j\phi_0 = 0$ for all j. Then, letting $\{\phi_m : m = 0, 1, 2, \ldots\}$ be a complete orthonormal set in $\mathcal{H}$, we have

$$0 = \langle\phi_0|L(|\phi_m\rangle\langle\phi_n|)\phi_0\rangle = \langle\phi_0|K^*\phi_m\rangle\,\delta_{n0} + \langle\phi_n|K\phi_0\rangle\,\delta_{m0} , \tag{3.4}$$

and $K\phi_0 = 0$. □

<u>Example 3.3</u> (thermal relaxation). Let ρ be an invertible density operator on $\mathcal{H}$, and let L be given by

$$L(X) = i[H, X] + V^*XV - \tfrac{1}{2}[V^*V, X]_+ + e^{-\beta}(VXV^* - \tfrac{1}{2}[VV^*, X]_+) \tag{3.5}$$

for all X in $\mathcal{B}(\mathcal{H})$, where $H = H^*$, V in $\mathcal{B}(\mathcal{H})$ satisfy

$$\rho H \rho^{-1} = H , \quad \rho V \rho^{-1} = e^{\beta}\,V \quad (\beta\in\mathbb{R}^+) . \tag{3.6}$$

Then a dilation of $T_t = \exp[L\,t]$ can be obtained by solving the stochastic differential equation

$$dU(t) = U(t)\Big[i(V^*\, dB_\beta(t) + V\, dB^*_\beta(t)) + (i\,H - \frac{1}{2}\,V^*V - \frac{1}{2}\,e^{-\beta}V\,V^*)\,dt\Big], \tag{3.7}$$

where

$$B_\beta(t) = A_1(t) + e^{-\beta/2}A^*_2(t) \quad : \quad t \in \mathbb{R}^+ , \tag{3.8}$$

$A_1(t)$, $A_2(t)$ being two independent annihilation processes, and the state $\varphi(.) = \mathrm{Tr}\,[\rho\, E(.)]$ on $\mathcal{A}$ is faithful and stationary under α_t .

Proof (Sketch). $\mathcal{A}$ is contained in the von Neumann algebra $\mathcal{M}$ generated by $\mathcal{B}(\mathcal{H}) \otimes \mathbf{1}_3$ and the (time translated) increments of $B_\beta(t)$, $B^*_\beta(t)$. The restriction of φ to $\mathcal{M}$ is a faithful normal state, with modular automorphism group $\{\sigma_t : t \in \mathbb{R}\}$ such that

$$\sigma_t(X \otimes \mathbf{1}_3) = \rho^{it}\, X\, \rho^{-it} \otimes \mathbf{1}_3 , \quad \sigma_t(B_\beta(s)) = e^{i\beta t}\, B_\beta(s) . \tag{3.9}$$

Taking into account (3.6), we see that the solution $U(t)$ of (3.7) is invariant under the modular automorphism group (3.9) , whence

$$\begin{aligned}\varphi(\alpha_t(A)) &= \varphi(U(t)\,\alpha^0_t(A)\,U(t)^*) = \varphi(\alpha^0_t(A)\,U(t)^*U(t)) \\ &= \varphi(\alpha^0_t(A)) = \varphi(A) \quad : \quad A \in \mathcal{M} ,\ t \in \mathbb{R}^+ . \square\end{aligned}$$

Remark 3.4. For $\beta > 0$, the pair of processes $\{ B_\beta(t) , B^*_\beta(t) : t \in \mathbb{R}^+\}$ is a multiple of the finite temperature quantum Brownian motion of Hudson and Lindsay ([15] and this volume); for $\beta = 0$ it is a realization of the classical complex Brownian motion.

A generalization of Example 3.3 may be stated as follows:

Theorem 3.5. Any norm continuous dynamical semigroup of $\mathcal{B}(\mathcal{H})$ satisfying the quantum detailed balance condition of [16] with respect to a faithful normal state ρ possesses a stationary Markov dilation.

The proof is a straightforward extension of the reasoning in Example 3.3. See also [5], where the same result is obtained through a different construction.

In the converse direction, we have

Theorem 3.6 (Kümmerer [2]). Suppose that the state $\varphi(.) = \mathrm{Tr}\left[\rho E(.)\right]$ is stationary under $\{\alpha_t : t \in \mathbb{R}\}$. Then

$$\mathrm{Tr}\left[\rho X T_t(Y)\right] = \mathrm{Tr}\left[\rho T_{-t}(X) Y\right] \quad : X , Y \in \mathcal{B}(\mathcal{H}) , t \in \mathbb{R}^+ . \quad (3.10)$$

In particular:

(1) ρ is stationary under both T_t and $\hat{T}_{-t}$;

(2) if ρ is faithful, then T_t commutes with the modular automorphism group σ_t ($= \rho^{it} (.) \rho^{-it}$) associated with it;

(3) if α_t is constructed through quantum Brownian motion, so that $L - L_- = 2i\left[H , .\right]$, then T_t satisfies the quantum detailed balance condition with respect to ρ .

Remark 3.7. As a by-product of Theorem 3.2 , we have that any norm continuous dynamical semigroup on $\mathcal{B}(\mathcal{H})$ with a pure stationary state $\rho = |\phi_0\rangle\langle\phi_0|$ may be said to satisfy the quantum detailed balance condition with respect to ρ (however, the "ρ-adjoint" dynamical semigroup T_{-t} satisfying (3.10) is not uniquely determined; see (2.14)). Since detailed balance characterizes (in some sense) relaxation to equilibrium in isothermal surroundings [16], this was to be expected on physical grounds: the only way to obtain relaxation of an open system to its ground state is to let it interact with external reservoirs all at absolute zero temperature.

4. Stationary Markov processes: general case.

It is easy to construct examples of dilations where T_t has a stationary state ρ , but $\varphi = \mathrm{Tr}\left[\rho E(.)\right]$ is not stationary under α_t , by considering situations in which T_t and ρ fail to satisfy some of the conditions (1)—(3) of Theorem 3.6. In Example 4.n below (n = 1 , 2 , 3) conditions (1) to (n - 1) hold and condition (n) fails.

Example 4.1 (laser model [17]). The system under consideration consists of two-level atoms interacting with one mode of the radiation field. For the sake of notational simplicity, we shall consider only one atom. The generator L of T_t is given by

$$
\begin{aligned}
L(X) = \; & i\,\omega\left[a^*a\, ,\, X\right] + \kappa\left(a^* X a - \tfrac{1}{2}\left[a^*a\, ,\, X\right]_+\right) \\
& + i\,\varepsilon\left[S^z, X\right] + \gamma_\downarrow\left(S^+ X S^- - \tfrac{1}{2}\left[S^+S^-, X\right]_+\right) + \gamma_\uparrow\left(S^- X S^+ - \tfrac{1}{2}\left[S^-S^+, X\right]_+\right) \\
& + i\,g\left[a S^+ + a^* S^-, X\right] \quad , \qquad (4.1)
\end{aligned}
$$

where, as usual,

$$
a\,|n\rangle = \sqrt{n}\,|n-1\rangle \quad , \quad a^*|n\rangle = \sqrt{n+1}\,|n+1\rangle \quad : n = 0, 1, 2, .. \qquad (4.2)
$$

and

$$
S^+ = \begin{pmatrix} 0 & 1 \\ 0 & 0 \end{pmatrix}, \quad S^- = \begin{pmatrix} 0 & 0 \\ 1 & 0 \end{pmatrix}, \quad S^z = \frac{1}{2}\begin{pmatrix} 1 & 0 \\ 0 & -1 \end{pmatrix}. \qquad (4.3)
$$

If a dilation of T_t is constructed through quantum Brownian motion (or, equivalently, through singular couplings to quasi-free reservoirs, as in [18]), then the generator L_- of the time reversed semigroup T_{-t} is obtained from L by changing the sign of the Hamiltonian part ($i \longmapsto -i$ in (4.1)). If a state ρ were stationary under both T_t and T_{-t}, then one would simultaneously have

$$
\kappa\left(a\,\rho\, a^* - \tfrac{1}{2}\left[a^*a\, ,\, \rho\right]_+\right) + \gamma_\downarrow\left(S^-\rho\, S^+ - \tfrac{1}{2}\left[S^+S^-, \rho\right]_+\right) + \gamma_\uparrow\left(S^+\rho\, S^- - \tfrac{1}{2}\left[S^-S^+, \rho\right]_+\right) \qquad (4.4)
$$

and

$$
\omega\left[a^*a\, ,\, \rho\right] + \varepsilon\left[S^z, \rho\right] + g\left[a S^+ + a^* S^-, \rho\right] = 0\ . \qquad (4.5)
$$

However, the only density operator ρ satisfying (4.4) is

$$
\rho = |0\rangle\langle 0| \otimes \frac{1}{1 + \gamma_\uparrow/\gamma_\downarrow}\begin{pmatrix} 1 & 0 \\ 0 & \gamma_\uparrow/\gamma_\downarrow \end{pmatrix}, \qquad (4.6)
$$

which does not satisfy (4.5) for $g \neq 0$ [19].

<u>Example 4.2</u> (heat conduction [20]). Davies' model of a heat conducting bar consists of a chain of two-level atoms interacting through intermediate reservoirs and coupled at its ends to two thermal reservoirs at different inverse temperatures β_L and β_R. In the simplest case of only two atoms, the generator L of the resulting dynamical semigroup T_t is given by

$$
\begin{aligned}
L(X) = \; & i\,\varepsilon\left[S_1^z + S_2^z\, ,\, X\right] \\
& + \gamma_L\left\{\left(S_1^+ X S_1^- - \tfrac{1}{2}\left[S_1^+S_1^-, X\right]_+\right) + e^{-\beta_L \varepsilon}\left(S_1^- X S_1^+ - \tfrac{1}{2}\left[S_1^-S_1^+, X\right]_+\right)\right\} \\
& + \gamma_R\left\{\left(S_2^+ X S_2^- - \tfrac{1}{2}\left[S_2^+S_2^-, X\right]_+\right) + e^{-\beta_R \varepsilon}\left(S_2^- X S_2^+ - \tfrac{1}{2}\left[S_2^-S_2^+, X\right]_+\right)\right\} +
\end{aligned}
$$

$$+ \gamma\left\{ S_1^+ S_2^- X\, S_1^- S_2^+ - \tfrac{1}{2}\left[S_1^+ S_1^- S_2^- S_2^+ ,\, X\right]_+ + \quad S_1^- S_2^+ X\, S_1^+ S_2^- - \tfrac{1}{2}\left[S_1^- S_1^+ S_2^+ S_2^- ,\, X\right]_+ \right\} , \tag{4.7}$$

where

$$S_1^{\#} = S^{\#} \otimes \mathbb{1} = \begin{pmatrix} S^{\#} & 0 \\ 0 & S^{\#} \end{pmatrix} , \quad S_2^{\#} = \mathbb{1} \otimes S^{\#} \; : \; \# = + , - , z . \tag{4.8}$$

Supposing

$$\gamma_L = (1 + e^{-\beta_L \varepsilon})^{-1} \gamma \qquad , \; \gamma_R = (1 + e^{-\beta_R \varepsilon})^{-1} \gamma , \tag{4.9}$$

and letting

$$\vartheta_L = \tanh(\beta_L \varepsilon/2) \qquad , \; \vartheta_R = \tanh(\beta_R \varepsilon/2) , \tag{4.10}$$

the unique stationary state ρ is found to be

$$\rho = \frac{1}{4}\mathbb{1} - \frac{1}{6}\left[(2\vartheta_L + \vartheta_R)S_1^z + (\vartheta_L + 2\vartheta_R)S_2^z - (\vartheta_L^2 + \vartheta_R^2 + 4\vartheta_L\vartheta_R)S_1^z S_2^z\right] . \tag{4.11}$$

Now ρ is invariant also under T_{-t} , but T_t does not commute with the modular automorphism $\sigma_t = \rho^{it}(.)\,\rho^{-it}$, unless $\beta_L = \beta_R$; hence it does not have a stationary Markov dilation.

<u>Example 4.3</u> (heat conduction, modified). With the same assumptions and notations as in Example 4.2, let

$$L^{\natural}(X) = \lim_{a \to \infty} \frac{1}{2a} \int_{-a}^{a} \rho^{it}\, L(\rho^{-it}\, X\, \rho^{it})\, \rho^{-it}\, dt \; ; \tag{4.12}$$

then $T_t^{\natural} = \exp\left[L^{\natural} t\right]$ commutes with σ_t and has the same action as T_t on the algebra of "classical" observables (diagonal operators). However, also $T_t^{\natural}$ does not satisfy the quantum detailed balance condition with respect to its stationary state (4.11); hence the state $\varphi = \mathrm{Tr}\left[\rho\, E(.)\right]$ is not stationary under the dilation α_t constructed through quantum Brownian motion. It remains an open problem whether it is possible to find a stationary Markov dilation of $T_t^{\natural}$ by different means. Since the explicit form of $L^{\natural}$ might possibly be useful for this purpose, we give it below, in terms of the matrix units in $M(4, \mathbb{C}) = M(2, \mathbb{C}) \otimes M(2, \mathbb{C})$, denoted by D_{ij} : $i , j = 1,\dots,4$,

$$(D_{ij})_{kl} = \delta_{ki}\, \delta_{lj} \; :$$

$$
\begin{aligned}
L^{\natural}(X) \;=\;& i\,\frac{\varepsilon}{2}\left[D_{11}+D_{44}\,,\,X\right]\\
&+\gamma_L\Big\{\big(D_{12}\,X\,D_{21}-\tfrac{1}{2}\left[D_{11},X\right]_+ + D_{34}\,X\,D_{43}-\tfrac{1}{2}\left[D_{33},X\right]_+\big)\\
&\qquad + e^{-\beta_L\varepsilon}\big(D_{21}\,X\,D_{12}-\tfrac{1}{2}\left[D_{22},X\right]_+ + D_{43}\,X\,D_{34}-\tfrac{1}{2}\left[D_{44},X\right]_+\big)\Big\}\\
&+\gamma_R\Big\{\big(D_{13}\,X\,D_{31}-\tfrac{1}{2}\left[D_{11},X\right]_+ + D_{24}\,X\,D_{42}-\tfrac{1}{2}\left[D_{22},X\right]_+\big)\\
&\qquad + e^{-\beta_R\varepsilon}\big(D_{31}\,X\,D_{13}-\tfrac{1}{2}\left[D_{33},X\right]_+ + D_{42}\,X\,D_{24}-\tfrac{1}{2}\left[D_{44},X\right]_+\big)\Big\}\\
&+\gamma\,\big(D_{32}\,X\,D_{23}-\tfrac{1}{2}\left[D_{33},X\right]_+ + D_{23}\,X\,D_{32}-\tfrac{1}{2}\left[D_{22},X\right]_+\big)\;.
\end{aligned}
\tag{4.13}
$$

When ρ is a stationary state for T_t, but $\varphi = \mathrm{Tr}\left[\rho\, E(.)\right]$ is not a stationary state for α_t, it might be expected that a stationary state φ_∞ is approached by $\varphi\circ\alpha_t$ in the limit as $t\longmapsto\infty$. Here we show that this is indeed the case, and that $(\mathcal{A},\{j_t : t\in\mathbb{R}\},\varphi_\infty)$ is a stationary Markov process.

<u>Theorem 4.4.</u> Let $(\mathcal{A},\{j_t=\alpha_t\, j_0 : t\in\mathbb{R}\},\varphi = \mathrm{Tr}[\rho\, E(.)])$ be a covariant Markov process over $\mathcal{B}(\mathcal{H})$, where ρ is a stationary state for the associated dynamical semigroup T_t. Then, for all "local observables" A, the limit

$$
\varphi_\infty(A) = \lim_{t\to+\infty}\varphi(\alpha_t(A)) \;:\; A\in\mathcal{A}_I\;,\; I \text{ bounded} \subseteq \mathbb{R}, \tag{4.14}
$$

exists, and defines a locally normal state φ_∞ on $\mathcal{A}$, which is stationary under $\{\alpha_t : t\in\mathbb{R}\}$ and compatible with the conditional expectations $\{E_{s]} : s\in\mathbb{R}^+\}$. Then $(\mathcal{A},\{j_t : t\in\mathbb{R}\},\varphi_\infty)$ is a stationary Markov W*-stochastic process. Explicitly, we have, for all $X_1,\dots,X_n$ in $\mathcal{B}(\mathcal{H})$, $t_1,\dots,t_n$ in $\mathbb{R}$, and n in $\mathbb{N}$,

$$
\varphi_\infty(j_{t_1}(X_1)\dots j_{t_n}(X_n)) = \mathrm{Tr}\left[\rho\, E(j_{t_1+t}(X_1)\dots j_{t_n+t}(X_n))\right]\;, \tag{4.15}
$$

where t is any real number such that $t_1+t,\dots,t_n+t$ are in $\mathbb{R}^+$.

<u>Proof</u>(Sketch). Let $A\in\mathcal{A}_{[-t_1,t_2]} : t_1,\, t_2\in\mathbb{R}^+$. For $t>t_1$, we have

$$
\begin{aligned}
\varphi(\alpha_t(A)) &= \varphi(\alpha_{t-t_1}\,\alpha_{t_1}(A)) = \varphi(E_{t-t_1]}\alpha_{t-t_1}\,\alpha_{t_1}(A))\\
&= \varphi(\alpha_{t-t_1}E_{0]}\alpha_{t_1}(A))\;,
\end{aligned}
\tag{4.16}
$$

by (2.9) and (2.7). Now $\alpha_{t_1}(A)$ is in $\mathcal{A}_{[0,t_1+t_2]}$, and by the Markov property (2.8) there is $X(A,t_1)$ in $\mathcal{B}(\mathcal{H})$ such that

$$E_{0]}\alpha_{t_1}(A) = j_0(X(A,t_1)) . \tag{4.17}$$

Inserting (4.17) into (4.16) and using (2.12), we obtain

$$\begin{aligned}\varphi(\alpha_t(A)) &= \mathrm{Tr}\left[\rho\, E(\alpha_{t-t_1}\, j_0(X(A,t_1)))\right] \\ &= \mathrm{Tr}\left[\rho\, T_{t-t_1}(X(A,t_1))\right] = \mathrm{Tr}\left[\rho\, X(A,t_1)\right] : t > t_1,\end{aligned} \tag{4.18}$$

which is independent of t for $t > t_1$. Then the limit (4.14) exists, and defines a state φ_∞ on $\mathcal{A}$ which is stationary under α_t. Clearly φ_∞ and φ have the same restriction to $\mathcal{A}_{[0,+\infty)}$; hence (4.15) holds, and φ_∞ is locally normal as φ is. We have also, for s, t in $\mathbb{R}^+$ and A in $\mathcal{A}$,

$$\varphi(\alpha_t E_{s]}(A)) = \varphi(E_{s+t]}\alpha_t(A)) = \varphi(\alpha_t(A)) , \tag{4.19}$$

so that φ_∞ is compatible with $E_{s]}$. □

<u>Remark 4.5</u>. Another stationary state $\varphi_{-\infty}$ for α_t can be constructed by taking ρ to be stationary under T_{-t} and letting

$$\varphi_{-\infty}(A) = \lim_{t\to-\infty} \varphi(\alpha_t(A)) : \quad A \in \mathcal{A}_I \quad , \text{ I bounded} \subseteq \mathbb{R} . \tag{4.20}$$

Then $\varphi_{-\infty}$ is compatible with $\{E_{[-s} : s \in \mathbb{R}^+\}$. If T_t does not have a stationary Markov dilation, then necessarily $\varphi_\infty \neq \varphi_{-\infty}$, since otherwise it would be compatible with $E_0 = E_{0]}E_{[0}$. This is, for instance, the situation of Example 4.2 (and also of Example 4.3, if α_t is constructed through quantum Brownian motion), although the same state ρ is stationary under T_t and T_{-t}. This asymmetry between past and future is linked with the lack of "microreversibility"[19].

References.

1. Evans, D.E., and Lewis, J.T.: Dilations of irreversible evolutions in algebraic quantum theory. Commun. Dublin Institute for Advanced Studies, Ser. A, No. 24, 1977.
2. Kümmerer, B.: A dilation theory for completely positive operators on W*-algebras. Thesis, Tübingen, 1982;
   ___: Markov dilations on W*-algebras. J. Funct. Anal. (to appear).
3. Kümmerer, B., and Schröder, W.: A Markov dilation of a non-quasifree Bloch evolution. Commun. Math. Phys. 90, 251-262 (1983).
4. Kümmerer, B.: Examples of Markov dilations over the 2 × 2 matrices. In: Accardi, L., Frigerio, A., and Gorini, V.(Eds.): Quantum Probability and Applications to the Quantum Theory of Irreversible Processes; Proceedings, Villa Mondragone, 1982. Lecture Notes in Mathematics 1055, pp. 228-244. Berlin Heidelberg New York Tokyo, Springer-Verlag, 1984.
5. Frigerio, A., and Gorini, V.: Markov dilations and quantum detailed balance. Commun. Math. Phys. 93, 517-532 (1984).
6. Hudson, R.L., and Parthasarathy, K.R.: Quantum Ito's formula and stochastic evolutions. Commun. Math. Phys. 93, 301-323 (1984).
7. Applebaum, D.B., and Hudson, R.L.: Fermion Ito's formula and stochastic evolutions. Commun. Math. Phys. (to appear).
8. Hudson, R.L., and Parthasarathy, K.R.: Stochastic dilations of uniformly continuous completely positive semigroups. Acta Math. Applicandae (to appear).
9. Frigerio, A.: Covariant Markov dilations of quantum dynamical semigroups. Preprint, 1984.
10. Maassen H.: The construction of continuous dilations by solving quantum stochastic differential equations. Semesterbericht Funktionalanalysis Tübingen, Sommersemester 1984, 183-204 (1984).
11. Lindblad, G.: On the generators of quantum dynamical semigroups. Commun. Math. Phys. 48, 119-130 (1976).
12. Hudson, R.L., and Streater, R.F.: Itô's formula is the chain rule with Wick ordering. Phys. Lett. 86 A , 277-279 (1981).
13. Accardi, L., Frigerio, A., and Lewis, J.T.: Quantum stochastic processes. Publ. RIMS Kyoto Univ. 18, 97-113 (1982).
14. Accardi, L.: On the quantum Feynman-Kac formula. Rend. Sem. Mat. Fis. Milano 48, 135-180 (1980).
15. Hudson, R.L., and Lindsay, J.M.: A non-commutative martingale representation theorem for non-Fock quantum Brownian motion. J. Funct. Anal. (to appear).
16. Kossakowski, A., Frigerio, A., Gorini, V., and Verri, M.: Quantum detailed balance and KMS condition. Commun. Math. Phys. 57, 97-110 (1977).
17. Haken, H.: Laser Theory. Handbuch der Physik., vol. XXV/2c. Berlin Heidelberg New York, Springer-Verlag, 1970.
18. Hepp, K., and Lieb, E.H.: Phase transitions in reservoir-driven open systems, with applications to superconductors and lasers. Helv. Phys. Acta 46, 575-603 (1973).
19. Agarwal, G.S.: Open quantum Markovian systems and the microreversibility. Z. Phys. 258, 409-422 (1973).
20. Davies, E.B.: A model of heat conduction. J. Stat. Phys. 18, 161-170 (1978).

A MODEL FOR A UNIFIED QUANTUM DESCRIPTION OF MACROSCOPIC AND MICROSCOPIC SYSTEMS

G.C.Ghirardi[*], A. Rimini[**], T. Weber[***]

1. Introductory considerations.

As is well known crucial conceptual problems in Quantum Theory arise in connection with the description of the behaviour of macroscopic objects and of their interactions with microscopic ones. Even though most features of the behaviour of macroscopic objects are accounted for by quantum mechanics in a natural way due to the irrelevant spreads of wave packets for macroscopic masses, one of the basic principles of quantum theory, i.e. the superposition principle, becomes puzzling when it involves macroscopically distinguishable states of a macro-object. This occurs for instance in the measurement process.

Various solutions for these difficulties have been proposed,which can be schematically fitted into one of the two following conceptual frameworks:

a) One accepts two principles of evolution yielding a different dynamical behaviour for micro and macro-objects.

b) One limits <u>in principle</u> the set of observables of a macrosystem to an Abelian set.

We do not want to enter here into the delicate question of whether these attitudes can lead to a satisfactory solution of the above mentioned conceptual difficulties. At any rate we want to stress that, in our opinion, to keep the standard quantum dynamics and to abandon or to make ineffective for some systems the superposition principle amounts to accepting (at least to a certain extent) a dualistic description of natural phenomena. This means to give up the program of a unified derivation of the behaviour of all objects from the basic dynamics of the microscopic world.

We present here an attempt of such a unified description through the discussion of a dynamical model in which linear superpositions of states describing systems localized in far apart spatial regions are naturally suppressed for macroscopic bodies.Let us sketch the line of thought we will follow to obtain this result.We start by considering the dynamics of macro-objects. We accept a modification of the dynamics of these objects with respect to the standard one implied by quantum mechanics, keeping

---

* Istituto di Fisica Teorica, Universita' di Trieste and ICTP, Trieste Italy.
** Dipartimento di Fisica Nucleare e Teorica, Universita' di Pavia , Italy.
*** Istituto di Fisica Teorica, Universita' di Trieste, Italy.

in mind the requirement of suppressing linear combinations of far apart localized states (as an example one can think of the elimination of states involving linear superpositions of different pointer positions of a measuring instrument). To satisfy the above requirement the dynamical equation must induce transitions from pure states to statistical mixtures. A way of obtaining this result is to add in the dynamical equation a term corresponding to some measurement process to the one describing the Hamiltonian evolution. With reference to our programme we are then naturally led to consider localization measurements.

Up to now, as we are dealing with a macrosystem, the introduction of an irreversible dynamics could be justified by recalling that, as properly pointed out by various authors(1), a macro-object can never be considered as isolated. Our equation could then be considered as describing the reduced dynamics of an object interacting with some external environment. Whether these interactions with the rest of the world can be accounted for as measurements is obviously open to debate. Here we are not interested in discussing this point, since we want to take a very different attitude, i.e. we will postulate that also the microscopic objects are governed by a dynamical equation of the previously introduced type. We are induced to make this assumption by the fact that(as we shall show below):

i) One can choose the parameters in the equation in such a way that the dynamics of microsystems coincides for all practical purposes with the standard Hamiltonian quantum dynamics.

ii) The dynamics of a macro object can be consistently deduced from that of its microscopic components and turns out to forbid linear superpositions of far away states, and to give an evolution compatible with classical mechanics.

Our starting point, i.e. the introduction of a non-Hamiltonian dynamical equation for a macro object, has been inspired by the important works of A. Barchielli, L. Lanz and G.M. Prosperi(2) and constitutes a generalization of the equations they have used.

2. The Evolution Equation

We deal with a macroscopic particle in one dimension. As already stated we will consider it as subjected to appropriate, obviously approximate, localization measurements. If one wants to introduce such processes, one either uses projection operators on definite space intervals (and introduces therefore an arbitrary discretization of space), or resorts to the concept of operation valued measures(3).

Following ref. (2) we consider the operation valued measure

$$T_I[\varrho] = \sqrt{\frac{\alpha}{\pi}} \int_I dx\, e^{-\frac{\alpha}{2}(\hat{q}-x)^2} \varrho\, e^{-\frac{\alpha}{2}(\hat{q}-x)^2} \quad , \tag{1}$$

where I is a Borel set in $\mathbb{R}$ and $\hat{q}$ is the position operator. In connection with the process $T_I[\cdot]$ one defines the probability $P(q\in I|\varrho)$ that the system in the state ϱ be found in the Borel set I in the position measurement, according to $P(q\in I|\varrho) = \mathrm{Tr}\, T_I[\varrho]$.In ref. 2 the system was considered as evolving by pure Hamiltonian dynamics and to be subjected to the process described by (1) at definite equally spaced instants. This discretization of the time axis was then eliminated by taking in a suitable way the infinite frequency limit, (i.e. if the time interval between two measurements is denoted by $1/\lambda$, one takes $\lambda\to\infty, \lambda\alpha = const$). The aim of the authors of ref. (2) was to define a functional probability distribution on an appropriate σ-algebra of the subsets of the space of the continuous functions of t (trajectories). The infinite frequency limit raises problems since it forbids the direct use of the process $T_I[\cdot]$ to perform selections on the statistical ensemble in order to define the probability distributions. The way used in ref. (2) to overcome this difficulty consists in adding to the infinite frequency measurement process a purely selective process, the selection being based on the mean values of the results of the measurements occurring between two selections.

To eliminate the arbitrary discretization of time we follow another way, i.e. we assume that the process described by $T_I[\cdot]$ occurs at random times[(4)]. In this way there is no need to take the infinite frequency limit for the process.If no selection on the basis of the results of the measurements is performed the evolution equation for the statistical operator ϱ is

$$\frac{d\varrho}{dt} = -\frac{i}{\hbar}[H,\varrho] - \lambda(\varrho - T[\varrho]) \quad , \tag{2}$$

where

$$T[\varrho] = \sqrt{\frac{\alpha}{\pi}} \int_{-\infty}^{+\infty} dx\, e^{-\frac{\alpha}{2}(\hat{q}-x)^2} \varrho\, e^{-\frac{\alpha}{2}(\hat{q}-x)^2} \, . \tag{3}$$

We note that in the infinite frequency limit $\lambda\to\infty, \lambda\alpha =$ const. equation (2) becomes the basic equation considered in ref.(2).

3. Many Particle Systems.

In the previous section Eq. (2) has been introduced as referring to a macroscopic system. We shall later investigate its implications.

In the spirit of obtaining a unified description of micro- and macro-systems, we tentatively assume that the approximate measurement process

$T[\cdot]$ acts individually on each constituent of a many particle system.

The evolution equation for such a system is then

$$\frac{d\varrho}{dt} = -\frac{i}{\hbar}[H,\varrho] - \sum_i \lambda_i \left(\varrho - T_i[\varrho]\right) \quad , \tag{4}$$

where $T_i[\cdot]$ is defined as in eq. (3) with $\hat{q}_i$ replacing $\hat{q}$. The parameters λ_i characterize the frequencies of the processes suffered by the constituents of the system; we have taken the parameter α which is related to the accuracy of the localizations, to be the same for all constituents.

We now want to show that the presence of the non-Hamiltonian terms for all constituents gives rise to an analogous localization process taking place for the centre of mass motion, but occurring with a much larger frequency. The simplest way to see this is obtained by deriving the reduced dynamical equation for the centre of mass motion implied by eq. (4). To this purpose we introduce the centre of mass and relative motion position operators $\hat{Q}$ and $\hat{r}_j$ respectively, which are connected to the operators $\hat{q}_i$ by: $\hat{q}_i = \hat{Q} + \sum_j t_{ij} \hat{r}_j$. If we assume that the Hamiltonian H of the system can be split into the sum of the centre of mass and internal motion parts $H = H_Q + H_r$, which act in the respective Hilbert spaces, taking the partial trace on the internal degrees of freedom and defining the statistical operator $\varrho_Q = Tr^{(r)}\varrho$ one obtains from eq. (4):

$$\frac{d\varrho_Q}{dt} = -\frac{i}{\hbar}[H_Q, \varrho_Q] - \left(\sum_i \lambda_i\right)\left(\varrho_Q - T_Q[\varrho_Q]\right) \quad , \tag{5}$$

where $T_Q[\cdot]$ is again given by eq. (3) with the operator $\hat{Q}$ replacing $\hat{q}$. Thus, when the constituents are subjected to localizations with frequencies λ_i, the centre of mass suffers analogous localizations with frequency $\sum_i \lambda_i$.

It is worthwhile stressing that the non-Hamiltonian term in eq. (5) is a direct consequence of the presence of the analogous terms of eq. (4) and is not due to the elimination of the internal degrees of freedom. In fact, if one starts with a composite system with Hamiltonian dynamics, the reduced dynamics for the centre of mass motion is always Hamiltonian and therefore allows the occurrence of linear superpositions of far away states of the centre of mass. To forbid this, one has to couple the system to some other system whose dynamics is then eliminated. This, however, gives rise to a chain procedure when larger and larger external parts are included. If one wants to reach a point where linear superpositions of far away states cannot occur one has to break in an arbitrary way this chain. On the contrary, in our case, even when the localization processes occur extremely seldom for the constituents, their effects sum up for the centre of mass. This makes more and more rapid with increasing number of constituents the suppression of the off diagonal elements which, as we shall see, is induced by the non-Hamiltonian term of eq. (5).

The mechanism by which any localization process on one of the constituents becomes a localization process for the reduced dynamics of the centre of mass is easily understood. In fact, suppose the statistical operator for the system in coordinate representation $(q_1, \ldots q_N|\rho|q_1 \ldots q_N)$ be practically zero when $|q_k - q_k'| > \Delta$. By going to the centre of mass one has that the matrix element of the reduced statistical operator ρ_Q :

$$\langle Q|\rho_Q|Q'\rangle = \int \langle Q+\sum_j t_{1j} r_j, \ldots |\rho| Q'+\sum_j t_{1j} r_j, \ldots\rangle dr_1 \ldots dr_{N-1}$$

vanishes for $|Q-Q'| > \Delta$. There follows that any process which decomposes the state of a constituent of the system into a statistical mixture of well localized states, does the same for the reduced statistical operator for the centre of mass. From eq. (5) we see that, if one assumes that the test frequencies λ_i for all microscopic (e.g. atomic) constituents of a macro object are of the same magnitude, $(\lambda_i = \lambda_{micro})$, the centre of mass is affected by the same process with a frequency $\lambda_{macro} = \mathcal{N} \lambda_{micro}$, where $\mathcal{N}$ is of the order of Avogadro's number. As we shall see this allows us to choose the parameters λ_{micro} and α in such a way that standard quantum mechanics holds exactly for extremely long times for microscopic systems, while for macrosystems possible linear superpositions of far away states are almost immediately suppressed by the dynamical equation.

One can also see that, in the case in which the internal motion Hamiltonian gives rise to a sharp (with respect to $1/\sqrt{\alpha}$) localization of the internal coordinates (as it happens, for appropriate choice of α, e.g. in a crystal) the internal and centre of mass motions decouple almost exactly and the internal motion is not affected (for suitable choices of λ_i) by the non-Hamiltonian process.

4. The Free Particle

It is quite evident that, for sufficiently large values of α and λ, eqs. (2),(3) lead to a rapid suppression of the off-diagonal elements of the statistical operator in the coordinate representation. Linear superpositions of far away states are therefore forbidden. It is, however, necessary to check that the modification of the Hamiltonian evolution of the system induced by the localizations, does not give rise to appreciable deviations from the classical behaviour of a macro-object. We limit our discussion to the case in which the operator H is the Hamiltonian of a free particle. In this Section we solve the evolution equation and we derive some relations whose physical implications will be discussed later.

A. Solution of the evolution equation

In coordinate representation eq. (2) becomes

$$\frac{\partial \rho(q,q',t)}{\partial t} = \frac{i\hbar}{2m}\left\{\frac{\partial^2\rho(q,q',t)}{\partial q^2} - \frac{\partial^2\rho(q,q',t)}{\partial q'^2}\right\} - \lambda\left\{1 - e^{-\frac{\alpha}{4}(q-q')^2}\right\}\rho(q,q',t). \tag{6}$$

One can express the solution of the above equation satisfying certain initial conditions in terms of the solution $\rho_{Sch}(q, q',t)$ of the same equation with $\lambda = 0$ satisfying the same initial conditions according to

$$\rho(q,q',t) = \frac{1}{2\pi}\int_{-\infty}^{+\infty} d\mu \int_{-\infty}^{+\infty} dy\, e^{-i\mu y} F(\lambda,\mu,q-q',t)\,\rho_{Sch}\left(q+\frac{y}{2}, q'+\frac{y}{2}, t\right), \tag{7}$$

where

$$F(\lambda,\mu,q-q',t) = e^{-\lambda t}\, e^{\lambda\int_0^t e^{-\frac{\alpha}{4}\left[\frac{2\mu\hbar}{m}y - (q-q')\right]^2} dy}. \tag{8}$$

The fact that expression (7) satisfies eq. (6), under appropriate regularity properties of $\rho_{Sch}(q,q',t)$, is easily checked. We note that $F(0,\mu,q-q',t)=1$. This implies $\rho(q,q',t) = \rho_{Sch}(q,q',t)$ as it must be. We list here some properties of the function F which will be useful in what follows:

$$F(\lambda,0,0,t) = 1, \quad \left.\frac{\partial F(\lambda,\mu,0,t)}{\partial \mu}\right|_{\mu=0} = 0, \quad \left.\frac{\partial^2 F(\lambda,\mu,0,t)}{\partial \mu^2}\right|_{\mu=0} = -\frac{2\hbar^2\alpha\lambda}{3m^2}t^3, \tag{9a}$$

$$\left.\frac{\partial F(\lambda,\mu,q-q',t)}{\partial q}\right|_{\substack{q=q'\\ \mu=0}} = 0, \quad \left.\frac{\partial^2 F(\lambda,\mu,q-q',t)}{\partial q^2}\right|_{\substack{q=q'\\ \mu=0}} = -\frac{\lambda\alpha}{2}t, \tag{9b}$$

$$1 - F(\lambda,\mu,0,t) \leq \frac{\lambda\alpha\mu^2\hbar^2}{m^2}t^3. \tag{9c}$$

Note that this inequality is meaningful only when its r.h.s. is smaller than 1.

We shall also use a bound on F for $(q-q') > 2\sqrt{\frac{\pi}{\alpha}}$. To this purpose we remark that the function F can be written as $F(\lambda,\mu,q-q',t) = e^{-\lambda t}\, e^{\lambda t\, h(A\mu, \frac{\sqrt{\alpha}}{2}(q-q'))}$, where $A = \hbar\sqrt{\alpha}\,t/m$ and $h(x,\omega) = \frac{1}{x}\int_0^x e^{-(y-\omega)^2} dy$.

Since $h(x,\omega)$, as a function of x is a positive function with only one maximum at a point x_M satisfying $\omega \leq x_M \leq 2\omega$ there immediately follows

$$h(x_M,\omega) = \frac{1}{x_M}\int_0^{x_M} e^{-(y-\omega)^2} dy < \frac{\sqrt{\pi}\,\mathrm{erf}(\omega)}{\omega}.$$

From the above expression for F we then get for $q > q'$ the inequality

$$F(\lambda,\mu,q-q',t) < e^{-\lambda\beta t} \quad , \quad \beta = 1 - \frac{\sqrt{\pi}}{\frac{\sqrt{\alpha}}{2}(q-q')} \operatorname{erf}\left[\sqrt{\frac{\alpha}{2}}(q-q')\right] . \tag{9d}$$

B. Mean values and spreads of position and momentum.

We now evaluate the mean values and spreads of the position and momentum operators induced by the general solution (7) of eq.(6). We have

$$\langle\hat{q}\rangle = \mathrm{Tr}[\hat{q}\rho] = \frac{1}{2\pi}\int_{-\infty}^{+\infty} d\mu \int_{-\infty}^{+\infty} dy \int_{-\infty}^{+\infty} dq \left(q-\frac{y}{2}\right) e^{-i\mu y} F(\lambda,\mu,0,t)\rho_{Sch}(q,q,t) =$$

$$= \int_{-\infty}^{+\infty} dq\, q\, \rho_{Sch}(q,q,t)\, F(\lambda,0,0,t) + \frac{i}{2}\int_{-\infty}^{+\infty} dq\, \rho_{Sch}(q,q,t) \left.\frac{\partial F(\lambda,\mu,0,t)}{\partial\mu}\right|_{\mu=0} = \langle\hat{q}\rangle_{Sch} . \tag{10a}$$

In deriving the result we have assumed appropriate regularity properties of $\rho_{Sch}(q,q,t)$, and we have used eq.(9a). We have indicated by $\langle\hat{q}\rangle_{Sch}$ the mean value associated to $\rho_{Sch}(q,q',t)$. Analogously we get

$$\langle\hat{q}^2\rangle = \langle\hat{q}^2\rangle_{Sch} F(\lambda,0,0,t) + i\langle\hat{q}\rangle_{Sch} \left.\frac{\partial F(\lambda,\mu,0,t)}{\partial\mu}\right|_{\mu=0} -$$

$$- \frac{1}{4}\left\{\mathrm{Tr}\,\rho_{Sch}(t)\right\} \left.\frac{\partial^2 F(\lambda,\mu,0,t)}{\partial\mu^2}\right|_{\mu=0} = \langle\hat{q}^2\rangle_{Sch} + \frac{\hbar^2\alpha\lambda}{6m^2}t^3 \quad , \tag{10b}$$

where we have used eq.(9a) and the fact that the pure Schroedinger evolution is trace preserving. With an analogous procedure one gets for the mean values of $\hat{p}$ and $\hat{p}^2$

$$\langle\hat{p}\rangle = \langle\hat{p}\rangle_{Sch} - i\hbar \left.\frac{\partial F(\lambda,\mu,q-q',t)}{\partial q}\right|_{\substack{q'=q\\ \mu=0}} \mathrm{Tr}\,\rho_{Sch}(t) = \langle\hat{p}\rangle_{Sch} \tag{11a}$$

and

$$\langle\hat{p}^2\rangle = \langle\hat{p}^2\rangle_{Sch} - \hbar^2 \left.\frac{\partial^2 F(\lambda,\mu,q-q',t)}{\partial q^2}\right|_{\substack{q'=q\\ \mu=0}} \mathrm{Tr}\,\rho_{Sch}(t) -$$

$$- 2i\hbar\langle\hat{p}\rangle_{Sch} \left.\frac{\partial F(\lambda,\mu,q-q',t)}{\partial q}\right|_{\substack{q'=q\\ \mu=0}} = \langle\hat{p}^2\rangle_{Sch} + \frac{\hbar^2\lambda\alpha}{2}t \quad , \tag{11b}$$

where we have used eqs. (9b).

Summarizing we have for the mean values and spreads:

$$\langle\hat{q}\rangle = \langle\hat{q}\rangle_{Sch} \quad , \quad \langle\hat{p}\rangle = \langle\hat{p}\rangle_{Sch} \quad , \tag{12a}$$

$$\Delta q^2 = (\Delta q^2)_{Sch} + \frac{\hbar^2 \alpha\lambda}{6m^2} t^3 \quad , \quad \Delta p^2 = (\Delta p^2)_{Sch} + \frac{\lambda\alpha\hbar^2}{2} t \quad . \tag{12b}$$

We note that the mean values are not affected by the non-Hamiltonian term in eq.(2), while, as far as the spreads are concerned, this term induces a kind of stochasticity which gives rise to the modification, with respect to the pure Schroedinger case, exhibited in eq.(12b).

C. Comments on the relations between $\rho(t)$ and $\rho_{Sch}(t)$.

Let us recall that $\lambda = 0$ implies $F(\lambda, \mu, q-q', t) = 1$ and $\rho(q,q',t) = \rho_{Sch}(q,q',t)$. We further note that the properties of the function F are remarkably different in the two cases $q = q'$ and $q \neq q'$.

In the first case, according to eq.(9c), for not too large times F is near to one. The scale of time for $F \simeq 1$ is determined by the coefficient of t^3 at the r.h.s. of eq.(9c). If the Fourier transform of ρ_{Sch} is strongly damped in μ (as it is usually the case for reasonable intial conditions), eq.(9c) gives a bound on the difference $\rho - \rho_{Sch}$. This fact can be used to prove that the diagonal matrix elements of ρ_{Sch} which are appreciably different from zero, coincide with the same elements of ρ up to an appropriate time. This time turns out to be the same as the one for which the modification to Δq^2 given in eq.(12b) can be neglected, i.e.

$$t \simeq \sqrt[3]{\frac{m^2 (\Delta q^2)_{Sch}}{\lambda\alpha\hbar^2}} \quad . \tag{13}$$

In the case $q \neq q'$ and $q - q' > 2\sqrt{\frac{\pi}{\alpha}}$, eq.(9d) gives a bound on F independent of μ. This shows that the expression (7) for $\rho(q,q',t)$ contains an exponentially damped factor whose lifetime τ equals $1/\lambda\beta$.

5. Particle Trajectories

Within our formalism since we keep λ finite we can identify trajectories by using the selective form of eq. (2):

$$\frac{d\rho(t)}{dt} = -\frac{i}{\hbar}[H,\rho(t)] - \lambda\rho + \lambda\sqrt{\frac{\alpha}{\pi}}\int_{I(t)} dx\, e^{-\frac{\alpha}{2}(\hat{q}-x)^2}\rho(t)\, e^{-\frac{\alpha}{2}(\hat{q}-x)^2} , \tag{14}$$

where I(t) for any fixed t is an interval in $\mathbb{R}$. Taking the trace we get

$$\frac{d\,\mathrm{Tr}\,\rho(t)}{dt} = -\lambda\,\mathrm{Tr}\,\rho(t) + \lambda\int_{-\infty}^{+\infty} dq\,\rho(q,q,t)\sqrt{\frac{\alpha}{\pi}}\int_{I(t)} dx\, e^{-\alpha(q-x)^2} . \tag{15}$$

If we call M the maximum extent of the interval I(t) we obtain from eq. (15)

$$\frac{d\,\mathrm{Tr}\,\rho(t)}{dt} < -\lambda\gamma\,\mathrm{Tr}\,\rho(t) \quad , \tag{16}$$

with $\gamma = 1-\mathrm{erf}(M\sqrt{\alpha}/2)$ .Taking into account that $\mathrm{Tr}\,\rho(0) = 1$ we then get

$$\mathrm{Tr}\,\varrho(t) < e^{-\lambda\gamma t} \tag{17}$$

Since γ remains finite for $\alpha \to 0$, one gets that in the limit $\lambda \to \infty$ (even when $\alpha\lambda$ is kept constant) $\mathrm{Tr}\varrho(t) \to 0$ whatever family of intervals I(t) has been chosen, provided I(t) is not the whole real line, which amounts to no selection. This explains why in ref.(2) one cannot use the process itself to select sets of trajectories.

For $t \to \infty$, $\mathrm{Tr}\,\varrho(t) \to 0$ for any choice of the family I(t). This gives rise to a disagreement with the classical case in which the probability of having the particles in a given tube remains constant if the tube is made of possible physical trajectories. However it has to be remarked that the idea of using the process to define trajectories can work only if two conditions are satisfied,i.e., the amplitudes of the intervals I(t) must always be much larger than the localization distance $1/\sqrt{\alpha}$ (because this parameter represents the distance beyond which the ensemble is decomposed in subensembles), i.e. $M\sqrt{\alpha} \gg 1$. Moreover the ensemble, in the time interval which we are interested in, must be subjected to many tests, otherwise the trajectory would not have been identified. This second condition implies $\lambda t \gg 1$. A combined use of the above inequalities shows that the damping factor $\exp(-\gamma\lambda t)$ is ineffective up to times extremely long with respect to $1/\lambda$.

6. Numerical Example and Conclusions.

To illustrate the physical significance of our model we can now choose explicit values for the parameters appearing in it. We take the attitude that the localization distance $1/\sqrt{\alpha}$ is of the order of the precision attainable in the determination of the position of a macro-object. We also assume that the mean time $1/\lambda$ elapsing between two successive localizations for a macro-object is such that the suppression of the off-diagonal elements for $|q-q'|$ larger than the localization distance takes place in a very small fraction of a second. We then choose $1/\sqrt{\alpha} = 10^{-5}$ cm, $\lambda = 10^{7}\,\mathrm{sec}^{-1}$.

According to the analysis of Section 3, this implies $\lambda_{micro} \simeq 10^{-16}\,\mathrm{sec}^{-1}$, which means that a microscopic system is localized once every 10^{8}-10^{9} years. Therefore standard quantum mechanics remains fully valid for this type of systems.

Coming back to macro-objects, let us take, for the sake of definiteness $m \simeq 1g$ and an initial spread of the position Δq_0 again of the order of 10^{-5} cm. For such a system, as well known, the quantum increase of the spread in position is negligible for extremely long times ($\sim 10^{10}$ years), so that the quantum evolution is practically the same as the classical one. The additional term appearing in Δq^2 (see eq.(12b)) becomes of the same order of magnitude of Δq_0^2 for times of the order of 100 years. This is a very long time for keeping isolated a macro-object. A much larger time is required in order that the additional term in Δp^2 become competitive with any reasonably chosen initial spread of the momentum.

As far as the off-diagonal elements are concerned, we have seen that they are exponentially suppressed with a life time $\tau = 1/\lambda\beta$. For $|q-q'| = 4\cdot 10^{-5}$ cm, we have $\tau = 10^{-6}$ sec. Therefore, after times of this order, the linear superpositions of states separated by distances larger than 10^{-4} cm are transformed into statistical mixtures.

Considerations of this type are important for the quantum theory of measurement. In fact, at least in the case in which the interaction leading to the triggering of the apparatus takes place in a very short time, we can apply our considerations to the macroscopic parts of the apparatus itself, obtaining in this way a consistent solution of the difficulties related to the final positions of the pointer.

Concluding, we have here introduced a model having the following features: it is a generalization of quantum mechanics which reduces to standard quantum mechanics for microscopic systems, and implies for macro-objects, when their dynamics is consistently deduced from that of their microscopic constituents,a stochastic behaviour having classical features. In particular linear combinations of far away states are naturally suppressed for a macro-object, a fact which is relevant for the conceptual problems of the quantum theory of measurement.

Finally we want to point out that, taking advantage of the fact that the considered dynamical equation maps the set of statistical mixtures of gaussian states into itself,one can introduce a description of the evolution in terms of a classical phase space density obeying to a Markov diffusion process. This will be discussed elsewhere.

References

1. See e.g. H.D. Zeh in Foundations of Quantum Mechanics (B. d'Espagnat ed.) Academic Press, N.Y. 1971.
2. A. Barchielli, L. Lanz and G.M. Prosperi, Nuovo Cim. B72, 79 (1982); Found. of Phys. 13, 779 (1983); Proc. Int. Symp. Foundations of Quantum Mechanics (S. Kamefuchi ed.) 165, Tokyo 1983.
3. K. Kraus, States, Effects and Operations, Springer-Verlag, Berlin 1983.
4. A. Rimini, Proc. Meeting of Theoretical Physics, Amalfi 1983.

CONDITIONAL EXPECTATIONS IN L^p- SPACES OVER VON NEUMANN ALGEBRAS

S. Goldstein (Łódź)

<u>Introduction</u>. The aim of the paper is to clarify the behaviour of conditional expectations in L^p-spaces over a von Neumann algebra M. We consider only interpolation L^p-spaces associated with a continuous embedding η of M into M_*. The spaces $L^p(M,\eta)$ are obtained by the complex interpolation method of Calderon (see [3], [7]) applied to the compatible pair of Banach spaces $(\eta(M), M_*)$. We restrict our attention to the family η_α, $0 \leq \alpha \leq 1$, of embeddings given by $\eta_\alpha(x) = \sigma_{-i\alpha}(x)\varphi_0$, where φ_0 is a faithful normal state on M. For more detail see Kosaki [7].

Two questions will be examined. The first concerns norm convergence of the sequence $E_n x$ for $x \in L^p(M,\eta_\alpha)$,where E_n are conditonal expectations associated with an increasing sequence M_n of von Neumann subalgebras of M. The answer for $\alpha = 0$ and E_n - projections of norm one was given by the author [5], while for $\alpha = 1/2$ and E_n being conditional expectations in the sense of Accardi and Cecchini [1] by Cecchini and Petz [4]. Theorem 8 contains both the results as special cases.

The second question concerns existence of operators that would act as conditional expectations in $L^p(M,\eta_\alpha)$ with $\alpha \neq 1/2$ (the conditional expectation of Accardi and Cecchini was intended to suit the case $\alpha = 1/2$). Construction of the appropriate "α- conditional expectations" (including $\alpha = 1/2$ as its special case) is described in detail in section "Conditional expectations".

We believe it makes good sense to use the symbols #, ♮ and ♭ whenever the cases $\alpha = 0$, $1/2$ and 1 are considered, and we shall use the convention throughout the paper.

<u>Preliminaries</u>. Let M be a von Neumann algebra acting in a Hilbert space $H = H_M$ with a cyclic and separating vector ξ_0. We write φ_0

for ω_{ξ_0}. The symbols $J = J_M$, $\Delta = \Delta_M$ and $\sigma = \sigma_M$ denote the usual objects in the Tomita - Takesaki theory associated with the triple (M, H, ξ_0), and $L^p(M)$, $p \in [1,\infty]$ are the L^p - spaces of Haagerup (see [6], [11]). We identify M with $L^\infty(M)$ and M_* with $L^1(M)$. Let $h_0 \in L^1(M)$ denote the operator corresponding to φ_0 through this identification, and let η_α, $0 \leq \alpha \leq 1$, be the embedding given by $\eta_\alpha(k) = h_0^\alpha k h_0^{1-\alpha} \in L^1(M)$ for $k \in M$. We put

$$L^\infty(M, \varphi_0, \alpha) \overset{df}{=} \eta_\alpha(M), \quad L^1(M, \varphi_0, \alpha) \overset{df}{=} L^1(M),$$

and define $L^p(M, \varphi_0, \alpha)$ to be the interpolation spaces of Kosaki [7] between $L^\infty(M, \varphi_0, \alpha)$ and $L^1(M, \varphi_0, \alpha)$, so that

$$L^p(M, \varphi_0, \alpha) = h_0^{\alpha/q} L^p(M) h_0^{(1-\alpha)/q}, \text{ where } 1/p + 1/q = 1.$$

We recall that the functional tr on $L^1(M)$ is given by $tr(h_\varphi) = \varphi(1)$, where h_φ is the operator corresponding to $\varphi \in M_*$. The norm $\| \ \|_{p,\alpha}$ in $L^p(M, \varphi_0, \alpha)$ is given by

$$\|h_0^{\alpha/q} x h_0^{(1-\alpha)/q}\|_{p,\alpha} = tr(|x|^p)^{1/p} \quad \text{for } x \in L^p(M),\ 1 \leq p < \infty.$$

Hence, if $x = \eta_\alpha(k) \in L^\infty(M, \varphi_0, \alpha)$ for some $k \in M$, then

$$\|x\|_{p,\alpha} = tr(|h_0^{\alpha/p} k h_0^{(1-\alpha)/p}|^p)^{1/p}, \quad \|x\|_{\infty,\alpha} = \|k\|.$$

We use the norm $\| \ \|_{p,\alpha}$ on M with the meaning $\|k\|_{p,\alpha} \overset{df}{=} \|\eta_\alpha(k)\|_{p,\alpha}$. The following facts will be needed later:

(1) $\|\Delta^{\alpha/2} k\xi_0\| \leq \|k^* \xi_0\|^\alpha \|k \xi_0\|^{1-\alpha}$ for $k \in M$,

which is the Hölder inequality (see [2, (C. 7)]);

(2) $\Delta^{\alpha/2} k\xi_0 \mapsto \eta_\alpha(k)$, $k \in M$, extends to an isometry $\Phi^{(\alpha)}$ of H onto $L^2(M, \varphi_0, \alpha)$ (see [8] and the definition of $\| \ \|_{2,\alpha}$).

<u>Basic inequality</u>.

1. THEOREM. For an arbitrary $x \in L^\infty(M, \varphi_0, \alpha)$, $0 \leq \alpha \leq 1$, and $s = 1, 2, \ldots,$

$$\|x\|_{2^s,\alpha} \leq \|x\|_{1,\alpha}^{1/2^s} \|x\|_{\infty,\alpha}^{1-1/2^s}$$

<u>Proof</u>. By repeated use of the Hölder inequality in Haagerup's L^p - spaces $L^p(M)$ with p-th norm $\|a\|_p = tr(|a|^p)^{1/p}$ we get

$$\|a^{2^s}\|_1 \leq \|a\|_{2^s}^{2^s} \quad \text{for } a \in L^{2^s}(M) \text{ and } s = 1, 2, \ldots .$$

Hence, for $x = \eta_\alpha(k)$ with $k \in M$,

$$\|x\|_{2^{s+1},\alpha}^{2^{s+1}} = \mathrm{tr}((k^* h_o^{\alpha/2^s} k h_o^{(1-\alpha)/2^s})^{2^s})$$
$$\leq \|(k^* h_o^{\alpha/2^s} k h_o^{(1-\alpha)/2^s})^{2^s}\|_1$$
$$\leq \|k^* h_o^{\alpha/2^s} k h_o^{(1-\alpha)/2^s}\|_{2^s}^{2^s}$$
$$\leq \|k\|^{2^s} \|h_o^{\alpha/2^s} k h_o^{(1-\alpha)/2^s}\|_{2^s}^{2^s} = \|x\|_{\infty,\alpha}^{2^s} \|x\|_{2^s,\alpha}^{2^s},$$

and the conclusion of the theorem follows.

2. PROPOSITION. If $k_n \in M$ and $k_n \to 0$ strongly*, then $\|\eta_\alpha(k_n)\|_{2,\alpha} \to 0$.

Proof. Observe that $\|\eta_\#(k_n)\|_{2,\#} = \|k_n \xi_o\|$ and $\|\eta_\flat(k_n)\|_{2,\flat} = \|k_n^* \xi_o\|$, so that it is enough to apply (1) and (2) from Preliminaries.

Conditional expectations.

Let N be a von Neumann subalgebra of M. Put $H_N = [N \xi_o]$. Since $N \ni x \mapsto x|H_N$ is a von Neumann algebra isomorphism, we can (and do) treat N as an algebra acting in H_N with a cyclic and separating vector ξ_o. We denote by N' the commutant of N in the Hilbert space H_N, by P the projection of H_M onto H_N, by Q the inclusion mapping of H_N into H_M, by $\mathcal{A}_M$ and $\mathcal{A}_N$ the sets of analytic elements of M and N, and by $J^{(\alpha)}$ the operator $J\Delta^{(\alpha-1)/2}$. Note that $Px'Q \in N'$ for any $x' \in M'$.

3. PROPOSITION. For a fixed $\alpha \in [0, 1]$,

(i) $\mathcal{A}_M \xi_o \subset \mathcal{D}(\Delta_N^{-\alpha/2} P \Delta_M^{\alpha/2}) \cap \mathcal{D}(J_N^{(\alpha)} P J_M^{(\alpha)})$;

(ii) $\Delta_N^{-\alpha/2} P \Delta_M^{\alpha/2} | \mathcal{A}_M \xi_o = J_N^{(\alpha)} P J_M^{(\alpha)} | \mathcal{A}_M \xi_o$;

(iii) the operator $E^{(\alpha)} = E_{N,M}^{(\alpha)} : \mathcal{A}_M \xi_o \to H_N$ given by (ii) is bounded and of norm one.

Proof. Of course, $\mathcal{A}_M \xi_o \subset \mathcal{D}(\Delta_M^{\alpha/2}) \cap \mathcal{D}(J_M^{(\alpha)})$ and

$$\Delta_M^{\alpha/2}(\mathcal{A}_M \xi_o) \cup J_M^{(\alpha)}(\mathcal{A}_M \xi_o) \subset \mathcal{A}_M \xi_o \subset M' \xi_o .$$

Hence,

$$P\Delta_M^{\alpha/2}(\mathcal{A}_M\xi_o)\cup PJ_M^{(\alpha)}(\mathcal{A}_M\xi_o)\subset N'\xi_o$$

and (i) follows. Let now $x\in\mathcal{A}_M$ and $y\in\mathcal{A}_N$. Then

$$(J_N\Delta_N^{(\alpha-1)/2}PJ_M\Delta_M^{(\alpha-1)/2}x\xi_o\,,\,y\xi_o)=$$

$$=(\sigma^N_{i\alpha/2}(y)^*\xi_o\,,\,J_M\Delta_M^{-1/2}\sigma^M_{-i\alpha/2}(x)\xi_o)=$$

$$=(\sigma^M_{-i\alpha/2}(x)\xi_o\,,\,\sigma^N_{i\alpha/2}(y)\xi_o)=(P\Delta_M^{\alpha/2}x\xi_o\,,\,y\xi_o),$$

which yields (ii). To show (iii), fix $x\in\mathcal{A}_M$ and $y\in\mathcal{A}_N$ and define the function

$$f(z)=(P\Delta_M^{z/2}x\xi_o\,,\,\Delta_N^{-\bar z/2}y\xi_o)$$

on the strip $0\leq \mathrm{re}\, z\leq 1/2$. It is easy to check that f is well--defined and continuous on the strip, and holomorphic in its interior. Moreover, by the Hölder inequality (1),

$$|f(z)|\leq \|\Delta_M^{z/2}x\xi_o\|\,\|\Delta_N^{-\bar z/2}y\xi_o\|$$

$$\leq \|x\xi_o\|^{1-\mathrm{re}\,z}\|x^*\xi_o\|^{\mathrm{re}\,z}\|y\xi_o\|^{1-\mathrm{re}\,z}\|\Delta_N^{-1/2}y\xi_o\|^{\mathrm{re}\,z},$$

so that f is bounded on the strip. Since

$$|f(it)|\leq \|x\xi_o\|\,\|y\xi_o\| \quad\text{and}\quad |f(it+1)|\leq\|x\xi_o\|\,\|y\xi_o\|,$$

we get $\|E^{(\alpha)}\|\leq 1$ by the three lines theorem. $\|E^{(\alpha)}\|=1$ follows from $E^{(\alpha)}\xi_o=\xi_o$.

4. DEFINITION. The continuous extension of $E^{(\alpha)}$ to H_M is still denoted by $E^{(\alpha)}$ and called (by abuse of language) the α- conditional expectation of M onto N. We use the same terminology and notation when speaking about the operators $\Phi_N^{(\alpha)}E_{N,M}^{(\alpha)}(\Phi_M^{(\alpha)})^{-1}$ from $L^2(M,\varphi_o,\alpha)$ into $L^2(N,\varphi_o|N,\alpha)$.

5. REMARKS. 1° The "natural" conditonal expectation $E^{\natural}$ is exactly the conditional expectation of Accardi and Cecchini. (cf. [9] and Proposition 7).

2° If there is a projection of norm one E of M onto N, then $E^{(\alpha)}=E$ for every α, $0\leq\alpha\leq 1$.

3° The chain rule $E^{(\alpha)}_{K,N}\, E^{(\alpha)}_{N,M} = E^{(\alpha)}_{K,M}$ is satisfied as for the usual conditional expectation.

6. PROPOSITION. For each $\alpha \in [0,1]$,
$E^{(\alpha)} \mid \mathcal{D}(S) = S$ for $S = \Delta_N^{-\alpha/2} P \Delta_M^{\alpha/2}$ and $S = J_N^{(\alpha)} P J_M^{(\alpha)}$.
In consequence, $\Delta_N^{-\alpha/2} P \Delta_M^{\alpha/2}$ and $J_N^{(\alpha)} P J_M^{(\alpha)}$ coincide on their common domain.

<u>Proof</u>. Let $\xi \in \mathcal{D}(\Delta_N^{-\alpha/2} P \Delta_M^{\alpha/2})$. We have to show that $\Delta_N^{-\alpha/2} P \Delta_M^{\alpha/2} = E^{(\alpha)}\xi$. Choose $x_n \in \mathcal{A}_M$ so that $x_n \xi_o \to \xi$. Then $\Delta_N^{-\alpha/2} P \Delta_M^{\alpha/2} x_n \xi_o \to E^{(\alpha)}\xi$. For $y \in N$ we get

$$(E^{(\alpha)}\xi\,,\, \Delta_N^{\alpha/2} y\, \xi_o) = \lim\, (\Delta_M^{\alpha/2} x_n \xi_o\,,\, y\,\xi_o) =$$
$$= \lim\, (x_n \xi_o\,,\, \Delta_M^{\alpha/2} y\, \xi_o) = (\Delta_M^{\alpha/2}\xi\,,\, y\,\xi_o)\,.$$

Therefore $E^{(\alpha)}\xi \in \mathcal{D}(\Delta_N^{\alpha/2})$ ($N\xi_o$ is a core for $\Delta_N^{\alpha/2}$) and $\Delta_N^{\alpha/2} E^{(\alpha)}\xi = P\Delta_M^{\alpha/2}\xi$, which gives the conclusion. A similar proof yields the result for $J_N^{(\alpha)} P J_M^{(\alpha)}$.

7. PROPOSITION. The operators $E^{(\alpha)} : L^2(M, \varphi_o, \alpha) \to L^2(N, \varphi_o|N, \alpha)$, $0 \le \alpha \le 1$, can be extended to bounded operators of norm one (still denoted $E^{(\alpha)}$ and called α-conditional expectations) from $L^p(M, \varphi_o, \alpha)$ into $L^p(N, \varphi_o|N, \alpha)$ for any $p \in [1,2]$. Moreover, on $L^1(M) = M_*$ the α-conditional expectation is given by $E^{(\alpha)}(\varphi) = \varphi\,|\,N$.

<u>Proof</u>. It suffices to prove the second assertion of the proposition, as then the first one follows by the non-commutative Riesz - Thorin theorem (see [3], [7]). It is enough to show that

$$(\eta_\alpha^M(k)\,,\, \eta_\alpha^M(y^*)) = (E^{(\alpha)}_{N,M}\, \eta_\alpha^M(k)\,,\, \eta_\alpha^N(y^*)) \quad \text{for } k \in \mathcal{A}_M\,,\ y \in N,$$

which (by Preliminaries (2)) amounts to

$$(\Delta_N^{-\alpha/2} P \Delta_M^{\alpha/2} \Delta_M^{\alpha/2} k\,\xi_o\,,\, \Delta_N^{\alpha/2} y^*\xi_o) = (\Delta_M^{\alpha/2} k\,\xi_o\,,\, \Delta_M^{\alpha/2} y^*\xi_o)\,.$$

But the last equality is evident.

<u>Convergence of conditional expectations</u>

Suppose there exists a (φ_o - invariant) projection of norm one

E of M onto its von Neumann subalgebra N (cf. [10]). Then

$$\|\eta_\alpha^N(Ek)\|_{p,\alpha} \leq \|\eta_\alpha^M(k)\|_{p,\alpha} \quad \text{for } p = 1,\infty \text{ and } k \in M$$

(use the equality $Ex = PxQ$). Hence, the operator $\eta_\alpha^M(k) \mapsto \eta_\alpha^N(Ek)$ from $L^\infty(M, \varphi_o, \alpha)$ into $L^\infty(N, \varphi_o|N, \alpha)$ can be extended by continuity to an operator of norm one from $L^p(M, \varphi_o, \alpha)$ into $L^p(N, \varphi_o|N, \alpha)$ for any $p \geq 1$. It is still denoted by E. Now, let $i_\alpha : L^\infty(N, \varphi_o|N, \alpha) \to L^\infty(M, \varphi_o, \alpha)$ be given by $i_\alpha(\eta_\alpha^N(k)) = \eta_\alpha^M(k)$ for $k \in N$. We check easily that

$$\|\eta_\alpha^M(k)\|_{p,\alpha} \leq \|\eta_\alpha^N(k)\|_{p,\alpha} \quad \text{for } p = 1,\infty \text{ and } k \in N,$$

so that i_α can be extended to an operator (still denoted by i_α) from $L^p(N, \varphi_o|N, \alpha)$ into $L^p(M, \varphi_o, \alpha)$ for any $p \geq 1$. Moreover, $i_\alpha(\varphi) = \varphi \circ E$ for each α and $\varphi \in M_*$ (cf. [12]). Therefore, the subscript α in i_α is redundant. We put $\mathcal{E} = i \circ E$, so that $\mathcal{E}$ is a bounded operator on $L^p(M, \varphi_o, \alpha)$ for any p and α.

Even if the operator E does not exist, the above results rest true for $\alpha = 1/2$, with E replaced by $E^\natural$ (here, $E^\natural : M \to N$ is given by $E^\natural k = J_N P J_M k J_M Q J_N$). In particular, we obtain a bounded operator $\mathcal{E}^\natural = i^\natural \circ E^\natural$ on $L^p_\natural(M, \varphi_o)$ for any $p \geq 1$.

8. THEOREM. Let (M_n) be an increasing sequence of von Neumann subalgebras of M such that $M = \bigvee_{n=1}^\infty M_n$. Then:

(i) if $x \in L^p(M, \varphi_o, \alpha)$, then $\|\mathcal{E}_n x - x\|_{p,\alpha} \to 0$;

(ii) if $x \in L^p_\natural(M, \varphi_o)$, then $\|\mathcal{E}_n^\natural x - x\|_{p,\natural} \to 0$;

(iii) if $k \in M$, then $\|E_n^\natural k - k\|_{p,\alpha} \to 0$.

<u>Proof</u>. Since $\|\mathcal{E}_n x\|_{p,\alpha} \leq \|x\|_{p,\alpha}$ and $\|\mathcal{E}_n^\natural x\|_{p,\natural} \leq \|x\|_{p,\natural}$, we may approximate x by $\eta_\alpha(k)$ in the norm $\|\ \|_{p,\alpha}$ and infer (i) and (ii) from (iii). To prove (iii), use (1) from Preliminaries to obtain

$$\|E_n^\natural k - k\|_{1,\alpha} \leq \|E_n^\natural k - k\|_{2,\alpha} \leq \|(E_n^\natural k^* - k^*)\xi_o\|^\alpha \|(E_n^\natural k - k)\xi_o\|^{1-\alpha}.$$

Since $E_n^\natural k \to k$ strongly for any $k \in M$ (see [9]), we get $\|E_n^\natural k - k\|_{1,\alpha} \to 0$, which implies $\|E_n k - k\|_{p,\alpha} \to 0$ by Theorem 1.

R E F E R E N C E S

[1] L. Accardi, C. Cecchini, Conditional Expectations in von Neumann Algebras and a Theorem of Takesaki, J. Func. Anal. 45(1982), 245 - 273.

[2] H. Araki, T. Masuda, Positive Cones and L^p-Spaces for von Neumann Algebras, Publ. Res. Inst. Math. Sci. 18(1982), 339 - 411.

[3] J. Bergh, J. Löfström, Interpolation Spaces, Springer - V. 1976.

[4] C. Cecchini, D. Petz, Norm convergence of generalized martingales in L^p-spaces over von Neumann algebras, preprint 1984.

[5] S. Goldstein, Norm convergence of martingales in L^p-spaces over von Neumann algebras, preprint 1983.

[6] U. Haagerup, L^p-spaces associated with an arbitrary von Neumann algebra, Colloq. Internat. CNRS, No. 274, 1979, pp. 175 - 184.

[7] H. Kosaki, Applications of the complex interpolation method to a von Neumann algebra, J. Func. Anal. 56(1984), 29 - 78.

[8] H. Kosaki, Positive cones associated with a von Neumann algebra, Math. Scand. 47(1980), 295 - 307.

[9] D. Petz, A Dual in von Neumann Algebras with Weights, to appear in Quart. J. Math. Oxford(2).

[10] M. Takesaki, Conditional Expectations in von Neumann Algebras, J. Func. Anal. 9(1972), 306 - 321.

[11] M. Terp, L^p-spaces associated with von Neumann algebras, preprint 1981.

[12] M. Tsukada, Strong convergence of martingales in von Neumann algebras, Proc. Amer. Math. Soc. 88(1983), 537 - 540.

Acknowledgement. I am grateful to C. Cecchini and D. Petz for making me acquainted with their recent papers [4], [9].

Institute of Mathematics, Łódź University,
ul. Banacha 22, 90-238 Łódź, Poland

QUANTUM GIBBS STATES AND THE ZEROTH LAW OF THERMODYNAMICS

Vittorio Gorini(*), Alberto Frigerio(*), and Maurizio Verri(**)

(*) Dipartimento di Fisica, Sezione Fisica Teorica
Università di Milano, Milano, Italy;
and INFN, Sezione di Milano.

(**) Dipartimento di Matematica,
Politecnico di Milano, Milano, Italy;
and INFN, Sezione di Milano.

Abstract. We show, without the use of ad hoc hypotheses and employing only elementary mathematical techniques, that an equilibrium state of a spatially confined quantum system is described by the Gibbs canonical ensemble, under a stability assumption which amounts essentially to the zeroth law of thermodynamics.

1. Introduction.

Within the framework of algebraic quantum statistical mechanics [1,2], a characterization of equilibrium states has emerged, that regards dynamical stability [3] as the fundamental property of an equilibrium state: an equilibrium state for a system with Hamiltonian H should not be drastically altered when H is modified by a small localized perturbation λV. Alternatively, an equilibrium state can be characterized by the property of (complete) passivity [4] : no cyclic perturbation should be able to extract work from a (finite collection of identically prepared) system(s) in an equilibrium state. Passivity expresses the most important empirically familiar property of equilibrium: the second law of thermodynamics; however, it is not so obvious a priori why Nature should chiefly produce passive states (cf. [2] p. 192). On the other hand, the characterization of equilibrium states via stability conditions has an obvious physical motivation. However, the existing proofs that stability implies the KMS condition ([3] , see also [5, 1, 2]) require some assumptions of asymptotic abelianness in time, which can hold only for infinitely extended systems, and have been shown to hold only for the free (Bose or Fermi) gas [1] and for a weakly, locally perturbed Fermi gas [6]. In particular, for an isolated, spatially confined quantum system, stability implies only that the density operator describing the state is a function of the Hamiltonian H [3, 5].

The purpose of the present note is to show that, if a finite system S with Hamiltonian H is allowed to interact weakly (to be in thermal contact) with some other (finite) systems in its surroundings, a stable state ρ for S must be of the form

$$\rho = \exp[-\beta H] / \mathrm{Tr}\{\exp[-\beta H]\} \qquad (1.1)$$

under the sole condition that $\mathrm{Tr}\{\exp[-\beta H]\} < +\infty$ for all positive β. To this end, we define and require, in addition to the usual stability condition, two additional stability conditions, that we call stability of order two and stability of order three (stability of order one being the usual stability condition for a system in isolation). Stability of order two requires, roughly speaking, that nothing dramatic should happen when the system S in the state ρ is allowed to interact weakly with another (finite) system S', provided S' is in a suitable state ρ' : it is a condition of mutual equilibrium of the two systems. Stability of order three is essentially a statement of the zeroth law of thermodynamics (transitivity of mutual equilibrium): if also another system S'' in a state ρ'' is in "equilibrium" with S in the state ρ, then it must be in "equilibrium" also with S' in the state ρ', so that the product state $\rho \otimes \rho' \otimes \rho''$ of the composite system S+S'+S'' must be stable of order one.

In Section 2, we prove that stability of order two implies, among other things, that ρ is a monotonically nonincreasing function of H, hence it describes a passive state [4,7]. In Section 3, we prove that stability of order three implies that ρ is the canonical state (1.1) for some positive inverse temperature β (possibly, $\beta = +\infty$). For details, see [8].

Here we fix the notations and collect some preliminary results. The word "system" will be used as an abbreviation for the expression "spatially confined quantum system with finitely many degrees of freedom". A system S is identified by the Hilbert space $\mathcal{H}$ of its wave-functions and by its Hamiltonian H, which is assumed to be a self-adjoint operator with

$$\mathrm{Tr}\{\exp[-\beta H]\} < +\infty \quad \text{for all positive } \beta. \qquad (1.2)$$

In particular, H is bounded from below and its spectrum consists of isolated eigenvalues E_n ($E_0 < E_1 < E_2 <$ ......) with finite multiplicities d(n). We shall denote by ϱ a density operator (positive trace class operator with $\mathrm{Tr}[\varrho]$ =1). If ϱ commutes with H, we shall denote by $\{u_{nj} : j=1,\dots,d(n); n=0,1,2.....\}$ a complete orthonormal set of simultaneous eigenvectors of H and ϱ , with

$$H u_{nj} = E_n u_{nj} \tag{1.3}$$

$$\varrho u_{nj} = p_{nj} u_{nj} \quad \left(p_{nj} \geqslant 0, \sum_{nj} p_{nj} = 1 \right). \tag{1.4}$$

Such a density operator represents a stationary state for the system S.

A state ϱ for the system S will be said to be stable of order one if, for each H-bounded self-adjoint operator V in $\mathcal{H}$ and for all sufficiently small positive constant λ , there exists a state $\varrho^{\lambda V}$ satisfying

$$\exp\left[-\tfrac{i}{\hbar}(H+\lambda V)t\right]\varrho^{\lambda V}\exp\left[\tfrac{i}{\hbar}(H+\lambda V)t\right]=\varrho^{\lambda V} : t\in\mathbb{R} \tag{1.5}$$

and

$$\lim_{\lambda\to 0} \varrho^{\lambda V} = \varrho \tag{1.6}$$

in the trace norm topology.

If ϱ is stable of order one, then ϱ is a function of the Hamiltonian H [3, 5] . The converse is obvious when $\mathcal{H}$ is finite-dimensional; more generally, if f is a function such that

$$0 \leq f(x) \leq C\exp[-\beta x] : x\in\mathbb{R} \tag{1.7}$$

then $\varrho = f(H)/\mathrm{Tr}[f(H)]$ is a stable state, with $\varrho^{\lambda V} = f(H+\lambda V)/\mathrm{Tr}[f(H+\lambda V)]$ [8] .

2. Mutual equilibrium

Let S and S' be two systems, with associated Hilbert spaces $\mathcal{H}$ and $\mathcal{H}'$ and with Hamiltonians H and H' respectively. The Hilbert space of the composite system S+S' is $\mathcal{H} \otimes \mathcal{H}'$. The condition of thermal contact between S and S' is expressed by the requirement that the time evolution of S+S' is generated by a small perturbation $H \otimes \mathbb{1}' + \mathbb{1} \otimes H' + \lambda V$ of the total Hamiltonian $H \otimes \mathbb{1}' + \mathbb{1} \otimes H'$ of the decoupled composite system.

A state ϱ for the system S with Hamiltonian H is said to be <u>stable of order two</u> if, for any system S' with Hamiltonian H', there exists a state ϱ' for S' such that the product state $\varrho \otimes \varrho'$ of the composite system S+S' is stable under small perturbations of H+H'. For example, the canonical ensemble (1.1) is stable of order two. Indeed, it suffices to take $\varrho' = \exp(-\beta H')/\mathrm{Tr}\{\exp(-\beta H')\}$, thus obtaining

$$\varrho \otimes \varrho' = \exp[-\beta(H \otimes \mathbb{1}' + \mathbb{1} \otimes H')]/\mathrm{Tr}\{\exp[-\beta(H \otimes \mathbb{1}' + \mathbb{1} \otimes H')]\}, \quad (2.1)$$

which is indeed stable of order one.

In general, let ϱ be stable of order two, and let ϱ' be a state for S' such that $\varrho \otimes \varrho'$ is stable of order one; then it follows that ϱ and ϱ' must be separately stable under small perturbations of H and of H' respectively. Then we have

$$\varrho u_{nj} = p_n u_{nj}, \quad \varrho' u'_{rk} = p'_r u'_{rk}.$$

Since $\varrho \otimes \varrho'$ is stable of order one, it must be a function of $H \otimes \mathbb{1}' + \mathbb{1} \otimes H'$, so that

$$p_n p'_r = p_m p'_s \quad \text{whenever } E_n - E_m = E'_s - E'_r. \quad (2.2)$$

For fixed S and S', condition (2.2) may be void; think e.g. of two harmonic oscillators with mutually irrational frequencies (cf. [9], p. 877). But since S' is arbitrary, (2.2) entails some consequences for ϱ.

<u>Theorem.</u> If the state ϱ is stable of order two, then it is a monotonic nonincreasing function of H, i.e.

$$p_n \leq p_m \quad \text{if} \quad E_n > E_m; \quad (2.3)$$

moreover, one has $p_n < p_m$ if $p_m \neq 0$.

<u>Proof.</u> Let $E_n > E_m$ be two eigenvalues of H. Choose S' to be a harmonic oscillator with frequency

$$\omega = (E_n - E_m)/\hbar \ ;$$

then $E'_r = (r + 1/2)(E_n - E_m)$, and (2.2) implies that

$$p_n p'_r = p_m p'_{r+1} \quad \text{for all } r = 0,1,2,\ldots \tag{2.4}$$

If $p_m = 0$, it follows that also $p_n = 0$, otherwise one would have $p'_r = 0$ for all r. If $p_m \neq 0$, one has

$$p'_{r+1} = (p_n/p_m)\, p'_r \quad \text{for all } r = 0,1,2,\ldots,$$

and $p_n < p_m$; otherwise ρ' would not be trace class ■

Consider in particular the case when S is a harmonic oscillator with frequency ω ; then it follows from (2.2) that either $p_n = 0$ for $n > 1$, and ρ is the ground state, or $p_n \neq 0$ for all n and p_{n+1}/p_n is independent of n. In the latter case, we may set

$$\beta = -\log(p_1/p_0)/\hbar\omega \quad (0 < \beta < +\infty) \tag{2.5}$$

and obtain

$$p_n = \exp(-\beta n \hbar \omega)\, p_0 \ : n = 1,2,\ldots \ . \tag{2.6}$$

Also the former case may be regarded as a special case of (2.6), with $\beta = +\infty$ and with the convention $e^{-\infty} = 0$. For a general system, we can define

$$\beta_n = -\log(p_n/p_0)/(E_n - E_0) \ : n = 1,2,\ldots, \tag{2.7}$$

($\log 0 = -\infty$), so that $0 < \beta_n \leq +\infty$ and

$$p_n = \exp[-\beta_n(E_n - E_0)]\, p_0 \ : n = 1,2,\ldots \ . \tag{2.8}$$

We have not been able to ascertain whether or not stability of order two implies (in general) that all β_n are equal.

3. The zeroth law and the canonical ensemble.

As we have already discussed, it is natural to require that the equilibrium states should satisfy the following stability condition: A state ρ for the system S with Hamiltonian H is said to be stable of order three if, for any system S' with Hamiltonian H', there exists a state ρ' for S' such that the product state $\rho\otimes\rho'$ of the composite system S+S' is stable of order two; i.e., for any system S'' with Hamiltonian H'', there exists a state ρ'' for S'' such that the product state $\rho\otimes\rho'\otimes\rho''$ of the composite system S+S'+S'' is stable of order one (note that the state ρ' for S' is independent of the choice of S'').

It is obvious from Eq. (2.1) that the canonical ensemble (1.1) is stable of order three, with $\rho'' = \exp(-\beta H'')/\mathrm{Tr}\{\exp(-\beta H'')\}$.

Theorem. If the state ρ is stable of order three, then it is either the canonical ensemble

$$\rho = \exp[-\beta H]/\mathrm{Tr}\{\exp[-\beta H]\} \tag{3.1}$$

for some positive real β, or it is the "ground state"

$$\rho = \frac{1}{d(0)} \sum_{j=1}^{d(0)} |u_{oj}\rangle\langle u_{oj}| \,. \tag{3.2}$$

Proof. Taking into account (2.8), it suffices to prove that, for any pair E_m, E_n of eigenvalues of H, one has $\beta_m = \beta_n$. We study the two alternative possibilities separately:

Case 1: $(E_m - E_o)/(E_n - E_o) = r/s$, with r, s positive integers. Then let

$$\omega_o = (E_m - E_o)/r\hbar = (E_n - E_o)/s\hbar \ , \tag{3.3}$$

and choose for S' a harmonic oscillator with frequency ω_o. Since ρ is stable of order three, the state ρ' for S' is stable of order two, hence, by Eq. (2.6), we have

$$P'_k = \exp[-\beta k\hbar\omega_o]\, P'_o \quad : k = 1, 2, \dots \tag{3.4}$$

for some positive β (possibly $\beta = +\infty$). Then

$$P_m P'_o = P_o P'_r \ ; \quad P_n P'_o = P_o P'_s \tag{3.5}$$

since $E_m + \hbar\omega_o/2 = E_o + (r+1/2)\hbar\omega_o$, $E_n + \hbar\omega_o/2 = E_o + (s+1/2)\hbar\omega_o$, by (3.3). Upon inserting (2.8), (3.3) and (3.4) into (3.5), we get

$$\beta_m = \beta = \beta_n \; . \tag{3.6}$$

<u>Case 2:</u> $(E_m - E_o)/(E_n - E_o)$ irrational. Then, for each positive integer s, there exists a (unique) non-negative integer r, depending on s, E_o, E_m, E_n, such that

$$\frac{r}{s}(E_n - E_o) < E_m - E_o < \frac{r+1}{s}(E_n - E_o) . \tag{3.7}$$

Fix s, and choose for S' a harmonic oscillator with frequency

$$\omega_s = (E_n - E_o)/s\hbar \; . \tag{3.8}$$

By the same argument as in Case 1, we have

$$p'_k = \exp[-\beta_n k\hbar\omega_s]\, p'_o \; : \; k = 1, 2, \ldots \; . \tag{3.9}$$

Now we use crucially the assumption that $\rho \otimes \rho'$ is stable of order two. Since $E_o + (r+1/2)\hbar\omega_s < E_m + \hbar\omega_s/2 < E_o + (r+1+1/2)\hbar\omega_s$ by (3.7), we must have

$$p_o p'_r \geqslant p_m p'_o \geqslant p_o p'_{r+1} \; , \tag{3.10}$$

or, dividing by $p_o\, p'_o$ and taking into account (3.9)

$$\exp[-\beta_n r\hbar\omega_s] \geqslant \exp[-\beta_m(E_m - E_o)] \geqslant \exp[-\beta_n(r+1)\hbar\omega_s] . \tag{3.11}$$

Upon inserting (3.8) into (3.11) and taking logarithms, we have

$$\beta_n \frac{r}{s}(E_n - E_o) \leq \beta_m (E_m - E_o) \leq \beta_n \frac{r+1}{s}(E_n - E_o) , \tag{3.12}$$

where s is arbitrary, and r is the unique integer satisfying (3.7). In the limit as $s \to \infty$, both $\frac{r}{s}$ $(E_n - E_o)$ and $\frac{r+1}{s}$ $(E_n - E_o)$ converge to $E_m - E_o$, so that we obtain $\beta_m = \beta_n$ once again ■

References

1. O. Bratteli and D.W. Robinson: Operator Algebras and Quantum Statistical Mechanics I, II. Springer-Verlag, New York Heidelberg Berlin, 1979, 1981.

2. W. Thirring: A Course in Mathematical Physics 4: Quantum Mechanics of Large Systems. Springer-Verlag, New York Wien, 1983.

3. R. Haag, D. Kastler and E. Trych-Pohlmeyer: Stability and equilibrium states. Commun. Math. Phys. 38 (1974), 173-193.

4. W. Pusz and S.L. Woronowicz: Passive states and KMS states for general quantum systems. Commun. Math. Phys. 58 (1978), 273-290.

5. F. Hoekman: On stability and symmetries in quantum statistical mechanics. Ph.D. Thesis, University of Groningen, 1977.

6. D.D. Botvich and V.A. Malyshev: Unitary equivalence of temperature dynamics for ideal and locally perturbed Fermi-gas. Commun. Math. Phys. 91 (1983), 301-312.

7. A. Lenard: Thermodynamical proof of the Gibbs formula for elementary quantum systems. J. Stat. Phys. 19 (1978), 575-586.

8. A. Frigerio, V. Gorini and M. Verri: The zeroth law of thermodynamics. Preprint 1984, submitted to Ann. Phys. (N.Y.).

9. P. Caldirola, R. Cirelli and G.M. Prosperi: Introduzione alla Fisica Teorica. UTET, Torino, 1982.

DISSIPATIVE QUANTUM TUNNELING

H. Grabert, Institut für Theoretische Physik
Universität Stuttgart, D 7000 Stuttgart 80

The talk reviews some recent developments in the theory of quantum mechanical tunneling. The new phenomena discussed arise from the coupling of the tunneling system to a heat bath of temperature T. The question is how dissipation and thermal fluctuation affect the tunneling rate. Quantum tunneling proves to be very sensitive to these environmental influences. The results are not only important for the analysis of tunneling experiments in macroscopic systems but also provide a crucial test of various approaches put forward in the theory of quantum stochastic processes.

1. Introduction

A great variety of phenomena in physical and chemical sciences is caused by transitions out of states that would be stable if there were no thermal and quantum fluctuations. Since several decades, a popular model of such systems has been a Brownian particle of mass m moving in a multistable potential $V(q)$ [Fig.1] while coupled to a heat bath at temperature T. At sufficiently high temperatures the particle can overcome the barriers separating the potential wells in a purely classical fashion by thermal hopping. This process was studied in great detail by Kramers in 1940 /1/. On the other hand, at very low temperatures thermal fluctuations die out, and, in particular at zero temperature, the particle can escape from a metastable well only by quantum mechanical tunneling. In 1981, Caldeira and Leggett /2/ have shown that the zero temperature decay rate is strongly affected by the frictional influence of the environment. This gave

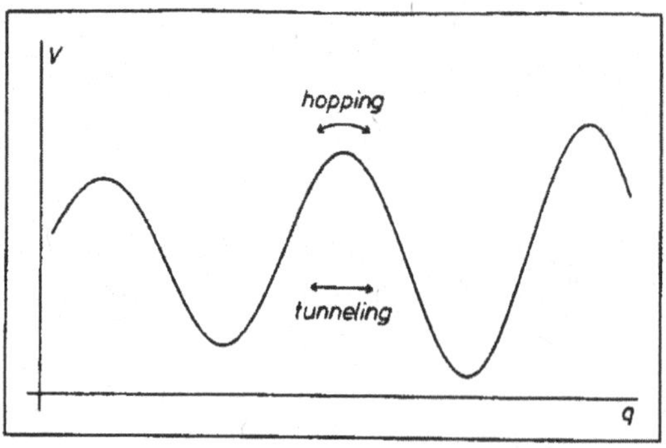

Fig. 1: A multistable potential

se to an intense investigation of dissipative quantum tunneling and a number of nteresting results were found meanwhile. In this talk I will discuss some of these ecent developments.

One of the reasons why dissipative quantum tunneling is interesting is that it an actually be observed in various systems of low temperature physics, and I will ortly mention a few. There are centers in solids /3/, like small Li^+ ions bstituting K^+ ions in a KCl crystal, which can occupy several equivalent quilibrium positions between which they tunnel back and forth at low temperatures. e dominant environmental influence comes from the coupling to the phonons of the ystal. The low temperature mobility of muons in metals /4/ is also due to quantum nneling. In this case, both the couplings to phonons and to electrons are significant. ile for these systems the connection with the model of a Brownian tunneling particle obvious, such a relation is not so apparent for the phenomenon of macroscopic antum tunneling in Josephson systems /5-8/, a problem which has attracted a great al of interest recently. In one type of experiments one observes the tunneling of e magnetic flux trapped in a SQUID /7/, that is a superconducting ring interrupted a Josephson junction. Here the coordinate q of the tunneling particle corresponds the trapped flux, and its mass is related to the capacitance of the junction. The tential $V(q)$ is the sum of the Josephson phase locking energy and the magnetic ergy /9/. Dissipation arises from the Ohmic part of the current through the junction. nally, I mention that the anomalous electric conductivity of charge density wave stems has also been explained by means of a tunneling model /10/. The problem can ain be recast in the language of Brownian motion where q corresponds to the phase the charge density wave. Further, frictional influence was found to be substantial 1/ for these systems too.

Another root for the renewed interest in quantum tunneling originates from its treme sensitivity to dissipation /2, 12-15/. Since Pauli's seminal work in 1928 /16/, great variety of approaches aiming at a consistentquantum mechanical description of ssipation were developed /17/. Probably the as yet most evolved theory is the -called dynamical semi group approach /18/ which was so successfully used to describe mping phenomena, e.g., in quantum optics /19/ and spin relaxation theory /20/. wever, since this approach treats the environmental coupling perturbatively, it is stricted to weakly damped systems where the largest damping constant γ (inverse laxation time) satisfies

$$\gamma \ll \omega_0 \quad , \quad \hbar\gamma \ll k_B T \tag{1}$$

ere ω_0 is the smallest frequency of the reversible motion. For a tunneling stem ω_0 may differ from zero just by a tunnel splitting so that the first equality is violated even for very weak damping. Furthermore, to observe tunneling enomena, one often has to go to very low temperatures where the second inequality ases to hold. Clearly, we have to treat the frictional influence of the heat bath ry carefully if we want to study phenomena involving tunneling. That is why dissipative quantum tunneling has strongly stimulated the further development of a theory of quantum stochastic processes beyond the weak coupling regime /21-25/.

2. Functional Integrals for Quantum Brownian Motion

Within classical physics we all know how to describe friction and thermal fluctuation. We write down Langevin equations, that is stochastic differential equations, for the variables of interest, or, alternatively, we start from a master equation for the probability distribution of these variables /26/. For continuous processes there is still another approach going back to Onsager and Machlup /27/ which characterizes the process in terms of a functional integral /28/. The relation between these approaches is well understood and one can pass over from one description of the process to another. For damped quantum systems there is presently only one approach that works at arbitrary low temperatures and for arbitrary damping. This is the functional integral representation of the process pioneered by Feynman and Vernon /29/. Recent extensions of the Langevin equation approach /22/ and the master equation approach /23/, which are both reasonably well understood in the weak coupling regime /19/, are only approximately valid for nonlinear systems so that the calculation of tunneling rates cannot reliably be based on these methods at present.

Even a crude presentation of the Feynman-Vernon theory is naturally not feasible here and I will confine myself to a broad outline. Let me consider a particle with Hamiltonian $H = p^2/2m + V(q)$. In the undamped case, its partition function reads

$$Z_\beta = \mathrm{tr}\, \exp(-\beta H). \tag{2}$$

Following Feynman /30/ this can be represented as a functional integral

$$Z_\beta = \int D[q]\, \exp(-\tfrac{1}{\hbar} S_E[q]) \tag{3}$$

where the integral is over all periodic paths with period $\hbar\beta$, that is $q(o) = q(\hbar\beta)$, and where every path is weighted according to a path probability determined by the Euclidean action

$$S_E[q] = \int_0^{\hbar\beta} d\tau \left(\tfrac{1}{2} m\dot{q}^2 + V(q)\right) . \tag{4}$$

Here the kinetic and potential energies are added in contrast to the familiar action functional of Hamilton's principle where $V(q)$ is subtracted from the kinetic term. This difference is obvious since the canonical operator $\exp(-\beta H)$ is just the time evolution operator $\exp[-(i/\hbar)Ht]$ for an imaginary time $\tau = -i\hbar\beta$, and a rotation to imaginary times changes the relative sign of kinetic and potential terms. Incidentally, it is worth mentioning that this fundamental connection between the canonical operator and the time evolution operator, which is the basis, for instance, of the fluctuation-dissipation theorem /31/, is destroyed by the weak coupling limit /18-20/ because terms of order γt are kept while terms of order $\gamma\hbar\beta$ are disregarded /32/.

In the damped case, the particle is coupled to a heat bath, and we may start out from a functional integral for the partition function of the entire system including the environmental degrees of freedom. Now, if every single degree of freedom of the heat bath responds only linearly to the particles motion, which is mostly the case and even strictly so for a reservoir consisting of harmonic oscillators /5,29/, the environmental variables can be integrated out explicitly and one is left again with an expression of the form (3) but the Euclidean action now contains an additional term /5,33/

$$S_E[q] = \int_0^{\hbar\beta} d\tau\left(\tfrac{1}{2}m\dot{q}^2 + V(q)\right) + \frac{1}{2}\int_0^{\hbar\beta} d\tau \int_0^{\hbar\beta} d\tau'\; k(\tau-\tau')q(\tau)q(\tau') \tag{5}$$

describing the frictional influence of the environment. In the case of frequency independent damping γ, which corresponds to linear Ohmic dissipation arising from a frictional force $-m\gamma\dot{q}$ in the classical regime, the kernel $k(\tau)$ is given by

$$k(\tau) = \frac{m\gamma}{\hbar\beta} \sum_{n=-\infty}^{+\infty} |\nu_n| e^{i\nu_n\tau} \tag{6}$$

where the $\nu_n = 2\pi n/\hbar\beta$ are Matsubara frequencies. This shows that in the functional integral representation of the process, dissipation is conveniently included as a nonlocal term in the effective action.

A corresponding formulation can also be given for the real time dynamics of the quantum particle. Let $w(q,o)$ be the probability distribution of the coordinate of the Brownian particle measured at $t = o$. Then the probability $w(q,t)$ at time t may be written as a double functional integral /24,29/

$$w(q_f,t) = \int D[q]D[q']e^{\frac{i}{\hbar}\{S[q] - S[q']\} + \frac{1}{\hbar}\Phi[q,q']\}} w(q_i,0) \tag{7}$$

where the integral is over all paths $q(s)$, $q'(s)$, $0 \leq s \leq t$ with $q(o) = q'(o) = q_i$, $q(t) = q'(t) = q_f$, and where q_i is integrated over. Here

$$S[q] = \int_0^t ds\left(\tfrac{1}{2}m\dot{q}^2 - V(q)\right) \tag{8}$$

is the action of the undamped particle, and $\Phi[q,q']$ is the Feynman-Vernon influence functional /29,34/ describing again the frictional influence of the heat bath. For Ohmic dissipation $\Phi[q,q']$ can be written as /35/

$$\Phi[q,q'] = \int_0^t ds \int_0^s ds'[\dot{q}(s) - \dot{q}'(s)][Q(s-s')\dot{q}(s') - Q^*(s-s')\dot{q}'(s')] \tag{9}$$

where

$$Q(s) = \frac{m\gamma}{\pi}\ln\,\sinh\left(\frac{\pi s}{\hbar\beta}\right) + \frac{i}{2}m\gamma \tag{10}$$

is a damping kernel. Both, the imaginary and the real time functional integral representations of quantum Brownian motion are convenient starting points for the study of tunneling phenomena.

3. Dissipative Quantum Decay

The first tunneling problem I shall consider is the quantum decay of a metastable state. In the Brownian motion picture, this corresponds to the escape of a particle from a metastable potential well of the form depicted in Fig. 2. Thus, initially the particle sits in the well near $q = o$, and we ask for the decay rate Γ of the occupation probability of this well. It is assumed that the slope of the potential at the zero temperature exit point q_0 [Fig.2] is not too small, so that a particle which has penetrated the barrier is rapidly removed to the region of lower potential. Then coherent return tunneling is negligible.

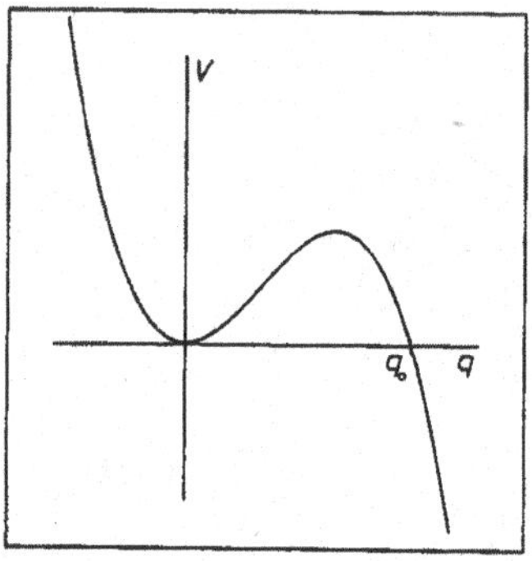

Fig. 2: A metastable well

The free energy $F = - k_B T \ln Z_\beta$ of the particle in the metastable well is given by the functional integral

$$F = - \frac{1}{\beta} \ln \int D[q] \exp(-\frac{1}{\hbar} S_E[q]) \tag{11}$$

where $S_E[q]$ has been defined in (5). Now, the state $q = 0$ is only metastable if its lifetime Γ^{-1} is long compared to other characteristic time scales of the problem as, for instance, the period $2\pi/\omega_0$ of small undamped oscillations about the metastable minimum. This means that the barrier must be high and wide in order that we may bring the concept of metastability to bear upon our problem. However, under these circumstances the functional integral is dominated by those paths for which the action $S_E[q]$ is smallest.

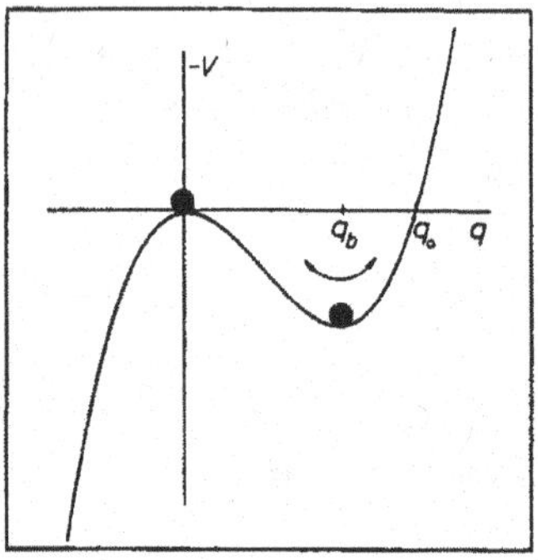

Fig. 3: The inverted potential

As we have already noticed, the motion in imaginary time τ corresponds to a real time motion in the potential $-V(q)$. In this inverted potential [Fig.3] there is a trivial periodic solution $q(\tau) \equiv 0$ where the particle just sits on top of the potential barrier, and another solution, $q(\tau) \equiv q_b$, where it sits at the bottom of the well. However, for temperatures T below a certain crossover temperature T_c, the period $\hbar\beta = \hbar/k_B T$ is large enough that the particle may oscillate along a periodic orbit in the classically forbidden region $0<q<q_0$. Coleman /36/ has coined the name "bounce" for this trajectory. The temperature T_c where the bounce ceases to exist is a characteristic temperature for the crossover between quantum tunneling and thermal hopping /37,38/

elow T_c the trivial solution $q(\tau)\equiv q_b$ may be disregarded.

The detailed analysis shows that the bounce is actually not a minimum of the ction but a saddlepoint, which means that there is one fluctuation mode in function pace with respect to which the bounce is a maximum of the action. Because of this eculiarity the functional integral (11) is in fact not defined. This should come as ot too big a surprise, after all, we are trying to compute the free energy of an nstable system. Langer /39/ has explained that in such a situation the free energy an still be defined by an analytical continuation from a stable to the unstable ituation. This leads to an exponentially small imaginary part of the free energy hich is proportional to the decay rate /36,40/

$$\Gamma = \frac{2}{\hbar} \text{Im } F. \tag{12}$$

he explicit evaluation of this formula is still complicated in the presence of issipation because a nonlinear potential and a nonlocal term in the action must be reated simultaneously. A detailed analysis of the zero temperature rate and analytical esults both for weak and strong damping are due to Caldeira and Leggett /5/. They ound a strong suppression of the decay as compared to undamped systems. For inter-ediate damping one has to resort to numerical methods and an appropriate treatment as given by Chang and Chakravarty /41/.

For finite temperatures the rate is enhanced by thermal fluctuations. Together ith Weiss and Hanggi, I have shown /15,33/ that this enhancement is very pronounced n the presence of dissipation as distinguished from undamped systems where Γ is lmost constant at low temperatures /40,42/. As $T = 0$ is approached, the decay rate f a dissipative system follows a power law /15,33/

$$\ell n[\Gamma(T)/\Gamma(0)] \propto T^n \tag{13}$$

here the exponent n is independent of the particular form of the metastable otential and is a distinctive feature of the dissipative mechanism. For Ohmic issipation this exponent equals 2.

The formula /12/ for the decay rate ceases to hold as the crossover temperature $_c$ is approached from below. In the case of frequency independent damping γ, the rossover temperature is given by /15/

$$T_c = (\hbar\omega_b/2\pi k_B)[(1+\varkappa^2)^{1/2}-\varkappa] \tag{14}$$

here $\omega_b = [-V''(q_b)/m]^{1/2}$ is the curvature of the potential at the barrier top nd $\varkappa = \gamma/2\omega_o$. Hence, damping leads to a reduction of T_c and one has to go to ower temperatures in order to observe quantum decay in a dissipative system.

For temperatures above T_c there is no bounce trajectory and the imaginary art of the free energy arises from the trivial saddle point, $q(\tau)\equiv q_b$, of the unctional integral (11). In this region the decay rate reads /38,43/

$$\Gamma(T) = f(T)\exp(-V_b/k_BT) \tag{15}$$

where $V_b = V(q_b) - V(o)$ is the barrier height and $f(T)$ is a prefactor. This is the familiar Arrhenius formula for thermally activated hopping. However, quantum corrections contribute to $f(T)$ and the classical attempt frequency $f_{c\ell}$ /1/ is only reached well above T_c. These quantum corrections are in fact important for a precise analysis of decay experiments /44/.

The calcualtion of the decay rate in the vicinity of T_c needs special care because in this region the functional integral (11) cannot be done by steepest descents. An appropriate treatment was given independently by Larkin and Ovchinnikov /45/ and by myself in collaboration with Ulrich Weiss. We also found that the transition between quantum tunneling and thermal hopping is described by a smooth crossover function which has a universal form /38/.

Very recently, Olschowski et al. /46/ have performed a numerical calculation of the decay rate for the whole range of temperature and damping coefficients of interest. Because of the classical Kramers formula $\Gamma = f_{c\ell}\exp(-V_b/k_BT)$, it is convenient to plot the logarithm of the rate as a function of the inverse temperature 1/T. In such a diagram the classical result is represented by a falling straight line. The upper line in Fig. 4 shows the rate of an undamped system. Because of quantum effects the

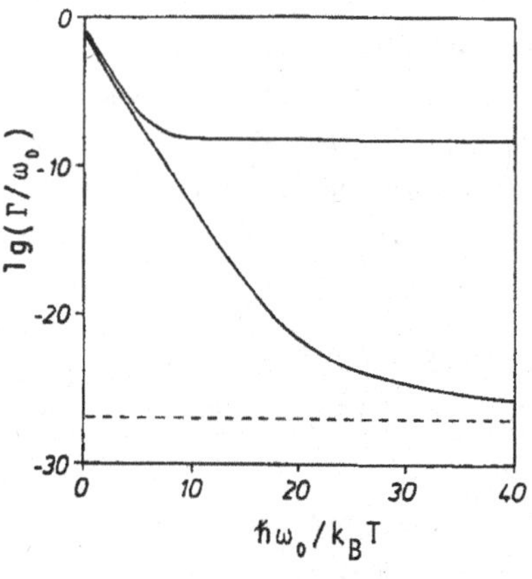

Fig. 4: $\ell g(\Gamma/\omega_0)$ is plotted as a function of $\hbar\omega_0/k_BT$ for a system with a cubic potential with barrier height $V_b = 3\hbar\omega_0$. The upper line shows the decay rate of an undamped system and the lower line of a damped system with an Ohmic damping coefficient $\gamma = 2\omega_0$. The dashed line shows the T = 0 rate of the damped system. After Ref. 46

rate does not decrease continuously as T is lowered but levels off at a certain value which is determined by the WKB decay rate of the ground state in the metastable well. For the undamped system there is a rather sharp transition between the classical regime of thermal hopping and the quantum regime of tunneling.

The lower line in Fig. 4 shows the decay rate of a damped system. The classical rate is slightly reduced because the classical attempt frequency $f_{c\ell}$ is diminished by dissipation as calculated by Kramers /1/. The zero temperature decay rate, however, is strongly reduced by an exponential suppression factor which was calculated by Caldeira and Leggett /2/. As a function of temperature the rate now rather gradually changes from classical hopping into pure zero temperature quantum tunneling and there a large crossover region where thermal and quantum fluctuations interplay. Due to e recent progress in finite temperature tunneling /15,37,38,43-47/, which I have etched roughly, this behavior at intermediate temperatures is now also well understood.

Dissipative Double Well Systems

e "bounce" technique used so far is no longer applicable if the potential has a all slope at the exit point q_0. This is the case whenever there is no, or only small potential drop between the metastable minimum and the region beyond the barrier. a simple model, I shall consider a slightly asymmetric double well system where the pths of the potential minima at $\pm q_0/2$ differ by a small bias energy $\hbar\sigma$ [Fig.5]. r large asymmetry σ we recover the old decay problem. However, as σ is lowered, turn jumps become feasible and a more careful treatment is required.

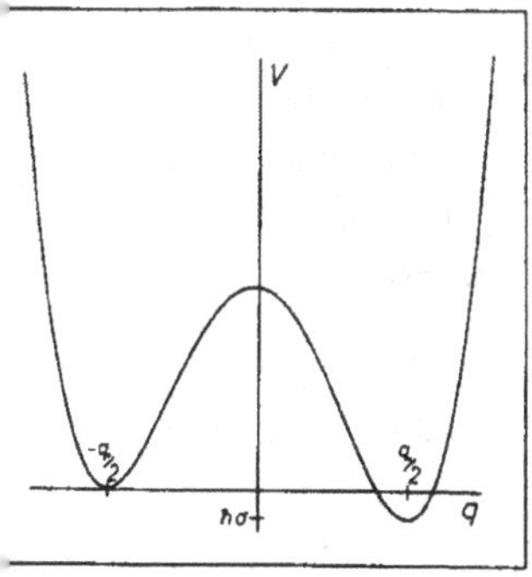

g. 5: The biased double well

Within the functional integral approach this change manifests itself in the following way. The interesting region is $\sigma \ll \omega_0$, $k_B T \ll \hbar\omega_0$, where ω_0 is the small oscillation frequency in either well. Now, a bounce trajectory which starts from the bottom of the left well may, for a little additional action, stay for a long time in the right well before returning to the starting point. Hence, the sojourn time in the right well now becomes a strongly fluctuating quantity which must be treated as a collective variable of the functional integral /48/. Loosely speaking, the bounce breaks up into a pair interacting instantons.

Furthermore, for very small dissipation, the system tunnels in a coherent "clocklike" manner and e time evolution of the occupation probability of a well is no longer determined simply a decay rate. Hence, the imaginary time functional integral (11) is not sufficient the whole range of parameters and a real time calculation is required. A convenient arting point is the functional integral (7) discussed earlier.

For an ensemble of systems initially in the left well, the probability distribu- on $w(q,o)$ is centered around $-q_0/2$. Now, since the probability distribution $w(q,t)$ time t will be centered around $+q_0/2$ and $-q_0/2$, it is sufficient to consider e occupation probabilities $P_{+-}(t) = \int_0^\infty dq\, w(q,t)$, $P_{--}(t) = \int_{-\infty}^0 dq\, w(q,t)$ of the two lls. The sum of these probabilities being 1, the dynamics of tunneling transitions characterized conveniently by their difference

$$P(t) = P_{--}(t) - P_{-+}(t). \tag{16}$$

e time evolution of $P(t)$ arises from multi instanton contributions to the nctional integral (7). In the dissipative case the interaction between these stantons must be treated carefully. For an unbiased system, an appropriate technique was explained by Chakravarty and Leggett /49,50/. Extending this method to biased systems we found /35/ for a wide range of parameters incoherent relaxation

$$P(t) = -\tanh(\tfrac{1}{2}\beta\hbar\sigma) + [1 + \tanh(\tfrac{1}{2}\beta\hbar\sigma)]\exp(-\Gamma t) \tag{17}$$

with the tunneling rate

$$\Gamma = \frac{\Delta^2}{2\omega_0}\left(\frac{\hbar\beta\omega_0}{2\pi}\right)^{1-2\alpha}\cosh(\tfrac{1}{2}\beta\hbar\sigma)\,\frac{|\Gamma(\alpha+i\beta\hbar\sigma/2\pi)|^2}{\Gamma(2\alpha)} \tag{18}$$

where Δ is the a bare tunneling frequency depending on the form of the potential, and where

$$\alpha = m\gamma q_0^2/2\pi\hbar \tag{19}$$

is a dimensionless Ohmic dissipation coefficient. In the region $\alpha>1$, the result (17) holds for all k_BT and $\hbar\sigma$ small compared with $\hbar\omega_0$. It also extends to the region $0<\alpha<1$ for systems sufficiently far from the $T = 0$, $\sigma = 0$ line /35/. Note that at zero temperature $P(t) = 2e^{-\Gamma t}-1$ if the particle starts out from the upper well $(\sigma>0)$ and $P(t) = 1$ if it starts out from the lower well $(\sigma<0)$. Thus, at $T = 0$ and for $\alpha>1$ there are only transitions from the upper to the lower well. Furthermore, for an unbiased system the rate (18) reduces to /49/

$$\Gamma = \frac{\sqrt{\pi}\Delta^2}{2\omega_0}\,\frac{\Gamma(\alpha)}{\Gamma(\alpha+\frac{1}{2})}\left(\frac{\pi k_BT}{\hbar\omega_0}\right)^{2\alpha-1} \tag{20}$$

so that the tunneling rate vanishes at $T = 0$ for $\alpha>1$. This remarkable fact was first noticed by Chakravarty /12/ and by Bray and Moore /13/. For the unbiased system there is a transition at $T = 0$, $\alpha = 1$ leading to spontaneous symmetry breaking for $\alpha>1$ /12-14/. The region of parameters where (17) fails to hold is of particular interest. For $0<\alpha<\frac{1}{2}$ and k_BT, $\hbar\sigma$ both of order $\hbar\Delta\ll\hbar\omega_0$, a damped oscillatory part of $P(t)$ is found. For $\sigma = 0$, $T = 0$ this coherent component was calculated explicitly by Chakravarty and Leggett /49/. Further work on this problem is presently in progress.

Finally I would like to mention that the methods I have presented can also be applied to dissipative quantum tunneling in periodic potentials. In fact, several interesting results appear in recent papers /4,51/, which, unfortunately, I will not be able to discuss here, as it is the case with the numerous important experiments on dissipative quantum tunneling.

Acknowledgements

I would like to thank Ulrich Weiss for the enjoyable collaboration in tunneling problems and M. Büttiker, S. Chakravarty, J. Clarke, U. Eckern, P. Hanggi, A.J. Leggett, A. Muramatsu, P. Olschowski, Yu.N. Ovchinnikov, A. Schmid, G. Schön, R.F. Voss, and W. Zwerger for helpful discussions and/or preprints.

eferences

H.A. Kramers, Physica 7,284(1940)

A.O. Caldeira and A.J. Leggett, Phys.Rev.Lett.46,211(1981)

V. Narayanamurti, R.O. Pohl, Rev.Mod.Phys. 42,201(1970)

F. Bridges, CRC Crit.Rev.Solid State Sci. 5,1(1975)

J. Kondo, Physica 126B,377(1984)

A.O. Caldeira and A.J. Leggett, Ann.Phys.(N.Y.)149,374(1983)

R.F. Voss and R.A. Webb, Phys.Rev.Lett.47,265(1981)

L.D. Jackel et al. Phys.Rev.Lett.47,697(1981)

M.H. Devoret, J.M. Martinis, D. Esteve, and J. Clarke, Phys.Rev.Lett.53,1260(1984)

R. de Bruyn Ouboter, Physica 126B,423(1984)

A.I. Larkin, K.K. Likharev, and Yu.N. Ovchinnikov, Physica 126B,414(1984)

J. Kurkijärvi, Phys.Rev.B6,832(1972)

0 J. Bardeen, Phys.Rev.Lett.42,1498(1979, 45,1978(1980)

1 W. Wonneberger, Z.Phys.B50,23(1983), K. Hida and U. Eckern (preprint)

2 S. Chakravarty, Phys.Rev.Lett.49,681(1982)

3 A.J. Bray and M.A. Moore, Phys.Rev.Lett.49,1545(1982)

4 W. Zwerger, Z.Phys.B53,(1983), B54,87(1983)
V. Hakim, A. Muramatsu, and F. Guinea, Phys.Rev.B30,464(1984)

5 H. Grabert, U. Weiss, and P. Hanggi, Phys.Rev.Lett.52,2193(1984)

6 W. Pauli, in Festschrift zum 60. Geburtstage A. Sommerfeld, (Hirzel, Leipzig 1928) p. 30

7 H. Dekker, Phys.Rep.80,1(1981) and references therein

8 V. Gorini, A. Kossakowski, and E.C.G. Sudarshan, J.Math.Phys.17,821(1976)

9 H. Haken, Rev.Mod.Phys.47,67(1975)

0 P.N. Agyres and P.L. Kelley, Phys.Rev.134,A98(1964)

1 G.W. Ford, M. Kac, and P. Mazur, J.Math.Phys.6,504(1965)

2 R.H. Koch, D.J. van Harlingen, and J. Clarke, Phys.Rev.Lett.45,2132(1980)
A. Schmid, J. Low Temp.Phys.49,609(1982)
H. Metiu and G. Schön, Phys.Rev.Lett.53,13(1984)

3 H. Grabert, Z.Phys.B49,161(1982)
H. Grabert and P. Talkner, Phys.Rev.Lett.50,1335(1983)

4 A.O. Caldeira and A.J. Leggett, Physica 121A,587(1983)
A. Schmid, J. Low. Temp.Phys.49,609(1982)

5 H. Grabert, U. Weiss, and P. Talkner, Z.Phys.B55,87(1984)

6 R.L. Stratonovich, Topics in the Theory of Random Noise (Gordon and Breach, N.Y. 1963), P. Hanggi and H. Thomas, Phys.Rep.88,207(1982)
H. Risken, The Fokker-Planck Equation, Springer Series in Synergetics, Vol. 18, (Springer 1984)

27 L. Onsager and S. Machlup, Phys.Rev.91,1505(1953) ; 91,1512(1953)
28 H. Grabert, R. Graham, and M.S. Green, Phys.Rev.A21,2136(1980), and references therein
29 R.P. Feynman and F.L. Vernon, Ann.Phys.(N.Y.)24,118(1963)
30 R.P. Feynman, Statistical Mechanics (Benjamin, N.Y. 1972)
31 H. B. Callen and T.A. Welton, Phys.Rev.83,34(1951)
32 P. Talkner, thesis, Univ.Stuttgart (1979)
33 H. Grabert and U. Weiss, Z.Phys.B56,171(1984)
34 R.P. Feynman and A.R. Hibbs, Quantum Mechanics and Path Integrals, (Mc Graw-Hill, N.Y. 1965)
35 H. Grabert and U. Weiss (in preparation)
A.Dorsey and M. Fisher (in preparation)
36 S. Coleman, in The Whys of Subnuclear Physics, ed. by A. Zichichi (Plenum N.Y.1979)
37 A.I. Larkin and Yu.N. Ovchinnikov JEPT Lett.37,382(1983)
38 H. Grabert and U. Weiss, Phys.Rev.Lett.53,1787(1984)
39 J.S. Langer, Ann.Phys.(N.Y.)41,108(1967)
40 I. Affleck, Phys.Rev.Lett.46,388(1981)
41 L.D. Chang and S. Chakravarty, Phys.Rev.B29,130(1984), E B30,1566(1984)
42 U. Weiss and W. Häffner, Phys.Rev.D27,2916(1983)
43 P.G. Wolynes, Phys.Rev.Lett.47,968(1981)
V.I. Mel'nikov and S.V. Meschkov, JEPT Lett.38,130(1983)
44 H. Grabert, P. Olschowski, and U. Weiss, post-deadline paper, LT 17 Karlsruhe (1984)
45 A.I. Larkin and Yu.N. Ovchinnikov, Zh.Eksp.Teor.Fiz.86,719(1984)
46 P. Olschowski, H. Grabert, and U. Weiss (in preparation)
47 W. Zwerger, (preprint)
48 U. Weiss, P. Riseborough, P. Hanggi, and H. Grabert, Phys.Lett.104A,10(1984)
49 S. Chakravarty and A.J. Leggett, Phys.Rev.Lett.52,5(1984)
50 S. Chakravarty, Physica 126B,385(1984)
51 A. Schmid, Phys.Rev.Lett.51,1506(1983); S.A. Bulgadaev, JETP Lett.39,315(1984)
F. Guinea, V. Hakim, and A. Muramatsu (preprint)
U. Weiss and H. Grabert (in preparation)

RLEN PROCESSES: A NEW CLASS OF DIFFUSIONS WITH SINGULAR DRIFTS.

ancesco Guerra
partimento di Matematica, Istituto "Guido Castelnuovo",
iversità di Roma "La Sapienza",
azzale Aldo Moro, I-00185 Roma, Italy.

INTRODUCTION.

the frame of Nelson's stochastic mechanics[1,2,3], recently Carlen[4] s shown that it is possible to associate, in a rigorous mathematical y, a Markov stochastic process to any solution of the quantum Schroed- ger equation, satisfying mild regularity assumptions, related to the niteness of the action on some bounded interval of time. These processes y have very singular drifts, more singular than those usually con- dered in the mathematical literature about the solutions of stochastic fferential equations. Carlen's proof exploits only techniques of para- lic differential equations and is based on a clever control of the proximation of the singular processes through the regular ones, for ich the standard theory applies.

turns out that, by a sligth generalization of Carlen's treatment, it possible to introduce a new class of stochastic processes, with very ngular drifts, not necessarily related to solutions of the Schroedinger uation. These processes, therefore, may have applications beyond the ame of stochastic mechanics. The purpose of this note is to give a incise and selfcontained presentation of the strategy leading to the nstruction of this new class of processes. This strategy is based on e following steps. We start from the class of regular diffusions on me bounded time interval, then introduce a suitable metric on this ass and take the completion, proving that in the limit we still get me stochastic processes, even though, in the limiting procedure, drifts y become very singular.

e organization of the paper is as follows. In section 2 we introduce e class of regular diffusions and state their basic properties, easily rived from the standard classical theory of parabolic differential uations. In section 3 we consider the problem of constructing new ochastic processes, by approximating them with regular diffusions. en we introduce a suitable metric on the class of regular diffusions. ction 4 is devoted to the proof that the completion, with respect to is metric, still defines some stochastic processes. The core of the

proof is based on methods introduced by Carlen in Ref. 4. Therefore, we call Carlen processes the class of stochastic processes obtained in the completion procedure. Finally, section 5 is devoted to some conclusion and outlook for new possible developments, especially in the direction of systems with an infinite number of degrees of freedom, of interest in the construction of interacting relativistic quantum field theories.
In conclusion, the author would like to thank the Organizers, in particular Luigi Accardi and Wilhelm von Waldenfels, for the kind hospitality and the excellent organization of the meeting.

2. REGULAR DIFFUSIONS.

for the sake of simplicity, we consider stochastic processes taking values in a configuration space R^n, but our considerations could be easily extended to the more general case where the configuration space is a generic smooth manifold. In some bounded time interval (t_0, t_1), regular diffusions are completely defined by their smooth everywhere positive density $\rho(x,t)$ and their smooth transition probability densities $p(x',t';x,t)$, $t'>t$. These can be easily constructed, through standard methods of parabolic partial differential equations. Let us start, for example, from a given time dependent vector field $v_{(+)}(\cdot,t)$, which we assume C^∞ and bounded. Introduce the operator

$$(1)\qquad D_{(+)} = \partial_t + v_{(+)} \cdot \nabla + \nu \Delta \quad ,$$

where ∇ and Δ are, respectively, the gradient and the Laplacian in R^n and ν is some fixed diffusion constant. Then the transition probability density satisfies the forward equation

$$(2)\qquad D_{(+)} p(x',t';x,t) = 0 \quad ,$$

where $D_{(+)}$ acts on the initial x,t variables $(t'>t)$, and the boundary condition, written in symbolic form,

$$(3)\qquad \lim_{t \uparrow t'} p(x',t';x,t) = \delta(x'-x) \quad .$$

It is very well known (see for example Ref. 5) that (2) and (3) can be translated into a unique integral equation, by introducing the transition probability density for the Brownian motion with fixed diffusion coefficient

$$(4)\qquad p_0(x',t';x,t) = \left[4\pi\nu(t'-t)\right]^{-\frac{n}{2}} \exp\left[-\|x'-x\|^2/4\nu(t'-t)\right] \quad .$$

The integral equation reads

$$(5)\qquad p(x_1,s_1;x_0,s_0) = p_0(x_1,s_1;x_0,s_0) + \int_{s_0}^{s_1} dt \int v_{(+)}(x,t) \cdot \; p_0(x_1,s_1;x,t)\, p(x,t;x_0,s_0)\, dx$$

d can be easily solved by iteration, under the stated regularity con-
tions for $v_{(+)}$. As a result of the control of the iteration the following
del'man[6] estimates hold

6) $$|p(x',t';x,t) - p_o(x',t';x,t)| \leq c\sqrt{t'-t}\ p_o'(x',t';x,t) ,$$

7) $$|\nabla p(x',t';x,t) - \nabla p_o(x',t';x,t)| \leq c'\ p_o'(x',t';x,t) ,$$

here c,c' are constants, depending on $v_{(+)}$, and p_o' is defined as in (4),
ut with ν replaced by a different conveniently chosen ν', also depending
n $v_{(+)}$. For $t' > t$ the solution p of (5) is C^∞ everywhere positive,
oreover the Markov property is expressed through the Kolmogorov con-
ition

8) $$p(x_1,s_1;x_o,s_o) = \int p(x_1,s_1;x,t)\ p(x,t;x_o,s_o)\ dx\ ,\quad t_o \leq t \leq t_1 ,$$

hile the conservation of probability gives

9) $$\int p(x',t';x,t)\ dx' = 1 .$$

e assume some given C^∞ everywhere positive initial density $\rho(\cdot,t_o)$
hen we have at each time

10) $$\rho(x,t) = \int p(x,t;x_o,t_o)\ \rho(x_o,t_o)\ dx_o .$$

lso $\rho(\cdot,t)$ will be C^∞ everywhere positive.
et us call q the process associated to (ρ,p). Then for any choice of
mooth functions f_i, $i=1,2,\dots,n$, $n=1,2,\dots$, we have the following
xpressions for the averages

11) $$E(f_1(q(s_1))\ f_2(q(s_2)) \dots f_n(q(s_n))) =$$
$$= \int \dots \int f_n(x_n)\ p(x_n,s_n;x_{n-1},s_{n-1})\ f_{n-1}(x_{n-1})\ p(x_{n-1},s_{n-1};x_{n-2},s_{n-2})$$
$$\dots f_2(x_2) p(x_2,s_2;x_1,s_1)\ f_1(x_1)\ \rho(x_1,s_1)\ dx_n \dots dx_1 ,$$

here $s_1 \leq s_2 \leq \dots \leq s_n$.
or fixed t' and for a smooth $\hat{f}$ we define the conditional expectations

12) $$f(x,t) = E(\hat{f}(q(t'))/q(t)=x) = \int \hat{f}(x')p(x',t';x,t)dx' ,\quad t' \geq t ,$$

nd notice

13) $$(D_{(+)}f)(x,t) = 0 .$$

Moreover

(14) $\left|f(x,t)\right| \leq \sup|\hat{f}|$, $\left|\nabla f(x,t)\right| \leq c' \sup|\hat{f}| + \sup|\nabla\hat{f}|$,

as a consequence of (7) and a simple integration by parts.

Let us recall also the following basic transport formula, valid for any smooth function F(.,t)

$$(15) \quad E(F(q(s_1),s_1)) = E(F(q(s_o),s_o)) + \\ + \int_{s_0}^{s_1} E((D_{(+)}F)(q(t),t))\,dt \quad .$$

Regular diffusions can be easily time inverted, according to the rules

$$(16) \quad t \to t'=-t \ , \ q(t) \to q'(t')=q(t) \ , \\ v_{(\pm)}(x,t) \to v'_{(\pm)}(x,t')=-v_{(\mp)}(x,t) \quad ,$$

where

$$(17) \quad v_{(-)}=v_{(+)}-2u \ , \ u = \tfrac{1}{2}(v_{(+)}-v_{(-)}) = \nu\nabla\rho/\rho \quad .$$

It is also useful to introduce

$$(18) \quad v = \tfrac{1}{2}(v_{(+)}+v_{(-)}) \ .$$

For other kinematical properties of regular diffusions we refer for example to Ref. 2.

3. A METRIC ON REGULAR DIFFUSIONS.

Let us consider the problem of taking suitable limits of regular diffusions, in order to construct new more general processes, also with singular drifts. A good starting point is offered by (11), since these averages completely define the process. Therefore, let us consider (11) for two different regular diffusions ($v'_{(+)}$, ρ') and ($v''_{(+)}$, ρ'') . Call E' and E" the l.h.s. of (11) in the two cases, respectively. Let F be a common upper bound for all f_i's. By changing one p at a time and finally ρ , we have

$$(19) \quad |E''-E'| \leq F^{n-1}\left|\iint f_n(x_n)(p''-p')(x_n,s_n;x_{n-1},s_{n-1})\cdot \right. \\ \left. \cdot\,\rho''(x_{n-1},s_{n-1})dx_n dx_{n-1}\right| + \ldots + \\ + F^n\int\left|(\rho''-\rho')(x_1,s_1)\right| dx_1 \quad ,$$

ere the dots ... denote other n-1 terms similar to the first one but volving $f_{n-1},\dots,f_1$, respectively. Let us introduce in general f" and defined as in (12), for p=p" and p=p', respectively, and $\hat{f}=f_i$, i=1,. ,n. Then it is clear that, in order to be able to approximate general erages through those related to regular diffusions, we should be able control terms of the type

0) $\int |(\rho''-\rho')(x,t)|\, dx$,

1) $\int |f''-f'|\, \rho''(x,t)\, dx$.

e following simple estimates hold

2) $|\rho''-\rho'| \leq |\sqrt{\rho''}-\sqrt{\rho'}|\,(\sqrt{\rho''}+\sqrt{\rho'})$,

3) $|f''-f'|\rho'' \leq |f''\sqrt{\rho''}-f'\sqrt{\rho'}|\sqrt{\rho''} + F|\sqrt{\rho''}-\sqrt{\rho'}|\sqrt{\rho''}$,

en Schwartz inequality and the normalization condition for ρ', ρ'' ow that, in order to have control on (20),(21), it is sufficient to ntrol the $L^2(dx)$ norms

4) $\|\sqrt{\rho''}-\sqrt{\rho'}\|_2$, $\|f''\sqrt{\rho''}-f'\sqrt{\rho'}\|_2$,

each fixed time.

is control can be obtained, through a long sequence of steps, by introcing the following metric on the class of regular diffusions in the xed time interval (t_0,t_1)

5) $d^2(q'',q') = \sup \|\sqrt{\rho''_t}-\sqrt{\rho'_t}\|^2 + (2\nu)^{-1}\left[\|v''_{(+)}\sqrt{\rho''}-v'_{(+)}\sqrt{\rho'}\|^2 + \dots\right]$,

ere ... is the term with $v_{(-)}$ replacing $v_{(+)}$, $\|\varphi_t\|$ is the norm in (dx) and the sup is taken for $t_0 \leq t \leq t_1$, $\|\cdot\|$ is the norm in $(dxdt)\equiv L^2(R^n \times [t_0,t_1], dxdt)$. We have introduced the constant in 5) so that d is adimensional.

e metric can also be written in equivalent form, by the substitution the term in square bracket in (25) with

6) $[\ \ ] = \|v''\sqrt{\rho''}-v'\sqrt{\rho'}\|^2 + (2\nu)^2\,\|\nabla\sqrt{\rho''}-\nabla\sqrt{\rho'}\|^2$,

cording to (17).

th the definition (25) the class of regular diffusions becomes a tric space. Let us now take the completion. We get some abstract metric space $\mathcal{D}$, whose elements are Cauchy sequences of regular diffusions, with respect to the given metric. Of course, there is no a priori guarantee that elements of $\mathcal{D}$ can still be interpreted as diffusions. The objective of the next section will be to show that this is indeed the case.

4. THE COMPLETION DEFINES A NEW CLASS OF DIFFUSIONS: THE CARLEN PROCESSES.

In order to show that Cauchy sequences of regular diffusions define stochastic processes, we firstly prove a very simple result, which shows the existence of a density ρ and drifts $v_{(+)}, v_{(-)}$, together with the convergence of the basic transport equation (15). Then we can rely on the ingenious strategy found by Carlen and based on methods of parabolic partial differential equations.

The basic result, connecting the method based on the completion for the metric (25) with Carlen's treatment, is given by the following.

Theorem 1. Let q' be regular diffusions, characterized by $(\rho', v'_{(\pm)})$, moving along a Cauchy sequence for the given metric (25). Then there exist a density $\rho(.,t) \in L^1(dx)$ and drifts $v_{(\pm)} \in L^2(\rho\, dxdt) \equiv L^2(R^n \times [t_0,t_1], \rho\, dxdt)$ such that

a) $\rho'(.,t) \to \rho(.,t)$ in $L^1(dx)$ uniformly in t, $t_0 \leq t \leq t_1$,

b) $\int_{t_0}^{t_1} dt \int v'_{(\pm)} \sqrt{\rho'}\, f \sqrt{\rho}\, dx \to \int_{t_0}^{t_1} dt \int v_{(\pm)}\, f\, \rho\, dx$, for any $f \in L^2(\rho dxdt)$

c) $v'_{(\pm)} \rho' \to v_{(\pm)} \rho$ in $L^1(dxdt)$.

The proof is completely elementary. In fact, if q' is Cauchy for d, then $\sqrt{\rho'}(.,t)$ is Cauchy in $L^2(dx)$, uniformly in t. But we also have, from (22),

$$\|\rho'' - \rho'\|_1 \leq 2\, \|\sqrt{\rho''} - \sqrt{\rho'}\|_2 \quad . \tag{27}$$

Therefore $\rho'(.,t)$ is uniformly Cauchy in $L^1(dx)$. The density $\rho(.,t)$ is defined by the L^1 limit. This shows a). In order to prove the existence of the drifts, let us notice that $v'_{(\pm)} \sqrt{\rho'}$ are Cauchy in $L^2(dxdt)$. Call $V_{(\pm)}$ their limits. If $f \in L^2(\rho\, dxdt)$ then $f\sqrt{\rho} \in L^2(dxdt)$. Therefore, we can define the functionals

$$\ell_{(\pm)}(f) = \int_{t_0}^{t_1} dt \int V_{(\pm)}\, f \sqrt{\rho}\, dx \;, \tag{28}$$

d obtain the bounds

9) $$\left| \ell_{(\pm)}(f) \right| \leq \| V_{(\pm)} \|_2 \, \| f\sqrt{\rho} \|_2 \; .$$

t we have also $\| f\sqrt{\rho} \|_2 = \| f \|_\rho$, where $\| \; \|_\rho$ denotes the norm $L^2(\rho\, dxdt)$. Therefore $\ell_{(\pm)}(f)$ are uniformly bounded in $L^2(\rho\, dxdt)$ d, by Fisher-Riesz, there are $v_{(\pm)} \in L^2(\rho\, dxdt)$ such that

0) $$\ell_{(\pm)}(f) = \int_{t_0}^{t_1} dt \int v_{(\pm)} \, f \rho \, dx \quad ,$$

d b) follows.

nally, let us write

1) $$v'_{(\pm)}\rho' - v_{(\pm)}\rho = v'_{(\pm)}\sqrt{\rho'}\left(\sqrt{\rho'} - \sqrt{\rho}\right) + \left(v'_{(\pm)}\sqrt{\rho'} - V_{(\pm)}\right)\sqrt{\rho} \quad ,$$

exploiting the fact that $v_{(\pm)}\rho = V_{(\pm)}\sqrt{\rho}$ in $L^1(dxdt)$ (see (28) d (30)). Then c) follows from Schwartz inequality. Q.E.D.

w we can also define the current velocity and the osmotic velocity

2) $$v = \tfrac{1}{2}(v_{(+)} + v_{(-)}) \quad , \quad u = \tfrac{1}{2}(v_{(+)} - v_{(-)}) \quad ,$$

th $v_{(\pm)}$ given by Theorem 1. It must be noticed that, while in the gular case we have $u = \nu \nabla\rho / \rho$ (see (17)) , no such relation holds general in the limiting case. In fact we have only

3) $$2\nu \nabla\sqrt{\rho'} \to u\sqrt{\rho} \quad \text{in } L^2(dxdt) \; .$$

e next result shows that the system $(\rho, v_{(\pm)})$ has all expected ansport properties.

eorem 2. For any smooth $F(.,.)$ we have

4) $$\int F(x_1,t_1)\,\rho(x_1,t_1)\,dx_1 - \int F(x_0,t_0)\,\rho(x_0,t_0)\,dx_0 = \\ = \int_{t_0}^{t_1} dt \int \left(D_{(\pm)} F \right)(x,t)\, \rho(x,t)\, dx \quad ,$$

ere $D_{(\pm)} = \partial_t + v_{(\pm)} \cdot \nabla \pm \nu \Delta$.

oof. Formula (34) is true for regular diffusions $(\rho', v'_{(\pm)})$ and has a aning also for the limiting $(\rho, v_{(\pm)})$. It holds also in the limit, as consequence of a) and c) of Theorem 1 and the smoothness of F.

tice that $D_{(\pm)}$ in (34) can be substituted by

5) $$D = \tfrac{1}{2}\left(D_{(+)} + D_{(-)} \right) = \partial_t + v \cdot \nabla \; .$$

eorems 1 and 2 give all basic ingredients, which allow us to apply rlen's strategy, without any reference to solutions of the Schroedinger equation. The main results are collected in the following.

<u>Theorem 3</u> (Carlen). For a regular diffusion q' define f' as in (12), with p=p', for some smooth $\hat{f}$. Let q' move along a Cauchy sequence, with density ρ in the limit. Then there exist a function $f(.,t) \in L^2(\rho_t dx) \equiv L^2(R^n, \rho_t d$ such that

(36) $\quad f'(.,t) \rightarrow f(.,t)$ in $L^2(\rho_t dx)$ uniformly in t .

Moreover, the map

(37) $\quad P_{tt'} : \hat{f} \rightarrow f \quad , t \leq t'$,

can be extended to $L^2(\rho_{t'} dx') \rightarrow L^2(\rho_t dx)$, by removing all regularity assumptions on $\hat{f}$. $P_{tt'}$ is positivity preserving and also preserves the averages.

For the proof we refer to Ref. 4 (see also ref. 7).

The results of Theorem 3 allow us to define a stochastic process, by letting the map P play the same role as played explicitely by the transition probability densities in (11) and (12) (see ref. 4). For example, it is very simple to show that, if q',q" vary along a Cauchy sequence, then also the corresponding averages E',E" in (11) are Cauchy and define expectations with respect to the limiting process. This follows from (19)-(24) and the simple estimate at a generic time t

(38) $$\|f''\sqrt{\rho''} - f'\sqrt{\rho'}\|_2 = \|f''(\sqrt{\rho''}-\sqrt{\rho}) - f'(\sqrt{\rho'}-\sqrt{\rho}) + (f''-f')\sqrt{\rho}\|_2 \leq$$
$$\leq F\left(\|\sqrt{\rho''}-\sqrt{\rho}\|_2 + \|\sqrt{\rho'}-\sqrt{\rho}\|_2\right) + \|f''-f'\|_{\rho_t} ,$$

where $\| \; \|_{\rho_t}$ here denotes the norm in $L^2(\rho_t dx)$.

Therefore, we have shown that the introduction of the norm (25) allows us to find, through completion, a large class of processes, possibly with very singular drifts, not necessarily related to solutions of the Schroedinger equation.

5. CONCLUSIONS AND OUTLOOK.

A careful consideration of the method outlined here shows that its success is strongly connected with the introduction of the L^2 spaces with respect to the measures $\rho\, dxdt$ or $\rho_t dx$. In this way the density acts as a counterbalance to possible singularities of the drifts, as foreshadowed the appearance of the square roots multiplying the drifts in the etric (25). This corresponds to the physical idea that flows, in part- ular stochastic flows, do not have a physical meaning in themselves, t only if there is matter attached to them. In fact, in the transport uations of the type (34), the velocity fields appear always multiplied the density.

reover, Carlen method, based on these L^2 estimates, appears extremely werful and surely it would be very interesting to try to extend its ope, for example in the direction of systems with an infinite number degrees of freedom, as in the stochastic or Euclidean version of lativistic quantum field theory. This would provide new methods for e study of infrared and ultraviolet limits. Work on these subjects is progress and will be reported elsewhere.

FERENCES.

E.Nelson, Quantum Fluctuations, Princeton University Press, 1984.

F.Guerra, Phys. Rep. 77, 263 (1981).

F.Guerra and L.M.Morato, Phys. Rev. 27, 1774 (1983).

E.Carlen, Commun. Math. Phys. 94, 293 (1984).

A.M.Il'in,A.S.Kalasnikov and O.A.Oleinik, Uspehi Mat. Nauk 17,3(1962).

S.D.Eidel'man, Parabolic Systems, North-Holland,Amsterdam, 1969.

E.Carlen, Existence and Sample Path Properties of the Diffusions in Nelson's Stochastic Mechanics, M.I.T. Preprint, 1984.

ADIABATIC DRAG AND INITIAL SLIPS FOR RANDOM PROCESSES WITH SLOW AND FAST VARIABLES

F. Haake, M. Lewenstein*, and R. Reibold
Fachbereich Physik, Universität-GHS Essen
4300 Essen, Deutschland

*Permanent address: Institute for Theoretical Physics, Polish Academy of Sciences, 02-668 Warsaw, Poland

Introduction

The adiabatic elimination [1,2,3] of fast variables from a multidimensional process with fast and slow time scales is a twofold problem. One task is the construction of asymptotic equations of motion for the slow variables, valid for times larger than the typical life time τ_f of fast transients. The second, independent part of the problem is the calculation of effective initial values for the slow variables which one must complement the asymptotic equations of motion with. These effective initial values may be defined as the values taken on by the slow variables after the death of all fast transients while the slow transients have not yet begun their decay to a noticeable degree. The difference between the initial values imposed at $t = 0$ and the effective initial values of the slow transients, the so-called initial slips, need not be small [4].

Similar considerations apply to random and quantum mechanical processes involving different time scales [2]. A full description requires a probability density and a density operator $W(t)$, respectively, together with an appropriate generator L of infinitesimal time translations,

$$\dot{W}(t) = L\, W(t) \; . \tag{1}$$

In exploiting the smallness of the time scale ratio τ_f/τ_s it is often possible and convenient to determine an asymptotic, reduced, generator ℓ for a suitably reduced probability distribution or density operator $\rho(t)$. These quantities describe the dynamics of the slow part of the process once all fast transients have decayed,

$$\dot{\rho}(t) = \ell\, \rho(t) \; , \; t \gg \tau_f \; . \tag{2}$$

An initial state at t = 0 may be specified by W(0) which implies a reduced version ρ(0). Of course, this ρ(0) must be distinguished from the effective reduced initial distribution $\rho_{eff}(0)$ to be used in conjunction with the asymptotic equation of motion (2). Formally, we may define $\rho_{eff}(0)$ through

$$\rho(t) = \int_{fast} W(t) = \int_{fast} e^{Lt}\, W(0) = e^{\ell t} \rho_{eff}(0). \tag{3}$$

The difference

$$\rho(0) - \rho_{eff}(0) = \int_{fast} W(0) - \rho \tag{4}$$

describes all initial slips.

We propose to illustrate the above considerations for two exactly solvable processes which have both played important roles in statistical physics, the classical Ornstein Uhlenbeck process in the limit of strong damping [1] and a quantum mechanical harmonic oscillator interacting with a heat bath [5,6,7].

The Ornstein-Uhlenbeck Process

One dimensional Brownian motion of a particle in a harmonic potential can be described by the Fokker Planck equation [1,8]

$$\begin{aligned} \dot{W}(p,q,t) &= LW\,(p,q,t) \\ L &= - \frac{\partial}{\partial q}\, p + \frac{\partial}{\partial p}\,(\gamma p + q) + \frac{\partial^2}{\partial p^2}\,\gamma d\,. \end{aligned} \tag{5}$$

Note that we have set the mass of the particle and the spring constant equal to unity. The damping constant and the diffusion coefficient are denoted by γ and γd, respectively. Due to the linear dependence of the drift coefficients on the displacement q and the momentum p and since the diffusion constant is independent of p and q the general solution of (5) is easily obtained [9]. We shall be interested in the special case of strong damping,

$$\gamma \gg 1.$$

The drift matrix ($D_{qq} = 0$, $D_{qp} = -1$, $D_{pp} = \gamma$, $D_{pq} = 1$) in L has the two eigenvalues

$$\gamma_{\pm} = \gamma/_2 \pm \sqrt{\gamma^2/_4 - 1} \equiv \gamma/_2 \pm \Gamma \tag{7}$$

The larger one is of order γ and the smaller one of order $^1/\gamma$, the time scale ratio for the fast and slow mode thus being $\tau_f/\tau_s \approx {}^1/\gamma^2$. The fast eigenvector turns out to have a p component much larger than its q component while the slow eigenvector is dominated by its q component. It is therefore most natural to try an adiabatic elimination of the momentum, i.e. to inquire about the reduced distribution of the displacement alone

$$\rho(q,t) = \int_{-\infty}^{+\infty} dp\; W(p,q,t) \; . \tag{8}$$

For the sake of concreteness we assume an initial state

$$W(p,q,0) = \frac{e^{-p^2/2d}}{\sqrt{2\pi d}}\; \rho(q,0) \; . \tag{9}$$

There are then no initial correlations between p and q and the momentum has the equilibrium distribution. For this special initial condition it is possible to show the reduced distribution (8) to obey the equation of motion

$$\dot\rho(t) = \ell(t)\; \rho(t) \tag{10}$$

which the time dependent generator

$$\begin{aligned}\ell(t) = {} & \frac{\partial}{\partial q}\, q\, \frac{1}{\Gamma}\, \frac{\sinh \Gamma t}{\cosh \Gamma t + \frac{\gamma}{2\Gamma} \sinh \Gamma t} \\ & + \frac{\partial^2}{\partial q^2}\, \frac{d}{\Gamma} \left\{ \frac{\left(1-e^{-\Gamma t}\right) \sinh \Gamma t}{\cosh \Gamma t + \frac{\gamma}{2\Gamma} \sinh \Gamma t} + \frac{e^{-\gamma t/2} \sin \Gamma t}{\left(\cosh \Gamma t + \frac{\gamma}{2\Gamma} \sinh \Gamma t\right)^2} \right\} .\end{aligned} \tag{11}$$

Obviously, the displacement q undergoes a <u>non</u>stationary Gaussian random process. However, the generator $\ell(t)$ approaches the asymptotic form

$$\ell(\infty) = \frac{\partial}{\partial q}\, q \left(\gamma/_2 - \Gamma\right) + \frac{\partial^2}{\partial q^2}\, d \left(\gamma/_2 - \Gamma\right) \tag{12}$$

on the fast time scale $^1/\gamma_+ \sim {}^1/\gamma$. This limiting generator describes a stationary Gaussian process on a time scale $\tau_s = {}^1/\gamma_-$ given by the smaller one of the eigenvalues of the original drift matrix. In contrast to the time dependent generator

(11) the asymptotic one can be shown to be independent of the initial distribution $W(p,q,0)$.

The effective initial distribution $\rho_{eff}(0)$ is most easily obtained from Chandrasekhar's exact result [9] (which solves (10,11)),

$$\rho(q,t) = \frac{1}{\sqrt{2\pi d\,\sigma(t)}} \int dq_0 \rho(q_0,0)\, e^{-\frac{(q-q_0 a(t))^2}{2d\sigma(t)}} ,$$

$$a(t) = e^{-\frac{1}{2}\gamma t}\left(\cosh \Gamma t + \frac{\gamma}{2\Gamma} \sinh \Gamma t\right) , \tag{13}$$

$$\sigma(t) = 1 - e^{-\gamma t}\left\{1 + \frac{\gamma}{2\Gamma} \sinh \Gamma t + \frac{\gamma^2-2}{2\Gamma^2} \sinh^2 \Gamma t\right\} ,$$

by dropping all fast transients $e^{-\gamma_+ t}$ and by extrapolating the slow transients $e^{-\gamma_- t}$ back to $t = 0$. The resulting solution of the asymptotic initial value problem can be written in the form

$$\rho(q,t) = e^{\ell(\infty)\cdot(t-t_0)}\rho(q,0)$$

$$= \int dq_0\, \rho(q_0,0)\, \frac{\exp\left[-\frac{(q-q_0 e^{-\gamma_-(t-t_0)})^2}{2d\left(1-e^{-2\gamma_-(t-t_0)}\right)}\right]}{\sqrt{2\pi d\left(1-e^{-2\gamma_-(t-t_0)}\right)}} . \tag{14}$$

with

$$t_0 = \frac{1}{\gamma_-} \ln \frac{\gamma_+}{\gamma_+-\gamma_-} .$$

The initial slip simply amounts to a time shift of the order $^1/\gamma_+$. Due to the assumed smallness of γ_-/γ_+ the slip effects appear to be of little practical importance in this case.

Ullersma's Process

We now consider a quantum mechanical model often used to illustrate the origin of irreversibility in Hamiltonian systems [1,5,6]. An harmonic oscillator with unperturbed eigenfrequency ω_0 is coupled to N other harmonic oscillators with eigenfrequencies ω_n according to the Hamiltonian

$$H = \frac{1}{2} (P_0{}^2 + \omega_0{}^2 Q_0{}^2) + \frac{1}{2} \sum_{n=1}^{N} (P_n^2 + \omega_n^2 Q_n^2) + \sum_{n=1}^{N} \varepsilon_n Q_0 Q_n . \tag{15}$$

The N+1 pairs of canonical variables obey the commutation rules

$$[P_\nu, Q_\mu] = \delta_{\nu\mu} \frac{\hbar}{i} , \quad [P_\nu, P_\mu] = [Q_\nu, Q_\mu] = 0 \tag{16}$$

For the eigenvalues of H to have a lower bound the coupling constants ε_n and the frequencies ω_ν have to fulfill

$$\omega_0{}^2 - \sum_{n=1}^{N} \frac{\varepsilon_n^2}{\omega_n^2} \geq 0 . \tag{17}$$

Due to the harmonicity of the Hamiltonian it is possible to give an explicit solution to the general initial value problem [1]. It suffices for our present purpose, however, to assume an initial state of partial thermal equilibrium such that the N "heat bath" oscillators are represented by a canonical density operator. The central oscillator, on the other hand, may initially be in an arbitrary state, represented by the density operator $\rho(0)$. The complete density operator then is the product

$$W(0) = \rho(0) \; Z^{-1} \exp \{- \sum_{n=1}^{N} (P_n^2 + \omega_n^2 Q_n^2)/2kT\} . \tag{18}$$

We shall be interested in the temporal behavior of the reduced Wigner function $\rho(p,q,t)$ of the central oscillator. Because of the harmonicity of H and the effectively Gaussian initial statistics for the bath implied by (18) we can construct an exact generator of infinitesimal time translations for the reduced Wigner function,

$$\begin{aligned} \ell(t) = & - \frac{\partial}{\partial q} p - \frac{\partial}{\partial p} \{f_{pq}(t) q + f_{pp}(t) p\} \\ & + \frac{\partial^2}{\partial p^2} d_{pp}(t) + \frac{\partial^2}{\partial p \partial q} d_{pq}(t) . \end{aligned} \tag{19}$$

The drift and diffusion coefficients occuring here are quasiperiodic functions of the time. They can be expressed in terms of the orthogonal matrix which canonically transforms the N+1 pairs P_ν, Q_ν into the momenta and displacements of the eigenmodes of H and the corresponding eigenvalues [1]. We shall not need these rather lengthy expressions here.

We want the N "bath oscillators" to constitute a heat reservoir for the central oscillator. Therefore, we let their eigenfrequencies ω_n and the coupling constants be densely spaced such that sums may be replaced by integrals,

$$\sum_n \varepsilon_n^2(\ldots) \to \int_0^\infty d\omega\, \gamma(\omega)(\ldots) . \tag{20}$$

Formally, this replacement turns the quasiperiodic behavior of the drift and diffusion coefficients in (19) into an aperiodic one. Even though the final results are universal to a considerable degree, i.e. rather insensitive to the choice of the spectral density $\gamma(\omega)$ we here adopt Ullersma's strength function

$$\gamma(\omega) = \frac{2}{\pi} \frac{\kappa\alpha^2\omega^2}{\alpha^2+\omega^2} . \tag{21}$$

This particular choice for $\gamma(\omega)$ allows to evaluate all frequency integrals in closed form.

The parameters κ and α appearing in (21) are both frequencies by dimension. While κ measures the strength of the coupling between the central oscillator and the reservoir, we can identify α as the inverse response time of the reservoir variable to which the central oscillator is coupled in the Hamiltonian, $\sum_n \varepsilon_n Q_n$. Moreover, the drift coefficients $f_{pq}(t)$ and $f_{pp}(t)$ turn out to relax to stationary values in a time of the order $1/\alpha$,

$$\left.\begin{aligned} f_{pp}(t) &\to 2\Gamma \\ f_{pq}(t) &\to \Omega_1{}^2 = \omega_0{}^2 - \alpha\kappa \end{aligned}\right\} \text{ for } t \gg 1/\alpha . \tag{22}$$

The parameter 2Γ which can be expressed in terms of the "microscopic" quantities ω_0, κ, and α must obviously be interpreted as a damping constant for the central oscillator. The quantity Ω_1, on the other hand, describes the renormalization of the unperturbed frequency ω_0 of the central oscillator by the heat bath. It follows from the positivity condition (17) and the definition (20) that Ω_1 is real. The shifted frequency Ω_1 may but need not be larger than the damping constant Γ. Both Ω_1 and Γ must be small compared to the cut-off frequency α, however, if the heat bath is to deserve its name,

$$\Gamma, \Omega_1 \ll \alpha . \tag{23}$$

The diffusion coefficients $d_{pp}(t)$ and $d_{pq}(t)$ also assume stationary values as

$t \to \infty$. However, beyond the bath response time α^{-1} there is an additional time scale in these coefficients, the thermal time $\hbar/kT$. For the central oscillator to be describable by a time independent generator $\ell(\infty)$ we must therefore require

$$\Gamma \ll \nu \equiv kT/\hbar \quad \text{and} \quad \Gamma \ll \alpha \,. \tag{24}$$

For times larger than $1/\nu$ and $1/\alpha$ the diffusion coefficients are then given in terms of the equilibrium mean squares of the displacement and the momentum of the central oscillator [7],

$$\begin{aligned}\langle Q_0{}^2\rangle_{eq} &= kT/\Omega_1{}^2 + \frac{\hbar}{\pi\Omega}\,\mathrm{Im}\,\psi\left(1+\frac{\Gamma+i\Omega}{\nu}\right),\\ \langle P_0{}^2\rangle_{eq} &= kT \quad + \frac{2}{\pi}\hbar\,\Gamma\,\mathrm{Re}\left[\ln\frac{\alpha}{\nu} - \psi\left(1+\frac{\gamma+i\omega}{\nu}\right)\right]\\ &\qquad + \frac{\hbar(\Omega^2-\Gamma^2)}{\pi\Omega}\,\mathrm{Im}\,\psi\left(1+\frac{\Gamma+i\Omega}{\nu}\right)\end{aligned} \tag{25}$$

with $\Omega^2 = \Omega_1{}^2 - \Gamma^2$ as

$$\left.\begin{aligned} d_{pp}(t) &\to 2\Gamma\,\langle P_0{}^2\rangle_{eq}\\ d_{pq}(t) &\to \Omega_1{}^2\,\langle Q_0{}^2\rangle_{eq} - \langle P_0{}^2\rangle_{eq}\end{aligned}\right\}\quad \text{for } t \gg 1/\alpha,\ 1/\nu \,. \tag{26}$$

The resulting asymptotic generator $\ell(\infty)$ takes its simplest form in the classical limit. As is obvious from (25,26) the offdiagonal diffusion coefficient d_{pq} vanishes while the diagonal one has the well-known value $d_{pp}(\infty) = 2\Gamma kT$. Apart from notational differencies the generator $\ell(\infty)$ in that case is precisely of the Ornstein Uhlenbeck form (5). We would like to point out that the heat bath can impart strong damping to the central oscillator in the limit considered [10].

Similarly simple as in the classical case is the behavior of the central oscillator in the limit of weak damping. We then have, beyond (24), $\Gamma \ll \Omega_1$. By dropping Γ/Ω and $\hbar\Gamma/kT$ from (25) but allowing $\hbar\Omega/kT$ to remain finite we again find the generator to be of the Ornstein Uhlenbeck form (5), with the quantum energy instead of kT in the diffusion coefficient, $d_{pp}(\infty)/2\Gamma = (\hbar\Omega/2)\coth(\hbar\Omega/2kT)$.

In the general case when (25) allows for no further implification the diffusion matrix has a negative eigenvalue. The asymptotic generator $\ell(\infty)$ thus bears no analogy with any classical random process.

We now turn to the effective initial Wigner function $\rho_{as}(p,q,0)$ which is to be used in conjunction with the asymptotic generator $\ell(\infty)$ [7]. One would expect and indeed finds $\rho_{as}(p,q,0)$ to coincide with the initial Wigner function $\rho(p,q,0)$ in zeroth order in Γ, i.e. in the weak damping limit. The central oscillator then undergoes a Gaussian Markov process which is, in fact, the Ornstein Uhlenbeck one.

For strong damping, however, there is a nontrivial initial slip, $\rho_{as}(p,q,0) - \rho(p,q,0) \neq 0$. Such a slip indicates a non-Markovian nature of the process, the existence of a time independent asymptotic generator $\ell(\infty)$ notwithstanding. By again exploiting the diagonalizability of H it is easy to calculate the slip in question. The rather simple result

$$\rho_{as}(p,q,t) = \rho(p + 2\Gamma q,q,0) \tag{27}$$

is obtained in the classical limit. It is most interesting to see the slip to be large for strong damping. For a discussion of the generalization of (27) to low temperatures we refer to ref. [7].

Concluding Remarks

The exactly solvable models treated above yield nice illustrations of the concepts of adiabatic elimination and initial slip. These concepts are of greater practical importance, however, for systems which can be treated rigorously only in zeroth order in the time scale ratio τ_f/τ_s. In such cases the asymptotic generator $\ell(\infty)$ and the effective initial distribution (or density operator) can be calculated as perturbation series in τ_f/τ_s. Examples are the single-mode laser [4] or near-critical systems [11].

References

[1] F. Haake, Z. Phys. B48, 31 (1982).
[2] F. Haake, M. Lewenstein, Phys. Rev. A28, 3606 (1983).
[3] V. Geigenmüller, U. M. Titulaer, B. U. Felderhof, Physica 119A, (1983).
[4] F. Haake, M. Lewenstein, Phys. Rev. A27, 1013 (1982).
[5] P. Ullersma, Physica 32, 27 (1966); 32, 56 (1966) 32, 74 (1966); 32, 90 (1966).
[6] F. Haake, R. Reibold, Acta Physica Austriaca 56, 37 (1984).
[7] F. Haake, R. Reibold, submitted to Phys. Rev. A.
[8a] G. E. Uhlenbeck, L. S. Ornstein, Phys. Rev. 36, 823 (1930) reprinted in N. Wax "Selected Papers in Noise and Stochastic Processes", Dover, New York (1954).
[8b] H. A. Kramers, Physica 7, 284 (1940).
[9] S. Chandrasekhar, Rev. Mod. Phys. 15, 1 (1943), represented in N. Wax ed. "Selected Papers on Noise and Stochastic Processes", Dover, New York (1954).
[10] Φ is the Digamma function.
[11] F. Haake, M. Lewenstein, M. Wilkens, Z. Phys. B54, 333 (1984), and 55, 211 (1984).

Uses of non-Fock quantum Brownian motion and a quantum martingale representation theorem

by

R L Hudson
Mathematics Department
Nottingham University
Nottingham NG7 2RD
England

and

J M Lindsay
School of Mathematics
Bristol University
Bristol BS8 1TN
England

Abstract After reviewing theories of stochastic integration against Fock and non-Fock quantum Brownian motion, we prove a martingale representation theorem for the latter, extending the main result of [12] by incorporating an initial space. We construct unitary processes adapted to the filtration of non-Fock quantum Brownian motion and use the martingale representation theorem to characterise such processes in terms of covariantly adapted unitary evolutions [9] with a continuity property. The classical limits of the quantum dynamical semigroups associated with these processes are contrasted with those arising in the Fock case.

1. Introduction

Quantum Brownian motion [3] has come of age. The Boson quantum stochastic calculus based on Fock space creation and annihilation and number operators [11,14, 16,17], as well as being a natural yet far-reaching noncommutative generalisation of the classical theory [14], has found significant applications in mathematical physics, in dilations of quantum dynamical semigroups [15], in integration of Schrödinger evolutions [21], in the constructive theory of quantum Markov processes [5], in the description of the measurement process [2], and elsewhere. Fermion theories show similar promise [1].

However our purpose in this work is to show that the quantum Brownian motion based on Fock space may be advantageously replaced by one based on certain non-Fock representations of the CCR, namely those corresponding to so-called extremal universally invariant states of the CCR algebra [23]. The resulting stochastic calculus [20] is in many respects simpler, permitting a more advanced mathematical development and in particular the formulation of a martingale representation theorem as well as an interesting physical interpretation in the classical limit, than the Fock construction which from our point of view may be regarded as a mathematically degenerate and physically extreme case.

The mathematical degeneracy arises from the following circumstances.

(1) The Fock vacuum vector is cyclic but not separating for the von Neumann algebra generated by the Fock quantum Brownian motion (which consists of all bounded operators in Fock space). The analogs of L^2-spaces of random variables are thus mathematically unwieldy objects comprising equivalence classes of operators with coincident actions on the vacuum.

(2) The Fock vacuum, as the ground state of an infinite-degree-of-freedom

oscillator, is Gaussian; in particular each canonical pair of Fock field operators is distributed in the vacuum state as in a limit state of the Boson central limit theorem [4]. However this limit state is one of minimal uncertainty and arises as a central limit only in the case when each of the summands of the theorem is already distributed as in the limit state. From this view-point the Fock vacuum emerges as the analog of a classical degenerate (δ-function) distribution.

(3) There is no analog in the Fock theory of the Kunita-Watanabe martingale representation theorem, that every centred square-integrable martingale is a stochastic integral against Brownian motion; the gauge process [14] is a martingale which cannot be so represented. As we shall show, the existence of such a theorem is the key to the characterisation of unitary processes.

Corresponding to these mathematical aspects of the Fock theory, particularly to (2), we interpret the Fock vacuum physically as a state of zero temperature, and expect that, since quantum effects predominate at low temperatures, the classical limit of theories based on Fock space will be somewhat singular.

The plan of this work is as follows. In §2 we review all aspects of the Fock quantum stochastic calculus which are relevant to what follows, so that the paper is self-contained. The formulation given here is simpler than that of [14]. In §3 we discuss stochastic integration against the non-Fock quantum Brownian motion, and in §4 prove the martingale representation theorem for this theory, thus generalising [12] which treats only the case when there is no initial space. In §5 we construct unitary processes which are the solutions of certain stochastic differential equations, and we show in our main result, Theorem 5.2, that these processes are characterised by their adaptedness and covariance properties, together with a certain uniform continuity condition. If the latter could be relaxed to strong continuity we would obtain a "stochastic Stone's theorem [22]"; some progress has been made in this direction [20]. Finally in §6 we discuss the quantum dynamical semigroups, and their classical limits, of which the unitary processes give rise to dilations.

We use the following notations and conventions. Inner products $\langle \, , \, \rangle$ are linear on the right. B(h) denotes the Banach algebra of bounded operators on a Banach space h. I is the identity operator. We use the usual notations $\subseteq$, * as applied to densely defined operators in a Hilbert space and their extensions and adjoints, A pair $T^{\#} = (T,T^{\dagger})$ of such operators are mutually adjoint if $T^{\dagger} \subseteq T^*$. Algebraic tensor products of vector spaces and of operators in them are indicated by $\underline{\otimes}$; Hilbert space extensions and completions are indicated by $\otimes$, which is also used in both cases for product vectors. χ_S denotes the indicator function of a set S which is 1 on S and 0 on its complement S^c. $\int_s^t f$ denotes the Riemann-Lebesgue integral of the complex-valued function f over the interval [s,t].

Conversations with K R Parthasarathy are gratefully acknowledged.

2. Stochastic calculus and the construction of unitary processes in Fock space

Let h be a Hilbert space. The (Boson) Fock space over h and the family of exponential vectors (or coherent states) is characterised [8] to within unitary equivalence exchanging the exponential vectors, as a pair $(\Gamma(h), (\psi(f): f \in h))$, where $\Gamma(h)$ is a Hilbert space and the $\psi(f)$ are vectors in $\Gamma(h)$ satisfying

(a) $\langle\psi(f),\psi(g)\rangle = \exp\langle f,g\rangle, \quad f,g \in h$ (2.1)

(b) $\mathcal{E} = \text{span}\{\psi(f): f \in h\}$ is dense in $\Gamma(h)$. (2.2)

Wedenote by ψ_0 the Fock vacuum $\psi(0)$. When $h = h_1 \oplus h_2$ is a direct sum we can set

$$\Gamma(h_1 \oplus h_2) = \Gamma(h_1)\otimes\Gamma(h_2), \quad \psi(f_1,f_2) = \psi(f_1)\otimes\psi(f_2), \quad f_1 \in h_1,\ f_2 \in h_2. \tag{2.3}$$

Since the exponential vectors are linearly independent [4] an operator with domain $\mathcal{E}$ is well defined by specifying an arbitrary action on each $\psi(g)$, $g \in h$. In particular we may define, as operators on $\mathcal{E}$, the Fock annihilation operator $a(f)$, the Fock creation operator $a^\dagger(f)$ and the Fock Weyl operator $W(f)$ corresponding to $f \in h$, and the Fock second quantisation $\Gamma(T)$ of $T \in B(h)$ by the actions

$$a(f)\psi(g) = \langle f,g\rangle\psi(g), \quad a^\dagger(f)\psi(g) = \frac{d}{dt}\psi(g+tf)\Big|_{t=0} \tag{2.4}$$

$$W(f)\psi(g) = \exp(-\tfrac{1}{2}\|f\|^2 + \langle f,g\rangle)\psi(g+f) \tag{2.5}$$

$$\Gamma(T)\psi(g) = \psi(Tg). \tag{2.6}$$

Then $a(f)$ and $a^\dagger(f)$ are mutually adjoint and respectively conjugate-linear and linear in f. For arbitrary $d,e,f,g \in h$

$$\langle a^\dagger(f)\,\psi(d), a^\dagger(g)\psi(e)\rangle = \langle a(g)\psi(d), a(f)\psi(e)\rangle + \langle f,g\rangle\langle\psi(d),\psi(e)\rangle \tag{2.7}$$

(that is, formally, or on an appropriately enlarged domain, $[a(f),a^\dagger(g)] = \langle f,g\rangle I$). $W(f)$ is isometric and thus extends uniquely to an isometry on $\Gamma(h)$. The same is true of $\Gamma(T)$ if T is isometric or coisometric, or more generally a contraction since a contraction is the product of isometries and coisometries, and we have the functorial rules of second quantisation for $T_1,T_2,T \in B(h)$

$$\Gamma(T_1T_2) = \Gamma(T_1)\Gamma(T_2), \quad \Gamma(T^*) \subseteq \Gamma(T)^*, \quad \Gamma(I) = I. \tag{2.8}$$

We denote the extensions to $\Gamma(h)$ by the same symbols. The Weyl operators satisfy the Weyl relation (CCR)

$$W(f)W(g) = \exp(-i\,\text{Im}\langle f,g\rangle)W(f+g), \quad f,g \in h, \tag{2.9}$$

in particular since $W(0) = I$, each $W(f)$ is unitary with inverse $W(-f)$. The vacuum expectation functional for the Fock Weyl operators is

$$\langle\psi_0, W(f)\psi_0\rangle = \exp(-\tfrac{1}{2}\|f\|^2). \tag{2.10}$$

As operators on $\mathcal{E}$, for $f,g \in h$,

$$a(f)W(g) = W(g)(a(f) + \langle f,g\rangle I), \quad a^\dagger(f)W(g) = W(g)(a^\dagger(f) + \langle g,f\rangle I), \tag{2.11}$$

while if $T \in B(h)$ is isometric

$$W(f) = \Gamma(T)^*W(Tf)\Gamma(T). \tag{2.12}$$

The Weyl operators $(W(f): f\in h)$ generate $B(\Gamma(h))$ as a von Neumann algebra. The map

$f \mapsto W(f)$ is strongly continuous. We have

$$W(f) = \exp(a^{\dagger}(f) - a(f)) \quad (f \in h) \tag{2.13}$$

in the sense that

$$\frac{d}{dt}W(tf)\Big|_{t=0} = a^{\dagger}(f) - a(f) \tag{2.14}$$

on $\mathcal{E}$, so that the infinitesimal generator of the one-parameter group $t \to W(tf)$ extends $i^{-1}(a^{\dagger}(f) - a(f))$.

Now let $h = L^2(\mathbb{R}^+)$. For $s,t \in \mathbb{R}^+$ with $s \le t$ let $h^t = L^2[0,t]$, $h^{(t} = L^2(t,\infty)$ and $h^{(s,t]} = L^2(s,t]$. Let

$$h = h^t \oplus h^{(t}, \quad f = (f^t, f^{(t}) \quad (f \in h) \tag{2.15}$$

be the natural projective decompositions. Correspondingly, we write

$$\Gamma(h) = \Gamma(h^t) \otimes \Gamma(h^{(t}), \quad \psi_0 = \psi_0^t \otimes \psi_0^{(t}, \quad \mathcal{E} = \mathcal{E}^t \underline{\otimes} \mathcal{E}^{(t}. \tag{2.16}$$

Let there be given an initial Hilbert space H^0, equipped with a dense subspace $\mathcal{E}^0$ called the <u>initial domain</u>. We write

$$\tilde{\Gamma}(h) = H^0 \otimes \Gamma(h), \quad \tilde{\Gamma}(h^t) = H^0 \otimes \Gamma(h^t), \quad \tilde{\mathcal{E}} = \mathcal{E}^0 \underline{\otimes} \mathcal{E}, \quad \tilde{\mathcal{E}}^t = \mathcal{E}^0 \underline{\otimes} \mathcal{E}^t \tag{2.17}$$

so that

$$\tilde{\Gamma}(h) = \tilde{\Gamma}(h^t) \otimes \Gamma(h^{(t}), \quad \tilde{\mathcal{E}} = \tilde{\mathcal{E}}^t \underline{\otimes} \mathcal{E}^{(t}. \tag{2.18}$$

<u>Definition 2.1</u> A <u>(Fock) adapted process</u> is a family of operators $F = (F(t): t \in \mathbb{R}^+)$ such that for each $t \in \mathbb{R}^+$ a) $F(t)$ is an operator in $\tilde{\Gamma}(h)$ with domain $\tilde{\mathcal{E}}$, the domain of whose adjoint includes $\tilde{\mathcal{E}}$ b) $F(t)$ and $F^{\dagger}(t)$ are the ampliations to $\tilde{\mathcal{E}} = \tilde{\mathcal{E}}^t \otimes \mathcal{E}^{(t}$ of operators F^t and $F^{t\dagger}$ in $\tilde{\Gamma}(h^t)$ with domain $\tilde{\mathcal{E}}^t$, where $F^{\dagger}(t)$ is the restriction to $\tilde{\mathcal{E}}$ of the adjoint of $F(t)$. The adapted process $F^{\dagger} = (F^{\dagger}(t): t \in \mathbb{R}^+)$ is called the <u>adjoint process</u> to F.

An adapted process F is called <u>simple</u> if there exists an increasing sequence t_n, $n = 0,1,\ldots$ with $t_0 = 0$ and $t_n \underset{n}{\to} \infty$ such that for each $n \ge 0$ $F(t) = F(t_n)$ for $t \in [t_n, t_{n+1})$, <u>measurable</u> (resp. <u>continuous</u>) if $t \mapsto F(t)\phi$ is strongly measurable (resp. continuous) for each $\phi \in \tilde{\mathcal{E}}$, <u>locally square integrable</u> if it is measurable and, for each $t \in \mathbb{R}^+$ and $\phi \in \tilde{\mathcal{E}}$,

$$\|F\|_{t,\phi}^{\#2} = \int_0^t \|F^{\#}(s)\phi\|^2 \, ds \tag{2.19}$$

is finite, and a <u>martingale</u> if, for arbitrary $s,t \in \mathbb{R}^+$ with $s < t$ and $\phi_1, \phi_2 \in \tilde{\mathcal{E}}^s$

$$\langle \phi_1 \otimes \psi_0^{(s}, F(t)\phi_2 \otimes \psi_0^{(s} \rangle = \langle \phi_1 \otimes \psi_0^{(s}, F(s)\phi_2 \otimes \psi_0^{(s} \rangle. \tag{2.20}$$

We denote by $\mathcal{A}, \mathcal{A}_0, \mathcal{A}_c, \mathcal{A}_m, \mathcal{L}^2$ and $\mathcal{M}$ respectively the complex vector spaces of adapted, simple, continuous, measurable and locally square integrable processes and of martingales. As in [14], every locally square integrable process can be approximated by a sequence of simple processes in the sense of the seminorms $\| \; \|_{t,\phi}^{\#}$ defined by (2.19).

An adapted process is <u>bounded</u> (resp. <u>unitary</u>) if it consists of the restrictions to $\tilde{\mathcal{E}}$ of bounded (resp. unitary) operators on $\tilde{\Gamma}(h)$. Identifying the restrictions with the extensions, we see that for a bounded process F, each $F(t) \in \tilde{N}_t$, where $\tilde{N}_t$ is the von Neumann algebra in $\tilde{\Gamma}(h) = \tilde{\Gamma}(h^t) \otimes \Gamma(h^{(t})$,

$$\tilde{N}_t = B(\tilde{\Gamma}(h^t)) \otimes I. \tag{2.21}$$

Such a process F is a martingale if and only if, for all $s,t \in \mathbb{R}^+$ with $s \le t$

$$\mathbb{E}_s[F(t)] = F(s) \tag{2.22}$$

where $\mathbb{E}_s$ is the vacuum conditional expectation map from $\tilde{N} = B(\tilde{\Gamma}(h))$ onto $\tilde{N}_s$ defined by

$$\mathbb{E}_s[T] = \mathbb{E}^s[T] \otimes I, \quad \langle \phi_1, \mathbb{E}^s[T]\phi_2 \rangle = \langle \phi_1 \otimes \psi_0^{(s}, T\phi_2 \otimes \psi_0^{(s} \rangle. \tag{2.23}$$

We define Fock quantum Brownian motion to be the pair of mutually adjoint adapted processes

$$A^{\#}(t) = I \underline{\otimes} a^{\#}(\chi_{[0,t]}), \quad t \in \mathbb{R}^+. \tag{2.24}$$

In view of (2.13) and the strong continuity of the Weyl operators we may regard $\tilde{N}_t$ as generated by the initial von Neumann algebra $B(H^0) \otimes I$ together with $\{A^{\#}(s): s \le t\}$. With this understanding, Fock adapted processes are adapted to Fock quantum Brownian motion.

For $s,t \in \mathbb{R}^+$ with $s < t$, $A^{\#}(t) - A^{\#}(s)$ is the ampliation to $\tilde{\mathcal{E}} = \tilde{\mathcal{E}}^s \underline{\otimes} \mathcal{E}^{(s}$ of an operator $A^{(s,t]\#}$ in $H^{(s}$ with domain $\mathcal{E}^{(s}$. Given an adapted process F we may then define the products

$$F(s)(A^{\dagger}(t) - A^{\dagger}(s)) = F^s \underline{\otimes} A^{(s,t]\dagger}, \quad F^{\dagger}(s)(A(t) - A(s)) = F^{s\dagger} \underline{\otimes} A^{(s,t]} \tag{2.25}$$

as mutually adjoint operators on $\tilde{\mathcal{E}}$ which are ampliations of operators on $\tilde{\mathcal{E}}^t$. This permits the following definition of the stochastic integral of simple processes.

Definition 2.2 Let $F, G, \mathfrak{H}$ be simple processes, assumed to have common intervals of constancy bounded by $0 = t_0 < t_1 < \ldots t_n \underset{n}{\to} \infty$. The adapted process M defined inductively by $M(0) = 0$ and

$$M(t) = M(t_n) + F(t_n)(A^{\dagger}(t) - A^{\dagger}(t_n)) + G^{\dagger}(t_n)(A(t) - A(t_n)) + \mathfrak{H}(t_n)(t - t_n), \quad t_n < t \le t_{n+1} \tag{2.26}$$

is called the stochastic integral of $(F, G^{\dagger}, \mathfrak{H})$ against the Brownian motion $A^{\#}$ and time t.

We write

$$M(t) = \int_0^t (F dA^{\dagger} + G^{\dagger} dA + \mathfrak{H} dt). \tag{2.27}$$

Clearly then

$$M^{\dagger}(t) = \int_0^t (G dA^{\dagger} + F^{\dagger} dA + \mathfrak{H}^{\dagger} dt). \tag{2.28}$$

The differential notation $dM = F dA^{\dagger} + G^{\dagger} dA + \mathfrak{H} dt$ indicates that M differs from the stochastic integral (2.27) by the ampliation to $\tilde{\mathcal{E}}$ of an operator in H^0 with domain $\mathcal{E}^0$.

<u>Theorem 2.1</u> Let $F, G, \mathfrak{H} \in \mathfrak{a}_0$, $M(t) = \int_0^t (F dA^\dagger + G^\dagger dA + \mathfrak{H} ds)$. Then for $u \in \mathcal{E}^0$, $f, g \in h$

$$\langle M(t)u \otimes \psi(f),\ M(t)u \otimes \psi(g)\rangle = \int_0^t \{\langle M(s)u \otimes \psi(f),\ (\bar{f}(s)F(s) + g(s)G^\dagger(s) + \mathfrak{H}(s))u \otimes \psi(g)\rangle$$

$$+ \langle(\overline{g(s)}F(s) + f(s)G^\dagger(s) + \mathfrak{H}(s))u \otimes \psi(f),\ M(s)u \otimes \psi(g)\rangle + \langle F(s)u \otimes \psi(f),\ F(s)u \otimes \psi(g)\rangle\}\, ds. \tag{2.29}$$

In particular

$$\|M(t)u \otimes \psi(f)\|^2 = \int_0^t \{2\,\mathrm{Re}\,\langle M(s)u \otimes \psi(f),\ (\bar{f}(s)F(s) + f(s)G^\dagger(s) + \mathfrak{H}(s))u \otimes \psi(f)\rangle$$

$$+ \|F(s)u \otimes \psi(f)\|^2\}\, ds. \tag{2.30}$$

<u>Proof</u> For brevity we prove only (2.30); the proof of (2.29) is similar (notice that nonlinearity of $\psi(f)$ in f prevents polarisation of (2.30) to get (2.29), but that both identities may be polarised in M and u). Assume M(t) given by (2.26) so that

$$\|M(t)u \otimes \psi(f)\|^2 = \|[M(t_n) + F(t_n)(A^\dagger(t) - A^\dagger(t_n)) + G^\dagger(t_n)(A(t) - A(t_n))$$
$$+ (t-t_n)\mathfrak{H}(t_n)]u \otimes \psi(f)\|^2.$$

Using the action $(A(t)-A(t_n))u \otimes \psi(f) = \int_{t_n}^t f u \otimes \psi(f)$ and the corresponding adjoint action of $A^\dagger(t) - A^\dagger(t_n)$ on $u \otimes \psi(f)$, we replace $A(t) - A(t_n)$ by $\int_{t_n}^t f$ and $A^\dagger(t) - A^\dagger(t_n)$ by $\int_{t_n}^t \bar{f}$ in (2.33), the term $\|F(t_n)(A^\dagger(t) - A^\dagger(t_n))u \otimes \psi(f)\|^2$ giving rise to the Ito correction $(t-t_n)\|F(t_n)u \otimes \psi(f)\|^2$ in view of (2.7). Differentiating the resulting expression,

$$\|M(t)u \otimes \psi(f)\|^2 = \|[M(t_n) + \int_{t_n}^t \bar{f}F(t_n) + \int_{t_n}^t fG^\dagger(t_n) + (t-t_n)\mathfrak{H}(t_n)]u \otimes \psi(f)\|^2$$
$$+ (t-t_n)\|F(t_n)u \otimes \psi(f)\|^2,$$

we obtain

$$\frac{d}{dt}\|M(t)u \otimes \psi(f)\|^2 = 2\,\mathrm{Re}\langle[M(t_n) + \int_{t_n}^t \bar{f}F(t_n) + \int_{t_n}^t fG^\dagger(t_n) + (t-t_n)]u \otimes \psi(f),$$
$$[\bar{f}(t)F(t_n) + f(t)G^\dagger(t_n) + \mathfrak{H}(t_n)]u \otimes \psi(f)\rangle + \|F(t_n)u \otimes \psi(f)\|^2$$
$$= 2\,\mathrm{Re}\langle[M(t_n) + F(t_n)(A^\dagger(t) - A^\dagger(t_n)) + G^\dagger(t_n)(A(t) - A^\dagger(t_n)) + (t-t_n)$$
$$H(t_n)]u \otimes \psi(f),\ [\overline{f(t)}F(t) + f(t)G^\dagger(t) + \mathfrak{H}(t)]u \otimes \psi(f)\rangle + \|F(t)u \otimes \psi(f)\|^2$$

reversing the replacing of the A (t) - A (t_n) by their (adjoint) actions and using the constancy of F , G and $\mathfrak{H}$ on $[t_n, t_{n+1})$. Using (2.26) again we see that this is the differential form of (2.30). □

Corollary 1 Under the hypotheses of the theorem, for arbitrary $u,v \in \mathcal{E}^0$, $f,g \in h$ and $t > 0$

$$\langle u\otimes\psi(f),\ M(t)v\otimes\psi(g)\rangle = \int_0^t \langle u\otimes\psi(f),\ (\bar{f}(s)F(s)+g(s)G^\dagger(s)+\mathfrak{H}(s))v\otimes\psi(g)\rangle\, ds. \tag{2.31}$$

Proof We apply the polarised form of (2.29) to $M(t)$ and to $t = \int_0^t dt$ to obtain

$$t\langle u\otimes\psi(f),\ M(t)v\otimes\psi(g)\rangle = \int_0^t \{\langle u\otimes\psi(f),\ M(s)v\otimes\psi(g)\rangle + s\langle u\otimes\psi(f),\ (\bar{f}(s)F(s)+g(s)G^\dagger(s)+H(s))v\otimes\psi(g)\rangle\}\, ds.$$

Differentiating with respect to t gives the differential version of (2.31). □

Corollary 2 If $dM = F\,dA^\dagger + G^\dagger dA + \mathfrak{H}\,dt$ then, for arbitrary $s,t \in \mathbb{R}^+$ with $s \le t$, $\phi \in \tilde{\Gamma}(h^s)$, $f,g \in h$, $v \in \mathcal{E}^0$,

$$\langle \phi\otimes\psi(f^{(s)}),\ (M(t)-M(s))v\otimes\psi(g)\rangle = \int_s^t \langle \phi\otimes\psi(f^{(s)}),\ (\overline{f(\tau)}F(\tau)+g(\tau)G^\dagger(\tau)+\mathfrak{H}(\tau))v\otimes\psi(g)\rangle\, d\tau \tag{2.32}$$

Proof Replacing f by $d\chi_{[0,s]} + f\chi_{(s,\infty)}$ and M by the stochastic integral $M - M(0)$ in (2.31) and subtracting the corresponding equation with t replaced by s, we see that (2.32) holds when ϕ is of form $u\otimes\psi(d^s)$, $d \in h$. But vectors of this form are total in $\tilde{\Gamma}(h^s)$. □

Corollary 3 Under the hypotheses of the theorem

$$\|M(t)u\otimes\psi(f)\|^2 \le \int_0^t \exp(\|\chi_{(s,t]}f\|^2+t-s)[2\|F(s)u\otimes\psi(f)\|^2 + \|G^\dagger(s)u\otimes\psi(f)\|^2 + \|\mathfrak{H}(s)u\otimes\psi(f)\|^2]\, ds. \tag{2.33}$$

Proof Differentiating (2.30) and making several uses of the inequality $2\,\mathrm{Re}\langle\phi_1,\phi_2\rangle \le \|\phi_1\|^2 + \|\phi_2\|^2$ we obtain

$$\frac{d}{dt}\|M(t)u\otimes\psi(f)\|^2 \le (|f(t)|^2+1)\|M(t)u\otimes\psi(f)\|^2 + 2\|F(t)u\otimes\psi(f)\|^2 + \|G^\dagger(t)u\otimes\psi(f)\|^2 + \|\mathfrak{H}(t)u\otimes\psi(f)\|^2. \tag{2.34}$$

Multiplying by the integrating factor $\exp(-\|\chi_{[0,t]}f\|^2 - t)$ and integrating we get (2.33). □

Now let $F, G^\dagger, \mathfrak{H} \in \mathcal{L}^2$, and let F_n, $G_n^\dagger$, $\mathfrak{H}_n$, $n = 1,2,\ldots$ be simple processes approximating F, G, $\mathfrak{H}$ in the sense of the seminorms (2.19) whose stochastic integrals are M_n, $n = 1,2,\ldots$. Applying the estimate (2.34) to differences $M_n - M_m$ we see that $M_n(t)u\otimes\psi(f)$ is Cauchy hence convergent. Moreover the limit $M(t)u\otimes\psi(f)$ is independent of the choice of sequence of simple approximants. The operators $M(t)$ on $\tilde{\mathcal{E}}$ so defined, together with the corresponding limits $M^\dagger(t)$ of the adjoints $M_n^\dagger$ form mutually adjoint adapted processes which we define to be the stochastic integrals of $(F,G^\dagger,\mathfrak{H})$ and $(F^\dagger,G,\mathfrak{H}^\dagger)$ against Fock quantum Brownian motion. Since the convergence is uniform in t on finite intervals, as is clear from (2.33), we may pass to the limit of simple approximants on both sides to obtain that (2.29), (2.30), (2.31), (2.32) and the estimate (2.33) hold for arbitrary $F,G,\mathfrak{H} \in \mathcal{L}^2$. Replacing $F^\#$, $G^\#$ and $\mathfrak{H}^\#$ by $\chi_{(s,t]}F^\#$, $\chi_{(s,t]}G^\#$, $\chi_{(s,t]}\mathfrak{H}^\#$ in (2.34) shows that stochastic integrals are continuous processes.

It is easy to see from (2.31) that a stochastic integral in which there is no time integrand, $\mathfrak{H} \equiv 0$, is a martingale. One might conjecture that conversely every martingale M satisfies

$$dM = F\,dA^\dagger + G^\dagger\,dA \tag{2.35}$$

for some locally square integrable processes F, $G^\dagger$. For the Fock theory we have developed so far this conjecture is false; the gauge process $\Lambda = \Lambda^\dagger$ defined on $\tilde{\mathcal{E}}$ by

$$\Lambda(t) = I \otimes \frac{1}{i}\frac{d}{d\varepsilon}\Gamma(M_{\exp(i\varepsilon\chi_{[0,t]})})\Big|_{\varepsilon=0}, \tag{2.36}$$

where M_f is multiplication by f on h, is a martingale which does not satisfy (2.35). In [14] stochastic integration is developed in Fock space including Λ as an integrator. The natural conjecture, that every martingale M satisfies

$$dM = E\,d\Lambda + F\,dA^\dagger + G^\dagger\,dA \tag{2.37}$$

for locally square integrable E ,F ,G in the sense of [14], is open.

In the remainder of this section we take $\mathcal{E}^0 = H^0$. Then by an extension of the Hellinger-Toeplitz theorem [15], for arbitrary $F^\# \in \mathcal{Q}$, $t > 0$ and $f \in h$, the linear maps $u \to F^\#(t)u \otimes \psi(f)$ are bounded from H^0 to $\tilde{H}$; we denote their bounds by $\|F^\#(t)\|_f$.

We turn to the construction of unitary processes. Let $L^0, L^{0\dagger} = L^{0*}$ and $\mathfrak{H}^0 = \mathfrak{H}^{0*}$ be bounded operators on H^0, and denote by L_0, $L_0^\dagger$ and $\mathfrak{H}_0$ their Hilbert space ampliations to $\tilde{\Gamma}(h) = H^0 \otimes \Gamma(h)$.

Theorem 2.3 The stochastic differential equation

$$dU = U(L_0 dA^\dagger - L_0^\dagger dA + (i\mathfrak{H}_0 - \tfrac{1}{2}L_0^\dagger L_0)dt), \quad U(0) = I \tag{2.38}$$

has a unique solution.

Proof We establish existence by the iterative method, defining processes U_n, $n = 0,1,\ldots$ by

$$U_0 \equiv I$$

$$dU_n = U_{n-1}(L_0 dA^\dagger - L_0^\dagger dA + (i\mathfrak{H}_0 - \tfrac{1}{2}L_0^\dagger L_0)dt), \quad U_n(0) = I. \tag{2.39}$$

Clearly $U_0 \in \mathcal{Q}_c$; assuming $U_{n-1} \in \mathcal{Q}_c$ then $U_{n-1}L_0$, $U_{n-1}L_0^\dagger$, $U_{n-1}(i\mathfrak{H}_0 - \frac{1}{2}L_0^\dagger L_0) \in \mathcal{Q}_c$, with adjoint processes $L_0^\dagger U_{n-1}^\dagger$, $L_0 U_{n-1}^\dagger$, $-(i\mathfrak{H}_0 + \frac{1}{2}L_0^\dagger L_0)U_{n-1}^\dagger$ respectively, so that their stochastic integral is defined. Since it differs from I by this stochastic integral $U_n \in \mathcal{Q}_c$. Hence U_n is well defined for all n.

By iterating (2.33) we have, for $0 \le t \le s \in \mathbb{R}^+$, $u \in H_0$, $f \in h$,

$$\|(U_n^\dagger(t) - U_{n-1}^\dagger(t))u \otimes \psi(f)\|^2 \le \exp(\|f\|^2 + s)(n!)^{-1}[4C^2 s]^n \|u \otimes \psi(f)\|^2$$

where $C = \max\{\|L_0\|, \|i\mathcal{H}_0 - \frac{1}{2}L_0^\dagger L_0\|\}$, whence the sequence $(U_n(t) - U_{n-1}(t))u \otimes \psi(f)$ is uniformly Cauchy for $t \in [0,s]$. We define $U^\dagger \in \mathcal{Q}$ by setting $U(t)u \otimes \psi(f) = \lim_n U_n(t) u \otimes \psi(f)$; since the convergence is uniform on finite intervals we may pass to the limit on both sides of the integrated form of (2.39) to obtain that U solves (2.38).

Suppose V is a second solution. Then $W = U^\dagger - V^\dagger$ satisfies

$$dW = (-L_0 W dA^\dagger + L_0^\dagger W dA - (i\mathcal{H}_0 + \tfrac{1}{2}L_0^\dagger L_0)W dt), \qquad W(0) = 0.$$

From (2.34), for $t > 0$, $u \in H_0$, $f \in h$

$$\|W(t)u \otimes \psi(f)\|^2 \le 3C^2 \exp(\|f\|^2 + t) \int_0^t \|W(s)u \otimes \psi(f)\|^2 \, ds. \tag{2.40}$$

Being a stochastic integral, W is continuous, so that the map $s \to W(s)u \otimes \psi(f)$ is continuous, hence bounded on $[0,t]$. Hence by iterating the estimate (2.40) we find that $W(t)u \otimes \psi(f) = 0$, hence $W \equiv 0$. □

<u>Theorem 2.4</u> The process U defined by (2.38) is unitary.

<u>Proof</u> Applying (2.29) and (2.31) to the stochastic integral $U^\dagger - I$, we have

$$\langle (U^\dagger(t) - I)u \otimes \psi(f), (U^\dagger(t) - I)u \otimes \psi(g)\rangle = -\langle u \otimes \psi(f), (U^\dagger(t) - I)u \otimes \psi(g)\rangle - \langle (U^\dagger(t) - I)u \otimes \psi(f), u \otimes \psi(g)\rangle$$

for arbitrary $u \in H^\circ$, $f,g \in h$ and $t \ge 0$, whence

$$\langle U^\dagger(t)u \otimes \psi(f), U^\dagger(t)u \otimes \psi(g)\rangle = \langle u \otimes \psi(f), u \otimes \psi(g)\rangle.$$

Polarising u we see that $U^\dagger(t)$ is isometric. In particular $U^\dagger$ and hence U is bounded. Applying the differential version of (2.29), polarised in u, together with (2.30) to $U - I$, we find that, for fixed $f,g \in h$, the operators $K(t) \in B(H^\circ)$ defined by

$$\langle u, K(t)v\rangle = \langle U(t)u \otimes \psi(f), U(t)v \otimes \psi(g)\rangle \qquad (u,v \in H^\circ)$$

satisfy the differential equation

$$\frac{dK}{dt} = [K, \bar{f}L_0 - gL_0^\dagger + i\mathcal{H}_0] - \tfrac{1}{2}(L_0^\dagger L_0 K - 2L_0^\dagger K L_0 + K L_0^\dagger L_0), \qquad K(0) = \langle \psi(f), \psi(g)\rangle I, \tag{2.41}$$

of which $K \equiv \langle \psi(f), \psi(g)\rangle I$ is a solution. If f and g are restricted to be piecewise constant the right hand side of (2.41) is of form $\mathcal{L}(K)$ where $\mathcal{L}$ is a piecewise constant $B(B(H^\circ))$-valued function, so that the solution is unique and consequently $\langle U(t)u \otimes \psi(f), U(t)v \otimes \psi(g)\rangle = \langle u \otimes \psi(f), v \otimes \psi(g)\rangle$ in this case. Since by continuity of the map $f \to \psi(f)$, the exponential vectors corresponding to such f and g are total, we conclude that U is isometric as required. □

For $s,t \in \mathbb{R}^+$ with $s \le t$ we denote by $\tilde{N}_{(s,t]}$ the von Neumann algebra $B(H_0) \otimes I \otimes B(\Gamma(h^{[s,t]}) \otimes I$ generated by $\{T^\circ \otimes W(\chi_{(s,t]}f) : T^\circ \in B(H^\circ), f \in h\}$. For $r \in \mathbb{R}^+$ denote by Γ_r the operator $I\otimes\Gamma(S_r)$ on $\Gamma(h)$, where S_r is the shift

$$S_r\, f(t) = \begin{cases} 0 & t < r, \\ f(t-r), & t \geq r. \end{cases} \tag{2.42}$$

Let $\mathbb{E}^\circ$ be the vacuum conditional expectation from $\tilde{\Gamma}(h)$ onto H° defined by

$$\langle u, \mathbb{E}^\circ(T)v\rangle = \langle u\otimes\psi_0, Tv\otimes\psi_0\rangle, \qquad (u,v\in H_0,\ T\in B(\tilde{\Gamma}(h))). \tag{2.43}$$

Theorem 2.5 [15] The unitary process U defined by (2.38) satisfies

a) $U^\dagger(s)U(t)\in\tilde{N}_{(s,t]} \qquad (s,t\in\mathbb{R}^+,\ s\leq t)$ (2.44)

b) $U(t) = \Gamma_r^\dagger U_r^\dagger U_{r+t}\Gamma_r \qquad (r,t\in\mathbb{R}^+)$ (2.45)

c) $t\mapsto\mathbb{E}^\circ(U(t))$ is uniformly continuous from $\mathbb{R}^+$ to $B(H^\circ)$. (2.46)

Proof a) Since $U^\dagger(s)U(t)\in\tilde{N}_t = B(H\ )\otimes B(\Gamma(h^s))\otimes B(\Gamma(L^2(s,t]))\otimes I$ by adaptedness, it suffices to prove that $U^\dagger(s)U(t)$ commutes with every operator of form $T = I\otimes T_1\otimes I$ with $T_1\in B(\Gamma(h^s))$. To do this regard s as fixed, vary t and define bounded processes J and K by

$$J(t) = \begin{cases} 0 & \text{if} \quad t < s \\ U^\dagger(s)U(t) & \text{if} \quad t \geqslant s \end{cases} \qquad\qquad K(t) = [T,J(t)].$$

From (2.38) we deduce that

$$J(t) = I\chi_{[s,\infty)}(t) + \int_0^t J(\tau)(LdA^\dagger - L^\dagger dA + (iH - \tfrac{1}{2}L^\dagger L)d\tau).$$

From Corollary 2 to Theorem 2.1 we deduce that, for $u,v\in H^\circ$, $f,g\in h$ and $t\geq s$

$$\begin{aligned}\langle u\otimes\psi(f), K(t)v\otimes\psi(g)\rangle &= \langle T^\dagger u\otimes\psi(f), J(t)v\otimes\psi(g)\rangle - \langle Tv\otimes\psi(g), J^\dagger(t)u\otimes\psi(g)\rangle^- \\ &= \int_s^t \langle u\otimes\psi(f), K(\tau)(\bar{f}(\tau)L - g(\tau)L^\dagger + iH - \tfrac{1}{2}L^\dagger L)v\otimes\psi(g)\rangle d\tau \\ &= \int_0^t \langle u\otimes\psi(f), K(\tau)(\bar{f}(\tau)L - g(\tau)L^\dagger + iH - \tfrac{1}{2}L^\dagger L)v\otimes\psi(g)\rangle d\tau.\end{aligned} \tag{2.46}$$

Since (2.46) holds trivially when $t < s$, we deduce from (2.30) that K satisfies

$$dK = K(LdA^\dagger - L^\dagger dA + (iH - \tfrac{1}{2}L^\dagger L)dt), \qquad K(0) = 0,$$

that is addition of K to a solution of (2.38) yields a new solution, violating the uniqueness, unless $K\equiv 0$ as was to be proved.

b) Fix r and define

$$V(t) = \Gamma_r^\dagger U^\dagger(r)U(r+t)\Gamma_r \qquad (t\in\mathbb{R}).$$

Since $U^\dagger(r)U(r+t)\in\tilde{N}_{(r,r+t]}$ and conjugation by Γ_r maps $\tilde{N}_{(r,r+t]}$ into $\tilde{N}_{(0,t]}$, as follows from (2.12), V is an adapted process, inheriting continuity from U. From Theorem 2.1 and its Corollary, for arbitrary $u,v\in H$, $f,g\in h$ and $t\geq 0$

$$\left\langle u\otimes\psi(f),\ \int_0^t V(\tau)(LdA^\dagger - L^\dagger dA + (iH - \tfrac{1}{2}L^\dagger L)d\tau)v\otimes\psi(g)\right\rangle$$

$$= \int_0^t \left\langle U(r)u\otimes\psi(S_r f),\ U(r+\tau)(\overline{f(\tau)}L - g(\tau)L^\dagger + iH - \tfrac{1}{2}L^\dagger L)v\otimes\psi(S_r g)\right\rangle d\tau$$

$$= \int_r^{r+t} \left\langle U(r)u\otimes\psi(S_r f),\ U(\tau)(\overline{S_r f(\tau)}L - S_r g(\tau)L^\dagger + iH - \tfrac{1}{2}L^\dagger L)v\otimes\psi(S_r g)\right\rangle d\tau$$

$$= \left\langle U(r)u\otimes\psi(S_r f),\ (U(r+t) - U(r))v\otimes\psi(S_r g)\right\rangle$$

$$= \left\langle u\otimes\psi(f),\ (V(t) - I)v\otimes\psi(g)\right\rangle.$$

Hence V satisfies (2.38). Hence $V \equiv U$ as required.

c) From (2.38) and (2.30), in which we set $M = U$ and $f = g = 0$, we have

$$\left\langle u,\ (\mathbb{E}^\circ(U(t)) - I)v\right\rangle = \int_0^t \left\langle u,\ \mathbb{E}^\circ(U(s))(iH^\circ - \tfrac{1}{2}L^{\circ\dagger}L^\circ)v\right\rangle ds \tag{2.47}$$

for arbitrary $u,v \in H^\circ$, $t \in \mathbb{R}^+$. Thus $(\mathbb{E}^\circ[U(t)]:\ t \geq 0)$ is a contraction semigroup $(\exp t(iH^\circ - \tfrac{1}{2}L^{\circ\dagger}L^\circ):\ t \geq 0)$ with bounded infinitesimal generator. Hence $t \mapsto \mathbb{E}^\circ(U(t))$ is uniformly continuous. □

We conclude this review of the Fock theory by sketching the generalisation to finitely many independent quantum Brownian motions (in [15] the non-trivial further generalisation to infinitely many is given). Thus, let h now be the Hilbert space $L^2(\mathbb{R}^+;\mathbb{C}^N) = L^2(\mathbb{R}^+)\otimes\mathbb{C}^N$. With corresponding definitions of h^t and $h^{(t}$, (2.15)...(2.18) hold and the space of adapted processes and its various subsets are defined exactly as before. We now have N component Fock quantum Brownian motions

$$A_j^\#(t) = I\otimes a^\#(\chi_{[0,t]}\varepsilon_j), \quad t \in \mathbb{R}^+ \tag{2.48}$$

where $(\varepsilon_1,\dots,\varepsilon_N)$ is the natural basis for $\mathbb{C}^N$. We define stochastic integrals

$$M(t) = \int_0^t \left(\sum_{j=1}^N \left\{F_j dA_j^\dagger + G_j^\dagger dA_j\right\} + H ds\right), \quad M^\dagger(t) = \int_0^t \left(\sum_{j=1}^N \left\{G_j dA_j^\dagger + F_j^\dagger dA_j\right\} + H ds\right) \tag{2.49}$$

initially of (2N+1)-tuples of simple processes $F_1,\dots,G_N,H$, extending the definition to locally square integrable processes through the many-dimensional analog of (2.31)

$$\begin{aligned}\left\langle M(t)u\otimes\psi(f),\ M(t)u\otimes\psi(g)\right\rangle &= \int_0^t \Bigg[\left\langle M(s)u\otimes\psi(f),\ \left[\sum_{j=1}^N \{\overline{f_j(s)}F_j(s) + g_j(s)G_j^\dagger(s)\} + H(s)\right]u\otimes\psi(g)\right\rangle \\ &+ \left\langle\left\{\sum_{j=1}^N \{\overline{g_j(s)}F_j(s) + f_j(s)G_j^\dagger(s)\} + H(s)\right\}u\otimes\psi(f),\ H(s)u\otimes\psi(g)\right\rangle \\ &+ \sum_{j=1}^N \left\langle F_j(s)u\otimes\psi(f),\ F_j(s)u\otimes\psi(g)\right\rangle\Bigg] ds\end{aligned} \tag{2.50}$$

where $u \in \mathcal{E}^\circ$, $f = (f_1,\dots,f_N)$, $g = (g_1,\dots,g_N) \in h$ and $t \geq 0$. With the appropriate modification to the definitions of $\tilde{N}_{[s,t]}$, Γ_r and $\mathbb{E}_0$ we then have the following theore which is proved similarly to Theorems 2.3, 2.4 and 2.5.

Theorem 2.6 Let $L_1^\circ,\dots,L_N^\circ$, $\mathcal{H}^\circ \in B(H^\circ)$ with $\mathcal{H}^\circ = \mathcal{H}^{\circ *}$, and let $L_1,\dots,L_N$, $L_1^\dagger,\dots,L_N^\dagger$, $\mathcal{H}$ be the Hilbert space ampliations of $L_1^\circ,\dots,L_N^\circ$, $L_1^{\circ *},\dots,L_N^{\circ *}$, $\mathcal{H}^\circ$ to $H^\circ \otimes \Gamma(h)$. Then the stochastic differential equation

$$dU = U\left[\sum_{j=1}^{N}(L_j dA_j^\dagger - L_j^\dagger dA_j) + (i\mathcal{H} - \tfrac{1}{2}\sum_{j=1}^{N} L_j^\dagger L_j)\, dt\right], \quad U(0) = I \tag{2.51}$$

has a unique solution U. The process U is unitary and satisfies

a) $U^\dagger(s)U(t) \in \tilde{N}_{(s,t}$ $\quad (s,t \in \mathbb{R}^+,\ s \le t)$ (2.52)

b) $U(t) = \Gamma_r^\dagger U^\dagger(r)U(r+t)\Gamma_r$ $\quad (r,t \in \mathbb{R}^+)$ (2.53)

c) $t \mapsto \mathbb{E}^\circ(U(t))$ is uniformly continuous from $\mathbb{R}^+$ to $B(H_0)$. (2.54)

§3. Stochastic integration against quantum Brownian motion of variance $\sigma^2 > 1$

Fix a real number $\sigma^2 > 1$ called the variance and define positive numbers λ,μ by $\sigma^2 = \lambda^2 + \mu^2$, $\lambda^2 - \mu^2 = 1$.

For the Hilbert space $h = L^2(\mathbb{R}^+)$, we construct a triple $(H, (W(f): f \in h), \Omega)$ comprising a Hilbert space H, a strongly continuous map $f \to W(f)$ from h to unitary operators on H satisfying the Weyl relation (2.9), and a unit vector Ω, cyclic for the operators $W(f)$, $f \in h$, for which the expectation functional is

$$\langle \Omega, W(f)\Omega\rangle = \exp(-\tfrac{1}{2}\sigma^2\|f\|^2) \tag{3.2}$$

as follows:

$$H = \Gamma(h)\otimes\Gamma(h) = \Gamma(h\oplus h) = \Gamma(L^2(\mathbb{R}^+ : \mathbb{C}^2)) \tag{3.3}$$

$$W(f) = W_F(\lambda f)\otimes W_F(-\mu\bar{f}), \quad f \in h \tag{3.4}$$

where now the $W_F(f)$ are the Fock Weyl operators defined by (2.5) and $f \to \bar{f}$ the natural conjugation in $L^2(\mathbb{R}^+)$

$$\Omega = \psi_0 \otimes \psi_0. \tag{3.5}$$

The commutant N' of the von Neumann algebra N generated by the $W(f)$, $f \in h$ is generated by the operators

$$W'(f) = W_F(-\mu f)\otimes W_F(\lambda\bar{f}), \quad f \in h. \tag{3.6}$$

Ω is cyclic for N' also, hence cyclic and separating for both N and N'.

For a contraction $T \in B(h)$ we define its second quantisation $\Gamma(T)$ by

$$\Gamma(T) = \Gamma_F(T)\otimes\Gamma_F(\bar{T}) \tag{3.7}$$

where now $\Gamma_F(T)$ is the Fock second quantisation defined by (2.6) and $\bar{T}$ is defined by

$\overline{T}\overline{f} = (Tf)^{-}$. Then the functorial rules (2.8) and the relation (2.12) hold; in particular if T is unitary, conjugation by $\Gamma(T)$ implements the automorphism of N which maps each W(f) to W(Tf). An extension of Shale's theorem shows that this automorphism is inner, equivalently that $\Gamma(T)$ can be factorised as a product of elements N and N', if and only if T differs from I by a Hilbert-Schmidt operator on h [10].

We now denote by $\mathfrak{E}$ the dense subspace of H spanned by the vectors $W'(f)\Omega$, $f \in h$. Note that in view of (3.6), (3.5) and (2.5)

$$\mathfrak{E} \subseteq \mathfrak{E}_F \underline{\otimes} \mathfrak{E}_F \tag{3.8}$$

where now $\mathfrak{E}_F$ is the span (2.2) of the exponential vectors in the Fock space.

Corresponding to the decomposition (2.15) we write

$$H = H^t \otimes H^{(t}, \qquad \Omega = \Omega^t \otimes \Omega^{(t} \tag{3.9}$$

$$W(f) = W(f^t) \otimes W(f^{(t}), \qquad W'(f) = W'(f^t) \otimes W'(f^{(t}) \tag{3.10}$$

using (3.3) and (3.4) and corresponding Fock space decompositions. In particular we can write

$$N = N^t \otimes N^{(t}, \qquad N' = N^{t'} \otimes N^{(t'}, \qquad \mathfrak{E} = \mathfrak{E}^t \underline{\otimes} \mathfrak{E}^{(t}. \tag{3.11}$$

Let there be given an initial space H° carrying an <u>initial von Neumann algebra</u> N° with commutant $N^{\circ\prime}$ and cyclic separating vector Ω°. We take the <u>initial domain</u> $\mathfrak{E}^{\circ}$ to be $N^{\circ\prime}\Omega^{\circ}$. We write

$$\tilde{H} = H^{\circ} \otimes H, \qquad \tilde{N} = N^{\circ} \otimes N, \qquad \tilde{\Omega} = \Omega^{\circ} \otimes \Omega, \qquad \tilde{\mathfrak{E}} = \mathfrak{E}^{\circ} \underline{\otimes} \mathfrak{E} \tag{3.12}$$

with similar definitions of $\tilde{H}^t$, $\tilde{N}^t$, $\tilde{\Omega}^t$ and $\tilde{\mathfrak{E}}^t$, so that

$$\tilde{H} = \tilde{H}^t \otimes H^{(t}, \qquad \tilde{N} = \tilde{N}^t \otimes N^{(t}, \qquad \tilde{\Omega} = \tilde{\Omega}^t \otimes \Omega^{(t}, \qquad \tilde{\mathfrak{E}} = \tilde{\mathfrak{E}}^t \underline{\otimes} \mathfrak{E}^{(t}. \tag{3.13}$$

We set $\tilde{N}_t = \tilde{N}^t \otimes I$.

A pair $T^{\#} = (T, T^{\dagger})$ of operators in $\tilde{H}$ with common domain $\tilde{\mathfrak{E}}$ is said to be <u>weakly affiliated to</u> $\tilde{N}$ if, for arbitrary $S \in \tilde{N}'$, $S^*T^{\dagger} \subseteq (ST)^*$. Taking S = I we see that such operators are necessarily mutually adjoint. The equations

$$\psi^{\#} = T^{\#}\tilde{\Omega}, \qquad T^{\#}S^{\circ} \otimes W'(f)\tilde{\Omega} = S^{\circ} \otimes W'(f)\psi^{\#} \qquad (S^{\circ} \in N^{\circ\prime},\ f \in h) \tag{3.14}$$

establish a one-one correspondence between such pairs and pairs $\psi^{\#} = (\psi, \psi^{\dagger})$ of vectors in $\tilde{H}$ for which

$$\langle S^{\circ} \otimes W'(f)\tilde{\Omega}, \psi^{\dagger} \rangle = \langle \psi, S^{\circ *} \otimes W'(f)^*\tilde{\Omega} \rangle \qquad (S^{\circ} \in N^{\circ\prime},\ f \in h); \tag{3.15}$$

we say such pairs of vectors are <u>weakly affiliated</u> to $\tilde{\mathfrak{E}}$. By passing to the weak limit of finite linear combinations in (3.15), we see that a weakly affiliated vector pair is precisely an element of the graph of the adjoint of the conjugate-linear operator $\mathfrak{F}_0$: $\tilde{S}\tilde{\Omega} \to \tilde{S}^*\tilde{\Omega}$, $\tilde{S} \in \tilde{N}'$. Correspondingly, weakly affiliated pairs of operators are precisely the restrictions to $\tilde{\mathfrak{E}}$ of pairs of mutually adjoint operators with domain $\tilde{N}'\tilde{\Omega}$ affiliated to $\tilde{N}$ in the ordinary sense.

We shall need the following in the proof of Theorem 4.2.

Theorem 3.1 Let vectors $\psi^{\#}$ weakly affiliated to $\tilde{\mathcal{E}}$ satisfy the condition

$$\langle T^{\circ}\otimes W(f)\tilde{\Omega},\psi^{\dagger}\rangle = -\langle\psi, T^{\circ *}\otimes W(-f)\tilde{\Omega}\rangle \quad (T^{\circ}\in N^{\circ},\ f\in h) \tag{3.16}$$

in addition to (3.15). Then $\psi = \psi^{\dagger} = 0$.

Proof Passing to weak limits of finite linear combinations, we obtain

$$\langle T^*\tilde{\Omega},\psi^{\dagger}\rangle = -\langle\psi, T\tilde{\Omega}\rangle \quad (T\in\tilde{N})$$

from (3.16), that is the pair $(\psi^{\dagger}, -\psi)$ belongs to the graph of the adjoint of the conjugate linear operator $\mathcal{S}_0$: $\tilde{T}\tilde{\Omega}\to\tilde{T}^*\tilde{\Omega}$, $\tilde{T}\in\tilde{N}$. But $\mathcal{S}_0$ and $\mathcal{F}_0$, and hence their adjoints, are mutually adjoint. From this it follows that $\psi = \psi^{\dagger} = 0$. □

We define a pair of mutually adjoint adapted processes $F^{\#}$ to be a family $(F^{\#}(t): t\in\mathbb{R}^+)$ of pairs of operators in $\tilde{H}$ with domain $\tilde{\mathcal{E}}$ weakly affiliated to $\tilde{N}$ such that, for each $t\in\mathbb{R}^+$, the $F^{\#}(t)$ are the ampliations $F^t \underline{\otimes} I$ to $\tilde{\mathcal{E}} = \tilde{\mathcal{E}}^t \underline{\otimes} \mathcal{E}^{(t}$ of operators F^t in $\tilde{H}^t$ with domain $\tilde{\mathcal{E}}^t$. The subspaces $\mathcal{Q}_0$, $\mathcal{Q}_c$, $\mathcal{Q}_m$ and $\mathcal{L}^2$ of the complex vector space $\mathcal{Q}$ of all adapted processes of simple, continuous, measurable and locally square integrable processes are then defined exactly as in the Fock case, Definition 2.1. For the present theory the family of seminorms (2.19) can be replaced by the equivalent family

$$\|F\|_t^{\#2} = \int_0^t \|F^{\#}(s)\tilde{\Omega}\|^2, \quad t\geq 0. \tag{3.17}$$

The equations

$$\psi^{\#}(t) = F^{\#}(t)\Omega, \quad F^{\#}(t)S^{\circ}\otimes W'(f)\tilde{\Omega} = S^{\circ}\otimes W'(f)\psi \quad (S^{\circ}\in N^{\circ\prime},\ f\in h) \tag{3.18}$$

establish a one-one correspondence between pairs of mutually adjoint adapted processes $F^{\#}$ and families $(\psi^{\#}(t): t\in\mathbb{R}')$ of pairs of vectors weakly affiliated to $\tilde{\mathcal{E}}$ such that, for each $t\in\mathbb{R}^+$,

$$\psi^{\#}(t) = \psi^{t\#}\otimes\Omega^{(t} \tag{3.19}$$

for $\psi^t\in\tilde{\mathcal{E}}^t$. We call ψ the associated vector process of F and denote it by ψ_F. Simplicity, continuity and measurability of F are characterised by simplicity, continuity and measurability of ψ_F while for $F\in\mathcal{Q}$, $F\in\mathcal{L}^2$ if and only if $\int_0^t \|\psi_F^{\#}(s)\|^2 < \infty$ for all $t\geq 0$.

We say that $M\in\mathcal{Q}$ is a martingale if, whenever $s\leq t$

$$E_s\psi_M(t) = \psi_M(s) \tag{3.20}$$

where E_s is the projector onto $\tilde{H}^s\otimes\Omega^{(s}$ in $B(\tilde{H})$, and denote by $\mathfrak{M}$ the subspace of martingales. Since $E_s = I\otimes\Gamma(M_{\chi_{[0,s]}})$ martingales are continuous processes.

The restrictions to $\tilde{\mathcal{E}}$ of a family $(F(t),\ t\in\mathbb{R}^+)$ of bounded operators on $\tilde{H}$ forms an adapted process (then said to be a <u>bounded</u> process, and to be <u>unitary</u> if each F(t) is unitary) if and only if each $F(t) \in N_t$, and a martingale if and only if, for all $s,t \in \mathbb{R}^+$, $s \le t$,

$$\mathbb{E}_s F(t) = F(s) \tag{3.21}$$

where $\mathbb{E}_s$ is the conditional expectation from $\tilde{N}$ onto $\tilde{N}_s$ defined by

$$\mathbb{E}_s[T] = \mathbb{E}^s[T]\otimes I, \quad \langle\phi_1, \mathbb{E}^s[T]\phi_2\rangle = \langle\phi_1\otimes\Omega^{(s}, T\phi_2\otimes\Omega^{(s}\rangle, \quad \phi_1,\phi_2\in\tilde{H}^s, \quad T\in\tilde{N}. \tag{3.22}$$

That $\mathbb{E}_s$ does indeed map $\tilde{N}$ onto $\tilde{N}_s$ follows from the facts that, for $T^\circ \in N^\circ$, $f \in h$,

$$\mathbb{E}_s[T^\circ \otimes W(f)] = e^{-\frac{1}{2}\sigma^2\|\chi_{(s,\infty)}f\|^2}\, T^\circ \otimes W(\chi_{[0,s]}f) \in \tilde{N}_s \tag{3.23}$$

and that $\tilde{N}$ is the weak closure of the linear span of the operators $T^\circ \otimes W(f)$.

Now consider the vectors in $\tilde{H} = H^\circ \otimes \Gamma(h) \otimes \Gamma(h)$

$$\psi_A(t) = \mu\Omega^\circ \otimes \psi_0 \otimes a^\dagger(\chi_{[0,t]})\psi_0, \qquad \psi_A^\dagger(t) = \lambda\Omega^\circ \otimes a^\dagger(\chi_{[0,t]})\psi_0 \otimes \psi_0, \qquad t \in \mathbb{R}^+. \tag{3.24}$$

Using (2.11) and (3.6) it may be verified that these pairs satisfy the condition (3.15) for weak affiliation to $\tilde{\mathfrak{E}}$ as well as (3.19), and are thus the associated vector process of adapted processes $A^\#$, which we call <u>quantum Brownian motion of variance</u> σ^2.

In accordance with (3.8) we have

$$A(t) \subseteq \lambda A_{1F}(t) + \mu A_{2F}^\dagger(t), \qquad A^\dagger(t) \subseteq \lambda A_{2F}^\dagger(t) + \mu A_{2F}(t) \tag{3.25}$$

where $A_{1F}^\#, A_{2F}^\#$ are the components of two dimensional Fock quantum Brownian motion. From this we may infer that each $\tilde{N}_t$ is generated by $N^\circ \otimes I$ together with the $A^\#(s)$ with $s \le t$ in the same sense as in the Fock case.

Let $s,t \in \mathbb{R}^+$, $s \le t$. We write the vectors $\psi_A^\#(t) - \psi_A^\#(s)$ in $\tilde{H} = \tilde{H}^s \otimes \Gamma(h^{(s}) \otimes \Gamma(h^{(s})$ as

$$\psi_A(t) - \psi_A(s) = \mu\tilde{\Omega}^s \otimes \psi_0^{(s} \otimes \psi_{s,t}, \qquad \psi_A^\dagger(t) - \psi_A^\dagger(s) = \lambda\tilde{\Omega}^s \otimes \psi_{s,t} \otimes \psi_0^{(s} \tag{3.26}$$

where $\psi_{s,t} \in \Gamma(h^{(s})$, and satisfies

$$\psi_{s,t} \perp \psi_0^{(s}, \qquad \|\psi_{s,t}\|^2 = t - s, \tag{3.27}$$

as may be seen using (2.4) and (2.7). Now let $F \in \mathfrak{A}$. The vectors

$$\lambda\psi_F^s \otimes \psi_{s,t} \otimes \psi_0^{(s}, \qquad \mu\psi_{F^\dagger}^s \otimes \psi_0^{(s} \otimes \psi_{s,t} \tag{3.28}$$

are weakly affiliated to $\tilde{\mathfrak{E}}$; we define the products

$$F(s)(A^\dagger(t) - A^\dagger(s)), \qquad F^\dagger(s)(A(t) - A(s)) \tag{3.29}$$

to be the corresponding operators weakly affiliated to $\tilde{N}$; clearly these are ampliations to $\tilde{\mathfrak{E}}$ of operators in $\tilde{H}^t$ with domain $\tilde{\mathfrak{E}}^t$. We note also that the vectors (3.28) are orthogonal to each other and to any vector of the form $\phi \otimes \Omega^{(s}$ with $\phi \in H^s$, and that

$$\|\lambda\psi_F^s \otimes \psi_{s,t} \otimes \psi_0^{(s}\|^2 = \lambda^2(t-s)\|\psi_F^s\|^2, \qquad \|\mu\psi_{F^\dagger}^s \otimes \psi_0^{(s} \otimes \psi_{s,t}\|^2 = \mu^2(t-s)\|\psi_{F^\dagger}^s\|^2 \tag{3.30}$$

in view of (3.27).

We may now define the stochastic integral, in the first place of simple processes exactly as in Definition 2.2. Assuming that the simple process F, $G^\dagger$, $\mathfrak{H}$ have common intervals of constancy bounded by $t_0 = 0$, $t_1, \ldots, t_n \underset{n}{\to} \infty$ we have, for their stochastic integral M, if $t \in (t_n, t_{n+1}]$

$$\|\psi_M(t)\|^2 = \|\psi_M(t_n) + (t-t_n)\psi_{\mathfrak{H}}(t_n)\|^2 + \lambda^2(t-t_n)\|\psi_F(t_n)\|^2 + \mu^2(t-t_n)\|\psi_G^\dagger(t_n)\|^2$$

in view of (3.29). Differentiating, using the orthogonality properties of vectors of the form (3.28), and integrating again we find inductively that

$$\|\psi_M(t)\|^2 = \int_0^t \{2\,\mathrm{Re}\langle\psi_M(s),\psi_{\mathfrak{H}}(s)\rangle + \lambda^2\|\psi_F(s)\|^2 + \mu^2\|\psi_G^\dagger(s)\|^2\}\,ds. \tag{3.31}$$

More generally, for arbitrary $S_1, S_2 \in N^{\circ\prime}$, $f, g \in h$ and $t \geq 0$, we have by a similar argument

$$\langle M(t)S_1 \otimes W'(f)\tilde{\Omega},\ M(t)S_2 \otimes W'(g)\tilde{\Omega}\rangle = \int_0^t \langle\{\lambda\mu((-\bar{f}+\bar{g})F + (f-g)G^\dagger) + \mathfrak{H}\}S_1 \otimes W'(f)\tilde{\Omega},$$

$$MS_2 \otimes W'(g)\tilde{\Omega}\rangle$$

$$+ \langle MS_1 \otimes W'(f)\tilde{\Omega},\ \{\lambda\mu((-\bar{f}+\bar{g})F + (f-g)G^\dagger + \mathfrak{H}\}S_2 \otimes W'(g)\tilde{\Omega}\rangle + \lambda^2\langle FS_1 \otimes W'(f)\tilde{\Omega},\ FS_2 \otimes W'(f)\tilde{\Omega}\rangle$$

$$+ \mu^2\langle G^\dagger S_1 \otimes W'(f)\tilde{\Omega},\ G^\dagger S_2 \otimes W'(g)\tilde{\Omega}\rangle. \tag{3.32}$$

From (3.31) we obtain in the case $\mathfrak{H} \equiv 0$

$$\|\psi_M(t)\|^2 = \int_0^t \{\lambda^2\|\psi_F\|^2 + \mu^2\|\psi_G^\dagger\|^2\}$$

and hence

$$\|\psi_M(t)\|^2 + \|\psi_M^\dagger(t)\|^2 = \int_0^t \{\lambda^2(\|\psi_F\|^2 + \|\psi_G\|^2) + \mu^2(\|\psi_F^\dagger\|^2 + \|\psi_G^\dagger\|^2)\}, \tag{3.33}$$

from which it is clear that stochastic integration can be extended to arbitrary locally square integrable processes F and $G^\dagger$. If $\mathfrak{H} \not\equiv 0$ we use the estimate

$$\|\psi_M(t)\|^2 \leq \int_0^t e^{t-s}\{\lambda^2\|\psi_F(s)\|^2 + \mu^2\|\psi_{G^\dagger}(s)\|^2 + \|\psi_{\mathfrak{H}}(s)\|^2\}\,ds \tag{3.34}$$

which is derived from (3.31) as (2.33) is from (2.30) using the integrating factor e^{-t}, together with its companion estimate for $\psi_M^\dagger$ to effect the extension. We note that (3.32), (3.33) and (3.34) remain valid for the extended integral by uniformity of the convergence on finite intervals. The differential notation $dM = FdA + G^\dagger dA^\dagger + \mathfrak{H}dt$ means that the adapted process M differs from the stochastic integral of $(F, G^\dagger, \mathfrak{H})$ by $M(0) = M^\circ \otimes I$, when M°, $M^{\circ\dagger}$ are mutually adjoint operators with domain $\mathcal{E}^\circ$ affiliated to N°.

We prove a partial analog of Corollary 1 to Theorem 2.1.

Theorem 3.2 Let M be the stochastic integral of $F, G^\dagger, \mathfrak{H} \in \mathcal{L}^2$. Then for arbitrary $u \in H^\circ$, $f \in h$, $t \geq 0$

$$\langle u\otimes W(f)\Omega,\ \psi_M(t)\rangle = \int_0^t \langle u\otimes W(f)\Omega,\ \{\lambda^2\overline{f(s)}\psi_F(s) + \mu^2 f(s)\psi_G^\dagger(s) + \psi_{\mathfrak{H}}(s)\}\rangle\, ds. \qquad (3.35)$$

In particular, if $f = 0$, $\mathfrak{H} = 0$ we have

$$\langle u\otimes\Omega,\ \psi_M(t)\rangle = 0. \qquad (3.36)$$

Proof Using (3.25) and (3.4), we have, for $s,t \in \mathbb{R}^+$ with $s \le t$,

$$\langle u\otimes W(f)\Omega,\ F(s)(A^\dagger(t) - A^\dagger(s))\tilde{\Omega}\rangle$$

$$= \langle\{\lambda(A_{1F}(t) - A_{1F}(s)) + \mu(A_{2F}^\dagger(t) - A_{2F}^\dagger(s))u\otimes W_F(\lambda f)\psi_0\otimes W_F(-\mu\tilde{f})\psi_0,\ F(s)\tilde{\Omega}\rangle$$

$$= \langle u\otimes W(f)\Omega,\ \lambda^2\int_s^t \bar{f}.F(s)\tilde{\Omega}\rangle$$

using (2.11) and the fact that $a^\dagger(f)\psi_0 \perp \psi_0$. From this and the similarly proved identity

$$\langle u\otimes W(f)\Omega,\ G^\dagger(s)(A(t) - A(s))\tilde{\Omega}\rangle = \langle u\otimes W(f)\Omega,\ \mu^2\int_s^t f.G(s)\tilde{\Omega}\rangle$$

it is clear that (3.35) holds when $F,G,\mathfrak{H} \in \mathcal{A}_0$. More generally we approximate $F,G,\mathfrak{H} \in \mathcal{L}^2$ in the sense of the seminorms (3.17) by $F_n, G_n, \mathfrak{H}_n \in \mathcal{A}_0$, $n = 1,2,\ldots$ and use the inequality

$$\left|\int_0^t \langle u\otimes W(f)\Omega,\ \{\lambda^2\overline{f(s)}(\psi_F(s) - \psi_{F_n}(s)) + \mu^2 f(s)(\psi_G^\dagger(s) - \psi_{G_n}^\dagger(s)) + \psi_{\mathfrak{H}}(s) - \psi_{\mathfrak{H}_n}(s)\}\rangle\, ds\right|$$

$$\le \|u\|\int_0^t \{\lambda^2|f(s)|\,\|\psi_F(s) - \psi_{F_n}(s)\| + \mu^2|f(s)|\,\|\psi_G^\dagger(s) - \psi_{G_n}^\dagger(s)\| + \|\psi_{\mathfrak{H}}(s) - \psi_{\mathfrak{H}_n}(s)\|\}\, ds$$

$$\le \|u\|\Big\{\sqrt{\int_0^t |f|^2}\Big[\lambda^2\sqrt{\int_0^t \|\psi_F(s) - \psi_{F_n}(s)\|^2\, ds} + \mu^2\sqrt{\int_0^t \|\psi_G^\dagger(s) - \psi_{G_n}^\dagger(s)\|^2\, ds}\Big]$$
$$+ \sqrt{t}\sqrt{\int_0^t \|\psi_{\mathfrak{H}}(s) - \psi_{\mathfrak{H}_n}(s)\|^2\, ds}\Big\}$$

to pass to the limit of simple approximants on both sides of (3.35). □

Theorem 3.3 Let $T \in N^\circ$, $f \in h$. Then the restrictions to $\tilde{\mathcal{E}}$ of the bounded operators

$$W_{T,f}(t) = \exp(\tfrac{1}{2}\sigma^2\|f_t\|^2)T\otimes W(f_t), \qquad f_t = \chi_{[0,t]}f, \qquad t \ge 0 \qquad (3.37)$$

together with their adjoints form a martingale; moreover for each $t \ge 0$

$$W_{T,f}(t) = T\underline{\otimes} I + \int_0^t (fW_{T,f}dA^\dagger - \bar{f}W_{T,f}dA). \qquad (3.38)$$

Proof (3.23) implies that $W_{T,f}$ is a martingale, hence continuous. Hence the right hand side of (3.38) is well defined. To prove (3.38) we evaluate

$$\|\{W_{T,f}(t) - T\otimes I - \int_0^t (fW_{T,f}dA^\dagger - \bar{f}W_{T,f}dA)\}\tilde{\Omega}\|^2$$

$$= \|W_{T,f}(t)\tilde{\Omega}\|^2 + \|T\otimes I\tilde{\Omega}\|^2 + \|\int_0^t (fW_{T,f}dA^\dagger - \bar{f}W_{T,f}dA)\tilde{\Omega}\|^2$$

$$- 2\ \mathrm{Re}\Big\langle W_{T,f}\tilde{\Omega},\ T\otimes I\tilde{\Omega}\Big\rangle - 2\ \mathrm{Re}\Big\langle W_{T,f}\tilde{\Omega},\ \int_0^t (fW_{T,f}dA^\dagger - \bar{f}W_{T,f}dA)\tilde{\Omega}\Big\rangle$$

$$+ 2\ \mathrm{Re}\Big\langle T\otimes I\tilde{\Omega},\ \int_0^t (fW_{T,f}dA^\dagger - \bar{f}W_{T,f}dA)\tilde{\Omega}\Big\rangle$$

$$= \|T\Omega_0\|^2\Big\{e^{\sigma^2\|f_t\|^2} + 1 + \int_0^t \sigma^2|f(s)|^2 e^{\sigma^2\|f_s\|^2}\ ds - 2$$

$$- 2e^{\frac{1}{2}\sigma^2\|f_t\|^2}\int_0^t W(f_t)\Omega,\ \sigma^2|f(s)|^2 e^{\frac{1}{2}\sigma^2\|f_s\|^2}W(f_s)\Omega\ \ ds + 0\Big\}$$

using (3.37), (3.2), (3.32), (3.35) and (3.36)

$$= \|T\Omega_0\|^2\Big\{2e^{\sigma^2\|f_t\|^2} - 2 - 2e^{\frac{1}{2}\sigma^2\|f_t\|^2}\int_0^t \sigma^2|f(s)|^2 e^{\frac{1}{2}\sigma^2(\|f_s\|^2 - \|f_t - f_s\|^2)}\ ds\Big\}$$

using (2.9) and (3.2). Writing

$$\sigma^2|f(s)|^2 e^{\frac{1}{2}\sigma^2(\|f_s\|^2 - \|f_t-f_s\|^2)} = \frac{d}{ds}(e^{\frac{1}{2}\sigma^2(\|f_s\|^2 - \|f_t-f_s\|^2)})$$

and using the fundamental theorem of calculus we see that this vanishes as required.

□

§4. The martingale representation theorem

Because the unitary operators infinitesimally generated by multiplication by the function $\chi_{[0,t]}$ in h do not differ from the identity by a Hilbert-Schmidt operator, the corresponding automorphisms of N are not inner and no analog of the gauge process Λ can be defined comprising an adapted process. This circumstance permits a martingale representation theorem in the present theory. We first show that stochastic integrals against Brownian motion are martingales.

Theorem 4.1 Let $F,G \in \mathcal{L}^2$ and $dM = FdA^\dagger + G^\dagger dA$. Then M is a martingale.

Proof We may assume that $M(0) = 0$. For $s,t \in \mathbb{R}^+$, $s \le t$, $T \in N^\circ$, $f \in h$, by (3.23) and (3.35)

$$\Big\langle T\otimes W(f)\tilde{\Omega},\ E_s\psi_M(t)\Big\rangle = \Big\langle E_sT\otimes W(f)\tilde{\Omega},\ \psi_M(t)\Big\rangle = \Big\langle(\mathbb{E}_sT\otimes W(f))\tilde{\Omega},\ \psi_M(t)\Big\rangle$$

$$= \exp\{-\tfrac{1}{2}\sigma^2\|\chi_{(s,\infty)}f\|^2\}\Big\langle T\otimes W(\chi_{[0,s]}f)\tilde{\Omega},\ \psi_M(t)\Big\rangle$$

$$= \exp\{-\tfrac{1}{2}\sigma^2\|\chi_{(s,\infty)}f\|^2\}\int_0^t\Big\langle T\otimes W(\chi_{[0,s]}f)\tilde{\Omega},\ \lambda^2\chi_{[0,s]}(\tau)\bar{f}(\tau)\psi_F(\tau) + \mu^2\chi_{[0,s]}(\tau)f(\tau)\psi_G^\dagger(\tau)\Big\rangle d\tau$$

$$= \exp\{-\tfrac{1}{2}\sigma^2\|\chi_{(s,\infty)}f\|^2\}\int_0^s\Big\langle T\otimes W(\chi_{[0,s]}f)\tilde{\Omega},\ \lambda^2\overline{f(\tau)}\psi_F(\tau) + \mu^2 f(\tau)\psi_G(\tau)\Big\rangle d\tau$$

$$= \Big\langle T\otimes W(f)\tilde{\Omega},\ \psi_M(s)\Big\rangle.$$

Since vectors of form $T\otimes W(f)\tilde{\Omega}$ are total we conclude that $E_s\psi_M(t) = \psi_M(s)$ as required. □

Theorem 4.2 Let M be a martingale. Then there exist unique $F,G \in \mathcal{L}^2$ such that

$$dM = FdA^\dagger + G^\dagger dA. \tag{4.1}$$

Proof For $t > 0$, denote by $\mathcal{L}^2_t$ the subspace of $\mathcal{L}^2$ of processes $F^\# \in \mathcal{L}^2$ such that $F(s) = 0$ for all $s > t$, equipped with the norm defined by

$$\|F\|^2 = \int_0^t (\lambda^2\|\psi_F(s)\|^2 + \mu^2\|\psi_F^\dagger(s)\|^2)\, ds, \tag{4.2}$$

and by $\mathcal{M}_t$ the subspace of $\mathcal{M}$ of martingales M for which $M(s) = M(t)$ for all $s > t$, equipped with the norm defined by

$$\|M\|^2 = \|\psi_M(t)\|^2 + \|\psi_M^\dagger(t)\|^2. \tag{4.3}$$

$\mathcal{L}^2_t$ is a Hilbert space; indeed if F_n is a Cauchy sequence then, setting $\psi_n(s) \equiv \psi_{F_n}(s)$, the pairs $(\lambda\psi_n, \mu\psi_n^\dagger)$, $n = 1,2,\ldots$ are Cauchy in $L^2(\mathbb{R}^+; \tilde{H} \oplus \tilde{H})$, hence convergent to $(\lambda\psi, \mu\psi^\dagger)$ say. Choose a subsequence $(\lambda\psi_{n_j}, \mu\psi^\dagger_{n_j})$ which converges pointwise almost everywhere on $[0,\infty)$ to $(\lambda\psi, \mu\psi^\dagger)$. Replacing $\psi^\#(s)$ by 0 at points of nonconvergence we see that the pair $\psi^\#(s)$ is weakly affiliated to $\tilde{\mathcal{E}}$ and inherits from the $\psi^\#_{n_j}$ the factorisation property (3.19) for all $s \in \mathbb{R}^+$, as well as vanishing for $s \geq t$. It follows that F_n converges in $\mathcal{L}^2_t$ to F, the operator process associated with the pair $\psi^\#$. $\mathcal{M}_t$ is also a Hilbert space; indeed writing $\tilde{H}_t = E_t\tilde{H}$, we see that the map $M \to (\psi_M(t), \psi_M^\dagger(t))$ is isometric from $\mathcal{M}_t$ onto the orthogonal complement in $\tilde{H}_t \oplus \tilde{H}_t$ of $\{(S^\circ \otimes W'(f_t)\tilde{\Omega}, -S^{\circ *} \otimes W'(f_t)\tilde{\Omega}) : f \in h,\ S^\circ \in N^{\circ\prime}\}$.

Consider the real-linear map

$$V: (F,G) \mapsto \int_0^t (FdA^\dagger + G^\dagger dA)$$

from $\mathcal{L}^2_t \oplus \mathcal{L}^2_t$ to $\mathcal{M}_t$. The range of V is a complex-linear subspace of $\mathcal{M}_t$. Furthermore, since from (4.2) and (3.32) V is isometric, its range is closed. If $M \equiv M(0) = T^\circ \otimes I$, where $(T^\circ, T^{\circ\dagger})$ are mutually adjoint operators affiliated to N° is a constant martingale, then, for $F, G \in \mathcal{L}^2_t$

$$\langle M, V(F,G)\rangle = \left\langle T^\circ\Omega^\circ \otimes \Omega, \int_0^t (FdA + G^\dagger dA)\tilde{\Omega}\right\rangle + \left\langle \int_0^t (GdA^\dagger + F^\dagger dA)\tilde{\Omega}, T^{\circ\dagger}\Omega^\circ \otimes \Omega\right\rangle = 0$$

by (3.36). Thus the orthogonal complement of the range V contains the subspace $\mathcal{M}_0$ of constant martingales; we prove it consists precisely of $\mathcal{M}_0$. $\mathcal{M}_0$ is closed, being as in the case of $\mathcal{M}_t$ the isometric preimage of an orthogonal complement. Hence it suffices to show that if $M \in \mathcal{M}_t$ is orthogonal to both the range of V and to $\mathcal{M}_0$, then M is necessarily 0. But such M is then orthogonal to each of the martingales $W_{T,f}$ of Theorem 3.3. Thus, for arbitrary $T \in N^\circ$, $f \in h$,

$$\langle \psi_M(t), T \otimes W(f_t)\tilde{\Omega}\rangle + \langle T^* \otimes W(-f_t)\tilde{\Omega}, \psi_M^\dagger(t)\rangle = 0. \tag{4.5}$$

Applying the factorisation (3.19) to the vectors $\psi_M(t)$, we see that f_t may be replaced by f in (4.5).. But then (4.5) is seen to be equivalent to (3.16). Hence by Theorem 3.1, $\psi_M(t) = 0$ as required. It follows that every $M \in \mathcal{M}_t$ can be uniquely represented in the form $M = M_0 + V(F,G)$, where $M_0 \in \mathcal{M}_0$, or equivalently

$$M(s) = T^{\circ} \underline{\otimes} I + \int_0^s \left(F dA^{\dagger} + G^{\dagger} dA\right), \quad s \geq 0 \tag{4.6}$$

where $T^{\circ} \eta N^{\circ}$ and $F, G \in \mathcal{L}_t^2$ are unique.

For general $M \in \mathfrak{M}$, we define $M_t \in \mathfrak{M}_t$ for each $t > 0$ by

$$M_t(s) = M(s) \quad \text{if} \quad s \leq t, \qquad M_t(s) = M(t) \quad \text{if} \quad s > t$$

and represent M_t in the form (4.6) as

$$M_t(s) = T_t \underline{\otimes} I + \int_0^s (F_t dA^{\dagger} + G_t^{\dagger} dA). \tag{4.7}$$

Using the uniqueness of the representation it is easily seen that, for $t \leq s$

$$T_t = T_s \quad \text{and} \quad F_t = F_s \chi_{[0,t]}, \quad G_t = G_s \chi_{[0,t]}. \tag{4.8}$$

From this it is clear that there exist T weakly affiliated to N° and $F, G \in \mathcal{L}^2$ such that, for all $t \in \mathbb{R}^+$

$$T_t = T, \quad F_t = F \chi_{[0,t]}, \quad G_t = G \chi_{[0,t]}.$$

But then M satisfies (4.1). Uniqueness is clear. □

§5. Unitary processes

Adapting the corresponding concepts of §2 to the case of non-unit variance, for $s,t \in \mathbb{R}^+$ with $s \leq t$ we denote by $\tilde{N}_{(s,t]}$ the von Neumann algebra generated by the operators $T^{\circ} \otimes W(\chi_{[s,t]} f)$, $T^{\circ} \in N^{\circ}$, $f \in h$ (where now $W(f)$ is defined by (3.4)), and for $r \in \mathbb{R}^+$ we denote by Γ_r, the ampliation to H of the second quantisation, now defined by (3.7), of the shift S_r given by (2.42). As in (3.22), $\mathbb{E}^{\circ}$ denotes the conditional expectation from N onto N° defined by

$$\langle u, \mathbb{E}^{\circ}[T] v \rangle = \langle u \otimes \Omega, T v \otimes \Omega \rangle \quad (u, v \in H^{\circ}, \ T \in \tilde{N}).$$

Let $L^{\circ}, L^{\circ\dagger} = L^{\circ *}$ and $\mathcal{H}^{\circ} = \mathcal{H}^{\circ *}$ be elements of N°. Denote by L, $L^{\dagger}$, $\mathcal{H}$ their Hilbert space ampliations in $\tilde{N}$.

Theorem 5.1 The stochastic differential equation

$$dU = U(L dA^{\dagger} - L^{\dagger} dA + (i\mathcal{H} - \tfrac{1}{2}\lambda^2 L^{\dagger} L - \tfrac{1}{2}\mu^2 L L^{\dagger}) dt), \qquad U(0) = I \tag{5.1}$$

has a unique solution. Moreover the process U is unitary and satisfies

a) $U^{\dagger}(s) U(t) \in \tilde{N}_{(s,t]} \quad (s, t \in \mathbb{R}^+, \ s \leq t)$ (5.2)

b) $U(t) = \Gamma_r^{\dagger} U^{\dagger}(r) U(r+t) \Gamma_r \quad (r, t \in \mathbb{R}^+)$ (5.3)

c) $t \mapsto \mathbb{E}^{\circ}[U(t)]$ is uniformly continuous from $\mathbb{R}^+$ to N_0. (5.4)

Proof To prove the existence of the solution, we use (3.25) and replace (5.1) by the stochastic differential equation against two-dimensional Fock quantum Brownian motion

$$dU = U(\lambda L dA_{1F}^{\dagger} - \mu L^{\dagger} dA_{2F}^{\dagger} - \lambda L^{\dagger} dA_{1F} + \mu L dA_{2F} + (i\mathcal{H} - \tfrac{1}{2}\lambda^2 L^{\dagger} L - \tfrac{1}{2}\mu^2 L L^{\dagger}) dt), \quad U(0) = I \tag{5.5}$$

which has a unique unitary solution by Theorem 2.6. To show that the solution U of (5.5) is an adapted process satisfying (5.1), it suffices to prove that each $U(t) \in \tilde{N}_t$. By the commutation theorem of tensor products and the fact that U is adapted to Fock quantum Brownian motion, so that $U(t) \in B(H^{\circ} \otimes H^{t}) \otimes I$, this will be so if and only if, for each $S \in N^{\circ\prime}$ and $\alpha \in h$, U(t) commutes with $W'_{s,\alpha}(t)$, where

$$W'_{s,\alpha}(t) = S \otimes W'(\alpha_t) = S \otimes W_F(-\mu\alpha_t) \otimes W_F(\lambda\bar{\alpha}_t)$$

is the solution of the stochastic differential equation

$$dW'_{s,\alpha} = W'_{s,\alpha}(-\mu\alpha dA_{1F}^{\dagger} + \lambda\bar{\alpha} dA_{2F}^{\dagger} + \mu\bar{f} dA_{1F} - \lambda f dA_{2F} - \tfrac{1}{2}\sigma^2 |\alpha|^2 dt), \tag{5.6}$$

$$W_{s,\alpha}(0) = S \otimes I.$$

But, using the differential form of the polarisation of the Fock Ito formula (2.50), we find that, from (5.5) and (5.6), for arbitrary $u, v \in H^{\circ}$, $f = (f_1, f_2)$, $g = (g_1, g_2)$ $\in h \oplus h$ and $t > 0$,

$$\frac{d}{dt}\{\langle U(t)u \otimes \psi(f),\ W'^{*}_{s,\alpha}(t)v \otimes \psi(g)\rangle - \langle W'_{s,\alpha}(t)u \otimes \psi(f),\ U^{*}(t)v \otimes \psi(g)\rangle\} = 0.$$

Since U(0) commutes with $W'^{*}_{s,f}(0)$, it follows that U(t) commutes with $W'_{s,\alpha}$ as required.

To prove the uniqueness of the solution, we apply the estimate (3.34) with $M = W^{\dagger}$, where W is the difference of two solutions of (5.1), obtaining

$$\|\psi_{W^{\dagger}}(t)\|^2 \le \int_0^t e^{t-s}\{\lambda^2\|L\psi_{W^{\dagger}}(s)\|^2 + \mu^2\|L^{\dagger}\psi_{W^{\dagger}}(s)\|^2 + \|(i\mathcal{H} - \tfrac{1}{2}\lambda^2 L^{\dagger}L - \tfrac{1}{2}\mu^2 LL^{\dagger})\psi_{W^{\dagger}}(s)\|^2\}\, ds$$

$$\le 3\lambda^2 e^{t} C \int_0^t \|\psi_{W^{\dagger}}(s)\|^2\, ds$$

where C is the larger of $\|L\|$, $\|i\mathcal{H} - \tfrac{1}{2}\lambda^2 L^{\dagger}L - \tfrac{1}{2}\mu^2 LL^{\dagger}\|$. Iterating, and using the continuity of the stochastic integral $W^{\dagger}$ to infer boundedness of $\|\psi_{W^{\dagger}}(s)\|$ on [0,t], we obtain that $\psi_{W^{\dagger}} \equiv 0$ as required.

To verify (5.2) we note that, by virtue of Theorem 2.6,

$$U^{\dagger}(s)U(t) \in B(H^{\circ}) \otimes I \otimes B(\Gamma(L^2(s,t]) \otimes \Gamma(L^2(s,t])) \otimes I \tag{5.7}$$

and in particular commutes with elements of $I \otimes B(H^{s}) \otimes I \otimes B(H^{(t})$. But, being adapted, $U^{\dagger}(s)U(t)$ also belongs to $\tilde{N}_t$. (5.2) now follows from the commutation theorem of tensor products.

(5.3) and (5.4) also follow from the corresponding clauses of Theorem 2.6, together with the observations that, for the shift S_r, $\bar{S}_r = S_r$, so that the present Γ_r coincides with the Fock Γ_r, and that the present conditional expectation is the restriction to $\tilde{N}$ of that of the Fock theory. □

We now establish our main result, that the three properties of Theorem 5.1 characterise unitary processes.

Theorem 5.2 Let $U = (U(t): t \geq 0)$ be a family of unitary operators in $\tilde{N}$ satisfying condition a), b) and c) of Theorem 5.1. Then there exist $L^\circ, L^{\circ\dagger} = L^\circ{}^*$, $\mathfrak{H}^\circ = \mathfrak{H}^\circ{}^* \in N^\circ$ with Hilbert space ampliations L, $L^\dagger$, $\mathfrak{H}$ in $\tilde{N}$ such that U is the solution of the stochastic differential equation (5.1).

For the proof we need three lemmas. We note the following properties of the conditional expectations $\mathbb{E}_s$ defined by (3.22).

$$\mathbb{E}_0\mathbb{E}_s = \mathbb{E}_0 \tag{5.8}$$

$$\mathbb{E}_s[S_1TS_2] = S_1\mathbb{E}_s[T]S_2 \quad (T \in \tilde{N},\ S_1,S_2 \in \tilde{N}_s). \tag{5.9}$$

Also, for arbitrary $T \in \tilde{N}_{(s,t]}$,

$$\mathbb{E}_s[T] = \mathbb{E}_0[T]. \tag{5.10}$$

We note also that, since second quantisations map the vacuum to itself

$$\mathbb{E}_0[\Gamma_s^\dagger T\Gamma_s] = \mathbb{E}_0[T]. \tag{5.11}$$

Lemma 1 [9] $(\mathbb{E}^\circ[U(t)]: t \geq 0)$ is a uniformly continuous contraction semigroup.

Proof For $s,t \in \mathbb{R}^+$, using (5.8), (5.9), (5.10), (5.11) and (5.3) respectively,

$$\begin{aligned}
\mathbb{E}_0[U(s+t)] &= \mathbb{E}_0[U(s)U^\dagger(s)U(s+t)] = \mathbb{E}_0\mathbb{E}_s[U(s)U^\dagger(s)U(s+t)] \\
&= \mathbb{E}_0[U(s)\mathbb{E}_s[U^\dagger(s)U(s+t)]] = \mathbb{E}_0[U(s) \qquad \mathbb{E}_0[U^\dagger(s)U(s+t)]] \\
&= \mathbb{E}_0[U(s)\mathbb{E}_0[\Gamma_s U^\dagger(s)U(s+t)\Gamma_s]] = \mathbb{E}_0[U(s)]\mathbb{E}_0[U(t)] \\
&= \mathbb{E}_0[U(s)\mathbb{E}_0[U(t)]].
\end{aligned} \tag{5.12}$$

Since $\|U(t)\| \leq 1$ implies that $\mathbb{E}_0[U(t)] = \sup\{|\langle u\otimes\Omega,\ U(t)v\otimes\Omega\rangle| : u,v \in H_0,\ \|u\|,\|v\| \leq 1\} \leq 1$, $(\mathbb{E}_0[U(t)])$ and hence $(\mathbb{E}^\circ[U(t)])$ is a contraction semigroup, uniformly continuous by hypothesis. □

We write $\mathbb{E}^\circ[U(t)] = e^{tZ^\circ}$ where $Z^\circ \in N_0$. We denote by Z the Hilbert space ampliation of Z° in $\tilde{N}$.

Lemma 2 $$M(t) = U(t) - I - \int_0^t U(\tau)Zd\tau, \quad t \geq 0 \tag{5.13}$$

defines a martingale M.

Proof For $s,t \in \mathbb{R}^+$, $s \leq t$, by arguments like those leading to (5.12),

$$\begin{aligned}
\mathbb{E}_s[M(t)] - M(s) &= \mathbb{E}_s\left[U(t) - I - \int_0^t U(\tau)Zd\tau\right] - M(s) \\
&= \mathbb{E}_s[U(s)U^\dagger(s)U(t)] - I - \int_0^s U(\tau)Zd\tau - \int_s^t \mathbb{E}_s[U(s)U^\dagger(s)U(\tau)Z]d\tau - M(s)
\end{aligned}$$

$$= U(s)\mathbb{E}_0[U\ (s)U(t)] - U(s)\int_{s'}^{t}\mathbb{E}_0[U^\dagger(s)U(\tau)Z]d\tau - U(s)$$

$$= U(s)\mathbb{E}_0[U(t-s)] - U(s)\int_s^t \mathbb{E}_0[U(\tau-s)Z]d\tau - U(s)$$

$$= U(s)\left\{e^{(t-s)Z} - \int_s^t e^{(\tau-s)Z}Z\ d\ - I\right\} = 0. \qquad \square$$

Since $M(0) = 0$, it follows from Theorem 4.2 that there exist unique $F,G \in \mathcal{L}^2$ such that

$$M(t) = \int_0^t (FdA^\dagger + G^\dagger dA), \quad M^\dagger(t) = \int_0^t (GdA^\dagger + F^\dagger dA) \quad (t \geq 0). \tag{5.14}$$

Lemma 3 For arbitrary $s,\tau \in \mathbb{R}^+$

$$\Gamma_s^\dagger U^\dagger(s)F(s+\tau)\Gamma_s = F(\tau), \quad \Gamma_s^\dagger U^\dagger(s)G^\dagger(s+\tau)\Gamma_s = G^\dagger(\tau). \tag{5.15}$$

Proof We note first that, for fixed s, the left handsides of (5.15) define adapted processes. For $s \leq t$, subtracting $M(s)$ from $M(t)$, we have

$$U(t) - U(s) - \int_s^t U(\tau)Z\ d\tau = \int_0^t (\chi_{(s,t]}FdA^\dagger + \chi_{(s,t]}G^\dagger dA).$$

Multiplying this equation on the left by $\Gamma_s^\dagger U^\dagger(s)$ and on the right by $U(s)$, and using (5.3) we get

$$M(t-s) = \Gamma_s^\dagger \int_0^t (\chi_{(s,t]}U^\dagger(s)FdA^\dagger + \chi_{(s,t]}U^\dagger(s)G^\dagger dA)\Gamma_s. \tag{5.16}$$

Using (3.35), we have, for arbitrary $u \in H_0$, $f \in h$

$$\left\langle u\otimes W(f)\Omega,\ \Gamma_s^\dagger \int_0^t (\chi_{(s,t]}U^\dagger(s)FdA^\dagger + \chi_{(s,t]}U^\dagger(s)G^\dagger dA)\Gamma_s\tilde{\Omega}\right\rangle$$

$$= \left\langle u\otimes W(S_sf)\Omega,\ \int_0^t (\chi_{(s,t]}U^\dagger(s)FdA^\dagger + \chi_{(s,t]}U^\dagger(s)G^\dagger dA)\tilde{\Omega}\right\rangle$$

$$= \int_s^t \left\langle u\otimes W(S_sf)\Omega,\ (\lambda^2\overline{f(\tau-s)}U^\dagger(s)F(\tau) + \mu^2 f(\tau-s)U^\dagger(s)G^\dagger(\tau))\tilde{\Omega}\right\rangle d\tau$$

$$= \int_0^{t-s} \left\langle u\otimes W(S_sf)\Omega,\ (\lambda^2\overline{f(\tau)}U^\dagger(s)F(s+\tau) + \mu^2 f(\tau)U^\dagger(s)G^\dagger(s+\tau)\tilde{\Omega}\right\rangle d\tau$$

$$= \int_0^{t-s} \left\langle u\otimes W(f)\Omega,\ (\lambda^2\overline{f(\tau)}\Gamma_s^\dagger U^\dagger(s)F(s+\tau)\Gamma_s + \mu^2 f(\tau)\Gamma_s^\dagger U^\dagger(s)G^\dagger(s+\tau)\Gamma_s\tilde{\Omega}\right\rangle d\tau$$

$$= \left\langle u\otimes W(f)\Omega,\ \int_0^{t-s} (\Gamma_s^\dagger U^\dagger(s)F(s+.)\Gamma_s dA^\dagger + \Gamma_s^\dagger U^\dagger(s)G(s+.)\Gamma_s dA)\tilde{\Omega}\right\rangle.$$

Since vectors of the form $u\otimes W(f)\Omega$ are total in $\tilde{H}$ and processes are determined by their actions on $\tilde{\Omega}$ it follows that

$$\Gamma_s^\dagger \int_0^t (\chi_{(s,t]}U^\dagger(s)FdA^\dagger + \chi_{(s,t]}U^\dagger(s)G^\dagger dA)\Gamma_s = \int_0^{t-s} (\Gamma_s^\dagger U^\dagger(s)F(s+.)\Gamma_s dA^\dagger + \Gamma_s^\dagger U^\dagger(s)G^\dagger(s+.)\Gamma_s dA).$$

Substituting in (5.16) and comparing with that obtained from (5.14) the resulting expression for $M(t-s)$ establishes the Lemma. □

<u>Proof of Theorem 5.2</u> For $t \in \mathbb{R}^+$ we set

$$J(t) = U^\dagger(t)F(t), \quad K^\dagger(t) = U^\dagger(t)G(t) \tag{5.17}$$

so that

$$dU = U(JdA^\dagger + K^\dagger dA + Zdt). \tag{5.18}$$

Setting $\tau = 0$ in (5.15) and recalling that $U(0) = I$ we see that, for arbitrary $s \in \mathbb{R}^+$,

$$\Gamma_s^\dagger J(s)\Gamma_s = J(0), \quad \Gamma_s^\dagger K^\dagger(s)\Gamma_s = K^\dagger(0). \tag{5.19}$$

We denote by P_s the projector $\Gamma_s\Gamma_s^\dagger$ onto $H^\circ \otimes \Omega^s \otimes H^{(s}$, and note that, by adaptedness,

$$P_sJ(0) = J(0)P_s, \quad P_sK^\dagger(0) = K^\dagger(0)P_s. \tag{5.20}$$

From (5.19) and (5.20) we have

$$P_sJ(s)P_s = J(0)P_s, \quad P_sK^\dagger(s)P_s = K^\dagger(0)P_s. \tag{5.21}$$

From (5.18), (3.31) and the unitarity of U we have that

$$2\,\mathrm{Re}\langle\tilde{\Omega}, Z\tilde{\Omega}\rangle + \lambda^2\|J(s)\tilde{\Omega}\|^2 + \mu^2\|K^\dagger(s)\tilde{\Omega}\|^2 = 0 \tag{5.22}$$

and hence that, for all $s \in \mathbb{R}^+$,

$$\lambda^2\|J(s)\tilde{\Omega}\|^2 + \mu^2\|K^\dagger(s)\tilde{\Omega}\|^2 = \lambda^2\|J(0)\tilde{\Omega}\|^2 + \mu^2\|K^\dagger(0)\tilde{\Omega}\|^2. \tag{5.23}$$

By Pythagoras's theorem we have

$$\lambda^2\|(1-P_s)J(s)\tilde{\Omega}\|^2 + \mu^2\|(1-P_s)K^\dagger(s)\tilde{\Omega}\|^2$$

$$= \lambda^2\|J(s)\tilde{\Omega}\|^2 + \mu^2\|K^\dagger(s)\tilde{\Omega}\|^2 - \lambda^2\|P_sJ(s)\tilde{\Omega}\|^2 - \mu^2\|P_sK^\dagger(s)\tilde{\Omega}\|^2$$

$$= \lambda^2\|J(0)\tilde{\Omega}\|^2 + \mu^2\|K^\dagger(0)\tilde{\Omega}\|^2 - \lambda^2\|P_sJ(s)P_s\tilde{\Omega}\|^2 - \mu^2\|P_sK^\dagger(s)P_s\tilde{\Omega}\|^2,$$

using (5.23) together with the fact that $P_s\tilde{\Omega} = \tilde{\Omega}$,

$$= \lambda^2\|J(0)\tilde{\Omega}\|^2 + \mu^2\|K^\dagger(0)\tilde{\Omega}\|^2 - \lambda^2\|J(0)P_s\tilde{\Omega}\|^2 - \mu^2\|K^\dagger(0)\tilde{\Omega}\|^2 = 0.$$

Hence, since both summands are nonnegative, both are zero and we have

$$J(s)\tilde{\Omega} = P_sJ(s)\tilde{\Omega} = P_sJ(s)P_s\tilde{\Omega} = J(0)P_s\tilde{\Omega} = J(0)\tilde{\Omega},$$

and similarly $K^\dagger(s)\tilde{\Omega} = K^\dagger(0)\tilde{\Omega}$. Since operators weakly affiliated to $\tilde{N}$ are determined by their actions on $\tilde{\Omega}$ it follows that

$$J(s) = J(0), \quad K^\dagger(s) = K^\dagger(0). \tag{5.23}$$

We write $J(0) = J^\circ \underline{\otimes} I$, $K(0) = K^\circ \underline{\otimes} I$ where $J^{\circ\#}, K^{\circ\#}\, \eta\, N^\circ$. Putting $r = t = 0$ in (3.3) show that $U(0) = I$. Hence, using (5.18) and (5.23) we can write

$$U(t) = I + \int_0^t U(s)(J^\circ \underline{\otimes} I dA^\dagger + K^{\circ\dagger} \underline{\otimes} I dA + Z^\circ \underline{\otimes} I ds). \tag{5.24}$$

Applying the differential form of (3.32) to the stochastic integral $U(t) - I$ we find that, for arbitrary $S_1, S_2 \in N^{\circ\prime}$, $f,g \in h$, since each $U(t)$ is unitary

$$0 = \frac{d}{dt}\langle U(t)S_1 \otimes W'(f)\tilde{\Omega},\ U(t)S_2 \otimes W'(g)\tilde{\Omega}\rangle$$

$$= \langle\{\lambda\mu((-\bar{f}+\bar{g})J^{\circ} + (f-g)K^{\circ\dagger} + Z^{\circ}\}S_1\Omega^{\circ} \otimes W'(f)\Omega,\ S_2 \otimes W'(g)\tilde{\Omega}\rangle$$

$$+ \langle S_1 \otimes W'(f)\tilde{\Omega},\ \lambda\mu((-\bar{f}+\bar{g})J^{\circ} + (f-g)K^{\circ\dagger}) + Z^{\circ}\}S_2\Omega^{\circ} \otimes W'(g)\Omega\rangle$$

$$+ \lambda^2\langle J^{\circ}S_1\Omega^{\circ} \otimes W'(f)\Omega,\ J^{\circ}S_2\Omega^{\circ} \otimes W'(g)\Omega\rangle$$

$$+ \mu^2\langle K^{\circ\dagger}S_1\Omega^{\circ} \otimes W'(f)\Omega,\ K^{\circ\dagger}S_2\Omega^{\circ} \otimes W'(g)\Omega\rangle. \tag{5.25}$$

Setting $f = g = 0$, $S_2 = S_1$ in (5.25)

$$0 = \langle Z^{\circ}S_1\Omega^{\circ}, S_1\Omega^{\circ}\rangle + \langle S_1\Omega^{\circ}, Z^{\circ}S_1\Omega^{\circ}\rangle + \lambda^2\|J^{\circ}S_1\Omega^{\circ}\|^2 + \mu^2\|K^{\circ\dagger}S_1\Omega^{\circ}\|^2, \tag{5.26}$$

from which it follows that

$$\|J^{\circ}S_1\Omega^{\circ}\|^2 \le -\lambda^{-2}\langle S_1\Omega^{\circ}, (Z^{\circ}+Z^{\circ}*)S_1\Omega\rangle \le 2\lambda^{-2}\|Z^{\circ}\|\|S_1\Omega\|.$$

Since the vectors $S_1\Omega^{\circ}, S_1 \in N^{\circ\prime}$ are dense in H° it follows that J° is bounded. Similarly $K^{\circ\dagger}$ is bounded. (5.26) now implies that

$$0 = Z^{\circ} + Z^{\circ}* + \lambda^2 J^{\circ}*J^{\circ} + \mu^2 K^{\circ}K^{\circ}*. \tag{5.27}$$

Using (5.27), (5.25) now gives

$$0 = \langle S_1\Omega^{\circ} \otimes W'(f)\Omega,\ (\lambda\mu(-\bar{f}+\bar{g})(J^{\circ}+K^{\circ}) + (f-g)(J^{\circ}*+K^{\circ}*))S_2\Omega^{\circ} \otimes W'(g)\Omega\rangle.$$

From this it is clear that $J^{\circ} = -K^{\circ}$. Combining this with (5.27) we write the triple $(J^{\circ}, K^{\circ\dagger}, Z^{\circ})$ in the form $(L^{\circ}, -L^{\circ}*, i\mathcal{H}^{\circ} - \frac{1}{2}\lambda^2 L^{\circ\dagger}L^{\circ} - \frac{1}{2}\mu^2 L^{\circ}L^{\circ\dagger})$ for $L^{\circ}, L^{\circ\dagger} = L^{\circ}*$ and $\mathcal{H}^{\circ} = \mathcal{H}^{\circ}* \in N^{\circ}$, and the theorem is proved. □

§6. The classical limit of reduced quantum stochastic evolutions [6,7,13]

Theorem 6.1 Let $L_1^{\circ}, \ldots, L_N^{\circ}$, $\mathcal{H}^{\circ} \in B(H^{\circ})$ and let U be the solution of the stochastic differential equation (2.51) against N-dimensional Fock quantum Brownian motion. Then the formula

$$\mathcal{T}_t(X) = \mathbb{E}^{\circ}[U(t)X \otimes IU(t)^{-1}], \quad X \in B(H^{\circ}), \quad t \ge 0 \tag{6.1}$$

defines a uniformly continuous completely positive semigroup $(\mathcal{T}_t : t \ge 0)$ on $B(H^{\circ})$, of which the infinitesimal generator $\mathcal{L}$ is given by

$$\mathcal{L}(X) = i[H^{\circ}, X] - \tfrac{1}{2}\sum_{j=1}^{N}(L_j^{\circ\dagger}L_j^{\circ}X - 2L_j^{\circ\dagger}XL_j^{\circ} + XL_j^{\circ\dagger}L_j^{\circ}). \tag{6.2}$$

Proof We apply the differential version of the polarised form of (2.50), in which we set $f = g = 0$, to the processes $X^{\dagger} \otimes IU^{\dagger}$, $U^{\dagger}$ to obtain that

$$\frac{d}{dt}\langle u,\mathcal{J}_t(X)v\rangle = \langle u,\mathcal{J}_t(\mathcal{L}(X))v\rangle \qquad (t \ge 0,\ u,v \in H^\circ)$$

where $\mathcal{L}$ is given by (6.2), from which the theorem follows. □

<u>Theorem 6.2</u> Let $L^\circ, L^{\circ\dagger} = L^\circ{}^*$, $\mathcal{H}^\circ = \mathcal{H}^\circ{}^* \in B(H^\circ)$ and let U be the solution of the stochastic differential equation (5.1) against quantum Brownian motion of variance $\sigma^2 > 1$. Then the formula

$$\mathcal{J}_t(X) = \mathbb{E}^\circ[U(t)X\otimes IU(t)^{-1}] \tag{6.3}$$

defines a uniformly continuous completely positive semigroup $(\mathcal{J}_t : t \ge 0)$ on N° of which the infinitesimal generator $\mathcal{L}$ is given by

$$\mathcal{L}(X) = i[\mathcal{H}^\circ,X] - \tfrac{1}{2}\{\lambda^2(L^{\circ\dagger}L^\circ X - 2L^{\circ\dagger}XL^\circ + XL^{\circ\dagger}L^\circ) + \mu^2(L^\circ L^{\circ\dagger}X - 2L^\circ XL^{\circ\dagger} + XL^\circ L^{\circ\dagger})\}. \tag{6.4}$$

<u>Proof</u> We recall that the unique solution of (5.1) is at the same time the solution of the equation (5.2) against two-dimensional Fock quantum Brownian motion, and that the conditional expectation coincides with the restriction to $\tilde{N}$ of its two-dimensional Fock counterpart. The Theorem thus follows from Theorem 6.1, together with the observation that the generator (6.3) clearly maps N° to itself. □

For typographical convenience we now drop the superscript $^\circ$ on operators in $B(H^\circ)$.

By Lindblad's theorem [19] the general form of infinitesimal generator $\mathcal{L}$ of a uniformly continuous completely positive semigroup on H° is

$$\mathcal{L}(X) = i[H,X] - \tfrac{1}{2}\sum_j (L_j^\dagger L_j X - 2L_j^\dagger XL_j + XL_j^\dagger L_j) \tag{6.5}$$

where $\sum_j L^\dagger L$ converges strongly in $B(H^\circ)$. A generalisation [15] of Theorem 6.1 constructs a stochastic dilation (6.1) of the semigroup $(\mathcal{J}_t)$ with generator (6.5) using a unitary process U formally satisfying a stochastic differential equation against infinite-dimensional Fock quantum Brownian motion. From this generalisation we may deduce, as Theorem 6.2 is deduced from Theorem 6.1, that there exists a similar stochastic dilation, now involving an infinite dimensional quantum Brownian motion of variance $\sigma^2 > 1$, of the uniformly continuous completely positive semigroup with generator

$$\mathcal{L}(X) = i[H,X] - \tfrac{1}{2}\sum_j \{\lambda^2(L_j^\dagger L_j X - 2L_j^\dagger XL_j + XL_j^\dagger L_j) + \mu^2(L_jL_j^\dagger X - 2L_jXL_j^\dagger + XL_jL_j^\dagger)\}. \tag{6.6}$$

However not every such semigroup has infinitesimal generator of form (6.6).

We obtain a physical interpretation of the difference between (6.5) and (6.6) by considering the classical limit of the generators as differential operators in a phase space. To discuss the classical limit (as $\hbar \to 0$) we need to relax the convention that Planck's constant h = 6 [18]. For simplicity we consider only the case when there is only one term L_j, and consider the generator

$$\mathcal{L}_{\hbar}(X) = i\hbar^{-1}[H,X] - \tfrac{1}{2}\hbar^{-1}\{\lambda^2(L^\dagger LX - 2L^\dagger XL + XL^\dagger L) + \mu^2(LL^\dagger X - 2LXL^\dagger + XLL^\dagger)\}. \tag{6.7}$$

We relate the variance σ^2 to a reciprocal temperature β by writing

$$\sigma^2 = \coth\left(\frac{\beta\hbar}{2}\right), \qquad \lambda^2 = \frac{1}{1-e^{-\beta\hbar}}, \qquad \mu^2 = \frac{e^{-\beta\hbar}}{1-e^{-\beta\hbar}} \tag{6.8}$$

and note that the Fock case $\sigma = \lambda = 1$, $\mu = 0$ is obtained in the zero temperature limit $\beta \to \infty$. Making the substitutions (6.8) in (6.7) we find that

$$\mathcal{L}_{\hbar}(X) = i\hbar^{-1}[H,X] - \frac{1}{\beta}\cdot\frac{1}{2\hbar^2}\,\frac{\beta\hbar}{1-e^{-\beta\hbar}}([L^\dagger,[L,X]] + [L,[L^\dagger,X]]) + \frac{1}{2\hbar}(L[L^\dagger,X] - [L,X]L^\dagger). \tag{6.9}$$

Now consider the space M_K of complex-valued Borel measures on $\mathbb{R}^2$ of compact support. Given a locally bounded Borel function ω on $\mathbb{R}^4$, and regarding $M_K(\mathbb{R}^2)$ as a subspace of the dual space of $C_0(\mathbb{R}^2)$, the formula

$$\int f\,d(\mu *_\omega \nu) = \int f(x+y)\omega(x,y)\; d\mu \times \nu(x,y) \qquad (f \in C_0(\mathbb{R}^2),\ \ \mu,\nu \in M_K) \tag{6.10}$$

defines a bilinear composition $*_\omega$, called the <u>twisted convolution</u> with <u>twist</u> ω, on M_K. The formula

$$\hat{\mu}\,o_\omega\,\hat{\nu} = (\mu *_\omega \nu)^\wedge \tag{6.11}$$

where $\hat{\mu}$ is the Fourier transform of μ, defines the corresponding <u>twisted multiplication</u> o_ω on the space $\hat{M}_K$ of Fourier transforms of measures of compact support. The twists

$$\pi \equiv 1, \qquad \gamma(x,y,x',y') = xy' - x'y \tag{6.12}$$

give rise in this way to ordinary multiplication and to the Poisson bracket respectively.

Let $(W_{x,y} : x,y \in \mathbb{R})$ be a <u>Weyl system of one degree of freedom</u>, that is, a strongly continuous family of unitary operators satisfying the Weyl relation

$$W_{x,y}W_{x',y'} = \exp(\tfrac{1}{2}i\hbar(xy'-x'y))W_{x+x',y+y'}. \tag{6.13}$$

The <u>Weyl quantisation</u> Q maps each $\hat{\mu} \in \hat{M}_K$ to the <u>Fourier-Weyl transform</u> $\tilde{\mu}$ of μ, defined as the operator-valued integral

$$\tilde{\mu} = \int W_{x,y}\mu(dx,dy). \tag{6.14}$$

Under Fourier-Weyl transformation,

$$\tilde{\mu}.\tilde{\nu} = (\mu *_{\pi(\hbar)} \nu)^\sim, \qquad (ih)^{-1}[\tilde{\mu},\tilde{\nu}] = (\mu *_{\gamma(\hbar)} \nu)^\sim \tag{6.15}$$

where the twists $\pi(h)$ and $\gamma(h)$ are defined by

$$\pi(\hbar)(x,y,x',y') = \exp(\tfrac{1}{2}i\hbar(xy'-x'y)), \qquad \gamma(\hbar)(x,y,x',y') = \frac{2}{\hbar}\sin\left(\frac{\hbar}{2}(xy'-x'y)\right). \tag{6.16}$$

The conventional wisdom that in the classical limit of quantum mechanics operator products go over to pointwise products of functions on phase space, whereas

$(i\hbar)^{-1}\times$ commutators go over to Poisson brackets, is made rigorous by the observation that the twists (6.16) converge to the twists (6.12) uniformly on compact sets, together with the following

<u>Theorem 6.3</u> Let $\omega_1(\hbar),\ldots,\omega_n(\hbar)$ be twists depending on the parameter $\hbar > 0$ and suppose that as $\hbar \to 0$ each $\omega_j(\hbar)$ converges to a twist ω_j uniformly on compact subsets of $\mathbb{R}^4$. Then, for arbitrary $\hat{\mu}_1,\ldots,\hat{\mu}_{n+1} \in \hat{M}_K$ and an arbitrary choice of bracketing of the non-associative product $\hat{\mu}\, o_{\omega_1(\hbar)} \ldots o_{\omega_n(\hbar)} \hat{\mu}_{n+1}$ converges to $\hat{\mu}\, o_{\omega_1} \ldots o_{\omega_n} \hat{\mu}_{n+1}$ in the weak* sense in $L^\infty(\mathbb{R}^2) = L^1(\mathbb{R}^2)^*$. In other words, for arbitrary $f \in L^1(\mathbb{R}^2)$,

$$\int f\, \hat{\mu}_1 o_{\omega_1(\hbar)} \ldots o_{\omega_n(\hbar)} \hat{\mu}_{n+1} \underset{\hbar}{\to} \int f\, \hat{\mu}_1 o_{\omega_1} \ldots o_{\omega_n} \hat{\mu}_{n+1}. \tag{6.17}$$

<u>Proof</u> Since the Fourier transformation is continuous from the space M of bounded Borel measures, regarded as the dual of $C_0(\mathbb{R}^2)$ and equipped with the corresponding weak*-topology, into $L^\infty(\mathbb{R}^2)$, it is sufficient to prove that, for arbitrary $f \in C_0(\mathbb{R}^2)$,

$$\int f\, d(\mu_1 *_{\omega_1(\hbar)} \ldots *_{\omega_n(\hbar)} \mu_{n+1}) \underset{\hbar}{\to} \int f\, d(\mu_1 *_{\omega_1} \ldots *_{\omega_n} \mu_{n+1}). \tag{6.18}$$

Writing

$$\int f\, d(\mu_1 *_{\omega_1(\hbar)} \ldots *_{\omega_n(\hbar)} \mu_{n+1}) = \int_{(x_1,\ldots,x_{n+1})} f(x_1+\ldots+x_{n+1}) \prod_{j=1}^{n} \omega_j(\hbar)(\xi_j,\eta_j)\, d(\mu_1 \times \ldots \times \mu_{n+1})$$

where each ξ_j and each η_j is the sum of a subset of $\{x_1,\ldots,x_{n+1}\}$ depending on the bracketing, together with a corresponding expression for the right hand side of (6.18), it is clear that (6.18) holds. □

Returning to (6.9) we assume that the operators $\mathcal{H}$, X, L and $L^\dagger$ are the Weyl quantisations of elements h, x, ℓ and $\bar{\ell}$ of $\hat{M}_K$. Then $\mathcal{L}_\hbar(X)$ is the Weyl quantisation of an element $\mathcal{G}_\hbar(x)$ of $\hat{M}_K$, and we have

$$\lim_{\hbar\to 0} \mathcal{G}_\hbar(x) = \mathcal{G}(x) \tag{6.19}$$

in the weak* sense of Theorem 6.3, where

$$\mathcal{G}(x) = \{h,x\} + \beta^{-1}\tfrac{1}{2}(\{\ell^\dagger,\{\ell,x\}\} + \{\ell\{\ell^\dagger,x\}\}) + \tfrac{1}{2}i(\ell\{\ell^\dagger,x\} - \ell^\dagger\{\ell,x\}).$$

Writing $\ell = c + id$ where c and d are real-valued, we find that

$$\mathcal{G}(x) = \{h,x\} + \beta^{-1}(\{c\{c,x\}\} + \{d\{d,x\}\}) + c\{d,x\} - d\{c,x\}. \tag{6.20}$$

It can be verified that $x \mapsto \{c,\{c,x\}\} + \{d,\{d,x\}\}$ is a strictly elliptic differential operator, except when $\{c,d\} = 0$ in which case it is semi-elliptic (then L is normal). Thus $\mathcal{G}$ is the generator of a Markov diffusion. Note however that in the Fock case, which is the limit $\beta^{-1} \to 0$, the elliptic terms disappear from (6.24) leaving only the drift.

References

[1] D B Applebaum and R L Hudson, Fermion Ito's formula and stochastic evolutions, Commun. Math. Phys. 96, 473-96 (1984).

[2] A Barchielli and G Lupieri, Quantum stochastic calculus, operation valued stochastic processes and continual measurements in quantum theory, preprint.

[3] A M Cockroft and R L Hudson, Quantum mechanical Wiener processes, J. Multivariate Anal. 7, 107-24 (1977).

[4] C D Cushen and R L Hudson, A quantum mechanical central limit theorem, J. Appl. Prob. 8, 454-69 (1971).

[5] A Frigerio, Covariant Markov dilations of quantum dynamical semigroups, preprint.

[6] A Frigerio and V Gorini, Diffusion processes, quantum dynamical semigroups and the classical KMS condition, J. Math. Phys. 25, 1050-65 (1984).

[7] A Frigerio and V Gorini, Markov dilations and quantum detailed balance, Commun. Math. Phys. 93, 517-32 (1984).

[8] A Guichardet, Symmetric Hilbert spaces and related topics, Springer LNM 261, Berlin (1972).

[9] R L Hudson, P D F Ion and K R Parthasarathy, Time-orthogonal unitary dilations and noncommutative Feynman-Kac formulae I, Commun. Math. Phys. 83, 761-80 (1982).

[10] R L Hudson, P D F Ion and K R Parthasarathy, Time orthogonal unitary dilations and noncommutative Feynman-Kac formulae II, Publ. RIMS 20, 607-33 (1984).

[11] R L Hudson, R L Karandikar and K R Parthasarathy, Towards a theory of non-commutative semimartingales adapted to Brownian motion and a quantum Ito's formula, in Theory and applications of random fields, Proceedings 1982, ed. Kallianpur, Springer LN Control Theory and Information Sciences 49, 96-110 (1983).

[12] R L Hudson and J M Lindsay, Stochastic integration and a martingale representation theorem for non-Fock quantum Brownian motion, to appear in J. Functional Anal.

[13] R L Hudson and J M Lindsay, The classical limit of reduced quantum stochastic evolutions, to appear in Ann. Inst. H Poincaré.

[14] R L Hudson and K R Parthasarathy, Quantum Ito's formula and stochastic evolutions, Commun. Math. Phys. 93, 301-23 (1984).

[15] R L Hudson and K R Parthasarathy, Stochastic dilations of uniformly continuous completely positive semigroups, Acta Applicandae Math. 2, 353-78 (1984).

[16] R L Hudson and K R Parthasarathy, Quantum diffusions, in Theory and applications of random fields, Proceedings 1982, ed. Kallianpur, Springer LN Control Theory and Information Sciences 49, 111-21 (1983).

[17] R L Hudson and R F Streater, Noncommutative martingales and stochastic integrals in Fock space, in Stochastic processes in quantum theory and statistical physics, proceedings 1981, Springer LNP 173, 216-22 (1982).

[18] Kings I, ch. 7, v. 23.

[19] G Lindblad, On the generators of quantum dynamical semigroups, Commun. Math. Phys. 48, 119-30 (1976).

[20] J M Lindsay, Nottingham thesis (1985).

[21] K R Parthasarathy, A remark on the integration of Schrödinger equation using quantum Ito's formula, Lett. Math. Phys. 8, 227-32 (1984).

[22] K R Parthasarathy, private communication.

[23] I E Segal, Mathematical characterisation of the physical vacuum, Ill. J. Math. 6, 500-23 (1962).

SUPERSYMMETRY AND A TWO-DIMENSIONAL REDUCTION IN RANDOM PHENOMENA

Abel Klein*
Department of Mathematics
University of California
Irvine, California 92717
U.S.A.

In the theory of random phenomena certain quantities of interest have been expressed as expectations of a field theory in two fewer dimensions. This dimensional reduction has been explained by a hidden supersymmetry.

This was done for

i) Some functional integrals related to the average correlation functions of a classical field theory in the presence of a random external source (Parisi and Sourlas [1]).

ii) The average density of states of electrons in the presence of a local random potential and of a uniform magnetic field, restricted to the lowest Landau level (Wegner [2], Brezin, Gross and Itzykson [3]).

In both cases the quantities of interest were re-written as n-point functions of supersymmetric field theories. These n-point functions, restricted to a hyperplane of codimension 2, were shown to be equal, order by order in perturbation theory, to the corresponding n-point functions of a scalar field theory in two fewer dimensions with the same Lagrangian (up to multiplication by a constant).

In this article I will show that the phenomenum of dimensional reduction is the same in both cases and will give a nonperturbative proof. I am reporting on joint work with L. J. Landau and J. F. Perez [4] for case (i) and on joint work with J. F. Perez [5] for case (ii).

This article is organized as follows:

1. Supersymmetries
2. Superfields and Supersymmetric Field Theories
3. The Parisi-Sourlas Dimensional Reduction
4. The Wegner-Brezin-Gross-Itzykson Dimensional Reduction
5. The Dimensional Reduction of Supersymmetric Field Theories

---

*Partially supported by the N.S.F. under grant MCS-830189

1. SUPERSYMMETRIES

Supersymmetries are rotations of the "superspace" with "coordinates" $(z,\theta,\bar{\theta})$, where $z \in \mathbb{R}^D$ and θ and $\bar{\theta}$ are anticommuting "variables" (i.e., $\theta^2 = \bar{\theta}^2 = \theta\bar{\theta} + \bar{\theta}\theta = 0$), which preserve the supermetric $z^2 + (4/\gamma)\bar{\theta}\theta$. Here $z^2 = z \cdot z$ and $\gamma \neq 0$ is a fixed constant.

In addition to the usual rotations in $\mathbb{R}^D$ and symplectic transformations of θ and $\bar{\theta}$, they include transformations of the type

$$\begin{aligned} z &\to z + 2\bar{b}\xi\theta + 2b\xi\bar{\theta} \\ \theta &\to \theta + \gamma b \cdot z\xi \\ \bar{\theta} &\to \bar{\theta} - \gamma\bar{b} \cdot z\xi \end{aligned}$$

where $b,\bar{b} \in \mathbb{R}^D$ and ξ is an anticommuting "c-number" $(\xi^2 = \xi\theta + \theta\xi = \xi\theta + \bar{\theta}\xi = 0)$. We will fix ξ and denote the above transformation by $\tau(b,\bar{b})$.

This discussion can be made rigorous by considering Grassman algebras $\mathcal{G}_2$ and $\mathcal{G}_3$, where $\mathcal{G}_2$ is a Grassman algebra with two generators θ and $\bar{\theta}$, and $\mathcal{G}_3$ is the Grassman algebra generated by $\mathcal{G}_2$ and ξ. Supersymmetries can be defined by their action on functions $F:\mathbb{R}^D \to \mathcal{G}_2$. Such a function can be written in a unique way as

$$F(z) = F_o(z) + F_1(z)\theta + F_2(z)\bar{\theta} + F_3(z)\bar{\theta}\theta ,$$

where $F_i:\mathbb{R}^D \to \mathbb{C}$ for $i = 0,1,2,3$. To emphasize that F takes values in the Grassman algebra generated by θ and $\bar{\theta}$ we will use the notation $F(z,\theta,\bar{\theta})$ for F. Notice also that if $H:\mathbb{R}^D \to \mathcal{G}_3$, then

$$H(z) = F(z,\theta,\bar{\theta}) + G(z,\theta,\bar{\theta})\xi$$

is a unique way, where $F,G:\mathbb{R}^D \to \mathcal{G}_2$.

We will say that $F:\mathbb{R}^D \to \mathcal{G}_2$ is of class $C^{1,0}$ if F_0,F_1,F_2 are of class C^1 and F_3 is of class C^0.

The action of supersymmetries on functions of class $C^{1,0}$ can now be rigorously defined. The action of space rotations and pure symplectic transformations is obvious. The action of the supersymmetry $\tau(b,\bar{b})$ is given by

$$(\tau(b,\bar{b})F)(z) = F(z,\theta,\bar{\theta}) + [(\gamma b\cdot z\ F_1(z) - \gamma\bar{b}\cdot zF_2(z))$$
$$+ (-2\nabla F_0(z)\cdot\bar{b} + \gamma\bar{b}\cdot z\ F_3(z))\theta + (-2\nabla F_0(z)\cdot b + \gamma b\cdot zF_3(z))\bar{\theta}$$
$$+ 2(\nabla F_1(z)\cdot b - \nabla F_2(z)\cdot\bar{b})\bar{\theta}\theta]\xi\ .$$

We say that $F(z,\theta,\bar{\theta})$ is supersymmetric if it is left invariant by all supersymmetries.

The following characterization of supersymmetric functions is now easy to prove:

<u>Proposition</u>. Let $F(z,\theta,\bar{\theta})$ be of class $C^{1,0}$. The following are equivalent:

(i) $F(z,\theta,\bar{\theta})$ is supersymmetric.

(ii) $F_1(z) = F_2(z) = 0$, and $(2/\gamma)\nabla F_0(z) = zF_3(z)$.

(iii) There exists a function $f:[0,\infty) \to \mathbb{C}$ of class C^1 such that $F(z,\theta,\bar{\theta}) = f(z^2 + (4/\gamma)\bar{\theta}\theta) \equiv f(z^2) + (4/\gamma)f'(z^2)\bar{\theta}\theta$. □

Following Berezin [6], we define integration over anticommuting variables by:

$$\int F(z,\theta,\bar{\theta})d\bar{\theta}d\theta = -F_3(z)$$
$$\int F(z,\theta,\bar{\theta})\xi d\bar{\theta}d\theta = -F_3(z)\xi\ .$$

Thus integration is defined as a linear functional on $\mathcal{S}_2$ and $\mathcal{S}_3$.

Notice that integration over $z,\theta,\bar{\theta}$ is supersymmetric, i.e., if $F(z,\theta,\bar{\theta})$ is integrable of class $C^{1,0}$ and τ is a supersymmetry, then

$$\int \tau F(z,\theta,\bar{\theta})d\bar{\theta}d\theta dz = \int F(z,\theta,\bar{\theta})d\theta d\theta dz\ .$$

Here $F(z,\theta,\bar{\theta})$ integrable means $F_0,\nabla F_0,F_1,\nabla F_1,F_2,\nabla F_2,F_3$ are integrable.

Distributions $T(z,\theta,\bar{\theta})$ can be defined in the usual way be the formal formula

$$T(F) = \int T(z,\theta,\bar{\theta})F(z,\theta,\bar{\theta})d\bar{\theta}d\theta dz\ .$$

We will write

$$T(z,\theta,\bar{\theta}) = T_o(z) + T_1(z)\theta + T_2(z)\theta + T_3(z)\theta\bar{\theta}\ ,$$

where T_0,T_1,T_2,T_3 are distributions in z. Then

$$T(F) = T_3(F_0) - T_0(F_3) + T_1(F_2) - T_2(F_1) \ .$$

The action of supersymmetries on distributions can be defined in the usual way. We will say that $T(z,\theta,\bar{\theta})$ is a supersymmetric distribution if for all supersymmetries τ we have $T(\tau F) = T(F)$ for all functions $F(z,\theta,\bar{\theta})$ in the domain of definition of $T(z,\theta,\bar{\theta})$

Proposition. A distribution $T(z,\theta,\bar{\theta})$ is supersymmetric if and only if

$$T(z,\theta,\bar{\theta}) = T_0(z) + T_3(z)\theta\bar{\theta} \ ,$$

where $T_o(z)$ and $T_3(z)$ are distributions invariant under rotations of z such that

$$(2/\gamma)\nabla T_0 = - zT_3 \ . \quad \square$$

2. SUPERFIELDS AND SUPERSYMMETRIC FIELD THEORIES

Superfields are fields whose test functions take values in the Grassman algebra $\mathcal{G}_2$. A superfield $\Phi(z,\theta,\bar{\theta})$ can be written as $\Phi(z,\theta,\bar{\theta}) = \varphi(z) + \psi_1(z)\theta + \psi_2(z)\bar{\theta} + \lambda(z)\theta\bar{\theta}$. In the cases we will be interested in, φ and λ will be commuting fields (possibly complex-valued) and ψ_1 and ψ_2 will be anticommuting fields; anticommuting fields are taken to anticommute with θ and $\bar{\theta}$.

As for ordinary fields, we can define n-point functions for superfields:

$$S_n(z_1,\theta_1,\bar{\theta}_1;z_2,\theta_2,\bar{\theta}_2;\cdots;z_n,\theta_n,\bar{\theta}_n) =$$

$$= \langle\Phi(z_1,\theta_1,\bar{\theta}_1)\Phi(z_2,\theta_2,\bar{\theta}_2)\cdots\Phi(z_n,\theta_n,\bar{\theta}_n)\rangle \ .$$

Here the $\theta_i,\bar{\theta}_i, i = 1,\cdots n$, are independent anticommuting variables. They are needed since S_n must be smoothed with n functions taking values in $\mathcal{G}_2$.

The superfield $\Phi(z,\theta,\bar{\theta})$ is supersymmetric if all its n-point functions are invariant under supersymmetries.

3. THE PARISI-SOULAS DIMENSIONAL REDUCTION

Let us consider a classical field theory in D dimensions with

$$\mathcal{L}(\varphi) = \tfrac{1}{2}(\nabla\varphi)^2 + \tfrac{1}{2}m^2\varphi^2 + V(\varphi) \ , \qquad (3.1)$$

where V is a bounded below polynomial, and let $\mathcal{L}_h(\varphi) = \mathcal{L}(\varphi) + h\varphi$ be the Lagrangian is the presence of an external source h.

The external sources are taken to be random, distributed as white-noise: $\{h(z); z \in \mathbb{R}^D\}$ form a generalized Gaussian system with mean zero and covariance $\overline{h(z)h(z')} = \gamma\delta(z-z')$, $\gamma > 0$.

The classical equation of motion is

$$(-\Delta + m^2)\varphi + V'(\varphi) = -h . \tag{3.2}$$

Proceeding formally, let φ_h denote the formal solution to (3.2) given by perturbation theory, and define the average correlation functions

$$R(z_1, \cdots, z_n) = \overline{\varphi_h(z_1)\cdots\varphi_h(z_n)}$$

Let us now consider a Euclidean field theory ϕ in $D-2$ dimensions with Lagrangian $(4\pi/\gamma)\mathcal{L}(\phi)$, where $\mathcal{L}$ is the same Lagrangian given by (3.1). Let

$$S(x_1, \cdots, x_n) = \frac{\int \phi(x_1)\cdots\phi(x_n)\exp\{-(4\pi/\gamma)\int \mathcal{L}(\phi(x))dx\}\mathcal{D}\phi}{\int \exp\{-(4\pi/\gamma)\int \mathcal{L}(\phi(x))dx\}\mathcal{D}\phi} .$$

We use the notation $z \in \mathbb{R}^D$, $x \in \mathbb{R}^{D-2}$, $y \in \mathbb{R}^2$, and also write $z = (x,y)$.

The Parisi-Sourlas dimensional reduction is

$$R((x_1,0), \cdots, (x_n,0)) = S(x_1, \cdots, x_n) . \tag{3.3}$$

To explain this dimensional reduction, Parisi and Sourlas [1] re-expressed $R(z_1, \cdots, z_n)$ as expectations of a supersymmetric field theory and showed (3.3) order by order in perturbation theory.

Formally,

$$R(z_1, \cdots, z_n) = \frac{\int \varphi_h(z_1)\cdots\varphi_h(z_n)\exp\{-(1/2\gamma)\int h(z)^2 dz\}\mathcal{D}h}{\int \exp\{-(1/2\gamma)\int h(z)^2 dz\}\mathcal{D}h} .$$

Changing the integration from h to φ by the change of variables given by (3.2), we get

$$R(z_1, \cdots, z_n) = \tag{3.4}$$

$$\frac{\int \varphi(z_1)\cdots\varphi(z_n)\exp\{-(1/2\gamma)\int [(-\Delta + m^2)\varphi(z) + V'(\varphi(z))]^2 dz\}\det(-\Delta + m^2 + V''(\varphi))\mathcal{D}\varphi}{\int \exp\{-(1/2\gamma)\int [(-\Delta + m^2)\varphi(z) + V'(\varphi(z))]^2 dz\}\det(-\Delta + m^2 + V''(\varphi))\mathcal{D}\varphi}$$

Even at the formal level there is need here to assume that V is convex so (3.2) actually gives a change of variables and the determinant is positive. But convexity is not needed if (3.4) is taken as the definition of $R(z_1,\cdots,z_n)$, which we do from now on.

In [4] we show that, with the introduction of appropriate cutoffs, $R(z_1,\cdots,z_n)$ can be rigorously constructed by (3.4).

To simplify (3.4), use

$$\exp\{-(1/2\gamma)\int V'(\varphi(z))^2 dz =$$

$$\frac{\int \exp\{-(i/\sqrt{\gamma})\int V'(\varphi(z))\omega(z)dz - (1/2)\int \omega(z)^2 dz\}\mathcal{D}\omega}{\int \exp\{-(1/2)\int \omega(z)^2 dz\}\mathcal{D}\omega}$$

and

$$\det(-\Delta + m^2 + V''(\varphi)) =$$

$$\int \exp\{-\int \bar{\psi}(z)[(-\Delta + m^2 + V''(\varphi(z)))\psi](z)dz\}\mathcal{D}\bar{\psi}\mathcal{D}\psi$$

where ω is a commuting field and $\psi,\bar{\psi}$ are anticommuting fields.

If we define the superfield

$$\Phi(z,\theta,\bar{\theta}) = \varphi(z) + \bar{\psi}(z)\theta + \bar{\theta}\psi(z) +$$

$$+ \theta\bar{\theta}[(1/\gamma)(-\Delta + m^2)\varphi(z) + (i/\sqrt{\gamma})\omega(z)] ,$$

and the super-Lagrangian

$$\mathcal{L}_{ss}(\Phi) = \tfrac{1}{2}\Phi(-\Delta_{ss} + m^2)\Phi = V(\Phi) ,$$

where $\Delta_{ss} = \Delta + \gamma^2 \dfrac{\partial^2}{\partial\bar{\theta}\partial\theta}$, with

$$\frac{\partial^2}{\partial\bar{\theta}\partial\theta}F(z,\theta,\bar{\theta}) = - F_3(z) ,$$

an explicit computation shows that

$$R(z_1,\cdots,z_n) = \frac{\int \varphi(z_1)\cdots\varphi(z_n)\exp\{-\mathcal{L}_{ss}(\Phi(z,\theta,\bar{\theta}))d\bar{\theta}d\theta dz\}\mathcal{D}\Phi}{\int \exp\{-\mathcal{L}_{ss}(\Phi(z,\theta,\bar{\theta}))d\bar{\theta}d\theta dz\}\mathcal{D}\Phi} .$$

If we write

$$\langle\cdot\rangle = \frac{\int\cdot\exp\{-(1/2)\int\Phi(-\Delta_{ss}+m^2)\Phi\, d\bar{\theta}d\theta dz\}\mathcal{D}\Phi}{\int\exp\{-(1/2)\int\Phi(-\Delta_{ss}+m^2)\Phi\, d\bar{\theta}d\theta dz\}\mathcal{D}\Phi}$$

the expectations $\langle\cdot\rangle$ are supersymmetric and we have

$$R(z_1,\cdots,z_n) = \frac{\langle\varphi(z_1)\cdots\varphi(z_n)\exp\{-\int V(\Phi(z,\theta,\bar{\theta}))d\bar{\theta}d\theta dz\}\rangle}{\langle\exp\{-\int V(\Phi(z,\theta,\bar{\theta}))d\bar{\theta}d\theta dz\}\rangle} \tag{3.5}$$

Introducing the appropriate cutoffs the passage from (3.4) to (3.5) is rigorous [4].

The dimensional reduction of (3.3) can now be written as

$$\frac{\langle\varphi(x_1,0)\cdots\varphi(x_n,0)\exp\{-\int V(\Phi(z,\theta,\bar{\theta}))d\bar{\theta}d\theta dz\}\rangle}{\langle\exp\{-\int V(\Phi(z,\theta,\bar{\theta}))d\bar{\theta}d\theta dz\}\rangle} \tag{3.6}$$

$$\frac{\langle\varphi(x_1,0)\cdots\varphi(x_n,0)\exp\{-(4\pi/\gamma)\int V(\varphi(x,0))dx\}\rangle}{\langle\exp\{-(4\pi/\gamma)\int V(\varphi(x,0))dx\}\rangle}$$

4. THE WEGNER-BREZIN-GROSS-ITZYKSON DIMENSIONAL REDUCTION

Let us consider electrons in the presence of a random potential and of a uniform magnetic field. The Hamiltonian is

$$H = H_0 + V\ , \quad \text{where} \quad H_0 = \frac{1}{2m}(\vec{p} - e\vec{A})^2\ ,$$

with $\vec{A} = \frac{1}{2}B(-x_2,x_1,0)$.

Since the magnetic field does not affect the direction orthogonal to the (x_1,x_2)-plane, we will only consider the two-dimensional case (see [3]).

H_0 has pure point spectrum with eigenvalues $\mathcal{E}_n = (\kappa^2/m)(n+\frac{1}{2})$, $n = 0,1,2,\cdots$, where $\kappa^2 = eB$; the corresponding eigenspaces $\mathcal{K}_n$, called the Landau levels, are infinitely degenerate (e.g., [3,5]). We will denote by P_0 the projection onto $\mathcal{K}_0$. P_0 has kernel $P_0(z,z') = (\kappa^2/2m)\exp\{-(\kappa^2/4)(z\bar{z}+z'\bar{z}'-2z\bar{z}')\}$, where $z = x_1+ix_2$.

The random potential V is required to be "local" in the sense that its characteristic functional has the form

$$\overline{\exp\{i\int V(\vec{x})\alpha(\vec{x})d\vec{x}\}} = \exp\{g(\alpha(\vec{x}))d\vec{x}\}$$

In a strong magnetic field the gap between the Landau levels is large so it is reasonable to assume that the electrons are confined within the subspace of the lowest level $\mathcal{H}_0$.

The average density of states with the electrons confined to the 0 - th Landau level is given by

$$\rho_0(E) = -\frac{1}{\pi}\lim_{\eta\downarrow 0} \overline{\mathrm{Im}\langle \vec{0}| P_0(E+i\eta - P_0HP_0)^{-1}P_0|\vec{0}\rangle} \ .$$

The Wegner-Brezin-Gross-Itzykson dimensional reduction is

$$\rho_0(E) = -\frac{1}{\pi}\,\mathrm{Im}\frac{\int \varphi\bar{\varphi}\exp\{(2\pi/\kappa^2)(i(E-\mathcal{E}_0)\varphi\bar{\varphi} + h(\varphi\bar{\varphi}))\}\, d\mathrm{Re}\varphi\, d\mathrm{Im}\varphi}{\int \exp\{(2\pi/\kappa^2)(i(E-\mathcal{E}_0)\varphi\bar{\varphi} + h(\varphi\bar{\varphi}))\}\, d\mathrm{Re}\varphi\, d\mathrm{Im}\varphi}$$

where $h(t) = \int_0^t \frac{g(s)}{s}\, ds$.

This was originally shown by Wegner [2] order by order in perturbation theory, when the random potential is distributed as white-noise. Wegner's argument relies on topological identities relating the number of Euler trails of a Feynman diagram to a determinant.

Brezin, Gross and Itzykson [3] derived (4.1), order by order in perturbation theory, from a hidden supersymmetry.

To uncover the supersymmetry, notice that

$$G(E+i\eta) = \langle \vec{0}| P_0(E+i\eta - P_0HP_0)^{-1}P_0|\vec{0}\rangle = \langle \vec{0}| P_0(E-\mathcal{E}_0+i\eta - P_0VP_0)^{-1}P_0|\vec{0}\rangle$$

$$= -i\langle \varphi_0(0)\bar{\varphi}_0(0)\exp\{i(E-\mathcal{E}_0)\int[\bar{\varphi}_0(z)\varphi_0(z) + \bar{\psi}_0(z)\psi_0(z)]d\vec{x} -$$

$$- i\int V(\vec{x})[\bar{\varphi}_0(z)\varphi_0(z) + \bar{\psi}_0(z)\psi_0(z)]d\vec{x}]\rangle$$

where $\varphi_0(z)$ is a complex commuting free field and $\psi_0, \bar{\psi}_0$ are anti-commuting free fields such that

$$\langle \varphi_0(z)\bar{\varphi}_0(z')\rangle = \langle \psi_0(z)\bar{\psi}_0(z')\rangle = \frac{1}{\eta}P_0(z,z')$$

and all other two-point functions are zero.

Averaging with respect to the random potential we get

$$\overline{G(E+i\eta)} = -i\langle \varphi_0(0)\bar{\varphi}_0(0)\exp\{i(E-\mathcal{E}_0)\int[\bar{\varphi}_0(z)\varphi_0(z) + \bar{\psi}_0(z)\psi(z)]d\vec{x}$$

$$+ \int g(\bar{\varphi}_0(z)\varphi_0(z) + \bar{\psi}_0(z)\psi_0(z))d\vec{x}\}\rangle \ .$$

Let us introduce the superfield

$$\Phi_0(z,\theta,\bar\theta) = \varphi_0(z) + \sqrt{\tfrac{1}{2}}\,\kappa\theta\psi_0(z) + \frac{\kappa^2}{4}\varphi_0(z)\bar\theta\theta$$

and its conjugate

$$\bar\Phi_0(z,\theta,\bar\theta) = \overline{\varphi_0}(z) + \sqrt{\tfrac{1}{2}}\,\kappa\bar\psi_0(z)\bar\theta + \frac{\kappa^2}{4}\bar\varphi_0(z)\bar\theta\theta \quad .$$

Notice that $\Phi_0(z,\theta,\bar\theta)$ is a free superfield and

$$\langle\Phi_0(z,\theta,\bar\theta)\bar\Phi_0(z',\theta',\bar\theta')\rangle = \frac{1}{\eta}\exp\{-(\kappa^2/4)(z\bar z + \theta\bar\theta + z'\bar z' + \theta'\bar\theta' - 2(z\bar z' + \theta\bar\theta'))\}$$

is manifestly supersymmetric with respect to the supermetric $z\bar z + \theta\bar\theta$.

This supermetric is different from the one studied in section 1. Here the supersymmetries consist of unitary transformations in z , symplectic transformations in $\theta,\bar\theta$, plus transformations of the type

$$z \to z + \bar a\xi\theta \ , \ \bar z \to \bar z + a\bar\theta\xi$$

$$\theta \to \theta + az\xi \ , \ \bar\theta \to \bar\theta + \bar a\bar z\xi$$

where $a \in \mathbb{C}$. But we still have similar results to the ones in sections 1, 2 and 5, the modifications being obvious. For expository purposes we will ignore the difference from the supersymmetries in section 1.

Let us now define

$$\Phi(z,\theta,\bar\theta) = \bar\Phi_0(z,\theta,\bar\theta)\Phi_0(z,\theta,\bar\theta) \quad .$$

It follows $\Phi(z,\theta,\bar\theta)$ is supersymmetric and its scalar component is $\varphi(z) = \bar\varphi_0(z)\varphi_0(z)$.

By an explicit computation,

$$G(E + i\eta) = -\,i\langle\varphi(0)\exp\{\frac{2}{\kappa^2}[i(E - \mathcal{E}_0)\int\Phi(z,\theta,\bar\theta)d\theta d\bar\theta d\vec{x} + \int h(\Phi(z,\theta,\bar\theta))d\theta d\bar\theta d\vec{x}]\}\rangle \quad .$$

Notice that

$$<\exp\{\frac{2}{\kappa^2}[i(E-\mathcal{E}_0)\int\Phi(z,\theta,\bar{\theta})d\theta d\bar{\theta}d\vec{x} + \int h(\Phi(z,\theta,\bar{\theta}))d\theta d\bar{\theta}d\vec{x}]\}> = 1 .$$

Thus (4.1) is equivalent to the following dimensional reduction

$$\frac{<\varphi(0)\exp\{\int H(\Phi(z,\theta,\bar{\theta}))d\theta d\bar{\theta}d\vec{x}\}>}{<\exp\{\int H(\Phi(z,\theta,\bar{\theta}))\}d\theta d\bar{\theta}d\vec{x}\}>} = \frac{<\varphi(0)\exp\{\pi H(\varphi(0))\}>}{<\exp\{\pi H(\varphi(0))\}>} \tag{4.2}$$

where $H(y) = \frac{2}{\kappa^2}[i(E-\mathcal{E}_0)y + h(y)]$.

5. THE DIMENSIONAL REDUCTION OF SUPERSYMMETRIC FIELD THEORIES

We will now state a dimensional reduction for supersymmetric field theories that includes the dimensional reduction of (3.6) and (4.2). As in section 3, we will use the notation $z \in \mathbb{R}^D$, $x \in \mathbb{R}^{D-2}$, $y \in \mathbb{R}^2$, and $z = (x,y)$.

FORMAL THEOREM. Let $\Phi(x,y,\theta,\bar{\theta})$ be a superfield, supersymmetric in $y,\theta,\bar{\theta}$ with respect to the supermetric $y^2 + (4/\gamma)\bar{\theta}\theta$, $\gamma \neq 0$, and let $\varphi(x,y)$ be its scalar part. Then

$$<\Gamma(\varphi_0)e^{-V(\Phi)}> = <\Gamma(\varphi_0)e^{-(4\pi/\gamma)V(\varphi_0)}> ,$$

where $\Gamma(\varphi_0)$ is a function of the field $\varphi_0(x) = \varphi(x,0)$, V is a function, and

$$V(\Phi) = \int V(\Phi(z,\theta,\bar{\theta}))d\bar{\theta}d\theta dz$$

$$V(\varphi_0) = \int V(\varphi_0(x))dx . \quad \square$$

We called it a formal theorem because clearly one needs technical conditions to define all the terms. One needs to introduce appropriate momentum cutoffs and space cutoffs to define $V(\Phi)$, one must show $e^{-V(\Phi)}$ is integrable, etc. In particular one must take care so the cutoffs are supersymmetric (see [4] for how it can be done for the theory of section 3).

The proof we will give is rigorous under the necessary technical assumptions. The basic idea behind the dimensional reduction is

LEMMA [1,4]. Let $F(y,\theta,\bar{\theta})$ be a supersymmetric integrable function of class $C^{1,0}$, $y \in \mathbb{R}^2$. Then

$$\int F(y,\theta,\bar{\theta})d\bar{\theta}d\theta dy = (4\pi/\gamma)F_0(0) . \quad \square$$

Since the expectation values of fields are distributions, we need an extension to distributions:

LEMMA [4]. Let $T(x,y,\theta,\bar{\theta})$ be a distribution supersymmetric in $y,\theta,\bar{\theta}$, $y \in \mathbb{R}^2$, such that

(i) $T_0(x,y)$ is a bounded continuous function.

(ii) $|T_3(g)| \leq C(\|g\|_1 + \|g\|_2)$ for some $C < \infty$ and all functions $g:\mathbb{R}^D \to \mathbb{C}$ of class C^1.

Then, if $F(x,y,\theta,\bar{\theta})$ is a function of class $C^{1,0}$, supersymmetric in $y,\theta,\bar{\theta}$, with $F_0 \in L^1 \cap L^2(\mathbb{R}^D,dz)$ and $F_3 \in L^1(\mathbb{R}^D,dz)$, we have

$$T(F) = (4\pi/\gamma)\int T_0(x,0)F_0(x,0)dx . \quad \square$$

We will now apply the Lemma to prove the Theorem. Under the appropriate technical assumptions the conditions in the last Lemma will be satisfied. If $G(\Phi)$ is some function of the superfield, let

$$\{G(\Phi)\}_s = \langle G(\Phi)\Gamma(\varphi_0)\exp\{-(1-s)V(\Phi) - s(4\pi/\gamma)V(\varphi_0)\}\rangle ,$$

for $0 \leq s \leq 1$.

Let $g(s) = \{1\}_s$. The statement of the Theorem is just $g(0) = g(1)$. We will show $g'(s) = 0$ for all $0 \leq s \leq 1$.

We have

$$g'(s) = \{V(\Phi)\}_s - (4\pi/\gamma)\{V(\varphi_0)\}_s .$$

But

$$\{V(\Phi)\}_s = \{\int V(\Phi(z,\theta,\bar{\theta}))d\bar{\theta}d\theta dz\}_s =$$

$$= \int\{V(\Phi(z,\theta,\bar{\theta}))\}_s d\bar{\theta}d\theta dz .$$

Since $\Gamma(\varphi_0)$, $V(\Phi)$ and $V(\varphi_0)$ are supersymmetric, $\{V(\Phi(z,\theta,\bar{\theta}))\}_s$ is a supersymmetric distribution and hence by the Lemma

$$\{V(\Phi)\}_s = (4\pi/\gamma)\int\{V(\varphi(x,0))\}_s dx = (4\pi/\gamma)\{V(\varphi_0)\}_s .$$

Thus $g'(s) \equiv 0$.

REFERECES

[1] G. Parisi and N. Sourlas, Random magnetic fields, supersymmetry, and negative dimensions. Phys. Rev. Lett. 43, 744-745 (1979).

[2] F. Wegner, Exact density of states for lowest Landau level in white noise potential, superfield representation for interacting systems. Z. Phys. B51, 279-285 (1983).

[3] E. Brezin, D. Gross and C. Itzykson, Density of states in the presence of a strong magnetic field and random impurities. Nucl. Phys. B235 [FS11], 24-44 (1984).

[4] A. Klein, L. J. Landau and J. F. Perez, Supersymmetry and the Parisi-Sourlas dimensional reduction: a rigorous proof. Comm. Math. Math. Phys. 94, 459-482 (1984).

[5] A. Klein and J. F. Perez, On the density of states for random potentials in the presence of a uniform magnetic field, Nucl. Phys. B251 [FS13], 199-211 (1985).

[6] F. A. Berezin, The method of second quantization. Academic Press, New York, 1966.

ON THE STRUCTURE OF MARKOV DILATIONS ON W*-ALGEBRAS

by

Burkhard Kümmerer
Mathematisches Institut
Universität Tübingen
Auf der Morgenstelle 10
7400 Tübingen
Germany

__Abstract__: In the first part of this paper we discuss the difficulties arising if one tries to immitate the Kolmogorov-Daniell reconstruction theorem in order to construct Markov dilations on non-commutative W*-algebras (which may be interpreted as non-commutative Markov processes). After reviewing a different procedure developed in an earlier paper we generalize this construction and show that in certain situations any Markov dilation is necessarily of this type.

This paper is part of a research project supported by the Deutsche Forschungsgemeinschaft.

Introduction

Dilations of completely positive operators on W*-algebras are interesting for various reasons:
First of all it is a natural mathematical procedure to establish relations between general completely positive operators on W*-algebras on one hand and the much better known automorphisms on the other hand. The analogous philosophy led to the extremely successful theory of unitary dilations on Hilbert spaces (cf. [Sz]).
Secondly, dilations on commutative W*-algebras are usually called "Markov processes corresponding to a transition operator", so dilations have a natural interpretation as non-commutative (quantum) stochastic processes.
Finally, in quantum statistical mechanics dilations appear in the context of coupling an irreversible thermodynamical system to a heat bath ([He], [Fo], [Kü 4], [Kü 3]).
In many situations it is desirable to take into account a distinguished (faithful normal) invariant state which is respected by the dilation. In the context of a thermodynamical interpretation such a state has the meaning of an equilibrium state and a general non-commutative stochastic process now becomes an identically distributed process. These considerations led to the problem of constructing and investigating dilations as they are defined in 1.1. below.
Without bothering about invariant states, dilations have been constructed for any completely positive operator in [Ev 1], [Ev 2], [Da]. However, it turned out that the consideration of an invariant state complicates the construction of a dilation considerably:
If the W*-algebra is commutative, then the Kolmogorov-Daniell reconstruction of a Markov process from its transition probabilities yields a dilation in the sense of the definition in 1.1 (cf. 2.1). However, various attempts to generalize this construction to the non-commutative situation led to solutions which did not satisfy all required properties (see, e.g.,[Ac 1], [Em 2], [Vi]).
In [Kü 1] we have presented a different construction scheme which might be paraphrased as coupling a (tensor) dilation of first order to the zero component of a (non-commutative) Bernoulli shift. Continuous analogues have been presented in [Kü 3], [Fr]. This procedure is quite different from a Kolmogorov-Daniell construction. However, as is shown in [Kü 5], it has a natural analogue in the construction of unitary dilations on Hilbert space. Moreover, it has a natural interpretation as describing a system interacting with completely stochastic surroundings (cf. [Kü 2]).

It is the aim of the present paper to show in section 3 that in certain situations any dilation is of this type, i.e., it can be viewed as a coupling of a dilation of first order to a (generalized) Bernoulli shift. Before doing that we discuss in section 2 some questions concerning the construction of dilations.

1 Preliminaries and Notation

1.1 Definitions.
As the objects of a category we consider pairs $(\mathfrak{A},\phi)$ consisting of a W*-algebra $\mathfrak{A}$ and a faithful normal state ϕ on $\mathfrak{A}$.
A morphism $T:(\mathfrak{A}_1,\phi_1)\longrightarrow(\mathfrak{A}_2,\phi_2)$ is a completely positive operator $T:\mathfrak{A}_1\longrightarrow\mathfrak{A}_2$ satisfying $T(1) = 1$ and $\phi_2\circ T = \phi_1$. In particular, a morphism is a normal operator.
If T is a morphism of $(\mathfrak{A},\phi)$ into itself we call it a morphism of $(\mathfrak{A},\phi)$ and $(\mathfrak{A},\phi,T)$ a dynamical system. If, moreover, T is a *-automorphism we call $(\mathfrak{A},\phi,T)$ a reversible dynamical system.

Definition. For a dynamical system $(\mathfrak{A},\phi,T)$ consider the following diagram

$$\begin{array}{ccc} (\mathfrak{A},\phi) & \xrightarrow{\quad T^n \quad} & (\mathfrak{A},\phi) \\ \Big\downarrow i & & \Big\uparrow P \\ (\hat{\mathfrak{A}},\hat{\phi}) & \xrightarrow[\quad \hat{T}^n \quad]{} & (\hat{\mathfrak{A}},\hat{\phi}) \end{array}$$

where $(\hat{\mathfrak{A}},\hat{\phi},\hat{T})$ is a reversible dynamical system and $i:(\mathfrak{A},\phi)\longrightarrow(\hat{\mathfrak{A}},\hat{\phi})$, $P:(\hat{\mathfrak{A}},\hat{\phi})\longrightarrow(\mathfrak{A},\phi)$ are morphisms.
The quadruple $(\hat{\mathfrak{A}},\hat{\phi},\hat{T};P)$ is called a
(i) (stationary) process over $(\mathfrak{A},\phi)$ if the diagram commutes for $n = 0$,
(ii) dilation of first order of $(\mathfrak{A},\phi,T)$ if the diagram commutes for $n = 0$ and $n = 1$,
(iii) dilation of $(\mathfrak{A},\phi,T)$ if the diagram commutes for $n \in \mathbb{N} \cup \{0\}$.

Remark: Let $(\hat{\mathfrak{A}},\hat{\phi},\hat{T};P)$ be a process over $(\mathfrak{A},\phi)$.
(i) The morphism i is an injective *-homomorphism and $i\circ P$ is a faithful normal conditional expectation of $\hat{\mathfrak{A}}$ onto $i(\mathfrak{A})$ leaving $\hat{\phi}$ invariant.

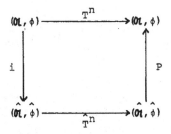

Note that if P is given then i is uniquely determined and conversely. Sometimes it will be more convenient to define i instead of P. We call i (P) the <u>injection (projection) corresponding to P (i)</u>.
(ii) Define the injective *-homomorphisms $j_k := \hat{T}^k \circ i : \mathfrak{A} \longrightarrow \hat{\mathfrak{A}}$ $(k \in \mathbb{Z})$. Then $(\hat{\mathfrak{A}}, \{j_k : k \in \mathbb{Z}\}, \hat{\phi})$ is a W*-stochatic process in the sense of [Ac 2].
(iii) $(\hat{\mathfrak{A}}, \hat{\phi}, \hat{T}; P)$ is a dilation of first order of $(\mathfrak{A}, \phi, P \circ i)$.

For the rest of this paper we assume that $(\mathfrak{A}, \phi, T)$ is a dynamical system.

1.2 <u>Markov Property</u>.
Let $(\hat{\mathfrak{A}}, \hat{\phi}, \hat{T}; P)$ be a process over $(\mathfrak{A}, \phi)$. For $I \subset \mathbb{Z}$ we denote by $\mathfrak{A}_I$ the W*-subalgebra of $\hat{\mathfrak{A}}$ generated by $\bigcup_{k \in I} \hat{T}^k \circ i(\mathfrak{A})$. In ([Kü 1],2.1.3) it is shown that there exists a conditional expectation P_I of $\hat{\mathfrak{A}}$ onto $\mathfrak{A}_I$ leaving $\hat{\phi}$ invariant.

<u>Definition</u>. A process $(\hat{\mathfrak{A}}, \hat{\phi}, \hat{T}; P)$ over $(\mathfrak{A}, \phi)$ is called <u>minimal</u> if $\hat{\mathfrak{A}} = \mathfrak{A}_{\mathbb{Z}}$. It is called a <u>Markov process</u> if for all $x \in \mathfrak{A}_{[0,\infty)}$: $P_{\{0\}}(x) = P_{(-\infty,0]}(x)$.

Note that from any process $(\hat{\mathfrak{A}}, \hat{\phi}, \hat{T}; P)$ one obtains a minimal process by restricting to $\mathfrak{A}_{\mathbb{Z}}$.

The following result is useful for establishing the dilation property.

<u>Proposition</u> ([Kü 1],2.2.7). Let $(\hat{\mathfrak{A}}, \hat{\phi}, \hat{T}; P)$ be a dilation of first order of $(\mathfrak{A}, \phi, T)$. If $(\hat{\mathfrak{A}}, \hat{\phi}, \hat{T}; P)$ satisfies the Markov property then it is a (Markov) dilation of $(\mathfrak{A}, \phi, T)$.

1.3 <u>Tensor Products</u>.
In order to simplify some of the forthcoming considerations we introduce the following notation.
Given $(\mathfrak{A}, \phi)$ and $(\mathfrak{B}, \psi)$ we define a state on the algebraic tensor product $\mathfrak{A} \odot \mathfrak{B}$ of $\mathfrak{A}$ and $\mathfrak{B}$ by $x \otimes y \longmapsto \phi(x) \cdot \psi(y)$. The GNS-construction with respect to this state leads to a representation of $\mathfrak{A} \odot \mathfrak{B}$ on some Hilbert space. The state extends to a faithful normal state $\phi \otimes \psi$ on the weak closure $\mathfrak{A} \otimes \mathfrak{B}$ of $\mathfrak{A} \odot \mathfrak{B}$ and we are led to a new pair $(\mathfrak{A} \otimes \mathfrak{B}, \phi \otimes \psi)$ which we sometimes denote by $(\mathfrak{A}, \phi) \otimes (\mathfrak{B}, \psi)$.

Similarily, if on the algebraic infinite tensor product $\odot_{\mathbb{Z}}\mathfrak{A}$ we have a state χ then χ extends to a normal state on the weak closure of the GNS-representation of $\odot_{\mathbb{Z}}\mathfrak{A}$ with respect to this state still denoted by χ. Thus we are led to a pair denoted by $(\otimes_{\mathbb{Z}}\mathfrak{A},\chi)$ (in the situations considered in the following χ will be faithful).
If, in particular, the state χ is a product state given by
$\chi(\ldots 1\otimes x_{-n}\otimes \ldots \otimes x_n\otimes 1 \ldots) = \phi(x_{-n})\cdots\phi(x_n)$ ($x_i \in \mathfrak{A}$ for $-n \leq i \leq n$) we write $(\otimes_{\mathbb{Z}}\mathfrak{A},\otimes_{\mathbb{Z}}\phi)$ or $\otimes_{\mathbb{Z}}(\mathfrak{A},\phi)$ for $(\otimes_{\mathbb{Z}}\mathfrak{A},\chi)$.

2 Construction of Dilations

2.1 The Kolmogorov-Daniell Construction.

We consider the dynamical system $(\mathfrak{A},\phi,T)$ and assume that $\mathfrak{A}$ is a commutative W*-algebra. Then the Kolmogorov-Daniell construction may be reviewed as follows:
On the infinite algebraic tensor product $\odot_{\mathbb{Z}}\mathfrak{A}$ we define a state χ by

$$\chi(\ldots 1\otimes x_{-n}\otimes x_{-n+1}\otimes \ldots \otimes x_{m-1}\otimes x_m\otimes 1\ldots) := \phi(x_{-n}\cdot T(x_{-n+1}\cdot T(\ldots x_{m-1}\cdot T(x_m)\ldots)))$$

and put $(\hat{\mathfrak{A}},\hat{\phi}) := (\otimes_{\mathbb{Z}}\mathfrak{A},\chi)$.
The tensor right shift on $\odot_{\mathbb{Z}}\mathfrak{A}$ extends to an automorphism $\hat{T}$ of $(\hat{\mathfrak{A}},\hat{\phi})$. Finally, we define an injection i of $(\mathfrak{A},\phi)$ into $(\hat{\mathfrak{A}},\hat{\phi})$ as the embedding into the 0-component, i.e.,
$i(x) := \ldots\otimes 1\otimes \underset{\widehat{0}}{x}\otimes 1\otimes\ldots$, and we denote by P the corresponding projection. Now the following is easily verified:

<u>Proposition</u>. $(\hat{\mathfrak{A}},\hat{\phi},\hat{T};P)$ is a minimal Markov dilation of $(\mathfrak{A},\phi,T)$.

2.2 The Failure of the Kolmogorov-Daniell Construction for Non-Commutative W*-Algebras.

Now we consider the opposite situation and assume that $\mathfrak{A}$ is a type I factor, i.e., $\mathfrak{A}$ is isomorphic to $\mathcal{B}(\mathcal{H})$ for some Hilbert space $\mathcal{H}$.

<u>Proposition</u>. Let $(\hat{\mathfrak{A}},\hat{\phi},\hat{T};P)$ be a dilation of first order of $(\mathfrak{A},\phi,T)$ and assume that $\mathfrak{A}$ is a type I factor. Then there exists $(\mathcal{C},\psi)$ such that
(i) $\hat{\mathfrak{A}} = \mathfrak{A}\otimes\mathcal{C}$,
(ii) $i(x) = x\otimes 1$ for $x \in \mathfrak{A}$,
(iii) $\hat{\phi} = \phi\otimes\psi$,
i.e., $(\hat{\mathfrak{A}},\hat{\phi},\hat{T};P)$ is a <u>tensor dilation</u> of first order of $(\mathfrak{A},\phi,T)$.

Proof: Conditions (i) and (ii) are verified by defining $\mathcal{C}$ as the relative commutant of $i(\mathfrak{A})$ in $\hat{\mathfrak{A}}$. Now define $\psi(y) := \hat{\phi}(1\otimes y)$ for $y \in \mathcal{C}$. For $x \in \mathfrak{A}$, $y \in \mathcal{C}$ we obtain by the modul property of conditional expectations: $x\cdot P(1\otimes y) = P(x\otimes 1\cdot 1\otimes y) = P(1\otimes y\cdot x\otimes 1) = P(1\otimes y)\cdot x$, hence $P(1\otimes\mathcal{C}) \subset \mathbb{C}\cdot 1$ since $\mathfrak{A}$ is a factor. It follows that
$P(1\otimes y) = \hat{\phi}(1\otimes y)\cdot 1 = \psi(y)\cdot 1$ and
$\hat{\phi}(x\otimes y) = \phi(P(x\otimes y)) = \phi(x\cdot\psi(y)) = \phi(x)\cdot\psi(y)$ for $x \in \mathfrak{A}$, $y \in \mathcal{C}$, hence
$\hat{\phi} = \phi\otimes\psi$.

Corollary. Let $(\hat{\mathfrak{A}},\hat{\phi},\hat{T};P)$ be a dilation of $(\mathfrak{A},\phi,T)$ where $\mathfrak{A}$ is a type I factor, and assume that $(\hat{\mathfrak{A}},\hat{\phi},\hat{T};P)$ is constructed as described in 2.1. Then $T(x) = \phi(x)\cdot 1$ for $x \in \mathfrak{A}$.

Proof: Since the dilation has the properties described in the above proposition, we obtain for $x_0, x_1 \in \mathfrak{A}$:

$$\begin{aligned}\hat{\phi}(\ldots 1\otimes x_0\otimes x_1\otimes 1\ldots) &= \phi(x_0)\cdot\psi(\ldots 1\otimes 1\otimes x_1\otimes 1\ldots)\\ &= \phi(x_0)\cdot\hat{\phi}(\hat{T}\circ i(x_1))\\ &= \phi(x_0)\cdot\hat{\phi}(i(x_1))\\ &= \phi(x_0)\cdot\phi(x_1)\ .\end{aligned}$$

On the other side, by the construction procedure, we have

$$\begin{aligned}\phi(x_0\cdot T(x_1)) &= \hat{\phi}(\ldots 1\otimes x_0\otimes x_1\otimes 1\ldots)\\ &= \phi(x_0)\cdot\phi(x_1)\\ &= \phi(x_o\cdot\phi(x_1)\cdot 1)\ ,\end{aligned}$$

hence $T(x_1) = \phi(x_1)\cdot 1$.

.3 Discussion.

The above corollary shows that the Kolmogorov-Daniell construction has no immediate non-commutative analogue. This is mainly due to the fact that in general there are not enough states on $\hat{\mathfrak{A}}$ allowing a conditional expectation onto $i(\mathfrak{A})$ leaving this state invariant whereas on a commutative W*-algebra such an expectation exists for every state on $\mathfrak{A}$. While we discussed only the most naive generalization of this construction, also the more elaborated generalizations meet similar difficulties.

Thus the problems arising in generalizing the Kolmogov-Daniell construction are to a considerable extend due to the fact that this construction puts all information on the dilation into the state $\hat{\phi}$ rendering the automorphism $\hat{T}$ very simple.

This observation leads to the idea that one could try to put more information into the automorphism while simplifying the structure of the state. Now it might become more difficult to find the automorphism of the whole dilation in one single step, so one could try to divide this problem into two steps. First construct a dilation of first order and secondly try to extend this dilation of first order to a full dilation, and if possible, do that in a canonical way. It turned out that following this strategy did indeed lead to new dilations and, moreover, these two steps pose quite different problems.

In the following we will be concerned exclusively with the second problem of extending a dilation of first order to a dilation.

2.4 Another Method for Constructing Dilations.

We now review the construction described in [Kü 1].

Theorem ([Kü 1], 4.2.1). If $(\mathfrak{A}, \phi, T)$ has a tensor dilation of first order then it has a Markov dilation.

Construction. By assumption, $(\mathfrak{A}, \phi, T)$ has a tensor dilation of first order $(\mathfrak{A} \otimes \mathfrak{C}, \phi \otimes \psi, T_1; Q)$ for some pair $(\mathfrak{C}, \psi)$. Now put
$(\hat{\mathfrak{A}}, \hat{\phi}) := (\mathfrak{A}, \phi) \otimes (\otimes_\mathbb{Z} (\mathfrak{C}, \psi))$.
Defining σ as the tensor right shift of $\otimes_\mathbb{Z} (\mathfrak{C}, \psi)$ we put
$\hat{\sigma} := \mathrm{Id}_\mathfrak{A} \otimes \sigma$ on $\hat{\mathfrak{A}}$.
Identifying $\mathfrak{A}_{\{0,1\}}$ as the subalgebra of $\hat{\mathfrak{A}}$ which is the tensor product of $\mathfrak{A}$ and the zero component of $\otimes_\mathbb{Z} \mathfrak{C}$ we can extend T_1 to an automorphism $\hat{T}_1$ of $(\hat{\mathfrak{A}}, \hat{\phi})$ by taking the identity on all other factors of $\otimes_\mathbb{Z} (\mathfrak{C}, \psi)$. Now put $\hat{T} := \hat{T}_1 \circ \hat{\sigma}$.
These definitions may be illustrated by the following sketch:

$$(\mathfrak{A}, \phi)$$
$$\otimes \quad \Big\} \, T_1$$
$$\ldots \otimes (\mathfrak{C}, \psi) \otimes (\mathfrak{C}, \psi) \otimes (\mathfrak{C}, \psi) \underbrace{\Big] \otimes (\mathfrak{C}, \psi) \otimes (\mathfrak{C}, \psi) \otimes \ldots}_{\sigma}$$

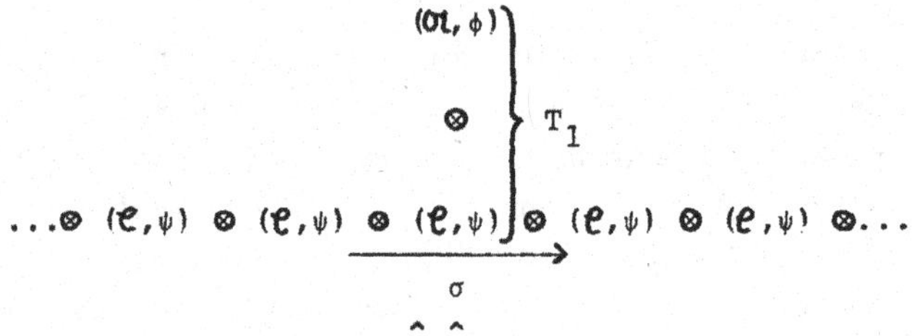

Finally, define $i: (\mathfrak{A}, \phi) \longrightarrow (\hat{\mathfrak{A}}, \hat{\phi}): x \mapsto x \otimes 1$ and P as the corresponding projection.
With the help of 1.2 it is easy to verify that $(\hat{\mathfrak{A}}, \hat{\phi}, \hat{T}; P)$ is a Markov dilation of $(\mathfrak{A}, \phi, T)$.

3 On the Structure of Dilations

3.1 Proposition. Let $(\hat{\mathfrak{A}},\hat{\phi},\hat{T};P)$ be a minimal process over $(\mathfrak{A},\phi)$. The following conditions are equivalent:

(a) For all $I,J \subset \mathbb{Z}$ such that $I < J$ (i.e., $n < m$ for all $n\in I$, $m\in J$) we have: $\hat{\phi}(x\cdot y) = \hat{\phi}(x)\cdot\hat{\phi}(y)$ for $x \in \mathfrak{A}_I$, $y \in \mathfrak{A}_J$.

(b) $P_I(y) = \hat{\phi}(y)\cdot 1$ for $y \in \mathfrak{A}_J$.

Proof: For all $x \in \hat{\mathfrak{A}}_I$, $y \in \mathfrak{A}_J$ we obtain
$\hat{\phi}(x\cdot y) = \hat{\phi}(P_I(x\cdot y)) = \hat{\phi}(x\cdot P_I(y))$.
Hence $\hat{\phi}(x\cdot y) = \hat{\phi}(x)\cdot\hat{\phi}(y)$ $(= \hat{\phi}(x\cdot\hat{\phi}(y)\cdot 1)$) if and only if $P_I(y) = \hat{\phi}(y)\cdot 1$ which proves the desired equivalence.

Remarks. (i) If $I_1, I_2, \ldots I_n$ are subsets of $\mathbb{Z}$ such that $I_1 < I_2 < \ldots < I_n$ and if $x_i \in \mathfrak{A}_{I_i}$ for $1 \le i \le n$, we immediately obtain from condition (a) that $\hat{\phi}(x_1\cdot x_2\cdots x_n) = \hat{\phi}(x_1)\cdot\hat{\phi}(x_2)\cdots\hat{\phi}(x_n)$. In particular, the family $\{\mathfrak{A}_{\{n\}}: n \in \mathbb{Z}\}$ of subalgebras of $\hat{\mathfrak{A}}$ is "weakly independent" in the sense of [Ba]. In terms of a probabilistic interpretation these subalgebras are stochastically independent.

(ii) Given $n \in \mathbb{N}$, condition (b) gives
$P\circ\hat{T}^n\circ i(x) = P\circ P_{\{0\}}\circ\hat{T}^n\circ i(x) = \phi(x)\cdot 1$, hence under this condition $(\hat{\mathfrak{A}},\hat{\phi},\hat{T};P)$ is a dilation of $(\mathfrak{A},\phi,T)$ with $T = \phi\otimes 1$, i.e., $T(x) = \phi(x)\cdot 1$ for $x \in \mathfrak{A}$. Condition (b) may be viewed as a weakened Markov property and a Markov dilation $(\hat{\mathfrak{A}},\hat{\phi},\hat{T};P)$ of $(\mathfrak{A},\phi,\phi\otimes 1)$ fulfils the above conditions.

3.2 Definition. If a minimal Markov process $(\hat{\mathfrak{A}},\hat{\phi},\hat{T};P)$ over $(\mathfrak{A},\phi)$ satisfies the equivalent conditions of 3.1 then we call $(\hat{\mathfrak{A}},\hat{\phi},\hat{T};P)$ a generalized Bernoulli shift over $(\mathfrak{A},\phi)$.

3.3 Remark. If $\hat{\mathfrak{A}}$ is a commutative W*-algebra, then every generalized Bernoulli shift is a Bernoulli shift as is easily derived from well known results in probability theory.
If $\mathfrak{A}$ is any W*-algebra, then $(\hat{\mathfrak{A}},\hat{\phi}) = \otimes_{\mathbb{Z}}(\mathfrak{A},\phi)$ with the tensor right shift $\hat{T}$ naturally leads to a generalized Bernoulli shift over $(\mathfrak{A},\phi)$. However, there exist other generalized Bernoulli shifts over $(\mathfrak{A},\phi)$.

3.4 In 2.4 we have constructed a dilation by "coupling" a tensor dilation of first order to a non-commutative Bernoulli shift. We are now going to show that under an additional assumption we can couple the tensor dilation of first order to any generalized Bernoulli shift in order to obtain a Markov dilation (3.6). Moreover, this result has a converse, which is proved in 3.8: If a dilation contains a certain tensor dilation of first order then it consists of a coupling of this dilation of first order to a generalized Bernoulli shift.

3.5 <u>Definition</u>. Let $(\mathcal{A},\phi,T)$ be a dynamical system.

(i) A dilation of first order $(\hat{\mathcal{A}},\hat{\phi},\hat{T};P)$ is an <u>inner tensor dilation of first order</u> if it is a tensor dilation of first order and if $\hat{T}$ is an inner automorphism.

(ii) A dilation $(\hat{\mathcal{A}},\hat{\phi},\hat{T};P)$ of $(\mathcal{A},\phi,T)$ <u>contains an inner tensor dilation of first order</u> if there exists a unitary $U \in \mathcal{A}_{\{0,1\}}$ which is in the centralizer of $(\hat{\mathcal{A}},\hat{\phi})$ such that $\hat{T}\circ i(x) = U^*\cdot i(x)\cdot U$ for all $x \in \mathcal{A}$ and if $(\mathcal{A}_{\{0,1\}},\hat{\phi}|\mathcal{A}_{\{0,1\}},\mathrm{Ad}\,U|\mathcal{A}_{\{0,1\}};P|\mathcal{A}_{\{0,1\}})$ is a tensor dilation of first order.

3.6 Let $(\mathcal{A},\phi,T)$ be a dynamical system and $(\mathcal{A}_1,\phi_1,T_1;P_1) = (\mathcal{A}\otimes\mathcal{C}_1,\phi\otimes\psi_1,\mathrm{Ad}\,U;P_1)$ be an inner tensor dilation of first order of $(\mathcal{A},\phi,T)$. Moreover, let $(\mathcal{C},\psi,\sigma;Q)$ be any generalized Bernoulli shift over $(\mathcal{C}_1,\psi_1)$.

If j denotes the injection corresponding to Q then $\mathrm{Id}_{\mathcal{A}}\otimes j$ is an injection of $(\mathcal{A}\otimes\mathcal{C}_1,\phi\otimes\psi_1)$ into $(\mathcal{A}\otimes\mathcal{C},\phi\otimes\psi)$ and we put

$\hat{U} := (\mathrm{Id}_{\mathcal{A}}\otimes j)(U)$. Moreover, with

$\mathrm{Id}_{\mathcal{A}}\otimes Q\colon (\mathcal{A}\otimes\mathcal{C},\phi\otimes\psi)\longrightarrow(\mathcal{A}\otimes\mathcal{C}_1,\phi\otimes\psi_1)$ we put

$P := P_1\circ\mathrm{Id}_{\mathcal{A}}\otimes Q\colon(\mathcal{A}\otimes\mathcal{C},\phi\otimes\psi)\longrightarrow(\mathcal{A},\phi)$ and denote by

$i := (\mathrm{Id}_{\mathcal{A}}\otimes j)\circ i_1\colon(\mathcal{A},\phi)\longrightarrow(\mathcal{A}\otimes\mathcal{C},\phi\otimes\psi)$ the corresponding injection.

The following diagram may illustrate this situation:

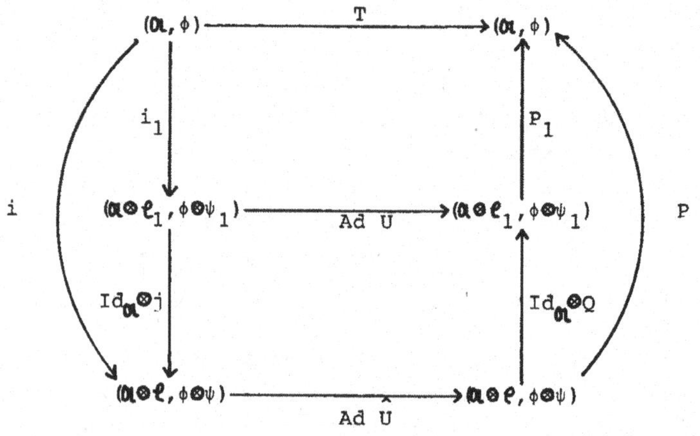

Finally we introduce the abbreviations $(\hat{\mathfrak{A}}, \hat{\phi}) := (\mathfrak{A}\otimes\mathfrak{C}, \phi\otimes\psi)$, $\hat{T} := \mathrm{Ad}\,\hat{U}\circ(\mathrm{Id}_{\mathfrak{A}}\otimes\sigma)$.

Theorem. Using the notation introduced above $(\hat{\mathfrak{A}}, \hat{\phi}, \hat{T}; P)$ is a Markov dilation of $(\mathfrak{A}, \phi, T)$.

Proof: For $x \in \mathfrak{A}$ we obtain

$$
\begin{aligned}
P\circ\hat{T}\circ i(x) &= P\circ(\mathrm{Ad}\,\hat{U}\circ \mathrm{Id}_{\mathfrak{A}}\otimes\sigma)(x\otimes 1_{\mathfrak{C}}) \\
&= P\circ \mathrm{Ad}\,\hat{U}(x\otimes 1_{\mathfrak{C}}) \\
&= P(\hat{U}^*\cdot(x\otimes 1_{\mathfrak{C}})\cdot\hat{U}) \\
&= P_1(U^*\cdot(x\otimes 1_{\mathfrak{C}_1})\cdot U) \\
&= P_1\circ \mathrm{Ad}\,U\circ i_1(x) \\
&= T(x),
\end{aligned}
$$

hence $(\hat{\mathfrak{A}}, \hat{\phi}, \hat{T}; P)$ is a dilation of first order of $(\mathfrak{A}, \phi, T)$.

Obviously, we have the inclusion

$\mathfrak{A}_{[0,\infty)} \subset \mathfrak{A}\otimes\mathfrak{C}_{[0,\infty)}$.

Moreover, for $x \in \mathfrak{A}$ we have

$$
\begin{aligned}
\hat{T}^{-1}\circ i(x) &= (\mathrm{Ad}\,\hat{U}\circ \mathrm{Id}_{\mathfrak{A}}\otimes\sigma)^{-1}\circ i(x) \\
&= (\mathrm{Id}_{\mathfrak{A}}\otimes\sigma)^{-1}\circ(\mathrm{Ad}\,\hat{U})^{-1}\circ i(x) \\
&\in \mathrm{Id}_{\mathfrak{A}}\otimes\sigma^{-1}(\mathfrak{A}_{\{0,1\}}) \\
&\subset \mathrm{Id}_{\mathfrak{A}}\otimes\sigma^{-1}(\mathfrak{A}\otimes j(\mathfrak{C}_1)) \\
&= \mathrm{Id}_{\mathfrak{A}}\otimes\sigma^{-1}(\mathfrak{A}\otimes\mathfrak{C}_{\{0\}}) \\
&= \mathfrak{A}\otimes\mathfrak{C}_{\{-1\}}
\end{aligned}
$$

and similarily

$\hat{T}^{-n}\circ i(x) \subset \mathfrak{A}\otimes\mathfrak{C}_{[-n,-1]}$ for $n \geq 0$, hence

$\mathfrak{A}_{(-\infty,-0]} \subset \mathfrak{A}\otimes\mathfrak{C}_{(-\infty,-1]}$.

Given $x \in \mathfrak{A}_{[0,\infty)}$ with $x = y\otimes w$, $y \in \mathfrak{A}$, $w \in \mathfrak{C}_{[0,\infty)}$ we obtain

$$
\begin{aligned}
P_{(-\infty,0]}(x) &= P_{(-\infty,0]}(y\otimes w) \\
&= y\otimes 1\cdot P_{(-\infty,0]}(1\otimes w) \\
&= y\otimes 1\cdot P_{(-\infty,0]}\circ(\mathrm{Id}_{\mathfrak{A}}\otimes Q_{(-\infty,-1]})(1\otimes w) \\
&= y\otimes 1\cdot P_{(-\infty,0]}(1\otimes Q_{(-\infty,-1]}(w)) \\
&= y\otimes 1\cdot P_{(-\infty,0]}(1\otimes\psi(w)\cdot 1) \qquad \text{(by 3.1.b)} \\
&= y\otimes 1\cdot 1\otimes\psi(w)\cdot 1 \\
&= y\otimes 1\cdot P_{\{0\}}(1\otimes w) \\
&= P_{\{0\}}(y\otimes w) \\
&= P_{\{0\}}(x) .
\end{aligned}
$$

Clearly, this relation extends to all $x \in \mathfrak{A}_{[0,\infty)} \subset \mathfrak{A}\otimes\mathfrak{C}_{[0,\infty)}$, thus we have proven the Markov property of $(\hat{\mathfrak{A}}, \hat{\phi}, \hat{T}; P)$. Now the assertion follows from the proposition in 1.2.

3.7 Definition. Let $(\mathfrak{A},\phi,T)$ be a dynamical system and $(\hat{\mathfrak{A}},\hat{\phi},\hat{T};P)$ be a minimal Markov dilation of $(\mathfrak{A},\phi,T)$. We say that $(\hat{\mathfrak{A}},\hat{\phi},\hat{T};P)$ can be decomposed into an inner tensor dilation of first order and a generalized Bernoulli shift if it can be constructed from an inner tensor dilation of first order and a generalized Bernoulli shift as in 3.6.

3.8 Theorem. Let $(\hat{\mathfrak{A}},\hat{\phi},\hat{T};P)$ be a minimal Markov tensor dilation of $(\mathfrak{A},\phi,T)$. If $(\hat{\mathfrak{A}},\hat{\phi},\hat{T};P)$ contains an inner tensor dilation of first order then it can be decomposed into an inner tensor dilation of first order and a generalized Bernoulli shift.

Proof: By assumption there exist $(\mathcal{C},\psi)$ and $(\mathcal{C}_1,\psi_1)$ such that $(\hat{\mathfrak{A}},\hat{\phi}) = (\mathfrak{A}\otimes\mathcal{C},\phi\otimes\psi)$, $(\mathfrak{A}_{\{0,1\}},\hat{\phi}|_{\mathfrak{A}_{\{0,1\}}}) = (\mathfrak{A}\otimes\mathcal{C}_1,\phi\otimes\psi_1)$ and $U \in \mathfrak{A}_{\{0,1\}}$ with $T\circ i(x) = \hat{T}(x\otimes 1) = U^*\cdot(x\otimes 1)\cdot U$ for $x \in \mathfrak{A}$.
Hence $\mathrm{Ad}\, U^*\circ\hat{T}(x\otimes 1) = x\otimes 1$ for $x \in \mathfrak{A}$, and there exists an automorphism σ of $(\mathcal{C},\psi)$ such that $\mathrm{Ad}\, U^*\circ\hat{T} = \mathrm{Id}_{\mathfrak{A}}\otimes\sigma$ hence $\hat{T} = \mathrm{Ad}\, U\circ(\mathrm{Id}_{\mathfrak{A}}\otimes\sigma)$.
If Q denotes the conditional expectation of $(\mathcal{C},\psi)$ onto the subalgebra $\mathcal{C}_1$ then $(\mathcal{C},\psi,\sigma;Q)$ is a process over $(\mathcal{C}_1,\psi_1)$, and it remains to show that it is a generalized Bernoulli shift. Clearly, the minimality of $(\hat{\mathfrak{A}},\hat{\phi},\hat{T};P)$ implies the minimality of $(\mathcal{C},\psi,\sigma;Q)$.
The same reasoning as in the proof of Theorem 3.6 yields

$$\mathfrak{A}_{[0,\infty)} = \mathfrak{A}\otimes\mathcal{C}_{[0,\infty)} ,$$
$$\mathfrak{A}_{(-\infty,0]} = \mathfrak{A}\otimes\mathcal{C}_{(-\infty,-1]} .$$

Therefore, given $y \in \mathcal{C}_{[0,\infty)}$ then $1\otimes y \in \mathfrak{A}_{[0,\infty)}$ and we obtain

$$\begin{aligned} 1\otimes Q_{(-\infty,-1]}(y) &= \mathrm{Id}_{\mathfrak{A}}\otimes Q_{(-\infty,-1]}(1\otimes y) \\ &= P_{(-\infty,0]}(1\otimes y) \\ &= P_{\{0\}}(1\otimes y) \\ &= \hat{\phi}(1\otimes y)\cdot 1\otimes 1 \\ &= 1\otimes\psi(y)\cdot 1, \end{aligned}$$

hence $Q_{(-\infty,-1]}(y) = \psi(y)\cdot 1$ for all $y \in \mathcal{C}_{[0,\infty)}$. This is easily seen to be equivalent to 3.1.b.

Thus we can rephrase 3.4: Given an inner tensor dilation of first order and a generalized Bernoulli shift these can be composed to obtain a Markov dilation. Conversely, if a Markov dilation contains an inner tensor dilation of first order, it is composed of this inner tensor dilation and a generalized Bernoulli shift.
This raises the question whether there are situations where a dilation contains such an inner tensor dilation of first order and hence is decomposable into this inner tensor dilation and a generalized Bernoulli shift. A first example is given by the following result.

3.9 Theorem. If $\mathfrak{A}$ is isomorphic to the n×n-matrices M_n, ϕ is the normalized trace tr on M_n, and T is any morphism of (M_n, tr) then any minimal Markov dilation $(\hat{\mathfrak{A}}, \hat{\phi}, \hat{T}; P)$ of (M_n, tr, T) can be decomposed into an inner tensor dilation of first order and a generalized Bernoulli shift.

Proof: By 2.2 $(\hat{\mathfrak{A}}, \hat{\phi}, \hat{T}; P)$ is a tensor dilation and in view of 3.8 we show that it has an inner tensor dilation of first order.
Since T has an invariant trace we know from [Kü 1],2.3.3 that $\hat{T}$ has an invariant trace. Moreover, by [Kü 1],3.1.4 every normal state $\hat{\psi}$ on $\hat{\mathfrak{A}}$ which is invariant under $\hat{T}$ is given by $\hat{\psi} = \psi \circ P$ for some T-invariant state ψ. By the uniqueness of the trace on M_n there is only one $\hat{T}$-invariant trace on $\hat{\mathfrak{A}}$ given by $tr \circ P = \hat{\phi}$. Thus it remains to find a unitary $U \in \mathfrak{A}_{\{0,1\}}$ such that $\hat{T} \circ i(x) = U^* \cdot i(x) \cdot U$ for $x \in \mathfrak{A}$.
Since $\hat{\phi}$ is a trace, $\hat{\mathfrak{A}}$ and hence $\mathfrak{A}_{\{0,1\}}$ are finite W*-algebras and we denote by $P^\natural$ the center valued trace on $\mathfrak{A}_{\{0,1\}}$ (cf. [Ta]).
Since $\mathfrak{A}_{\{0,1\}} = M_n \otimes \mathcal{C}_1$ for some W*-algebra $\mathcal{C}_1$, the center of $\mathfrak{A}_{\{0,1\}}$ is contained in $1 \otimes \mathcal{C}_1$, hence $P^\natural(M_n \otimes 1) \subset \mathbb{C} \cdot 1 \otimes 1$ and $P^\natural(\hat{T}(M_n \otimes 1)) \subset \mathbb{C} \cdot 1 \otimes 1$.
If we denote by e_{ik}, $1 \le i,k \le n$ a system of matrix units in M_n, it follows that
$P^\natural(e_{11} \otimes 1) = \frac{1}{n} \cdot 1 \otimes 1 = P^\natural(\hat{T}(e_{11} \otimes 1))$,
hence there exists a partial isometry $v \in \mathfrak{A}_{\{0,1\}}$ such that
$v \cdot v^* = e_{11} \otimes 1$ and $v^* \cdot v = \hat{T}(e_{11} \otimes 1)$.
Now put $U := \sum_{i=1}^{n} e_{i1} \cdot v \cdot \hat{T}(e_{1i})$.
Obviously, U is a unitary in $\mathfrak{A}_{\{0,1\}}$ and an easy computation shows that it fulfils our requirement.

For the case that $\mathfrak{A} = M_2$ this result has been proven before in [Kü 2], [Kü 4].

3.10 Remark. If $\mathfrak{A}$ is isomorphic to M_n and $(\hat{\mathfrak{A}}, \hat{\phi}, \hat{T}; P)$ is a dilation of some dynamical system (M_n, ϕ, T) then it is generally true that there exists a unitary $U \in \mathfrak{A}_{\{0,1\}}$ such that $\hat{T} \circ i(x) = U^* \cdot i(x) \cdot U$ for $x \in \mathfrak{A}$. However, if ϕ (hence $\hat{\phi}$) is not a trace it is more difficult to show that there exists a unitary in the centralizer of $(\hat{\mathfrak{A}}, \hat{\phi})$ with this property (U being in the centralizer is equivalent to Ad U, hence $Id_{\mathfrak{A}} \otimes \sigma$, leaving $\hat{\phi}$ invariant).
In [Kü 4] we showed that for certain morphisms on M_2 having only a non-tracial invariant state every minimal Markov dilation has an inner tensor dilation of first order and hence can be decomposed.

3.11 Concluding Remark. The construction of Markov dilations as reviewed in 2.4 has been invented in order to overcome the difficulties found in various attempts to immitate the Kolmogorov-Daniell construction in the non-commutative situation. While this construction might look quite artificial at first glance we showed in the present paper that in certain situations a slight generalization of this construction is necessary, i.e., every possible dilation looks that way. Moreover, in [Kü 2] we showed exemplarily that in a concrete physical situation such a dilation has a natural physical interpretation.

References.

[Ac 1] L. Accardi: Nonrelativistic quantum mechanics as a noncommutative Markof process, Advances in Math. 20 (1976), 329-366.

[Ac 2] L. Accardi, A. Frigerio, J.T. Lewis: Quantum stochastic processes, Publ. RIMS, Kyoto Univ. 18 (1982), 97-133.

[Ba] C. Batty: The strong law of large numbers for states and traces of a W*-algebra, Z. Wahrscheinlichkeitstheorie verw. Gebiete 48 (1979), 177-191.

[Da] E.B. Davies: Dilations of completely positive maps, J. London Math. Soc. (2), 17 (1978), 330-338.

[Em 1] G.G. Emch, S. Albeverio, J.P. Eckmann: Quasi-free generalized K-flows, Reports Math. Phys. 13 (1978), 73-85.

[Em 2] G.G. Emch: Minimal dilations of CP-flows, in "C*-Algebras and Applications to Physics", Proc. Los Angeles 1977, Lecture Notes in Mathematics 650, Springer-Verlag, Heidelberg 1978, 156-159.

[Ev 1] D.E. Evans: Positive linear maps on operator algebras, Comm. Math. Phys. 48 (1976), 15-22.

[Ev 2] D.E. Evans, J.T. Lewis: Dilations of dynamical semigroups, Comm. Math. Phys. 50 (1976), 219-227.

[Ev 3] D.E. Evans: Completely positive quasi-free maps on the CAR algebra, Comm. Math. Phys. 70 (1979), 53-68.

[Fo] G.W. Ford, M. Kac, P. Mazur: Statistical mechanics of assemblies of coupled oscillators, J. Math. Phys. 6 (1965), 504-515.

Fr] A. Frigerio, V. Gorini: Markov dilations and quantum detailed balance, Comm. Math. Phys. 93 (1984), 517-532.

He] K. Hepp, E.H. Lieb: Phase transition in reservoir-driven open systems with applications to lasers and superconductors. Helv. Phys. Acta 46 (1973), 573-603.

Kü 1] B. Kümmerer: Markov dilations on W*-algebras. To appear in J. Funct. Anal..
See also: B. Kümmerer: Markov dilations of completely positive operators on W*-algebras, in Gr. Arsene (Ed.), "Dilation Theory, Toeplitz Operators, and Other Topics"Timisoara and Herculane (Romania), 1982, Birkhäuser Verlag, Basel-Boston-Stuttgart 1983, 251-259.

Kü 2] B. Kümmerer, W. Schröder: A survey of Markov dilations for the spin-$\frac{1}{2}$-relaxation and physical interpretation, Semesterbericht Funktionalanalysis, Tübingen, Wintersemester 1981/82, 187-213.

Kü 3] B. Kümmerer: Examples of Markov dilations over the 2×2-matrices. In L. Accardi, A. Frigerio, V. Gorini (Eds.), "Quantum Probability and Applications to the Quantum Theory of Irreversible Processes", Villa Mondragone 1982, Lecture Notes in Mathematics 1055, Springer-Verlag, Heidelberg 1984, 228-244.

Kü 4] B. Kümmerer: Markov dilations on the 2×2-matrices. To appear in the Proceedings of the "Conference on operator algebras and ergodic theory" Busteni, Romania 1983.

Kü 5] B. Kümmerer, W. Schröder: On the structure of unitary dilations. Semesterbericht Funktionalanalysis. Tübingen, Wintersemester 1983/84, 177-225.

Sz] B. Sz.-Nagy, C. Foias:"Harmonic Analysis of Operators on Hilbert Spaces", Noth Holland, Amsterdam 1970.

Ta] M. Takesaki:"Theory of Operator Algebras I", Springer-Verlag, Berlin-Heidelberg-New York 1979.

Vi] G.F. Vincent-Smith: Dilations of a dissipative quantum dynamical system to a quantum Markov process, Proc. London Math. Soc. (3), 49 (1984), 58-72.

A NEW CONSTRUCTION OF UNITARY DILATIONS : SINGULAR COUPLING TO WHITE NOISE

B. Kümmerer and W. Schröder
Mathematisches Institut
Universität Tübingen
Auf der Morgenstelle 10
D-7400 Tübingen
Germany

Abstract: For any one-parameter semigroup of contractions on a Hilbert space we describe a new construction of a unitary dilation. Our method rests on a "singular coupling" of the original Hilbert space to an L^2-shift with appropriate multiplicity, i.e. to a Hilbert space version of white noise.

If the semigroup is self-adjoint then by a so-called "singular coupling limit" the dilation can be obtained from physically meaningful interactions between the dissipative system and a heat bath.

Moreover, we show that for any contraction semigroup our dilation satisfies a Hilbert space version of the Langevin equation.

From a mathematical point of view our construction clearly exhibits the Markovian structure inherent in any unitary dilation. On the other hand it permits an appealing physical interpretation.

This paper is part of a research project supported by the Deutsche Forschungsgemeinschaft.

. Preliminaries

.1 Notations and Definitions. Throughout this paper we denote by $(\mathcal{H}, T_t)$ a pair consisting of a separable Hilbert space $\mathcal{H}$ with scalar product $\langle\,\cdot\,,\,\cdot\,\rangle$ and a strongly continuous one-parameter contraction semigroup $(T_t)_{t\in\mathbb{R}^+}$ on $\mathcal{H}$ with $T_0 = 1$. (Denoting by $\mathbb{R}$ the set of real numbers we use the conventions $\mathbb{R}^+ := \{r \in \mathbb{R}: r \geq 0\}$ and $\mathbb{R}^- := \{r \in \mathbb{R}: r \leq 0\}$). The generator of $(T_t)_{t\in\mathbb{R}^+}$ will be denoted by $-A$, i.e. $T_t = \exp(-At)$ for all $t \in \mathbb{R}^+$.

If an arbitrary Hilbert space $\mathcal{K}$ is given we denote its scalar product by $\langle\,\cdot\,,\,\cdot\,\rangle_{\mathcal{K}}$ or simply by $\langle\,\cdot\,,\,\cdot\,\rangle$ when no confusion is likely. The corresponding norm will always be written as $\|\cdot\|$.

For any operator B on $\mathcal{K}$ we write $D(B)$ for its domain.

The algebra of all bounded linear operators on $\mathcal{K}$ will be denoted by $\mathcal{B}(\mathcal{K})$.

If $\mathcal{K}$ has a decomposition $\mathcal{K} = \mathcal{K}_1 \oplus \mathcal{K}_2$ into an orthogonal direct sum then an operator $S \in \mathcal{B}(\mathcal{K})$ can be written as a 2×2-matrix

$$\begin{pmatrix} S_{11} & S_{12} \\ S_{21} & S_{22} \end{pmatrix}$$

where S_{ij}, $1 \leq i, j \leq 2$, is a bounded operator from $\mathcal{K}_j$ into $\mathcal{K}_i$. Analogously, a decomposition of $\mathcal{K}$ into n orthogonal subspaces leads to a representation of S as $n\times n$-matrix.

Given an interval $I \subset \mathbb{R}$ we denote by $L^2(I)$ the Hilbert space of Borel measurable complex functions on I which are square integrable with respect to the Lebesgue measure on I.

The characteristic function of I will be written as χ_I.

Furthermore, if $\mathcal{K}$ is a Hilbert space, we write $L^2(I, \mathcal{K})$ for the Hilbert space of $\mathcal{K}$-valued square integrable Borel measurable functions on I. This space is canonically isomorphic to the Hilbert space tensor product $L^2(I) \otimes \mathcal{K}$.

By $W(\mathbb{R}, \mathcal{K})$ we denote the space of $\mathcal{K}$-valued Sobolev functions on $\mathbb{R}$, i.e. the domain of the generator $\frac{d}{dx} \otimes 1$ of the right shift on $L^2(\mathbb{R}, \mathcal{K})$. Finally, we define $W(I, \mathcal{K})$ as consisting of those $\psi \in L^2(I, \mathcal{K})$ for which there exists an element $\phi \in W(\mathbb{R}, \mathcal{K})$ such that $\psi(s) = \phi(s)$ for all $s \in I$.

Definitions. Let $(\mathcal{H}, T_t)$ be given. If $\hat{\mathcal{H}}$ is another Hilbert space, $(\hat{T}_t)_{t\in\mathbb{R}}$ a strongly continuous one-parameter group of unitaries on $\hat{\mathcal{H}}$, and $i: \mathcal{H} \longrightarrow \hat{\mathcal{H}}$ an isometry with adjoint $P := i^*$ such that the diagram

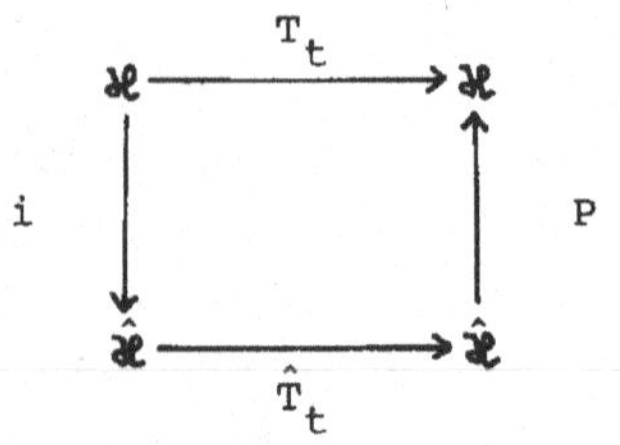

commutes for all $t \in \mathbb{R}^+$ then we call $(\hat{\mathcal{H}}, \hat{T}_t; i)$ a <u>unitary dilation</u> of $(\mathcal{H}, T_t)$. It is said to be <u>minimal</u> if $\hat{\mathcal{H}}$ is the closed linear span of $\cup\{\hat{T}_t i(\mathcal{H}) : t \in \mathbb{R}\}$.

1.2 Existence and Construction of Dilations.

1.2.1 The fundamental result on unitary dilations is due to Sz.-Nagy.

<u>Theorem</u> ([Sz], I.8.1). There exists a minimal unitary dilation of $(\mathcal{H}, T_t)$ which is unique up to unitary equivalence.

Let us indicate two widely used procedures for constructing dilations.

1.2.2 Given $(\mathcal{H}, T_t)$ we define a function $T^\dagger : \mathbb{R} \longrightarrow \mathcal{B}(\mathcal{H})$ by

$$T_t^\dagger := \begin{cases} T_t & \text{if } t \geq 0 \\ T_t^* & \text{if } t \leq 0 . \end{cases}$$

Let $\mathcal{M}$ be the vector space of functions $f : \mathbb{R} \longrightarrow \mathcal{H}$ such that $f(t) = 0$ except for finitely many points $t \in \mathbb{R}$. On $\mathcal{M}$ the function $T^\dagger$ gives rise to a positive semidefinite sesquilinear form

$(f, g) \longmapsto \sum_{s,t \in \mathbb{R}} \langle T_{t-s}^\dagger f(t), g(s) \rangle$.

If we denote its kernel by $\mathcal{M}_0$ then on the quotient $\mathcal{M}/\mathcal{M}_0$ this form induces a scalar product. The completion of the quotient with respect to this scalar product leads to a Hilbert space $\hat{\mathcal{H}}$.

One defines an isometry $i : \mathcal{H} \longrightarrow \hat{\mathcal{H}}$, $\xi \longmapsto f_\xi$ by

$$f_\xi(t) := \begin{cases} \xi & \text{if } t = 0 \\ 0 & \text{if } t \neq 0 . \end{cases}$$

The right shift on $\mathcal{M}$ induces a unitary $\hat{T}_t$ on $\hat{\mathcal{H}}$.

Altogether $(\hat{\mathcal{H}}, \hat{T}_t; i)$ is a unitary dilation of $(\mathcal{H}, T_t)$.

This construction relies on the fact that $T^\dagger$ is a positive definite $\mathcal{B}(\mathcal{H})$-valued function on the additive group $\mathbb{R}$.

1.2.3 There is another way of constructing dilations (cf. [Sz]).

Since $(T_t)_{t\geq 0}$ consists of contractions, the generator $-A$ is dissipative, i.e. the quadratic form F defined on $D(A)$ by

$$F(\xi,\eta) = \langle A\xi, \eta \rangle + \langle \xi, A\eta \rangle \qquad \text{for all } \xi, \eta \in D(A) \text{ is positive.}$$

If ker F denotes the kernel of the form F then the completion of the quotient $D(A)/\ker F$ with respect to the norm induced by F leads to a Hilbert space $\mathcal{N}$.

One puts $\mathcal{K} := L^2(\mathbb{R}, \mathcal{N})$.

By V_t we denote the right shift on $L^2(\mathbb{R},\mathcal{N})$, i.e.

$$(V_t f)(s) = f(s-t) \qquad \text{for } f \in L^2(\mathbb{R},\mathcal{N}) \text{, } s,t \in \mathbb{R}.$$

Since $s \longmapsto T_s^* T_s$ is a monotonically decreasing function of positive operators, we can define an operator Q by $Q := (\lim_{s\to\infty} T_s^* T_s)^{1/2}$ and then the Hilbert space $\mathcal{L} := \overline{Q\mathcal{H}}$.

For each $t \in \mathbb{R}^+$ the mapping $Q\xi \longmapsto QT_t\xi$ ($\xi \in \mathcal{H}$) has a continuous extension to an isometry W_t on $\mathcal{L}$.

Next put $\hat{\mathcal{H}} := \mathcal{L} \oplus \mathcal{K}$,

$$i: \mathcal{H} \longrightarrow \hat{\mathcal{H}} \text{ , } \xi \longmapsto Q\xi \oplus f_\xi \qquad \text{where}$$

$$f_\xi(s) := \begin{cases} T_{-s}\,\xi & \text{for } s \leq 0 \\ 0 & \text{for } s > 0 \end{cases},$$

and for $t \in \mathbb{R}^+$ define an isometry $\hat{T}_t$ on $\mathcal{L} \oplus \mathcal{K}$ as 2×2-matrix

$$\hat{T}_t := \begin{pmatrix} W_t & 0 \\ 0 & V_t \end{pmatrix} \qquad \text{(cf. 1.1).}$$

In general $\hat{T}_t$ is not unitary. Nevertheless the triple $(\hat{\mathcal{H}}, \hat{T}_t; i)$ makes the diagram of 1.1 commutative.

1.2.4   In the special case that $\lim_{t\to\infty} T_t = 0$ in the strong operator topology, we have $Q = 0$, $\mathcal{L} = \{0\}$, and hence $\hat{\mathcal{H}} = L^2(\mathbb{R},\mathcal{N})$; $\hat{T}_t = V_t$; $i: \mathcal{H} \longrightarrow \hat{\mathcal{H}}$, $\xi \longmapsto f_\xi$.

Now this gives a unitary dilation of $(\mathcal{H}, T_t)$.

Such dilations have in particular been used by Lax and Phillips [La] in their approach to scattering theory.

In [Ev] one can find a Langevin type equation for this type of unitary dilations.

2 A New Construction of a Unitary Dilation

The usual construction schemes reviewed above are in two respects not altogether satisfactory:

(1) In view of applications in the physical theory of open systems one would like to be able to interpret a dilation $(\hat{\mathcal{H}}, \hat{T}_t; i)$ of $(\mathcal{H}, T_t)$ as a coupling of the dissipative system to a heat bath.
In particular one wants $\hat{\mathcal{H}}$ to be of the form $\hat{\mathcal{H}} = \mathcal{H} \oplus \mathcal{A}$ with a certain auxilliary Hilbert space $\mathcal{A}$. The injection should be $i: \mathcal{H} \longrightarrow \mathcal{H} \oplus \mathcal{A}$, $\xi \longmapsto (\xi, 0)$. Moreover, the dynamics $\hat{T}_t$ should describe a coupling between $\mathcal{H}$ and $\mathcal{A}$ as well as an extremely random behaviour on the heat bath $\mathcal{A}$.

(2) On the other hand, one would like to have a dilation scheme which makes evident the stochastic features inherent in any unitary dilation, namely the linear Markov property or its equivalent formulation as orthogonality relations.
The intention to find a dilation scheme meeting these demands led to [Kü 1] and [Kü 2].

2.1 First we describe the linear Markov property. Let us consider a strongly continuous one-parameter group $(\hat{T}_t)_{t\in\mathbb{R}}$ of unitaries on a Hilbert space $\hat{\mathcal{H}}$ and let $i: \mathcal{H} \longrightarrow \hat{\mathcal{H}}$ be an isometry with adjoint $P := i^*$. If we define $S_t := P\cdot\hat{T}_t\cdot i$ for $t \geq 0$ then $(S_t)_{t\geq 0}$ is a one-parameter family of contractions, however, in general it is not a semigroup.
For a subset $I \subset \mathbb{R}$ we define $\mathcal{H}_I$ as the closed linear span of $\cup\{ \hat{T}_t i(\mathcal{H}): t \in I \} \subset \hat{\mathcal{H}}$ and P_I as the orthogonal projection of $\hat{\mathcal{H}}$ onto $\mathcal{H}_I$, while $P_I^{\perp}$ denotes the projection onto the orthogonal complement of $\mathcal{H}_I$.

<u>Proposition</u> ([Kü 1]). The following conditions are equivalent:

(a) $S_{s+t} = S_s S_t$ for $s, t \geq 0$, i.e., $(S_t)_{t\geq 0}$ is a one-parameter semigroup and $(\hat{\mathcal{H}}, \hat{T}_t; i)$ is a unitary dilation of $(\mathcal{H}, S_t)$.

(b) For all $\xi \in \mathcal{H}$ the vector $\hat{T}_s\cdot P_0^{\perp}\cdot\hat{T}_t\cdot i(\xi)$ is orthogonal to $i(\mathcal{H}) = P_0(\hat{\mathcal{H}})$.

(c) For all $\xi \in \mathcal{H}_{[0,+\infty)}$: $P_0(\xi) = P_{(-\infty,0]}(\xi)$.

Condition (b) means the following: Start with a vector $\xi \in i(\mathcal{H}) \subset \hat{\mathcal{H}}$ and consider $\hat{T}_t\xi$, i.e. look at ξ after t seconds. Decompose it into a part which still belongs to $i(\mathcal{H})$, i.e. $P_0\cdot\hat{T}_t\xi$, and a part which lies completely outside of $i(\mathcal{H})$, i.e. $P_0^{\perp}\cdot\hat{T}_t\xi$. Then condi-

tion (b) tells us that the part outside of $i(\mathcal{H})$ remains outside for all future times. In other words, the parts of ξ which are once lost from $i(\mathcal{H})$ remain lost until eternity: They never come back!
Condition (c) is a kind of Markov property. Here $\mathcal{H}_{[0,+\infty)}$ means the future while $\mathcal{H}_{(-\infty,0]}$ describes the past of some process. In this sense condition (c) says that given ξ in the future, the information gained from the presence $P_0\xi$ is the same as the information that can be deduced from the whole past $P_{(-\infty,0]}(\xi)$.
Summing up, we may say that in one way or another all three conditions describe the absence of memory.

2.2 The two basic ideas behind our new construction aim to meet the demands (1) and (2) stated in the beginning of this section.

2.2.1 We first find a certain "coupling operator" C_+ from $\mathcal{H}$ into another Hilbert space $\mathcal{K}_+$ such that C_+ infinitesimally compensates the dissipativity of T_t by transporting into $\mathcal{K}_+$ what is metrically lost by T_t in each moment. To be precise, we introduce an operator $C_+: D(A) \longrightarrow \mathcal{K}_+$ which satisfies

$$\begin{aligned}\| C_+\xi \|^2 &= \lim_{\varepsilon\downarrow 0} 1/\varepsilon(\|\xi\|^2 - \|T_\varepsilon\xi\|^2) \\ &= \langle A\xi, \xi \rangle + \langle \xi, A\xi \rangle \\ &= F(\xi, \xi) \qquad \text{for all } \xi \in D(A) .\end{aligned}$$

Moreover, we assume that $\mathcal{K}_+$ is the closure of $C_+D(A)$.
The existence of such a "coupling" $(C_+, \mathcal{K}_+)$ is shown in [Kü 1].
In order to construct the dilation also for negative times we replace T_t by $T_t{}^*$, resp. A by A^* in the preceding definition and thus obtain another "coupling" $(C_-, \mathcal{K}_-)$.

2.2.2 The second basic ingredient of our construction takes account of the Markovian structure inherent in any dilation. Property (b) of the proposition in 2.1 suggests to couple $(\mathcal{H}, T_t)$ to the right shift S_t^+ , $t \in \mathbb{R}^+$, on $L^2(\mathbb{R}^+, \mathcal{K}_+)$ for transporting away what has been brought into $\mathcal{K}_+$ by C_+ .
Correspondingly, for constructing the "past" we consider the right shift S_t^- on $L^2(\mathbb{R}^-, \mathcal{K}_-)$.
In a physical interpretation these shifts describe the random behaviour of the heat bath we looked for.

2.2.3 Combinig these ideas we put $\hat{\mathcal{H}} := L^2(\mathbb{R}^-, \mathcal{K}_-) \oplus \mathcal{H} \oplus L^2(\mathbb{R}^+, \mathcal{K}_+)$.
and define the injection by

$$i : \mathcal{H} \longrightarrow \hat{\mathcal{H}}, \ \xi \longmapsto \begin{pmatrix} 0 \\ \xi \\ 0 \end{pmatrix} .$$

On $\hat{\mathcal{H}}$ we introduce densely defined operators $\hat{T}_t$, $t \in \mathbb{R}^+$, as

$$\hat{T}_t = \begin{pmatrix} S_t^- & 0 & 0 \\ X_t & T_t & 0 \\ Z_t & Y_t & S_t^+ \end{pmatrix}$$

with the following components X_t, Y_t, Z_t :

The densely defined operator X_t from $L^2(\mathbb{R}^-, \mathcal{K}_-)$ into $\mathcal{H}$ is given by
$X_t\phi = \int_0^t T_s A^*\eta f(s-t)ds + A\int_0^t T_s\eta f(s-t)ds$
for any vector $\phi = \chi_I \otimes C_-\eta$ with I a finite interval and $\eta \in D(A^*) = D(C_-)$.

For the densely defined operator Y_t from $\mathcal{H}$ into $L^2(\mathbb{R}^+, \mathcal{K}_+)$ we have the formula

$$(Y_t\xi)(s) := \begin{cases} - C_+ \cdot T_{t-s}\xi & \text{if} \quad s \in [0,t] \\ 0 & \text{if} \quad s \notin [0,t] \end{cases} \qquad \text{provided} \quad \xi \in D(A) .$$

In the sequel we shall frequently use the short hand notation
$(Y_t\xi)(s) := \chi_{[0,t]}(s) \cdot (-C_+) \cdot T_{t-s}\xi$.

Finally, the densely defined operator Z_t from $L^2(\mathbb{R}^-, \mathcal{K}_-)$ into $L^2(\mathbb{R}^+, \mathcal{K}_+)$ is given by
$\chi_{[-t,-r]} \otimes C_-\eta \longmapsto \int_0^{t-r} Y_s A^*\eta \, ds - Y_{t-r}\eta$
with $r \in [0,t]$ and $\eta \in D(A^*) = D(C_-)$.

Remark. If each T_t, $t \in \mathbb{R}^+$, is self-adjoint, then the operators X_t, Y_t, Z_t take a more transparent form (cf. 4.1) which allows a physical interpretation (cf. [Kü 1]).

Theorem ([Kü 1]). The operators X_t, Y_t, and Z_t are contractions and extend to everywhere defined operators. Moreover, each $\hat{T}_t$, $t \in \mathbb{R}^+$, is a unitary and $(\hat{\mathcal{H}}, \hat{T}_t; i)$ is a minimal unitary dilation of $(\mathcal{H}, T_t)$.

This dilation may be visualized by the following sketch:

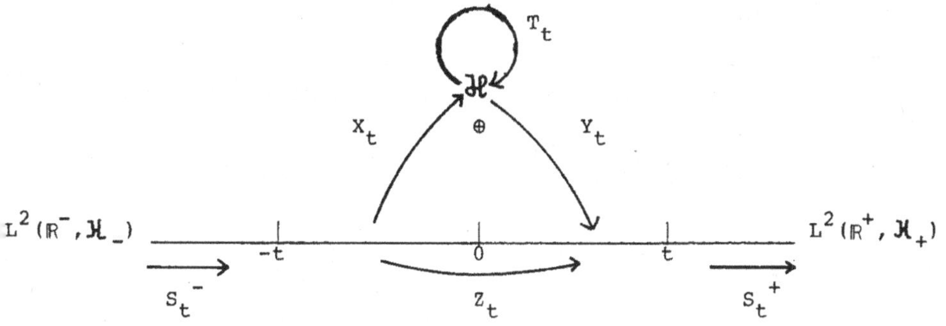

In the spirit of the motivation outlined above we have adopted the term "singular coupling to white noise" for the new construction.
In the following we want to show in more detail that our dilation indeed deserves this name.

3 A Hilbert Space Langevin Equation

3.1 Hilbert Space White Noise. In stochastics as well as in physics "white noise" means in essence the derivative of Brownian motion. We want to take this heuristic meaning of white noise as a starting point for our approach. So let us begin with considering Brownian motion and see how it can be reformulated in a Hilbert space theory.

Let (Ω, Σ, μ) be a probability space and $(b_t)_{t \in \mathbb{R}^+}$ a realization of Brownian motion on (Ω, Σ, μ). Then $b_t \in L^2(\Omega, \Sigma, \mu)$ for all $t \in \mathbb{R}^+$.
As the setting of the present work is the theory of Hilbert spaces it is enough for our purposes to consider just the closed linear hull of $(b_t)_{t \in \mathbb{R}^+}$. Since for all $s, t \in \mathbb{R}^+$ we have

$$< b_s, b_t >_{L^2(\Omega, \Sigma, \mu)} = \min \{s,t\} = < \chi_{[o,s]}, \chi_{[0,t]} >_{L^2(\mathbb{R}^+)}$$

the subspace generated by $(b_t)_{t \in \mathbb{R}^+}$ is isomorphic to $L^2(\mathbb{R}^+)$ via the identification $b_t = \chi_{[0,t]}$ for all $t \in \mathbb{R}^+$.
From now on we adopt this identification. (It is correct as far as two-point correlations are concerned.)

In this situation the flow of Brownian motion $(E_t)_{t \in \mathbb{R}^+}$ defined by

$$E_t(b_r - b_s) = b_{r+t} - b_{s+t} \quad \text{for all } r, s, t \in \mathbb{R}^+$$

is simply the right shift on $L^2(\mathbb{R}^+)$.

Furthermore, white noise, heuristically viewed as derivative of Brownian motion, should be given as

$db_t := \lim_{\varepsilon \downarrow 0} \frac{1}{\varepsilon} (b_{t+\varepsilon} - b_t)$, $t \in \mathbb{R}^+$,
which means something like the Dirac function δ_t at $t \in \mathbb{R}^+$. Of course, this limit does not exist in $L^2(\mathbb{R}^+)$. But formally it matches with an essential idea behind our construction scheme, namely the feature that the coupling is instantaneous: After an infinitesimal moment dt an initial vector

$$\begin{pmatrix} \xi \\ 0 \end{pmatrix} \in \begin{matrix} D(A) \\ \oplus_2 \\ L^2(\mathbb{R}^+, \mathcal{H}_+) \end{matrix} \quad \text{is transformed into} \quad \begin{pmatrix} -A\xi \; dt \\ db_t \otimes (-C_+\xi) \end{pmatrix}$$

where $db_t \otimes (-C_+\xi)$ describes a weighted multiplicity of white noises.

The occurence of this multiplicity is related to the famous rule (see e.g. [He]) that "each degree of freedom has its private (heat) bath". If one wants the dilation to be minimal one ought to provide a bath just for those degrees of freedom for which there is dissipation. In our setting this means that we have to set up white noise with multiplicity $\dim (D(A)/\ker C_+) = \dim \mathcal{H}_+$.
Correspondingly, for the forward time evolution of the reservoir we have to employ the flow of Brownian motion with multiplicity $\dim \mathcal{H}_+$, i.e. the restriction of $E_t \otimes 1$ to $L^2(\mathbb{R}^+, \mathcal{H}_+)$ which is precisely the right shift.

3.2 A Hilbert Space Version of the Langevin Equation

3.2.1 We now turn to the second complex of questions relating to the notion of singular coupling, i.e. the idea that the exchange between the original Hilbert space and the reservoir takes place during an infinitesimal time interval. One way of fixing this idea mathematically seems to be a differential equation.

Of particular interest is a description of the time evolution of a fixed vector $i(\xi)$, $\xi \in D(A)$, under the unitaries $\hat{T}_t$, $t \in \mathbb{R}^+$, i.e. one wants to describe the function

$$j(\xi) : \mathbb{R}^+ \longrightarrow \begin{matrix} \mathcal{H} \\ \oplus_2 \\ L^2(\mathbb{R}^+, \mathcal{H}_+) \end{matrix} , \quad t \longmapsto j_t(\xi) := \hat{T}_t \cdot i(\xi) = \begin{pmatrix} T_t\xi \\ Y_t\xi \end{pmatrix} .$$

Using the heuristic language developed in the preceding paragraph one may write down the following "stochastic differential equation"
$dj_t(\xi) = j_t(-A\xi)\, dt + db_t \otimes (-C_+\xi)$

which is a Hilbert space analogue of the famous Langevin equation in thermodynamics. This formal expression can be made precise in the following way:

3.2.2 Theorem [Kü2]. For $\xi \in D(A)$ the function $j(\xi)$ satisfies the following integral equation:

$$j_t(\xi) - j_s(\xi) = \int_s^t j_r(-A\xi)\,dr + (b_t - b_s)\otimes(-C_+\xi) \quad \text{for } s, t \in \mathbb{R}^+ .$$

Remarks. (1) We consider the function $(b_t - b_s)\otimes(-C_+\xi) \in L^2(\mathbb{R}^+, \mathcal{H}_+)$ canonically as an element of $0 \oplus L^2(\mathbb{R}^+, \mathcal{H}_+) \subseteq \mathcal{H} \oplus L^2(\mathbb{R}^+, \mathcal{H}_+)$.

(2) For the case that $\lim_{t\uparrow\infty} T_t = 0$ in the strong operator topology a result of this kind can already be found in [Ev].

For proving the Langevin equation we prepare a technical result.

Lemma. For all $\xi \in D(A^2)$ and $\eta \in D(A)$ we have

$$\int_s^t < C_+T_{t-r}A\xi,\ C_+\eta > dr = < C_+(1 - T_{t-s})\xi,\ C_+\eta > .$$

Proof: Recalling the definition of C_+ one computes

$$\int_s^t [< A\cdot T_{t-r}A\xi,\ \eta > + < T_{t-r}A\xi,\ A\eta >]dr$$
$$= < (1 - T_{t-s})\cdot A\xi,\ \eta > + < (1 - T_{t-s})\xi,\ A\eta >$$
$$= < C_+(1 - T_{t-s})\xi,\ C_+\eta > .$$

Proof of the Langevin equation:

The Langevin equation for $\xi \in D(A)$ and $s \leq t$ consists of two components:

$$T_t\xi - T_s\xi = \int_s^t T_r(-A\xi)\,dr \qquad \text{and}$$
$$Y_t\xi - Y_s\xi = \int_s^t Y_r(-A\xi)\,dr + \chi_{[s,t]}\otimes(-C_+\xi) .$$

While the first equation is obvious, we approach the second one by splitting the integral term as

$$\int_s^t Y_r(-A\xi)\ dr = \chi_{[0,s]}\cdot\int_s^t Y_r(-A\xi)\,dr + \chi_{[s,t]}\cdot\int_s^t Y_r(-A\xi)\,dr .$$

For dealing with the first summand we use the fact that the function

$$[s,t] \longrightarrow L^2(\mathbb{R}^+, \mathcal{H}_+)\ ,\ r \longmapsto \chi_{[0,s]}\cdot Y_r\xi$$

is differentiable and for all $r \in [s,t]$ we have

$$\frac{d}{dx}\,\chi_{[0,s]}\cdot Y_x\xi\ \Big|_r = \chi_{[0,s]}\cdot Y_r(-A\xi) .$$

Indeed, for $\varepsilon > 0$ we can estimate

$$\Big\| \frac{1}{\varepsilon}(\chi_{[0,s]}\cdot Y_{r+\varepsilon}\xi - \chi_{[0,s]}\cdot Y_r\xi) - \chi_{[0,s]}\cdot Y_r(-A\xi) \Big\|$$

$$\leq \| \chi_{[0,s]} \cdot Y_r \cdot (\tfrac{1}{\varepsilon}(T_\varepsilon - 1)\xi) - \chi_{[0,s]} \cdot Y_r(-A\xi) \|$$

$$+ \| \tfrac{1}{\varepsilon}\chi_{[0,s]} \cdot \chi_{[r,r+\varepsilon]} \cdot Y_{r+\varepsilon}\xi \|$$

$$\leq \| \chi_{[0,s]} \cdot Y_r \| \cdot \| \tfrac{1}{\varepsilon}(T_\varepsilon - 1)\xi + A\xi \| .$$

Hence the fundamental theorem of calculus yields

$$\chi_{[0,s]} \cdot \int_s^t Y_r(-A\xi)\,dr = \chi_{[0,s]} \cdot (Y_t\xi - Y_s\xi) = (\chi_{[0,s]} \cdot Y_t\xi) - Y_s\xi .$$

Thus it remains to verify

$$\chi_{[s,t]} \cdot Y_t\xi = \chi_{[s,t]} \cdot \int_s^t Y_r(-A\xi)\,dr + \chi_{[s,t]} \otimes (-C_+\xi) , \qquad \text{i.e.}$$

$$< \chi_{[s,t]} \cdot \int_s^t Y_r(-A\xi)\,dr, \ \chi_{[u,v]} \otimes C_+\eta > = < \chi_{[s,t]} \cdot (Y_t\xi + C_+\xi), \ \chi_{[u,v]} \otimes C_+\eta >$$

for all $u, v \in [s,t]$, $u \leq v$, and $\eta \in D(A)$.

For the moment let us suppose that $\xi \in D(A^2)$. Then we have

$$< \chi_{[s,t]} \cdot \int_s^t Y_r(-A\xi)\,dr, \ \chi_{[u,v]} \otimes C_+\eta >$$

$$= \int_s^t dr < Y_r(-A\xi), \ \chi_{[u,v]} \otimes C_+\eta >$$

$$= \int_u^t dr \int_u^{\min\{r,v\}} dx < C_+T_{r-x}A\xi, \ C_+\eta >$$

$$= \int_u^t dr < C_+(T_{r-\min\{r,v\}} - T_{r-u})\xi, \ C_+\eta > \quad \text{by the lemma above}$$

$$= \int_u^v < C_+(1 - T_{r-u})\xi, \ C_+\eta > dr + \int_v^t < C_+(T_{r-v} - T_{r-u})\xi, \ C_+\eta > dr$$

$$= |u-v| \cdot < C_+\xi, \ C_+\eta > - \int_u^t < C_+ \cdot T_{r-u}\xi, \ C_+\eta > dr$$

$$+ \int_v^t < C_+ \cdot T_{r-v}\xi, \ C_+\eta > dr$$

$$= |u-v| \cdot < C_+\xi, \ C_+\eta > + \int_u^t < C_+ \cdot T_{t-s}\xi, \ C_+\eta > ds$$

$$- \int_v^t < C_+ \cdot T_{t-s}\xi, \ C_+\eta > ds$$

$$= |u-v| \cdot < C_+\xi, \ C_+\eta > + \int_u^v < C_+ \cdot T_{t-s}\xi, \ C_+\eta > ds$$

$$= < \chi_{[s,t]} \cdot (Y_t\xi + C_+\xi), \ \chi_{[u,v]} \otimes C_+\eta > .$$

Finally, it is not difficult to extend the validity of the equation

$$\chi_{[s,t]} \cdot Y_t\xi = \chi_{[s,t]} \cdot \int_s^t Y_r(-A\xi)\,dr + \chi_{[s,t]} \otimes (-C_+\xi)$$

from $\xi \in D(A^2)$ to $\xi \in D(A)$: Given $\xi \in D(A)$ choose a sequence

$(\xi_n)_{n\in\mathbb{N}}$ in $D(A^2)$ such that $\lim_n \xi_n = \xi$ and $\lim_n A\xi_n = A\xi$.
Then one concludes $\lim_n C_+\xi_n = C_+\xi$ so that both sides of the equation
$\chi_{[s,t]}\cdot Y_t\xi_n = \chi_{[s,t]}\cdot\int_s^t Y_r(-A\xi_n)dr + \chi_{[s,t]}\otimes(-C_+\xi_n)$
converge to the corresponding expression with $\xi \in D(A)$.
This finishes the proof of the Langevin equation.

Dilations as Singular Coupling Limits

There is another way of interpreting a dilation as coupling to white noise. This approach relies on a construction usually called "singular coupling limit" (see [Fo], [Go], [He]). In this paragraph we show - at least for self-adjoint semigroups - that a unitary dilation can indeed be obtained by a singular coupling limit from physically meaningful interactions of the irreversible system with a heat bath whose evolution is governed by white noise.

.1 In order to simplify our considerations we assume that from now on $T_t = \exp(-At)$, $t \in \mathbb{R}^+$, is a self-adjoint semigroup, i.e. A is a positive self-adjoint operator. It is worthwhile to allow A to be unbounded since this reveals some phenomena which cannot be seen in the bounded case.
Without restriction we assume that A has no kernel, i.e. T_t is not the identity on any invariant subspace of $\mathcal{H}$. In this situation the unitary dilation can be written in a more transparent form than in 2.2 (see also [Kü1], [Kü2]):
The Hilbert space of the dilation is $\hat{\mathcal{H}} := \mathcal{H} \oplus L^2(\mathbb{R}, \mathcal{H})$.
It will often be convenient to decompose a vector $\phi \in L^2(\mathbb{R}, \mathcal{H})$ as $\phi = \phi_- + \phi_+$ with $\phi_- \in L^2(\mathbb{R}^-, \mathcal{H})$, $\phi_+ \in L^2(\mathbb{R}^+, \mathcal{H})$.
An element $\hat{\xi} \in \hat{\mathcal{H}}$ will be written as a column vector

$$\hat{\xi} = \begin{pmatrix} \xi \\ \phi \end{pmatrix} = \begin{pmatrix} \xi \\ \phi_- + \phi_+ \end{pmatrix} \quad \text{with } \xi \in \mathcal{H} , \ \phi \in L^2(\mathbb{R}, \mathcal{H}) .$$

Then $\hat{T}_t$ can be written as a 2×2-matrix

$$\hat{T}_t = \begin{pmatrix} T_t & X_t \\ Y_t & S_t + Z_t \end{pmatrix} \qquad \text{for } t \in \mathbb{R}^+$$

where $S_t : L^2(\mathbb{R}, \mathcal{H}) \longrightarrow L^2(\mathbb{R}, \mathcal{H})$ is the right shift and the other components are given by the following densely defined operators:

$X_t: L^2(\mathbb{R}, D(A^{1/2})) \longrightarrow \mathcal{H}$, $X_t\phi = (2A)^{1/2} \int_0^t T_r\phi(r-t)dr$;

$Y_t: D(A^{1/2}) \longrightarrow L^2(\mathbb{R}, \mathcal{H})$, $(Y_t\xi)(s) = \chi_{[0,t]}(s)\cdot(-(2A)^{1/2})\cdot T_{t-s}\xi$;

$Z_t: L^2(\mathbb{R}, D(A)) \longrightarrow L^2(\mathbb{R}, \mathcal{H})$, $(Z_t\phi)(s) = \chi_{[0,t]}(s)\cdot(-(2A)^{1/2})\cdot X_{t-s}\phi$.

Here we have written $L^2(\mathbb{R}, \mathcal{N})$ for the $\mathcal{N}$-valued functions in $L^2(\mathbb{R}, \mathcal{H})$ if $\mathcal{N}$ is any subspace of $\mathcal{H}$. For a physical interpretation of the operators X_t, Y_t, Z_t we refer to [Kül] . There it is also shown that these densely defined operators are contractions and have continuous extensions (again denoted by the same symbols) to operators

$X_t: L^2(\mathbb{R}, \mathcal{H}) \longrightarrow \mathcal{H}$; $\qquad Y_t: \mathcal{H} \longrightarrow L^2(\mathbb{R}, \mathcal{H})$;

$Z_t: L^2(\mathbb{R}, \mathcal{H}) \longrightarrow L^2(\mathbb{R}, \mathcal{H})$.

4.2 For describing the dilation as a coupling of the small system to a reservoir $L^2(\mathbb{R}, \mathcal{H})$ let us consider the Hamiltonian generator H of $\hat{T}_t = \exp(iHt)$, $t \in \mathbb{R}$. An interpretation as a coupling should correspond to a decomposition of H as

$H = H_S + H_R + H_C$

where H_S is the Hamiltonian of the free evolution of the small system, H_R is the Hamiltonian of the reservoir and H_C describes the coupling between both systems.

Since we chose the time evolution on $\mathcal{H}$ to be self-adjoint, it does not contain a unitary part and we expect H_S to be zero. This means that there are no exterior fields acting on $\mathcal{H}$ and the whole development is exclusively due to the energy transfer between the system and its heat bath.

The time evolution of the heat bath is assumed to be the flow of Brownian motion, i.e. on $L^2(\mathbb{R}, \mathcal{H}) = L^2(\mathbb{R}) \otimes \mathcal{H}$ we expect

$H_R = i\cdot\frac{d}{dx} \otimes 1$ (cf. 3.1).

Finally, the coupling should heuristically be given by an Hamiltonian of the form

$$H_C = \begin{pmatrix} 0 & -iC^* \\ iC & 0 \end{pmatrix} \qquad \text{with}$$

$C: D(A^{1/2}) \ni \xi \longmapsto \delta_0 \otimes (2A)^{1/2}\xi$ and $C^*: L^2(\mathbb{R}, \mathcal{H}) \ni \phi \longmapsto (2A)^{1/2}\phi(0)$.

While this expression for H is only formal it describes exactly what we had in mind: The dissipative system $\mathcal{H}$ is coupled to the Hilbert space $\mathcal{H}$ sitting at point zero of $L^2(\mathbb{R}, \mathcal{H})$ via the coupling operator $(2A)^{1/2}$. Comparing this with our considerations in 3.1 we may say that the dissipative system is coupled via $(2A)^{1/2}$ to the zero component of white noise. This also supports the terminology in [Kül]

where we introduced the term "coupling operator" for $(2A)^{1/2}$.
Rigorously, H is given as follows:

Theorem ([Kü 2]). The domain $D(H)$ of H consists of those vectors $\begin{pmatrix} \xi \\ \phi_- + \phi_+ \end{pmatrix} \in \hat{\mathcal{H}}$ which satisfy the following five conditions:

$\xi \in D(A^{1/2})$, $\phi_- \in W(\mathbb{R}^-, \mathcal{H})$, $\phi_+ \in W(\mathbb{R}^+, \mathcal{H})$,

$\phi_-(0) + \phi_+(0) \in D(A^{1/2})$, $\phi_+(0) = \phi_-(0) - (2A)^{1/2}\xi$.

Furthermore, for such a vector we have

$$H \begin{pmatrix} \xi \\ \phi_- + \phi_+ \end{pmatrix} = i \cdot \begin{pmatrix} -(2A)^{1/2}\, 1/2 \cdot (\phi_-(0) + \phi_+(0)) \\ \phi_-' + \phi_+' \end{pmatrix}.$$

1.3 In this final section we shall make precise the idea of the singular coupling limit: Guided by our formal expression for H we shall approximate H by well defined Hamiltonians H_n describing a coupling betweeen $\mathcal{H}$ and $L^2(\mathbb{R}, \mathcal{H})$ in the usual manner.
Our starting point is an approximation of the Dirac function.

Definition. For each $n \in \mathbb{N}$ let $f_n : \mathbb{R} \longrightarrow \mathbb{R}$ be a positive function such that

(i) $\int_{\mathbb{R}} f_n(s)ds = 1$,
(ii) $f_n \in L^2(\mathbb{R})$,
(iii) $f_n(x) = f_n(-x)$ for all $x \in \mathbb{R}$,
(iv) $\lim_n F_n \cdot \chi_{(-\infty,0]} = 0$ in $L^2(\mathbb{R})$ for $F_n(x) := \int_{-\infty}^{x} f_n(s)ds$.

Then we call the family $(f_n)_{n\in\mathbb{N}}$ a <u>symmetric approximation of the δ-function</u> (in $L^2(\mathbb{R})$).

Our terminology may be justified by the following result:

Lemma. If $\phi = \phi_- + \phi_+ \in W(\mathbb{R}^-, \mathcal{H}) \oplus W(\mathbb{R}^+, \mathcal{H})$ and $(f_n)_{n\in\mathbb{N}}$ is a symmetric approximation of the δ-function then
$\lim_n \int_{\mathbb{R}} f_n(s) \cdot \phi(s)\, ds = (\phi_-(0) + \phi_+(0))/2$.

Remark. There are many possibilities of defining an "approximation of the δ-function". The definition above, however, seems most suitable for our purposes and covers all relevant examples.

Frequently used examples of approximations of the δ-function are the following:

$f_n(s) = n \cdot \chi_{[-1/2n,1/2n]}(s)$, $f_n(s) = (n/2\pi)^{1/2} \cdot \exp(-ns^2/2)$,

or the normalized Fourier transform of the Lorentz kernel

$s \longmapsto n \cdot (n^2 + s^2)^{-1/2}$, $s \in \mathbb{R}$, $n \in \mathbb{N}$.

Next we define an approximation of the coupling.

For the following let $(f_n)_{n \in \mathbb{N}}$ be a fixed approximation of the δ-function. For each $n \in \mathbb{N}$ we introduce a densely defined linear operator

$C_n: D(A^{1/2}) \longrightarrow L^2(\mathbb{R}, \mathcal{H})$, $\eta \longmapsto f_n \otimes (2A)^{1/2}\eta$.

Clearly, C_n is a closed operator and thus has a densely defined adjoint $C_n{}^*$ which is given by

$D(C_n{}^*) := \{ \zeta \in L^2(\mathbb{R}, \mathcal{H}) : \int_{\mathbb{R}} f_n(s)\zeta(s)\, ds \in D(A^{1/2}) \}$ and

$C_n{}^*\zeta := (2A)^{1/2}\int_{\mathbb{R}} f_n(s)\zeta(s)\, ds$ for $\zeta \in D(C_n{}^*)$.

After these preparations we are in the position to define a sequence of approximating Hamiltonians: For every $n \in \mathbb{N}$ put

$$D(H_n) := \{ \begin{pmatrix} \eta \\ \zeta \end{pmatrix} \in \hat{\mathcal{H}} : \eta \in D(A^{1/2}), \ \zeta \in W(\mathbb{R}, \mathcal{H}) \cap D(C_n{}^*) \} \quad \text{and}$$

$$H_n := i \begin{pmatrix} 0 & -C_n{}^* \\ C_n & \frac{d}{dx} \otimes 1 \end{pmatrix} .$$

<u>Proposition</u> ([Kü 2]). The operators H_n , $n \in \mathbb{N}$, are self-adjoint.

Finally we are ready for the main result of this section.

<u>Theorem</u> ([Kü 2]). The Hamiltonians $(H_n)_{n \in \mathbb{N}}$ converge to H in the strong graph limit, i.e. the generated unitary groups converge uniformly on compact time intervals.

4.3 In conclusion we have fully verified our intuition of the new dilation scheme. The Hamiltonians H_n describe a physically interpretable interaction of $\mathcal{H}$ with $L^2(\mathbb{R}, \mathcal{H})$ localized around the time $0 \in \mathbb{R}$. As $n \in \mathbb{N}$ tends to infinity the interactions concentrate on a decreasing time interval converging to the point $0 \in \mathbb{R}$. Now the above theorem shows that H is indeed the limit of the $(H_n)_{n \in \mathbb{N}}$ in any reasonable sense and thus justifies the proposed interpretation of H . In the literature, this limit is usually called "singular coupling limit" (cf. [Fo], [Go], [He]) .

References

[Da] Davies, E.B.: One-parameter semigroups. Academic Press, London - New York - San Francisco 1980.

[Ev] Evans, D.E.; Lewis, J.T.: Dilations of irreversible evolutions in algebraic quantum theory. Commun. DIAS, Series A, 24 (1977).

[Fo] Ford, G.W.; Kac, M.; Mazur, P.: Statistical mechanics of assemblies of coupled oscillators, J. Math. Phys. 6 (1965), 504-515.

[Go] Gorini, V.; Frigerio, A.; Verri, M.; Kossakowski, A.; Sudarshan, E.C.G.: Properities of quantum Markovian master equations. Rep. Math. Phys. 13 (1978), 149-173.

[He] Hepp, K.; Lieb, E.H.: Phase transition in reservoir driven open systems with applications to lasers and superconductors. Helv. Phys. Acta 46 (1973), 573-603.

[Kü1] Kümmerer, B.; Schröder, W.: On the structure of unitary dilations. Semesterbericht Funktionalanalysis. Tübingen, Wintersemester 1983/84, 177-225.

[Kü2] Kümmerer, B.; Schröder, W.: Unitary dilations as singular couplings to white noise. Semesterbericht Funktionalanalysis. Tübingen, Sommersemester 1984, 137-181.

[La] Lax, P.D.; Phillips, R.S.: Scattering theory. Academic Press, New York - San Francisco - London 1967.

[Sz] Sz.-Nagy, B.; Foias, C.: Harmonic analysis of operators on Hilbert space. North Holland P.C. 1970.

A NEW APPROACH TO QUANTUM ERGODICITY AND CHAOS

Göran Lindblad

Department of Theoretical Physics
Royal Institute of Technology
S-100 44 Stockholm, Sweden

Abstract. The random properties of the Hamiltonian dynamics of a finite quantum system are described in terms of the set of quantum correlation functions defined by sequences of incomplete observations of the system. The amount of information (hence the degree of randomness) is measured by entropy quantities. Ergodicity is the property of the system of giving the maximal amount of information allowed by the finite size of it, while chaos is characterized by a high, but transient, rate of information gain. A chaotic time scale is found which increases at most linearly with the size of the system and which restricts quantum chaos to be a transient phenomenon.

1. Introduction

This talk deals with the concepts of 'randomness' and 'chaos' in the context of the Hamiltonian dynamics of finite quantum systems. The expression 'quantum chaos' has been used rather extensively in the last few years without a proper definition. Here is outlined an attempt to introduce a definition inspired by classical ergodic theory. A more detailed treatment is given in ref. 1.

Developments in experimental physics during the last decade, especially in the field called 'laser chemistry' [2,3], seem to demand the introduction of a theoretical concept like quantum chaos. A typical example is provided by the vibrational relaxation after laser excitation of a relatively small molecule like SF_6 or C_6H_6. The molecule can be considered to be a finite, closed quantum system with a line spectrum (below the dissociation limit) under the relevant experimental conditions. The initial state is created from a very 'cold' molecule through the selective excitation of one of the vibrational modes by a pulse of resonant IR laser light. The experimentalist claim to see, in general, a quasiperiodic behaviour for low levels of excitation, but an apparently aperiodic motion for higher energies which effectively leads to the equipartition of the excitation energy among the normal modes [4]. It is believed that no interaction with the world outside the molecule is necessary in order to explain the relaxation, it is a purely intramolecular phenomenon.

. Classical chaos

he theoretical interest in this type of physics lies largely in the behaviour on the orderline between periodic and aperiodic motion, with a possible transition between hem as some parameter is varied. Such properties are seen frequently in the dynamics f classical non-linear systems, especially in numerical simulations of simple models, ike the Hénon-Heiles model. It is thus possible to construct a classical model of he vibrating molecule as a set of harmonic oscillators (normal modes) coupled by anarmonic forces, and this model may show a transition of the type described, although ot a very sharp one.

The KAM theory gives an analytic treatment of the limit of weak coupling or low nergy, where the 'constants of motion' associated with the normal modes are still onserved to a large extent [5,6]. The resulting 'regular' motion shows at most a linar divergence of the trajectories [7]. The destruction of the constants of motion as he strength of the coupling or the energy increases leads to exponentially diverging rajectories and a sensitive dependence on the initial conditions. This 'hyperbolic' roperty is often described by introducing the Liapounov exponents, which then take n non-zero values. The motion of a phase space point becomes essentially unpredictable over long time intervals. The global behaviour is highly complex due to the presnce of both types of motion in the same energy hypersurface, being interleaved in an nseparable Cantor-set-like structure.

In classical ergodic theory, which deals with a larger class of measure-preservng transformations T of a measure space (Ω,μ), the notion of randomness or chaos is onveniently described in terms of the Kolmogorov-Sinai entropy [5]. This quantity easures the unpredictability of the dynamics in terms of the additional information eeded to predict the evolution for one more unit of time, or, equivalently, the additional information obtained through the observation of this evolution. For some ice systems there is a relation (Pesin's formula) expressing the KS entropy as a hase space average of the sum of the positive Liapounov exponents, thus providing a ink between the geometric and the probabilistic aspects of the dynamics [8].

Considering simple examples, like the Bernoulli shifts, one can realize that he randomness has a dual origin in this picture [9]

) Each observation of the system (specified by a generating partition) is incomplete, nd it takes an infinite amount of information to specify one phase space point.

) The dynamics leads to a generation of new information in each repetition of the ame observation. The asymptotic rate of information gain is unchanged by a choice of finer partition. Thus the KS entropy $h(T,\mu)$ is defined by the dynamics T and the nvariant measure μ only, it is an intrinsic' property of the system.

The beautiful simplicity of the picture given by ergodic theory hides some funamental problems which turn up in the physical interpretation of any particular odel.

a) There is no physical scale of the size of the system. An infinite amount of information demands an infinite thermodynamic entropy, and without this there can be no asymptotic information rate $h(T,\mu) > 0$,
b) Due to the use of asymptotic quantities there is no intrinsic time scale. As a result a change in time scale gives a trivial rescaling of the KS entropy for a flow T_t

$$h(T_t,\mu) = |t|\cdot h(T_1,\mu).$$

3. Problems of quantum chaos

The two properties a) and b) above are in contradiction with a quantum description of a finite system. Only a finite information is needed to specify a pure quantum state as the total equilibrium entropy is finite if the energy is. Thus there can be no asymptotic rate of information gain coming solely from the initial state of the system itself (similar conclusions have been drawn by several authors using various arguments). We will be dealing with transient phenomena and the time scale is an essential ingredient. In quantum theory there is a measure of size (entropy) and a time scale given by $\hbar$ and the energy scale. Classical ergodic theory is based on the neglect of any effect of the finite value of $\hbar$.

In the quantum dynamics of a finite system there are in fact several time scales which depend on the size of the system in different ways. The shortest time scale (the dephasing time) is given by the total energy spread in the relevant set of states

$$\tau_D = \hbar/\Delta E$$

One can take τ_D, which depends rather weakly on system size, as a unit of time. There is a time scale associated with the average level spacing in the discrete spectrum

$$\tau_M = \hbar/\delta E$$

This will increase exponentially with system size if there are no large scale degeneracies. For $t > \tau_M$ the discrete nature of the spectrum will become evident and there is no further systematic decay of the correlation functions. This implies a deviation from the behaviour of a mixing classical system [5]. The discreteness of the spectrum is also reflected in the almost periodic property of the dynamics and the existence of recurrence periods τ_P. However, τ_P increases much faster than exponentially with the size of the system and it is larger than the age of the universe even for small molecules [10].

There is a time scale shorter than τ_M which is more relevant for the chaotic property of the dynamics. If the total entropy of the system is S (taken to be dimensionless by dividing with Boltzmann's constant) and the information gain per unit time is h, then there is a time scale

$$\tau_C = S/h$$

which grows at most linearly with system size. τ_C is an upper bound on the time scale over which a transient information rate h is possible for a finite quantum system. A more rigorous form of this argument will be sketched below.

The quantum measurement process introduces additional problems. There are in principle complete measurements which tell us all about the system without considering the dynamics at all. In order to have a rate of information gain coming from the dynamics, it is necessary to let incomplete quantum measurements correspond to the classical partitions. Furthermore, the quantum measurement process introduces an unpredictability coming from the interaction between system and apparatus and which exists even when the observed system has a trivial dynamics. It is thus necessary to separate this contribution from that which is due to the dynamics.

The following conclusions are drawn from the arguments given above:

) Quantum chaos must be a transient phenomenon, and it takes place on a time scale which increases at most linearly with the size of the system. This seems to be confirmed in some work on the 'kicked quantum rotator' [11].

) It is necessary to specify which part of the system we consider to be observable. This subsystem (called S below) corresponds to the generating partition of classical ergodic theory. In the context of laser chemistry we can identify S with the 'active mode' which couples to the laser field (called M), while the rest of the vibrational degrees of freedom of the molecule act as a reservoir (R).

. Mathematical formalism

In order to model the physical picture described above, the following mathematical setup is used. The Hilbert space of the system $S+R$ (which can be assumed to be finite-dimensional for simplicity) is a tensor product $H_{S+R} = H_S \otimes H_R$, and so is the operator algebra $A_{S+R} = B(H_{S+R}) = A_S \otimes A_R$, where we can identify $A_S = A_S \otimes 1$, etc. The dynamics of $S+R$ is given by a group of unitaries defined by the Hamiltonian

$$H_{S+R} = H_S \otimes 1 + 1 \otimes H_R + H_I$$

$$U(t) = \exp(-itH_{S+R}/\hbar),$$

or, expressed as automorphisms of the operator algebra

$$X \to T(t)[X] = U(t)^+ X\, U(t).$$

We also introduce the space of states represented by density operators: $\rho(X) = \mathrm{Tr}(\rho X)$, $X \in A$, $E(A) = \{\rho;\ \rho \geq 0,\ \mathrm{Tr}\,\rho = 1\}$, and the tracial state of maximal entropy $\tau = (\dim H)^{-1}\cdot 1$. The partial states are defined in the standard way

$$\rho_S = \mathrm{Tr}_R \rho_{S+R} \qquad\qquad \rho_R = \mathrm{Tr}_S \rho_{S+R}\ .$$

To each map of the operator algebra into itself the dual map of the states is denoted by a $*$: $T(t)^*[\rho] = \rho\cdot T(t)$.

We assume that only the 'observables' in A_S are actually observable, as the apparatus (or laser field) M couples only to this subsystem. This picture makes it necessary to discuss the possible distinction between a truly chaotic system and a regular system weakly perturbed by a heat bath (say, Brownian motion or spontaneous radiation). For a classical dynamical system this difference seems clearcut, but it is not so in the quantum case. When the S-R interaction is weak compared to the level spacings in H_S, then we have the latter situation. Every line in the discrete spectrum of H_S acquires a linewidth due to the interaction with R, but the spectrum is still recognizable. On the other hand, when the interaction is of the same order of magnitude as the rest of the Hamiltonian then the dynamics does not distinguish the subsystem S in any way except for the coupling to M. No proper spectrum for S is likely to be seen, only the broad and dense spectrum of the total system S+R. It is only in this case we expect to have a quantum counterpart of classical chaos, but the dividing line must necessarily be fluid.

Consider observations of the system S or more general instantaneous interactions of S with M. They are modelled by operations generated by the simple type of completely positive (CP) maps

$$X \to V^+ X V , \qquad V \in A_S , \qquad ||V|| \leq 1, \qquad X \in A_{S+R} ,$$

or the dual action on the states $\rho \to V \rho V^+$. To each such map and given initial state ρ is associated a probability $\rho(V^+V)$. Successive operations and the dynamics are (by definition) composed in a time-ordered semigroup fashion. If we discretize the time parameter by introducing the time scale τ_D and put $U = U(\tau_D)$, then a sequence of operations on S combined with the dynamics of S+R give rise to a set of quantum correlation functions (QCFs), one for each $n = 1,2,..$

$$R(\underline{X}_n^+,\underline{X}_n) = \rho(X_1^+U^+X_2^+ \ldots U^+X_n^+X_nU \ldots X_2U\, X_1)$$

$$\rho \in E(A_{S+R}), \qquad \underline{X}_n = (X_n, X_{n-1}, \ldots, X_1), \qquad X_i \in A_S .$$

These QCFs define all the operationally accessible probabilities associated with the choice of S as the observed subsystem, the time scale and the initial state of S+R. In the following it is assumed that ρ is stationary, i.e. $U \rho U^+ = \rho$, and that it is faithful: $\rho(X^+X) = 0 \Rightarrow X = 0$.

The set of QCFs have the following more or less evident properties:

(1) *Positivity*: $R(\underline{X}_n^+,\underline{Y}_n)$, which is defined by polarization, is positive semidefinite in the sense that for all $\underline{X}_n(i) \in \overset{n}{\times} A_S$, $\lambda_i \in \mathbb{C}$, all n

$$\sum_{i,j} \bar{\lambda}_i \lambda_j \, R(\underline{X}_n(i)^+, \underline{X}_n(j)) \geq 0 .$$

(2) *Compatibility*: For $\underline{Y}_n = (1, X_{n-1}, \ldots, X_1)$

$$R(\underline{Y}_n^+,\underline{Y}_n) = R(\underline{X}_{n-1}^+,\underline{X}_{n-1}).$$

When ρ is stationary and $\underline{Y}_n = (X_{n-1},\ldots,X_1,1)$ the same relation holds.

(3) *Time homogeneity*: When ρ is stationary then the QCFs are invariant under the simultaneous translation of all time parameters.

In view of the Kolmogorov construction of classical probability theory, it seems natural to ask if the set of time-ordered QCFs suffice to define the system S+R completely. In order for this to be true an ergodicity property must be assumed.

5. Quantum ergodicity

Given A_S , A_R and U (or H_{S+R} in the continuous time case), we can introduce a property which replaces the condition that a classical partition shall be generating

$$\{A_S \cup U\}' = \mathbb{C}\,1 \qquad \text{(QE)}$$

or, in the continuous case, $\{A_S \cup H_{S+R}\}' = \mathbb{C}\,1$ (' denotes the commutant). The QE condition is equivalent to

$$\{T^n[X],\ X \in A_S\ ,\ n \in Z\}'' = A_{S+R}\ ,$$

which means that A_{S+R} is spanned by products of time translated elements in A_S .

The almost periodic property of the dynamics can be written in the following way for finite dimensional Hilbert spaces and a discretized time parameter. For every $\delta > 0$ there is an infinite sequence of integers $m \to \infty$ such that

$$\|T^m - I\| < \delta\ .$$

In conjunction with the QE property this means that the time ordered products

$$T^n[X_n]T^{n-1}[X_{n-1}] \ldots T[X_1]X_0, \quad \text{all } X_i \in A_S\ ,\ \text{all } n$$

suffice to span A_{S+R} . As a consequence the following result holds:

Reconstruction theorem: From a compatible set of time-ordered stationary QCFs can be constructed a quantum dynamical system with the given QCFs. If the QE condition holds then this system is unitarily equivalent with the original one.

For a proof see ref. 1. The restriction to finite dimensions is not essential. Note that the similar reconstruction theorem of Accardi, Frigerio and Lewis [12] uses a larger set of not necessarily time-ordered QCFs. The proof of the theorem involves a standard GNS construction where one obtains a Hilbert space $\hat{H}$, a unitary *-representation $\hat{\pi}$ of A_S in $\hat{H}$, a unitary operator $\hat{U}$ in $\hat{H}$ and a $\hat{U}$-invariant vector $\hat{\Omega} \in \hat{H}$. In order to define the unitary equivalence with the original system we note that if we choose a CON basis $\{|k\rangle\}$ in $H = H_{S+R}$ where ρ is diagonal $\rho = \sum_k p_k\ |k\rangle\langle k|$, then we can define a pure state $\omega = |\Omega\rangle\langle\Omega|$ in the following way:

$$|\Omega\rangle = \sum_k p_k^{\frac{1}{2}}\ |k\rangle \otimes |k\rangle \in H \otimes H$$

$$\omega(X \otimes 1) = \omega(1 \otimes X) = \rho(X), \quad \text{all } X \in A_{S+R}\ .$$

The unitary equivalence now identifies

$$\hat{H} \simeq H \otimes H\ , \qquad \hat{\pi}(X) \simeq X \otimes 1,\ \ X \in A_S$$

$$\hat{\Omega} \simeq \Omega, \qquad \hat{U} \simeq U \otimes \tilde{U}^+,$$

where ~ denotes the matrix transposition in the chosen basis.

The fact that the statistics of all sequences of measurements on S allows us to reconstruct $S+R$ indicates that there is a maximal randomness in these sequences as a whole, as it is not possible to have more information about the system. If the QE property does not hold, then the selfadjoint elements of

$$C \equiv \{A_S \cup U\}' = A_R \cap \{U\}'$$

can be interpreted as 'constants of motion'. It is obvious that the QCFs do not give any information on the relative phases between different eigenspaces of any of these operators. The relation to the classical concept of a constant of motion is not clear except for their property of reducing the randomness of the dynamics.

A relation between a slightly stronger form of QE and a more familiar physical property is the following non-crossing rule [1]. Let $H_1 = H_1^+ \in A_S$ be such that $\{H_1 \cup H_{S+R}\}' = \mathbb{C}\,1$ (which means that QE holds), then the family of Hamiltonians

$$H(\lambda) = H_{S+R} + \lambda H_1$$

will generally have avoided crossings, i.e. the energy levels as functions of λ do not cross. $H(\lambda)$ may represent the coupling of a variable external field to S. Such avoided crossings seem to be generic in the quantization of classically ergodic systems [13].

7. Subdynamics

The QCFs always exist but they do not automatically define a dynamics of the subsystem S. In order to define a measure of randomness in the present scheme it seems to be necessary to have a subdynamics for S in the following very weak sense (also called a 'quantum stochastic process' = QSP [14]): Let there exist a set of sesquilinear functions $\hat{R}_n$ with values in A_S

$$\hat{R}_n(\underline{X}_n^+, \underline{X}_n) \in A_S\,, \qquad \underline{X}_n \in \overset{n}{\times} A_S\,, \qquad n = 1,2,..$$

such that the set of QCFs are given by

$$R(\underline{X}_{n+1}^+, \underline{X}_{n+1}) = \rho_S(X_0^+ \hat{R}_n(\underline{X}_n^+, \underline{X}_n) X_0)$$

for all $\underline{X}_n = (X_n, \ldots, X_1)$, $\underline{X}_{n+1} = (X_n, \ldots, X_1, X_0)$, all n. It is not difficult to check that when the QE property holds then the set of QCFs is of this form if and only if the invariant state is of the product form $\rho_{S+R} = \rho_S \otimes \rho_R$ and then

$$\hat{R}_n(\underline{X}_n^+, \underline{X}_n) = Tr_R\{\rho_R U^+ X_1^+ U^+ \ldots U^+ X_n^+ X_n U \ldots U\, X_1 U\}\ .$$

In finite dimension it is convenient to choose for ρ the tracial state $\bar{\rho}$ which is always of product form. Trouble comes if we want to describe thermal equilibrium QCFs at a finite temperature in this way, as the Gibbs states do not factorize for an interacting system. Thus the QCFs derived from a genuinely non-unitary subdynamics of S will not satisfy the KMS condition for finite temperature.

Given that the invariant state has a tensor product form the resulting QSP can be written in a different but equivalent fashion [1,14]. To $\hat{R}_n$ is associated a CP map

$$T_n \in CP(\overset{n}{\otimes} A_S), \qquad n = 1,2,..$$

To every set $\{T_n\}$ of such CP maps satisfying certain compatibility conditions we can find a set $\{\hat{R}_n\}$ and hence a complete set of QCFs. The correspondence between $\{T_n\}$ and the QCFs is an affine bijection of the natural convex structures. The description $\{T_n\}$ has some very nice properties. The subdynamics is Markovian in the sense that the 'quantum regression theorem'

$$\hat{R}_n(\underline{X}_n^+,\underline{X}_n) = T[X_1^+T[X_2^+\ldots T[X_n^+X_n]\ldots]X_1]$$

holds for all n, some $T \in CP(A_S)$, if and only if $T_n = \overset{n}{\otimes} T$. The subdynamics is unitary (no S-R interaction) if and only if all the T_n are unitary (in which case the Markov condition holds). The Markov condition holds, by assumption, for the action of the measuring instrument on S. Let this action be described by elementary operations

$$E_k[X] = V_k^+ X V_k, \qquad V_k \, , X \in A_S$$

at time $t_k = k\,\tau_D$. The total action of M in n consecutive instants $\{t_k\}_1^n$ is given by the map $\overset{n}{\otimes} E_k \in CP(\overset{n}{\otimes} A_S)$, and it turns out that the combined effect of the subdynamics due to the interaction with R and the action of M in this time interval is described by the composed CP map (see ref. 1)

$$T_n \cdot (\overset{n}{\otimes} E_k) \in CP(\overset{n}{\otimes} A_S).$$

In this way the subdynamics, as defined from the total set of QCFs, is separated from the quantum measurement process itself, and the two parts can be recombined by a composition of CP maps. This construction makes it possible to associate to the QCFs a notion of randomness which is an effect of the subdynamics alone, not of the measurement process. In order to do this the QSP is mapped in an affine and bijective way into a state of a quantum lattice system.

First note that the trick described in connection with the reconstruction theorem can be repeated here for each n. From the faithful state ρ_S we can construct a pure state $\omega \in E(A_S \otimes A_S)$ and hence a 'purification' of $\rho(n) \equiv \overset{n}{\otimes} \rho_S$

$$\omega(n) = \overset{n}{\otimes} \omega \in E(\overset{2n}{\otimes} A_S).$$

Then T_n and $\omega(n)$ define the state

$$\sigma(n) = (T_n^* \otimes I)[\omega(n)] \in E(\overset{2n}{\otimes} A_S).$$

From the compatibility and stationarity properties of the QCFs follows that $\{\sigma(n)\}_1^\infty$ define a translation invariant state of a 1D quantum lattice system where each lattice point corresponds to a unit time interval (t_k,t_{k+1}) and carries the Hilbert space $H_S \otimes H_S$. This construction has the following properties [1]:

(1) A translation invariant state of the lattice system which satisfies a couple of

subsidiary conditions conversely defines a QSP and hence a full set of QCFs. The correspondence between the QCFs and lattice states is an affine bijection of the convex structures.

(2) The QSP is Markovian if and only if the lattice state is of product form $\sigma(n) = \overset{n}{\otimes}\sigma(1)$, all n. The QSP is similar to a Bernoulli shift, in that the outcomes of successive observations are statistically independent, if in addition $\sigma(1) = \rho_S \otimes \rho_S$.

(3) The QSP is unitary if and only if all the $\sigma(n)$ are pure (vector) states, in which case they are of the Markovian product form.

This construction has a lot of similarity with the classical case where a stochastic process is a 1D random field, but there is a significant difference, and the classical case is not a special case of the quantum construction.

8. Quantum chaos

In order to have a measure of the randomness of the QSP we use an idea borrowed from the corresponding classical problem. There the KS entropy of a stationary shift is equal to the specific entropy of the associated random field in the thermodynamic limit. For a quantum state the entropy is defined in the standard way (dimensionless)

$$S(\rho) = -\,\mathrm{Tr}(\rho \ln \rho)\ .$$

The entropy of the n-point lattice state is

$$S(n) = S(\sigma(n)), \qquad S(0) = 0.$$

I claim that S(n) is a suitable measure of the randomness or unpredictability associated with the QCF of order n+1, but this statement can only be justified through applications to particular examples [1]. In the unitary case S(n) = 0 for all n. This means that {S(n)} measures only the unpredictability due to the *S*-*R* interaction, while that due to the quantum measurement process is left out.

From the strong subadditivity property of the quantum entropy and the translation invariance of the state we obtain [15]

$$h(n) \equiv S(n) - S(n-1) \geq 0,$$

$$h(n+1) \leq h(n).$$

Consequently the following limits both exist, though the first may be $+\infty$, the second may be 0:

$$S(\infty) = \lim_{n\to\infty} S(n),$$

$$h(\infty) = \lim_{n\to\infty} h(n).$$

For a finite system (with finite energy and hence finite entropy) $S(\infty)$ must be finite, in fact the following bound holds [1]

$$S(\infty) \leq S_{max} \equiv 2\,S(\rho_R).$$

ote that when the state $\bar{\rho}_R$ is used, then the dimension of H_R must be chosen such hat S_{max} has the correct physical value. It is shown in ref. 1 that for the choice $_R$ the QE property holds if and only if

$$S(\infty) = S_{max}$$

nd this is true also for a general ρ_R, at least in finite dimension. This fact indi-ates that $S(\infty)$ measures the total information content in the set of QCFs.

When $S(\infty) < \infty$ then $h(\infty) = 0$. Now $h(\infty)$ is the quantity corresponding most close-y to the classical KS entropy [14], and it must thus be zero for a finite quantum sy-tem. The relation

$$h(n) \leq h(1) = S(1) \leq 2\, S(\rho_S), \quad \text{all } n,$$

ives a bound on the transient information rate. For a Markov QSP it holds that $(\infty) = S(1)$, in fact $S(n) = n\, S(1)$, which shows that the Markov property can only old for an infinite R when $S(1) > 0$. For a finite R there can be a subdynamics look-ng like a (non-unitary) Markov process when we consider the QCFs of order at most $_C+1$, where

$$n_C = [S(1)^{-1} S_{max}] ,$$

nd the corresponding transient rate of information gain is

$$h_C = n_C^{-1} S(n_C) .$$

he notion of quantum chaos can now be introduced in a rather vague way as the prop-rty that the QSP defined by S+R looks as much like a Markov process as is allowed by he finite size of R.

The time scale limit for quantum chaos is given by the following multiple of he dephasing time

$$\tau_C = n_C\, \tau_D$$

hich increases at most linearly with the size of the system R. The size of R comes n rather than that of S+R as the chaos is due only to the S-R interaction. The fac-or 2 in S_{max} is a quantum feature which may be seen as a reflection of the possible PR correlations between the system and the environment.

There seems to be no unique way of measuring the degree of chaos in the tran-ient sense. The index

$$0 \leq \chi_0 = S_{max}^{-1}\; S(\infty) \leq 1$$

easures the degree of ergodicity and QE holds if and only if $\chi_0 = 1$. The index

$$0 \leq \chi_1 = S_{max}^{-1}\; S(n_C) \leq \chi_0$$

as the property that $\chi_1 = 1$ implies that the QSP looks precisely like a Markov pro-ess for QCFs of order $\leq n_C+1$, only showing the finite nature of the system for igher orders. Even in cases where $\chi_1 \simeq 1$, if $S(1)$ is very small compared to the aximal value we will see the evolution of S as governed mainly by H_S while the influence of R is a small perturbation (as in the case of spontaneous radiation). In order that the dynamics shall be dominated by the S-R interaction it seems necessary to assume that the index χ_2

$$0 \leq \chi_2 = (2\, S(\rho_S))^{-1} S(1) \leq 1$$

has a value of the order of 1. Another desirable feature for the observability of the chaotic property is that

$$n_C >> 1 \; .$$

Together with $\chi_1 \simeq 1$, $\chi_2 \simeq 1$, this implies that $S(\rho_S) << S(\rho_R)$ is necessary for quantum chaos in its most extreme form.

9. Chaotic properties

It is generally believed that quantum chaos, whatever the way in which it is defined, should be reflected in the fine scale spectral properties of the Hamiltonian, like the statistics of level spacings. With the present definition, however, it is clear that for large systems the details of the level spacings can not effect the evolution of the system on the chaotic time scale τ_C. The average level spacing δE in a non-degenerate spectrum is expected to decrease exponentially with system size and we need a time $\tau_M = \hbar/\delta E >> \tau_C$ to see this detail.

To illustrate this and other points it is possible to construct a class of models of a maximally chaotic nature where $\chi_1 = \chi_2 = 1$ and n_C can be chosen arbitrarily [1]. These models look like quantum counterparts of Bernoulli shifts for the QCFs of order $\leq n_C+1$,

$$R(\underline{X}_n^+, \underline{X}_n) = \prod^n \bar{\rho}_S(X_i^+ X_i), \qquad n \leq n_C+1.$$

The algebra describing R is taken to be $A_R = \overset{N-1}{\otimes} A_S$ $(N = n_C+1)$ and the dynamics is given by a cyclic shift and a unitary V acting in H_S:

$$U(\varphi_1 \otimes \varphi_2 \ldots \otimes \varphi_N) = V\varphi_2 \otimes \varphi_3 \ldots \otimes \varphi_1 \quad .$$

The spectrum of U is easily calculated from that of V and it is found to be highly degenerated by the symmetric nature of the model. The eigenstates of U show strong 'mode mixing': most of them will not factorize but are non-trivial linear combinations

$$\psi = \sum \xi_k \otimes \eta_k \in H_S \otimes H_R$$

Strong mode mixing is often taken to be typical of quantum chaos in the vague sense, and it is the result of a sufficiently strong interaction of the subsystems, an effect which is enhanced by resonances in the uncoupled system. A related property is that operators in A_S will connect the different eigenstates of U in an efficient way: It takes at most n_C matrix elements of the form

$$\langle\psi, (X \otimes 1)\psi'\rangle, \qquad X \in A_S \; ,$$

o connect any two eigenstates by a chain of non-zero elements in the model above. his property is reminiscient of some aspects of the 'irregular spectrum' introduced y Percival [16]. It is clear that it will make the system $S+R$ highly sensitive to erturbations which act on S.

It can be shown quite generally that the chaotic property as defined here is losely related to such a sensitivity property [1]. In fact there is an isometric quivalence

$$\sigma(n) \simeq \sigma_R(n) \in E(A_R \otimes A_R)$$

$$\sigma_R(n) \equiv (T_R^* \otimes I)^n[\omega_R], \qquad T_R^*[\rho] \equiv Tr_S\{U(\rho_S \otimes \rho)U^+\},$$

here ω_R is a 'purification' of ρ_R of the type described above. This means that the on-zero eigenvalues of $\sigma(n)$ and $\sigma_R(n)$ are the same (with multiplicities), hence the ntropies are the same

$$S(n) = S(\sigma_R(n)) ,$$

hich shows that $S(n) \leq 2\, S(\rho_R)$ and leads to the bound on n_C. But $\{\sigma_R(n)\}$ represent semigroup evolution of R when the operation

$$E[X] = \rho_S(X)\cdot 1$$

cts on S in each instant of the (discrete) time. This operation represents a trongly dissipative action of M driving S instantaneously into the 'equilibrium' tate ρ_S . In a chaotic system $S+R$ this results in a dissipative evolution of the tate of R . In fact, the sequence $\{S(n)\}$ gives the increase of the entropy of a ystem $R+R$ living in the Hilbert space $H_R \otimes H_R$, starting from the pure initial state $_R$, and where S interacts with one of the R-systems only. The sequence of states $_R(n)$ for $R+R$ also defines the evolution of any initial state of R in $S+R$, as such state can be written in the form $\rho(X) = \omega_R(X \otimes Y)$ for a suitable positive Y.

The representation $\sigma_R(n)$ makes clear the role of the 'constants of motion' in $= \{A_S \cup U\}'$. A non-trivial projection $P \in C$ satisfies, for every n

$$\sigma_R(n)[P \otimes (1-P)] = \omega_R[P \otimes (1-P)] = 0.$$

onsequently the sequence $\{\sigma_R(n)\}$ can not approach the limit $\rho_R \otimes \rho_R$ as $n \to \infty$ which s characteristic of the QE property, and the inequality $S(\infty) < S_{max}$ is strict. A rojection P for which there is a slow increase (for $n >> n_C$, say) to the limit

$$\sigma_R(n)\,[P \otimes (1-P)] \to \rho_R(P)\rho_R(1-P)$$

an be interpreted as an 'approximate constant of motion' for the system.

Acknowledgement. The research behind this note was supported by the Swedish Natural Science Research Council.

References

1. G. Lindblad: Quantum ergodicity and chaos. Preprint TRITA-TFY-84-12, Stockholm 1984
2. A.H. Zewail: Phys. Today 33(11), 27 (1980)
3. J. Jortner, R.D. Levine, S.A. Rice (eds): Adv. Chem. Phys. 47, New York: Wiley-Interscience 1981
4. D.W. Noid, M.L. Koszykowski, R.A. Marcus: Ann. Rev. Phys. Chem. 32, 267 (1981)
5. V.I. Arnold, A. Avez: Ergodic problems in classical mechanics. New York: Benjamin 1968
6. A.J. Lichtenberg, M.A. Lieberman: Regular and stochastic motion. New York: Springer 1983
7. G. Casati, B.V. Chirikov, J. Ford: Phys. Lett. 77A, 91 (1980)
8. Ya.B. Pesin, Ya.G. Sinai: Sov. Sci. Rev. C2, 53 (1981)
9. R. Shaw: Z. Naturforschung 36a, 81 (1981)
10. A. Peres: Phys. Rev. Lett. 49, 1118 (1982)
11. B.V. Chirikov, F.M. Izrailev, D.L. Shepelyanski: Sov. Sci. Rev. C2, 209 (1981)
12. L. Accardi, A. Frigerio, J.T. Lewis: Publ. RIMS (Kyoto U.) 18, 97 (1982)
13. M.V. Berry: Ann. Phys. 131, 163 (1981)
14. G. Lindblad: Commun. Math. Phys. 65, 281 (1979)
15. A. Wehrl: Rev. Mod. Phys. 50, 221 (1978)
16. I.C. Percival: J. Phys. B6, L229 (1973)

Quantum Markov processes on Fock space described by integral kernels.

Hans Maassen, Dep. of Mathematics and Computer Science,
Technical University Delft, the Netherlands.

Abstract: A description is introduced of operators on Fock space by way of integral kernels. In terms of these kernels, the quantum stochastic differential equation for a Markov process over the n×n matrices can be explicitly solved. As an example the Wigner-Weisskopf atom is treated.

§ 1 Introduction and summary

1.1. The dilation problem is the problem of finding all stationary Markov processes which correspond to a given semigroup of transition probabilities with an invariant probability distribution. The problem was proposed in the early seventies by Davies [Dav], Lewis and Evans [EvL1], [EvL2] in a non-commutative setting.

In commutative probability theory it had been long solved completely by Kolmogorov's reconstruction theorem: there exists a unique Markov dilation for every semigroup of transition probabilities. In non-commutative probability theory, however, the question of existence has yet to be answered in general, whereas uniqueness does not hold [AFL]. Although in recent years Kümmerer and Schröder [KüS], [Küm 1], [Küm 2] have begun to develop a systematic theory of Markovian dilations, the field is still in a stage of orientation, and in need of examples.

One way to find a Markov process is to solve a stochastic differential equation. The solutions of ordinary stochastic differential equations, however, necessarily yield dilations of trace-preserving semigroups [KüM].For more general semigroups non-commutative stochastic differential equations (i,e, with non-commuting noise terms) are needed. These are in fact available due to the work of Hudson, Parthasarathy and co-workers [HuP], [ApH].

In this paper it will be shown that these quantum stochastic differential equations can be used to construct Markov dilations of a wide class of semigroups of transition probabilities, at least on the $n \times n$ matrices. This was realised independently by Frigerio [Fri]. Here we shall explicitly solve the stochastic differential equations by writing operators on Fock space in terms of integral kernels on Guichardet's symmetric space of the real line [Gui]. The germ of the method used here is already present in the work of Evans and Lewis ([EvL2], § 16,17)

1.2 Summary

In § 2 the symmetric Fock space of $L^2(\mathbb{R})$ is introduced, and the action of integral kernels on this space is described. In § 3 stochastic integrals of kernel processes are defined. In § 4 the conditions are formulated for a kernel process to satisfy a quantum stochastic differential equation (QSDE).

An Itô formula for products of kernel processes is proved. A certain QSDE serves to form dilations of semigroups of transition probabilities in § 5. Finally, in § 6 we treat an example of a semigroup on the 2 × 2 matrices, the evolution of the Wigner-Weisskopf atom. In a different fashion, the latter has been treated by von Waldenfels [vWa].

§ 2 Fock space

2.1

Let I be a closed interval on the real line, and let $\Omega(I)$ be the symmetric space of I [Gui], i.e. the set of all finite subsets of I.

$$\Omega(I) = \bigcup_{n=0}^{\infty} \Omega_n(I), \tag{2.1}$$

$$\Omega_0(I) = \{\emptyset\}; \qquad \Omega_n(I) = \{\{t_1,\cdots,t_n\} \mid t_1,\cdots,t_n \in I\}.$$

By the total ordering of the interval I, a finite subset of I corresponds to an ordered sequence (say, in increasing order). Thus $\Omega_n(I)$ corresponds to $\{\ulcorner t_1,\cdots,t_n \urcorner \in I^n \mid t_1<t_2<\cdots<t_n\}$, and we transpose the Lebesgue measure on I^n to $\Omega_n(I)$. Let $d\omega$ denote the measure on $\Omega(I)$ which has $\emptyset$ as an atom of weight 1, and which equals this Lebesgue measure on $\Omega_n(I)$ for n=1,2,···. By the *symmetric Fock space* $\mathcal{F}(I)$ we shall mean

$$\mathcal{F}(I) = L^2(\Omega(I), d\omega).$$

Corresponding to (2.1) we have the orthogonal decomposition

$$\mathcal{F}(I) = \bigoplus_{n=0}^{\infty} \mathcal{F}_n(I) \tag{2.2}$$

with $\mathcal{F}_n(I) = L^2(\Omega_n(I), d\omega)$. Note that

$$\mathcal{F}(I) \cong \mathbb{C} \oplus \bigoplus_{n=1}^{\infty} L^2_{sym}(I^n).$$

Let $C_0(I)$ be the set of all continuous functions $I \to \mathbb{C}$ with compact support. For $f \in C_0(I)$, define $\pi(f) \in \mathcal{F}(I)$ by

$$\pi(f)(\omega) = \prod_{t \in \omega} f(t).$$

The functions $\pi(f)$ are known as the *coherent states*, or *exponential vectors*. They generate $\mathcal{F}(I)$ as a Hilbert space.

2.2 Operators on Fock space, given by integral kernels

We consider operators X on $\mathcal{F}(\mathrm{I})$ of the form

$$(X\xi)(\omega) = \sum_{\sigma\subset\omega} \int_{\tau\in\Omega(\mathrm{I})} x(\sigma,\tau)\xi((\omega\setminus\sigma)\cup\tau)\,d\tau \ , \tag{2.3}$$

where x is a function $\Omega(\mathrm{I})\times\Omega(\mathrm{I}) \to \mathbb{C}$, which we shall call the *integral kernel* of X. In order that (2.3) makes sense, we must make certain assumptions on ξ and x. Let $\mathrm{K}_1(\mathrm{I})$ be the class of functions ξ: $\Omega(\mathrm{I}) \to \mathbb{C}$ with the properties

(i) $\exists_{C>0}$: $|\xi(\{t_1,\cdots,t_n\})| \leq C^n$,

(ii) ξ is continuous except for a finite number of jumps, i.e. points $t\in\mathrm{I}$ for which the limits

$$\lim_{s\downarrow t} \xi(\omega\cup\{s\}) \qquad \text{and} \qquad \lim_{s\uparrow t} \xi(\omega\cup\{s\})$$

do exist, but are not equal for all ω.

(iii) There is a bounded set $\mathrm{J}\subset\mathrm{I}$ such that $\xi(\omega)$ vanishes for $\omega \not\subset \mathrm{J}$.

By $\mathrm{K}_2(\mathrm{I})$ we shall denote the class of functions x: $\Omega(\mathrm{I})\times\Omega(\mathrm{I}) \to \mathbb{C}$ with the same properties. (In *(i)* the total number of arguments must be taken.) Then every $x\in\mathrm{K}_2(\mathrm{I})$ determines a linear map $\mathrm{K}_1(\mathrm{I}) \to \mathrm{K}_1(\mathrm{I})$ via (2.3). Notice that X is not necessarily a bounded operator, even when x is bounded. The problem of extending X to an operator on all of $\mathcal{F}(\mathrm{I})$ must be considered separately.

2.3 Examples and properties

For $f\in C_0(\mathrm{I})$ the *annihilation operator* $A(f)$ is given by

$$(A(f)\xi)(\omega) = \int_{\mathrm{I}} \overline{f(t)}\ \xi(\omega\cup\{t\})\ dt \ ;$$

the *creation operator* $A^*(f)$ by

$$(A^*(f)\xi)(\omega) = \sum_{s\in\omega} f(s)\ \xi(\omega\setminus\{s\}).$$

These operators have kernels $a(f)$ and $\tilde{a}(f)$ respectively, given by

$$a(f)(\sigma,\tau) = \begin{cases} \overline{f(t)} \text{ if } \sigma=\emptyset \text{ and } \tau=\{t\} \\ 0 \text{ otherwise} \end{cases} ; \quad \tilde{a}(f)(\sigma,\tau) = \overline{a(f)(\tau,\sigma)}$$

The *Weyl operators* $W(f)$ have integral kernels $w(f)$, given by

$$w(f)(\sigma,\tau) = \pi(f)(\sigma)\pi(-\overline{f})(\tau)\ \exp(-\tfrac{1}{2}\|f\|^2).$$

<u>Proof.</u> One checks that the $W(f)$, thus defined have the properties

$$W(-f) = W(f)^* \quad \text{and} \quad W(f)W(g) = \exp(-i\ \mathrm{Im}\langle f,g\rangle)W(f+g).$$

These are the defining properties of Weyl operators. ∎

To the transpose $\tilde{X}$ of an operator X (given by $\langle\eta,\tilde{X}\xi\rangle=\langle X\eta,\xi\rangle$), corresponds the transpose $\tilde{x}$ of the kernel x of X (given by $\tilde{x}(\sigma,\tau)=\overline{x(\tau,\sigma)}$).

The composition XY has integral kernel $x*y$, given by

$$(x*y)(\sigma,\tau) = \sum_{\alpha\subset\sigma}\ \sum_{\beta\subset\tau}\ \int_{\gamma\in\Omega(I)} x(\alpha,\beta\cup\gamma)\ y((\sigma\setminus\alpha)\cup\gamma,(\tau\setminus\beta))\ d\gamma. \qquad (2.4)$$

Operators of the form (2.3) with $x\in K_2(I)$ are strongly dense in $\mathcal{L}(\mathcal{F}(I))$, because finite linear combinations of the Weyl operators are.

§3. Quantum stochastic integrals.

It is convenient to define stochastic integrals not only for those families $\{x_t\}$ where x_t is the kernel of some operator on $F(I)$, $0 \le t \le T$ but simply for all $x_t \in L^2(\Omega[0,T]^2)$ which are non-anticipating in the sense that

$$x_t(\sigma,\tau) = 0 \quad \text{unless } \sigma\cup\tau\subset[0,t].$$

The space of all such non-anticipating kernel processes form a Hilbert space $L[0,T]$ with the norm

$$\|x\|^2 = \int_0^T \|x_t\|^2\,dt = \int_0^T \Big(\int_{\Omega[0,t]^2} |x_t(\sigma,\tau)|^2\, d\sigma d\tau\Big)dt. \qquad (3.1)$$

We define the maps I_T and $\tilde{I}_T$: $L[0,T] \to L^2(\Omega[0,T]^2)$ by

$$(I_T x)(\sigma,\tau) = \begin{cases} 0 & \text{if } \tau=\emptyset \\ x_{\max(\tau)}(\sigma,\tau\setminus\{\max(\tau)\}) & \text{if } \tau\neq\emptyset; \end{cases}$$

$$(\tilde{I}_T\ )(\sigma,\tau) = \begin{cases} 0 & \text{if } \sigma=\emptyset, \\ x_{\max(\sigma)}(\sigma\setminus\{\max(\sigma)\},\tau) & \text{if } \sigma\neq\emptyset. \end{cases}$$

I_T and $\tilde{I}_T$ will be considered as stochastic integration w.r.t. a and a respectively. The following lemmas motivate this.

Lemma 3.1. Let $t_0,\dots,t_n \in [0,T]$ be such that $0 = t_0 < t_1 < \dots < t_n = T$. Suppose $\{x_t\}_{0\le t\le T}$ is a non-anticipating kernel process which is constant on each of the intervals $[t_k,t_{k+1})$. Then

$$\sum_{k=0}^{n-1} x_{t_k} * (a_{t_{k+1}} - a_{t_k}) = I_T x \qquad (3.2)$$

and

$$\sum_{k=0}^{n-1} x_{t_k} * (\tilde{a}_{t_{k+1}} - \tilde{a}_{t_k}) = \tilde{I}_T x . \tag{3.3}$$

Proof. For all σ and τ in $\Omega[0,T]$ we have

$$\left(\sum_{k=0}^{n-1} x_{t_k} * (a_{t_{k+1}} - a_{t_k})\right)(\sigma,\tau) =$$

$$= \sum_{k=0}^{n-1} \sum_{\alpha\subset\sigma} \sum_{\beta\subset\tau} \int_{\gamma\in\Omega[0,T]} x_{t_k}(\alpha,\beta\cup\gamma)a(\chi_{[t_k,t_{k+1})})((\sigma\setminus\alpha)\cup\gamma,\tau\setminus\beta)d\gamma.$$

The integrand is zero unless $\gamma = \emptyset$, $\alpha = \sigma$ and $\tau = \beta\cup\{t\}$ with $t\in[t_k,t_{k+1})$. So the above is equal to

$$\sum_{k=0}^{n-1} \sum_{t\in\tau\cap[t_k,t_{k+1})} x_{t_k}(\sigma,\tau\setminus\{t\}) = \sum_{t\in\tau} x_t(\sigma,\tau\setminus\{t\}).$$

Now, we must have $t = \max(\tau)$ for $x_t(\sigma,\tau\setminus\{t\})$ not to vanish. Hence the above sum is equal to $(Ix)(\sigma,\tau)$. To prove (3.3) we note that x_{t_k} commutes with $a_{t_{k+1}} - a_{t_k}$ and that $\tilde{I}x = (I\tilde{x})^{\sim}$.

□

Lemma 3.2. I and $\tilde{I}$ are isometries $L[0,T] \to L^2(\Omega[0,T]^2)$.

Proof. Let $x\in L[0,T]$. Then

$$\|x\|^2 = \int_0^T dt \int_{\Omega[0,t]} d\rho \left(\int_{\Omega[0,t]} d\sigma\, |x_t(\sigma,\rho)|^2\right) =$$

$$= \int_{\Omega[0,T]} d\tau \int_{\Omega[0,\max(\tau)]} d\sigma |x_{\max(\tau)}(\sigma,\tau\setminus\{\max(\tau)\})|^2 =$$

$$= \int_{\Omega[0,T]} d\tau \int_{\Omega[0,T]} d\sigma |(Ix)(\sigma,\tau)|^2 = \|Ix\|^2 .$$

The proof for $\tilde{I}$ is obtained analogously.

□

This shows that I and $\tilde{I}$ are the isometric extensions of the left hand sides of (3.2) and (3.3). We may write symbolically

$$I_T x = \int_0^T x_t * da_t \quad \text{and} \quad \tilde{I}_T x = \int_0^T x_t * d\tilde{a}_t .$$

§ 4. Quantum Stochastic Differential Equations

Let x, f, g and h be non-anticipating kernel processes. We say that the quantum stochastic differential equation (QSDE)

$$dx_t = f_t * d\tilde{a}_t + g_t * da_t + h_t dt \tag{4.1}$$

is satisfied if for all $T > 0$,

$$x_T - x_0 = \int_0^T f_t * da_t + \int_0^T g_t * da_t + \int_0^T h_t dt.$$

From lemma 3.1 it follows that (4.1) holds if and only if for almost all σ and τ in $\Omega[0,\infty)$!

$$\frac{d}{dt} x_t(\sigma,\tau) = h_t(\sigma,\tau) \quad \text{if } \tau \notin \sigma \cup \tau \tag{4.2}$$

$$(\lim_{s \downarrow t} - \lim_{s \uparrow t}) x_s(\sigma,\tau) = h_t(\sigma\backslash\{t\},\tau) \quad \text{if} \quad t \in \sigma, \tag{4.3}$$

$$(\lim_{s \downarrow t} - \lim_{s \uparrow t}) x_s(\sigma,\tau) = h_t(\sigma,\tau \backslash \{t\}) \text{ if } \quad t \in \tau. \tag{4.4}$$

We shall prove an Itô formula (theorem 4.2) for products of kernel processes. First a lemma is needed.

Lemma 4.1. Let $\{\zeta_t \in F[0,\infty)\}_{t \geq 0}$ be such that for some $\phi_t, \psi_t \in [0,\infty)$,

$$\frac{d}{dt} \zeta_t(\omega) = \psi_t(\omega) \quad \text{if } t \notin \omega,$$

$$(\lim_{s \downarrow t} - \lim_{s \uparrow t}) \zeta_s(\omega) = \phi_t(\omega\backslash\{t\}) \quad \text{if } t \in \omega.$$

Then

$$\frac{d}{dt} \int_{\Omega[0,\infty)} \zeta_t(\omega) d\omega = \int_{\Omega[0,\infty)} (\psi_t(\omega) + \phi_t(\omega)) d\omega.$$

Proof.

$$\frac{d}{dt} \int_{\Omega[0,\infty)} \zeta_t(\omega) d\omega =$$

$$= \lim_{\varepsilon \downarrow 0} (2\varepsilon)^{-1} \int_{\Omega([0,\infty)\backslash(t-\varepsilon,t+\varepsilon))} d\omega \left(\int_{\Omega(t-\varepsilon,t+\varepsilon)} d\sigma (\zeta_{t+\varepsilon}(\omega \cup \sigma) - \zeta_{t-\varepsilon}(\omega \cup \sigma)) \right) =$$

$$=\lim_{\varepsilon\downarrow 0}(2\varepsilon)^{-1}\int_{\Omega([0,\infty)\setminus(t-\varepsilon,t+\varepsilon))}(\zeta_{t+\varepsilon}(\omega)-\zeta_{t-\varepsilon}(\omega)+\int_{t-\varepsilon}^{t+\varepsilon}\zeta_{t+\varepsilon}(\omega\cup\{s\})-\zeta_{t-\varepsilon}(\omega\cup\{s\})+0(\varepsilon))$$

$$=\int_{\Omega[0,\infty)}(\psi_t(\omega)+\phi_t(\omega))d\omega. \qquad \square$$

heorem 4.2. (Itô product formula for operator kernels). Let $x^{(1)}$, $x^{(2)}$, $f^{(1)}$, $f^{(2)}$, $g^{(1)}$, $g^{(2)}$, $h^{(1)}$ and $h^{(2)}$ be non-anticipating kernel processes with values in $K_2[0,\infty)$. uppose that for $j=1,2$:

$$dx_t^{(j)}=f_t^{(j)}*d\tilde{a}_t+g_t^{(j)}*da_t+h_t^{(j)}dt,$$

then

$$\begin{aligned}d(x_t^{(1)}*x_t^{(2)}) &= \left(f_t^{(1)}*x_t^{(2)}+x_t^{(1)}*f_t^{(2)}\right)*d\tilde{a}_t\\ &+\left(g_t^{(1)}*x_t^{(1)}+x_t^{(1)}*g_t^{(2)}\right)*da_t\\ &+\left(h_t^{(1)}*x_t^{(1)}+x_t^{(1)}*h_t^{(2)}+g_t^{(1)}*f_t^{(2)}\right)dt.\end{aligned}$$

This may be paraphrased as follows: all differentials commute with the processes and all products of differentials are zero, except

$$da_t * d\tilde{a}_t = dt\ .$$

Proof. Let $\sigma,\tau\in\Omega([0,\infty))$ be such that $\sigma\cap\tau=\emptyset$. Take $t\in\sigma$ and put $\sigma'=\sigma\setminus\{t\}$. Then

$$\begin{aligned}&(\lim_{s\downarrow t}-\lim_{s\uparrow t})(x_s^{(1)}*x_s^{(2)})(\sigma,\tau)=\\ &=\sum_{\alpha\subset\sigma'}\sum_{\beta\subset\tau}\int_{\gamma\in\Omega}(\lim_{s\downarrow t}-\lim_{s\uparrow t})\Big(x_s^{(1)}(\alpha\cup\{s\},\beta\cup\gamma)x_s^{(2)}((\sigma'\setminus\alpha)\cup\gamma,\tau\setminus\beta)+\\ &\qquad+x_s^{(1)}(\alpha,\beta\cup\gamma)x_s^{(2)}((\sigma'\setminus\alpha)\cup\{s\},\tau\setminus\beta)\Big)\,d\gamma\\ &=(f_t^{(1)}*x_t^{(2)})(\sigma',\tau)+(x_t^{(1)}*f_t^{(2)})(\sigma',\tau).\end{aligned}$$

In the same way one calculates that for $t\in\tau$:

$$(\lim_{s\downarrow t}-\lim_{s\uparrow t})(x_s^{(1)}*x_s^{(2)})(\sigma,\tau)=(g_t^{(1)}*x_t^{(2)}+x_t^{(1)}*g_t^{(2)})(\sigma,\tau\setminus\{t\}).$$

Finally, let $t\notin\sigma\cup\tau$. Then

$$\frac{d}{dt}(x_t^{(1)}*x_t^{(2)})(\sigma,\tau)=\sum_{\alpha\subset\sigma}\sum_{\beta\subset\tau}\frac{d}{dt}\int_{\gamma\in\Omega}x_t^{(1)}(\alpha,\beta\cup\gamma)x_t^{(2)}((\sigma\setminus\alpha)\cup\gamma,\tau\setminus\beta)\,d\gamma$$

Now, for fixed α and β, call the integrand $\xi_t(\gamma)$, and apply lemma 4.1. One finds

$$\frac{d}{dt}(x_t^{(1)}*x_t^{(2)})=x_t^{(1)}*h_t^{(2)}+h_t^{(1)}*x_t^{(2)}+g_t^{(1)}*f_t^{(2)}\ . \qquad \blacksquare$$

§ 5. Markovian tensor dilations of dynamical systems

5.1 Definitions

We shall conform to the terminology in [KüS] and [Küm 1], except for a few explicitly stated deviations.

If $\mathbb{A}$ is a W*-algebra, and $\{T_t\}_{t\geq 0}$ is a semigroup of completely positive, identity preserving normal maps $\mathbb{A} \to \mathbb{A}$, we shall call $\{\mathbb{A},T\}$ a *dynamical system*. (In contrast to [KüS], no invariant state on $\mathbb{A}$ is assumed to exist.) If the semigroup T consists of *-automorphisms, it can be naturally extended to a group by putting $T_t = (T_{-t})^{-1}$ for $t<0$. In this case $\{\mathbb{A},T\}$ will be called a *reversible dynamical system*.

Below we shall only consider dilations of dynamical systems on the W*-algebra M_n of all complex n×n matrices. As M_n is a factor of type I, the only possible dilations are of tensor form (see [Küm 1]). For this reason we restrict ourselves to this type of dilations in the present paper.

Definition 5.1. Let $\{\mathbb{A},T\}$ be a dynamical system. Suppose that a W*-algebra $\mathbb{C}$, a normal state γ on $\mathbb{C}$, and a group $\{\hat{T}_t\}_{t\in\mathbb{R}}$ of *-automorphisms of $\mathbb{A}\otimes\mathbb{C}$ exists, such that the diagram

$$\begin{array}{ccc} \mathbb{A} & \xrightarrow{\ T_t\ } & \mathbb{A} \\ {\scriptstyle \mathrm{Id}\otimes\mathbb{1}}\downarrow & & \uparrow{\scriptstyle \mathrm{Id}\otimes\gamma} \\ \mathbb{A}\otimes\mathbb{C} & \xrightarrow[\ \hat{T}_t\ ]{} & \mathbb{A}\otimes\mathbb{C} \end{array} \tag{5.1}$$

commutes for all $t\geq 0$. Then we shall call $\{\mathbb{A}\otimes\mathbb{C},\hat{T},\mathrm{Id}\otimes\gamma\}$ a *tensor dilation* of $\{\mathbb{A},T\}$.

Again, we have refrained from requiring the existence of a $\hat{T}$- invariant state on $\mathbb{A}\otimes\mathbb{C}$.

Let $\{\mathbb{A}\otimes\mathbb{C},\hat{T},\mathrm{Id}\otimes\gamma\}$ be a tensor dilation of $\{\mathbb{A},T\}$. For $I\in\mathbb{R}$, let $\mathbb{A}_I$ denote the subalgebra of $\mathbb{A}\otimes\mathbb{C}$, generated by

$$\bigcup_{t\in I} T_t(\mathbb{A}\otimes\mathbb{1}).$$

There exists a conditional expectation P_I onto $\mathbb{A}_I$.

Definition 5.2. Let $\{\mathbb{A}\otimes\mathbb{C},\hat{T},\mathrm{Id}\otimes\gamma\}$ be a tensor dilation of $\{\mathbb{A},T\}$. It is called a *Markovian* tensor dilation if for all $X\in\mathbb{A}_{[0,\infty)}$:

$$P_{(-\infty,0]}(X) = P_{\{0\}}(X). \tag{5.2}$$

5.2. Construction

Let $\{A,T\}$ be a dynamical system with $A = M_n$. Then T is necessarily of the form $T_t = \exp(tL)$, where $L:M_n \to M_n$ is given by [Lin]

$$L(X) = \sum_{j=1}^{k} V_j^* X V_j - \tfrac{1}{2}\{V_j^* V_j, X\} + i[H,X] \ . \tag{5.3}$$

Here, $0 \leq k \leq n^2$ and $V_j \in M_n$ for $j=1,\cdots,k$, and $H = H^* \in M_n$. By $\{A,B\}$ the sum $AB + BA$ is meant.

For simplicity we shall assume henceforth that L is extremal, i.e. $k=1$, and omit the Hamiltonian term. So we assume that

$$L(X) = V^* X V - \tfrac{1}{2}\{V^* V, X\} \ , \tag{5.4}$$

for some $V \in M_n$. The convex combination (5.3) can be treated by taking a tensor product of k copies of Fock space instead of one in the construction below. (See, for instance, [Fri]). The inclusion of the Hamiltonian term poses no difficulties either (although it does have consequences for the question of the existence of a stationary state [Fri].)

Let $\mathbb{C}$ be the W*-algebra of all bounded operators on $\mathcal{F}(\mathbb{R})$. The space $\mathbb{C}^n \otimes \mathcal{F}(\mathbb{R})$ can be identified with the space $L^2(\Omega(\mathbb{R}) \to \mathbb{C}^n)$. The W*-algebra $A \otimes \mathbb{C}$ consists of all bounded operators on this space. Their integral kernels are functions $\Omega(\mathbb{R}) \times \Omega(\mathbb{R}) \to M_n$. Definition 4.2 applies to these kernels as well.

Let $S_t: \mathcal{F}(\mathbb{R}) \to \mathcal{F}(\mathbb{R})$ denote the left shift, given by

$$(S_t \xi)(\omega) = \xi(\omega + t),$$

where $\omega + t = \{s+t \mid s \in \omega\}$. Define $\sigma_t: \mathbb{C} \to \mathbb{C}: C \to S_t^{-1} C S_t$. Let γ be the vacuum state on $\mathbb{C}$: $C \mapsto \langle \delta_\emptyset, C\delta_\emptyset \rangle$.

Now consider the QSDE

$$du_t = Vu_t * d\tilde{a}_t - V^* u_t da_t - \tfrac{1}{2} V^* V u_t dt \ , \tag{5.5}$$

with initial condition $u_0(\sigma,\tau) = \delta_\emptyset(\sigma)\delta_\emptyset(\tau)1\!\!1_{M_n}$.

We define a process u solving this QSDE as follows. For $\sigma,\tau \subset \mathbb{R}$ with $\sigma \cap \tau = \emptyset$, put $\sigma \cup \tau = \{t_1,\cdots,t_k\}$ with $t_1 < t_2 < \cdots < t_k$. Let $u_t(\sigma,\tau)$ be the $n \times n$ matrix

$$\begin{aligned} u_t(\sigma,\tau) = \pi(\chi_{[0,t]})(\sigma\cup\tau)\cdot\exp(-\tfrac{1}{2}(t-t_k)V^*V)\,V_k \exp(-\tfrac{1}{2}(t_k - t_{k-1})V^*V) \times \\ \times V_{k-1} \times \cdots \times V_2 \exp(-\tfrac{1}{2}(t_2-t_1)V^*V)\,V_1 \exp(-\tfrac{1}{2}t_1 V^*V)\cdot 1\!\!1 . \end{aligned} \tag{5.6}$$

where

$$V_j = \begin{cases} V & \text{if } t_j \in \sigma \ , \\ -V^* & \text{if } t_j \in \tau \ . \end{cases}$$

Theorem 5.3. *The process u defined by (5.6) solves the QSDE (5.5). It defines a family* $\{U_t\}_{t\geq 0}$ *of unitary operators on* $\mathbb{C}^n \otimes \mathcal{F}(\mathbb{R})$. *The *-automorphisms* $\{\hat{T}_t\}_{t\in\mathbb{R}}$ *of* $M_n \otimes \mathcal{L}(\mathcal{F}(\mathbb{R}))$, *given by*

$$\hat{T}_t(X) = \begin{cases} U_t^{-1}\sigma_t(X)U_t & \text{if } t\geq 0, \\ \sigma_t(U_{-t}XU_{-t}^{-1}) & \text{if } t<0, \end{cases}$$

determine a Markovian tensor dilation $\{M_n \otimes \mathcal{L}(\mathcal{F}(\mathbb{R})), \hat{T}, \mathrm{Id}\otimes\gamma\}$ *of* $\{M_n, T\}$.

Proof. Clearly, u is non-anticipating, and for $t \notin \sigma\cup\tau$ we have

$$\frac{d}{dt}u_t(\sigma,\tau) = -\tfrac{1}{2}V^*Vu_t(\sigma,\tau)$$

and

$$\lim_{s\uparrow t} u_s(\sigma\cup\{t\},\tau) = 0, \qquad \lim_{s\downarrow t} u_s(\sigma\cup\{t\},\tau) = Vu_t(\sigma,\tau),$$

$$\lim_{s\uparrow t} u_s(\sigma,\tau\cup\{t\}) = 0, \qquad \lim_{s\downarrow t} u_s(\sigma,\tau\cup\{t\}) = -V^*u_t(\sigma,\tau).$$

So u is a solution of the QSDE; note that $u_0(\sigma,\tau)=0$ unless $\sigma = \tau = \emptyset$, and that $u_0(\emptyset,\emptyset) = \mathbb{1}$. Now, the process $\{\tilde{u}_t\}$ solves the QSDE

$$d\tilde{u}_t = \tilde{u}_t*(-Vd\tilde{a}_t + V^*da_t - \tfrac{1}{2}V^*Vdt).$$

Hence, by the Itô product formula (theorem 4.3),

$$d(\tilde{u}_t*u_t) = 0 ,$$

and

$$d(u_t*\tilde{u}_t) = [V,u_t*\tilde{u}_t]*d\tilde{a}_t - [V^*,u_t*\tilde{u}_t]*da_t + (-\tfrac{1}{2}\{V^*V,u_t*\tilde{u}_t\}+V^*u_t*\tilde{u}_tV)dt.$$

Therefore $(\tilde{u}_t*u_t)(\sigma,\tau) = (u_t*\tilde{u}_t)(\sigma,\tau) = \delta_\emptyset(\sigma)\delta_\emptyset(\tau)\cdot\mathbb{1}$, the integral kernel of the unit operator, for all $t\geq 0$, so that the operator U_t with kernel $\mathbf{u}_t$ extends to a unique unitary operator on $\mathbb{C}^n \otimes \mathcal{F}(\mathbb{R})$ for all $t\geq 0$.

To see that $\hat{T}$ is a group, it suffices to prove the cocycle property $U_{t+s} = \sigma_t(U_s)U_t$. Indeed, for (almost) all $\sigma,\tau \in \Omega(\mathbb{R})$:

$$(u_s(\cdot-t,\cdot-t)*u_t)(\sigma,\tau) = \sum_{\alpha\subset\sigma}\sum_{\beta\subset\tau}\int_{\gamma\in\Omega} u_s(\alpha-t,\beta\cup\gamma-t)u_t((\sigma\setminus\alpha)\cup\gamma,\tau\setminus\beta)\,d\gamma =$$

$$= \pi(\chi_{[0,t+s]})(\sigma\cup\tau)\cdot u_s(\alpha'-t,\beta'-t)u_t(\alpha'',\beta''),$$ where α', β' are the parts of σ, τ in $[t,t+s]$ and α'', β'' those in $[0,t]$

$$= u_{t+s}(\sigma,\tau).$$

The first step follows from the fact that $u_t(\omega,\nu)\neq 0 \Rightarrow \omega,\nu\subset[0,t]$. The second is easily read off from the definition (5.6) of u_t.

We must show that we have a dilation, i.e. for all $X\in M_n$ and $t\geq 0$:

$$(\mathrm{Id}\otimes\gamma)\hat{T}_t(X\otimes\mathbb{1}) = T_t(X). \tag{5.7}$$

Because $X\otimes 1\!\!1$ is σ_t-invariant, we have

$$(\mathrm{Id}\otimes\gamma)\hat{T}_t(X\otimes 1\!\!1) = \mathrm{Id}\otimes\gamma(U_t^{-1}(X\otimes 1\!\!1)U_t) = \int_{\rho\in\Omega}\tilde{u}_t(\emptyset,\rho)Xu_t(\rho,\emptyset)\,d\rho .$$

An application of lemma 4.1 with $\xi_t(\rho) = \tilde{u}_t(\emptyset,\rho)Xu_t(\rho,\emptyset)$ leads to

$$\frac{d}{ds}(\mathrm{Id}\otimes\gamma)\hat{T}_s(X\otimes 1\!\!1)\Big|_{s=t} = \int_{\rho\in\Omega}\tilde{u}_t(\emptyset,\rho)\left(-\tfrac{1}{2}\{V^*V,X\} + V^*XV\right)u_t(\rho,\emptyset)\,d\rho =$$

$$= (\mathrm{Id}\otimes\gamma)\hat{T}_t(\mathrm{L}(X)\otimes 1\!\!1),$$

which proves (5.7). The Markov property follows from the fact that for any interval $\mathrm{I}\subset\mathbb{R}$: $A_{\mathrm{I}}\subset M_n\otimes\mathcal{L}(\mathcal{F}(\mathrm{I}))$ and that the projection of $M_n\otimes\mathcal{L}(\mathcal{F}([0,\infty)))$ on $M_n\otimes\mathcal{L}(\mathcal{F}((-\infty,0]))$ lies inside $M_n\otimes 1\!\!1$. ∎

§ 6 Example

Let us put

$$V = \begin{pmatrix} 0 & 0 \\ 1 & 0 \end{pmatrix} .$$

The associated dynamical semigroup is given by

$$T_t\begin{pmatrix} X_{11} & X_{12} \\ X_{21} & X_{22} \end{pmatrix} = \begin{pmatrix} e^{-t}(X_{11}-X_{22})+X_{22} & e^{-\frac{1}{2}t}X_{12} \\ e^{-\frac{1}{2}t}X_{21} & X_{22} \end{pmatrix} .$$

The reversible evolution $\hat{T}$ on $M_2\otimes\mathcal{L}(\mathcal{F}(\mathbb{R}))$ is of the form

$$\hat{T}_t(X) = R_t^{-1}XR_t ,$$

where $R_t = S_tU_t$ is the evolution operator on $\mathbb{C}^2\otimes\mathcal{F}(\mathbb{R})$, given by the formula

$$(R_t\xi)(\omega) = \sum_{\sigma\subset\omega+t}\int_{\tau\in\Omega} u_t(\sigma,\tau)\ \xi(((\omega+t)\setminus\sigma)\cup\tau)\,d\tau . \tag{6.1}$$

The kernel u_t is described in (5.6); it has the following behaviour in this case:

(i) $u_t(\emptyset,\emptyset) = \begin{pmatrix} e^{-\frac{1}{2}t} & 0 \\ 0 & 1 \end{pmatrix}$;

(ii) $u_t(\sigma,\tau) = 0$ unless σ and τ are included in $[0,t]$ and alternating;

(iii)

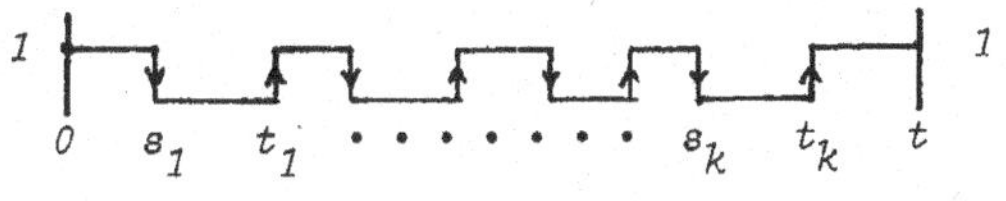

$$u_t(\{s_1,\cdots,s_k\},\{t_1,\cdots,t_l\})_{11} = \begin{cases} 0 \text{ unless } k=l \text{ and } s_1<t_1 , \\ \quad\text{in which case it equals:} \\ (-1)^l\exp(-\tfrac{1}{2}(s_1-t_1+\ \ +s_k-t_k+t)) \end{cases}$$

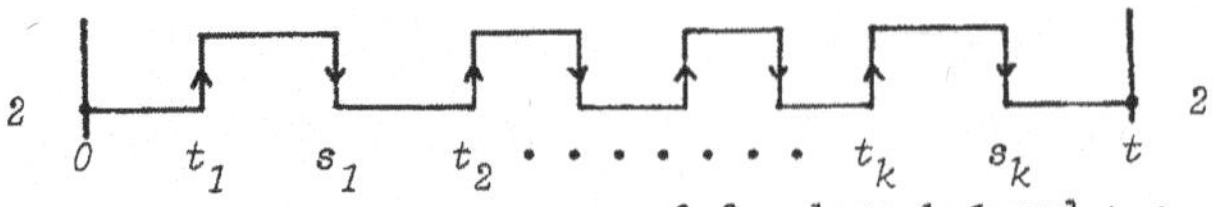

$$u_t(\{s_1,\cdots,s_k\},\{t_1,\cdots,t_l\})_{22} = \begin{cases} 0 \text{ unless } k=l \text{ and } t_1<s_1 \\ \text{in which case it is equal to} \\ (-1)^l \exp(-\tfrac{1}{2}(-t_1+s_1-\cdots-t_k+s_k)) \end{cases}$$

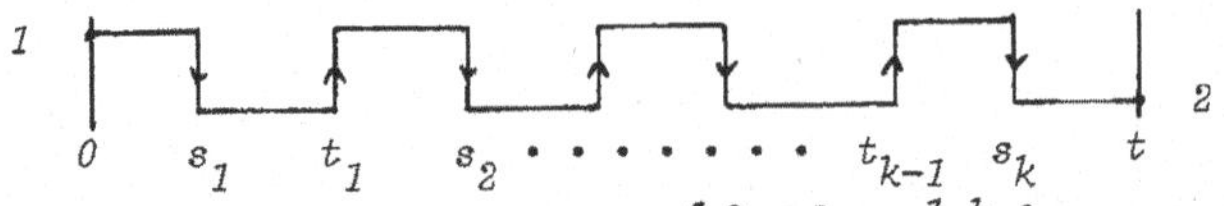

$$u_t(\{s_1,\cdots,s_k\},\{t_1,\cdots,t_l\})_{21} = \begin{cases} 0 \text{ unless } l=k-1, \\ \text{in which case it is equal to} \\ (-1)^l \exp(-\tfrac{1}{2}(s_1-t_1+\cdots-t_{k-1}+s_k)) \end{cases}$$

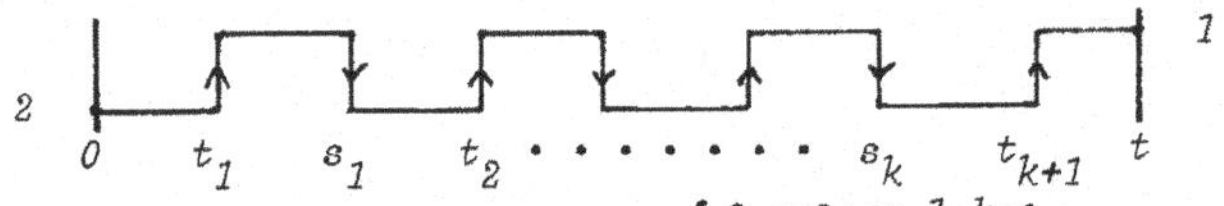

$$u_t(\{s_1,\cdots,s_k\},\{t_1,\cdots,t_l\})_{12} = \begin{cases} 0 \text{ unless } l=k+1, \\ \text{in which case it is equal to} \\ (-1)^l \exp(-\tfrac{1}{2}(-t_1+s_1-\cdots+s_k-t_{k+1}+t)). \end{cases}$$

The evolution $\{R_t\}$ leaves each of the spaces

$$\mathcal{D}_0 = \mathbb{C}\begin{pmatrix}0\\1\end{pmatrix}\otimes\delta_\emptyset; \qquad \mathcal{D}_k = \begin{pmatrix}0\\1\end{pmatrix}\otimes\mathcal{F}_k(\mathbb{R}) \;\oplus\; \begin{pmatrix}1\\0\end{pmatrix}\otimes\mathcal{F}_{k-1}(\mathbb{R}) \qquad (k=1,2,\cdots).$$

invariant. The restriction of R_t to each of these spaces can be read off from the above formulae. Let us only take a look at the space $\mathcal{D}_1$.

$$\left(R_t\begin{pmatrix}\delta_\emptyset\\0\end{pmatrix}\right)(\emptyset) = u_t(\emptyset,\emptyset)\begin{pmatrix}1\\0\end{pmatrix} = \begin{pmatrix}u_t(\emptyset,\emptyset)_{11}\\u_t(\emptyset,\emptyset)_{21}\end{pmatrix} = \begin{pmatrix}e^{-\frac{1}{2}t}\\0\end{pmatrix};$$

$$\left(R_t\begin{pmatrix}\delta_\emptyset\\0\end{pmatrix}\right)(\{s\}) = u_t(\{s\},\emptyset)\begin{pmatrix}1\\0\end{pmatrix} = \begin{pmatrix}u_t(\{s\},\emptyset)_{11}\\u_t(\{s\},\emptyset)_{21}\end{pmatrix} = \chi_{[0,t]}(s)e^{\frac{1}{2}(s-t)}\begin{pmatrix}0\\1\end{pmatrix}.$$

Therefore

$$R_t\begin{pmatrix}\delta_\emptyset\\0\end{pmatrix} = e^{-\frac{1}{2}t}\begin{pmatrix}\delta_\emptyset\\0\end{pmatrix} + \begin{pmatrix}0\\\chi_{[0,t]}e^{\frac{1}{2}(\cdot-t)}\end{pmatrix}.$$

In the same way we find that

$$R_t\begin{pmatrix}0\\f\delta_1\end{pmatrix} = \left(-\int_0^t e^{-\frac{1}{2}t}f(s)\,ds\right)\begin{pmatrix}\delta_\emptyset\\0\end{pmatrix} + \left(S_tf - \int_0^t f(s)e^{\frac{1}{2}(\cdot-s)}\,ds\right)\begin{pmatrix}0\\\delta_1\end{pmatrix}.$$

(where $\delta_1(\omega) = \delta_1(\#(\omega))$.) Thus $R_t\restriction\mathcal{D}_1$ is the linear dilation of the semigroup $\mathbb{C}\to\mathbb{C}$: $\lambda\mapsto e^{-\frac{1}{2}t}\lambda$. In the higher spaces $\mathcal{D}_k$, however, the behaviour gets more complicated, and no linear dilations are found.

Interpratation. If we interpret M_2 as the algebra of observables of an atom with two energy levels, a higher level (level 1) and a lower level (level 2), then the semigroup T describes the evolution associated with spontaneous emission of a quantum of radiation. The dilation turns out to have a physical interpretation as well: The algebra $\mathcal{L}(\mathcal{F}(\mathbb{R}))$ is the algebra of observables of a radiation field; the layers $\mathcal{F}_k(\mathbb{R})$ of Fock space corresponding to the possible numbers of quanta of radiation. The symmetric character of this Fock space reflects the boson nature of the quanta. $\mathbb{R}$ is the space in which the quanta can move. The algebra $M_2 \otimes \mathcal{L}(\mathcal{F}(\mathbb{R}))$ stands for the coupled system of atom and field. The Hilbert space of the system is $\mathbb{C}^2 \otimes \mathcal{F}(\mathbb{R})$, on which R describes a unitary time evolution, costomary in quantum mechanics. We read from formula (6.1) the following interpretation of the matrix elements of the kernel u_t:

$u_t(\sigma,\tau)_{jk}$ is the amplitude that, given the presence at time 0 of at least quanta at the positions τ, and the location of the atom in its level nr. k, at time t the atom will be in level j, and the quanta at τ will have been absorbed and re-emitted so that they get to the positions $\sigma - t$, all other quanta having moved to the left by t.

The diagrams on the previous pages are Feynman diagrams of this process.

REFERENCES:

[ApH] : Applebaum D., Hudson R.L.: Fermion Diffusions. Journ. Math. Phys. 25 (1984) 858-861.

[AFL] : Accardi L., Frigerio A., Lewis J.T.: Quantum Stochastic Processes, Publ. of the R.I.M.S., Kyoto, 18 (1982) 97-133.

[Dav] : Davies E.B., Quantum Theory of Open Systems. Academic Press 1976.

[EvL1] : Evans D.E., Lewis J.T.: Dilations of dynamical semigroups. Comm. Math. Phys. 50 (1976) 219-227.

[EvL2] : Evans D.E., Lewis J.T.: Dilations of Irreversible Evolutions in Algebraic Quantum Theory. Communications of the Dublin Institute for Advanced Studies, 24A (1977).

[Fri] : Frigerio A. : Covariant Markov Dilations of Quantum Dynamical Semigroups. (preprint, Milan).

[Gui] :Guichardet A.: Symmetric Hilbert Spaces and Related Topics. Lecture Notes in Mathematics 261, Springer 1972.

[HuP] :Hudson R.L., Parthasarathy K.R.: Quantum Itô's formula and Stochastic Evolutions. Comm. Math. Phys. 93(1984), 301-323.

[Küm 1]:Kümmerer B. : A dilation theory for completely positive operators on W*-algebra Thesis, Tübingen 1982.

[Küm 2]:Kümmerer B. : A non-commutative example of a continuous Markov dilation. Semesterbericht Funktionalanalysis Tübingen, Wintersemester 1982/83.

[KüM] :Kümmerer B., Maassen H.: The commutative dilations of dynamical semigroups on M_n In preparation.

[KüS] :Kümmerer B., Schröder W,: A Markov dilation of a non-quasifree Bloch evolution. Comm. Math. Phys. 90(1983) 251-262.

[Lin] :Lindblad G.: On the generators of quantum dynamical semigroups. Comm. Math. Phys 48(1976) 119.

[vWa] :von Waldenfels W.: Itô solution of the linear quantum stochastic differential equation describing light emission and absorption. Proceedings Villa Mondragone 1982; Lecture Notes in Mathematics 1055, 384-411, Springer.

Quantization of Brownian Motion Processes in Potential Fields

Hiroshi Nakazawa

Department of Physics, Kyoto University, Kyoto 606, Japan

1. Introduction

We discuss problems of the quantization of Brownian motion processes for particles in potential fields of force. Let $f(t)$ be a Gaussian, stationary, mean-square continuous process with mean zero and covariance function $\gamma(\tau)$ characterized by a bounded spectral measure $\nu(\omega)$,

$$\gamma(\tau) = E\{f(t)f(t+\tau)\} = \int_0^\infty \cos(\omega\tau)\,d\nu(\omega), \qquad \nu(0)=0, \qquad \nu(\omega+0)=\nu(\omega), \tag{1}$$

with the expectation $E\{\cdots\}$. The major role in this report is played by the following, stochastic integro-differential equation (SIDE),

$$dQ/dt = P/M, \qquad dP/dt = -V'(Q) - (2M)^{-1}\int_0^t \gamma(t-s)P(s)\,ds + \sigma f(t), \tag{2}$$

with generally nonlinear $V'(Q)$ and constants $M>0$, $\sigma=(2\beta)^{-1/2}$, $0<\beta<\infty$. The stochastic differential equation (SDE),

$$dQ/dt = P/M, \qquad dP/dt = -V'(Q) - (4M)^{-1}P + \sigma w(t), \tag{3}$$

with a standard, Gaussian white noise $w(t)$ plays the title role, though it will participate only in the final stage.

As we discuss in §2, the SIDE (2) has a stationary solution that is approximated arbitrarily well by the motion of some degrees of freedom in a class of classical, finite Hamiltonian systems in thermal equilibrium at inverse temperature β. This physical origin attributes on (2) some neat mathematical structures, and (2) approximates (3) in an obvious way. Thus (2) works as an able intermediary, enabling us to define and discuss the quantization of (3) starting from the mentioned, finite Hamiltonian systems and then taking limits. These are precisely what we undertake below.

The present subject is related most directly to Ford, Kac and Mazur [1], Benguria and Kac [2], Maassen [3] and Nakazawa [4,5] in regard to the quantization of (3). In a slightly different context it is also related deeply to Lewis and Thomas [6,7]. We shall have (§4) general in-ferences on the representation and its structure of quantal Gaussian processes in the class of Lewis and Thomas [7] by a quantized Gaussian white noise (Q-noise, for short) of [2~4]. Concerning (2) and (3), the feasible class of V(Q) shifts from case to case. In classical mechanics (§2) V(Q) need be sufficiently smooth, but boundedness is unnecessary if the following statility condition is satisfied:

$$V(Q) \geqq M\Omega^2 Q^2/2 - c, \qquad \exists \Omega > 0, \qquad \exists c > 0. \tag{4}$$

Quantum mechanics (§§3~5) introduces some tractability by its Hilbert space structure, but together with a difficult, domain problem. The final quantitative statements on the quantized (2) and (3) will be restricted to the case

$$V(Q) = M\Omega^2 Q^2/2 + V_1(Q), \qquad V_1(Q) \in L^\infty(\mathbb{R}) \cap C^0(\mathbb{R}). \tag{5}$$

We shall have (§5) statements for the global existence of solutions of the quantized SIDE (2) and SDE (3) (Q-SIDE and A-SDE, for short) with V(Q) of (5) in a version of initial value problems.

In his penetrating analysis [3] Maassen discussed the problem by a direct quantization of a class of infinite systems, and showed the existence of, and the approach to, a stationary solution for the quantized (3) with $V_1(Q)$ of (5) taken to be smooth with $V_1'(Q)$ in the class $L^\infty(\mathbb{R}) \cap C^0(\mathbb{R})$. Though there remain points to be inquired on the interrelation of the formalisms of [3] and the present work, the results of this work on (5) are suggested to be extended to problems of the approach to equilibrium and, conversely, the present analysis suggests the extensibility of [3] to a more general cases.

This report arose from [5] whose main topics were read at the Second Workshop on Quantum Probability and Applications. Large portion of the following attainments must be dated as late as December, 1984. The author would like to express his sincere gratitude to Professors T. Hida, J. R. Klauder and R. F. Streater for illuminating comments. He also thanks deeply to the Organizing Committee of the Workshop and many of the attendants for hearty hospitality.

2. Realizable SIDE's in classical statistical mechanics

We start with the following fact.

Lemma 1 *Let the Hamiltonian* $H^{(N)}$ *be defined by*

$$H^{(N)} = H_0(Q,P) + H_B^{(N)}, \qquad H_0(Q,P) = P^2/(2M) + V(Q),$$
$$H_B^{(N)} = \frac{1}{2}\sum_{n=1}^{N} [m_n^{(N)} \omega_n{}^2 (q_n - Q)^2 + p_n{}^2/m_n^{(N)} - \hbar\omega_n], \tag{6}$$

ith $m_n^{(N)}>0$ and $\omega_n>0$. Assume a sufficient regularity[1] *of $V(Q)$ together ith (4). There hold (2) and (1) with the replacement of $v(\omega)$ by*

$$v_N(\omega) = \sum_{\omega_n \leq \omega} 2m_n^{(N)}\omega_n{}^2, \qquad v_N'(\omega) = \sum_{n=1}^{N} 2m_n^{(N)}\omega_n{}^2\delta(\omega-\omega_n). \tag{7}$$

f the probability density $\propto exp(-\beta H^{(N)})$ with the expectation $E_N\{\cdots\}$ and $=(2\beta)^{-1/2}$ is posed, then $f(t)$ is a stationary Gaussian process independ-nt of $\{Q(0),P(0)\}$ with $E_N\{f(t)\}=0$ and

$$E_N\{f(t)f(t+\tau)\} = \gamma(\tau), \tag{8}$$

nd $\{Q(t),P(t)\}$ have the invariant probability density

$$\rho_{eq} \propto exp[-\beta H_0(Q,P)]. \tag{9}$$

Proof) Equations of motion for (6) read

$$dq_n/dt = p_n/m_n^{(N)}, \qquad dp_n/dt = -m_n^{(N)}\omega_n{}^2[q_n - Q(t)], \tag{10}$$

$$dQ/dt = P/M, \qquad dP/dt = -V'(Q) + \sum_{n=1}^{N} m_n^{(N)}\omega_n{}^2[q_n(t)-Q(t)]. \tag{11}$$

ince $Q(t)$ exists smoothly by (4), we have a solution of (10) as follows:

$$q_n(t) = q_n(0)\cos(\omega_n t) + p_n(0)\sin(\omega_n t)/(m_n^{(N)}\omega_n) + \omega_n\int_0^t \sin[\omega_n(t-s)]Q(s)ds.$$

artial integration of the last term gives (2) with

$$\gamma(\tau) = \sum_{n=1}^{N}\gamma_n(\tau), \qquad \gamma_n(\tau) = 2m_n^{(N)}\omega_n{}^2\cos(\omega_n\tau), \qquad f(t) = \sum_{n=1}^{N} f_n(t),$$
$$f_n(t) = (2\beta)^{1/2}\{m_n^{(N)}\omega_n{}^2[q_n(0)-Q(0)]\cos(\omega_n t) + \omega_n p_n(0)\sin(\omega_n t)\}. \tag{12}$$

he probability density $\propto\exp(-\beta H^{(N)})$ is invariant, and the initial data $Q(0),P(0),q_n(0)-Q(0),p_n(0);\ 1\leq n\leq N\}$ are independent. Specifically, $f_n(t)$ s Gaussian and independent of $\{Q(0),P(0)\}$ or of $f_m(s)$ for $m\neq n$. There old $E_N\{f_n(t)\}=0$, $E_N\{f_n(t)f_n(t+\tau)\}=\gamma_n(\tau)$, and (7)~(9) follow. □

After physical literatures (e.g. Mori [8]; cf. also [9] and [5]) we all (8), or (1) in (2) the fluctuation-dissipation relation.

Let $v(\omega)$ be any given, bounded spectral measure of (1), and consider he associated Gaussian process $f(t)$. A significant fact is that the tationary solution of (2) need not be unique for general $v(\omega)$.[2] In this egard we note the following which will enable us later to pick out cases

. Existence of locally Lipschitz continuous $V'(Q)$ is sufficient.

of unique stationary states.

Lemma 2 *Let the bounded measure $v(\omega)$ be absolutely continuous and $v'(\omega) = v'(-\omega)$ satisfy the condition* [10]

$$\int_{-\infty}^{\infty} d\omega |\log v'(\omega)| / (1+\omega^2) < \infty. \tag{13}$$

Let $V(Q)$ be smooth and satisfy (4). The ordinary IDE with $\gamma(\tau)$ of (1),

$$dQ/dt = P/M, \qquad dP/dt = -V'(Q) - (2M)^{-1}\int_0^t \gamma(t-s)P(s)ds, \tag{14}$$

gives $\lim_{t\to\infty}\{Q(t),P(t)\}=\{\bar{Q},0\}$, where $\bar{Q}$ satisfies $V'(\bar{Q})=0$.

(Proof) We define $h(t)=V[Q(t)]+P^2(t)/(2M)$. There holds

$$h(0) - h(t) = -(2M)^{-1}\int_0^t ds\int_0^t ds' P(s)\gamma(s-s')P(s') \leqq 0. \tag{15}$$

The condition (13) assures [10] that the integral above can vanish for all $t \geqq 0$ only when $P(s) \equiv 0$ holds. Since $h(t)$ is bounded from below by (4) and decreases as $t\to\infty$, it has a limit $h_\infty \equiv h(\infty)$. This implies $\lim_{t\to\infty} P(t) = 0$, which gives by (14) $\lim_{t\to\infty} V'[Q(t)]=0$. □

Our bounded spectral measure $v(\omega)$ is in the space NBV of right-continuous functions of bounded variation. The point spectra of the type of $v_N(\omega)$ of (7) are dense in NBV. Therefore, Lemma 1 suggests that some stationary solution of (2) for any Gaussian process $f(t)$ with the covariance function $\gamma(\tau)$ of (1), may be approximated by finite, Hamiltonian systems of Lemma 1. A precise version of this statement is given below. We use tentatively a phrase that a set of continuous, stationary processes are *mechanically realizable* when there exists a sequence of finite, classical Hamiltonian systems in thermal equilibrium in such a way that some degrees of freedom in these systems form continuous, stationary processes that converge samplewise to the mentioned processes, the convergence being uniform w.r.t. t in any finite interval.

Lemma 3 *Let $f(t)$ be any mean-square continuous, Gaussian stationary process with zero mean and covariance function $\gamma(\tau)$ of (1) for $v(\omega)$ NBV. Let $V(Q)$ be a polynomial*[3] *fulfilling (4). Define (2) samplewise as ordinary IDE's. (A) The SIDE (2) has a stationary solution (to be called normal) that is realizable mechanically by systems of Lemma 1 at inverse temperature β characterized by $v_N(\omega)$ of (7) that satisfies $v_N(\omega) \leqq v(\omega)$ and converges in NBV to $v(\omega)$ as $N\to\infty$. (B) As $v_N(\omega)\to v(\omega)$ in (A),*

---

2. If $V'(Q)$ is locally Lipschitz continuous, the local solution of (2) is unique. The mentioned non-uniqueness of the stationary solution may be seen typically with $V(Q)=M\Omega^2Q^2/2$, where $H^{(N)}$ decomposes into N+1 independent oscillators called normal modes.

the covariance functions of any polynomials of $\{Q(t),P(t),f(t)\}$ converge to those for the normal stationary state uniformly on any finite time interval. Specifically, $\{Q(t),P(t)\}$ of the normal stationary solution retain the probability density (9).

(Proof) Let $\{Q_N(t),P_N(t)\}$ be the solution of (2) for $\{f(t),\gamma(\tau)\}$ characterized by $v_N(\omega)$. By (9) we have $E_N\{P_N^2(0)\}\leq A$, $E_N\{Q_N^2(0)\}\leq A$, where A is independent of N. We show an estimate with a,b,B independent of $v_N(\omega)$ for $0\leq h\leq T<\infty$ and $F_N(t)=\int_0^t f(s)ds$,

$$\max[E_N\{|Q_N(t+h)-Q_N(t)|^a\},\ E_N\{|P_N(t+h)-P_N(t)|^a\},\ E_N\{|F_N(t+h)-F_N(t)|^a\}\leq Bh^{1+b}. \tag{16}$$

In fact (2) and (9) give, typically,

$$E_N\{[\int_t^{t+h}dt'\int_0^{t'}ds\gamma(t'-s)P_N(s)]^2\}=\int_0^h d\tau\int_0^h d\tau'\int_0^{t+\tau}d\sigma\int_0^{t+\tau'}d\sigma'$$

$$\times\gamma(\sigma)\gamma(\sigma')E_N\{P_N(t+\tau-\sigma)P_N(t+\tau'-\sigma')\}\leq E_N\{P_N^2(t)\}\gamma^2(0)h^2,$$

$$E_N\{[F_N(t+h)-F_N(t)]^2\}=\int_0^h d\tau\int_0^h d\tau'\gamma(\tau-\tau')\leq h^2\gamma(0).$$

Since $\gamma(0)=\int_0^\infty dv_N(\omega)\leq\int_0^\infty dv(\omega)$ holds, (16) is proved. Theorem 4.3 in §4, Chap. I of Ikeda and Watanabe [11] assures that a suitable subsequence of $\{Q_N(t),P_N(t),F_N(t)\}$ can be realized on one and the same probability space, and converges in the sense of Lemma 3. Once this representation is chosen, the limiting $\{Q(t),P(t)\}$ is readily seen to satisfy (2) for $v(\omega)$. Also, any polynomial of $\{Q_N(t),P_N(t),f(t)\}$ is seen to converge to the limit in the mean sense, assuring (9) for the limiting $\{Q(t),P(t)\}$, and also the convergence of covariance functions on any finite time interval. □

It is significant that, with a sufficiently stringent restrictions on $\{f(t),\gamma(\tau)\}$ we may have the uniqueness of the stationary solution of (2). If such a uniqueness holds, Lemma 3 stipulates that it is normal. One such case of sufficiency is given below for quadratic V(Q), which represents a generalization of Ornstein-Uhlenbeck problem and will be used later.

Lemma 4 *Let $v(\omega)$ be a bounded measure that is absolutely continuous, with $v'(\omega)$ satisfying (13) and the following two conditions: [a] There holds for $\gamma(\tau)$ of (1) $\int_0^\infty|\gamma(\tau)|d\tau<\infty$; [b] $v'(\omega)$ is continuous with $v'(\omega)\neq 0$ for any finite ω. Consider a SIDE for $V(Q)=M\Omega^2Q^2/2$,*

---

3. This simplifies the matter; what is needed is that $E\{[V'(Q)]^2\}$ under (9) exists finitely.

$$dQ/dt = P/M, \qquad dP/dt = -M\Omega^2 Q - (2M)^{-1}\int_a^t \gamma(t-s)P(s)ds + \sigma f(t), \tag{17}$$

where f(t) is a stationary Gaussian process with mean zero and covariance function $\gamma(\tau)$, and the initial data are assigned at t=a. The stationary solution of (17) for $t\to\infty$ or $a\to-\infty$ is in law unique, and given by the following.

$$\{Q(t),P(t)\} = \int_{-\infty}^{t} \{\phi(t-s),\psi(t-s)\}f(s)ds, \qquad \phi(t) = \psi(t) = 0 \text{ for } t<0,$$

$$\mathcal{F}\{\phi(t),\psi(t)\} \equiv (2\pi)^{-1/2}\int_{-\infty}^{\infty} \{\phi(t),\psi(t)\}e^{-i\omega t}dt \equiv \{\Phi(\omega),\Psi(\omega)\}$$

$$= \{-(\sigma/M)(2\pi)^{-1/2}J(i\omega+0),\ -i\omega\sigma(2\pi)^{-1/2}J(i\omega+0)\}, \tag{18}$$ [4]

$$J(i\omega+0) \equiv [\omega^2 - \Omega^2 - i\omega\tilde{\gamma}(i\omega+0)/(2M)]^{-1},$$

$$\tilde{\gamma}(i\omega+0) = -(\pi/2)v'(\omega) - (i/2)\mathcal{P}\int_{-\infty}^{\infty} d\bar{\omega}v'(\bar{\omega})/(\omega-\bar{\omega}).$$

(Proof) Introduce Laplace transforms,

$$\{\tilde{q}(z),\tilde{p}(z),\tilde{g}(z)\} \equiv (2\pi)^{-1/2}\int_a^{\infty} e^{-z(t-a)}\{Q(t),P(t),f(t)\}dt,$$

$$\tilde{\gamma}(z) \equiv \int_0^{\infty} e^{-z\tau}\gamma(\tau)d\tau = \int_0^{\infty} d\omega v'(\omega)z/(z^2+\omega^2).$$

We obtain the solution of (17) as follows.

$$\tilde{q}(z) = \{[z+\tilde{\gamma}(z)/(2M)]Q(a) + P(a)/M + (\sigma/M)\tilde{g}(z)\}J(z),$$

$$\tilde{p}(z) = [-M\Omega^2 Q(a) + zP(a) + \sigma z\tilde{g}(z)]J(z),$$

$$J(z) = [z^2+\Omega^2+z\tilde{\gamma}(z)/(2M)]^{-1}.$$

Lemma 2 assures that Laplace inverse transforms of the coefficients of Q(a) and P(a) in the above all exist uniformly bounded, and vanish as $a\to-\infty$. Therefore, J(z) and zJ(z) are analytic for Re $z>0$, and their boundary functions for $z\to i\omega+0$ are $J(i\omega+0) = -[\omega^2-\Omega^2-i\omega\tilde{\gamma}(i\omega+0)/(2M)]^{-1}$ and $i\omega J(i\omega+0)$ with $\tilde{\gamma}(i\omega+0)$ of (18). Assumption [a] implies, by $\gamma(0)\geqq|\gamma(t)|$, that $\gamma(\tau)\in L^p(\mathbb{R}_+)$ holds for any $p\geqq 1$, and $v'(\omega)\in L^2(\mathbb{R})$ holds in particular. The Hilbert transform in $\tilde{\gamma}(i\omega+0)$ thus vanishes as $|\omega|\to\infty$, and [b] assures that both of $J(i\omega+0)$ and $i\omega J(i\omega+0)$ are in $L^2(\mathbb{R})$. Up to some constants they give $\{\Phi(\omega),\Psi(\omega)\} = \{\phi(t),\psi(t)\}$, where $\{\phi(t),\psi(t)\}$ must vanish for t<0 by these facts. Examining coefficients, we have

---

4. Test (or smearing) functions on the real (t-) space are denoted by lower case Greek letters $\phi(t), \rho_k(t), \hat{\xi}_\ell(t), \cdots$, and their Fourier transforms (defined exclusively by (18)) are expressed, often without comments, by the corresponding capital letters $\Phi(\omega)=\mathcal{F}[\phi(t)]$, $R_k(\omega)=\mathcal{F}[\rho_k(t)]$, $\hat{\Xi}_\ell(\omega)=\mathcal{F}[\hat{\xi}_\ell(t)]$, $\cdots$.

:he assertion by letting $a \to -\infty$. □

3. Quantization

Hereafter we regard $\{Q(t), P(t), q_n(t), p_n(t), f(t)\}$ as operators, and start with the quantization of the finite Hamiltonian system (6). Heisenberg equations for (6) retain the same forms (10) and (11) as the classical case. The solution (12) is valid again, and a procedure now obvious gives us (2) as the reduced equation of motion for $\{Q(t), P(t)\}$ with identical expressions (7) and (12) for $v_N(\omega)$, $\gamma(\tau)$ and $f(t)$. A problem arises, however, with the density matrix $\propto \exp(-\beta H^{(N)})$. Since $q_n(0)-Q(0)$ does not commute with $P(0)$, $f(t)$ of (12) is non-Gaussian for non-harmonic $V(Q)$. To circumvent this difficulty we reformulate the problem by taking a time-dependent Hamiltonian,

$$H^{(N)}(t) = \begin{cases} H_1^{(N)} \equiv M\Omega^2 Q^2/2 + P^2/(2M) + H_B^{(N)}, & t \leq 0, \\ H_1^{(N)} + V_1(Q) \equiv P^2/(2M) + V(Q) + H_B^{(N)}, & t > 0. \end{cases} \qquad (19)$$

he system for $t<0$ is assumed to be in equilibrium with the density matrix $\rho_N \propto \exp(-\beta H_1^{(N)})$.

As a departure from [5] we adopt an interaction picture, taking $H_1^{(N)}$ as unperturbed. For clarity we denote the unperturbed solution s follows.

$$\{q(t), p(t)\} = K^*(t)\{Q(0), P(0)\}K(t), \qquad K(t) = \exp[-(it/\hbar)H_1^{(N)}]. \qquad (20)$$

ince $H_1^{(N)}$ is harmonic, $\{q(t), p(t)\}$ of (20) obey the linear, reduced eisenberg equations of the form of (17). The remaining Schrödinger roblem in the interaction picture is

$$i\hbar\partial\psi/\partial t = V_1[q(t)]\psi, \qquad t > 0. \qquad (21)$$

f we show the existence of a propagator [12] $L(t,s)$ for this generator $(i/\hbar)V_1[q(t)]$, the problem is through with the solution

$$\{Q(t), P(t)\} = L^*(t,0)\{Q(0), P(0)\}L(t,0), \qquad t > 0. \qquad (22)$$

This reformulation to an interaction picture (20) and (21) is awkard if $V_1(Q)$ is unbounded as (4); the original dynamics is well-defined y (4), but (21) is marred by the domain problem for $V_1[q(t)]$. However, ur destination is the limit $N \to \infty$. In (20) and (21) this limit need nly be discussed on the linear problem for $\{q(t), p(t)\}$, and the contruction of $L(t,s)$ may by done in a separated form. This is a great erit in view of the famous non-existence of $P(t)$ process for the quanized Ornstein-Uhlenbeck problem found by Ford, Kac and Mazur [1]. We herefore abandon generality, and proceed by adopting the restriction

[5] in full.

Our concern is in the observables formed with $\{Q(t),P(t)\}$. We need to specify commutation relations (CR's) and expectation functions (EF's) of $\{q(t),p(t)\}$, observe their limiting forms for $N\to\infty$, construct a minimal representation of these limiting CR's and EF's on a suitable Hilbert space $\mathcal{H}$ as an Araki-Woods construction, and finally discuss the existence and convergence of the propagator $L(t,s)$ for the generator $-(i/\hbar)V_1[q(t)]$ on $\mathcal{H}$. These will be undertaken in the following two sections.

4. Quantized Gaussian processes

We use the quantized, standard Gaussian white noise (Q-noise) of [2~4] as a basic tool for representation.

Definition 5 *Let Y be the real, linear space of test functions,*

$$Y \equiv \{\xi(t);\ t\in\mathbb{R},\ \ \Xi(\omega)=\Xi^*(-\omega)\in L^2\,[\mathbb{R},\ \ \mu(\omega)d\omega\,]\},$$
$$\mu(\omega) \equiv \tfrac{1}{2}\beta\hbar\omega\, coth(\tfrac{1}{2}\beta\hbar\omega). \tag{23}$$

Let $\mathcal{H}$ be a complex Hilbert space with the inner product (A,B), and $\Omega\in\mathcal{H}$ be a unit vector. A Q-noise $w(t)$ at inverse temperature β is a real linear mapping from $\xi\in Y$ to a self-adjoint operator $w(\xi)$ on $\mathcal{H}$, denoted formally $w(\xi)=\int_{-\infty}^{\infty}\xi(t)w(t)dt$, that gives the cyclic representation of the following:

$$e^{iw(\xi)}e^{iw(\eta)} = e^{iw(\xi+\eta)}e^{-i\sigma(\xi,\eta)},\qquad \sigma(\xi,\eta) = -i\int_{-\infty}^{\infty}\Xi^*(\omega)H(\omega)\tfrac{1}{2}\beta\hbar\omega d\omega, \tag{24}$$

$$\langle e^{iw(\xi)}\rangle = (\Omega, e^{iw(\xi)}\Omega) = e^{-s(\xi,\xi)/2},\qquad s(\xi,\eta) = \int_{-\infty}^{\infty}\Xi^*(\omega)H(\omega)\mu(\omega)d\omega. \tag{25}$$

We quote a few, relevant properties of the Q-noise [3,4], and add a statement (E) for later use.

Proposition 6 *(A) A Q-noise exists uniquely up to unitary equivalence. (B) The (cyclic) representation is primary. (C) Ω gives a β-KMS state w.r.t. the time shift τ_t: $w(s)\to w(s+t)$. (D) Ω is in the domain of any polynomials of $w(\xi)$'s. (E) Equip Y with the inner product $s(\eta,\zeta)$. If ξ_n converges in $\{Y,s(\eta,\zeta)\}$ to ξ, then $w(\xi_n)$ converges to $w(\xi)$ in the strong resolvent sense.*

(Proof of (E)) We note a basic inequality obtained from $\mu(\omega)\geq\beta\hbar|\omega|/2$,

$$|\sigma(\xi,\eta)|^2 \leq s(\xi,\xi)s(\eta,\eta). \tag{26}$$

By Trotter's theorem (cf. Theorem VIII.21 of [12]) we need only to prove

$$1 = \underset{n\to\infty}{\text{s-lim}}\, \exp[iw(\xi)]\exp[-iw(\xi_n)] = \underset{n\to\infty}{\text{s-lim}}\, \exp[iw(\xi-\xi_n)]\exp[i\sigma(\xi,\xi_n)].$$

ince $|\sigma(\xi,\xi_n)|^2=|\sigma(\xi,\xi_n-\xi)|^2\leqq s(\xi,\xi)s(\xi_n-\xi,\xi_n-\xi)$ hold, and since $\{e^{iw(\eta)}\Omega;\ \eta\in Y\}$ spans a linear space dense in $\mathcal{H}$, the above follows from $0=\lim_{n\to\infty}\|\{\exp[iw(\xi-\xi_n)]-1\}e^{iw(\eta)}\Omega\|$ with the norm $\|\cdots\|$ in $\mathcal{H}$. The latter is at once from (25). □

From (24) and (25) we also have useful formulas, the first being valid on a dense domain including Ω,

$$[w(\xi),w(\eta)]=2i\sigma(\xi,\eta),\quad \langle w(\xi)\rangle=0,\quad \langle w(\xi)\circ w(\eta)\rangle=s(\xi,\eta). \tag{27}$$

Here $A\circ B\equiv(AB+BA)/2$. After Itô [13] we summarize the following, which are obtained from these.

Lemma 7 *Let the bounded measure $\nu(\omega)$ of (1) be absolutely continuous. Let $\mathcal{S}$ denote the space of real, rapidly decreasing functions of Schwartz. Let $G=\{w(\xi);\ \xi\in Y\}$ be equipped with the inner product $\langle\langle A,B\rangle\rangle\equiv\langle A\circ B\rangle$. Define with $R(\omega)=[\nu'(\omega)]^{1/2}e^{i\chi(\omega)}$,*

$$f(t)=\int_{-\infty}^{\infty}\rho(t-s)w(s)ds=w\{\mathcal{F}^{-1}[e^{-i\omega t}R^*(\omega)]\}, \tag{28}$$

for any real function $\chi(\omega)=-\chi(-\omega)$. $f(t)$ gives a linear continuous mapping from $\xi(t)\in\mathcal{S}$ to a self-adjoint operator $f(\xi)\in G$ by

$$f(\xi)=\int_{-\infty}^{\infty}\xi(t)f(t)dt=w\{\mathcal{F}^{-1}[\Xi(\omega)R^*(\omega)]\}, \tag{29}$$

with the commutator- and the covariance tempered distributions

$$[f(s),f(t)]=i\beta\hbar\gamma'(s-t)=-i\beta\hbar\int_0^{\infty}\omega\sin[\omega(s-t)]\nu'(\omega)d\omega, \tag{30}$$

$$\langle f(s)\circ f(t)\rangle=\gamma_{\beta\hbar}(s-t)=\int_0^{\infty}\cos[\omega(s-t)]\nu'(\omega)\mu(\omega)d\omega. \tag{31}$$

Ω gives a β-KMS state for $f(t)$ w.r.t. the time shift $\tau_t f(s)=f(s+t)$.

If the condition

$$\int_0^{\infty}\omega\nu'(\omega)d\omega<\infty, \tag{32}$$

or equivalently, $\int_0^{\infty}\nu'(\omega)\mu(\omega)d\omega<\infty$ holds, then $f(t)$ of (28) is a well-defined operator in G, and by Proposition 6 (E) it is t-continuous in the strong resolvent sense. It gives a representation of a portion of the quantal Gaussian processes whose existence was shown by Lewis and Thomas [7].

Lemma 8 *Let the bounded measure $\nu(\omega)$ be absolutely continuous, with $\nu'(\omega)$ satisfying (13), [a] and [b] of Lemma 4 and also (32). Let $\gamma(\tau)$ be defined by (1), and take a Gaussian stationary operator process $f(t)$ of Lewis-Thomas class defined by (28) on $\mathcal{H}$ with the properties (30) and (31). The quantized SIDE (Q-SIDE, for short),*

$$dq/dt = p/M, \qquad dp/dt = -M\Omega^2 q - (2M)^{-1}\int_{-\infty}^{t} \gamma(t-s)p(s)ds + \sigma f(t), \tag{33}$$

defined by its solution as follows with $\{\phi(\tau),\psi(\tau)\}$ *of (18),*

$$q(t) = \int_{-\infty}^{t} \phi(t-s)f(s)ds = w\{\mathcal{F}^{-1}[e^{-i\omega t}\phi^*(\omega)R^*(\omega)]\},$$
$$p(t) = \int_{-\infty}^{t} \psi(t-s)f(s)ds = w\{\mathcal{F}^{-1}[e^{-i\omega t}\psi^*(\omega)R^*(\omega)]\}, \tag{34}$$

realize on $\mathcal{H}$ *the limiting CR's and EF's of (20) as* $v_N(\omega)$ *of (7) converges to* $v(\omega)$ *from below as* $N\to\infty$. *Specifically,* $[q(t),p(t)]=i\hbar$ *holds.*

For the proof we use the following.

Corollary 9 *Let* $\{a(t),b(t)\}$ *be the coordinate and the momentum of a quantal harmonic oscillator with Hamiltonian* $h=m\omega^2a^2/2+b^2/(2m)-\hbar\omega$. *Denote by* $\{\tilde{a}(t),\tilde{b}(t)\}$ *the corresponding, classical solution, and assume that both systems are in equilibrium at inverse temperature* β, $\langle\cdots\rangle$ *and* $E\{\cdots\}$ *being the quantal and the classical expectations. If* c_k,d_k *are real constants and if* $x_k(t)=c_ka(t)+d_kb(t)$ *fulfils KMS condition w. r.t.* $\tau_t:x_k(s)\to x_k(s+t)$, *then* $d_k=c_k/(m\omega)$ *must hold with*

$$E\{x_k(s)x_\ell(t)\} \equiv \tilde{\gamma}_{k\ell}(s-t;\omega) = c_kc_\ell[2/(m\omega^2\beta)]\cos[\omega(s-t)],$$
$$[x_k(s),x_\ell(t)] = i\beta\hbar\tilde{\gamma}'_{k\ell}(s-t;\omega), \qquad \langle x_k(s)\circ x_\ell(t)\rangle = \mu(\omega)\tilde{\gamma}_{k\ell}(s-t;\omega).$$

(Proof) With a suitable choice of the origin of t we have $y(t)=(c_k^2+d_k^2\omega^2)^{1/2}x_k(t)=a(0)\cos(\omega t)+[b(0)/(m\omega)]\sin(\omega t+\theta)$, $0\leq\theta<2\pi$. The density matrix $\propto\exp(-\beta h)$ gives $m\omega^2\beta\langle y(s)\circ y(t)\rangle/[2\mu(\omega)]=\cos[\omega(s-t)]+\sin(\omega\theta)\cos[\omega(s+t+\theta)]$. Thus $\theta=0$ or π must hold. The KMS condition $F(\tau)=\langle y(t)\times y(t+\tau)\rangle$, $F(\tau+i\beta\hbar)=\langle y(t+\tau)y(t)\rangle$ with $[y(s),y(t)]=-[i\hbar/(m\omega)]\sin[\omega(s-t)]$ shows that $\theta=0$ is the unique solution. Relations for $k\neq\ell$ are evident. □

(Proof of Lemma 8) By assumptions on $v(\omega)$ and $v_N(\omega)$ there hold

$$\lim_{N\to\infty}\int_0^\infty \cos(\omega\tau)dv_N(\omega) = \gamma(\tau), \qquad \lim_{N\to\infty}\int_0^\infty [-\omega\sin(\omega\tau)]dv_N(\omega) = \gamma'(\tau). \tag{35}$$

The finite Hamiltonian $H_1^{(N)}$ of (19) has N+1 normal modes in terms of which $H_1^{(N)}$ is decomposed into independent harmonic oscillators. Conversely, $\{q(t),p(t)\}$ of (20) are linear combinations of elements that satisfy (34) of Corollary 9. This stipulates, upon summation over normal modes, that $\{q(t),p(t)\}$ must be in Lewis-Thomas class whose CR's and EF's are determined by their classical covariance functions in thermal equilibrium. Lemma 3 assures that these covariance functions of $\{q(t),p(t)\}$ converge to those of $\{Q(t),P(t)\}$, the *unique* (hence normal) stationary solution of (17). It is also possible, using (35) and the equation for d^2P/dt^2, that the convergence also holds with derivatives of these covariance functions. Therefore, CR's and EF's of $\{q(t),p(t)\}$

onverge to those of operator Gaussian processes of Lewis-Thomas class hat is determined by the classical covariance functions of $\{Q(t),P(t)\}$ f (18). It is now at once to check that these CR's and EF's are re-roduced by (34), with the aid of (27). Classical (9) gives $E\{P^2(t)\}=$ $/\beta$, which implies $[q(t),p(t)]=i\hbar$ by (34). □

. Q-SIDE's, Q-SDE's and comments

We now introduce:

efinition 10 *Let $v(\omega)$ fulfil the requirements of Lemma 8, and $f(t)$ be he stationary, operator Gaussian process (28) defined by $v(\omega)$ with a -noise $w(t)$ on $\mathcal{H}$. The Q-SIDE with $\gamma(\tau)$ of (1) for $t>0$,*

$$dQ/dt = P/M, \qquad dP/dt = (i/\hbar)[V(Q),P] - (2M)^{-1}\int_0^t \gamma(t-s)P(s)ds + \sigma f(t), \tag{36}$$

s defined to give $\{Q(t),P(t)\}$ of (22) with the quantized, generalized rnstein-Uhlenbeck process $\{q(t),p(t)\}$ of (34) by the propagator $L(t,s)$ or the generator $-(i/\hbar)V_1[q(t)]$ on $\mathcal{H}$.

urther discussions on this definition will follow shortly. Lemma 8 ives the following.

emma 11 *The global solution $\{Q(t),P(t)\}$ of (36) exists for $V(Q)$ satis-ying (5).*

Proof) By (34) and Proposition 6 (E) $q(t)$ is t-continuous in the strong esolvent sense. Since $V_1(Q)$ is real, bounded and continuous, $V_1[q(t)]$ s a bounded, self-adjoint operator that is strongly continuous in t on $\mathcal{H}$. Existence of Dyson's series for $t\geq s$,

$$L(t,s) = 1 + \sum_{n=1}^{\infty} (-i/\hbar)^n \int_s^t dt_1 \int_s^{t_1} dt_2 \cdots \int_s^{t_{n-1}} dt_n V[q(t_1)] \times V[q(t_2)]\cdots V[q(t_n)], \tag{37}$$

s assured [12] as a unitary propagator on $\mathcal{H}$.

efinition 12 *The Q-SDE (3) for $V(Q)$ of (5),*

$$dQ/dt = P/M, \qquad dP/dt = (i/\hbar)[V(Q),P] - (4M)^{-1}P + \sigma w(t), \tag{38}$$

or the Q-noise $w(t)$, is defined by $Q(t)$ of (22) with the propagator (t,s) for the generator $-(i/\hbar)V_1[q(t)]$, where $q(t)$ is the quantized rnstein-Uhlenbeck coordinate process of Ford, Kac and Mazur [1],

$$q(t) = w\{\mathcal{F}^{-1}[2\sigma(2/\pi)^{1/2}e^{-i\omega t}(i\omega - 4M\omega^2 + 4M\Omega^2)^{-1}]\}. \tag{39}$$

heorem 13 *(A) For $V(Q)$ of (5) the Q-SDE (38) has a solution on $\mathcal{H}$. B) There exists a sequence $\{v_k(\omega);\ k=1,2,\cdots\}$ of measures that con-erge to ω/π as $k\to\infty$, and the sequence of the propagators $\{L_k(t,s)\}$ of*

Q-SIDE (36) for $v_k(\omega)$'s, converges strongly on $\mathcal{H}$ to $L(s,t)$ of (38) at any finite t and s.

(Proof) (A) $q(t)$ of (39) is a well-defined operator on $\mathcal{H}$ by

$$\Xi(\omega) = 2\sigma(2/\pi)^{1/2}[i\omega - 4M\omega^2 + 4M\Omega^2]^{-1} \in L^2[\mathbb{R}, \mu(\omega)d\omega].$$

Since $\mathcal{F}^{-1}[e^{-i\omega t}\Xi(\omega)]$ moves continuously in $\{Y, s(\eta,\zeta)\}$ with t, $V_1[q(t)]$ has again the necessary t-continuity. Dyson's series (37) gives a unitary propagator. (B) Let $\forall v(\omega)$ satisfy the requirements of Lemma 8, specifically (13). We normalize it so as for $v'(0)=\int_{-\infty}^{\infty}\gamma(\tau)d\tau/\pi=1/\pi$ to hold. Put $v_k'(\omega)\equiv v'(\omega/k)$ and $R_k(\omega)\equiv[v_k'(\omega)]^{1/2}$, and define $q_k(t)$ by substituting $R_k(\omega)$ into $R(\omega)$ of (34). $\mathcal{F}^{-1}\{[\Phi(\omega)R_k(\omega)]*e^{-i\omega s}\}$ is readily seen to converge to $\mathcal{F}^{-1}[\Phi^*(\omega)e^{-i\omega t}]$ in $\{Y, s(\eta,\zeta)\}$ as $k\to\infty$ and $s\to t$. By Proposition 6 (E) $V[q_k(t)]$ converges strongly on $\mathcal{H}$ to $V[q(t)]$. The principle of uniform boundedness gives a common bound for the operator norm of $\{V[q_k(t)];\ k=1,2,\cdots,\ 0\leq t\leq T\}$. Taking differences of Dyson's series for $V[q(t)]$ and $V[q_k(t)]$, and using dominated convergence theorem on a suitable, finite sum, the assertion follows. □

The feasible representation (34) of the generalized, Ornstein-Uhlenbeck process by a Q-noise is possible only when $f(t)$ has an absolutely continuous spectrum or only for infinite systems. A proof of convergence of the propagator $L(t,s)$ from finite systems remains as a problem for the future, though Theorem 13 (B) will be approving enough. Finally, we comment on the fact that the representation (34) of $q(t)$ by a Q-noise is just sufficient without redundance.

Let $\mathcal{S}_+$ denote the subspace of $\mathcal{S}$ composed of functions with supports in $\mathbb{R}_+$. Introduce subsets of Y,

$$\begin{aligned} Y_t^{(\pm)} &\equiv \{\xi(s)=\eta[\pm(s-t)];\ \eta(u)\in\mathcal{S}_+\}, \\ \bar{Y}_t^{(\pm)} &\equiv \{\xi(s)\in Y;\ \sigma(\xi,\eta)=0,\quad \forall\eta(s)\in Y_t^{(\mp)}\}. \end{aligned} \tag{40}$$

Lemma 14 *Denote $w(S)$ for $\{w(\xi);\ \xi\in S\}$. The sets $w(\bar{Y}_t^{(\pm)})$ $(\supset w(Y_t^{(\pm)}))$ are closed linear subspaces of $\{G,\langle\!\langle A,B\rangle\!\rangle\}$ with the properties*

$$w(\bar{Y}_s^{(-)})\subset w(\bar{Y}_t^{(-)}),\qquad w(\bar{Y}_s^{(+)})\supset w\bar{Y}_t^{(+)}),\qquad s<t, \tag{41}$$

$$\bigcap_{s=-\infty}^{\infty} w(\bar{Y}_s^{(\pm)}) = \{0\},\qquad \bigvee_{s=-\infty}^{\infty} w(\bar{Y}_s^{(\pm)}) = G. \tag{42}$$

Here $V_s A_s$ denotes the smallest closed subspace that includes all of A_s. Let the spectral measure $v(\omega)$ satisfy the requirements of Lemma 8, specifically (13). Denote $\rho_c(\tau)$ for the unique, optimal kernel [10] for the $\rho(\tau)$ of (28), and $f_c(t)$ be defined with $\rho_c(\tau)$ by (28). There holds $w(\bar{Y}_t^{(-)})=V_{s=-\infty}^{\infty}f_c(s)$.

(Proof) Manifestly, $\bar{Y}_t^{(\pm)}$ are linear subspaces of Y. Take $\{\xi_n(s)\in\bar{Y}_t^{(\pm)};\ n=1,2,\cdots\}$ that converges to ξ in $\{Y, s(\eta,\zeta)\}$. For any $\eta(s)\in Y_t^{(\mp)}$ there holds $\sigma(\xi,\eta)=0$ by $\sigma(\xi_n,\eta)=0$ and $|\sigma(\xi_n,\eta)-\sigma(\xi,\eta)|^2 \leq s(\xi_n-\xi,\xi_n-\xi)s(\eta,\eta)$. Thus $\bar{Y}_t^{(\pm)}$ so that $w(\bar{Y}_t^{(\pm)})$ are closed. A variant of (27),

$$[w(\xi),w(\eta)] = -i\beta\hbar\int_{-\infty}^{\infty}\xi'(t)\eta(t)dt, \qquad \forall\xi\in\mathcal{S}, \quad \forall\eta\in Y, \tag{43}$$

gives $\bar{Y}_t^{(\pm)} \supset Y_t^{(\pm)}$. Any $\xi(t)\in\bigcap_{s=-\infty}^{\infty}\bar{Y}_s^{(\pm)}$ fulfils $\sigma(\xi,\eta)=0$ for $\forall\eta\in\bigcup_{s=-\infty}^{\infty} Y_s^{(\mp)} \supset C_0^\infty(\mathbb{R})$. Since $C_0^\infty(\mathbb{R})$ is dense in $\{Y,s(\eta,\zeta)\}$, this and (43) prove the first of (42). The l.h.s. in the second of (42) contains $w[C_0^\infty(\mathbb{R})]$, and the assertion is manifest. Since $\mu(\omega)$ of (23) satisfies (13) by itself, and since $[v'(\omega)]^{1/2}$ as well as $[v'(\omega)\mu(\omega)]^{1/2}$ are in $L^2(\mathbb{R})$, the optimal kernel corresponding to $[v'(\omega)\mu(\omega)]^{1/2}$ has [10] the form $[R_c(\omega)M_c(\omega)]^*$, with $R_c(\omega)\equiv\mathcal{F}[\rho_c(\tau)]$. By the equivalence of $s(\xi,\eta)$ to the inner product of $L^2(\mathbb{R})$, $\int_{-\infty}^{\infty}[\Xi(\omega)M_c(\omega)]^*[H(\omega)M_c(\omega)]d\omega$, the assertions are proved. □

[1] G. W. Ford, M. Kac and P. Mazur, J. Math. Phys. 6 (1965), 504.

[2] R. Benguria and M. Kac, Phys. Rev. Lett. 46 (1981), 1.

[3] H. Maassen, thesis, Groningen University, 1982.

[4] H. Nakazawa, unpublished communication, 1982.

[5] H. Nakazawa, submitted to Prog. Theor. Phys., 1984.

[6] J. T. Lewis and L. C. Thomas, in Functional Integration and Its Applications (edited by Arthurs), Clarendon Press, Oxford, 1975.

[7] J. T. Lewis and L. C. Thomas, Ann. Inst. Henri Poincaré A22 (1975), 241.

[8] H. Mori, Prog. Theor. Phys. 33 (1965), 423.

[9] H. Nakazawa, Prog. Theor. Phys. Suppl. No. 36 (1966), 172.

[10] K. Karhunen, Arkiv för Mat. 1 (1950), 141.

[11] N. Ikeda and S. Watanabe, *Stochastic Differential Equations and Diffusion Processes*, North Holland, 1981.

[12] M. Reed and B. Simon, *Methods of Modern Mathematical Physics*, Vol. II, Academic Press, 1975.

[13] K. Itô, Mem. Coll. Sci. Kyoto Univ. A28 (1953~1954), 209.

GLEASON MEASURES AND QUANTUM COMPARATIVE PROBABILITY

Wilhelm Ochs

Fraunhofer-Institut INT, Appelsgarten 2
D-5350 Euskirchen, BRD

1. Introduction

Classical probability theory usually describes a random experiment by a probability space $(\Omega, \mathfrak{A}, \mu)$ where the measurable space $(\Omega, \mathfrak{A})$ models the event-structure of the experiment in question while the probability measure μ comprises all quantitative probability assessments of the form "$\mu(A)=q$" which quantify the chance event $A \in \mathfrak{A}$ will occur by a real number $q \in [0,1]$. Probability assessments, however, need not be made through numbers. As an alternative, one can determine, for any two events under consideration, which one has a greater chance of occurrence. Standardized in the form "B is at least as probable as A" with $A,B \in \mathfrak{A}$, these probability assessments can be represented by a binary relation on the algebra of events, $\mathfrak{A}$, and give rise to the concept of <u>comparative</u> (or <u>qualitative</u>) probability [1-4].

This paper introduces the concept of comparative probability into quantum theory and investigates its relation to quantitative quantum probability. Our investigation is confined to the simplest version of quantum probability theory where the event-structure of a quantum system is represented by the lattice $\mathsf{P}(H)$ of all orthogonal projection operators on a (usually separable) complex Hilbert space H. Quantitative probability assessments are then made by means of Gleason measures [5,6].

In the following we will use <u>Hilbert space</u> to mean a complex, not necessarily separable, Hilbert space of dimension greater than one. The least and greatest elements of the orthocomplemented lattice $\mathsf{P}(H)$ are denoted by $\mathbb{0}$ and $\mathbb{1}$, respectively, and $Q^{\perp}$ denotes the orthocomplement of Q in $\mathsf{P}(H)$. For $P \leqslant Q^{\perp}$ we write $P \perp Q$. For $\alpha \in H$, $\alpha \neq 0$, $P(\alpha)$ denotes the orthogonal projection operator (OPO) onto the 1-dimensional subspace spanned by α. The range of a mapping f is denoted by ran(f) and tr(A) denotes the trace of a trace class operator A. For further notations and details we refer the reader to [7,8].

Definition 1 Let H be a Hilbert space. A mapping $\phi:P(H)\to[0,1]$ is called a Gleason measure if ϕ is σ-orthoadditive and $\phi(1)=1$. By $G(H)$ we denote the set of all Gleason measures on $P(H)$.

Definition 2 Let H be a Hilbert space. A binary relation $\trianglelefteq$ on $P(H)$ is called a quantum comparative probability (QCP) if the following axioms are satisfied by all $P,Q,R\in P(H)$:

(i) $P\trianglelefteq Q$ or $Q\trianglelefteq P$;

(ii) If $P\trianglelefteq Q$ and $Q\trianglelefteq R$, then $P\trianglelefteq R$;

(iii) $0\trianglelefteq P\trianglelefteq 1$, $1\ntrianglelefteq 0$;

(iv) If $P\perp R$ and $Q\perp R$, then $P\trianglelefteq Q \iff (P+R)\trianglelefteq(Q+R)$;

(v) $P\trianglelefteq Q$ iff $Q^{\perp}\trianglelefteq P^{\perp}$.

As usual, we write $P\triangleleft Q$ for $Q\ntrianglelefteq P$, and $P\triangleq Q$ if $P\trianglelefteq Q$ and $Q\trianglelefteq P$. Axioms (i) to (iv) are almost literal translations of the axioms which form the generally accepted axiom system of the classical comparative probability relation [1-4]. In the classical context, Axiom (v) is a consequence of Axioms (i) to (iv) [2]. In the quantum context, however, Axiom (v) is not implied by Axioms (i) to (iv) as shown by the following example. Consider a 2-dimensional Hilbert space H and fix two OPOs $P,Q\in P(H)$ such that tr(P)=tr(Q)=1 and $PQ\neq QP$. On $P(H)$ we define a binary relation $\leqslant\cdot$ by setting $0<\cdot P<\cdot X_1\doteq X_2<\cdot Q^{\perp}<\cdot 1$ for all $X_1,X_2\in P(H)$ with $X_i\notin\{0,P,Q^{\perp},1\}$. One easily checks that $\leqslant\cdot$ satisfies Axioms (i) to (iv) but violates Axiom (v). Since property (v) seems to be indispensable for any reasonable concept of comparative probability, we postulate it as an additional axiom in the quantum case. Note in passing that the implication $P\leqq Q \Rightarrow P\trianglelefteq Q$ is a consequence of Axioms (i) to (iv) as in the classical case.

Lemma 1 Let H be a Hilbert space. Every $\phi\in G(H)$ defines a QCP $\trianglelefteq_{\phi}$ on $P(H)$ by $P\trianglelefteq_{\phi}Q :\iff \phi(P)\leq\phi(Q)$.

The proof of Lemma 1 is trivial. The inverse problem, on the other hand, seems rather difficult and gave rise to the present paper. We say that a QCP $\trianglelefteq$ on $P(H)$ can be represented by a Gleason measure $\phi_{\trianglelefteq}$ on $P(H)$ if $P\trianglelefteq Q \iff \phi_{\trianglelefteq}(P)\leq\phi_{\trianglelefteq}(Q)$ for all $P,Q\in P(H)$. So the inverse problem consists of two parts, first, which QCPs can be represented by Gleason measures, and secondly, when is such a representation unique? This paper treats mainly the uniqueness problem; the existence of representations is only considered in the case of $\dim H = 2$.

2. Results

Definition 3 Let $\trianglelefteq$ be a QCP on $\mathcal{P}(H)$. A subset $\mathcal{X}$ of $\mathcal{P}(H)$ is called $\trianglelefteq$-dense in $\mathcal{P}(H)$ if, for all $P,Q \in \mathcal{P}(H)$, $P \triangleleft Q$ implies that $P \trianglelefteq X \trianglelefteq Q$ for some $X \in \mathcal{X}$.

Theorem 1 Let H be a 2-dimensional Hilbert space and let $\trianglelefteq$ be a QCP on $\mathcal{P}(H)$. If $\mathcal{P}(H)$ contains a countable $\trianglelefteq$-dense subset, then $\trianglelefteq$ can be represented by a Gleason measure.

Proof: By Axioms (i) and (ii), $(\mathcal{P}(H), \trianglelefteq)$ is a connected preorder. Transferring the relation $\trianglelefteq$ to the quotient set $\tilde{\mathcal{P}} := \mathcal{P}(H)/\hat{=}$, we obtain a total order $(\tilde{\mathcal{P}}, \trianglelefteq)$. Assume that $\mathcal{P}(H)$ contains a $\trianglelefteq$-dense subset $\mathcal{X}$. Then the set $\tilde{\mathcal{X}} := \{[P] \in \tilde{\mathcal{P}} : P \in \mathcal{X}\}$ is $\trianglelefteq$-dense in $\tilde{\mathcal{P}}$ (where $[P]$ denotes the $\hat{=}$-equivalence class containing P), and so $(\tilde{\mathcal{P}}, \trianglelefteq)$ is a total order with a countable $\trianglelefteq$-dense subset. By a theorem of Debreu[3], there exists a function $f: \tilde{\mathcal{P}} \to \mathbb{R}$ such that $a \trianglelefteq b \iff f(a) \leq f(b)$ for all $a,b \in \tilde{\mathcal{P}}$. Moreover, this function can be chosen such that $\operatorname{ran}(f) \subseteq (0,\frac{1}{2})$. Next we define a function $\phi : \mathcal{P}(H) \to [0,1]$ by $\phi(P) := 0$ for all $P \hat{=} 0$, $\phi(P) := f([P])$ for all $0 \triangleleft P \triangleleft P^{\perp}$, $\phi(P) := \frac{1}{2}$ for all $P \hat{=} P^{\perp}$ and $\phi(P) := 1 - \phi(P^{\perp})$ for all $P^{\perp} \triangleleft P$. These equations define a function ϕ on $\mathcal{P}(H)$ which is a Gleason measure by construction. And one easily checks that $P \trianglelefteq Q$ iff $\phi(P) \leq \phi(Q)$ for all $P,Q \in \mathcal{P}(H)$. ∎

Definition 4 Two Gleason measures $\phi,\psi \in \mathcal{G}(H)$ are called QCP-equivalent (in symbol $\phi \sqsubset\!\sqsupset \psi$) iff $\trianglelefteq_{\phi} = \trianglelefteq_{\psi}$. A Gleason measure $\phi \in \mathcal{G}(H)$ is called exclusive iff $\trianglelefteq_{\phi} = \trianglelefteq_{\psi}$ implies that $\phi = \psi$ for all $\psi \in \mathcal{G}(H)$.

The relation $\sqsubset\!\sqsupset$ is in fact an equivalence relation, and one should note that $\phi \not\sqsubset\!\sqsupset \psi$ if $\phi(P) = \phi(Q)$ but $\psi(P) \neq \psi(Q)$ for some $P,Q \in \mathcal{P}(H)$. The uniqueness problem is solved if we know which Gleason measures are exclusive and which are not.

Theorem 2 Let H be a 2-dimensional Hilbert space and let $\phi \in \mathcal{G}(H)$. Then ϕ is exclusive iff $\operatorname{ran}(\phi) \subseteq \{0,\frac{1}{2},1\}$.

Proof: (I) Assume that $\mathrm{ran}(\phi)\subseteq\{0,\frac{1}{2},1\}$ and consider a $\psi\in G(H)$ with $\phi\neq\psi$. Then $P(H)$ contains an OPO P with $\mathrm{tr}(P)=1$ and $\phi(P)\neq\psi(P)$. If $\phi(P)=1$, then $\phi(P)=\phi(1)$ whereas $\psi(P)\neq\psi(1)$. If $\phi(P)=\frac{1}{2}$, then $\phi(P)=\phi(P^{\perp})$ whereas $\psi(P)\neq\psi(P^{\perp})$. If $\phi(P)=0$, then $\phi(P)=\phi(0)$ whereas $\psi(P)\neq\psi(0)$. Hence ϕ ⊄ ψ and ϕ is exclusive.

(II) Let $\mathrm{ran}(\phi)\nsubseteq\{0,\frac{1}{2},1\}$. Then $\phi(Q)\notin\{0,\frac{1}{2},1\}$ for some $Q\in P(H)$ with $\mathrm{tr}(Q)=1$. Consider the bijective mapping $f: [0,1]\to[0,1]$ given by $f(x) := x + 10^{-1}\sin(2\pi x)$. Then $\psi := f\circ\phi$ is also a Gleason measure and $\phi \sqsubset \psi$ since f is strictly monotone. $\phi(Q)\notin\{0,\frac{1}{2},1\}$ implies that $\phi\neq\psi$ and hence ϕ is not exclusive. ■

When we consider a separable Hilbert space H with $\dim H\geq 3$, we can rely on Gleason's theorem which establishes a bijective mapping $\phi\mapsto W_{\phi}$ from $G(H)$ onto the set of all positive trace class operators on H with trace one, $S(H)$, such that $\phi(P)=\mathrm{tr}(W_{\phi}P)$ for all $P\in P(H)$ [9]. An element of $S(H)$ will be called state operator (STO). A STO $W\in S(H)$ has a spectral representation of the form $W=\sum_{k=1}^{K} w_k R_k$ with $K\in\mathbb{N}$ or $K=\infty$, $w_1>w_2>\dots>0$, $0\neq R_k\in P(H)$, $R_iR_j=0$ for $i\neq j$ and $\sum_{k=1}^{K} w_k\mathrm{tr}(R_k)=1$. $P_W := \sum_{k=1}^{K} R_k$ is the OPO onto $\overline{\mathrm{ran}(W)}$. To the unique spectral representation of W, there correspond (in general many) diagonal representations of the form $W=\sum_{i=1}^{I} w_i' P(\alpha_i)$ with $I\in\mathbb{N}$ or $I=\infty$, $w_1'\geq w_2'\geq\dots>0$, $\{w_i': i=1,..,I\}=\{w_k: k=1,..,K\}$, $P(\alpha_i)\leq R_k$ iff $w_i'=w_k$, and $\{\alpha_i: i=1,..,I\}$ is an orthonormal system in H. Because of the bijection between $G(H)$ and $S(H)$, we will use the notions "QCP-equivalent" and "exclusive" also for STOs. So, $U,V\in S(H)$ are QCP-equivalent iff

$$(\forall P,Q\in P(H))\qquad \mathrm{tr}(UP)\leq\mathrm{tr}(UQ)\iff\mathrm{tr}(VP)\leq\mathrm{tr}(VQ)\ . \tag{1}$$

Lemma 2 Let H be a Hilbert space and let U,V be two QCP-equivalent STOs in $S(H)$. Then the spectral representations of U,V have the form

$$U=\sum_{k=1}^{K} u_k R_k\ ,\quad V=\sum_{k=1}^{K} v_k R_k\ ,$$

i.e. they differ at most in their eigenvalues.

Proof: As STOs, U and V have diagonal representations

$$U=\sum_{i=1}^{I} a(i)P(\alpha_i)\ ,\quad V=\sum_{j=1}^{J} b(j)P(\beta_j) \tag{2}$$

with $1 \geq a(1) \geq a(2) \geq \ldots > 0$, $1 \geq b(1) \geq b(2) \geq \ldots > 0$.

Step 1: Assume that $P_U \neq P_V$. Then there exists, without loss of generality, a vector $\gamma \in H$ such that $P_U\gamma = 0$, $P_V\gamma \neq 0$, and we have $\mathrm{tr}(UP(\gamma)) = \mathrm{tr}(U0) = 0$ but $\mathrm{tr}(VP(\gamma)) \nleq \mathrm{tr}(V0) = 0$, inconsistent with our assumption. Hence $P_U = P_V$ and $I = J$.

Step 2: Choosing $P = P(\alpha_1)$, $Q = P(\beta_1)$, we conclude from (1),(2) and Step 1 that

$$a(1) \leq \sum_{i=1}^{I} a(i)|\langle\alpha_i, \beta_1\rangle|^2 \iff \sum_{i=1}^{I} b(i)|\langle\alpha_1, \beta_i\rangle|^2 \leq b(1)$$

which implies that

$$\sum_{i=1}^{I} \{a(i) - a(1)\} |\langle\alpha_i, \beta_1\rangle|^2 \geq 0 \ . \tag{3}$$

A few simple arguments based on (3), Step 1 and on the ordering of the eigenvalues of U and V show that there is a natural number s_1 such that

$$a(1) = \ldots = a(s_1) =: u_1 > a(s_1+1),\ b(1) = \ldots = b(s_1) =: v_1 > b(s_1+1)$$

and

$$\sum_{i=1}^{s_1} P(\alpha_i) = \sum_{i=1}^{s_1} P(\beta_i) =: R_1 \ .$$

Step 3: In complete analogy to Step 2 one shows that, if the equations

$$a(s_{r-1}+1) = \ldots = a(s_r) =: u_r > a(s_r+1) \ ,\ b(s_{r-1}+1) = \ldots = b(s_r) =: v_r > b(s_r+1)$$

and (4)

$$\sum_{i=s_{r-1}+1}^{s_r} P(\alpha_i) = \sum_{i=s_{r-1}+1}^{s_r} P(\beta_i) =: R_r$$

hold for $r = 1, \ldots, m$ and some chain of integers $s_o := 0 < s_1 < \ldots < s_m$, then there exists an integer s_{m+1} with $s_{m+1} > s_m$ such that Eqs.(4) hold also for $r = m+1$. So the assertion follows by induction. ■

Theorem 3 Let H be a separable Hilbert space with $\dim H \geq 3$ and let ϕ be a Gleason measure on $\mathcal{P}(H)$ such that $\phi(X) = 0$ for some $X \in \mathcal{P}(H)$, $X \neq 0$. Then ϕ is exclusive.

Proof: Let W be the STO corresponding to ϕ and let $V \in \mathcal{S}(H)$ be any STO QCP-equivalent to W. Then, according to Lemma 2, W and V have the spectral representations

$$W = \sum_{k=1}^{K} w_k R_k \ , \quad V = \sum_{k=1}^{K} v_k R_k \ . \tag{5}$$

If $K = 1$, then $W = V = R_1/\mathrm{tr}(R_1)$. So we assume that $K > 1$. We choose a natual number $t \leq K$ and define $W' := W - w_t R_t$, $V' := V - v_t R_t$. It follows from 5) that $\overline{\mathrm{ran}(W')} = \overline{\mathrm{ran}(V')}$. Since the function $\alpha \mapsto \langle \alpha, (W'-V')\alpha \rangle$ is ontinuous on H, there is a unit vector $\gamma \in \overline{\mathrm{ran}(W')}$ such that $\langle \gamma, W'\gamma \rangle = \langle \gamma, V'\gamma \rangle > 0$. By assumption, there exists a unit vector ε ith $\varepsilon \perp \overline{\mathrm{ran}(W')}$. Consider the family $\{\eta_c : 0 \leq c \leq 1\}$ of unit vectors $_c := c\gamma + \sqrt{1-c^2}\,\varepsilon$. By construction, $P(\eta_c) \perp R_t$ and $\mathrm{tr}(W'P(\eta_c)) =$ $r(V'P(\eta_c)) = c^2 \langle \gamma, W'\gamma \rangle$. By assumption, U and V satisfy Eq.(1) for arbi- rary $P, Q \in P(H)$. If we choose $P = P(\eta_c)$ and restrict Q by the conditions $\leq R_t + (1 - P_W)$ and $\mathrm{tr}(R_t Q) \neq 0$, then Eq.(1) takes the form

$$F(c,Q) \leq w_t \iff F(c,Q) \leq v_t$$

ith $F(c,Q) := c^2 \langle \gamma, W'\gamma \rangle / \mathrm{tr}(R_t Q)$. Since, by construction, $\mathrm{ran}(F) = [0,\infty)$, e see that $w_t = v_t$. And since the index t has been picked out arbitra- ily, we conclude from Eq.(5) that W=V. ∎

Now we come to the main result of this paper.

heorem 4 Let H be an infinite-dimensional (not necessarily separa- le) Hilbert space. Then every $\phi \in G(H)$ is exclusive.

roof: (I) In part (I) of the proof we assume that H is separable and onsider an arbitrary $\phi \in G(H)$. In view of Theorem 3, we can then con- ine the proof to the case that $\phi(X) \neq 0$ for all $0 \neq X \in P(H)$. This implies hat $\overline{\mathrm{ran}(W)} = H$ if W is the STO associated with ϕ by Gleason's theorem. et V be any STO which is QCP-equivalent to W. Then there exists, by emma 2, a complete orthonormal system $\{\alpha_i : i \in \mathbb{N}\}$ in H such that

$$W = \sum_{i=1}^{\infty} a_i P(\alpha_i) \quad , \quad V = \sum_{i=1}^{\infty} b_i P(\alpha_i)$$

ith $a_i > 0$, $b_i > 0$ for all $i \in \mathbb{N}$, $a_i \searrow 0$, $b_i \searrow 0$ and $\sum_{i=1}^{\infty} a_i = \sum_{i=1}^{\infty} b_i = 1$. here is an index $k \in \mathbb{N}$ such that $|a_k - b_k| = \sup\{|a_i - b_i| : i \in \mathbb{N}\} =: \Delta$. If $=0$, then V=W. So we assume that $\Delta > 0$ and, without restriction of ge- erality, that $a_k > b_k$.

ase 1: $(\exists n > k)\ a_n \leq b_n$ In this case, we choose an index $z \in \mathbb{N}$ with $> n$, $3(a_z + b_z) \leq a_n \Delta / a_k$ and consider the unit vector $\chi := c\alpha_k + \sqrt{1-c^2}\,\alpha_z$ ith $c > 0$, $c^2 = a_n / a_k$. Setting $P = P(\alpha_n)$ and $Q = P(\chi)$, Eq.(1) yields

$$a_n \leq a_n + \left(1 - \frac{a_n}{a_k}\right) a_z \iff b_n \leq \frac{a_n}{a_k} b_k + \left(1 - \frac{a_n}{a_k}\right) b_z \ .$$

ere, the left inequality is satisfied, whereas the right inequality is violated since $b_n \geq a_n = (a_n b_k/a_k + a_n \Delta/a_k) > (a_n b_k/a_k + (1-c^2) b_z)$.

Case 2: $(\forall j>k)\ a_j > b_j$ In this case, there are indices $s,u \in \mathbb{N}$ such that $s<k<u$, $b_s > a_s$ and $3(a_u+b_u) \leq \Delta$. Consider the unit vector $\zeta := c\alpha_s + \sqrt{1-c^2}\alpha_u$ with $c>0$, $c^2=b_k/b_s$. Setting $P=P(\zeta)$ and $Q=P(\alpha_k)$, Eq.(1) yields

$$\frac{b_k}{b_s} a_s + (1 - \frac{b_k}{b_s}) a_u \leq a_k \iff b_k + (1 - \frac{b_k}{b_s}) b_u \leq b_k .$$

Here, the right inequality is violated, whereas the left inequality is satisfied since $a_k = b_k + \Delta > b_k a_s/b_s + (1-c^2) a_u$.

In both cases, the assumption $\Delta > 0$ leads to a contradiction and we conclude that $V=W$, hence that ϕ is exclusive.

(II) In the second part of the proof we assume that H is nonseparable and that $G(H)$ contains a nonexclusive Gleason measure. Then $G(H)$ contains two Gleason measures ϕ, ψ such that $\phi \sqsupset \psi$ and $\phi(Q) \neq \psi(Q)$ for some $Q \in P(H)$. In what follows, $|P|$ denotes the cardinality of the complete orthonormal systems in $\overline{\mathrm{ran}(P)}$; hence, $|P| = \mathrm{tr}(P)$ for finite $|P|$. Without restriction of generality, we assume that $|Q| \geq |Q^\perp|$. Hence we can choose a partition $Q=T_1+T_2$ of Q in OPOs T_i such that $|T_1|=|T_2|=|Q|$ and we define $R:=Q$ in case of $|Q|=|Q^\perp|$, $R:=T_1$ in case of $|Q|>|Q^\perp|$ and $\phi(T_1) \neq \psi(T_1)$, and $R:=T_1+Q$ in case of $|Q|>|Q^\perp|$ and $\phi(T_1)=\psi(T_1)$. In all cases, we have $|R|=|R^\perp|=|1|$ and $\phi(R) \neq \psi(R)$. Then there exists a tensor decomposition $H = K_1 \otimes K_2$ of H into an infinite-dimensional separable factor K_1 and a factor K_2 with $\dim K_2 = \dim H$ such that $R = S \otimes 1_2$ and $R^\perp = S^\perp \otimes 1_2$ for a suitable OPO $S \in P(K_1)$. We define the functions $\hat{\phi}: P(K_1) \to \mathbb{R}$ and $\hat{\psi}: P(K_1) \to \mathbb{R}$ by $\hat{\phi}(P) := \phi(P \otimes 1_2)$, $\hat{\psi}(P) := \psi(P \otimes 1_2)$. One easily checks that $\hat{\phi}$ and $\hat{\psi}$ are QCP-equivalent Gleason measures in $G(K_1)$ with $\hat{\phi}(S) \neq \hat{\psi}(S)$. But this contradicts part(I) of the proof and thus we conclude that $G(H)$ contains only exclusive Gleason measures. ■

The results so far show that "most" Gleason measures are exclusive. As illustrated by the following example, the situation is quite different for classical probability measures (for which the notions of exclusiveness and CP-equivalence can simply be adopted).

Example: Fix three natural numbers k,m,n with $k<m<n$. To every real number a with $0<a<(2n)^{-1}$, associate the probability measure μ_a on

$\mathbb{N}, \mathcal{P}(\mathbb{N}))$ defined by $\mu_a(X) := \frac{1}{n}|X \cap N| + a\hat{X}(k) - a\hat{X}(m)$ for all $X \subseteq \mathbb{N}$ here $N := \{1,..,n\}$, $|M|$ denotes the cardinality of the set M and $\hat{X}$ denotes the characteristic function of X on $\mathbb{N}$. Hence

$$\mu_a(X) \leq \mu_a(Y) \Longleftrightarrow na\left[\hat{X}(k)-\hat{X}(m)-\hat{Y}(k)+\hat{Y}(m)\right] \leq |Y \cap N|-|X \cap N| .$$

ince $0 \leq |na[...]| < 1$ for all $X,Y \in \mathcal{P}(\mathbb{N})$, we conclude that all probability measures μ_{xa} with $0 < x \leq 1$ are CP-equivalent and hence nonexclusive. On the other hand, $\mu_{xa}(M) = 0$ for all $M \in \mathcal{P}(\mathbb{N})$ with $M \cap N = \emptyset$, nd $\mathcal{P}(\mathbb{N})$ contains infinitely many orthogonal (or disjoint) events. o the classical counterparts of Theorems 3 and 4 do not hold.

In order to complete the characterization of exclusive Gleason measures, we quote (without proof) a result from a forthcoming paper.

emma 3 Let H be a Hilbert space with $3 \leq \dim H =: n < \infty$. Then the Gleason measure $\omega \in G(H)$, defined by $\omega(P) := n^{-1}\mathrm{tr}(P)$, is exclusive.

This lemma is a corollary to Lemma 2 since $n^{-1}\mathbb{1}$ is the STO associaed with ω.

heorem 5 Let H be a Hilbert space with $3 \leq \dim H =: n < \infty$ and let ϕ be Gleason measure in $G(H)$ such that $\phi \neq \omega$ and $\phi(P) > 0$ for all $P \neq 0$. Let W_ϕ be the STO corresponding to ϕ and let $w_1, \ldots, w_n$ be the eigenvalues f W_ϕ arranged in decreasing order and allowing for multiplicity. Then is exclusive iff there exist a $P \in P(H)$ and a $r \in \mathbb{N}$ with $r < \mathrm{tr}(P)$ uch that $\mathrm{tr}(W_\phi P) = \sum_{i=1}^{r} w_i$.

cknowledgements

e author is indebted to R.Werner for his idea to use tensor decompoition in the proof of Theorem 4, and would also like to thank D.Casrigiano for reading the manuscript.

eferences

] B.de Finetti, Ann.Inst.H.Poincaré 7(1937),1-68.

] L.J.Savage, The Foundation of Statistics, New York 1954

] D.H.Krantz,R.D.Luce,P.Suppes and A.Tversky, Foundations of Measurement(Vol.1), New York 1971

] T.Fine, Theories of Probability: An Examination of Foundations, New York 1973

[5] S.Gudder, Stochastic Methods in Quantum Mechanics, New York 1979

[6] E.Beltrametti and G.Cassinelli, The Logic of Quantum Mechanics (Encyclopedia of Mathematics,Vol.15), Reading(Mass.)1981

[7] M.Reed and B.Simon, Methods of Modern Mathematical Physics,Vol.1: Functional Analysis, New York 1972

[8] E.Pflaumann und H.Unger, Funktionalanalysis(2 Bände), Mannheim 1968(Bd.1) und 1974(Bd.2)

[9] A.Gleason, J.Rat.Mech.Anal.6(1957),885-894.

STATE CHANGE AND ENTROPIES IN QUANTUM DYNAMICAL SYSTEMS

Masanori Ohya

Department of Information Sciences
Science University of Tokyo
Noda City, Chiba 278, Japan

Introduction

Any physical system or most of more general dynamical systems can be described by using the concepts like state and observable associated with that system. In particular, a careful consideration of the dynamical change of states under some external or internal effects is important for studying physical properties of the system. Hence it is interesting from both mathematical and physical points of view to rigorously study the state change for noncommutative systems.

It is well-known that we have many different types of entropies of states in classical and quantum mechanical systems and these entropies play essential role to discuss dynamical properties of some physical systems.

In this paper, we introduce some entropies for states in C*-dynamical systems and discuss the dynamics of these entropies under the state change, based on our works [1,2,3,4,5], in order to study some irreversible processes and formulate quantum communication processes. The following problems are pertinent to our investigation:
(P1) Study dynamical transformations describing the state change. (We call this transformation a "channel" in the sequel.)
(P2) Define the entropy of a general state in C*-systems and consider its dynamical properties.
(P3) When a state φ changes to another state $\bar{\varphi}$, construct a so-called compound state expressing the correlation existing between φ and $\bar{\varphi}$.
(P4) Formulate the mutual entropy defining the amount of information

transmitted from φ to $\bar{\varphi}$.

(P5) Study the time development of the mutual entropy under a dynamical channel (semigroup).

§1 : Channels

In this section, we formulate a quantum mechanical channel as an extension of a classical channel. This concept was first introduced in Shannon's communication theory and has been extended in measure theoretic frameworks, which we will briefly review below.

Let X,Y be compact Hausdorff spaces, $\mathcal{F}_X$, $\mathcal{F}_Y$ be their Borel σ-fields and P(X), P(Y) be the set of all regular probability measures on X and Y respectively. We often call (X, $\mathcal{F}_X$, P(X)) an input space and (Y, $\mathcal{F}_Y$, P(Y)) an output space. A channel Λ^* is a mapping from P(X) to P(Y) defined by

$$\Lambda^*\varphi(Q) = \int_X \lambda(x,Q)\varphi(dx), \qquad \varphi \in P(X),\ Q \in \mathcal{F}_X, \tag{1.1}$$

where $\lambda : X \times \mathcal{F}_Y \to [0,1]$ with (i) $\lambda(x,\cdot) \in P(X)$ for each fixed $x \in X$ and (ii) $\lambda(\cdot,Q) \in B(X)$, the set of all bounded mesurable functions on X, for each fixed $Q \in \mathcal{F}_Y$. This mapping λ is often called a Markov kernel in the theory of stochastic processes. We then have the following theorem [6].

Theorem 1.1 (Umegaki): For the above channel Λ^*, there exists a mapping $\Lambda : B(Y) \to B(X)$ such that (i) $f \geqq 0 \to \Lambda(f) \geqq 0$, (ii) $f_j \downarrow 0 \to \Lambda(f_j) \downarrow 0$ and (iii) $\Lambda^*\varphi(f) = \varphi(\Lambda(f))$ for any $\varphi \in P(X)$.

Based on this theorem, we define a quantum mechanical channel as follows : Let $(\mathcal{A}, \mathfrak{S}, \alpha(R))$ be a C*-dynamical system, that is, $\mathcal{A}$ is a C*-algebra with unity I, $\mathfrak{S}$ is the set of all states on $\mathcal{A}$ and $\alpha(R)$ is a strongly continuous automorphism group on $\mathcal{A}$. We denote another

C*-dynamical system by ($\bar{\mathcal{A}}$, $\bar{\mathfrak{S}}$, $\bar{\alpha}(R)$). Then a mapping Λ^* from $\mathfrak{S}$ to $\bar{\mathfrak{S}}$ is called a channel if its dual map $\Lambda : \bar{\mathcal{A}} \to \mathcal{A}$ is completely positive [7]. Further if $\mathcal{A}$ and $\bar{\mathcal{A}}$ are von Neumann algebras, then Λ is assumed to be normal. Therefore the study of channels is strongly related to that of completely positive maps. The ergodic properties of channels have been studied in [1].

Let $I(\alpha)$ be the set of all α-invariant states, $K(\alpha)$ be the set of all KMS states w.r.t. α_t at $\beta = 1$ and $exI(\alpha)$ (resp. $exK(\alpha)$) is the set of all extreme points in $I(\alpha)$ (resp. $K(\alpha)$) in the sequel discussion.

From the next section, we will consider the problems (P2) - (P5) mentioned in Introduction when a state φ changes to another state $\bar{\varphi} = \Lambda^*\varphi$ under a channel Λ^*. We stand "CDS" for commutative (or classical) dynamical systems and "NDS" for noncommutative (or quantum) dynamical systems in the sequel sections.

§2 : Entropy in C*-Systems

The entropy of a state is a measure of the uncertainty of a system. The information obtained from a system carrying much uncertainty is more valuable than that obtained from a system carrying less uncertainty. Therefore we can regard the entropy of a state as a measure of the information carried by the state. Under this consideration, Shannon brought the notion of entropy used in thermodynamics and statistical mechanics into communication processes of information and constructed the so-called communication theroy [8,10]. The Shannon's entropy of a state (probability distribution) p $= \{p_k\}_{k=1}^{n}$ is given by

$$S(p) = - \Sigma_k p_k \log p_k .$$

On the other hand, the von Neumann entropy [9 ,11] for NDS is defined for a state expressed by a density operator ρ such as

$$S(\rho) = - \operatorname{tr} \rho \log \rho .$$

In this section, we formulate the entropy of a general state φ in C*-dynamical systems [2]: The problem (P2) stated in Introduction.

Let $\mathcal{S}$ be a weak* compact convex subset of $\mathfrak{S}$. For any $\varphi \in \mathcal{S}$, there exists [7] a maximal measure μ pseudosupported on ex $\mathcal{S}$ such that

$$\varphi = \int_{(\mathrm{ex}\,\mathcal{S})} \omega \, d\mu \quad (= \int_{\mathcal{S}} \omega \, d\mu) \tag{2.1}$$

This measure is not always unique, and we denote the set of such measures by $M_\varphi(\mathcal{S})$. Moreover put

$$D_\varphi(\mathcal{S}) = \{\mu \in M_\varphi(\mathcal{S}) \; ; \exists \{\mu_k\} \subset R^+ \text{ and } \{\varphi_k\} \subset \mathrm{ex}\,\mathcal{S} \text{ s.t. } \Sigma_k \mu_k = 1$$
$$\text{and } \mu = \Sigma_k \mu_k \delta(\varphi_k) \text{ with delta measure } \delta\},$$

$$H(\mu) = - \Sigma_k \mu_k \log \mu_k \quad \text{for any } \mu \in D_\varphi(\mathcal{S}).$$

Then the entropy of φ w.r.t. $\mathcal{S}$ is defined by

$$S^{\mathcal{S}}(\varphi) = \begin{cases} \inf \{H(\mu) \; ; \mu \in D_\varphi(\mathcal{S})\} \\ + \infty \qquad (\text{if } D_\varphi(\mathcal{S}) = \emptyset). \end{cases} \tag{2.2}$$

We introduced this entropy and studied its properties in [2]. Our entropy does depend on the set $\mathcal{S}$ chosen. Particularly the cases $\mathcal{S} = \mathfrak{S}$, $I(\alpha)$ and $K(\alpha)$ are interesting. Three entropies $S^{\mathfrak{S}}(\varphi)$ ($= S(\varphi)$ for simplicity), $S^{I(\alpha)}(\varphi)$ and $S^{K(\alpha)}(\varphi)$ are generally different even for $\varphi \in K(\alpha)$. The entropy $S^{\mathcal{S}}(\varphi)$ is the uncertainty of φ measured from the coordinate $\mathcal{S}$. We will discuss some fundamental properties of $S^{\mathcal{S}}(\varphi)$.

Our entropy is an extension of von Neumann entropy, indeed,

<u>Theorem</u> 2.1 : <u>Let $\mathcal{A} = B(\mathcal{H})$ and $\alpha_t = \mathrm{Ad}(U_t)$ with a unitary U_t, then for any state $\varphi \in \mathfrak{S}$ given by $\varphi(A) = \operatorname{tr} \rho A$ for any $A \in \mathcal{A}$ with a density operator $\rho \in T(\mathcal{H})_{+,1}$, the set of all positive trace class operators on $\mathcal{H}$ with trace = 1, we have the following</u> : (1) $S(\varphi) = -\operatorname{tr} \rho \log \rho$; (2) <u>if φ is an α-invariant faithful state and every eigenvalue of ρ is nondegenerate, then</u> $S^{I(\alpha)}(\varphi) = S(\varphi)$; (3) <u>if</u> $\varphi \in$

$K(\alpha)$, <u>then</u> $S^{K(\alpha)}(\varphi) = 0$.

<u>Sketch</u>: (1): Let $\rho = \Sigma_k \lambda_k \rho_k$ be a decomposition of ρ into extremal (pure) states ρ_k (i.e., $\rho_k^2 = \rho_k$ for each k). It is known that $-\Sigma_k \lambda_k \log \lambda_k$ attains to the minimum value when λ_k is the eigenvalue of ρ (the eigenvalue of multiplicity n is repeated precisely n times) and ρ_k is the one-dimensional projection from $\mathcal{H}$ to the subspace generated by a pairwise orthonormal eigenvector x_k associated with $\lambda_k : \rho_k = |x_k\rangle\langle x_k|$. Hence, $S(\varphi) = -\mathrm{tr}\rho\log\rho$. (2): Since φ is α-invariant, the equality $[u_t, \rho] = 0$ holds for all $t \in R$. From the assumptions, we have $[u_t, \rho_k] = 0$ for each ρ_k. Thus ρ_k is α-invariant for every k, by which we obtain $S(\varphi) \geqq S^{I(\alpha)}(\varphi)$. The converse inequality is proved by using the ergodic decomposition of φ. (Q.E.D.)

Some relations among $S(\varphi)$, $S^{I(\alpha)}(\varphi)$ and $S^{K(\alpha)}(\varphi)$ exist, for instance, we have

<u>Theorem</u> 2.2 : <u>For any</u> $\varphi \in K(\alpha)$, (1) $S^{I(\alpha)}(\varphi) \geqq S^{K(\alpha)}(\varphi)$; (2) $S(\varphi) \geqq S^{K(\alpha)}$; (3) <u>if our dynamical system</u> $(\mathcal{A}, \alpha(R))$ <u>is</u> G-<u>abelian on</u> φ, <u>then</u> $S(\varphi) \geqq S^{I(\alpha)}(\varphi) \geqq S^{K(\alpha)}(\varphi)$; (4) <u>if our dynamical system</u> $(\mathcal{A}, \alpha(R))$ <u>is</u> η-<u>abelian, then</u> $S^{I(\alpha)}(\varphi) = S^{K(\alpha)}(\varphi)$.

Most of properties (e.g., positivity, concavity and additivity) satisfied by von Neumann entropy also hold for our entropy $S^{\mathcal{S}}(\varphi)$ under some conditions [2]. For any state φ of a finite convex combination of extreme KMS states, $S^{K(\alpha)}(\varphi)$ will change w.r.t. the temperature, which might open the possibility to discuss some phase transition by our entropy.

See [2] for the complete proof of Theorem 2.1 and the proof of Theorem 2.2.

§3 : Compound State

When a state φ changes to another state $\bar{\varphi} = \Lambda^*\varphi$ under a channel Λ^*, there exists some correlation between φ and $\bar{\varphi}$. This correlation is expressed by the compound state (measure) in CDS, which is given by

$$\Phi(Q_1,Q_2) = \int_{Q_1} \lambda(x,Q_2)\varphi(dx), \quad Q_1 \in \mathcal{F}_X,\ Q_2 \in \mathcal{F}_Y. \tag{3.1}$$

As an example, when $\varphi = \{p_k\}_{k=1}^n$ and $\Lambda^* = (p_{i,j})_{i,j=1}^n$ (transition matrix), the compound state Φ is the joint probability distribution :

$$\Phi = \{p(i,j)\} \quad \text{with } p(i,j) = p_{i,j}p_j. \tag{3.2}$$

In order to define the quantum mechanical compound state expressing the correlation existing between an initial state φ and a final state $\Lambda^*\varphi$, we use the decomposition (2.1) in §2. When a state $\varphi \in \mathcal{S}$ changes to $\bar{\varphi} = \Lambda^*\varphi$, the compound state w.r.t. $\mathcal{S}$ and μ is a state on the tensor product C*-algebra $\mathcal{A} \otimes \bar{\mathcal{A}}$ and defined by ([3,4])

$$\Phi_\mu = \int_{(\mathrm{ex}\,\mathcal{S})} \omega \otimes \Lambda^*\omega \, d\mu. \tag{3.3}$$

When $\mathcal{A}$ is an abelian C*-algebra and $\mathcal{S} = \mathfrak{S}$, it is easily shown [3] that (3.3) is reduced to the classical compound state (3.1). Before closing this section, we give a useful example of the compound state. For $\mathcal{A} = B(\mathcal{H})$ and a density operator $\rho \in T(\mathcal{H})_{+,1} (\subset \mathfrak{S})$, if an extremal decomposition of ρ is given by $\rho = \Sigma_n \mu_n \rho_n$, then our compound state for $\mathfrak{S}$ is

$$\sigma = \Sigma_n \mu_n \rho_n \otimes \Lambda^*\rho_n.$$

More precisely the compound state Φ is defined by $\Phi(\cdot) = \mathrm{tr}(\sigma\ \cdot)$. In particular, let $\rho = \Sigma_n \mu_n E_n$ be a Schatten decomposition of ρ (i.e., $E_n = |x_n\rangle\langle x_n|$ with the eigenvector x_n of the eigenvalue μ_n of ρ). Then the compound state is

$$\sigma_E = \Sigma_n \mu_n E_n \otimes \Lambda^*E_n, \tag{3.4}$$

where E represents a Schatten decomposition $\{E_n\}$ of ρ because this decomposition is not unique unless every eigenvalue μ_n is nondegenerate. This compound state is useful in laser communication processes [4].

§4 : Mutual Entropy and its Dynamical Change

In this section we formulate the mutual entropy in quantum systems and discuss its time development. When a state φ of an input system dynamically changes to a state $\bar{\varphi}$ (= $\Lambda^*\varphi$) of an output system under a channel Λ^*, it is natual for us to ask how much information of φ can be transmitted to the output system. It is the mutual entropy that represents this amount of information transmitted from φ to $\bar{\varphi}$. The mutual information in CDS is given by the compound state (3.1) such as

$$I(\varphi;\Lambda^*) = S(\Phi|\Psi), \tag{4.1}$$

where $\Psi = \varphi \otimes \Lambda^*\varphi$ and $S(\Phi|\Psi)$ is the Kullback-Leibler information (relative entropy) of two measures Φ and Ψ defined by

$$S(\Phi|\Psi) = \begin{cases} \int \log(\frac{d\Phi}{d\Psi})d\Phi & (\Phi<<\Psi) \\ +\infty & (\text{otherwise}). \end{cases} \tag{4.2}$$

In the case of Shannon (i.e., the case of (3.2)), the mutual entropy is

$$I(\varphi;\Lambda^*) = \Sigma_{ij}p(i,j)\log(p(i,j)/p_j q_i) \tag{4.3}$$

with $q_i = \Sigma_j p_{i,j} p_j$. For this case (4.3), we have a fundamental theorem.

<u>Theorem</u> 4.1 (Shannon) : $I(\varphi;\Lambda^*) \leqq S(\varphi) = -\Sigma_j p_j \log p_j$.

Now we will discuss the quantum case. In order to formulate the mutual entropy, we first review the formulation of relative entropy. The relative entropy was first introduced by Umegaki [12] in σ-finite and semifinite von Neumann algebras and was extensively studied by Lindblad [13,14], which was extended in more general operator algebras by Araki [15,16] and Uhlmann [17]. In [18,19], we showed that the

Araki's relative entropy is equal to the Uhlmann's relative entropy for states in a C*-algebra, and we applied the relative entropy to sufficient statistics. In this section we assume that $\mathcal{A}$ and $\bar{\mathcal{A}}$ are σ-finite von Neumann algebras acting on Hilbert spaces $\mathcal{H}$ and $\bar{\mathcal{H}}$ resp..

Now let us define the relative entropy according to Araki [16] : For two normal states φ and ψ on a von Neumann algebra $\mathcal{N}$ represented by the vectors x and y such as $\varphi(A) = \langle x,Ax\rangle$ and $\psi(A) = \langle y,Ay\rangle$ for any $A \in \mathcal{N}$, the operator $S_{x,y}$ is defined by

$$S_{x,y}(Ay+z) = s^{\mathcal{N}}(y)A^*x,\ A \in \mathcal{N}\ ,\ s^{\mathcal{N}'}(y)z = 0, \tag{4.4}$$

on the domain $\mathcal{N}y + (I - s^{\mathcal{N}'}(y))\mathcal{K}$, where $s^{\mathcal{N}}(y)$ denotes the $\mathcal{N}$-support of the vector y and $\mathcal{K}$ is the Hilbert space on which $\mathcal{N}$ acts. By this operator $S_{x,y}$, the relative modular operator $\Delta_{x,y}$ is defined as $\Delta_{x,y} = (S_{x,y})^*\bar{S}_{x,y}$ and the relative entropy $S(\psi|\varphi)$ is given by

$$S(\psi|\varphi) = \begin{cases} -\langle y,(\log \Delta_{x,y})y\rangle & (\psi \ll \varphi) \\ +\infty & (\text{otherwise}) \end{cases} \tag{4.5}$$

In particular, if $\mathcal{N} = B(\mathcal{K})$ and φ, ψ are given by density operators σ, ρ such as $\varphi(A) = \mathrm{tr}\sigma A$ and $\psi(A) = \mathrm{tr}\rho A$ for any $A \in \mathcal{N}$, then

$$S(\psi|\varphi) = \mathrm{tr}\rho(\log\rho - \log\sigma). \tag{4.6}$$

We introduce the mutual entropy w.r.t. a normal state $\varphi \in \mathcal{S}$ (weak* compact convex subset of $\mathfrak{S}$) and a channel Λ^* as [5]

$$I^{\mathcal{S}}(\varphi;\Lambda^*) = \lim_{\varepsilon\to 0}\sup\{I_\mu^{\mathcal{S}}(\varphi;\Lambda^*)\ ;\ \mu \in F_\varphi^\varepsilon(\mathcal{S})\}, \tag{4.7}$$

$$I_\mu^{\mathcal{S}}(\varphi;\Lambda^*) = S(\Phi_\mu^{\mathcal{S}}|\Psi), \tag{4.8}$$

where $\Psi = \varphi \otimes \Lambda^*\varphi$ and $F_\mu^\varepsilon(\mathcal{S})$ is the set $\{\mu \in D_\varphi(\mathcal{S})\ ;\ S^{\mathcal{S}}(\varphi) \leqq H(\mu) < S^{\mathcal{S}}(\varphi) + \varepsilon < +\infty\}$ (remark: $F_\varphi^0(\mathcal{S}) = \{\mu \in D_\varphi(\mathcal{S})\ ;\ S^{\mathcal{S}}(\varphi) = H(\mu)\}$) or is $M_\varphi(\mathcal{S})$ if $S^{\mathcal{S}}(\varphi) = +\infty$. When $\mathcal{S} = \mathfrak{S}$, we simply write $I^{\mathcal{S}}(\varphi;\Lambda^*)$ as $I(\varphi;\Lambda^*)$ and $I_\mu^{\mathcal{S}}(\varphi;\Lambda^*)$ as $I_\mu(\varphi;\Lambda^*)$.

We call two states φ and ψ are orthogonal each other (denoted by $\varphi \perp \psi$) if their supports $s(\varphi)$ and $s(\psi)$ are orthogonal, and we call the measure $\mu \in M_\varphi(\mathcal{S})$ is orthogonal if $(\int_Q \omega d\mu) \perp (\int_{\mathcal{S}-Q}\omega d\mu)$ is satisfied for every Borel set Q in $\mathcal{S}$.

<u>Theorem</u> 4.2 : <u>For a normal state</u> φ <u>and a channel</u> Λ^*, <u>if a measure</u> μ <u>of</u> (2.1) <u>is in</u> $F_\varphi^\varepsilon(\mathfrak{S}) \cap D_\varphi(\mathfrak{S})$ <u>and is orthogonal</u>, <u>then</u> $I_\mu(\varphi;\Lambda^*) = \int S(\Lambda^*\omega|\Lambda^*\varphi)d\mu < S(\varphi) + \varepsilon$.

<u>Sketch</u>: It is enough to show this theorem for the case of $\varphi = \mu_1\varphi_1 + \mu_2\varphi_2$ with $\varphi_1 | \varphi_2$ because the general case can be shown by mathematical induction. Let x, x_k be in $\mathcal{H}$ and y, y_k be in $\bar{\mathcal{H}}$ (k=1,2) such that $\varphi(\cdot) = \langle x, \cdot x\rangle$, $\varphi_k(\cdot) = \langle x_k, \cdot x_k\rangle$ and $\Lambda^*\varphi(\cdot) = \langle y, \cdot y\rangle$ and $\Lambda^*\varphi_k(\cdot) = \langle y_k, \cdot y_k\rangle$. It is easily seen from the conditions that $\varphi_1 \otimes \Lambda^*\varphi_1 \perp \varphi_2 \otimes \Lambda^*\varphi_2$ and $\Delta_{x_k \otimes y_k, x\otimes y} = \Delta_{x_k,x} \otimes \Delta_{y_k,y}$ (k=1,2) hold, so we obtain

$$I_\mu(\varphi;\Lambda^*) = \mu_1 S(\Lambda^*\varphi_1|\Lambda^*\varphi) + \mu_2 S(\Lambda^*\varphi_2|\Lambda^*\varphi) \ (= \int S(\Lambda^*\omega|\Lambda^*\varphi)d\mu)$$

after some computation by using the theorem 3.6 of [16]. The inequality is obvious for the case of $S(\varphi) = +\infty$. When $S(\varphi) < +\infty$, we have

$$I_\mu(\varphi;\Lambda^*) = \Sigma_k \mu_k S(\Lambda^*\varphi_k|\Lambda^*\varphi) \ (= \int S(\Lambda^*\omega|\Lambda^*\varphi)dl)$$
$$\leqq \Sigma_k \mu_k S(\varphi_k|\varphi) = - \Sigma_k \mu_k \log \mu_k < S(\varphi) + \varepsilon,$$

where we used an inequality $S(\Lambda^*\varphi_k|\Lambda^*\varphi) \leqq S(\varphi_k|\varphi)$ proved in [16] and the assumption $\mu \in F_\varphi^\varepsilon(\mathfrak{S})$. (Q.E.D.)

If $\mathcal{A} = B(\mathcal{H})$ and $\bar{\mathcal{A}} = B(\bar{\mathcal{H}})$, then any normal state φ is represented by a density operator ρ such as $\varphi(A) = \mathrm{tr}\rho A$ for any $A \in \mathcal{A}$. In this case, we obtain

<u>Theorem</u> 4.3 : $I(\rho;\Lambda^*)$ $(= I(\varphi;\Lambda^*)) \leqq \min\{S(\rho), S(\Lambda^*\rho)\}$.

<u>Sketch</u>: According to Theorem 4.2, we have $I_E(\rho;\Lambda^*) \leqq S(\rho)$ for every Schatten decomposition E, which follows the inequality $I(\rho;\Lambda^*) \leqq S(\rho)$ by taking the supremum over E. Another inequality can be easily seen by routine calculation. (Q.E.D.)

The above two theorems correspond to the fundamental theorem 4.1 of Shannon.

We now consider the time development of the mutual entropy when the state change is caused by a time dependent channel Λ_t^*. Here we assume that $\bar{\mathcal{A}} = \mathcal{A}$ and $\Lambda(R^+) = \{\Lambda_t ;\ t \in R^+\}$ is a dynamical semigroup on $\mathcal{A}$ (i.e., $\Lambda(R^+)$ is a weakly* continuous semigroup and Λ_t^* is a channel) having at least one faithful normal stationary state ϑ (i.e., $\Lambda_t^*\vartheta = \vartheta$ for any $t \in R^+$). For $\Lambda(R^+)$, put

$$\mathcal{A}_\Lambda = \{A \in \mathcal{A} ;\ \Lambda_t(A) = A,\ t \in R^+\},$$
$$\mathcal{A}_C = \{A \in \mathcal{A} ;\ \Lambda_t(A^*A) = \Lambda_t(A^*)\Lambda_t(A),\ t \in R^+\}.$$

Then $\mathcal{A}_\Lambda$ is a von Neumann subalgebra of $\mathcal{A}$ and there exists a unique conditional expectation $\mathcal{E}$ from $\mathcal{A}$ to $\mathcal{A}_\Lambda$ [20].

<u>Theorem</u> 4.4 : <u>Under the same conditions of Theorem</u> 4.2, <u>if</u> $\mathcal{A}_\Lambda = \mathcal{A}_C$ <u>holds and</u> $\mathcal{A}$ <u>is type</u> I, <u>then</u> $I_\mu(\varphi;\Lambda_t^*)$ <u>decreases to</u> $I_\mu(\varphi;\mathcal{E}^*)$ <u>as</u> $t \to +\infty$.

<u>Sketch</u>: Since $\mathcal{A}_\Lambda = \mathcal{A}_C$ and $\mathcal{A}$ is type I, $\Lambda_t^*\omega$ converges to $\mathcal{E}^*\omega$ in norm for any normal states ω. Hence

$$\|\Lambda_t^*\varphi - \mathcal{E}^*\varphi\| \to 0 \text{ and } \|\Lambda_t^*\varphi_k - \mathcal{E}^*\varphi_k\| \to 0 \quad (t\to\infty);$$

where $\varphi = \Sigma_k \mu_k \varphi_k$ with $\varphi_i \perp \varphi_j$ $(i \neq j)$. As there exists a constant $\lambda_k \in R^+$ satisfying $\varphi_k \leqq \lambda_k \varphi$ for each k, the inequality $\Lambda_t^*\varphi_k \leqq \lambda_k \Lambda_t^*\varphi$ holds for all $t \in R^+$. Therefore the theorem 3.7 of [16] applies and we obtain

$$\lim_{t\to+\infty} S(\Lambda_t^*\varphi_k | \Lambda_t^*\varphi) = S(\mathcal{E}^*\varphi_k | \mathcal{E}^*\varphi).$$

This equality and the equality given in the proof of Theorem 4.2 concludes the existence of $\lim_{t\to\infty} I_\mu(\varphi;\Lambda_t^*)$. This limit is decreasing in time because of $S(\Lambda_{t+s}^*\varphi_k | \Lambda_{t+s}^*\varphi) \leqq S(\Lambda_t^*\varphi_k | \Lambda_t^*\varphi)$ for all $s \in R^+$. (Q.E.D.)

<u>Theorem</u> 4.5 : <u>If</u> $\mathcal{A} = B(\mathcal{H})$ <u>and</u> $\mathcal{A}_\Lambda = \mathcal{A}_C$ <u>holds, then we have the followings</u> : (1) $I(\rho;\Lambda_t^*)$ <u>decreases to</u> $I(\rho;\mathcal{E}^*)$ <u>as</u> $t \to +\infty$ <u>for any density operator</u> ρ ; <u>and</u> (2) <u>there exists only one stationary normal state</u> w.r.t. Λ_t^* <u>iff</u> $I(\rho;\mathcal{E}^*) = 0$ <u>for all</u> ρ.

Sketch: Every Schatten decomposition is discrete and orthogonal, so the convergence of $I_E(\rho;\Lambda_t^*)$ to $I_E(\rho;\mathcal{E}^*)$ as $t\to\infty$ is proved in Theorem 4.4. Hence we obtain (1) by taking the supremum over E. The statement (2) follows from some properties of the relative entropy. (Q.E.D.)

It is easy to see that $S(\Lambda_t^*\rho|\Lambda_t^*\sigma)$ is equal to $S(\rho|\sigma)$ for any density operators ρ and σ when $\Lambda(R)$ is a unitary implemented group, we immediately conclude

Theorem 4.6 : When $\Lambda(R)$ is a unitary implemented group, the mutual entropy $I(\rho;\Lambda_t^*)$ is equal to the entropy $S(\rho)$.

We apply our discussions in communication processes for Gaussian channels [21] and in quantum stochasic processes based on the work [22], which will be discussed elsewhere.

Finally we mention a few questions still unsolved: (1) Can we take off the orthogonality of μ in Theorem 4.2 ? (2) Find some suitable conditions under which Theorems 4.2 and 4.3 hold for $I(\varphi;\Lambda^*)$. (3) In the course of proof of Theorem 4.3, we obtain an equality $I_\mu(\varphi;\Lambda^*) = \int S(\Lambda^*\omega|\Lambda^*\varphi)d\mu$ for a measure μ satisfying the conditions given in Theorem 4.2. Under what conditions does this equality hold for a more general measure μ ?

References

[1] M.Ohya, J. Math. Anal. Appl., 84, 318 (1981).
[2] ______, J. Math. Anal. Appl., 100, 222 (1984).
[3] ______, L. Nuovo Cimento, 38, 402 (1983).
[4] ______, IEEE Inform. Theory, 29, 770 (1983).
[5] ______, Res. Rep at TIT, (1984).
[6] H.Umegaki, J. Math. Anal. Appl., 25, 41 (1969).
[7] M.Takesaki, "Theory of Operator Algebra I", Springer, (1981).
[8] H.Umegaki and M.Ohya, "Entropies in Probability Theory (in Japanese), Kyoritsu Shuppan, (1983).

[9] ________________, "Quantum Mechanical Entropies (in Japanese)", Kyoritsu Shuppan, (1984).
[10] C.R.Shannon, Bell System Tech. J., 27, 379 and 623 (1948).
[11] J.von Neumann, "Die Mathematischen Grundlagen der Quantenmechanik" Springer, (1932).
[12] H.Umegaki, Kodai Sem. Rep., 14, 59 (1962).
[13] G.Lindblad, Commun. Math. Phys., 39, 111 (1974).
[14] _________, Commun. Math. Phys., 40, 147 (1975).
[15] H.Araki, Publ. RIMS Kyoto Univ., 11, 809 (1976).
[16] ______, Publ. RIMS Kyoto Univ., 13, 173 (1977).
[17] A.Uhlmann, Commun. Math. Phys., 54, 21 (1977).
[18] F.Hiai, M.Ohya and M.Tsukada, Pacific J. Math., 96, 99 (1981).
[19] ____________________________, Pacific J. Math., 107, 117 (1983).
[20] A.Frigerio, Commun. Math. Phys., 63, 269 (1978).
[21] M.Ohya and N.Watanabe, Res. Rep. at Science Univ. of Tokyo, (1984)
[22] L.Accardi, A.Frigerio and J.T.Lewis, Publ. RIMS Kyoto Univ., 18, 97 (1982).

SOME REMARKS ON THE INTEGRATION OF SCHRÖDINGER EQUATION USING THE QUANTUM STOCHASTIC CALCULUS

by

K.R. Parthasarathy
Indian Statistical Institute
7 Sansanwal Marg
New Delhi 110016
India

§ 0. Introduction : In [2] a quantum stochastic calculus was developed on the basis of canonical commutation relations. This leads to a quantum Ito's formula which is useful in integrating certain irreversible equations of motion governed by semigroups of completely positive maps. Taking a hint from [1] Ito's formula was used in [3] to integrate the Schrödinger equation when the potential is the Fourier transform of a complex valued measure in $\mathbb{R}^n$. Pursuing the same line of thought we present here a formula for the Schrödinger one parameter group when the number of degrees of freedom is infinite and the potential is the Fourier transform of a complex valued measure in a Hilbert space. Quantum stochastic calculus enables us to examine the continuity of the Schrödinger group as a function of certain vector parameters.

§ 1. The Weyl Representation : Let $\mathfrak{h}$ be a complex separable Hilbert space with inner product $<\cdot,\cdot>$ which is antilinear in the first variable. We write

$$\Gamma_s(\mathfrak{h}) = \mathbb{C} \oplus \mathfrak{h} \oplus \mathfrak{h} \circledS \mathfrak{h} \oplus \cdots \oplus \mathfrak{h}^{\circledS n} \oplus \cdots$$

where $\mathfrak{h}^{\circledS n}$ denotes the n-fold symmetric tensor product of $\mathfrak{h}$ and call $\Gamma_s(\mathfrak{h})$ the symmetric or boson Fock space over $\mathfrak{h}$. For any $u \in \mathfrak{h}$, we denote by $\psi(u)$ the exponential or coherent vector defined by

$$\psi(u) = 1 \oplus u \oplus \frac{u \otimes u}{\sqrt{2!}} \oplus \cdots \oplus \frac{u^{\otimes n}}{\sqrt{n!}} \oplus \cdots$$

and observe that

$$<\psi(u), \psi(v)> = \exp <u,v>, \quad u,v \in \mathfrak{h}, \qquad (1.1)$$

the symbol $< \cdot , \cdot >$ denoting inner product in any Hilbert space. Let

$$\psi(0) = \Omega = 1 \oplus 0 \oplus 0 \oplus \ldots$$

denote the vacuum vector. Let $\mathcal{U}(\mathfrak{h})$ be the group of all unitary operators on $\mathfrak{h}$ with strong topology and let $\mathcal{E}(\mathfrak{h}) = \mathfrak{h} \odot \mathcal{U}(\mathfrak{h})$ be the semidirect product of the additive group $\mathfrak{h}$ with norm topology and $\mathcal{U}(\mathfrak{h})$. As a topological space $\mathcal{E}(\mathfrak{h})$ is the cartesian product of $\mathfrak{h}$ and $\mathcal{U}(\mathfrak{h})$ but the group multiplication is defined by

$$(u,U)\cdot(v,V) = (u + Uv,\ UV).$$

Then $\mathcal{E}(\mathfrak{h})$ is a topological group called the Euclidean group over $\mathfrak{h}$. With these notations we have the following result.

<u>Theorem 1.1</u> : For any $(u,U) \in \mathcal{E}(\mathfrak{h})$ there exists a <u>unique</u> unitary operator $W(u,U)$ on $\Gamma_s(\mathfrak{h})$ satisfying

$$W(u,U)\psi(v) = e^{-\frac{1}{2}||u||^2 - <u,Uv>}\ \psi(Uv + u) \tag{1.2}$$

for all $v \in \mathfrak{h}$. The map $(u,U) \to W(u,U)$ is strongly continuous and

$$W(u,U)W(v,V) = e^{-i\,\mathrm{Im}\,<u,Uv>}\ W(u + Uv,\ UV)\,. \tag{1.3}$$

<u>Proof</u> : The first part is immediate from the observation that the exponential vectors form a total family in $\Gamma_s(\mathfrak{h})$ and $W(u,U)$ defined by (1.2) is inner-product preserving in view of (1.1). Strong continuity is immediate from the continuity of inner products. Equation (1.3) follows from a routine verification.

<u>Remark</u> : The map $(u,U) \to W(u,U)$ is a continuous projective unitary representation of the topological group $\mathcal{E}(\mathfrak{h})$ in $\Gamma_s(\mathfrak{h})$ with multiplier σ defined by

$$\sigma((u,U),\ (v,V)) = \exp - i\ \mathrm{Im}\ <u,Uv>.$$

We call this the <u>Weyl representation</u> and $W(u,U)$ the <u>Weyl operator</u> corresponding to (u,U).

<u>Lemma 1.1</u> : Let $a(u)$, $a^{\dagger}(u)$ denote the annihilation and creation operators corresponding to $u \in \mathfrak{h}$ so that $a(u)$ is antilinear in u. Then

$$a(u)\psi(v) = \langle u,v\rangle \psi(v)$$

$$a^{\dagger}(u)\psi(v) = \frac{d}{d\varepsilon}\psi(v+\varepsilon u)\Big|_{\varepsilon=0}$$

the derivative being taken in the strong sense,

$$W(u,U)a(v)W(u,U)^{\dagger} = a(Uv) - \langle Uv,u\rangle ,$$

$$W(u,U)a^{\dagger}(v)W(u,U)^{\dagger} = a^{\dagger}(Uv) - \langle u,Uv\rangle .$$

Proof : This is immediate from the definitions of annihilation and creation operators.

Lemma 1.2 : Let $t \to u_t$ and $t \to U_t$ be strongly differentiable maps from $[0,\infty)$ into $\mathfrak{h}$ and $\mathcal{U}(\mathfrak{h})$ respectively such that

$$\sup_{0 \le t \le T} (||\dot{u}_t|| + ||\dot{U}_t||) < \infty \quad \text{for all } T > 0.$$

Let $\{e_j\}$ be a complete orthonormal basis in $\mathfrak{h}$. Then

$$\frac{d}{dt}W(u_t, U_t) = W(u_t, U_t)\{ \sum_j a^{\dagger}(U_t^{\dagger} e_j)a(\dot{U}_t^{\dagger} e_j)$$

$$+ a^{\dagger}(U_t^{\dagger}\dot{u}_t) - a(U_t^{\dagger}\dot{u}_t) + \tfrac{1}{2}(\langle u_t,\dot{u}_t\rangle - \langle \dot{u}_t,u_t\rangle)\}$$

in a domain which includes all the exponential vectors.

Proof : This is a straightforward computation using (1.2) and Lemma 1.1.

Corollary 1 : Let $\mathfrak{h}_0$ be a real Hilbert space with orthonormal basis $\{e_j\}$ and let L_0 be a bounded positive selfadjoint operator satisfying

$$L_0 e_j = \lambda_j e_j, \quad j = 1,2, \ldots .$$

Let $t \to X(t)$ be a locally bounded Borel map from $[0,\infty)$ into $\mathfrak{h}_0$. Suppose $\mathfrak{h} = \mathfrak{h}_0 + i\,\mathfrak{h}_0$ is the complexification of $\mathfrak{h}_0$,

$$U_t = e^{-itL_0}, \quad u_t = 2^{-\frac{1}{2}}\int_0^t L_0 e^{-isL_0} x(s)ds,$$

$$\phi(t) = -\tfrac{1}{2}\langle x(t), L_0 x(t)\rangle - \operatorname{Im}\langle u_t, \dot{u}_t\rangle ,$$

$$J(t,x,L_0) = e^{i\int_0^t \phi(s)ds} W(u_t, U_t) . \tag{1.4}$$

Then

$$\frac{dJ}{dt} = - i J \left(\sum_j \lambda_j\{\tfrac{1}{2}(p_j + x_j(t))^2 + \tfrac{1}{2} q_j^2 - 1\}\right) \tag{1.5}$$

where

$$a_j = a(e_j),\ a_j^{\dagger} = a^{\dagger}(e_j),$$

$$q_j = \frac{a_j + a_j^{\dagger}}{\sqrt{2}},\quad p_j = \frac{a_j - a_j^{\dagger}}{\sqrt{2}},$$

$$X_j(t) = \langle e_j, x(t)\rangle,\quad j = 1,2,\ \dots\ .$$

Equation (1.5) holds in a domain which includes all exponential vectors.

Corollary 2 : Let $\mathfrak{h}_o$ be a real separable Hilbert space and let S be a selfadjoint operator on $\mathfrak{h}_o$ such that for a complete orthonormal basis $\{e_j\}$ in $\mathfrak{h}_o$,

$$S\, e_j = \beta_j\, e_j,\quad j = 1,2,\ \dots\ ,\ \beta_j > 0 \text{ for all } j,$$

$$\sum_j (\beta_j^2 - 1)^2 < \infty\ .$$

Let $\tilde{\Lambda}(S)$ be the unitary operator defined by

$$\tilde{\Lambda}(S) = \exp - \frac{i}{2} \sum_j (\log \beta_j)(q_j p_j + p_j q_j)$$

so that

$$\tilde{\Lambda}(S)^{\dagger}\, p_j\, \tilde{\Lambda}(S) = \beta_j^{-1}\, p_j\ ,$$

$$\tilde{\Lambda}(S)^{\dagger} q_j\, \tilde{\Lambda}(S) = \beta_j\, q_j \quad \text{for all } j,$$

where p_j, q_j are defined as in Corollary 1. Define

$$J_1(t,x,L_o) = J(t,S^{-1}x,\ L_o)\tilde{\Lambda}(S) \tag{1.6}$$

where J is defined by (1.4). Then

$$\frac{dJ_1}{dt} = - iJ_1\left(\sum_j \lambda_j \beta_j^{-2}\ \{\tfrac{1}{2}(p_j + x_j(t))^2 + \tfrac{1}{2}\beta_j^4\ q_j^2 - \beta_j^2\}\right) \tag{1.7}$$

in a domain which includes all the exponential vectors of $\Gamma_s(\mathfrak{h})$, $\mathfrak{h} = \mathfrak{h}_o + i\, \mathfrak{h}_o$.

Proof : This is immediate from Corollary 1 and the fact that the unitary operator $\tilde{\Lambda}(S)$ implements the Bogoliubov automorphism $p_j \to \beta_j^{-1}\, p_j$, $q_j \to \beta_j\, q_j$ of the canonical variables p_j, q_j.

Remark : The generator of the unitary evolution defined by (1.6) is a time dependent Hamiltonian which is a superposition of independent Harmonic oscillators which are suitably shifted in the momenta and renormalised.

§ 2. Integration of Schrödinger Equations :

Combining the remarks of Section 1 and the methods of quantum stochastic calculus we shall obtain a formula for the one parameter unitary group generated by a Schrödinger operator. To this end we introduce some notation.

Let $\mathfrak{k} = L_2(\mathbb{R}^n)$, $\tilde{H} = \mathfrak{k} \otimes \Gamma_s(L_2[0,\infty) \otimes H)$ where H is a fixed complex separable Hilbert space. For any operator J on $\tilde{H}$ we define its vacuum expectation $\mathbb{E}_\Omega(J)$ as the operator on $\mathfrak{k}$ determined by the identity

$$\langle u, \mathbb{E}_\Omega(J)v\rangle = \langle u \otimes \Omega, J v \otimes \Omega\rangle \quad \text{for all} \quad u,v \in \mathfrak{k} \quad ,$$

Ω being the vacuum vector in $\Gamma_s(L_2[0,\infty) \otimes H)$.

Let P be a spectral measure on $\mathbb{R}^n$ with values in the lattice of orthogonal projections in H. Define

$$V_x = \int_{\mathbb{R}^n} e^{-ix.y} P(dy), \quad x \in \mathbb{R}^n , \tag{2.1}$$

$$\delta(x) = V_x u - u \tag{2.2}$$

where u is a fixed vector in H. Using the Weyl operators of Theorem 1.1 define the unitary operators

$$W_x(t) = e^{it \operatorname{Im} \langle u,V_x u\rangle} W(\chi_{[0,t]} \otimes \delta(x), \chi_{[0,t]} \otimes V_x + \chi_{(t,\infty)} \otimes 1)$$

in $\Gamma_s(L_2[0,\infty) \otimes H)$ where the symbol χ denotes indicator function as well as the canonical projection of multiplication by χ in $L_2[0,\infty)$. By Theorem 1.1 it follows that $\{W_x(t), t \geq 0, x \in \mathbb{R}^n\}$ is a commuting family of unitary operators and

$$W_x(t)\, W_y(t) = W_{x+y}(t),$$

$$\langle \Omega, W_x(t)\Omega\rangle = \exp t \langle u, V_x u-u\rangle .$$

Hence we can express

$$W_x(t) = \exp - i\, x \cdot \underline{X}(t) \tag{2.4}$$

where $\underline{X}(t) = (X_1(t),\ldots,X_n(t))$ is a commuting family of selfadjoint operators in $\Gamma_s(L_2[0,\infty) \otimes \mathcal{H})$. Indeed, in the vacuum state $\{\underline{X}(t), t \geq 0\}$ is a classical mixed Poisson process with 'intensity measure' $\langle u, P(\cdot)u\rangle$. Furthermore, in the language of the quantum stochastic calculus developed in [2] we have

$$dW_x = W_x(d\Lambda_{V_x-1} + dA_{\delta(-x)} + dA^{\dagger}_{\delta(x)} + \langle u, \delta(x)\rangle dt) \tag{2.5}$$

where Λ, A, $A^{\dagger}$ denote respectively the gauge, annihilation and creation processes with the suffixes indicating that their respective strengths are $\chi_{[0,t]} \otimes (V_x-1) + \chi_{(t,\infty)} \otimes 1$, $\chi_{[0,t]} \otimes \delta(-x)$ and $\chi_{[0,t]} \otimes \delta(x)$ respectively. Define the unitary operator valued adapted process $\{W(t), t \geq 0\}$ by

$$W(t) = \int_{\mathbb{R}^n} P(dx) \otimes W_x(t) \tag{2.6}$$

in the Hilbert space $\tilde{\mathcal{H}}$.

Let H_o be a real valued continuous function on $\mathbb{R}^n$. For any $w \in \mathcal{H}$, let

$$Q_w(t) = (1 \otimes \{a(\chi_{[0,t]} \otimes w) + a^{\dagger}(\chi_{[0,t]} \otimes w)\})^{\sim} \tag{2.7}$$

where $\sim$ indicates closure. Then $\{Q_w(t), t \geq 0\}$ is a selfadjoint operator valued commutative adapted process. We are now ready to state the first basic result.

<u>Theorem 2.1</u> For $u, v \in \mathcal{H}$, let

$$J_{u,v}(t) = e^{t/2(||u||^2 + ||v||^2) - i\{Q_{v+iu}(t) + t\,\mathrm{Re}\langle v,u\rangle\}} \cdot e^{-i\int_o^t H_o(\underline{p} \otimes 1 + 1 \otimes \underline{X}(s))ds} W(t) \tag{2.8}$$

where Q_w, $\underline{X}$ and W are adapted processes defined by (2.7), (2.4) and (2.6) respectively and $\underline{p}$ is the canonical vector of momentum operators in $\mathbb{R}^n$. Then

$$\mathbb{E}_{\Omega} J_{u,v}(t) = e^{-it(H_o(\underline{p}) + V(\underline{q}))} \tag{2.9}$$

where

$$V(x) = \langle v, V_x u\rangle, \quad x \in \mathbb{R}^n$$

and $\underline{q}$ is the canonical vector of position operators.

<u>Proof</u> : For a proof of this result using the quantum Ito's formula see [3].

Remark : Equation (2.9) can be looked upon as a separation of the noncommuting variables $\underline{p}$ and $\underline{q}$. It is interesting to note that $\{J_{u,v}(t)\exp-\frac{t}{2}(||u||^2+||v||^2)\}$ is a unitary operator valued processes. Furthermore formula (2.9) covers all potentials which are Fourier transforms of complex measures on $\mathbb{R}^n$. Instead of using Feynman integrals or generalised Poisson processes with complex intensity measures (see [1]) we have used noncommutative integration which is a linear and positivity preserving operation.

We shall now investigate the integration problem when the number of degrees of freedom is infinite. To this end we consider the Hilbert spaces

$$\mathfrak{k} = \Gamma_s(\mathfrak{h}), \quad \mathfrak{h} = \mathfrak{h}_o + i\,\mathfrak{h}_o$$

where $\mathfrak{h}_o$ is a real separable Hilbert space and $\mathfrak{h}$ is its complexification and

$$\tilde{H} = \mathfrak{k} \otimes \Gamma_s(L_2[0,\infty) \otimes H),$$

H being a complex separable Hilbert space. Let P be a spectral measure on the Borel σ-algebra of $\mathfrak{h}_o$ with values in the lattice of orthogonal projections in H. We define

$$V_x = \int_{\mathfrak{h}_o} e^{-i\langle x,y\rangle}\, P(dy) \tag{2.10}$$

$$\delta(x) = V_x u - u \tag{2.11}$$

$$W_x(t) = e^{it\,\mathrm{Im}\,\langle u, V_x u\rangle}\, W(\chi_{[0,t]} \otimes \delta(x),\, \chi_{[0,t]} \otimes V_x + \chi_{(t,\infty)} \otimes 1) \tag{2.12}$$

exactly as before. Then

$$W_x(t) = e^{-i\,x\cdot \underline{X}(t)}$$

where $\{\underline{X}(t),\ t \geq 0\}$ is a family of commutative selfadjoint operator valued adapted processes which can be interpreted as a mixed Poisson process with values in $\mathfrak{h}_o$ and intensity measure $\langle u, P(\cdot)u\rangle$. If $\mathfrak{h}_o = \ell_2$ over real scalars we can express

$$\underline{X}(t) = (X_1(t),\ X_2(t),\ \ldots).$$

The fact that $\Gamma_s(\mathfrak{h})$ can be identified with $L_2(\mathbb{R}^\infty, \mathcal{B}_{\mathbb{R}^\infty}, \mu)$ where μ is the probability measure of independent and identically distributed standard Gaussian random variables shows that

$$W(t) = e^{-i \sum_j q_j \otimes X_j(t)}, \quad t \geq 0 \tag{2.13}$$

is well defined as a unitary adapted process in $\tilde{H}$.

Viewing $(X_1(t), X_2(t), \ldots)$ as a sequence of random variables satisfying $\sum_j X_j(t)^2 < \infty$ we can construct the unitary operator valued adapted process $\{J_1(t, \underline{X}, L_o), t \geq 0\}$ by using (1.6) so that

$$dJ_1 = - iJ_1(\sum_j \lambda_j \beta_j^{-2}\{\tfrac{1}{2}(p_j+X_j(t))^2 + \tfrac{1}{2}\beta_j^4 q_j^2 - \beta_j^2\})dt \tag{2.14}$$

where $p_j, q_j, j = 1,2, \ldots$ is a sequence of independent canonical pairs of momentum and position operators.

Theorem 2.2 For $u, v \in H$, let

$$J_{u,v}(t) = e^{t/2(||u||^2 + ||v||^2)} e^{-i\{Q_{v+iu}(t)+t \operatorname{Re} \langle v,u\rangle\}} J_1(t,\underline{X},L_o)W(t). \tag{2.15}$$

Then

$$\mathbb{E}_\Omega[J_{u,v}(t)] = e^{-it\{\sum_{j=1}^{\infty} \lambda_j \beta_j^{-2}(\frac{1}{2} p_j^2 + \frac{1}{2}\beta_j^4 q_j^2 - \beta_j^2) + V(\underline{q})\}}$$

where

$$V(\alpha_1,\alpha_2,\ldots) = \int_{\mathfrak{h}_o} e^{-i \sum \alpha_j y_j} \langle v, P(d\underline{y})u\rangle$$

is defined almost everywhere when $(\alpha_1,\alpha_2,\ldots)$ is a sequence of independent and identically distributed standard Gaussian random variables.

Proof : Let

$$J_2(t) = e^{t/2(||u||^2 + ||v||^2)-i\{Q_{v+iu}(t) + t \operatorname{Re} \langle v,u\rangle\}}.$$

By the quantum Ito's formula we have

$$dJ_2 = -i J_2(dQ_{v+iu} + \langle v,u\rangle dt) \tag{2.16}$$

By the generalisation of (2.5) when $\mathbb{R}^n$ is replaced by $\mathcal{B}_o$ and (2.13) we have

$$dW = \int_{\mathcal{B}_o} P(dx) \otimes W(d\Lambda_{V_x - 1} + dA_{\delta(-x)} + dA^{\dagger}_{\delta(x)} + \langle u, \delta(x)\rangle dt) \quad (2.17)$$

Once again applying the quantum Ito's formula to the product $J_{u,v}(t) = J_2(t)J_1(t)W(t)$ we obtain from (2.14), (2.16) and (2.17)

$$\begin{aligned} d\,J_{u,v} = \; & J_{u,v}(-idQ_{v+iu} - i(v,u\rangle dt) \\ & - i\,J_2 J_1 (\sum_j \lambda_j \beta_j^{-2}\{\tfrac{1}{2}(p_j + X_j(t))^2 + \tfrac{1}{2}\beta_j^4 q_j^2 - \beta_j^2\}) W\,dt \\ & - J_{u,v}\int_{\mathcal{B}_o} P(dx) \otimes (d\Lambda_{V_x - 1} + dA_{\delta(-x)} + dA^{\dagger}_{\delta(x)} + \langle u, \delta(x)\rangle dt) \\ & - J_{u,v}\int_{\mathcal{B}_o} P(dx) \otimes (dA_{(V_{-x}-1)(v+iu)} + \langle v+iu, \delta(x)\rangle dt). \end{aligned} \quad (2.18)$$

Since for any canonical pair q_j, p_j satisfying $[q_j, p_j] = i$,

$$(p_j + \alpha)e^{-i\alpha q_j} = e^{-i\alpha q_j} p_j \quad \text{for all } \alpha \in \mathbb{R}$$

the second term on the right hand side of (2.18) is equal to

$$-i\,J_{u,v}\,(\sum_j \lambda_j\,\beta_j^{-2}\{\tfrac{1}{2}(p_j^2 + \beta_j^4\,q_j^2) - \beta_j^2\})dt\;.$$

Since the gauge, annihilation and creation processes have vacuum expectation zero we obtain from (2.18)

$$\begin{aligned} d\{\mathbb{E}_\Omega\,J_{u,v}(t)\} &= (\mathbb{E}_\Omega J_{u,v}(t))\{-i\sum_j \lambda_j\beta_j^{-2}(\tfrac{1}{2}\,p_j^2 + \tfrac{1}{2}\,\beta_j^4 q_j^2 \;-\; \beta_j^2) \\ &\quad + \int_{\mathcal{B}_o} \{-i\langle v,u\rangle - \langle u, \delta(x)) - i\langle v+iu, \delta(x)\rangle\}P(dx)\}dt \\ &= -\,i(\mathbb{E}_\Omega\,J_{u,v}(t))\{\sum_j \lambda_j\beta_j^{-2}(\tfrac{1}{2}\,p_j^2 + \tfrac{1}{2}\,\beta_j^4 q_j^2 - \beta_j^2) \\ &\quad + \int_{\mathcal{B}_o} \langle v, V_x u\rangle\;P(dx)\}dt\;. \end{aligned}$$

□

In Theorem 2.1 and Theorem 2.2 the adapted process $\{J_{u,v}(t),\ t \geq o\}$ has the special form

$$J_{u,v}(t) = S_{u,v}(t)R_u(t) \quad (2.19)$$

where $R_u(t)$ is unitary and

$$S_{u,v}(t) = e^{t/2(||u||^2 + ||v||^2) - i\{Q_{v+iu}(t) + t\ \mathrm{Re}\langle v,u\rangle\}}. \tag{2.20}$$

We write

$$\mathbb{E}_\Omega\, J_{u,v}(t) = \exp -it\, H_{u,v}\ .$$

<u>Theorem 2.3</u>

$$||e^{-it H_{u,v_1}} - e^{-it H_{u,v_2}}|| \le ||\phi_{u,v_1}(t) - \phi_{u,v_2}(t)||$$

where

$$\phi_{u,v}(t) = e^{it\ \mathrm{Re}\langle v,u\rangle}\ \psi(i\chi_{[0,t]} \otimes v + iu).$$

In particular, the map $(t,v) \to \exp -it\, H_{u,v}$ is continuous in operator norm.

<u>Proof</u> : Let $T_{u,v}(t) = \exp - it\, H_{u,v}$. Then for $\xi,\eta \in \mathcal{K}$

$$\langle \xi\ ,\ T^\dagger_{u,v}(t)\eta\rangle = \langle \xi \otimes \Omega,\ J^\dagger_{u,v}(t)\eta \otimes \Omega\rangle\ .$$

By (2.19) and (2.20) we have

$$\begin{aligned} ||T_{u,v_1}(t) - T_{u,v_2}(t)|| &= ||T^\dagger_{u,v_1}(t) - T^\dagger_{u,v_2}(t)|| \\ &\le ||(S^\dagger_{u,v_1}(t) - S^\dagger_{u,v_2}(t)\Omega||\ . \end{aligned} \tag{2.21}$$

Since

$$e^{iQ_w(t)} = W(\chi_{[0,t]} \otimes iw\ ,\ I)$$

we have

$$e^{iQ_w(t)}\Omega = e^{-t/2||w||^2}\ \psi(\chi_{[0,t]} \otimes iw),\quad w \in H\ .$$

Now (2.20) and (2.21) imply

$$||T_{u,v_1}(t) - T_{u,v_2}(t)|| \le ||\phi_{u,v_1}(t) - \phi_{u,v_2}(t)||\ .$$

□

References

[1] Ph. Combe, R. Rodriguez, R. Hoegh-Krohn, M. Sirigue, M. Sirigue-Collin : Generalised Poisson processes in quantum mechanics and field theory, Physics Reports Vol. 77, No. 3, 1981 (New Stochastic Methods in Physics, Ed. C. Dewitt - Morette and K.D. Elworthy, North-Holland Publishing Company, Amsterdam).

[2] R.L. Hudson and K.R. Parthasarathy : Quantum Ito's formula and stochastic evolution, Commun. Math. Phys. 93, 301-323 (1984).

[3] K.R. Parthasarathy : A remark on the integration of Schrödinger equation using quantum Ito's formula, Lett. Math. Phys 8(3), 227 - (1984).

CONVERGENCE ALMOST EVERYWHERE IN W^*- ALGEBRAS

by

Adam Paszkiewicz

§1. Introduction

1.1. In non-commutative probability theory a W^*-algebra $\mathcal{A}$ with a faithful normal state ϱ is treated as a generalization of the commutative W^*-algebra $L^\infty_\mu = L^\infty(\Omega, \mathcal{F}, \mu)$ of bounded random variables with the tracial state $\tau_\mu(f) = \int f \, d\mu$. By simple manipulations of the usual definition, it is possible to describe the notion of the almost everywhere convergence of a sequence of bounded random variables on Ω purely in terms of L^∞_μ. The generalization of this notion to the non-commutative case leads to the following three conditions which are not equivalent ([2] Theorem 1.2 d.):

(i) For any $\varepsilon > 0$, there are a projection e in $\mathcal{A}$ with $\varrho(e^\perp) < \varepsilon$ and an integer m, such that $\|(x_n - x)e\| < \varepsilon$ for all $n \geqslant m$.

(ii) For any $\varepsilon > 0$, there is a projection e in $\mathcal{A}$ with $\varrho(e^\perp) < \varepsilon$, such that $\|(x_n - x)e\| \to 0$.

(iii) For any projection $e \neq 0$ in $\mathcal{A}$, there is a projection $0 \neq f \leqslant e$, f in $\mathcal{A}$, such that $\|(x_n - x)f\| \to 0$.

If condition (i), (ii) or (iii) is fulfilled, then (x_n) is said to converge (in $\mathcal{A}$) to x (x_n, x are in $\mathcal{A}$) closely on large sets (c.l.s.), almost uniformly (a.u.) or quasi uniformly (q.u.), respectively [3], [4], [5], [6].

In the present paper the following theorem will be proved.

1.2. Theorem. For any operators x_n, x in $\mathcal{A}$, if $x_n \to x$ (c.l.s.), then $x_{n_k} \to x$ (a.u.) for some subsequence (x_{n_k}) of the sequence (x_n).

We shall also present an interesting Proposition 3.2 concerning sequence of projectors e_n in $\mathcal{A}$ satysfying $\rho(e_n^{\perp}) \to 0$.

§2

2.1. We shall start with some auxiliary properties of a sequence projectors in $\mathcal{A}$, resulting from the following Halmos's representation of two projections P, Q ([1] Theorem 2). The projections P, are said to have a generic position in a Hilbert space H if

$$P \wedge Q = P \wedge Q^{\perp} = P^{\perp} \wedge Q = P^{\perp} \wedge Q^{\perp} = 0.$$

it is the case, then $H = H' \oplus H'$,

$$P = \begin{bmatrix} 1_{H'} & 0 \\ 0 & 0 \end{bmatrix}, \qquad Q = \begin{bmatrix} C^2 & SC \\ SC & S^2 \end{bmatrix}$$

r some positive operators S, C in H', $\ker S = \ker C = 0$ and $+ C^2 = 1_{H'}$.

For any projectors P, Q in H, we can write

$$\begin{aligned} H &= H_1 \oplus H_2 \oplus H_3 \oplus H_4 \oplus H_5 , \\ P &= P_1 \oplus 1 \oplus 1 \oplus 0 \oplus 0 , \\ Q &= Q_1 \oplus 1 \oplus 0 \oplus 1 \oplus 0 , \end{aligned} \tag{1}$$

d P_1, Q_1 have a generic position in H_1.

2.2. Lemma ([2], 2.3(iii)). If P_1, Q_1 have a generic position, en

$$\|P_1 - Q_1\|^2 = \|P_1 - P_1Q_1P_1\|.$$

2.3 Proposition. For any projectors P, Q in H,

(i) $\|P - Q\| = \varepsilon < 1$ implies $\|P - PQP\| = \varepsilon^2$;

(ii) $\|P - PQP\| = \varepsilon^2$ implies $\|P - \tilde{Q}\| = \varepsilon$ for some projector $\tilde{Q} \leq Q$ ($\tilde{Q}$ belongs to $\mathcal{A}$ if P, Q belong to $\mathcal{A}$).

Proof. (i). If $\|P - Q\| = \varepsilon < 1$, then 2.1 (1) takes the form

$$H = H_1 \oplus H_2 \oplus H_5,$$
$$P = P_1 \oplus 1 \oplus 0,$$
$$Q = Q_1 \oplus 1 \oplus 0,$$

and thus, $\|P - PQP\| = \|P_1 - P_1Q_1P_1\|_{H_1} = \|P_1 - Q_1\|^2_{H_1} = \|P - Q\|^2$.

(ii). Assume $\|P - PQP\| = \varepsilon^2$. If $\varepsilon = 1$, then in 2.1 (1) we get $\|P_1 - P_1Q_1P_1\|_{H_1} = 1$ or $H_3 \neq 0$, and thus, $\|P - Q\| = 1$. If $\varepsilon < 1$, then in 2.1 (1) $H_3 = 0$, $\|P_1 - P_1Q_1P_1\|_{H_1} = \varepsilon^2$. It is enough to assume $\tilde{Q} = Q_1 \oplus 1 \oplus 0 \oplus 0$.

2.4. Proposition. If projectors e^s in $\mathcal{A}$ satisfy

$$\| e^{s+1} - e^{s+1} e^s e^{s+1} \| < \varepsilon_s^2, \qquad s \geq 1, \tag{1}$$

$\Sigma_{s=1}^{\infty} \varepsilon_s < \infty$, then there exist a strong limit $e = \lim e^s$ being a projector in $\mathcal{A}$ as well as subprojectors $\tilde{e}_s \leq e^s$ in $\mathcal{A}$ fulfilling

$$\| \tilde{e}_s - e \| \leq \Sigma_{t=s}^{\infty} \varepsilon_t, \qquad s \geq 1. \tag{2}$$

If, additionally, $\varrho(e^s - e^{s+1}) < \eta_s$, $s = 0, 1, \ldots$, $e^0 = 1$, then, obviously, $\varrho(e^{\perp}) \leq \Sigma_{s=0}^{\infty} \eta_s$.

Proof. We shall first prove the existence of a matrix of projectors from $\mathcal{A}$

$$\begin{aligned} e^1 = e_{11} &\geq e_{12} \geq \cdots \\ e^2 = \quad & e_{22} \geq \cdots \\ & \cdots\cdots\cdots\cdots \end{aligned} \qquad (3)$$

with the properties

$$\|e_{st} - e_{s+1,t}\| < \varepsilon_s, \qquad 1 \leq s < t. \qquad (4)$$

Let us assume that conditions (3), (4) are satisfied for e_{st} in $\mathcal{A}$, $1 \leq s \leq t \leq t_o - 1$, $t_o = 2, 3, \ldots$. The t_o-th column of matrix (3) will be obtained by backward induction. Let $e_{t_o,t_o} = e^{t_o}$ and we assume that there exist projectors e_{s,t_o} for $1 \leq s_o < s \leq t_o$, satisfying (4) and $e_{s,t_o} \leq e_{s,t_o-1}$ for $s_o < s < t_o$ (if $s_o < t_o - 1$). Then

$$\| e_{s_o+1,t_o} - e_{s_o+1,t_o} e_{s_o,t_o-1} e_{s_o+1,t_o} \| < \varepsilon_{s_o}^2. \quad (5)$$

Indeed, when $s_o = t_o - 1$, (5) is a consequence of (1). When $s_o < t_o - 1$, the left-hand side of (5) equals

$$\|e_{s_o+1,t_o}(e_{s_o+1,t_o-1} - e_{s_o+1,t_o-1} e_{s_o,t_o-1} e_{s_o+1,t_o-1}) e_{s_o+1,t_o}\|$$

$< \varepsilon_{t_o}^2$ (Proposition 2.3(i) and the assumption $\|e_{s_o,t_o-1} - e_{s_o+1,t_o-1}\| < \varepsilon_{s_o}$).

In view of Proposition 2.3(ii), inequality (5) implies $\| e_{s_o+1,t_o} - e_{s_o,t_o} \| < \varepsilon_{s_o}$ for some projector $e_{s_o,t_o} \leq e_{s_o,t_o-1}$ in $\mathcal{A}$. The existence of matrix (3) satisfying (4) is proved.

Let $\tilde{e}_s = \bigwedge_{t=s}^{\infty} e_{s\,t} = \lim_{t\to\infty} e_{s\,t}$ in strong operator (s.o.)

topology. Then $\tilde{e}_s$ is a projection in $\mathcal{A}$, $\tilde{e}_s \leqslant e^s$ $(s \geqslant 1)$, and, for ξ in H, $\varepsilon > 0$, an index $t \geqslant s$ can be chosen such that

$$\|(\tilde{e}_s - \tilde{e}_{s+1})\xi\| < \|(e_{st} - e_{s+1,t})\xi\| + 2\varepsilon$$

and then, by (4), $\|\tilde{e}_s - \tilde{e}_{s+1}\| \leqslant \varepsilon_s$. In consequence, the limit in norm $e = \lim \tilde{e}_s$ exists, e is a projection in $\mathcal{A}$ and (2) holds. Moreover, $e^t \rightarrow e$ in s.o. topology. Indeed,

$$\|\tilde{e}_s - e\| \leqslant \Sigma_{u=s}^{\infty} \varepsilon_u$$

and, by (4),

$$\|e_{st} - e^t\| < \Sigma_{u=s}^{t} \varepsilon_u < \Sigma_{u=s}^{\infty} \varepsilon_u, \qquad s < t.$$

For ξ in H, $\varepsilon > 0$, we can choose an index s satisfying

$$\Sigma_{u=s}^{\infty} \varepsilon_u < \varepsilon,$$

and $t_o > s$ satisfying $\|(e_{st} - \tilde{e}_s)\xi\| < \varepsilon$ for $t > t_o$; then

$$\|(e^t - e)\xi\| \leqslant \|e^t - e_{st}\| \, \|\xi\| + \|(e_{st} - \tilde{e}_s)\xi\| + \\ + \|\tilde{e}_s - e\| \, \|\xi\| < 2\varepsilon\|\xi\| + \varepsilon$$

for $t > t_o$.

§3. <u>Existence of a trapped projector</u>.

3.1 Definition. Let $\delta_m > 0$ for m in N. We shall say that a sequence of projectors e_m traps a projector e on the level (δ_m) if

$$\|\tilde{e}_s - e\| < \delta_{k_s}$$

for some projectors $\tilde{e}_s \leqslant e_{k_s}$ and numbers $k_s \nearrow \infty$.

3.2. Proposition. If $\varrho(e_m{}^{\perp}) \to 0$ for some projectors e_m in $\mathcal{A}$, then, for all numbers $\eta > 0$, $\delta_m > 0$, there exists a projector e in $\mathcal{A}$, $\varrho(e^{\perp}) < \eta$, trapped by (e_m) on the level (δ_m).

It is obvious that, for any numbers $\delta_m > 0$, there exist numbers $\varepsilon_m > 0$ such that

$$\delta_m > \Sigma_{n=m}^{\infty} \varepsilon_n, \qquad m \geqslant 1,$$

and Proposition 3.2 is equivalent to the following

3.3 Lemma. If $\varrho(e_m{}^{\perp}) \to 0$ for some projectors e_m in $\mathcal{A}$, then, for any numbers $\eta > 0$, $\varepsilon_s > 0$, $\Sigma \varepsilon_s < \infty$, there exists a projector e in $\mathcal{A}$ such that

$$\| \tilde{e}_s - e \| < \Sigma_{t=s}^{\infty} \varepsilon_t , \qquad \varrho(e^{\perp}) < \eta, \tag{1}$$

for some projectors

$$\tilde{e}_s \leqslant e_{k_s} \tag{2}$$

and numbers $k_s \nearrow \infty$.

Proof. Let us choose a number k_1 for which $\varrho(e_{k_1}{}^{\perp}) < \eta/2$ and put $e^1 = e_{k_1}$. Then we can assume that the projectors $e^1, \ldots,$ e^s in $\mathcal{A}$ and numbers $k_1 < \ldots < k_s$ in N, satisfying

$$\begin{gathered} e^t \leqslant e_{k_t}, \qquad t = 1, \ldots, s, \\ \| e^{t+1} - e^{t+1} e^t e^{t+1} \| < \varepsilon_t{}^2, \qquad \varrho(e^t - e^{t+1}) < 2^{-t-1}\eta, \end{gathered} \tag{3}$$

$t = 1, \dots, s-1$, are chosen. As $\rho(e_m^{\perp}) \to 0$, the strong convergence $e_m \to 1$ holds and $\rho(e^s - e^s e_m e^s) \to 0$ for $m \to \infty$. In consequence, for $\tilde{\varepsilon}_s = \min(\varepsilon_s, 2^{-s-2}\eta)$, the projector

$$f^s = e([0, \tilde{\varepsilon}_s^{\,2})) - e^{s\perp} \leq e^s$$

(where $e^s - e^s e_{k_{s+1}} e^s = \int \lambda\, e\,(d\lambda)$) satisfies $\rho(e^s - f^s) < 2^{-s-2}\eta$ for some $k_{s+1} > k_s$, and

$$\| f^s - f^s e_{k_{s+1}} f^s \| = \| f^s(e^s - e^s e_{k_{s+1}} e^s) f^s \| \leq \tilde{\varepsilon}_s^{\,2}.$$

Let us choose a projection $e^{s+1} \leq e_{k_{s+1}}$ in $\mathcal{A}$ for which $\| e^{s+1} - f^s \| < \tilde{\varepsilon}_s$ (according to Proposition 2.3(ii)). Then

$$\| e^{s+1} - e^{s+1} e^s e^{s+1} \| \leq \| e^{s+1} - e^{s+1} f^s e^{s+1} \| \leq \tilde{\varepsilon}_s^{\,2}$$

and $\rho(f^s - e^{s+1}) < \tilde{\varepsilon}_s \leq 2^{-s-2}\eta$. Consequently, $\rho(e^s - e^{s+1}) < 2^{-s-1}\eta$ and (3) holds for $s+1$ instead of s. By the use of projectors e^s defined by induction, the projectors $\tilde{e}_s$ and e in $\mathcal{A}$ can be formed analogically as in Proposition 2.4. Inequalities (1) and (2) are then satisfied.

§4. <u>Proof of Theorem 1.2.</u>

4.1 Lemma. If $x_n \to x$ (c.l.s.) (x_n, x in $\mathcal{A}$), then, for each $\eta > 0$, there exists a projector e in $\mathcal{A}$ such that $\|(x_{r_s} - x)e\| \to 0$ for some subsequence (x_{r_s}) of (x_n).

Proof. By assumption, there exist projectors e_m in $\mathcal{A}$ such that $\varrho(e_m^{\perp}) \to 0$, $\|(x_n - x)e_m\| < m^{-1}$ for $n \geq n_m$, n_m - an increasing sequence of indices. Let us denote by e a projector in $\mathcal{A}$, $\varrho(e^{\perp}) < \eta$, trapped by (e_m) on the level (δ_m) where $\delta_m = 1/m \, \|x_{n_m} - x\|$, according to Proposition 3.2. Then, for some sequences $k_s \nearrow \infty$, $\tilde{e}_s \leq e_{k_s}$, $\tilde{e}_s$ - projectors in $\mathcal{A}$,

$$\|(x_{n_{k_s}} - x)e\| \leq \|(x_{n_{k_s}} - x)\tilde{e}_s\| + \|x_{n_{k_s}} - x\| \, \|\tilde{e}_s - e\|$$
$$\leq k_s^{-1} + \|x_{n_{k_s}} - x\| \delta_{k_s} = 2k_s^{-1} .$$

4.2. Proof of Theorem 1.2. Let $x_n \to x$ (c.l.s.), x_n, x in $\mathcal{A}$. In virtue of Lemma 4.1, we define, by induction, subsequences $(x_n) \supset (x_{r_s^{(1)}}) \supset (x_{r_s^{(2)}}) \supset \ldots$ and projectors $e_1, e_2, \ldots$ in $\mathcal{A}$, satisfying $\varrho(e_\ell^{\perp}) \to 0$, $\|(x_{r_s^{(\ell)}} - x)e_\ell\| \to 0$ $(n \to \infty)$. Then it is enough to put $n_k = r_k^{(k)}$.

References

[1] Halmos, P. R.: Two subspaces, Trans. Amer. Math. Soc. 144, 381-389 (1969);

[2] Pazkiewicz, A.: Convergences in W* -algebras, to appear;

[3] Batty, C. J. K.: The Strong Law of Large Numbers for States and Traces of a W* -algebras, Z. Wahrscheinlichkeitstheorie verw. Gebiete 48, 177-191 (1979);

[4] Gol'dštein, M. Š.: Almost sure convergence theorems in von Neumann algebras (in Russian), J. Operator Theory, 6, 233-311 (1981);

[5] Petz, D.: Quasi-uniform ergodic theorems in von Neumann algebras, Bull. London Math. Soc. 16, 151-156 (1984);

[6] Segal, I. E.: A non-commutative extension of abstract integration, Ann. of Math. 57, 401-457 (1953).

PROPERTIES OF QUANTUM ENTROPY

Dénes Petz
Mathematical Institute, H.A.S.
H-1364, Budapest, P.O.Box 127

Introduction

Entropy is a crutial notion in thermodynamics and statistical mechanics. It is not an observable of the physical system but a functional on the states. In the mathematical formalism of quantum mechanics observables are described by selfadjoint operators affiliated with a von Neumann algebra and physical states are characterized by normal positive functionals on the algebra. When this algebra is $B(H)$ (that is, all bounded operators on a Hilbert space H) then normal positive functionals are given by density matrices.

Generalizing the classical expression of Boltzmann and Gibbs the entropy of a state ϕ was defined by J. von Neumann ([19]):

$$S(\phi) = -\mathrm{Tr}\ \rho \log \rho$$

where ρ is the density corresponding to ϕ. For convenience, we shall use the standard notation $-t \log t = \eta(t)$.

If $\{P_k\}$ is a set of mutually orthogonal projections satisfying $\sum P_k = I$ then the map $\alpha : A \mapsto \sum P_k A P_k$ describes the interaction of the quantum system with a classical apparatus. Then for every state ϕ we

$$S(\phi) \leq S(\phi \circ \alpha)$$

(monotonicity). On the other hand $S(\phi)$ is a concave function on the states. Monotonicity and concavity are important properties of the entropy and it turned out a long time ago ([19]) that both are due to the concavity of the function η. More precisely, if f is a continuous concave function then

$$S_f(\phi) = \mathrm{Tr}\ f(\rho)$$

will be concave and monotone in the sense above.

Von Neumann's entropy has a very natural extension to semifinite von Neumann algebras. On such algebras there is an abstract faithful semifinite normal trace τ which is not unique but reminds the usual trace on $B(H)$. The extension is due to I. E. Segal ([28]). Fixing τ we have a density operator A with spectral resolution $\int_0^\infty \lambda dE(\lambda)$ for every normal state ϕ and set

$$S(\phi) = \int_0^\infty \eta(\lambda)d\tau(E_\lambda)$$

where $\tau(E_\lambda)$ is a positive Borel measure on $\mathbb{R}^+$.

In the present lecture we are going to deal with relative entropy theory rather than entropy. As an analogue of the relative entropy of the information theory (referred also as I-divergence or Kullback-Leibler information for discrimination, see [14] and [7]) the relative entropy of two states on a semifinite von Neumann algebra was introduced by H. Umegaki ([31]):

$$S(\phi,\omega) = \tau(\rho_\omega(\log \rho_\omega - \log \rho_\phi))$$

where ρ_ϕ and ρ_ω are the densities of ϕ and ω , respectively. In fact, $S(\phi,\omega)$ does not depend on the chosen trace but we needed it in the definition very heavily.

Quantum mechanics is often not satisfied with semifinite von Neumann algebras so let us consider the general case when trace is not at our disposal. Assume that the algebra M acts on the Hilbert space H , moreover ϕ and ω are vectorstates:

$$\phi(a) = \langle a\Phi,\Phi\rangle \quad \text{and} \quad \omega(a) = \langle a\Lambda,\Lambda\rangle ..$$

For $\xi \in H$ we denote by $s^M(\xi)$ the smallest projection in M such that $s^M(\xi)\xi = \xi$. (It is easy to see that $s^M(\xi)H$ is the closure of $M'\xi$.) We can define a conjugate linear operator $S(\Phi,\Lambda)$ as follows.

$$S(\Phi,\Lambda)(a\Lambda+\xi) = s^M(\Lambda)a^*\Phi$$

where $a \in M$ and $\xi \in [M\Lambda]^\perp$. Since $S(\Phi,\Lambda)$ is closable and densely defined

$$\Delta(\Phi,\Lambda) = S(\Phi,\Lambda)^*\overline{S(\Phi,\Lambda)}$$

is a positive selfadjoint operator ([4], see also [5] and [23]) called relative modular operator. Now Araki's relative entropy is defined as

$$S(\phi,\omega) = \begin{cases} -\langle \log \Delta(\Phi,\Lambda)\Phi,\Phi\rangle & s(\omega) \geq s(\phi) \\ +\infty & \end{cases}$$

(Here $s(\cdot)$ denotes the support of a state.) We shall see later that

$S(\phi,\omega)$ is independent of the vector representatives Φ and Λ.

In accordance with the historical development of the entropy concept (in von Neumann algebras) we should distinguish three levels of technicalities in the rest of the paper. There are the finite dimensional, the semifinite and the general cases. The first one corresponds to finite quantum systems, in the second one the density operator technique is available and in the general case the relative modular operator is the main mathematical tool.

Concerning operator algebras our basic reference is the two volumes of [5] where all the involved mathematical objects are defined.

Quasi-entropies

Let M be a semifinite von Neumann algebra with a faithful normal semifinite trace τ. For a convex continuous function $f:[0,+\infty)\to\mathbb{R}$ a functional $S_f: M_*^+ \to \mathbb{R}\cup\{+\infty,-\infty\}$ can be defined in the following way. Assume that $\phi \in M_*^+$ has density operator ρ with spectral resolution $\int_0^\infty \lambda dE_\lambda$ then

$$S_f(\phi) = \int_0^\infty f(\lambda)d\tau(E_\lambda)$$

where the integral always exists but can be infinite. S_f is an entropy like functional and called quasi-entropy in [32]. Due to the convexity of f the functional S_f is convex on the states (see [24]).

THEOREM 1. Let M, τ and f be as above. Suppose that $f(0)=0$ and let $\alpha:M\to M$ be a linear mapping such that

(i) $\alpha(a)\geq 0$ if $a\geq 0$,

(ii) $\alpha(1)=1$,

(iii) $\tau(\alpha(a))\leq\tau(a)$ if $a\geq 0$,

then

$$S_f(\phi)\geq S_f(\phi\circ\alpha)$$

for every $\phi\in M_*^+$.

I give the proof only in the finite dimensional case. (In general one can apply an approximation argument through the spectral theorem, see [25].)

There exists a linear mapping $\alpha^*: M\to M$ satisfying the conditions (i)-(iii) such that

$$\tau(\alpha(a)b) = \tau(a\alpha^*(b))$$

for every $a,b\in M_+$. We shall show that

$$\tau(f(\alpha^*(a))) \le \tau(\alpha^*(f(a))) \quad (a\in M_+) .$$

f $\alpha^*(a) = \sum_{j=1}^n \mu_j p_j$ and $a = \sum_{k=1}^m \nu_k q_k$ are spectral resolutions then e have

$$\tau(f(\alpha^*(a))) = \sum_{j=1}^n f(\mu_j)\tau(p_j) .$$

ince $\mu_j = \tau(\alpha^*(a)p_j)/\tau(p_j) = \tau(a\alpha(p_j))/\tau(p_j) = \sum_{k=1}^m \tau(q_k a\alpha(p_j))/\tau(p_j) = \sum_{k=1}^m \nu_k \tau(q_k\alpha(p_j))/\tau(p_j)$ the Jensen inequality gives that

$$f(\mu_j) \le \sum_{k=1}^m f(\nu_k)\tau(q_k\alpha(p_j))/\tau(p_j)$$

nd we conclude

$$\sum_{j=1}^n f(\mu_j)\tau(p_j) \le \sum_{j=1}^n \sum_{k=1}^m f(\nu_k)\tau(\alpha^*(q_k)p_j) = \sum_{k=1}^n f(\nu_k)\tau(\alpha^*(q_k)) =$$

$$= \tau(\alpha^* f(a)) .$$

Now I am going to carry out a similar generalization for the re-ative entropy. To do so the convexity of the function f will not be ufficient, we need its operator convexity. I recall that a continuous unction $f\colon \mathbb{R}^+ \to \mathbb{R}$ is operator convex if

$$f(\lambda A+(1-\lambda)B) \le \lambda f(A)+(1-\lambda)f(B)$$

or any $0<\lambda<1$ and for any bounded positive selfadjoint operators A nd B acting on a Hilbert space. f is operator concave when $-f$ is perator convex. The function f is called operator monotone if $\le A\le B$ implies $f(A)\le f(B)$ for bounded positive selfadjoint operators and B . It is known that operator monotone and operator convex func-ions have nice integral representation but we do not need this. What e need is a C^*-version of the Jensen inequality. Namely, if f is n operator convex function then

$$f\left(\sum_{i=1}^n c_i A_i c_i^*\right) \le \sum_{i=1}^n c_i f(A_i) c_i^*$$

here all involved operators are bounded, $A_i\ge 0$ for $1\le i\le n$ and $\sum_{i=1}^n c_i c_i^* = I$.

The quasi-entropies $S_f^k(\phi,\omega)$ are defined for normal positive func-ionals ϕ and ω and depend on two parameters: an operator k from he algebra and a continuous function f . Namely I set

$$S_f^k(\phi,\omega) = \langle f(\Delta(\Phi,\Lambda))k\Lambda, k\Lambda\rangle$$

here Φ and Λ are vector representatives for the functionals ϕ and

ω , respectively. The concept is intimately related to Kosaki's form of the Lieb's convexity theorem (see [12]) and to Csiszár's paper in classical information theory ([7]). On the other hand, it enables us to recapture the main convexity properties of the relative entropy by quite natural application of the Jensen's operator inequality.

THEOREM 2. Let $f: \mathbb{R}^+ \to \mathbb{R}$ be an operator monotone function with $f(0)\geq 0$. Assume that M_0 and M are von Neumann algebras with positive normal functionals ϕ_0, ω_0 and ϕ, ω respectively. If $\alpha: M_0 \to M$ is a unit preserving two-positive mapping such that

$$\omega \circ \alpha \leq \omega_0 \quad \text{and} \quad \phi \circ \alpha \leq \phi_0$$

then for every $k \in M_0$ we have

$$S_f^k(\phi_0,\omega_0) \geq S_f^{\alpha(k)}(\phi,\omega) \ .$$

Proof. I suppose that M (M_0) acts on a finite dimensional Hilbert space H (H_0) and $\phi(\omega,\ \phi_0,\ \omega_0)$ is given by a cyclic and separating vector $\Phi(\Lambda,\Phi_0,\Lambda_0)$. (For the general case see [23]). The formulae

$$\begin{aligned} V_\omega(a_0\Lambda_0) &= \alpha(a_0)\Lambda \\ V_\phi(a_0\Phi_0) &= \alpha(a_0)\Phi \end{aligned} \qquad (a \in M_0)$$

define contractions $V_\phi, V_\omega: H_0 \to H$. Denote by S_0 and S the conjugate linear operators occuring in the definition of the relative modular operators $\Delta(\Phi_0,\Lambda_0)=\Delta_0$ and $\Delta(\Phi,\Lambda)=\Delta$. So we have

$$V_\phi S_0 = S V_\omega$$

and

$$V_\omega^* \Delta V_\omega = S_0^* V_\phi^* V_\phi S_0 \leq \Delta_0 \ .$$

The Jensen's operator inequality gives that

$$S_f^{\alpha(k)}(\phi,\omega) = \langle f(\Delta)\alpha(k)\Lambda,\alpha(k)\Lambda\rangle = \langle V_\omega^* f(\Delta)V_\omega k\Lambda_0,k\Lambda_0\rangle \leq$$

$$\leq \langle f(V_\omega^*\Delta V_\omega)k\Lambda,k\Lambda\rangle \leq \langle f(\Delta_0)k\Lambda,k\Lambda\rangle = S_f^k(\phi_0,\omega_0) \ .$$

The finite dimensional case has been proved and it follows from the theorem that $S_f^k(\phi,\omega)$ does not depend on the vector representatives Φ and Λ . Other consequences are contained in the following

COROLLARY 3. Let $f: \mathbb{R}^+ \to \mathbb{R}$ be an operator monotone function and let ϕ, ω be normal positive functionals on the von Neumann algebra M . Then

(i) When M_0 is a subalgebra of M and $k \in M_0$ then

$$S_f^k(\phi|M_0, \omega|M_0) \geq S_f^k(\phi,\omega) .$$

(ii) $S_f^1(\phi,\omega) \leq f(\phi(1)/\omega(1))\omega(1)$.

(iii) If ϕ and ω are states then

$$S_f^1(\phi,\omega) \leq f(1)$$

nd for non-linear f the equality holds if and only if $\phi=\omega$.

Here (i) is a direct consequence of Theorem 2. We get (ii) by utting $\mathbb{C}\cdot 1$ in the role of M_0 in statement (i). To prove (iii) we onsider the Jordan decomposition of $\phi-\omega$ and let M_0 be the commuative subalgebra generated by the support of the positive part. Then

$$\|\phi-\omega\| = \|\phi_0-\omega_0\|$$

here ϕ_0 and ω_0 are the restrictions of ϕ and ω , respectively. n M_0 we apply the classical result ([7]) since f is strictly conave. Hence $S_f^1(\phi_0,\omega_0)=f(1)$ implies that $\phi_0=\omega_0$ and we obtain $\phi=\omega$.

The next theorem is proved by Kosaki ([12]). Instead of interpolaion methods and the integral representation of operator monotone funcions it can be deduced from Theorem 2.

THEOREM 4. Assume that $f:\mathbb{R}^+ \to \mathbb{R}$ is operator monotone and (0)=0 . Let $\phi_1, \phi_2, \omega_1, \omega_2$ and ω be positive normal functionals n the von Neumann algebra M such that

$$\lambda\phi_1+\mu\phi_2 \leq \phi \quad \text{and} \quad \phi\omega_1+\mu\omega_2 \leq \omega .$$

f $k \in M$ and $\lambda,\mu>0$ then

$$\lambda S_f^k(\phi_1,\omega_1)+\mu S_f^k(\phi_2,\omega_2) \leq S_f^k(\phi,\omega) .$$

This proposition formulates a stronger form of concavity (may be alled Lieb-Kosaki concavity). I note that if ω is supposed to be aithful then both Theorem 2 and 4 can be proven under the weaker asumption that f is operator concave (see [23] for the details).

Now I consider some particular cases. Let $\eta(t)=-t \log t$ be as bove. Araki's relative entropy may be expressed as a quasi-entropy:

$$S(\omega,\phi) = -\langle\Delta(\phi,\omega)^{1/2}J \log \Delta(\omega,\phi)J\Delta(\phi,\omega)^{1/2}\Lambda, \Lambda\rangle =$$

$$= \langle\eta(\Delta(\phi,\omega))\Lambda,\Lambda\rangle = S_\eta^1(\phi,\omega) .$$

Another example is the case of $f(t)=\sqrt{t}$. Then

$$S_f^1(\phi,\omega)=\langle\Delta(\Phi,\Lambda)^{1/2}\Lambda,\Lambda\rangle=\langle\Phi,\Lambda\rangle$$

if Φ and Λ are the vector representatives from the natural positive cone. This quantity is interpreted as a kind of transition probability ([26], [27]) and denoted by $P_A(\phi,\omega)$. The basic properties of $P_A(\phi,\omega)$ follow from those of the quasi-entropies.

Assume now that M possesses a faithful normal semifinite trace τ and the states $\phi,\omega \in M$ correspond to densities ρ_ϕ and ρ_ω. Choosing $k=1$ and $f(t)=t^\alpha$ we have

$$S_f^k(\phi,\omega) = \tau(\rho_\omega^{1-\alpha}\rho_\phi^\alpha) .$$

Rényi's α-entropies are related to this example. Namely,

$$S_\alpha(\phi,\omega) = \frac{1}{\alpha-1} \log \tau(\rho_\omega^\alpha \rho_\phi^{1-\alpha}) .$$

Among the generalized relative entropies treated here α-entropies are the only additive ones (see [33] and [22]).

Sufficiency

Let M be a von Neumann algebra with faithful normal states ϕ and ω. The relative entropy $S(\phi,\omega)$ is regarded as a kind of measure for the mutual information between these two states. When a subalgebra $N\subset M$ corresponds to a measurement the observed states are $\phi|N$ and $\omega|N$. One can compare the information obtained by the measurement to discriminate between ϕ and ω with that for $\phi|N$ and $\omega|N$. So we say the subalgebra N to be weakly sufficient for ϕ and ω if $S(\phi,\omega)=S(\phi|N,\omega|N)$. The reason for using the term weak sufficiency is the paper [10] where another sufficiency was introduced. Namely, N is sufficient for ϕ and ω if there is a conditional expectation preserving both ϕ and ω. Since conditional expectations are Schwarz mappings, a sufficient subalgebra is weakly sufficient by Theorem 2. In the commutative case weak sufficiency implies sufficiency as it follows from the next result.

THEOREM 5. Let M be a von Neumann algebra with faithful normal states ϕ and ω and let N be a subalgebra of M. Assume that there is a conditional expectation $E: M\to N$ preserving ϕ.

$$S(\phi,\omega) = S(\phi|N,\omega|N)+S(\omega\circ E,\omega)$$

provided that all terms are finite.

Proof. We use the formula

$$S(\phi,\omega)=i \lim_{t\to 0} t^{-1}(\omega([D\phi,D\omega]_t)-1)$$

to compute the entropy from the Radom-Nikodym cocycle ([29], [20]). By the chain rule we have

$$[D\phi,D\omega]_t = [D\phi,D\omega\circ E]_t[D\omega\circ E,D\omega]_t \ .$$

Since $[D\phi,D\omega\circ E]_t=[D\phi|N,D\omega|N]_t$ the derivation gives the result.

In particular, under the conditions of Theorem 5 $S(\phi,\omega) = S(\phi|N,\omega|N)$ if and only if $\omega\circ E=\omega$, that is, E preserves ω. A less general result is Theorem 3.2 in [10] where it is assumed that N is included in the centralizer of ϕ. (In this case the conditional expectation preserving ϕ always exists.)

Now I study weak sufficiency for commutative subalgebras of finite dimensional algebras. The story of this subject is the following. Dealing with a possible extension of the Donsker and Varadhan's theory to a noncommutative setup I met a question with L. Accardi: When ϕ and ω are states on the algebra $M_2(\mathbb{C})$ and $N\subset M_2(\mathbb{C})$ is a commutative subalgebra then

$$S(\phi,\omega) \geq S(\phi|N, \omega|N)$$

by the monotonicity of the relative entropy. However, may the right hand side reach $S(\phi,\omega)$ for an appropriate N ? In other words, may a commutative subalgebra be weakly sufficient for non-commuting states? G. Lindblad showed us a counterexample. Since that time it has turned out that the counterexample is generic.

THEOREM 6. Let M be a finite dimensional algebra with faithful states ϕ_0 and ϕ_1. Assume that N is a commutative subalgebra of M. Then N is weakly sufficient for ϕ_0 and ϕ_1 if and only if ϕ_0 commutes with ϕ_1 and $[D\phi_0,D\phi_1]_t$ is in N.

Proof. The claim is strongly related to the paper [11] with minor changes.

For $0\leq t\leq 1$ we define

$$\phi(t) = (1-t)\phi_0 + t\phi_1$$

and

$$K(0,\phi_0,\phi_1) = tS(\phi_1, \phi(t)) + (1-t)S(\phi_0, \phi(t)) \ .$$

So $K(0,\phi_0,\phi_1)=K(1,\phi_0,\phi_1)=0$ and if ω_i is the restriction of ϕ_i to a subalgebra of M then we have

$$K(t,\omega_0,\omega_1) \le K(t,\phi_0,\phi_1) .$$

Using the formula $\frac{d}{dt}\tau(f(at+b)) = \tau(f'(at+b)a)$ we infer

$$\frac{d}{dt}K(t,\phi_0,\phi_1) = -\tau(\sigma\log(t\sigma+\rho_0))+\tau(\rho_1\log\rho_1 - \rho_0\log\rho_0)$$

where ρ_i is the density of ϕ_i and $\sigma=\rho_1-\rho_0$. The operator $t\sigma+\rho_0$ is the density of $\phi(t)$ and it will be denoted by $\rho(t)$. Since

$$\log x = \int_0^\infty (1+t)^{-1}-(x+t)^{-1}dt$$

we obtain

$$\frac{d^2}{dt^2} K(t,\phi_0,\phi_1) = -\tau\left[\int_0^\infty \sigma(\rho(t)+s)^{-1}\sigma(\rho(t)+s)^{-1}ds\right] .$$

Let $\sum_i \lambda_i^t p_i^t$ be the spectral resolution of $\rho(t)$. We can write

$$\frac{d^2}{dt^2}K(t,\phi_0,\phi_1) = -\sum_{i,j} \tau(\sigma p_i^t \sigma p_i^t) \int_0^\infty \frac{1}{(s+\lambda_i^t)(s+\lambda_j^t)} ds .$$

Here the coefficient $\int_0^\infty 1/(s+\lambda_i^t)(s+\lambda_j^t)ds = L(\lambda_i^t,\lambda_j^t)^{-1}$ equals to $1/\lambda_i^t$ when $i=j$ and to $(\log\lambda_i^t - \log\lambda_j^t)/(\lambda_i^t-\lambda_j^t)$ when $i\neq j$. $L(\alpha,\beta)$ is called the logarithmic mean of α and β , furthermore

$$(*) \qquad \sqrt{\alpha\beta} \le L(\alpha,\beta) \le \frac{\alpha+\beta}{2} .$$

The equalities hold if and only if $\alpha=\beta$ (see, for example, [6]). Let $w(t) = 2\sum_{i,j} p_i^t \sigma p_j^t/(\lambda_i^t+\lambda_j^t)$. Using the inequality (*) we have

$$\frac{d^2 K(t,\phi_0,\phi_1)}{dt^2} \le -\tau(\sigma w(t))$$

where the equality holds if and only if $\tau(p_i^t\sigma p_j^t\sigma) = 0$ for any $i\neq j$. Since

$$\tau(|\sigma\rho(t)-\rho(t)\sigma|) = \sum_{i,j} (\lambda_i^t-\lambda_j^t)^2\tau(p_i^t\sigma p_j^t\sigma)$$

the equality above is equivalent to the condition $[\sigma,\rho(t)]=0$. By the commutativity of N we have

$$\frac{d^2K(t,\omega_0,\omega_1)}{dt^2} = -\tau(E(\sigma)E(\rho(t))^{-1}E(\sigma)) = -\tau(\sigma v(t)) = -\tau(\rho(t)v(t)^2)$$

where $v(t)=E(\sigma)/E(\rho(t))$ and $E: M\to N$ is the trace preserving conditional expectation.

It is easy to check that $\rho(t)w(t)+w(t)\rho(t) = 2\sigma$. Therefore

$\tau(\sigma w(t))=\frac{1}{2}\tau(\rho(t)w(t)^2+w(t)\rho(t)w(t))=\tau(\rho(t)w(t)^2)$ and we estimate as follows:

$$-\tau(\sigma w(t))=-\tau(\rho(t)w(t)^2)=-\tau(\rho(t)[v(t)+(w(t)-v(t))]^2) =$$
$$=-\tau(\rho(t)v(t)^2)-\tau(\rho(t)[w(t)-v(t)]^2)+2\tau(\rho(t)v(t)^2)-$$
$$-\tau([w(t)\rho(t)+\rho(t)w(t)]v(t)) =$$
$$= -\tau(\rho(t)v(t)^2)-\tau(\rho(t)[w(t)-v(t)]^2) .$$

So we conclude that

$$\frac{d^2K(t,\phi_0,\phi_1)}{dt^2} \le \frac{d^2K(t,\omega_0,\omega_1)}{dt^2}$$

and the equality implies that $[\rho(t),\sigma]=0$ on the one hand and on the other hand $w(t)=v(t)$. The function

$$y(t) = K(t,\phi_0,\phi_1)-K(t,\omega_0,\omega_1)$$

satisfies the conditions

$$y(0) = y(1) = 0$$
$$y''(t) \le 0 , \ y(t) \ge 0 .$$

Consequently, $y'(0)$ is equivalent to $y(t)\equiv 0$. Since

$$y'(0) = S(\phi_0,\phi_1)-S(\omega_0,\omega_1)$$

we obtain that $S(\phi_0,\phi_1)=s(\omega_0,\omega_1)$ may happen in the only case when $[\rho_0,\rho_1]=0$ and $\rho_0\cdot\rho_1^{-1}=E(\rho_0)\cdot E(\rho_1)^{-1}$.

The theorem is proved. Weak sufficiency of non-commutative subalgebras seems to be an interesting and non-trivial problem even in the finite dimensional case.

Lower estimate in Donsker and Varadhan's theory

In this section we prove a lower estimate in a possible non-commutative Donsker and Varadhan's theory of stationary quantum Markov processes. The main theorem is joint work with L. Accardi and as far as we know it is the first result of this kind.

For convenience, I summarize the basic theory of quantum Markov chains with finite state space (see [1] and [2]). Let B_n be isomorphic to a fixed finite dimensional full matrix algebra M and form the

the C^*-tensor product $A=\bigotimes_{n=-\infty}^{\infty} B_n$. $\pi_n : M \to A$ will stand for the embedding into the n-th factor. The local algebra $A_{[k,n]}$ is spanned by $\bigcup_{i=k}^{n} \pi_i(M)$. The shift automorphism of A is defined by the formula $\alpha\pi_i(a)=\pi_{i+1}(a)$ $(a\in M)$. If ϕ_n is a state on $A_{[-n,n]}$ such that $\phi_n=\phi_{n+k}|A_{[-n,n]}$ then there is a unique state ϕ of A with $\phi|A_{[-n,n]}=\phi_n$.

We denote by E_1 the trace preserving conditional expectation from $A_{[0,1]}$ onto A_0 . If $K\in A_{[0,1]}$ is an operator such that $E_1(KK^*)=I$ and ρ_0 is a density matrix in A_0 then

$$\rho_{[0,n+1]} = \alpha^n(K^*)\dots K^*\rho_0 K\dots\alpha^n(K)$$

is a density matrix in $A_{[0,n+1]}$. In the sequel we assume that $[K,\alpha(K)]=0$, $[\rho_0,K]=0$ and $[\alpha(K^*),K]=0$. So

$$\rho_{[0,n+1]} = \rho_0 L\alpha(L)\alpha^2(L)\dots\alpha^n(L)$$

with $L=K^*K$. There is a state ϕ on A such that $\phi/A_{[0,n]}$ has density $\rho_{[0,n]}$ and this is called Markov state determined by the initial distribution ρ_0 and the conditional density amplitude K .

All the states on A what we consider will be stationary states, that is, invariant under α .

THEOREM 7. Let ω be an arbitrary stationary state and ϕ be a Markov state on A . Then the mean relative entropy

$$S_M(\phi,\omega) = \lim_{n\to\infty} \frac{1}{n} S(\phi|A_{[0,n]},\omega|A_{[0,n]})$$

exists.

Proof. Let f_n and w_n be the densities of ϕ and ω restricted to $A_{[0,n]}$, respectively. Then $f_n=\rho_0 L\alpha(L)\dots\alpha^{n-1}(L)$ and we have

$$S(\phi|A_{[0,n]},\omega|A_{[0,n]}) = \tau(w_n \log w_n)-\tau(w_n \log\rho_0)-\sum_{i=0}^{n-1} (w_n \log\alpha^i(L)) .$$

Here $\lim_{n\to\infty} \frac{1}{n}\tau(w_n \log w_n)$ exists and equals to $\sup n^{-1} \tau(w_n \log w_n)$ by the subadditivity of the entropy ([5]). Moreover, the stationarity gives

$$\tau(w_n \log \alpha^i(L)) = \tau(\alpha^{-i}(w_n)\log L) = \tau(w_1 \log L)$$

and we conclude

$$S_M(\phi,\omega) = \sup_n n^{-1} (w_n \log w_n) - \tau(w_1 \log L) .$$

I note that O. Besson has computed the Connes-Størmer entropy of quantum Markov chain and it equals to the mean relative entropy with espect to the tracial state (see his talk in this conference).

LEMMA 8. Let $\mathcal{B}$ be a finite dimensional C^*-algebra with faith- ul trace τ. Assume that ϕ and ω are states on $\mathcal{B}$ with densities and w, respectively. Then

$$\log \phi(\exp A) \geq -S(\phi,\omega)+\omega(A)$$

or every $A \in \mathcal{B}_{sa}$.

Proof. We estimate using the Golden-Thomson and Peierls-Bogoliubov nequalities:

$$\begin{aligned}\phi(\exp A) &= \tau(f \exp A) \geq \tau(\exp(\log \phi + A)) = \\ &= \tau(\exp[(\log f - \log w) + A + \log w]) \geq \\ &\geq \exp(\tau(f[\log w - \log f])+\tau(fA)) .\end{aligned}$$

Now I prove the lower estimate for quantum Markov chains. Concern- ng the probabilistic theory I refer to [8] and [3].

THEOREM 9. Let ϕ be a stationary Markov state on A. With the otation above we have

$$\underline{\lim}_{n \to +\infty} \frac{1}{n} \log \phi\left(\exp \sum_{i=0}^{n} \alpha^i(a)\right) \geq \sup_{\omega} - S_M(\phi,\omega)+\omega(a)$$

here $a \in \bigcup_n A_{[0,n]}$ and the sup is taken over all stationary states ω.

Proof. Let $a \in A_{[0,\ell]}$ and $u_n = \sum_{i=0}^{n-1} \alpha^i(a)$. The previous lemma ives

$$\log \phi(\exp u_n) \geq -S(\phi_{n+\ell}, \omega_{n+\ell}) + \omega(u_n)$$

here $\phi_{n+\ell}$ $(\omega_{n+\ell})$ is the restriction of ϕ (ω) to $A_{[0,n+\ell)}$. By The- em 7 we know that

$$\frac{1}{n} S(\phi_{n+\ell}, \omega_{n+\ell}) \to S_M(\phi,\omega)$$

s $n \to \infty$. On the other hand,

$$\frac{1}{n} \omega(u_n) = \frac{1}{n}\omega\left(\sum_{i=0}^{n-1} \alpha^i(a)\right) = \omega(a)$$

ince ω is stationary.

Acknowledgements

I would like to thank all organizers for the kind invitation to participate in a most stimulating conference. I am grateful to Professor Luigi Accardi for the invitation to the 2nd University of Rome and for his collaboration in the subject of the last part of the talk. Thanks are due to A. S. Holevo for fruitful discussions on his paper.

Bibliography

[1] L. Accardi, Topics in quantum probability, Physics Reports, 77 (1981), 169-192.

[2] L. Accardi and A Frigerio, Markovian cocyclies, Proc. Royal Irish Acad. 83A (1983), 251-263.

[3] L. Accardi and S. Olla, Donsker and Varadhan's theory for stationary processes, Preprint, Rome, 1983.

[4] H. Araki, Relative entropy of states of a von Neumann algebra, I and II, Publ. RIMS, Kyoto Univ. 11 (1976), 809-833 and 13 (1977), 173-192.

[5] O. Bratteli and D. V. Robinson, Operator algebras and quantum statistical mechanics I and II, Springer Verlag, Berlin, 1981.

[6] B. C. Carlson, The logarithmic mean, Amer. Math. Monthly, 79 (1972), 615-618.

[7] I. Csiszár, Information-type measures of difference of probability distributions and indirect observations, Studia Sci. Math. Hungar. 2 (1967), 299-318.

[8] M. Donsker and S. R. S. Varadhan, Asymptotic evaluation of certain Markov process expectations for large time, Comm. Pure Appl. Math. 28 (1975), 1-47.

[9] V. Ya. Golodets and G. N. Zholkevich, Markovian KMS-states (in Russian), Teoret. Mat. Fiz. 56 (1983), 80-86.

[10] F. Hiai, M. Ohya and M. Tsukuda, Sufficiency, KMS condition and relative entropy in von Neumann algebras, Pacific J. Math. 96 (1981) 99-109.

[11] A. S. Holevo, Some estimates for the amount of information transmittable by a quantum communication channel (in Russian), Problemy Peredaci Informacii, 9 (1973), 3-11.

[12] H. Kosaki, Interpolation theory and the Wigner-Yanase-Dyson-Lieb conjecture, Commun. Math. Phys. 87 (1982), 315-329.

[13] H. Kosaki, Variational expressions of relative entropy of states on W* -algebras, Preprint, 1984.

[14] S. Kullback and R. A. Leibler, On information and sufficiency, Ann. Math. Stat. 22 (1951), 79-86.

[15] E. H. Lieb, Some convexity and subadditivity properties of entropy, Bull. Amer. Math. Soc. 81 (1975), 1-14.

[16] G. Lindblad, Entropy, information and quantum measurements, Commun. Math. Phys. 33 (1973), 305-322.

[17] G. Lindblad, Expectations and entropy inequalities for finite quantum systems, Commun. Math. Phys. 39 (1974), 111-119.

[18] G. Lindblad, Letter to the author, 1984.

[19] J. von Neumann, Mathematische Grundlagen der Quantenmechanic, Springer Verlag, Berlin, 1932.

[20] D. Petz, The relative entropy of states of von Neumann algebras Proc. Second Intern. Conf. on Operator Algebras, Ideals and their Appl. in Theor. Physics, Teubner-Texte zur Math. 67, 112-117, Teubner Verlag, 1984.

[21] D. Petz, Properties of the relative entropy of states of a von Neumann algebra, to appear in Acta Math. Hungar.

[22] D. Petz, Quasi-entropies for finite quantum systems, to appear in Rep. Math. Phys.

[23] D. Petz, Quasi-entropies for states of a von Neumann algebra, Preprint, Budapest, 1984.

[24] D. Petz, Spectral scale of selfadjoint operators and trace inequalities, to appear in J. Math. Anal. Appl.

[25] D. Petz, Jensen's inequality for trace reducing positive mappings, in preparation.

[26] G. A. Raggio, Comparison of Uhlmann's transition probability with the one induced by the natural positive cone of a von Neumann algebra in standard form, Lett. Math. Phys. 6 (1982), 233-236.

[27] G. A. Raggio, Generalized transition probabilities and applications, Quantum Prob. and Appl. to the Quant. Theor. of Irrev. Processes (ed. by L. Accardi, A. Frigerio and V. Gorini), Lecture Notes in Math. 1055, 327-335, Springer Verlag, Berlin, 1984.

[28] I. E. Segal, A note on the concept of entropy, J. Math. Mech. 9 (1960), 623-629.

[29] S. Stratila, Modular theory of operator algebras, Abacus Press, Tunbridge Wells, 1981.

[30] A. Uhlmann, Relative entropy and the Wigner-Yanase-Dyson-Lieb concavity in an interpolation theory, Commun. Math. Phys. 54 (1977), 21-32.

[31] H. Umegaki, Conditional expectations in an operator algebra IV (entropy and information), Kodai Math. Sem. Rep. 14 (1962/, 59-85.

[32] A. Wehrl, A remark on the concavity of entropy, Found. Phys. 9 (1979), 939-946.

[33] A. Wehrl, General properties of entropy, Rev. Modern Phys. 50 (1978), 221-260.

SEMICLASSICAL DESCRIPTION OF N-LEVEL SYSTEMS INTERACTING WITH RADIATION FIELDS.

Guido A. Raggio, and Henri S. Zivi

Laboratorium für physikal. Chemie, ETH Zürich, CH-8092 Zürich.

The objective of our work is to try to understand the theoretical status of time-dependent hamiltonians for N-level systems, of the type

(I) $H(t) = F + f(t)V$, f real-valued, F and V selfadjoint,

widely used in the analysis of spectroscopic experiments and quantum optical phenomena. We consider a fully quantal model where the N-level system is coupled to the electromagnetic field. We look at the evolution of the full system when the field is in a coherent state, and evaluate the limiting case of *weak coupling and high field but constant coupling energy*. We show that in this asymptotic situation, the dynamics of the N-level system is governed by (I), and does not influence the field which is free. Detailed proofs of the results, as well as a more elaborate discussion of the underlying physics, will be published elsewhere [1].

1. Motivating the hamiltonian.

Consider a system of K non-relativistic particles with masses m_j, charges e_j, and charge distributions $e_j\rho(x_j)$, where ρ is positive, spherically symmetric, and satisfies $\int d^3x\, \rho(x) = 1$. If the particle system interacts with a radiation field, the hamiltonian in the Coulomb gauge is given by

$$H = \sum_{j=1}^{K} (2m_j)^{-1} :(p_j - e_jA(x_j))^2: + V + 1/2 \int dx : E_\perp^2 + B^2 : ,$$

where

$$V = \sum_{j=1}^{K} \sum_{\substack{k=1 \\ j<k}}^{K} e_je_k \int d^3r \int d^3r' \frac{\rho(x_j - r)\rho(x_k - r')}{|r-r'|} \; ; \quad A(x) = \int d^3y A_\perp(x+y)\rho(y) \; ;$$

$$A_\perp(x) = (2\pi)^{-3/2} \int d^3k (2|k|)^{-1/2} \sum_{s=1}^{2} e_s(k)\{a_s(k)e^{ikx} + a_s^*(k)e^{-ikx}\} .$$

Here, the usual commutation relations $[x_{j\alpha}, p_{m\beta}]=i\delta_{jm}\delta_{\alpha\beta}$, $[a_s(k), a^*_{s'}(k')]=\delta_{ss'}\delta(k-k')$ hold, and $e_s(k)$, s=1,2, are the polarization vectors satisfying $ke_s(k)=0$, and $e_s(k)e_{s'}(k)=\delta_{ss'}$. *Neglecting the* $A(x_j)^2$ *terms,* we have

$$H = H_0 + H' + \sum_{s=1}^{2} \int d^3k|k|\ a^*_s(k)a_s(k) \quad ,$$

where

$$H_0 = \sum_{j=1}^{K} (2m_j)^{-1}p_j^2 + V \quad ,$$

$$H' = \int d^3k(2|k|)^{-1/2}\tilde{\rho}(k) \sum_{s=1}^{2} e_s(k)\{a_s(k)G(p,x,k)+a^*_s(k)G(p,x,-k)\}$$

$$G(p,x,k) = -\sum_{j=1}^{K} (e_j/2m_j)\{p_j\exp(ikx_j)+\exp(ikx_j)p_j\}$$

$$\tilde{\rho}(k)= (2\pi)^{-3/2}\int d^3x\ e^{-ikx}\rho(x) \quad ^{1)} .$$

Assume H is an N-dimensional subspace of $L^2(\mathbb{R}^{3K})$ contained in the domains of definition of H_0 and the p_j's. Let P be the projection operator onto this subspace, and consider the projection PHP of H denoted by H_N. One has

$$H_N= F+\sum_{s=1}^{2} \int d^3k|k|a^*_s(k)a_s(k) +\int d^3k(2|k|)^{-1/2}\tilde{\rho}(k) \sum_{s=1}^{2} e_s(k) \cdot$$

$$\cdot\{a_s(k)G(k)+a^*_s(k)G(-k)\} \quad ,$$

where $F=PH_0P$, and $G(k)=PG(p,x,k)P$. In two special cases, one can show that $e_s(k)G(\mp k)=g_s(k)V_s$, for a selfadjoint operator V_s on H, and a real-valued function g_s:

a) *Special two-level approximation:* Assume φ_1, φ_2 are eigenfunctions of H_0, to different eigenvalues, that φ_1 is real-valued and even, and that φ_2 is real-valued and odd. Let P_j be the projection operator onto the ray spanned by φ_j, j=1,2, and $P=P_1+P_2$. Then a lengthy computation [2], using the properties of φ_1, φ_2 and the transversality of the vectors $e_s(k)$, gives $e_s(k)G(\mp k)=g_s(k)V$, with

$$g_s(k) = i\sum_{j=1}^{K} (e_j/m_j)\langle\varphi_1,\cos(kx_j)p_j\varphi_2\rangle_{L^2(\mathbb{R}^{3K})}\cdot e_s(k) \quad ,$$

1) $\tilde{\rho}$ is by assumptions on ρ real-valued and spherically symmetric.

$V = Q_- - Q_+$, $Q_\mp$ being the projection operator onto the ray spanned by $\varphi_\mp = 2^{-1/2}(\varphi_1 \mp i\varphi_2)$.

b) *Dipole approximation for linearly polarized field:* The dipole approximation of H is obtained by replacing $A(x)$ by $A(0)$. Then, $G(p,x,k)$ does not depend on k (and x). If furthermore the field is linearly polarized, $e_1(k)=(1,0,0)$, $e_2(k)=(0,1,0)$, then $e_s(k)G(\mp k)=g(k)V_s$, with

$$g(k)=1$$
$$V_s = -\sum_{j=}^{K} (e_j/m_j)Pp_{js}P \quad .$$

In these instances the projected hamiltonian reads (keeping only one of the two polarization components and introducing a coupling parameter $\lambda \geq 0$)

$$H_N = F + \int d^3k|k|a^*(k)a(k) + \lambda\int d^3k(2|k|)^{-1/2}\tilde{\rho}(k)g(k)\{a(k)+a^*(k)\}V$$

This operator will be given a precise mathematical meaning as a selfadjoint operator on $\mathfrak{F}\otimes H$, where $\mathfrak{F}$ is the symmetric Fock-space constructed over the Hilbert space $\mathfrak{H} = L^2(\mathbb{R}^3;|x|^{-1}d^3x)$ [2].

Semiclassical radiation theory is considered as a suitable approximation to quantum electrodynamics if the field is in a highly excited coherent state $\Omega(f)$,

$$\Omega(f)=W(f)\Phi \quad , f \in \mathfrak{H} \; ; \; W(f) = \exp\{2^{-1/2}(a^*(f)-a(f))\}, \; f \in \mathfrak{H} \; ;$$

$a(f)$, resp. $a^*(f)$, being the usual annhilation, resp. creation, operators on $\mathfrak{F}$ with $[a(f),a^*(g)]=(f,g)1$, and $\Phi = 1\oplus 0\oplus 0\oplus 0 \ldots$ being the Fock vacuum.

This leads us to consider the time-evolved observables of the N-level system H reduced with the state $\Omega(f)$, and the expectation values of the time-evolved field observables with respect to the state $|\Omega(f)\rangle\langle\Omega(f)|\otimes D$, D being the initial state

---

2) The scalar product in $\mathfrak{H}$ is given by $(f,g)= \int\overline{f(x)}g(x)|x|^{-1}d^3x$.

(density operator) of the N-level system, in the limit $\|f\| \to \infty$. For this purpose, it is convenient to replace $\|f\|$ by $\varepsilon^{-1}\|f\|$, and to let ε go to zero. In order that the expectation values of the field operators remain finite in this limit, one has to rescale $a(f) \to a^{\varepsilon}(f)=\varepsilon\, a(f)=a(\varepsilon f)$. Since the commutator then scales as ε^2, one expects, heuristically, that the $a^{\varepsilon}(f)$ become commuting (i.e. classical) variables in the limit $\varepsilon \to 0$. It will also be seen in section 2., that the coupling constant must be rescaled as $\lambda \to \varepsilon\lambda$ in order that the coupling energy remain at a finite prescribed value in the limit $\varepsilon \to 0$.

2. *The model, its semiclassical limit, and fluctuations.*

Let the one-particle hamiltonian h on $\mathfrak{H}$ be given by $\{hf\}(x)=|x|f(x)$, $\int d^3x|f(x)|^2|x| < \infty$; denote by $\{u_t=\exp\{-ith\}:t\in \mathbb{R}\}$ the generated unitary group on $\mathfrak{H}$. Denote by $\mathfrak{F}$ the symmetric Fock space built upon $\mathfrak{H}$, by Ω the second-quantization map, and write $\langle .,. \rangle$ for the scalar product on $\mathfrak{F}$. Let $U_0(t)=\exp\{-it\Omega(h)\}$, $t \in \mathbb{R}$. For $\lambda \geq 0$, F and V selfadjoint operators on $\mathbb{C}^N$, and $\xi \in \mathrm{Dom}(h^{-1})$ [3)], let

$$H^{\lambda} =\Omega(h)\otimes 1 + \lambda 2^{-1/2}\{a(\xi)+a^{*}(\xi)\}\otimes V + 1\otimes F \quad .$$

H^{λ} defines a selfadjoint operator on $\mathfrak{F}\otimes\mathbb{C}^N$; let $\{U^{\lambda}(t)=\exp\{-itH^{\lambda}\}:t \in \mathbb{R}\}$ be the generated unitary group.

If we compute the expectation value of the field-part of the coupling in H^{λ}, given by $\lambda 2^{-1/2}\{a(\xi)+a^{*}(\xi)\}$, in the coherent state $\Omega(f)$, we get $\lambda\mathrm{Re}(f,\xi)$. Thus, if f is replaced by $\varepsilon^{-1}f$, and the limit $\varepsilon \to 0$ is considered, the coupling remains constant only if λ is scaled as $\varepsilon\lambda$.

Let us now assume that the initial state is given by the density operator

---

3) $\xi = \tilde{\rho}g$, appears as the Fourier transform of $\rho * \mathfrak{g}$. The condition $\xi \in \mathfrak{H}$, implicit in $\xi \in \mathrm{Dom}(h^{-1})$, amounts to an ultraviolet cutoff in the hamiltonian H of *1.*; $\xi \in \mathrm{Dom}(h^{-1})$ corresponds to an infrared cutoff. H can still be defined as a selfadjoint operator when $\xi \in \mathrm{Dom}(h^{-1/2})$, [3].

$|\Omega(f)\rangle\langle\Omega(f)|\otimes D$, where $f\in\mathfrak{D}$, and D is any density operator on $\mathbb{C}^N$. If A is any observable of the N-level system (i.e., linear operator on $\mathbb{C}^N$), its dynamical evolution is given by $\mathcal{U}^\lambda(-t)(1\otimes A)\mathcal{U}^\lambda(t)$, and its dynamics reduced by the field state $\Omega(f)$ by

$$A(f,\lambda,t)= \mathrm{Tr}_{\mathfrak{F}}\{(|\Omega(f)\rangle\langle\Omega(f)|\otimes 1)\mathcal{U}^\lambda(-t)(1\otimes A)\mathcal{U}^\lambda(t)\} \quad .$$

If $P(g_1,g_2,\ldots,g_\mu;h_1,h_2,\ldots,h_\nu)= a^*(g_1)a^*(g_2)\ldots a^*(g_\mu)a(h_1)a(h_2)\ldots a(h_\nu)$, $g_j,h_k \in \mathfrak{D}$, is any (normaly ordered) polynomial in the field operators, its expectation value in time is given by ($\vec{g}_\mu=(g_1,g_2,\ldots,g_\mu)$, $\vec{h}_\nu=(h_1,h_2,\ldots,h_\nu)$)

$$p(\vec{g}_\mu;\vec{h}_\nu;f,\lambda,t)=\mathrm{Tr}_{\mathfrak{F}\otimes\mathbb{C}^N}\{(|\Omega(f)\rangle\langle\Omega(f)|\otimes D)\mathcal{U}^\lambda(-t)(P(\vec{g}_\mu;\vec{h}_\nu)\otimes 1)\mathcal{U}^\lambda(t)\} \quad .$$

Finally, the "state generating functional", is given by

$$\omega(g;f,\lambda,t)=\mathrm{Tr}_{\mathfrak{F}\otimes\mathbb{C}^N}\{(|\Omega(f)\rangle\langle\Omega(f)|\otimes D)\mathcal{U}^\lambda(-t)(W(g)\otimes 1)\mathcal{U}^\lambda(t)\} \quad ,$$

for $g \in \mathfrak{D}$.

THEOREM:

a) Semiclassical limit:

1. $A_{(0)}(f,\lambda,t) = \lim_{\varepsilon\to 0} A(\varepsilon^{-1}f,\varepsilon\lambda,t)$ exists and is the solution of

   $i\,\dot{A}_{(0)}(f,\lambda,t)=[A_{(0)}(f,\lambda,t),H(f,\lambda,t)]_-$, $A_{(0)}(f,\lambda,0)=A$,

   where the semiclassical hamiltonian $H(f,\lambda,t)$ is given by

   $H(f,\lambda,t)=F + \lambda\mathrm{Re}(u_t f,\xi)V$.

2. $\omega_{(0)}(g;f,\lambda,t) = \lim_{\varepsilon\to 0} \omega(\varepsilon g;\varepsilon^{-1}f,\varepsilon\lambda,t)$, and $p_{(0)}(\vec{g}_\mu;\vec{h}_\nu;f,\lambda,t) = \lim_{\varepsilon\to 0} p(\varepsilon\vec{g}_\mu;\varepsilon\vec{h}_\nu;\varepsilon^{-1}f,\varepsilon\lambda,t)$ both exist, and

   $$\omega_{(0)}(f,\lambda,t)=\exp\{i\mathrm{Im}(u_t f,g)\}$$

   $$p_{(0)}(\vec{g}_\mu;\vec{h}_\nu;f,\lambda,t)=\langle\Omega(u_t f),P(\vec{g}_\mu;\vec{h}_\nu)\Omega(u_t f)\rangle = 2^{-1/2(\mu+\nu)} \prod_{j=1}^{\mu} \{(u_t f,g_j)\} \prod_{k=1}^{\nu} \{(h_k,u_t f)\}$$

b) Fluctuations:

1. The fluctuation of order n, n=1,2,3,... , of $A(f,\lambda,t)$ defined (recursively) by

$$A_{(n)}(f,\lambda,t) = \lim_{\varepsilon\to 0} A^{\varepsilon}_{(n)}(f,\lambda,t) \text{ , where}$$
$$A^{\varepsilon}_{(n)}(f,\lambda,t)= \varepsilon^{-2n}\{A(\varepsilon^{-1}f,\varepsilon\lambda,t)-\sum_{m=0}^{n-1}\varepsilon^{2m}A_{(m)}(f,\lambda,t)\} \text{ ,}$$

exists.

2. The fluctuations of order n, n=1,2,3,..., of $p(\vec{g}_\mu;\vec{h}_\nu;f,\lambda,t)$ and $\omega(g;f,\lambda,t)$ defined (recursively) by

$$p_{(n)}(\vec{g}_\mu;\vec{h}_\nu;f,\lambda,t) = \lim_{\varepsilon\to 0} p^{\varepsilon}_{(n)}(\vec{g}_\mu;\vec{h}_\nu;f,\lambda,t) \text{ ,}$$
$$\omega_{(n)}(g;f,\lambda,t) = \lim_{\varepsilon\to 0} \omega^{\varepsilon}_{(n)}(g;f,\lambda,t) \text{ , where}$$
$$p^{\varepsilon}_{(n)}(\vec{g}_\mu;\vec{h}_\nu;f,\lambda,t)= \varepsilon^{-2n}\{p(\varepsilon\vec{g}_\mu;\varepsilon\vec{h}_\nu;\varepsilon^{-1}f,\varepsilon\lambda,t)-\sum_{m=0}^{n-1}\varepsilon^{2m}p_{(m)}(g_\mu;h_\nu;f,\lambda,t)\}$$
$$\omega^{\varepsilon}_{(n)}(g;f,\lambda,t)= \varepsilon^{-2n}\{\omega(\varepsilon g;\varepsilon^{-1}f,\varepsilon\lambda,t)-\sum_{m=0}^{n-1}\varepsilon^{2m}\omega_{(m)}(g;f,\lambda,t)\} \text{ ,}$$

exist.

REMARKS: Notice that $\lambda\mathrm{Re}(u_t f,\xi)$, which appears as the time-dependent potential in the semiclassical hamiltonian $H(f,\lambda,t)$, is nothing but $\langle U_0(t)\Omega(f),2^{-1/2}\{a(\xi)+a^*(\xi)\}U_0(t)\Omega(f)\rangle=\lambda\langle\Omega(u_t f),2^{-1/2}\{a(\xi)+a^*(\xi)\}\Omega(u_t f)\rangle$, which is the expectation value of the field part of the coupling in the coherent state $\Omega(f)$ evolved according to the free field-dynamics $U_0(t)$. The semiclassical limit in the field, described by a)2., is in fact a classical limit; in this limit, the field is equivalent to a system of (infinitely many) classical, uncoupled harmonic oscillators. The phase-space is given by $\mathfrak{H}$ and the dynamics by $f: \to u_t f$. If X_g, $g \in \mathfrak{H}$, is the characteristic function defined by $X_g(f)=\exp\{i\mathrm{Im}(f,g)\}$, $f \in \mathfrak{H}$, then $\omega_{(0)}(g;f,\lambda,t)=X_g(u_t f)$, which is the dispersion-free value of X_g at time t when the field is initially in the state f. By b), the fluctuations (around the semiclassical solution) appear as the terms of asymptotic series in ε (only even powers of ε appear) the sense of Poincaré:

$$A(\varepsilon^{-1}f,\varepsilon\lambda,t) \sim \sum_{n=0}^{\infty} \varepsilon^{2n} A_{(n)}(f,\lambda,t) \quad , \text{ etc.}$$

OUTLINE OF THE PROOF: Choosing an orthonormal basis of $\mathbb{C}^N$ which diagonalizes V, letting $\{v_n : n=1,2,..,N\}$ be the set of eigenvalues of V numbered according to their multiplicities, identifying $\mathfrak{F}\otimes\mathbb{C}^N$ with the direct sum of N copies of $\mathfrak{F}$, using a result of Cook [4], and letting $\lambda_n = \lambda v_n$, we have

$$H_o^\lambda = \Omega(h)\otimes 1 + 2^{-1/2}\{a(\xi)+a^*(\xi)\}\otimes V = \bigoplus_{n=1}^{N} \{W(-\lambda_n h^{-1}\xi)\Omega(h)W(\lambda_n h^{-1}\xi) +$$
$$-\lambda_n^2(\xi,h^{-1}\xi)1/2\}. \text{ [4)]}$$

Thus H_o is diagonal. With the operator-norm convergent Dyson-series for the propagator of $F(t)=\exp\{itH_o^\lambda\}(1\otimes F)\exp\{-itH_o^\lambda\}$, we have

(1) $$U^\lambda(t)=\exp\{-itH_o^\lambda\}\{1+\sum_{n=1}^{\infty}(-i)^n \, T\int_0^t dt^{(n)} F(t_1)F(t_2)..F(t_n)\} .$$

The following formulas are well-known in Fock-space calculus, and will permit us to obtain explicit expressions for $U^\lambda(t)$ when the partial state of the field is coherent:

$$W(f)W(g)=\exp\{-i\mathrm{Im}(f,g)/2\}W(f+g) \; ; \; W(-f)a(g)W(f)=a(g)+2^{-1/2}(g,f)1;$$
$$U_o(-t)a(g)U_o(t)=a(u_{-t}g) \; ; \; U_o(-t)W(f)U_o(t)=W(u_{-t}f) \; ;$$
$$\langle\Omega(f),W(g)\Omega(f)\rangle=\exp\{i\mathrm{Im}(f,g)-\|g\|^2/4\} .$$

By our identification of $\mathfrak{F}\otimes\mathbb{C}^N$ with $\bigoplus_{n=1}^{N}\mathfrak{F}$, every operator on $\mathfrak{F}\otimes\mathbb{C}^N$ corresponds to an NxN-matrix whose elements are operators on $\mathfrak{F}$. Then,

(2) $$\left[\exp\{-itH_o^\lambda\}\right]_{jk} = \delta_{jk}\exp\{i\lambda_j^2(\xi,h^{-1}\xi)t/2\}W(-\lambda_j h^{-1}\xi)U_o(t)W(\lambda_j h^{-1}\xi)$$
$$\left[F(t)\right]_{jk} = F_{jk}\exp\{i(\lambda_k^2-\lambda_j^2)(t(\xi,h^{-1}\xi)+\mathrm{Im}(h^{-1}\xi,u_t h^{-1}\xi))/2\}$$
$$W((\lambda_j-\lambda_k)\{u_t-1\}h^{-1}\xi) \; . \text{ [5)]}$$

We introduce the notation :

- $\gamma(f,t)=(f,h^{-1}f)t+\mathrm{Im}(h^{-1}f,u_t h^{-1}f) \in \mathbb{R}$, for $t\in\mathbb{R}$, and $f \in \mathfrak{H}$;
- $\zeta(f,t)=\{u_t-1\}h^{-1}f \in \mathfrak{H}$, for $t\in\mathbb{R}$, and $f \in \mathfrak{H}$;

4) This formula shows that H_o^λ is indeed selfadjoint; the selfadjointness of H^λ follows from the boundedness of $1\otimes F$.

5) F_{jk} is the j,k-th matrix element of F in the chosen basis.

- If $n=0,1,2,\ldots$, $\{m_{-1},m_0,m_1,m_2,\ldots,m_n\}$ is a set of $(n+2)$ indices with values in $\{1,2,\ldots,N\}$, and $\{t_0,t_1,t_2,\ldots,t_n\}$ are $(n+1)$reals, and if $\kappa=0$ or $\kappa=-1$, then let

$$\alpha(m_\kappa,m_{\kappa+1},\ldots,m_n;t_{\kappa+1},t_{\kappa+2},\ldots,t_n;\lambda;g)=(1/2)\sum_{\nu=\kappa+1}^{n}\{(\lambda^2_{m_\nu}-\lambda^2_{m_{\nu-1}})\gamma(g,t_\nu)\}+$$
$$(1/2)\sum_{\nu=\kappa+1}^{n}\sum_{r=0}^{n-\nu-1}\{(\lambda_{m_\nu}-\lambda_{m_{\nu-1}})(\lambda_{m_{\nu+r+1}}-\lambda_{m_{\nu+r}})(\gamma(g,t_{\nu+r+1}-t_\nu)+\gamma(g,t_\nu)+$$
$$-\gamma(g,t_{\nu+r+1}))\}\in\mathbb{R},\ \text{for } \lambda\in\mathbb{R},\ \text{and } g\in\mathfrak{H};$$

$$\psi(m_\kappa,m_{\kappa+1},\ldots,m_n;t_{\kappa+1},t_{\kappa+2},\ldots,t_n;\lambda;g)=\sum_{\nu=\kappa+1}^{n}\{(\lambda_{m_{\nu-1}}-\lambda_{m_\nu})\zeta(g,-t)\}\in\mathfrak{H},$$

for $\lambda\in\mathbb{R}$, and $g\in\mathfrak{H}$;

$$\beta(m_\kappa,m_{\kappa+1},\ldots,m_n;t_{\kappa+1},t_{\kappa+2},\ldots,t_n;f,\lambda;g)=\mathrm{Im}(f,\psi(m_\kappa,m_{\kappa+1},\ldots,m_n;t_{\kappa+1},t_{\kappa+2},$$
$$\ldots,t_n;\lambda;g))\in\mathbb{R},\ \text{for } \lambda\in\mathbb{R},\ \text{and } f,g\in\mathfrak{H}.$$

If $n=1,2,\ldots$, $\{l_1,l_2,\ldots,l_{n-1}\}$ are $(n-1)$ indices with values in $\{1,2,\ldots,N\}$, and $\{t,t_1,t_2,\ldots,t_n\}$ are $(n+1)$ reals, then we use the following shorthand notations:

- $\alpha(j,m,\vec{l}_{n-1},k;t,\vec{t}_n;\lambda;g)\equiv\alpha(j,m,l_1,l_2,\ldots,l_{n-1},k;t,t_1,t_2,\ldots,t_n;\lambda;g)$
  where $j,m,k\in\{1,2,\ldots,N\}$;
  analogously for ψ and β;
- $F(j,\vec{l}_{n-1},k)\equiv F_{jl_1}F_{l_1l_2}F_{l_2l_3}\cdots F_{l_{n-1}k}$, where $j,k\in\{1,2,\ldots,N\}$;
- $\displaystyle\sum_{\vec{l}_{n-1}=1}^{N}\equiv\sum_{l_1=1}^{N}\sum_{l_2=1}^{N}\cdots\sum_{l_{n-1}=1}^{N}$

Notice that $\gamma(f,t)$ is an odd function of t, that $\alpha(j,k,\vec{l}_{n-1},m;t,\vec{t}_n;\lambda;g)=\alpha(j,k;t;\lambda;g)+\alpha(k,\vec{l}_{n-1},m;\vec{t}_n;\lambda;g)-\mathrm{Im}(\psi(j,k;t;\lambda;g),\psi(k,\vec{l}_{n-1},m;\vec{t}_n;\lambda;g))/2$, and that if we scale $\lambda\to x\lambda$, $f\to x^y f$, with $x,y\in\mathbb{R}$, then $\alpha\to x^2\alpha$, $\beta\to x^{y+1}\beta$, and $\psi\to x\psi$. We can now rewrite our expression for the matrix elements of $F(t)$ as follows

$$\left[F(t)\right]_{jk}=F_{jk}\exp\{i\alpha(j,k;t;\lambda;\xi)\}W(\psi(j,k;t;\lambda;\xi))\quad.$$

Using the commutation relations for the Weyl operators $W(.)$, we have

$$[F(t_1)F(t_2)..F(t_n)]_{jk}=\sum_{\vec{1}_{n-1}=1}^{N} F(j,\vec{1}_{n-1},k)\exp\{i\alpha(j,\vec{1}_{n-1},k;\vec{t}_n;\lambda;\xi)\}\cdot \tag{3}$$

$$\cdot W(\psi(j,\vec{1}_{n-1},k;\vec{t}_n;\lambda;\xi)) \quad .$$

Now, for a linear operator A on $\mathbb{C}^N$, $f,g \in \mathfrak{D}$, and $t \in \mathbb{R}$, let the linear operator $\Phi_{A,g}(f,\lambda,t)$ on $\mathbb{C}^N$, be defined by

$$\Phi_{A,g}(f,\lambda,t)=\mathrm{Tr}_{\mathfrak{F}}\{(|\Omega(f)\rangle\langle\Omega(f)|\otimes 1)U^\lambda(-t)(W(g)\otimes A)U^\lambda(t)\} \ .$$

We have $A(f,\lambda,t)=\Phi_{A,0}(f,\lambda,t)$ and $\omega(g;f,\lambda,t)$ is obtained by tracing $\Phi_{1,g}(f,\lambda,t)$ with D; i.e., $\Phi_{A,g}(f,\lambda,t)$ contains all the dynamical information we want. The matrix elements of $\Phi_{A,g}(f,\lambda,t)$ are given by

$$\begin{aligned}\Phi_{A,g}(f,\lambda,t)_{jk}&=\langle\Omega(f),[U^\lambda(-t)(W(g)\otimes A)U^\lambda(t)]_{jk}\Omega(f)\rangle\\ &=\sum_{l=1}^{N}\sum_{m=1}^{N} A_{lm}\langle\Omega(f),[U^\lambda(t)]^*_{lj}W(g)[U^\lambda(t)]_{mk}\Omega(f)\rangle \quad .\end{aligned}$$

Using (1),(2),(3), we can obtain a formula for $[U^\lambda(t)]_{jk}$ as an infinite series of time-ordered repeated integrals of Weyl operators. This will then provide us (using the commutation relations for the W(.)) with a formula for $[U^\lambda(t)]^*_{lj}W(g)$ $[U^\lambda(t)]_{mk}$ as a sum of infinite series of time-ordered repeated integrals of Weyl operators. Taking the expectation value w.r.t the state $\Omega(f)$ we arrive at:

$$\begin{aligned}&\langle\Omega(f),[U^\lambda(t)]^*_{lj}W(g)[U^\lambda(t)]_{mk}\Omega(f)\rangle=\exp\{i\mathrm{Im}(u_t f,g)\}\Big\{\delta_{jl}\delta_{mk}\exp\{-i(\lambda_j+\lambda_k)\mathrm{Im}(g,\zeta(\xi,\\ &t))/2 +i\alpha(j,k;t;\lambda;\xi)+i\beta(j,k;t;\lambda;\xi)-\|\psi(j,k;t;\lambda;\xi)+u_{-t}g\|^2/4\}+\delta_{jl}\Big\{\exp\{-i(\lambda_j+\lambda_m)\\ &\mathrm{Im}(g,\zeta(\xi,t))/2\}\sum_{n=1}^{\infty}(-i)^n\sum_{\vec{1}_{n-1}=1}^{N}F(m,\vec{1}_{n-1},k)\ T\int_0^t dt^{(n)}\exp\{i\alpha(j,m,\vec{1}_{n-1},k;t,\vec{t}_n;\lambda;\\ &\xi)-i\beta(m,\vec{1}_{n-1},k;\vec{t}_n;u_{-t}g,\lambda;\xi)/2 +i\beta(j,m,\vec{1}_{n-1},k;t,\vec{t}_n;f,\lambda;\xi)-\|\psi(j,m,\vec{1}_{n-1},k;t,\vec{t}_n;\\ &\lambda;\xi)+u_{-t}g\|^2/4\}\Big\}+\delta_{mk}\Big\{\exp\{-i(\lambda_l+\lambda_k)\mathrm{Im}(g,\zeta(\xi,t))/2\}\sum_{n=1}^{\infty}i^n\sum_{\vec{1}_{n-1}=1}^{N}\overline{F(l,\vec{1}_{n-1},j)}\ T\int_0^t dt^{(n)}\\ &\exp\{-i\alpha(k,l,\vec{1}_{n-1},j;t,\vec{t}_n;\lambda;\xi)-i\beta(l,\vec{1}_{n-1},j;\vec{t}_n;u_{-t}g,\lambda;\xi)/2 -i\beta(k,l,\vec{1}_{n-1},j;t,\vec{t}_n;\\ &f,\lambda;\xi)-\|\psi(k,l,\vec{1}_{n-1},j;t,\vec{t}_n;\lambda;\xi)-u_{-t}g\|^2/4\}\Big\} + \exp\{-i(\lambda_l+\lambda_m)\mathrm{Im}(g,\zeta(\xi,t))/2\}\cdot\\ &\sum_{n=1}^{\infty}\sum_{p=1}^{\infty}(-i)^n i^p\sum_{\vec{1}_{n-1}=1}^{N}\sum_{\vec{m}_{p-1}=1}^{N}\overline{F(l,\vec{m}_{p-1},j)}F(m,\vec{1}_{n-1},k)\ T\int_0^t ds^{(p)}\ T\int_0^t dt^{(n)}\\ &\exp\{-i\alpha(m,l,\vec{m}_{p-1},j;t,\vec{s}_p;\lambda;\xi)+i\alpha(l,m,\vec{1}_{n-1},k;t,\vec{t}_n;\lambda;\xi)-i\alpha(l,m;t;\lambda;\xi)-i\beta(l,m;t,\end{aligned}$$

$f,\lambda;\xi)-i\beta(m,1,\vec{m}_{p-1},j;t,\vec{s}_p;f,\lambda;\xi)+i\beta(1,m,\vec{l}_{n-1},k;t,\vec{t}_n;f,\lambda;\xi)-i\beta(1,\vec{m}_{p-1},j;\vec{s}_p;$

$u_{-t}g,\lambda;\xi)/2 -i\beta(m,\vec{l}_{n-1},k;\vec{t}_n;u_{-t}g,\lambda;\xi)/2 +i\,\mathrm{Im}(\psi(1,\vec{m}_{p-1},j;\vec{s}_p;\lambda,\xi),\psi(m,\vec{l}_{n-1},k;$

$\vec{t}_n;\lambda;\xi))/2 -\|\psi(1,m,\vec{l}_{n-1},k;t,\vec{t}_n;\lambda;\xi)-\psi(m,1,\vec{m}_{p-1},j;t,\vec{s}_p;\lambda;\xi)-\psi(1,m,t;\lambda;\xi)$

$+u_{-t}g\|^2/4\}\Big\} = T^{jk}_{1m}(\xi;g;f,\lambda,t)$

One now proceeds to:

(1) establish that the above infinite series are absolutely convergent, uniformly in $\|f\|$, $\|g\|$, $\|\xi\|$, λ, and t for t in any bounded subset of $\mathbb{R}$.

(2) perform the limit $\varepsilon\to 0$ in $T^{jk}_{1m}(\xi;\varepsilon g;\varepsilon^{-1}f,\varepsilon\lambda,t)$ term-by-term as one may due to (1).

(3) in $T^{jk}_{1m}(\xi;\varepsilon g;\varepsilon^{-1}f,\varepsilon\lambda,t)$ expand the exponentials (they contain only ε^2 due to the scaling properties of α, β, ψ) by $\exp(x)=\sum_{n=0} x^n/n!$, interchange the orders of summation, and verify that in the formal power-series for $T^{jk}_{1m}(\xi;\varepsilon g;\varepsilon^{-1}f,\varepsilon\lambda,t)$ in powers of ε^2, each summand, which is given by an infinite series of repeated time-ordered integrals of polynomials in α's, β's and norms of ψ's , is in fact well-defined.

This leads to the proofs of existence in a) and b), and gives us explicit expressions for all the quantities. Finally, to prove that the equation of motion of $A_{(0)}(f,\lambda,t)$ is the one given in a]1., we proceed as follows:

(4) If $\dot{U}^\lambda_o(t)=-iH_o(f,\lambda,t)U^\lambda_o(t)$, $U^\lambda_o(0)=1$, where $H_o(f,\lambda,t)=\lambda\mathrm{Re}(u_t f,\xi)V$, then $U^\lambda_o(t)=\exp\{-i\int_0^t ds H_o(f,\lambda,s)\}$, with matrix elements $U^\lambda_o(t)_{jk}=\delta_{jk}\exp\{-i\lambda_j\int_0^t ds\ \mathrm{Re}(u_s f,\xi)\}$ (recall that V is diagonal in the chosen basis). But, since $\mathrm{Im}(f,\zeta(\xi,-t))=\int_0^t ds\mathrm{Re}(u_s f,\xi)$ as can be easily seen by differentiating, we have $U_o(t)_{jk}=\delta_{jk}\exp\{-i\lambda_j\mathrm{Im}(f,\zeta(\xi,-t))\}$.

(5) Letting, $\hat{F}(t)= U_o(t)^* F U_o(t)$, we have $\hat{F}(t)_{jk}=F_{jk}\exp\{i\beta(j,k;t;\lambda;\xi)\}$.

(6) If $\dot{U}^\lambda(t)=-iH(f,\lambda,t)U^\lambda(t)$, $U^\lambda(0)=1$, then

$$U^\lambda(t) = U^\lambda_o(t)\{1+\sum_{n=1}^{\infty}(-i)^n\ T\int_0^t dt^{(n)}\ \hat{F}(t_1)\hat{F}(t_2)..\hat{F}(t_n)\} .$$

With this and the above we obtain a formula for $U^\lambda(t)_{jk}$, and then one for $U^\lambda(t)^* A U^\lambda(t)\ _{jk}$. This last formula coincides with the one obtained for $A_{(0)}(f,\lambda,t)_{jk}$, completing the proof since $U^\lambda(t)^* A U^\lambda(t)$ is the solution of the differential equation we propose.

3. *Acknowledgements*

We gratefully acknowledge discussions with Dr. Peter Pfeifer. Part of this work was done while one of us (G.A.R.) was at the Naturwissenschaftlich-Theoretisches-Zentrum der Karl-Marx-Universität, Leipzig; it is a pleasure to thank the members of the NTZ for their hospitality. We are also grateful to the organizers of the II. Workshop on Quantum Probability and Applications for permitting us to present these results at that meeting.

4. *References*

[1] H.S. Zivi, and G.A. Raggio, *Semiclassical description on N-level systems interacting with radiation fields*, in preparation.

[2] P. Pfeifer, *Chiral molecules - a superselection rule induced by the radiation field.* Diss. ETH-Zürich No. 6551; DK Gotthard S+D A.G., Zürich 1980.

[3] E.B. Davies, Ann. Inst. H. Poincaré A35, 149-171 (1981)

[4] J.M. Cook, J. Math. Phys. 2, 33-45 (1961).

5. *Addendum of December 1984*

We have shown, in the meantime, that the unphysical infrared cutoff-condition (see footnote 3)) can be removed without altering the theorem; see [1] .

THE CHARGE CLASS OF THE VACUUM STATE IN A FREE MASSLESS DIRAC FIELD THEORY

S. Scarlatti*

Istituto Matematico
"Guido Castelnuovo"
Università di Roma
La Sapienza

M. Spera**

Dipartimento di Matematica
II Università di Roma
Torvergata

ABSTRACT - We comment upon the concept of charge class introduced by D. Buchholz in [1] and its modification suggested in [2] and we prove that in the Free Massless Dirac Field case the modified charge class of the Fock state does not contain any non trivial gauge invariant quasi free state.

§.0 - INTRODUCTION -

We shall adopt the algebraic approach to Quantum Field Theory proposed by R. Haag and D. Kastler in [3] . In their formalism, a physical system is described by assigning an isotonic net

$$\mathcal{O} \to \mathfrak{A}(\mathcal{O})$$

of C* algebras (called local observable algebras because their self adjoint elements represent the physical measurements which can be performed within $\mathcal{O}$) indexed by bounded regions of Minkowski space. For an unbounded region, say $\mathcal{Q}$, $\mathfrak{A}(\mathcal{Q})$ is defined through a C*

----

* Supported by Istituto Nazionale di Alta Matematica " F. Severi "

** Supported by Ministero della Pubblica Istruzione and CNR-GNAFA inductive limit:

$$\mathcal{O}\!\mathcal{L}(\mathcal{Q}) = \overline{\bigcup_{\sigma \subset \mathcal{Q}}}^{C^*} \mathcal{O}\!\mathcal{L}(\sigma)$$

$\mathcal{O}\!\mathcal{L} \equiv \mathcal{O}\!\mathcal{L}(\mathbb{R}^4)$ is called quasi local algebra. We briefly recall the properties that the net $\{\mathcal{O}\!\mathcal{L}(\sigma)\}$ should possess:

i) <u>causality</u> -

$$[\,\mathcal{O}\!\mathcal{L}(\sigma_1)\,,\ \mathcal{O}\!\mathcal{L}(\sigma_2)\,]_- = 0$$

if $\sigma_1 \subset \sigma_2'$. Here σ' is the space-like complement of σ and if A and B are elements of any C* algebra, we set $[A,B]_\pm \equiv AB \pm BA$ This postulate translates the requirement that experiments performed in causally disjoint regions should not interfere.

ii) <u>relativistic covariance</u> - The universal covering group $\tilde{\mathcal{P}}_+^\uparrow$ of the restricted Poincaré group $\mathcal{P}_+^\uparrow$ acts on $\mathcal{O}\!\mathcal{L}$ via a net structure preserving automorphism group $\alpha = \{\alpha_L\ ,\ L \in \tilde{\mathcal{P}}_+^\uparrow\}$ that is to say

$$\alpha_L \circ \mathcal{O}\!\mathcal{L}(\sigma) = \mathcal{O}\!\mathcal{L}(\Pi(L)\sigma)$$

where $L \in \tilde{\mathcal{P}}_+^\uparrow$ and Π is the covering map $\Pi : \hat{\mathcal{P}}_+^\uparrow \to \mathcal{P}_+^\uparrow$

iii) <u>existence of the vacuum representation</u> - Before stating this property let us recall some terminology. The quasi equivalence (unitary equivalence) class of a representation π of $\mathcal{O}\!\mathcal{L}$ will be denoted by $[\pi]_q$ (or $[\pi]$, respectively). Among all possible representations of $\mathcal{O}\!\mathcal{L}$ we shall consider henceforth those satisfying the so called Spectrum Condition, that is to say, those representations which are space time translation covariant($[\pi \circ \alpha_T] = [\pi]$ with $T \simeq \mathbb{R}^4$ the space time translation group and the unitary group

$\{U_\pi(a)\}_{a\in T}$ realizing the unitary equivalence on $\mathcal{H}_\pi$ being strongly continuous, and such that the joint spectrum of its generators is contained in $\overline{V^+}$, the closure of the forward light cone V^+. $V^\pm = \{x \in \mathbb{R}^4 : \langle x,x\rangle > 0 \, , \, x^0 \gtrless 0\}$. Such representations are also called positive energy representations or simply positive representations. Next axiom is then the following: there exists a distinguished positive representation $\{\pi_0 \, , \mathcal{H}_0 \, , \xi_0\}$ (called the vacuum or the Fock representation built up via the GNS theory from a (unique up to phase) Poincaré invariant vector ξ_0 . Any representation $\pi \in [\pi_0]_q$ will be called a trivial representation. In practice, all other positive representations are to be compared with π_0 . This is made precise through the important concept of normality.

0.1 - Definition - Let π be a positive representation of $\mathfrak{A}$

i) - π is said to be normal on a region $\mathcal{R}$ if

$$\left[\pi \upharpoonright \mathfrak{A}(\mathcal{R})\right]_q = \left[\pi_0 \upharpoonright \mathfrak{A}(\mathcal{R})\right]_q \qquad (*)$$

ii) - π is said to be locally normal if $(*)$ holds for any bounded region $\mathcal{O}$

iii) - Given a locally normal representation π, $[\pi]_c \equiv [\pi \upharpoonright \mathfrak{A}(V^+)]_q$ is called the Charge Class of π

Remarks - All definitions may be formulated in terms of states too, due to the GNS theory; we shall freely use the two versions.

All representations we use will be locally normal; the meaning of local normality is that local measurements cannot distinguish among the various representations, i.e. the superselection structure (the physical spectrum) of $\mathfrak{A}$ is dictated by the global features of the theory.

The concept of charge class is due to Buchholz [1]. In order to appreciate this notion we digress a little about the infrared problem in free massless theories. It is well known that in the case of an infinite number of degrees of freedom, the von Neumann uniqueness theorem for the Weyl Commutation Relations does no longer hold and we get uncountably many inequivalent representations of them. Among these we encounter those corresponding to generalized coherent states, [4], which are positive and are suitable for the description of the so called infrared catastrophe, [5], i.e. roughly speaking, the appearance with probability one of an infinite number of photons with total finite energy accompanying a charged particle treated classically as an exter nal field which vanishes adiabatically for $t \to \pm\infty$ Infrared represent ations for free massless Fermi fields have been introduced by S. Dopli cher [6] (see also[7],[2] f.i.) They will be discussed later.

Buchholz,[1], has started the algebraic approach to QED. He points out that, owing to Gauss law there are uncountably many superselection sectors in QED. In order to cope with this situation he first introduc es the notion of charge class of a representation, motivated by the fact that charged particles are massive and if they are future travelling they will definitely enter the cone V^+; then he proposes a candidate for the state space of QED introducing the so called infrared minimal states, which enjoy the best localization properties with respect to the vacuum state and admit scattering observables and he proves in particularthat within a single charge class, two represent ations may differ only by some radiation field ; we hope that these few remarks might give an idea of the usefulness of this concept.

Nevertheless, the analysis of [2] in a theory where univalence is the only superselection rule suggests the introduction of a new notion of charge class $\mathcal{C}'$, namely $\pi' \in [\pi]_{\mathcal{C}'}$ iff

$$i) \quad [\pi' \upharpoonright \mathcal{A}(V^+)]_q = [\pi \upharpoonright \mathcal{A}(V^+)]_q$$

$$ii) \quad [\pi \upharpoonright \mathcal{A}(V^-)]_q = [\pi \upharpoonright \mathcal{A}(V^-)]_q$$

However, this does not essentially modify the analysis of [1]. The main result of [2] may be formulated as follows: in the case of the free massless Dirac (or Majorana) field there exist representations π such that

$$[\pi]_{\mathcal{C}} = [\pi_0]_{\mathcal{C}}$$

but

$$[\pi]_{\mathcal{C}'} \neq [\pi_0]_{\mathcal{C}'}$$

The aim of this paper is a partial characterization of the $\mathcal{C}'$class of the Fock representation in the free massless Dirac field case: we prove the following theorem:

> The (modified) charge class of the vacuum state does not contain any non trivial pure gauge invariant quasi free state.

§.1 - REVIEW OF CAR ALGEBRA AND FREE STATES

We recall the definition of the CAR algebra (see [8] as a reference for the whole section) . Let $\mathcal{K}$ be a separable Hilbert space. The CAR algebra $\mathcal{A}(\mathcal{K})$ over $\mathcal{K}$ is the unital C* algebra generated by elements $a(f)$ depending linearly on $f \in \mathcal{K}$ and satisfying the CAR

$$[a(f), a(g)]_+ = 0$$
$$[a(f)^*, a(g)]_+ = \langle f, g\rangle_X I$$

The group $\mathbb{T}$ (the circle group) is called the gauge group of $\mathfrak{A}(X)$
It acts on X via the unitary representation

$$\mathcal{U}_\theta f = e^{i\theta} f \qquad , \theta \in \mathbb{T}$$

which in turn induces a unique * automorphism group $\{\alpha_\theta\}_{\theta\in\mathbb{T}}$ on $\mathfrak{A}(X)$. Set $\alpha_\pi = \gamma$ The even CAR algebra $\mathfrak{A}(X)_e$ is the fixed point algebra of γ . The map $\Gamma = 2^{-1}(1+\gamma)$ provides a normal conditional expectation from $\mathfrak{A}(X)$ onto $\mathfrak{A}(X)_e$. Now let $A \in \mathcal{B}(X)$, $0 \le A \le I$

1.1 - Definition - The gauge invariant quasi free state ω_A is defined through its n-point functions by the formula

$$\omega_A\left(a(f_n)^* \dots a(f_1)^* a(g_1) \dots a(g_m)\right) = \delta_{nm} \det\left(\langle f_i, A g_j\rangle\right)$$

$\{f_i\}_{i=1..n}$, $\{g_j\}_{j=1..m} \in X$. Gauge invariance means $\omega_A \circ \alpha_\theta = \omega_A$; quasi freedom means that the n-point functions depend only on the 2-point functions. Set $\pi_{\omega_A} \equiv \pi_A$. ω_0 is called the Fock state and π_0 the Fock representation. We recall the important theorem of [8]

Theorem (Powers - Størmer)

i) - ω_A is pure iff A is a projection $(A^2 = A = A^*)$

ii) - ω_A is a factor state for any A $(\pi_A'' \cap \pi_A' = \mathbb{C}I)$

iii) - $[\pi_A]_q = [\pi_B]_q$ iff $A^{\frac{1}{2}} - B^{\frac{1}{2}}$ and $(1-A)^{\frac{1}{2}} - (1-B)^{\frac{1}{2}} \in \mathcal{L}^2(X)$

Here $\mathcal{L}^p(X)$, $p \ge 1$ denotes the Shatten ideal of compact operators C in $\mathcal{B}(X)$ with eigenspectrum $\{c_i\}_{i=1,2...}$ such that $\sum_{i=1}^{\infty} |c_i|^p < \infty$ Then $\mathcal{L}^1 \equiv$ trace class operators ; $\mathcal{L}^2 \equiv$ Hilbert Schmidt operators.

Remark - We explicitly want to observe that $S = A^{\frac{1}{2}} - B^{\frac{1}{2}} \in \mathcal{L}^2$ implies $A - B \in \mathcal{L}^2$ In fact if $T = A^{\frac{1}{2}} + B^{\frac{1}{2}}$ then $A-B = 2^{-1}[T,S]_+$. We also note that if E is a projection then $\pi_E \in [\pi_0]_q$ iff E is finite dimensional.

§.2 - THE FREE MASSLESS DIRAC FIELD

See [2] for reference. The one particle space of this theory is the subspace $\mathcal{K}$ of $\mathcal{L}^2(\mathbb{R}^3, d^3k) \otimes \mathbb{C}^4$ consisting of the solutions of the massless Dirac equation

$$\not{k}\, \psi(\vec{k}) = 0$$

$k = (|\vec{k}|, \vec{k})$ or equivalently of the equation

$$(2|\vec{k}|)^{-1} \not{k} \gamma_0 \psi(\vec{k}) = \psi(\vec{k})$$

$\mathcal{K}$ carries a $[0, \frac{1}{2}] \oplus [0, -\frac{1}{2}]$ representation of $\tilde{\mathcal{P}}^1_+$. Let $\mathcal{S} \equiv \mathcal{S}(\mathbb{R}^4) \otimes \mathbb{C}^4$ be the test function space. Define two maps T and S from $\mathcal{S}$ to $\mathcal{K}$, the former linear, the latter antilinear, as follows

$$(Tf)(\vec{k}) = \sqrt{2\pi}\, \frac{\not{k}\gamma_0}{2|\vec{k}|}\, C_1 \hat{f}(|\vec{k}|, \vec{k})$$

$$(Sf)(\vec{k}) = \sqrt{2\pi}\, \frac{\not{k}\gamma_0}{2|\vec{k}|}\, \bar{\hat{f}}(|\vec{k}|, \vec{k})$$

where C_1 is the real Pauli matrix such that $C_1^2 = I$ and $C_1 \gamma_\mu C_1 = -g_{\mu\mu} \gamma_\mu^t$. We have $\overline{[T\mathcal{S}]} = \overline{[S\mathcal{S}]} = \mathcal{K}$. T and S intertwine the action of $\mathcal{P}^1_+$ and $\tilde{\mathcal{P}}^1_+$ on $\mathcal{S}$ and $\mathcal{K}$ respectively. The Dirac Field is defined as follows: consider the CAR algebra $\mathfrak{A}(\mathcal{K} \oplus \mathcal{K})$ (the Field Algebra), then

$$\psi(f) = a(Tf \oplus 0) \oplus a(0 \oplus Sf)^* \qquad f \in \mathcal{S}$$

ψ satisfies the right anticommutation relations, namely

$$[\, \psi(f), \psi(g)]_+ = 0 \qquad (*)$$

$$[\Psi(f)^*, \Psi(g)]_+ = (f, (-i S\gamma^\circ)^t g) I \qquad (**)$$

The right hand side of $(**)$ comes from the formula

$$(f, (-i S\gamma^\circ)^t g) = \langle Tf, Tg\rangle_{\mathcal{H}} + \langle Sf, Sg\rangle_{\overline{\mathcal{H}}} \qquad (***)$$

Remark - The use of $\overline{\mathcal{H}}$ (†) comes from the antilinearity of S ; hence we are led to the CAR algebra $\mathfrak{A}(\mathcal{H} \oplus \overline{\mathcal{H}})$ in order to interpret the anticommutation relations for the fields. From $(*)$, $(**)$ and $(***)$ we also get:

$$\| \Psi(f) \|^2 = \| Tf \|^2 + \| Sf \|^2$$

Here $\| \cdot \|$ denotes a C* norm in the left hand side and a $\mathcal{L}^2$ norm in the right hand side. $\Psi(F)$ makes sense for all $F \in \mathcal{S}'$ such that $TF \oplus SF \in$ $\in \mathcal{H} \oplus \overline{\mathcal{H}}$ Now we are ready to define the local structure of the Dirac field. Let $\mathcal{B}(\mathcal{O})$ be the C* algebra generated by polynomials in $\Psi(f)$ with supp $f \subset \mathcal{O}$. Then :

$$\mathfrak{A}(\mathcal{O}) = \Gamma \cdot \mathcal{B}(\mathcal{O})$$
$$\mathfrak{A} = \mathfrak{A}(\mathcal{H} \oplus \mathcal{H})_e$$

The Haag - Kastler axioms are readily verified. Moreover, time-like separated observables commute (field theoretic Huygens Principle). Now, for any region $\mathcal{R}$ define $\mathcal{G}_{\mathcal{R}} \equiv$ projector onto the closed sub space of $\mathcal{H} \oplus \overline{\mathcal{H}}$ spanned by $\{Tf \oplus Sf, \text{supp} f \subset \mathcal{R}\}$. Set $\mathcal{G}_{V^\pm} \equiv \mathcal{G}_\pm$

2.1 - Theorem - i) $\mathcal{G}_+ \mathcal{G}_- = \mathcal{G}_- \mathcal{G}_+ = 0$ ii) $\mathcal{G}_+ + \mathcal{G}_- = I$

Proof: i) is obvious because if $Tf \oplus Sf \in \mathcal{G}_+(\mathcal{H} \oplus \overline{\mathcal{H}})$, $Tg \oplus Sg \in \mathcal{G}_-(\mathcal{H} \oplus \overline{\mathcal{H}})$, then $\langle Tf, Tg\rangle_{\mathcal{H}} + \langle Sf, Sg\rangle_{\overline{\mathcal{H}}} = [\Psi(f)^*, \Psi(g)]_+ = 0$ due to Huygens Principle.

ii) - It is easily proved that $\overline{[TF \oplus SF]} = \mathcal{H} \oplus \overline{\mathcal{H}}$, where F is a distribution of the form $F = \mathcal{P}(\vec{\partial})\delta(\vec{x})f(t)$, $f \in \mathcal{S}(\mathbb{R})$, with $\mathcal{P}$ a

(†) $\overline{\mathcal{H}}$ is the dual of $\mathcal{H}$

polynomial, but the same result holds if f ranges across the functions of $\mathcal{S}(\mathbb{R}^4)$ whose support does not contain the origin. Any such function can be written as a sum $f_+ + f_-$ where $f_\pm$ have support in the interval $(0,+\infty)$ and $(-\infty, 0)$ respectively. But for this dense set

$$TF \oplus SF = TF_+ \oplus SF_+ + TF_- \oplus SF_-$$

with obvious notations and it is clear that $G_\pm(TF_\pm \oplus SF_\pm) = TF_\pm \oplus SF_\pm$. The result follows immediately. We also get the following :

2.2 - Corollary - If $F = \delta(\vec{\nabla})\delta(\vec{x})f(t)$, $f \in \mathcal{L}^2(\mathbb{R})$ and $TF \oplus SF$ makes sense then $G_\pm(TF \oplus SF) = T\chi_\pm F \oplus S\chi_\pm F$ with $\chi_+ = \chi_{[0,\infty)}, \chi_- = \chi_{(-\infty,0)}$.

§.3 - THE MAIN RESULT

We shall consider positive representations π_E of the CAR algebra of the Dirac field. Such representations were first considered by S. Doplicher (for the construction of relevant non pure - e.g. III_λ - examples see [7][2] for the fermion case,[9]for the boson case,[10] for the interacting case. For a brief review of these problems see also[11] The structure of E is the following: $E = \sum_i |\Psi_i\rangle\langle\Psi_i|$ with $\{\Psi_i\}$ an orthonormal system in $\mathcal{H}$ such that, if H is the one particle Hamiltonian, then $\sum_i \langle \Psi_i, H\Psi_i\rangle < \infty$ (see f.i.[6][7][12][2]etc)

We have to consider the state $\omega_{\tilde{E}}$ on $\mathfrak{A}(\mathcal{H}\oplus\mathcal{H})$ with $\tilde{E} = \left(\begin{smallmatrix} E & 0 \\ 0 & 0 \end{smallmatrix}\right)$. The 2-point functions of the field Ψ take the form :

$$\omega_{\tilde{E}}(\Psi(f)^*\Psi(g)) = \langle Tf, ETg\rangle_{\mathcal{H}} + \langle Sf, Sg\rangle_{\bar{\mathcal{H}}}$$

The right hand side can be interpreted as the two point function for the state $\omega_{\hat{E}}$, with $\hat{E} = \left(\begin{smallmatrix} E & 0 \\ 0 & I \end{smallmatrix}\right)$ on the CAR algebra $\mathfrak{A}(\mathcal{H}\oplus\bar{\mathcal{H}})$

The restriction of the representation $\tilde{\pi}_{\tilde{E}}$ of the field algebra on a region $\mathcal{Q}$ is accordingly interpreted as the representation $\tilde{\pi}_{G_{\mathcal{Q}} \hat{E} G_{\mathcal{Q}}}$ of $\mathfrak{A}(\mathcal{H} \oplus \bar{\mathcal{H}})$. The use of P-S theorem at this point yields the following:

3.1 - Proposition - $\tilde{\pi}_{\tilde{E}}$ is in the q' class of the Fock representation iff

$$(G_{\pm} \hat{E} G_{\pm})^{\frac{1}{2}} - (G_{\pm} \hat{O} G_{\pm})^{\frac{1}{2}} \in \mathcal{L}^2(\mathcal{H} \oplus \bar{\mathcal{H}})$$
$$(G_{\pm}(\hat{I} - \hat{E}) G_{\pm})^{\frac{1}{2}} - (G_{\pm} \hat{I} G_{\pm})^{\frac{1}{2}} \in \mathcal{L}^2(\mathcal{H} \oplus \bar{\mathcal{H}})$$

Owing to the remark just after P-S theorem, we have inparticular that if $\tilde{\pi}_E \in [\tilde{\pi}_0]q'$, then $G_{\pm} \left(\begin{array}{c|c} E & 0 \\ \hline 0 & 0 \end{array}\right) G_{\pm} \in \mathcal{L}^2$ But this implies $G_{\pm} \left(\begin{array}{c|c} E & 0 \\ \hline 0 & 0 \end{array}\right) \in \mathcal{L}^4$ and, due to theorem 2.1, $\left(\begin{array}{c|c} E & 0 \\ \hline 0 & 0 \end{array}\right) \in \mathcal{L}^4$ that is, E is finite dimensional. Thus we proved:

3.2 - Theorem - $[\tilde{\pi}_0]_{q'}$ does not contain any non trivial pure gauge invariant quasi free state.

Remark - We actually proved the theorem for the field algebra; the quasi local algebra case follows immediately. This theorem is the first instance of a phenomenon we expect in general: $[\pi_0]_{q'} = [\pi_0]_q$ for Fermi theories. Acknowledgements - We are indebted to Prof. S. Doplicher for his constant help and encouragement, and to the Organizers of the II Workshop on Quantum Probability and Applications for letting the second named author to present a communication on this subject.

REFERENCES

[1] - D. Buchholz, CMP 85, 49 1982
[2],[7] - S. Doplicher, M. Spera, CMP 89, 19 1983; 84, 505 1982
[3] - R. Haag, D. Kastler JMP 5, 848 1964
[4],[5] - G. Roepstorff, CMP 19, 301 1970; F.Bloch, A.Nordsieck Phys. Rev. 52, 54, ...
[6],[12] - S. Doplicher, CMP 3, 228 1966; K.Kraus et al, JMP 48, 2166 1977
[8] - R. Powers, E. Størmer, CMP 16, 1 1970
[9] - S. Doplicher, F. Figliolini, D. Guido AIHP 41, 49 1984
[10] - D. Buchholz, S. Doplicher, AIHP 1984
[11] - M. Spera, Springer LNM 1055 1984

DERIVATION OF CLASSICAL HYDRODYNAMICS OF A QUANTUM COULOMB SYSTEM

by Geoffrey L. Sewell

Department of Physics, Queen Mary College, London E1 4NS

ABSTRACT. We derive macroscopic electro-hydrodynamical equations of Euler and Maxwell from the many-particle Schrödinger equation of a quantum Coulomb system, subject to certain viable initial conditions.

1. Introduction and Discussion

A central problem in the theory of the relationship between microphysics and macrophysics is that of extracting the phenomenological laws of continuum mechanics, such as those of hydrodynamics or heat conduction, from the Schrödinger equation for large assemblies of particles. Here, the description of such systems by quantum, rather than classical, mechanics is dictated by the fact that the very stability of matter is a quantum phenomenon(1). The passage from a microscopic quantum description of matter to a classical, deterministic, phenomenological one is, of course, fraught with serious conceptual and mathematical problems. It is therefore not surprising that only in very limited contexts has this passage been acheived on a rigorous level. As positive examples, there are Davies's(2) derivation of Fourier's law of heat conduction for a rather rudimentary model and my derivation of the plasma hydrodynamics of a Coulomb system, sketched out in a research note(3). It is with this latter work that I shall be concerned in the present talk.

The system that I shall be concerned with is the so-called Jellium model, consisting of N electrons in a box with uniform neutralising background of positive charge. The interactions are taken to be the standard Coulomb electrostatic ones. My main result is that the Schrodinger evolution of a suitable class of pure states leads, in the limit where $N \to \infty$ at finite density, and in a certain macroscopic scaling, to the following Euler-cum-Maxwell equations for the normalised electron density, ρ, the drift velocity, u, and the electric field, E.

$$\frac{\partial \rho}{\partial t} + \mathrm{div}\,(\rho u) = 0 \; ; \; \frac{\partial u}{\partial t} + (u \cdot \nabla) u = E \; ; \; \mathrm{div}\, E = \rho - 1 \tag{1}$$

It will be seen that the classical character of these equations stems from the macroscopicality of the description, i.e. from the limit $N \to \infty$ and the chosen scaling, without recourse to any formal procedure of letting $\hbar \to 0$. On the debit side, these equations lack viscosity and pressure gradient terms. These deficiencies may be traced to the fact that the length scale we employ is the largest one available, given by the length of the box; and in this scaling, any viscosity and pressure gradient terms that might have appeared in the second equation of (1), are automatically removed by the limit $N \to \infty$. This means that a finer treatment, based on a shorter length scale, is needed to incorporate these terms.

I shall present the theory leading from the quantum dynamics of the Jellium model to its hydrodynamical equations (1) as follows. In section 2, I shall formulate the model. In section 3, I shall pass to its description in the chosen macroscopic scaling, in the limit $N \to \infty$. In section 4, I shall derive the classical Vlasov kinetic equation for its macroscopic evolution from its underlying quantum dynamics, by an adaptation of a method devised for a related model[4]. In section 5, I shall derive the hydrodynamical equations (1) from the Vlasov dynamics. The Appendix will be devoted to the proof of a technical point.

2. The Model

The Jellium model, Σ_N, consists of N electrons in a cube, Δ_L, of side L, with uniform neutralising background of positive charge. For simplicity, the particles are treated as spinless, and periodic boundary conditions are assumed. Thus, denoting spatial position and time by X and T, respectively, the Hamiltonian for Σ_N is[5]

$$H_N = -\frac{\hbar^2}{2m} \sum_{1}^{N} \Delta_{X_j} + e^2 \sum_{j,k=1}^{N} V^{(L)}(X_j - X_k) \tag{2}$$

where $V^{(L)}(X) = \frac{2\pi}{L^3} \sum_{q\neq 0}' \exp(iq\cdot X/L)/q^2$, (3)

the prime indicating that q runs over the values $2\pi(n_1, n_2, n_3)$, with the n's integers. The time-dependent state of Σ_N corresponds to a normalised antisymmetric wave-function $\Psi_T^{(N)}(X_1, \ldots, X_N)$, that evolves according to the Schrödinger equation

$$i\hbar\, \partial\Psi_T^{(N)}/\partial T = H_N \Psi_T^{(N)} \quad (4)$$

Macroscopic Description. We formulate the macroscopic properties of Σ_N on length and time scales whose units are L and ω^{-1}, respectively, ω being the classical plasma frequency[5],

$$\omega = (4\pi\bar{n}e^2/m)^{1/2} , \quad (5)$$

where

$$\bar{n} = N/L^3 . \quad (6)$$

Correspondingly, we take the unit of momentum per particle to be $mL\omega$. We centre our macroscopic description on intensive observables, given by N^{-1} times additive functions of positions and momenta on the specified scales. These then take the form

$$A^{(N)} = N^{-1} \sum_1^N A(X_j/L, P_j/mL\omega) , \quad P_j = -i\hbar\nabla_{X_j} \quad (7)$$

where $A(X/L,P/mL\omega) = \sum_\eta' \int d\xi \hat{A}(\xi,\eta)\exp(i\xi\cdot P/2mL\omega)\exp(i\eta\cdot X/L)\exp(i\xi\cdot P/2mL\omega)$, (8)

the prime over Σ having the same significance as in (3), and $\hat{A}(\xi,\eta)$ being continuous in ξ and satisfying the condition $\sum_\eta' \int d\xi |\hat{A}(\xi,\eta)| < \infty$, so that the R.H.S. of (8) is well-defined. The expectation values of products of the macro-observables, $A^{(N)}$, at time t on the ω^{-1}-scale are given by

$$\langle A_1^{(N)} \ldots A_n^{(N)}\rangle_t^{(N)} = \left(\Psi_{\omega^{-1}t}^{(N)} , A_1^{(N)} \ldots A_n^{(N)} \Psi_{\omega^{-1}t}^{(N)}\right) \quad (9)$$

The Rescaled Description. Since the macroscopic length scale is L, it is convenient to introduce a scale transformation $X \to x = X/L$, which maps Σ_N into a system, Σ_N', of N particles in a unit cube, Δ_1. It then maps the state $\Psi_{\omega^{-1}t}^{(N)}$ of Σ_N into the state, $\psi_t^{(N)}$, of Σ_N', defined by

$$\psi_t^{(N)}(x_1, \ldots, x_N) = L^{3N/2}\, \Psi^{(N)}_{\omega^{-1}t}\, (Lx_1 \ldots, Lx_N), \tag{10}$$

and transforms the Schrödinger equation (5) for Σ_N into the following one for Σ'_N :-

$$i\hbar_N\, \partial\psi_t^{(N)}/\partial t = H'_N\, \psi_t^{(N)} \tag{11}$$

where $$H'_N = -\frac{1}{2}\hbar_N^2 \sum_1^N \Delta_j + N^{-1} \sum_{j,k=1}^{N} V(x_j - x_k), \tag{12}$$

$$\hbar_N = \hbar/mL^2\omega \;\equiv\; \hbar\bar{n}^{2/3}/m\omega N^{2/3} \tag{13}$$

is a dimensionless, <u>effective Planck constant</u> and

$$V(x) = \sum_{q\neq 0}{}' \exp(iq\cdot x)/q^2 \tag{14}$$

the prime having the same significance as in (3). Two key properties of the model are that (a) $\hbar_N \to 0$ as $N \to \infty$ and (b) its potential energy takes the form $N^{-1}\sum V(x_j - x_k)$. As one might anticipate, it will be found that these properties lead to classical and mean field theoretic limits, respectively, and so to a Vlasov dynamics (cf. Ref. 4).

<u>Characteristic Functions</u>. We define the characteristic functions (CF's) for Σ'_N to be

$$\mu_t^{(N,n)}(\xi_1 \ldots, \xi_n; \eta_1, \ldots, \eta_n) =$$

$$\left(\psi_t^{(N)}, \exp\left(\frac{i}{2}\sum_1^n \xi_j\cdot p_j\right)\exp\left(i\sum_1^n \eta_j\cdot x_j\right)\exp\left(\frac{i}{2}\sum_1^n \xi_j\cdot p_j\right)\psi_t^{(N)}\right), \text{ with } p_j = -i\hbar\nabla_j,$$

where the ξ's run through R^3 and the η's through the values $2\pi(n_1, n_2, n_3)$, with the n's integers. Thus, by (7) - (9) and (15), the CF's of Σ'_N serve to determine the time-dependent expectation values of products of the intensive macro-observables of Σ_N.

<u>Initial Conditions</u>. Denoting the kinetic and potential energy observables of Σ_N by $T^{(N)}$ and $V^{(N)}$ and those of Σ'_N by $T'^{(N)}$ and $V'^{(N)}$, respectively, we assume the following initial conditions.

(I) The initial mean kinetic energy per particle of Σ_N is uniformly bounded w.r.t. N. Using the scale transformation $X \to x = X/L$ and the formulae (2), (12) to translate this condition to one for Σ'_N, this signifies that

$$\left(\psi_o^{(N)},\ T'^{(N)}\ \psi_o^{(N)}\right) = \frac{1}{2} N\left(\psi_o^{(N)}\ ,\ p_1^2\ \psi_o^{(N)}\right) < \text{const.}\ N^{1/3} \qquad (16)$$

(II) The initial mean potential energy of Σ_N is less than some constant times $N^{5/3}$ – this bound corresponds to the electrostatic energy of a charge distribution in Δ_L given by a smooth function of X/L. Correspondingly, for Σ_N',

$$\left(\psi_o^{(N)}\ ,\ V'^{(N)}\ \psi_o^{(N)}\right) < \text{const.}\ N \qquad (17)$$

(III) The macroscopic observables of Σ_N are sharply defined at $t = 0$, in that $\langle A_1^{(N)} \dots A_n^{(N)}\rangle_0^{(N)} - \prod_1^n \langle A_j^{(N)}\rangle_0^{(N)} \to 0$ as $N \to \infty$. Equivalently, by (7) – (9) and (15), the CF's of Σ_N' satisfy the condition

$$\mu_o^{(N,n)}\ (\xi_1,\ \dots,\ \xi_n;\ \eta_1 \dots,\ \eta_n) - \prod_1^n \mu_o^{(N,1)}\ (\xi_j,\eta_j) \to 0 \text{ as } N \to \infty \qquad (18)$$

Note. Conditions (I) – (III) are perfectly viable, as may be seen by checking that they are satisfied by states $\psi_o^{(N)}$, constructed as follows. Divide Δ_L into N' equal cubes $\{C_J\}$ of fixed size, centred at $\{X_J\}$. Distribute the electrons among these cubes so that n_J, the number in C_J, is a smooth function of X_J/L. For each C_J, construct a state Ψ_J of the n_J particles there, so that its kinetic and potential energies are bounded, uniformly w.r.t. J. Define $\Psi_o^{(N)}$ as the state of Σ_N given by the antisymmetrized product of the Ψ_J's and then define $\psi_o^{(N)}$ in terms of $\Psi_o^{(N)}$ according to (10).

Regularity Assumption. In view of the singularity in the Coulomb potential $V(x)$ $(\sim |x|^{-1})$, we need an assumption to the effect that, in the states under consideration, the repulsive character of the inter-electronic interactions keeps the particles apart at all times so that the effective force on a particle is always bounded. We express this assumption in terms of the two-particle density of Σ_N', namely

$$\rho_t^{(N,2)}\ (x_1,x_2) = \int dx_3\ \dots\ dx_N\ |\psi_t^{(N)}\ (x_1,\ \dots,\ x_N)|^2\ , \qquad (19)$$

as follows.

(A.1) <u>The two-particle probability density $\rho_t^{(N,2)}$ is uniformly bounded over finite time intervals, i.e. for any finite positive T, there is a finite K_T such that</u>

$$\rho_t^{(N,2)}(x_1,x_2) < K_T \text{ for } x_1,x_2 \in \Delta_1,\ 0 \le t \le T \text{ and all N.} \tag{20}$$

3. <u>The Limit $N \to \infty$</u>

To prove that the CF's of Σ_N' converge to limits as $N \to \infty$, we need to establish some bounds on the kinetic energy and on the derivatives of the CF's of that model.

<u>The K.E. Bound</u>. By the initial conditions (16), (17) and the conservation of energy,

$$\left(\psi_t^{(N)},\ T'^{(N)}\ \psi_t^{(N)}\right) + \left(\psi_t^{(N)},\ V'^{(N)}\ \psi_t^{(N)}\right) < \text{const. N} \tag{21}$$

Further, by the stability of neutral Coulomb systems of fermions[1], the ground energy level per particle of Σ_N has a finite lower bound. Equivalently, the ground energy level of Σ_N' exceeds a constant times $N^{1/3}$, and the same would be true if the particle mass were doubled. Hence

$$\frac{1}{2}\left(\psi_t^{(N)},\ T'^{(N)}\ \psi_t^{(N)}\right) + \left(\psi_t^{(N)},\ V'^{(N)}\ \psi_t^{(N)}\right) > \text{const. } N^{1/3}$$

and therefore, by (21),

$$\left(\psi_t^{(N)},\ T'^{(N)}\ \psi_t^{(N)}\right) = \frac{1}{2}\, N \left(\psi_t^{(N)},\ p_1^2\ \psi_t^{(N)}\right) < \text{const. N} \tag{22}$$

<u>Derivatives of CF's</u>. By (15) and (21),

$$\left|\frac{\partial\mu_t^{(N,1)}}{\partial\xi}(\xi,\eta)\right| \le \left(\psi_t^{(N)},\ p_1^2\ \psi_t^{(N)}\right)^{1/2} < \text{const. ;} \tag{23}$$

and, by (11) and (15),

$$\frac{\partial\mu_t^{(N,1)}}{\partial t} - \eta\cdot\frac{\partial\mu_t^{(N,1)}}{\partial\xi} =$$

$$-\frac{i}{\hbar_N}\left[\frac{N-1}{N}\right]\left(\psi_t^{(N)},\left[\exp(i\xi\cdot p_1/2)\exp(i\eta\cdot x_1)\exp(i\xi\cdot p_1/2),V(x_1-x_2)\right]\psi_t^{(N)}\right) =$$

$$-\,i\left[\frac{N-1}{N}\right]\int dx_1\ldots dx_N\,\psi^{(N)*}\left(x_1-\frac{1}{2}\hbar\,\xi,\ x_2,\ldots,x_N\right)\exp(i\eta\cdot x_1)\ \times$$

$$\hbar_N^{-1}\left(V(x_1-x_2+\tfrac{1}{2}\hbar_N\xi) - V(x_1-x_2-\tfrac{1}{2}\hbar_N\xi)\right)\ \times$$

$$\psi_t^{(N)}(x_1+\tfrac{1}{2}\hbar_N\xi,\ x_2,\ \ldots,\ x_N) \tag{24}$$

Using the Schwartz inequality $|\int AB|^2 \le \int |A|^2 \int |B|^2$, together with the formula (19) for $\rho_t^{(N,2)}$, one can majorise the RHS of this last equation by $\left(\Phi_t^{(N)}(\xi)\,\Phi_t^{(N)}(-\xi)\right)^{1/2}$, where

$$\Phi_t^{(N)}(\xi) = \int dx_1 dx_2 \rho_t^{(N,2)}\, \hbar_N^{-1} |V(x_1-x_2 + \hbar_N\xi) - V(x_1-x_2)|$$

Since $\hbar_N \to 0$ as $N \to \infty$ and $V(x) \sim |x|^{-1}$, it follows from (20) that $\Phi_t^{(N)}(\xi)$, and thus the RHS of (24), is uniformly bounded over the compacts in (ξ,η,t)-space. Hence, by (23) and (24), the same is true for $\partial\mu_t^{(N,1)}/\partial t$ and $\partial\mu_t^{(N,1)}/\partial\xi$. Similarly one may prove the same for $\partial\mu_t^{(N,n)}/\partial t$ and $\partial\mu_t^{(N,n)}/\partial\xi_j$.

<u>Convergence of CF's</u>. In view of this boundedness property, it follows from Ascoli's theorem that, for each n, $\mu_t^{(N,n)}$ converges pointwise to a limit, $\mu_t^{(n)}$, as $N \to \infty$ over some sequence of integers. Hence, by the diagonalisation process, one can pass to a subsequence that works for <u>all</u> n, i.e.

$$\mu_t^{(N,n)} \to \mu_t^{(n)} \text{ for all } n, \text{ as } N \to \infty \tag{25}$$

over some sequence of integers.

<u>The Classical Limit</u>. Since $\hbar_N \to 0$ as $N \to \infty$, it now follows from a standard argument(4), involving Bochner's theorem, that the limit functions $\mu_t^{(N)}$ are classical CF's, i.e. Fourier transforms of probability measures $m_t^{(n)}$:-

$$\mu_t^{(n)}(\xi_1,\ldots,\xi_n;\eta_1,\ldots,\eta_n) = \int \exp i\left(\sum_1^n \xi_j\cdot v_j + \eta_j\cdot x_j\right) dm_t^{(n)}(x_1,\ldots,x_n;v_1,\ldots,v_n) \tag{26}$$

In fact, $m_t^{(n)}$ is the restriction to $(\Delta_1 X\, R^3)^n$ of a probability measure m_t over $(\Delta_1 X\, R^3)^{\mathbb{N}}$. Denoting the two-particle position probability density induced by m_t (or $m_t^{(2)}$) by $\rho_t^{(2)}(x_1,x_2)$, we see from (25) and (26) that the regularity condition (20) implies that

$$\rho_t^{(N,2)}(x_1,x_2) < K_T \text{ for } 0 \le t \le T \tag{27}$$

We also note that, by (25) and (26), the condition (18), representing the sharp definition of the macroscopic observables at t = 0, reduces to the following form in the limit N → ∞.

$$dm_o^{(n)}(x_1,\ldots,x_n;\ v_1,\ldots,v_n) = \prod_1^n dm_o^{(1)}(x_j,v_j) \tag{28}$$

Further, since, by (16) and (231), $\partial\mu_o^{(N,1)}/\partial\xi = 0$ and hence that $dm_o^{(1)}(x,v)$ is of the form $d\sigma(x)\delta(v)dv$, for some probability σ. Therefore, as (27) implies that the spatial density is bounded,

$$dm_o^{(1)}(x,v) = \rho_o(x)\delta(v)\,dxdv \tag{29}$$

with ρ_o bounded. We shall further assume that this density is strictly positive.

4. The Vlasov Dynamics

To obtain the dynamics of m_t subject to the initial conditions (28) and (29), we start by multiplying equation (24) by an arbitrary function $\hat{\phi}(\xi,\eta)$ that is continuously differentiable w.r.t. ξ and has compact support. Applying $\Sigma'_\eta \int d\xi$ to the resultant formula and using equations (25) and (26), together with the bounds given by (20) and (27), to secure the passage to its limiting form as N → ∞, we find that

$$\frac{d}{dt}\int\phi(x,v)dm_t^{(1)}(x,v) = \int v.\frac{\partial\phi(x,v)}{\partial x}\,dm_t^{(1)}(x,v) - \int\nabla V(x-x').\frac{\partial\phi(x,v)}{\partial v}dm_t^{(2)}(x,x';v,v'),$$

ϕ being the Fourier transform of $\hat{\phi}$. This equation may be extended by continuity to C'_o-class functions $\phi(x,v)$, i.e. continuous ones with compact support w.r.t. v. Similarly, one may derive the following set of equations of motion, termed the Vlasov hierarchy, for $\{m_t^{(n)}\}$

$$\frac{d}{dt}\int\phi^{(n)}dm_t^{(n)} = \sum_{j=1}^{n}\left(\int v_j\cdot\frac{\partial\phi^{(n)}}{\partial x_j} - \int\nabla V(x_j-x_{n+1})\cdot\frac{\partial\phi^{(n)}}{\partial v_j}\,dm_t^{(n+1)}\right) \tag{30}$$

for C'_o-class functions $\phi^{(n)}$ on $(\Delta_1 \times R^3)^n$. This hierarchy has a solution that perpetuates the initial factorisation property (28), namely

$$dm_t^{(n)}(x_1,\ldots,x_n;\ v_1,\ldots,v_n) = \prod_1^n dm_t^{(1)}(x_j,v_j), \tag{31}$$

where $m_t^{(1)}$ satisfies the <u>Vlasov equation</u>

$$\frac{d}{dt}\int \phi(x,v)\ dm_t^{(1)}(x,v) = \int \left[v\cdot\frac{\partial}{\partial x} + E_t(x)\cdot\frac{\partial}{\partial v}\right]\phi(x,v)dm_t^{(1)}(x,v) \tag{32}$$

$$\text{with}\quad E_t(x) = -\int \nabla V(x-x')dm_t^{(1)}(x',v'). \tag{33}$$

We note, however, that although the <u>existence</u> of a solution of the Vlasov equation has been proved(6) for the case where V is the Coulomb potential, there is no corresponding uniqueness theorem. On the other hand, uniqueness has been proved, both for the Vlasov hierarchy(4,7) and for the Vlasov equation(8), for the case where V is suitably regular, e.g. C^2-class. Thus, in such cases, the solution of the hierarchy (30) is given by the factorisation property (31) and the Vlasov equation: hence the macroscopic determinacy condition (31) is satisfied at all times and

the field E_t and its spatial derivatives are bounded. We now introduce physical assumptions to the effect that, in the states under consideration of the Jellium model, the Coulomb singularity does not affect either the preservation of the determinacy condition (31) or the boundedness of the spatial derivatives of E_t, on the grounds that the repulsive character of the inter-electronic forces keeps the particles apart and so tames the singularity. Thus, we assume that

(A.2) <u>$dm_t^{(2)} = dm_t^{(1)}\ dm_t^{(1)}$ at all times, and</u>

<u>(A.3) the spatial derivatives of $E_t(x)$ are uniformly bounded over finite time intervals.</u>

Under assumption (A.2), it follows from (31) that $m_t^{(1)}$ satisfies the Vlasov equation (32). When considered as a Liouville equation for a particle in an external field, E_t, the condition (A.3) ensures that its solution is unique(8) and may be expressed in terms of that of the auxilliary equations

$$\frac{dx_t}{dt} = v_t \ ; \ \frac{dv_t}{dt} = E_t(x_t), \text{ with } x_o = x, \ v_o = v, \tag{34}$$

as follows. Denoting the solution of (34) by

$$x_t = \bar{X}_t(x,v) \ , \ v_t = \bar{V}_t \ (x,v) \ , \tag{35}$$

the solution of (32) is given by

$$\int \phi(x,v) \ dm_t^{(1)}(x,v) = \int \phi\Big(\bar{X}_t(x,v), \ \bar{V}_t(x,v)\Big) \ dm_o^{(1)}(x,v).$$

Hence, by (29),

$$\int \phi(x,v) \ dm_t^{(1)}(x,v) = \int \phi(X_t(x), \ V_t(x)) \ \rho_o(x) \ dx \tag{36}$$

$$\text{where } X_t(x) = \bar{X}_t(x,o) \text{ and } V_t(x) = \bar{V}_t(x,o) \tag{37}$$

By (A.3), (34), (35) and (37), $X_t(x)$ and $V_t(x)$ are C'-class w.r.t. x and t. Further, as we shall prove in the Appendix, X_t is an invertible function of x.

5. Hydrodynamical equations

We define

$$\rho_t(x) = \rho_o\left[X_t^{-1}(x)\right]/J_t\left[X_t^{-1}(x)\right] \text{ and } u_t(x) = V_t\left[X_t^{-1}(x)\right] \tag{38}$$

where $J_t(x)$ is the Jacobean for the transformation $x \to X_t(x)$. Thus the solution (36) for $m_t^{(1)}$ may be expressed in the form

$$dm_t(x,v) = \rho_t(x) \ \delta(v-u_t(x))dxdv \tag{39}$$

which signifies that ρ_t is the local density and u_t the drift velocity. By (38), these functions, like $X_t(x)$ and $V_t(x)$, are continuously differentiable w.r.t. x and t

To prove that u_t, ρ_t and E_t satisfy (1), we start by integrating the first equation of (38) over an arbitrary regular region K. Thus, denoting $X_t(K)$ by K_t,

$$\int_{K_t} \rho_t(x)dx = \int_K \rho_o(x)dx \ .$$

On differentiating this equation w.r.t. t and noting that, by (37) and (38), $\partial X_t(x)/\partial t = u_t(X_t(x))$, we see that

$$\int_{K_t} \frac{\partial\rho_t}{\partial t} + \int_{A_t} \rho_t u_t \cdot dS = 0,$$

A_t being the surface of K_t. Hence $\int_{K_t} ({}^{\partial\rho_t}/\partial t + \mathrm{div}(\rho_t u_t))dx = 0$,

which is equivalent to the first equation of (1), since K is arbitrarily chosen and $K_t = X_t(K)$, with X_t invertible

Next, we note that he second equation of (38) implies that

$$u_t(X_t(x)) = V_t(x) \; .$$

On differentiating this w.r.t. t and using (34), (35) and (37), we see that

$$\frac{\partial u_t}{\partial t} + (u_t \cdot \nabla)u_t = E_t \text{ at the point } X_t(x).$$

As X_t is invertible, this is equivalent to the second equation of (1).

To complete our derivation of (1), we simply note that the third equation there follows directly from (14), (33) and (39).

Appendix: Invertibility of X_t

By (34), (35) and (37), the tensor $\nabla X_t(x)$ satisfies the equation

$$\frac{\partial^2}{\partial t^2} \nabla X_t(x) = (\nabla E_t)(X_t(x)) \cdot \nabla X_t(x)$$

with $\nabla X_t(x) = I$ and $\frac{\partial}{\partial t} \nabla X_t(x) = 0$ at $t = 0$.

Hence, by (A.3), $\nabla X_t(x)$ and its time derivatives are uniformly bounded over finite time intervals. The Jacobean $J_t(x)$ for the transformation $x \to X_t(x)$, which takes the value unity at $t = 0$, therefore remains positive over at least a finite interval. It follows from (36) that, in that period, the one-particle probability density is $\rho_t(x)$, as defined by (38). Hence, $J_t(x)$ cannot fall below the value of $(\min \rho_o/\max \rho_t)$, a strictly positive quantity, by (27) and our assumption (following (29)) of the strict positivity of ρ_o. From this, it follows by continuity that $J_t(x)$ can never fall to zero and hence that x_t must be invertible.

REFERENCES

1. E.H. Lieb and W. Thirring : Phys.Rev.Lett. 35, 687 (1975)
2. E.B. Davies : J.Stat.Phys. 18, 161 (1978)
3. G.L. Sewell : Phys.Lett. 97A, 35 (1983)
4. H. Narnhofer and G.L. Sewell : Commun.Math.Phys. 79, 9 (1981)
5. D. Bohm and D. Pines : Phys.Rev. 92, 609 (1953)
6. R. Illner and H. Neunzert : Math.Meth. in the App.Sci. 1, 530 (1979)
7. H. Spohn : Math.Meth. in the App.Sci. 3, 445 (1981)
8. H. Neunzert : Fluid Dynamics Trans. 9, 229 (1978)

Positive and Conditionally Positive Linear Functionals on Coalgebras

Michael Schürmann

Abstract. For a linear functional φ on a complex coalgebra V the convolution exponential $\exp_{\star} \varphi$ can be defined. Let $C \subset V$ be a set. Under which assumptions on C and φ is $\exp_{\star}(t\varphi)(v)$ positive for all $v \in C$ and $t \geq 0$? We state a theorem for finite-dimensional coalgebras and apply it to two special cases. First we treat the case when V is the bialgebra of a semigroup G, and C is the convex cone in V of all positive functions on G. Then we formulate a theorem on sesquilinear forms on arbitrary complex coalgebras. In both cases $\exp_{\star}(t\varphi)$ is positive for all $t \geq 0$ if and only if φ is conditionally positive and hermitian. The theorem on sesquilinear forms on coalgebras is applied to linear functionals on certain graded bialgebras with an involution which we call skew graded $\star$-bialgebras. The theory developed in this paper covers several known theorems and has applications to quantum probability.

1. Introduction

Our results are based on two known theorems. The first one is a theorem on semigroups of matrices leaving invariant a convex cone [19]. The second one is a well-known theorem from the theory of coalgebras, the so-called Fundamental Theorem on Coalgebras [22]. Combining these two theorems we are able to develop a theory which covers

(a) a theorem on linear functionals on the coefficient algebra of a compact group ([14] Theorem 1.5.13)

(b) the formula for the generator of a finite-dimensional completely positive quantum dynamical semigroup by Lindblad [16] and Gorini, Kossakowski, Sudarshan [13]

(c) a theorem on sesquilinear forms on anticocommutative coalgebras by von Waldenfels [24] which includes a theorem on positive and negative definite kernels [3, 18, 20] and a theorem for free algebras [7]. The latter is related to a result of Mathon and Streater [17] who proved that an even representation of the CAR is infinitely divisible if and only if it is quasi-free.

Furthermore, we get a characterization of the generator of a positive convolution semigroup describing a quantum analogue of a stochastic process with independent and stationary increments [25].

A coalgebra over a field k is a vector space V together with a k-linear map $\Delta: V \to V \otimes V$ which is called a comultiplication [4]. A linear functional $\delta: V \to k$ is called a counit of (V, Δ) if $(\delta \circ \mathrm{Id}_V) \circ \Delta = (\mathrm{Id}_V \circ \delta) \circ \Delta = \mathrm{Id}_V$. The coalgebra (V, Δ) is said to be coassociative if $(\Delta \circ \mathrm{Id}_V) \circ \Delta = (\mathrm{Id}_V \circ \Delta) \circ \Delta$. Let (V, Δ) be a coalgebra. Suppose that a graduation $V = \bigoplus_{n=0}^{\infty} V_n$ of the vector space V is given. For a homogeneous element v set $\deg(v) = n$ if $v \in V_n \setminus \{0\}$. We regard $V \otimes V$ as a graded vector space with the usual tensor product graduation. If Δ is homogeneous of degree 0, the coalgebra (V, Δ) is called a graded coalgebra [4]. If δ is a counit for the graded coalgebra (V, Δ), it automatically vanishes on V_n for $n \geq 1$ (see [4] § 11.3). Of course every coalgebra (V, Δ) is a graded coalgebra with respect to the trivial graduation $V = \bigoplus_{n=0}^{\infty} V_n$ where $V_0 = V$ and $V_n = \{0\}$ for $n \geq 1$. Let (V, Δ, δ) be a graded coassociative coalgebra and let $(V, \mu, \mathbb{1})$ be a graded associative algebra with unit $\mathbb{1}$ where $\mu: V \otimes V \to V$ denotes the algebra multiplication map. Define the skew multiplication on $V \otimes V$ by

$$(v \otimes w) \cdot (v' \otimes w') = (-1)^{\deg(w) \deg(v')} (v \cdot v') \otimes (w \cdot w')$$

for v' and w homogeneous. Let Δ and δ be algebra homomorphisms (where the multiplication on $V \otimes V$ is the skew multiplication). Then $(V, \Delta, \delta, \mu, \mathbb{1})$ is called a skew graded bialgebra [4]. If the graduation of V is trivial, $(V, \Delta, \delta, \mu, \mathbb{1})$ simply is called a bialgebra. Let $(V, \Delta, \delta, \mu, \mathbb{1})$ be a skew graded bialgebra and suppose that $(V, \mu, \mathbb{1})$ also is a $*$-algebra such that $*$ is homogeneous of degree 0. It is easy to check that on $V \otimes V$ an involution is defined by setting

$$(v \otimes w)^* = (-1)^{\deg(v) \deg(w)} v^* \otimes w^*$$

for v and w homogeneous. If

$$\Delta(v^*) = (\Delta v)^* \quad \text{and} \quad \delta(v^*) = \overline{\delta(v)}$$

hold for all $v \in V$, we call $(V, \Delta, \delta, \mu, \mathbb{1}, *)$ a <u>skew graded $*$-bialgebra</u> (see also [24]). Again if the graduation is trivial, $(V, \Delta, \delta, \mu, \mathbb{1}, *)$ simply is called a $*$-bialgebra.

In the sequel all coalgebras are assumed to be coassociative coalgebras over $\mathbb{C}$ unless other assumptions are stated explicitly.

The dual algebra of a coalgebra (V,Δ) is defined as the algebraic dual space V^* of V with multiplication $*$ given by

$$\varphi * \psi = (\varphi \otimes \psi) \circ \Delta .$$

The linear functional $\varphi * \psi$ is called the convolution product of the linear functionals φ and ψ. The algebra V^* is associative and possesses the unit δ if δ is a counit of (V,Δ). There are associative algebras which do not appear as the dual of a coalgebra. However, there is a one-to-one duality correspondence between coalgebras (with a counit) and complex associative algebras (with a unit) if everything is restricted to finite dimensions [1].

The Fundamental Theorem on Coalgebras ([22] Theorem 2.2.1.) states that, given finitely many elements of an arbitrary coalgebra with a counit, the subcoalgebra generated by these elements is finite-dimensional. As a consequence of this theorem, for a coalgebra (V,Δ,δ) the series $\sum_{n=0}^{\infty} \frac{1}{n!}\varphi^{*n}(v)$ converges for all $v \in V$ and all $\varphi \in V^*$: The linear operator $T_\varphi = (\mathrm{Id}_V \otimes \varphi) \circ \Delta$ on V leaves invariant all subcoalgebras of V, so $\exp T_\varphi$ is well-defined on the (finite-dimensional) subcoalgebra generated by a single element $v \in V$. As $\varphi * \psi = \delta \circ T_\varphi \circ T_\psi$ for $\varphi, \psi \in V^*$ we get:

$$\sum_{n=0}^{\infty} \frac{1}{n!} \varphi^{*n}(v) = \delta \circ (\exp T_\varphi)(v) \tag{1}$$

We denote the linear functional on V given by (1) by $\exp_* \varphi$.

Let (V,W) be a dual pair of (real or complex) vector spaces. Let C be a subset of V. Then the subset $\{w \in W \mid \langle v|w\rangle \geq 0 \ \forall v \in C\}$ of W is denoted by $C^{+,W}$. If $W = V^*$ we write C^+ instead of C^{+,V^*}. The elements of C^+ are said to be positive on C. For a (real or complex) algebra A with unit $\mathbb{1}$ and $C \subset A^*$ we call $a \in A$ conditionally positive on C if $\varphi(a) \geq 0$ for all $\varphi \in C$ with $\varphi(\mathbb{1}) = 0$. For instance, a linear functional φ on a coalgebra (V,Δ,δ) is conditionally positive on $C \subset V \subset V^{**}$ if $\varphi(v) \geq 0$ for all $v \in C$ with $\delta(v) = 0$. A subset C of a complex vector space V is called a cone if $v \in C$ implies $\lambda v \in C$ for all $\lambda \geq 0$.

The plan of this paper is the following. Section 2 serves as a preparation for Sections 3 and 4. In Section 3 we start with a semigroup G. One can define a $*$-bialgebra $(\mathcal{R}(G),\Delta,\delta)$ where $\mathcal{R}(G)$ is a certain

$*$-subalgebra of the $*$-algebra of all complex-valued functions on G [1]. The $*$-bialgebra $\mathcal{R}(G)$ is called the $*$-bialgebra of G. We consider the convex cone C in $\mathcal{R}(G)$ of all positive functions and prove that $\exp_*(t\varphi)$ is positive on C for all $t \geq 0$, if and only if φ is conditionally positive on C and hermitian. In Section 4 we define the convolution $K \underset{g}{*} L$ of two sesquilinear forms K and L on a graded coalgebra. A sesquilinear form K on a complex vector space V is called positive if $K(v|v) \geq 0$ for all $v \in V$. We prove that for an even sesquilinear form on a graded coalgebra (V, Δ, δ) the convolution exponential $\exp_{\underset{g}{*}}(tK)$ is positive if and only if $K(v|v) \geq 0$ for all $v \in V$ with $\delta(v) = 0$. As a corollary of our theorem on sesquilinear forms we get a theorem for skew graded $*$-bialgebras which states that for a linear functional φ on a skew graded $*$-bialgebra $\exp_*(t\varphi)$ is positive for all $t \geq 0$, if and only if φ is hermitian and $\varphi(v \cdot v^*) \geq 0$ holds for all v with $\delta(v) = 0$. Section 5 is devoted to applications.

2. Finite-dimensional Coalgebras

Let C be a closed convex cone in $\mathbb{R}^N$. Schneider and Vidyasagar [19] proved that for an $N \times N$-matrix a the semigroup $\exp(ta)$ leaves C invariant for all $t \geq 0$, if and only if $\langle ax|y\rangle \geq 0$ for all $(x,y) \in C \times C^+$ with $\langle x|y\rangle = 0$. This theorem establishes a correspondence between positive and "cross-positive" matrices (in the terminology of [19]). We may call this kind of correspondence "Schoenberg correspondence", because an important special case goes back to Schoenberg [20]: the connection between positive and negative definite kernels (see application 5.5. of this paper). Evans and Hanche-Olsen [11] proved a generalization of the result of Schneider and Vidyasagar for bounded linear operators on a Banach space. See also [2] and [6] for a general treatment of positive semigroups on Banach spaces.

We want to prove the following version of Schoenberg correspondence for elements of a Banach algebra.

Lemma 2.1. Let A be a real Banach algebra with unit $\mathbb{1}$. Denote by A' the topological dual space of A. Let $C \subset A$ be a closed convex cone with non-empty interior such that $c \in C$ implies $c^n \in C$ for all $n \geq 0$ (where of course $c^0 = \mathbb{1}$). Then for an element a of A the following two statements are equivalent:

(i) a is conditionally positive on $C^{+,A'}$

(ii) $\exp(ta)$ is positive on $C^{+,A'}$ for all $t \geqq 0$, i.e. $\exp(ta) \in C$ for all $t \geqq 0$.

Proof: (ii)$\Rightarrow$(i) follows by differentiating $\exp(ta)$ at $t = 0$.

(i)$\Rightarrow$(ii): For obvious reasons it suffices to prove that $\exp a \in C$. Let b be an interior point of C. As a is conditionally positive, $a_\varepsilon = a + \varepsilon b$ is conditionally positive on $C^{+,A'}$ for all $\varepsilon \geqq 0$. We prove that $\exp a_\varepsilon \in C$ for all $\varepsilon > 0$. Then the lemma is established, for $\lim\limits_{\varepsilon \downarrow 0} \| a_\varepsilon - a \| = \lim\limits_{\varepsilon \downarrow 0} \varepsilon \| b \| = 0$, which implies $\lim \| \exp a_\varepsilon - \exp a \| = 0$.

Let us denote by $\mathcal{B} = \{ \varphi \in A' \mid \| \varphi \| \leqq 1 \}$ the unit ball and by $\mathcal{S} = \{ \varphi \in A' \mid \| \varphi \| = 1 \}$ the unit sphere in A'. We prove that a fulfills the following condition:

For every real number $\rho > 0$ there is a real number $\eta > 0$, such that $\varphi(a) \geqq -\rho$ for all $\varphi \in C^{+,A'} \cap \mathcal{B}$ with $\varphi(\mathbb{1}) < \eta$. (2)

Suppose this were wrong. Then there is a real number $\rho_0 > 0$ and a sequence (φ_n) in $C^{+,A'} \cap \mathcal{B}$ with the properties: $\lim\limits_{n\to\infty} \varphi_n(\mathbb{1}) = 0$ and $\varphi_n(a) < -\rho_0$ for all $n \in \mathbb{N}$. Since $C^{+,A'}$ is weak-$*$ closed the theorem of Banach-Alaoglu tells us that $C^{+,A'} \cap \mathcal{B}$ is weak-$*$ compact, so there is a cluster point $\varphi \in C^{+,A'}$ of (φ_n). Clearly φ satisfies $\varphi(\mathbb{1}) = 0$ and $\varphi(a) \leqq -\rho_0$ which contradicts the assumption that a is conditionally positive on $C^{+,A'}$.

We have: $\varphi(\exp a_\varepsilon) = \lim\limits_{n\to\infty} \varphi((\mathbb{1} + \frac{a_\varepsilon}{n})^n)$. We only must prove the existence of a number $n_0 \in \mathbb{N}$, such that

$$\varphi\left(\mathbb{1} + \frac{a_\varepsilon}{n}\right) \geqq 0 \tag{3}$$

for all $\varphi \in C^{+,A'}$ and all $n \geqq n_0$. As $C^{+,A'}$ is a cone, it suffices to prove the existence of a number $n_0 \in \mathbb{N}$ such that (3) holds for all $\varphi \in C^{+,A'} \cap \mathcal{S}$ and all $n \geqq n_0$.

Now b is an interior point of C, so there exists a real number $\delta > 0$ such that $b + \delta \frac{a}{\|a\|} \in C$ for all $a \in A \setminus \{0\}$. For $\varphi \in C^{+,A'}$ and $a \in A \setminus \{0\}$ we get $|\varphi(a)| = \frac{\|a\|}{\delta} |\varphi(\delta \frac{a}{\|a\|})|$ $= \frac{\|a\|}{\delta} |\varphi(\delta \frac{a}{\|a\|} + b) + \varphi(-b)| \leqq \frac{\|a\|}{\delta}(\varphi(\delta \frac{a}{\|a\|} + b) + \varphi(b))$. The same argument for $-a$ yields $|\varphi(a)| \leqq \frac{\|a\|}{\delta}(\varphi(-\delta \frac{a}{\|a\|} + b) + \varphi(b))$. Adding both inequalities we get $2|\varphi(a)| \leqq 4 \frac{\|a\|}{\delta} \varphi(b)$ or $\varphi(b) \geqq \frac{\delta}{2} \varphi(\frac{a}{\|a\|})$ which means $\varphi(b) \geqq \frac{\delta}{2} \|\varphi\|$. Since a fulfills the condition (2), there is a real number $\varrho > 0$ such that $\varphi(a) \geqq -\frac{\varepsilon\delta}{4}$ for all $\varphi \in C^{+,A'} \cap \mathcal{S}$ with $\varphi(\mathbb{1}) < \varrho$. Now we get $\varphi(\mathbb{1} + \frac{a_\varepsilon}{n})$ $= \varphi(\mathbb{1}) + \frac{1}{n}(\varphi(a) + \varepsilon\varphi(b)) \geqq \frac{\varepsilon\delta}{4} \geqq 0$ for all $\varphi \in C^{+,A'} \cap \mathcal{S}$ with $\varphi(\mathbb{1}) < \varrho$. Let $\varphi \in C^{+,A'} \cap \mathcal{S}$ with $\varphi(\mathbb{1}) \geqq \varrho$. Then $\varphi(\mathbb{1} + \frac{a_\varepsilon}{n})$ $\geqq \varrho + \frac{1}{n}\varphi(a_\varepsilon) \geqq \varrho - \frac{\|a_\varepsilon\|}{n} \geqq 0$ for $n \geqq \frac{\|a_\varepsilon\|}{\varrho}$ and we have proved that (3) holds for all $n \geqq \frac{\|a_\varepsilon\|}{\varrho}$ and all $\varphi \in C^{+,A'} \cap \mathcal{S}$. □

Now we want to apply the preceeding lemma to the dual algebra of a finite-dimensional coalgebra. First let V be a real finite-dimensional vector space. Let $C \subset V$ be a closed convex cone. C contains an interior point if and only if $C - C$ equals the whole of V. The orthogonal complement of $C - C$ in V^* is $C^+ \cap (-C^+)$, so $C - C = V$ if and only if $C^+ \cap (-C^+) = \{0\}$. A cone C in a (real or complex) vector space is said to be proper if $C \cap (-C) = \{0\}$. Using Lemma 2.1. we get:

<u>Lemma 2.2.</u> Let (V, Δ, δ) be a finite-dimensional real coalgebra and let $C \subset V$ be a closed, proper convex cone such that $\psi \in C^+$ implies $\psi^{*n} \in C^+$ for all $n \geqq 0$. Let φ be a linear functional on V. Then the following two statements are equivalent:

(i) φ is conditionally positive on C

(ii) $\exp_*(t\varphi)$ is positive on C for all $t \geqq 0$.

If we want to make use of Lemma 2.2. in the complex case, we have to assume the coalgebra (V, Δ, δ) to carry an additional structure which leads to the notion of hermitian elements and functionals. Let V be a complex vector space. Let $v \mapsto v^*$ be a selfinverse antilinear map on V. An element $v \in V$ is called hermitian if $v^* = v$. We denote the real vector space of all hermitian elements by V_h. A set $C \subset V$ is called hermitian if $C \subset V_h$. For $\varphi \in V^*$ let $\varphi^* \in V^*$ be given by $\varphi^*(v) = \overline{\varphi(v^*)}$. The map $\varphi \mapsto \varphi^*$ also is selfinverse and antilinear.

Let (V, Δ) be a (not necessarily coassociative) coalgebra. Let $*$ be a selfinverse antilinear map on V that fulfills $\Delta \circ * = (* \otimes *) \circ \Delta$ and $\delta^* = \delta$ (if there is a counit δ on V). We call $*$ a selfinverse antilinear automorphism on V. We have:

Theorem 2.3. Let (V, Δ, δ) be a finite-dimensional coalgebra with a selfinverse antilinear automorphism $*$. Let $C \subset V$ be a hermitian, closed, proper convex cone such that $\psi \in C^+ \cap (V^*)_h$ implies $\psi^{*n} \in C^+ \cap (V^*)_h$ for all $n \geq 0$. Then for a linear functional φ on V the following two statements are equivalent:

(i) φ is conditionally positive on C and hermitian

(ii) $\exp_*(t\varphi)$ is positive on C and hermitian for all $t \geq 0$.

Proof: V_h is a real subcoalgebra of V and $(V^*)_h$ is a real subalgebra of V^*. The map $\iota: (V^*)_h \rightarrow (V_h)^*$ with $(\iota\varphi)(v) = \varphi(v)$ for $v \in V_h$ is an algebra isomorphism. Moreover we get that $\exp_*(t\varphi) \in (V^*)_h$ implies $\varphi \in (V^*)_h$ by differentiating. Now Lemma 2.2. yields the theorem. □

3. The $*$-Bialgebra of a Semigroup

Let G be a semigroup with neutral element e. We consider the complex vector space $\mathcal{F}_{\mathbb{C}}(G)$ of all complex-valued functions on G and the linear map $j: \mathcal{F}_{\mathbb{C}}(G) \otimes \mathcal{F}_{\mathbb{C}}(G) \rightarrow \mathcal{F}_{\mathbb{C}}(G \times G)$ which is given by $j(f \otimes g)(x,y) = f(x)g(y)$. The map j is injective ([1] § 2.2). We define the linear map $\tilde{\Delta}: \mathcal{F}_{\mathbb{C}}(G) \rightarrow \mathcal{F}_{\mathbb{C}}(G \times G)$ by $\tilde{\Delta} f(x,y) = f(yx)$. Let $\mathcal{R}_{\mathbb{C}}(G) = \{f \in \mathcal{F}_{\mathbb{C}}(G) \mid \tilde{\Delta} f \in j(\mathcal{F}_{\mathbb{C}}(G) \otimes \mathcal{F}_{\mathbb{C}}(G))\}$. Define Δ: $\mathcal{R}_{\mathbb{C}}(G) \rightarrow \mathcal{F}_{\mathbb{C}}(G) \otimes \mathcal{F}_{\mathbb{C}}(G)$ by $\Delta = j^{-1} \circ \tilde{\Delta}$. We have $\Delta \mathcal{R}_{\mathbb{C}}(G) \subset \mathcal{R}_{\mathbb{C}}(G) \otimes \mathcal{R}_{\mathbb{C}}(G)$ ([1] § 2.2). Defining $\delta: \mathcal{R}_{\mathbb{C}}(G) \rightarrow \mathbb{C}$ by $\delta(f) = f(e)$, we get a coalgebra $(\mathcal{R}_{\mathbb{C}}(G), \Delta, \delta)$. In fact, every coalgebra can be considered as a subcoalgebra of a coalgebra of the type $(\mathcal{R}_{\mathbb{C}}(G), \Delta, \delta)$ ([1] § 2.3). By defining $f^*(x) = \overline{f(x)}$, the vector space $\mathcal{R}_{\mathbb{C}}(G)$ becomes a coalgebra with selfinverse antilinear automorphism. $\mathcal{R}_{\mathbb{C}}(G)$ also is a $*$-subalgebra of the commutative $*$-algebra $\mathcal{F}_{\mathbb{C}}(G)$. The $*$-algebra and the coalgebra structure turn $\mathcal{R}_{\mathbb{C}}(G)$ into a $*$-bialgebra. A linear functional φ on a linear subspace V of $\mathcal{F}_{\mathbb{C}}(G)$ is called positive if $\varphi(f) \geq 0$ for all $f \in V$ with $f \geq 0$, it is said to be conditionally positive if $\varphi(f) \geq 0$ for all $f \in V$ with $f \geq 0$ and $f(e) = 0$.

Theorem 3.1. Let G be a semigroup and let V be a $*$-subbialgebra of $\mathcal{R}_{\mathbb{C}}(G)$. Let φ be a linear functional on V. Then the following two statements are equivalent:

(i) φ is conditionally positive and hermitian,

(ii) $\exp_*(t\varphi)$ is positive for all $t \geq 0$.

Proof: Every positive linear functional φ on V is hermitian, because $(f|g) = \varphi(fg^*)$ defines a positive sesquilinear form on V and every positive sesquilinear form is hermitian.

Define $C = \{f \in \mathcal{R}_{\mathbb{C}}(G) \mid f \geq 0\}$. The set C is a hermitian, proper convex cone. To prove the theorem, we apply Theorem 2.3. to the following situation: We have the subcoalgebra V_f generated by a given function $f \in C$, the hermitian, closed, proper convex cone $C \subset V_f$, and the linear functional $\varphi | V_f$ on V_f. We prove that $(C \cap V_f)^+$ is a subsemigroup of V_f with respect to convolution and that V_f is left invariant by $*$. Let W be an arbitrary subcoalgebra of $\mathcal{R}_{\mathbb{C}}(G)$. For $\psi \in (C \cap W)^+ \subset W^*$ we prove that T_ψ leaves $C \cap W$ invariant. For $g \in W$ with $g(yx) = \sum g_i(x)g_i'(y)$ (where $g_i, g_i' \in W$) we have: $T_\psi g(x) = (\mathrm{Id}_V \otimes \psi)(\sum g_i \otimes g_i')(x) = \sum g_i(x)\psi(g_i') = \psi(\sum g_i(x)g_i')$ $= (\psi \circ L(x))(g)$ where L(x) denotes the left translation $f(\cdot)$ $f(x\cdot)$. Obviously, L(x) leaves $C \cap W$ invariant. As $\psi * \eta = \psi \circ T_\eta$ for $\psi, \eta \in W^*$, it follows that $(C \cap W)^+$ is a subsemigroup of W^*. The subcoalgebra V_f is equal to $\mathrm{lin}\{R(x)L(y)f \mid x, y \in G\}$ where R(x) denotes right translation ([1] § 2.2). The function f is hermitian, so $g \in V_f$ implies $g^* \in V_f$. □

4. Sesquilinear Forms on Coalgebras

Let V be a complex vector space. We introduce the complex conjugate vector space $\overline{V}$ of V ([5] § 1.1). As a set $\overline{V}$ consists of all formal elements $\overline{v}$ with $v \in V$. The set $\overline{V}$ becomes a complex vector space if we define the addition of vectors by $\overline{v} + \overline{w} = \overline{v + w}$ and the scalar multiplication by $\lambda \bar{v} = \overline{\bar{\lambda} v}$. The map $v \mapsto \bar{v}$ is an antilinear bijection from V to $\bar{V}$. If K is a sesquilinear form on V we define the linear functional $\tilde{K}$ on $V \otimes \bar{V}$ by $\tilde{K}(v \otimes w) = K(v|w)$. We get an identification of sesquilinear forms on V and linear functionals on $V \otimes \bar{V}$. We again write K for the functional $\tilde{K}$. Let (V, Δ) be a graded (not necessarily coassociative) coalgebra. $\bar{V}$ can be considered as a graded vector space with the graduation $\bar{V} = \bigoplus_{n=0}^{\infty} \bar{V}_n$. We regard $V \otimes \bar{V}$ as a graded vector space with the usual tensor product graduation. We define the linear map ${}^g\bar{\Delta}: \bar{V} \to \bar{V} \otimes \bar{V}$ by

$$ {}^g\bar{\Delta}\, v = \sum_i (-1)^{\deg(v_i)\, \deg(v_i')}\, v_i \otimes v_i' $$

where $\Delta v = \sum_i v_i \otimes v_i'$ with homogeneous elements $v_i, v_i' \in V$. The graded coalgebra $(\bar{V}, {}^g\bar{\Delta})$ is coassociative if and only if (V, Δ) is coassociative. If δ is a counit of (V, Δ) then a counit $\bar{\delta}$ of $(\bar{V}, {}^g\bar{\Delta})$ is given by $\bar{\delta}(\bar{v}) = \overline{(\delta v)}$. The flip operator ${}^g\tau : V \otimes \bar{V} \to \bar{V} \otimes V$ is defined as the linear map given by

$$ {}^g\tau\, (v \otimes \bar{w}) = (-1)^{\deg(v)\, \deg(w)}\, \bar{w} \otimes v $$

(with v,w homogeneous). We set ${}^g\Lambda = Id_V \otimes {}^g\tau \otimes Id_{\bar{V}}$. The pair $(V \otimes \bar{V}, {}^g\Lambda)$ again is a graded coalgebra which is coassociative if (V, Δ) is, and which possesses the counit $\delta \otimes \bar{\delta}$ if δ is a counit of (V, Δ). The convolution product of two sesquilinear forms K and L on V with respect to the comultiplication ${}^g\Lambda$ is denoted by $K \underset{g}{*} L$. There is always the trivial graduation on V. We write $\bar{\Delta}$, τ and Λ instead of ${}^g\bar{\Delta}$, ${}^g\tau$ and ${}^g\Lambda$ if the trivial graduation on V is considered. In this case we also write $K * L$ instead of $K \underset{g}{*} L$. The pair $(V \otimes \bar{V}, \Lambda)$ is the usual coalgebra tensor product of (V, Δ) and $(\bar{V}, \bar{\Delta})$. The antilinear map $*$ given by $(v \otimes \bar{w})^* = w \otimes \bar{v}$ defines a selfinverse antilinear automorphism on $(V \otimes \bar{V}, \Lambda)$. Now we prove a simple lemma which is crucial.

Lemma 4.1. Let (V, Δ) be a (not necessarily coassociative) coalgebra and let K and L be positive sesquilinear forms on V. Then the sesquilinear form $K * L$ is also positive.

Proof: The lemma is a direct consequence of a lemma by Schur [21]. We have $(K * L)(v|v) = (K \otimes L) \circ (Id_V \otimes \tau \otimes Id_{\bar{V}}) \circ (\Delta \otimes \bar{\Delta})(v \otimes \bar{v})$

$= \sum_{i,j} K(v_i|v_j)L(v_i'|v_j') \geq 0$ if the matrices $(k_{ij}) = (K(v_i|v_j))$ and $(l_{ij}) = (L(v_i'|v_j'))$ are positive definite, because this implies that their Schur product $(k_{ij}l_{ij})$ is positive definite, too. □

A linear functional φ on a graded vector space $V = \bigoplus_{n=0}^{\infty} V_n$ is called <u>even</u>, if $\varphi(v) = 0$ for all $v \in V_n$ with n odd. A sesquilinear form K on a coalgebra (V, Δ, δ) is called <u>conditionally positive</u> if $K(v|v) \geq 0$ for all $v \in V$ with $\delta(v) = 0$. The following theorem is a generalization of Theorem 2 in [24] where (V, Δ, δ) is assumed to be anticocommutative.

<u>Theorem 4.2.</u> Let (V, Δ, δ) be a graded coalgebra. Let K be an even sesquilinear form on V. Then the following two statements are equivalent:

(i) K is conditionally positive and hermitian

(ii) $\exp_{\underset{g}{*}}(tK)$ is positive for all $t \geq 0$.

<u>Proof</u>: First we prove that the theorem holds if the graduation of V is trivial. We consider the subset C of $V \otimes \bar{V}$ of all elements of the form $\sum_{i,j=1}^{n} a_{ij}\, v_i \otimes \bar{v}_j$ where $n \in \mathbb{N}$ and (a_{ij}) is a positive definite $n \times n$-matrix. C is the convex hull of the cone $\{v \otimes \bar{v} \mid v \in V\}$. The convex cone C is hermitian and proper. A sesquilinear form is positive (conditionally positive) if and only if it is positive (conditionally positive) on C. For a given $\tilde{v} = \sum_i v_i \otimes \bar{w}_i \in V \otimes \bar{V}$ let $V_{\tilde{v}}$ denote the subcoalgebra of V generated by v_i, w_i. The vector space $V_{\tilde{v}} \otimes \overline{V_{\tilde{v}}}$ is a finite-dimensional subcoalgebra of $V \otimes \bar{V}$ which contains $\tilde{v}$ and which is left invariant by $*$. The convex cone $C \cap V_{\tilde{v}} \otimes \overline{V_{\tilde{v}}}$ is closed, because the convex cone of all positive definite $n \times n$-matrices is closed. The previous lemma yields the result that $(C \cap V_{\tilde{v}} \otimes V_{\tilde{v}})^+$ is a subsemigroup of $(V_{\tilde{v}} \otimes \bar{V}_{\tilde{v}})^*$. Now apply Theorem 2.3.

Let (V, Δ, δ) be an arbitrary graded coalgebra. Let K and L be two sesquilinear forms on V, and let L be even. For $v, w \in V$ with $\Delta v = \sum_i v_i \otimes v_i'$ and $\Delta w = \sum_j w_j \otimes w_j'$ (where v_i, v_i', w_j, w_j' homogeneous) we get

$$(K \underset{g}{*} L)(v|w) = (K \otimes L) \circ (\mathrm{Id}_V \otimes {}^g\tau \otimes \mathrm{Id}_{\bar{V}})$$
$$\Big(\sum_{i,j} (-1)^{\deg(w_j)\deg(w_j')}\, v_i \otimes v_i' \otimes w_j \otimes w_j'\Big)$$
$$= \sum_{i,j} (-1)^{\deg(w_j)(\deg(w_j') + \deg(v_i'))}\, K(v_i|w_j)L(v_i'|w_j');$$

cf. [24] Lemma 1. Suppose that for a term of this sum the number $\deg(w_j') + \deg(v_i')$ is odd. As L is even, $L(v_i'|w_j') = 0$. This implies that the term vanishes and, hence, that the sum is equal to $\sum_{i,j} K(v_i|w_j)L(v_i'|w_j') = (K * L)(v|w)$. Now the theorem follows from the first part of the proof. □

Remark: Of course all sesquilinear forms are even in the trivial graduation case.

If there is some additional structure on (V, Δ, δ) Theorem 4.2. can be used to formulate a result for linear functionals on V. Let (V, Δ, δ) again be a graded coalgebra. Furthermore, let us assume that a (coalgebra) homomorphism ρ from $(V \otimes \overline{V}, {}^s\Lambda, \delta \otimes \overline{\delta})$ to (V, Δ, δ) is given. (ρ also is assumed to be homogeneous of degree 0.) For $\varphi \in V^*$ the map $K_\varphi = \varphi \circ \rho$ is a sesquilinear form on V. A linear functional φ is said to be ρ-positive (conditionally ρ-positive, ρ-hermitian) if K_φ is positive (conditionally positive, hermitian). $\varphi \mapsto K_\varphi$ is an algebra homomorphism. If φ is even K_φ is even, too. Let φ be an even linear functional on V. Then by Theorem 4.2. we have that $\exp_*(t\varphi)$ is ρ-positive for all $t \geq 0$ if and only if φ is conditionally ρ-positive and ρ-hermitian. Especially let $(V, \Delta, \delta, \mu, \mathbb{1}, *)$ be a skew graded $*$-bialgebra. Then $\rho(v \otimes \overline{w}) = \mu(v \otimes w^*) = v \cdot w^*$ defines a homomorphism from $(V \otimes \overline{V}, {}^s\Lambda, \delta \otimes \overline{\delta})$ to (V, Δ, δ). A linear functional φ on V is ρ-positive (ρ-hermitian) if and only if φ is positive (hermitian) in the usual sense as a linear functional on the $*$-algebra $(V, \mu, \mathbb{1}, *)$. We get:

Corollary 4.3. Let $(V, \Delta, \delta, \mu, \mathbb{1}, *)$ be a skew graded $*$-bialgebra and let φ be an even linear functional on V. Then the following two statements are equivalent:

(i) φ is conditionally positive (i.e. $\varphi(v \cdot v^*) \geq 0$ for all $v \in V$ with $\delta(v) = 0$) and hermitian

(ii) $\exp_*(t\varphi)$ is positive for all $t \geq 0$.

5. Applications

5.1. Let $M_k(N)$ denote the algebra with unit $\mathbb{1}$ of all $N \times N$-matrices over the field k. Let a be a hermitian element of $M_{\mathbb{C}}(N)$ and denote by A the real subalgebra (with unit $\mathbb{1}$) of $M_{\mathbb{C}}(N)$ generated by a. Consider the closed convex cone C of all positive definite matrices in A. Since $\mathbb{1} \in \mathring{C}$ we can apply Lemma 2.1. and get that exp a is positive definite.

5.2. Let k be $\mathbb{R}$ or $\mathbb{C}$. Let x_{ij} with $i,j \in \{1,2,\dots,N\}$ denote the matrix units in $M_k(N)$. We define a (real or complex) coassociative coalgebra $(M_k(N), \Delta, \delta)$ by setting

$$\Delta x_{ij} = \sum_{k=1}^{N} x_{ik} \otimes x_{kj} \text{ and } \delta = \text{trace}.$$

Consider the closed, proper convex cone C in $M_{\mathbb{R}}(N)$ of all matrices with nonnegative entries. We identify $M_{\mathbb{R}}(N)^*$ with $M_{\mathbb{R}}(N)$ by $c(b) = \text{trace}(cb)$ for $b,c \in M_{\mathbb{R}}(N)$. We have $C^+ = C$, and Lemma 2.2. yields the fact, widely used in probability theory, that for $a \in M_{\mathbb{R}}(N)$ the matrices $\exp(ta)$ have nonnegative entries for all $t \geq 0$ if and only if a has nonnegative off-diagonal entries ([9] VI § 1 p. 241).

5.3. Let G be a compact group. For a (continuous, unitary) representation π of G with representation space $\mathcal{H}(\pi)$ and for $\xi, \zeta \in \mathcal{H}(\pi)$ the continuous complex-valued function $d^{(\pi)}_{\xi,\zeta}$ on G given by $d^{(\pi)}_{\xi,\zeta}(x) = \langle \pi(x)\xi \mid \zeta \rangle$ is called a coefficient function of π. The space $\mathcal{K}(G)$ of all coefficient functions of all finite-dimensional representations of G is a $*$-subalgebra of $\mathcal{F}_{\mathbb{C}}(G)$ and is called the coefficient algebra of G [14]. Let $(\pi, \mathcal{H}(\pi))$ be a finite-dimensional representation of G and let $\{\xi_1, \xi_2, \dots, \xi_N\}$ be an orthonormal basis of $\mathcal{H}(\pi)$. We have:

$$\begin{aligned}(\tilde{\Delta} d^{(\pi)}_{\xi_i,\xi_j})(x,y) &= \langle \pi(yx)\xi_i \mid \xi_j \rangle \\ &= \sum_k \langle \pi(x)\xi_i \mid \xi_k \rangle \langle \pi(y)\xi_k \mid \xi_j \rangle \\ &= \sum_k d^{(\pi)}_{\xi_i,\xi_k}(x) d^{(\pi)}_{\xi_k,\xi_j}(y).\end{aligned}$$

We get $\mathcal{K}(G) \subset \mathcal{R}_{\mathbb{C}}(G)$ and $\Delta\mathcal{K}(G) \subset \mathcal{K}(G) \otimes \mathcal{K}(G)$, so $\mathcal{K}(G)$ is a $*$-subbialgebra of $\mathcal{R}_{\mathbb{C}}(G)$. Theorem 3.1. yields a well-known theorem ([14] Theorem 1.5.13.). We also can apply Corollary 4.3. to the (commutative) $*$-bialgebra $\mathcal{K}(G)$: We get for a linear functional φ on $\mathcal{K}(G)$ that $\exp_*(t\varphi)(|f|^2) \geq 0$ for all f and all $t \geq 0$ if and only if $\varphi(|f|^2) \geq 0$ for all f with $f(e) = 0$. However, it is known that $\varphi(|f|^2) \geq 0$ for all f if and only if $\varphi(f) \geq 0$ for all $f \geq 0$ ([14] Theorem 1.4.8.).

5.4. Let $(M_{\mathbb{C}}(N), \Delta, \delta)$ be the coalgebra of 5.2. We associate a linear operator $\hat{K}$ on $M(N) = M_{\mathbb{C}}(N)$ with every sesquilinear form K on M(N) by defining

$$\hat{K}(x_{ij}) = \sum_{k,l=1}^{N} K(x_{ki} | x_{lj}) x_{kl}.$$

The linear map $\wedge$ is bijective, the inverse map will be denoted by $\vee$.
If K and L are two sesquilinear forms on M(N) we have:

$$\begin{aligned}
&(K * L)(x_{ij} | x_{kl}) \\
&= (K \otimes L) \circ (Id_{M(N)} \otimes \tau \otimes Id_{\overline{M(N)}}) \circ (\Delta \otimes \bar{\Delta})(x_{ij} \otimes \bar{x}_{kl}) \\
&= \sum_{m,n=1}^{N} K(x_{im} | x_{kn}) L(x_{mj} | x_{nl})
\end{aligned}$$

which implies $\widehat{K * L} = \hat{K} \circ \hat{L}$. So $\wedge$ is an isomorphism between the algebras $(M(N) \otimes \overline{M(N)})^*$ and $\mathcal{B}(M(N))$ (= algebra of linear operators on M(N)).
A sesquilinear form K is hermitian if and only if $K(a^*) = K(a)^*$ for all $a \in M(N)$. A sesquilinear form K is positive if and only if K is completely positive [8]. An operator $\kappa \in \mathcal{B}(M(N))$ is called completely positive if the linear operators $\kappa \otimes 1$ on $M(N) \otimes M(N')$ given by $(\kappa \otimes 1)(a \otimes b) = (\kappa a) \otimes b$ are positive for all $N' \in \mathbb{N}$ (i.e., if they map positive definite matrices on positive definite matrices). The completely positive operators play an important role in the theory of quantum dynamical semigroups. A formula for the generator of a finite-dimensional quantum dynamical semigroup is given in [13] and in [16]. (In [16] the more general case of a norm continuous quantum dynamical semigroup is treated.) We get this formula as a consequence of Theorem 4.2. by proving that for $\kappa \in \mathcal{B}(M(N))$ the operators $\exp(t\kappa)$ are completely positive for all $t \geq 0$ if and only if κ is of the form

$$\kappa(a) = \nu(a) + ka + ak^* \tag{4}$$

where $\nu \in \mathcal{B}(M(N))$ is completely positive and $k \in M(N)$: We know by Theorem 4.2. that $\exp(t\kappa)$ is completely positive for all $t \geq 0$ if and only if $\check{\kappa}$ is conditionally positive and hermitian. We only have to prove that κ has the form (4) if and only if $\check{\kappa}$ is conditionally positive and hermitian.

To see this take a fixed $e_0 \in M(N)$ with $\delta(e_0) = 1$. Let $\check{\kappa}$ be conditionally positive and hermitian. We divide $\check{\kappa}$ into four parts:
$= \check{\kappa}_1 + \check{\kappa}_2 + \check{\kappa}_3 + \check{\kappa}_4$ where

$$\check{\kappa}_1(a|b) = \check{\kappa}(a - (\delta a)e_0|b - (\delta b)e_0),$$
$$\check{\kappa}_2(a|b) = \overline{(\delta b)}\,\check{\kappa}(a|e_0),$$
$$\check{\kappa}_3(a|b) = (\delta a)\check{\kappa}(e_0|b),$$
$$\check{\kappa}_4(a|b) = -(\delta a)\overline{(\delta b)}\,\check{\kappa}(e_0|e_0).$$

$\check{\kappa}_1$ is a positive sesquilinear form, so $\kappa_1 = \hat{\check{\kappa}}_1$ is completely positive. We have $\kappa_2(a) = l a$ where $l = (l_{ij}) = (\check{\kappa}(x_{ij}|e_0))$, $\kappa_3(a) = a l^*$ and $\kappa_4(a) = \gamma a$ where $\gamma = -\check{\kappa}(e_0|e_0)$. We get the representation (4) of κ with $\nu = \kappa_1$ and $k = l + \frac{\gamma}{2}\mathbb{1}$. Conversely, it can be seen easily that κ is conditionally positive and hermitian if it has the form (4).

An operator $\nu \in \mathcal{B}(M(N))$ is completely positive if and only if ν has the form $\nu(a) = \sum_{j=1}^{d} v_j^* a v_j$ with $v_j \in M(N)$; cf. [15]. The semigroup $\{\exp(t\kappa) \mid t \geq 0\}$ preserves the identity if and only if $\kappa(\mathbb{1}) = 0$. Inserting $k = ih - \frac{1}{2}\nu(\mathbb{1})$ into (4), we find that the identity-preserving semigroup $\{\exp(t\kappa)\}$ is completely positive if and only if κ admits the representation (cf. [10, 16])

$$\kappa(a) = i[h, a] + \sum_{j=1}^{d} (v_j^* a v_j - \tfrac{1}{2}(v_j^* v_j a + a v_j^* v_j)).$$

<u>5.5.</u> The following example goes back to Schoenberg [20]. Let X be a non-empty set and let $K: X \times X \to \mathbb{C}$ be a map. K is called a positive definite kernel if

$$\sum_{i,j=1}^{n} \bar{z}_i z_j K(x_j, x_k) \geq 0 \tag{5}$$

for all finite sequences of complex numbers $z_1, z_2, \ldots, z_n$ and elements $x_1, x_2, \ldots, x_n$ of X. The map K is called a conditionally positive definite kernel if (5) holds under the condition $\sum_{i=1}^{n} z_i = 0$ and K is said to be hermitian if $K(x,y) = \overline{K(y,x)}$. (Some authors call -K negative definite if K is conditionally positive definite and hermitian; cf. e.g. [3].) The following theorem holds ([3] Theorem 2.2., [18] Lemma 1.7.): The kernel exp(tK) is positive definite for all $t \geq 0$ if and only if K is conditionally positive definite and hermitian. We demonstrate that this theorem follows directly from Theorem 4.2.

Let $\mathbb{C}X$ be the free complex vector space spanned by X. Define the linear map $\Delta: \mathbb{C}X \to \mathbb{C}X \otimes \mathbb{C}X$ by $\Delta x = x \otimes x$ for $x \in X$, and the linear functional $\delta: \mathbb{C}X \to \mathbb{C}$ by $\delta(x) = 1$. It is easy to check that $(\mathbb{C}X, \Delta, \delta)$

forms a coalgebra, the so-called group-like coalgebra [22]. For two sesquilinear forms K and L on $\mathbb{C}X$ we get $(K \star L)(x|y) = K(x|y)L(x|y)$ for $x,y \in X$, that is, on $X \subset \mathbb{C}X$ the convolution becomes the pointwise multiplication. It is clear that K is positive (conditionally positive, hermitian) if and only if the kernel $K|X \times X$ on X is positive definite (conditionally positive definite, hermitian).

5.6. Let X be a nonempty set and let $\mathbb{C}\langle X\rangle$ be the free associative complex algebra with unit $\mathbb{1}$ generated by X. We divide X into two disjoint sets X_1 and X_2. For $x \in X$ set $\deg(x) = 1$ if $x \in X_1$ and $\deg(x) = 2$ if $x \in X_2$. For a monomial $x_1x_2\ldots x_n$ with $x_i \in X$ define $\deg(x_1x_2\ldots x_n) = \sum_{j=1}^{n} \deg(x_j)$ and $\deg(\mathbb{1}) = 0$. The direct sum $\mathbb{C}\langle X\rangle = \bigoplus_{n=0}^{\infty} V_n$ is a graduation of $\mathbb{C}\langle X\rangle$ (where V_n denotes the linear span of all monomials M with $\deg(M) = n$). We regard $\mathbb{C}\langle X\rangle \otimes \mathbb{C}\langle X\rangle$ as an algebra with respect to skew multiplication. $\Delta: \mathbb{C}\langle X\rangle \to \mathbb{C}\langle X\rangle \otimes \mathbb{C}\langle X\rangle$ is defined as the algebra homomorphism that maps $x \in X$ on $x \otimes \mathbb{1} + \mathbb{1} \otimes x$. Let $\delta: \mathbb{C}\langle X\rangle \to \mathbb{C}$ be the homomorphism given by $\delta(x) = 0$ for $x \in X$. We introduce the involution $\ast$ on $\mathbb{C}\langle X\rangle$ by setting $x^\ast = x$ for $x \in X$. The triplet $(\mathbb{C}\langle X\rangle, \Delta, \delta)$ is a coalgebra which turns $\mathbb{C}\langle X\rangle$ into a skew graded $\ast$-bialgebra [24]. The coalgebra $\mathbb{C}\langle X\rangle$ is anticocommutative. In the special case $X_2 = X$ we have $\mathbb{C}\langle X\rangle = \bigoplus_{n=0}^{\infty} V_{2n}$. The skew multiplication on $\mathbb{C}\langle X\rangle \otimes \mathbb{C}\langle X\rangle$ becomes the ordinary tensor algebra multiplication and $\mathbb{C}\langle X\rangle$ is a cocommutative $\ast$-bialgebra. Applying Theorem 4.2., we get a result which is stated in [7].

We consider the following example of a hermitian conditionally positive linear functional on the skew graded $\ast$-bialgebra $\mathbb{C}\langle X\rangle$ [24]: The linear functional δ_Q on $\mathbb{C}\langle X\rangle$ is defined by setting $\delta_Q(M) = Q(x,y)$ if M is a monomial of the form $M = x \cdot y$ with $x,y \in X$, and $\delta_Q(M) = 0$ if M is a monomial not of this form. A Gaussian functional γ_Q on $\mathbb{C}\langle X\rangle$ is a functional of the form $\gamma_Q = \exp_\ast(\delta_Q)$. In the case $X_2 = X$ the Gaussian functional γ_Q is a quasi-free state on the quantum-mechanical Boson algebra, in the case $X_1 = X$ it is a quasi-free Fermion state [12, 23]. Since δ_Q is even in both the Boson and the Fermion case, Corollary 4.3. can be applied and we get that the quasi-free Boson and the quasi-free Fermion states are positive if δ_Q is conditionally positive and hermitian. One easily sees that δ_Q is conditionally positive and hermitian if Q is a positive definite kernel. Conversely, if γ_Q is positive we have $\gamma_Q((\sum_{x\in X} v_x\, x)(\sum_{x\in X} v_x\, x)^*) = \sum_{x,y\in X} v_x\, \bar{v}_y\, Q(x,y) \geqq 0$, so Q is a positive definite kernel.

5.7. Let $(V,\Delta,\delta,\mu,\mathbb{1},*)$ be a $*$-bialgebra. A derivation d on V is a linear operator on V satisfying $d(v\cdot w) = (dv)w + v(dw)$. Let d_1 and d_2 be $*$-maps (i.e. $d_i(v^*) = (d_i v)^*$) and derivations on V. Let λ be a linear operator on V satisfying

$$\lambda(v\cdot w) = (\lambda v)w + v(\lambda w) + i((d_1 v)(d_2 w) - (d_2 v)(d_1 w)) \tag{6}$$

Then $\delta\circ\varkappa$ with $\varkappa = r(d_1^2 + d_2^2) + s\lambda$ is conditionally positive if $2r \geqq s \geqq 0$. Indeed, we have: $(\delta\varkappa)(v\cdot v^*) = 2r(|\delta d_1 v|^2 + |\delta d_2 v|^2 + is((\delta d_1 v)(\overline{\delta d_2 v}) - (\delta d_2 v)(\overline{\delta d_1 v})) \geqq s|(\delta d_1 v) - i(\delta d_2 v)|^2 \geqq 0$ for $v\in V$ with $\delta v = 0$. The cases $\varkappa = d$, where d is a derivation, and $\varkappa = d^2$, where d is a $*$-map and a derivation, are included. As a consequence of Corollary 4.3. we get that $\{\exp_*(t\,\delta(r(d_1^2 + d_2^2) + s\lambda)) \mid t \geqq 0\}$ is a positive semigroup of linear functionals on V if $2r \geqq s \geqq 0$.

For example, let $V = \mathbb{C}\langle x_{ij}, x_{ij}^*\rangle$ be the free associative complex $*$-algebra with unit $\mathbb{1}$ generated by the indeterminants x_{ij} where $i,j = 1,2,\dots,N$. Define $\Delta: V\to V\otimes V$ to be the $*$-homomorphism given by $\Delta x_{ij} = \sum_{k=1}^{N} x_{ik}\otimes x_{kj}$ and let $\delta: V\to\mathbb{C}$ be the hermitian homomorphism given by $\delta(x_{ij}) = \delta^i_j$ (Kronecker delta). It is straightforward to check that we defined a $*$-bialgebra. A quantum stochastic analogue of a process on the unitary group with independent and stationary increments is introduced in [25]. Such a process is characterized, in a way analogous to the classical case, by a positive convolution semigroup of states on $K(d) = V/I$ where I denotes the $*$-ideal of V generated by the elements of the form $\sum_{k=1}^{N} x_{ik}\, x_{jk}^* - \delta^i_j\mathbb{1}$ and $\sum_{k=1}^{N} x_{ki}^*\, x_{kj} - \delta^i_j\mathbb{1}$. The ideal I also is a coideal, so K(d) again can be considered as a $*$-bialgebra.

For a skew hermitian matrix $a\in M(N)$ we construct a derivation d on K(d): Define $\tilde{d}_a$ to be the derivation on V given by $\tilde{d}_a\, x_{ij} = \sum_{k=1}^{N} a_{ik}\, x_{kj}$ and $\tilde{d}_a\, x_{ij}^* = \sum_{k=1}^{N} \bar{a}_{ik}\, x_{kj}^*$. One easily checks that $\tilde{d}_a$ is a $*$-map and that it leaves invariant the ideal I, so $\tilde{d}_a$ gives rise to a $*$-map and derivation d_a on K(d). The process that describes the Ito solution of the linear quantum stochastic differential equation in [25] is given by the semigroup with generator $\kappa = \delta\,(r(d_p^2 + d_q^2) + s\lambda)$ where $r, s \in \mathbb{R}$ with $2r \geq s \geq 0$, the matrices $p, q \in M(N)$ are skew hermitian elements, and λ fulfills equation (6) with $d_1 = d_p$ and $d_2 = d_q$.

References

[1] Abe, E.: Hopf Algebras, Cambridge University Press (1980).

[2] Arendt, W. Chernoff, P.R., Kato, T.: A Generalization of Dissipativity and Positive Semigroups, J. Operator Theory 8, 167-180 (1982).

[3] Berg, C., Christensen, J.P.R., Ressel, P.: Harmonic Analysis on Semigroups, Graduate Texts in Math. Vol. 100, Springer, New York, Heidelberg, Berlin (1984).

[4] Bourbaki, N.: Elements of Mathematics, Algebra, Chap. III, Hermann, Paris (1973).

[5] Bourbaki, N.: Eléments de Mathématique, Algèbre, Chap. IX, Hermann, Paris (1959).

[6] Bratteli, O., Digernes, T., Robinson, D.W.: Positive Semigroups on Ordered Banach Spaces, J. Operator Theory 9, 371-400 (1983).

[7] Canisius, J.: Algebraische Grenzwertsätze und unbegrenzt teilbare Funktionale, Diplomarbeit, Heidelberg (1979).

[8] Choi, M.-D.: Completely Positive Linear Maps on Complex Matrices, Lin. Alg. Appl. 10, 285-290 (1975).

[9] Doob, J.L.: Stochastic Processes, John Wiley & Sons, New York (1953).

[10] Dümcke, R.: Über quantendynamische Halbgruppen und ihre Begründung aus der mikroskopischen Dynamik, Dissertation, München (1980).

[11] Evans, D.E., Hanche-Olsen, H.: The Generators of Positive Semigroups, J. Functional Analysis 32, 207-212 (1979).

[12] Giri, N., von Waldenfels, W.: An Algebraic Version of the Central Limit Theorem, Z. Wahrscheinlichkeitstheorie verw. Gebiete 42, 129-134 (1978).

[13] Gorini, V., Kossakowski, A., Sudarshan, E.C.G.: Completely Positive Dynamical Semigroups of N-Level Systems, J. Math. Phys. 17, 821-825 (1976).

[14] Heyer, H.: Probability Measures on Locally Compact Groups, Springer, New York, Heidelberg, Berlin (1977).

[15] Kraus, K.: General State Changes in Quantum Theory, Annals of Physics 64, 311-335 (1971).

[16] Lindblad, G.: On the Generators of Quantum Dynamical Semigroups, Comm. math. Phys. 48, 119-130 (1976).

[17] Mathon, D., Streater, R.F.: Infinitely Divisible Representations of Clifford Algebras, Z. Wahrscheinlichkeitstheorie verw. Gebiete 20, 308-316 (1971).

[18] Parthasarathy, K.R., Schmidt, K.: Positive Definite Kernels, Continuous Tensor Products, and Central Limit Theorems of Probability Theory, Lect. Notes in Math. 272, Springer, New York, Heidelberg, Berlin (1972).

[19] Schneider, H., Vidyasagar, M.: Cross-Positive Matrices, SIAM J. Numer. Anal. 7, 508-519 (1970).

[20] Schoenberg, I.J.: Metric Spaces and Positive Definite Functions, Trans. Amer. Math. Soc. 44, 522-530 (1938).

[21] Schur, I.: Bemerkungen zur Theorie der beschränkten Bilinearformen mit unendlich vielen Veränderlichen, J. Reine Angew. Math. 140, 1-29 (1911).

[22] Sweedler, M.E.: Hopf Algebras, Benjamin, New York (1969).

[23] von Waldenfels, W.: An Algebraic Central Limit Theorem in the Anticommuting Case, Z. Wahrscheinlichkeitstheorie verw. Gebiete 42, 135-140 (1978).

[24] von Waldenfels, W.: Positive and Conditionally Positive Sesquilinear Forms on Anticocommutative Coalgebras, in Lect. Notes in Math. 1064, Springer, New York, Heidelberg, Berlin, 450-466 (1983).

[25] von Waldenfels, W.: Ito Solution of the Linear Quantum Stochastic Differential Equation Describing Light Emission and Absorption, in Lect. Notes in Math. 1055, Springer, New York, Heidelberg, Berlin, 384-411 (1984).

Michael Schürmann
Institut f. Ang. Mathematik
Universität Heidelberg
Im Neuenheimer Feld 294
D-6900 Heidelberg

The Ito-Clifford Integral Part II

R.F. Streater
Dept. of Mathematics,
King's College,
London WC2R 2LS.

This article continues Part I, to appear in the proceedings of the 1st BiBos Symposium, under the title "Stochastic Processes - Mathematics and Physics" Ed. S. Albeverio, Ph. Blanchard and Ph. Combes.

Contents

Part I §1 Notation
§2 Three motivations
§3 The Clifford Process
§4 The Ito-Clifford Integral

Contents of the present article

§5 Martingales
§6 Stochastic differentiation
§7 Differential Equations
§8 Hitting times, stopping times

§5. Martingales

We saw that the Clifford process $(\Psi_t)_{t\geq 0}$ is a martingale; more precisely, it is a martingale with respect to the filtration $(\mathfrak{A}_t)$. It is easy to construct many examples of martingales by making use of the conditional expectation: let $A \in L^2(\mathfrak{A}, \mathbb{E})$ and define

$$M_t = \mathbb{E}(A|\mathfrak{A}_t)$$

Then M_t is a martingale, by property (4) of the conditional expectation. Barnett has proved a converse to this: if M_t is an L^2-martingale, with $\sup_t \|M_t\|_2 < \infty$, then [18] there exists $A \in L^2(\mathfrak{A}, \mathbb{E})$ such that $M_t = \mathbb{E}(A|\mathfrak{A}_t)$. A question arises whether such martingales are actually non-classical, i.e. do not commute; $[M_s, M_t]_- \neq 0$. The classical theory was shown [8] to be an example of continuous tensor products, a concept that led to the introduction of infinitely divisible group representations. Using non-abelian groups, we [9] were able to construct large classes of truly non-abelian martingales. These involved Wick ordering. Since the conditional expectation in the Clifford process also involves Wick ordering (see §3) it is easy to see that the Wick-ordered monomials

$$M_t = :\psi(\chi_{[0,t]}u_0) \ \cdots \ \psi(\chi_{[0,t]}u_n): \qquad (\neq)$$

are martingales [1].

In the classical theory of martingales (M_s) one can form stochastic integrals $\int f(s)dM_s$ where $f(s)$ is adapted. We show in [1] that such integrals exist for the operator martingales of the form $(\neq)$. The key to this construction as for the Ito-Clifford integral, lies in the remark that $M_t^*M_t$ is a multiple of the identity: $M_t^*M_t = a_t 1$, where a_t is a positive increasing continuous function. Writing $a_t = \int_0^t d\nu_t$ we then prove ((5.3) of [1]) that for simple adapted h:

$$\left\| \int_0^t h(s)dM_s \right\|_2^2 = \int_0^t \| h(s)\Omega_F \|_2^2 d\nu(s).$$

The proof follows that of the Clifford process, §3. The effective part of these proofs is the Clifford analogue of the Doob-Meyer decomposition: if M_t is a L^2 martingale, then

$$M_t^*M_t = \text{martingale} + \text{increasing process of bounded variation.}$$

In both the Clifford process and the Wick power $(\neq)$, the martingale part is zero (unlike Brownian motion) and, like Brownian motion, the increasing part is non-random (that is, is a multiple of the identity). Clearly, sums of Wick monomials also define martingales, but now the martingale part of the Doob-Meyer decomposition will not always be zero. The effective tool to control stochastic integrals is the martingale representation theorem (Fermion analogue of the Kunita-Watanabe theorem). This takes us to the next topic.

§6. Stochastic derivatives

We have seen (§3) that every stochastic integral relative to $d\Psi$ is a martingale; we have the following converse [1]: let $(X_t : t \in \mathbb{R}_+)$ be an L^2-martingale of mean 0, adapted to the family $(\mathfrak{A}_t)_{t\geq 0}$. Then there is $f \in \mathcal{H}$ such that $X_t = \int_0^t f(s)d\Psi_s$ for all $t \in \mathbb{R}_+$. Recall that $\mathcal{H}$ is the completion of the set of adapted simple functions in the norm $\left(\int_0^t \| h(s)\Omega_F \|_2^2 ds\right)^{\frac{1}{2}}$. For the full proof, see [1], theorem (4.1). But the idea is not too hard: X_t corresponds, under the duality map D, to a vector function $X_t\Omega$ in Fermion Fock space, and so has a strongly convergent orthogonal decomposition into n-particle parts. It is enough, then to prove the theorem for n-particle martingales, and indeed, for the dense set $\mathcal{S}_0$ of such (in $\Lambda^n L^2(0,t)$) with kernels in $\mathcal{S}(\mathbb{R}^n)$, vanishing when any two time-variables coincide. Then by Fock-space techniques, there is an antisymmetric $\hat{w}$:

$$X_t = \int_0^t \hat{w}(t_1,\ldots,t_n):\psi(t_1)\ldots\psi(t_n):dt_1\ldots dt_n$$

where $\hat{w} \in \mathcal{S}_0$. But now take the latest time to the right (use the anti-symmetry of $\hat{w}$) to get

$$X_t = n!\int_{\substack{0\\ t_1<\ldots<s}}^t \hat{w}(t_1,\ldots,t_{n-1};s):\psi(t_1)\ldots\psi(t_{n-1}):\psi(s)\,dt\ldots dt_{n-1}ds$$

$$= \int_0^t f(s)d\Psi(s)$$

(we have used the fact that if $s > t_1,\ldots,t_{n-1}$, then $\psi(t_1)\ldots\psi(t_{n-1})\psi(s): = :\psi(t_1)\ldots\psi(t_{n-1}):\psi(s)$). Thus $f(s)$ may be taken to be

$$n!\int_0^t \hat{w}(t_1,\ldots,t_{n-1};s):\psi(t_1)\ldots\psi(t_{n-1}):dt_1\ldots dt_{n-1},$$

which is clearly adapted and lies in L^2. We show it lies in $\mathfrak{h}$; [1]. Taking limits to the completion gives the theorem for any n-particle L^2-martingale.

We need the martingale to be 'centred' i.e. to have mean 0, so that the degree of the Wick polynomial is at least 1, and we can pull out one $\psi(s)ds = d\Psi_s$. In [1] we use the notation

$$X_t = \int_0^t \tilde{X}_s d\Psi_s$$

and call $(\tilde{X}_s)$ the stochastic derivative of (X_t): a physicist would write $\tilde{X}_t = \frac{\partial X_t}{\partial \Psi_t}$.

A submartingale is a process X_t (adapted to $(\mathfrak{A}_t)$) such that $X_s \leqslant \mathbb{E}(X_t|\mathfrak{A}_s)$ if $s \leqslant t$. Similarly we define supermartingale [1]. One shows that any submartingale is the sum of a martingale and a positive increasing process. This is our analogue of the Doob-Meyer decomposition. Conditions for the uniqueness, given in [1], have been verified for L^2 martingales by Barnett and Wilde (see Wilde's contribution to this volume).

The next problem is to form a stochastic integral $\int_0^t f_s dM_s$, where M_s is any martingale. Such integrals would occur in the solution of a stochastic differential equation driven by (M_s), instead of (Ψ_s), as noise. If (M_s) is square integrable, we can write $M_s = \int_0^t \tilde{M}_s d\Psi_s$, so $dM_s = \tilde{M}_s d\Psi_s$ and we could tentatively try the definition

$$\int_0^t f_s dM_s = \int_0^t f_s \tilde{M}_s d\Psi_s$$

which makes sense if $(f_s) \in L^\infty$ and is adapted and continuous in s.

There is an explicit formula for the Doob-Meyer decomposition which enables us to give a direct definition of $\int_0^t f_s dM_s$, and to prove this equation.

Define the automorphism β of $\mathcal{O}\!\!\mathcal{L}$ to be the second quantization of the operator -1 on $\mathcal{K}$. Thus, $\beta(A) = A$ if A is even, and $\beta(A) = -A$ if A is odd. Then we give [1] the following formula for the square of a stochastic integral

$$\left(\int_0^t fd\Psi_s\right)^*\left(\int_0^t fd\Psi_s\right) = Z_t + \int_0^t (\beta f)^*(\beta f)ds$$

where Z_t is a centred martingale. Indeed, this is the Doob-Meyer decomposition of the submartingale $X_t^* X_t$, where $X_t = \int_0^t f_s d\Psi_s$. By the martingale representation theorem, the same formula holds for any L^2-martingale X_t, with $f_s = \tilde{X}_s$.

<u>Theorem</u> [19]. Let f be a simple adapted L^∞ function, and let (X_t) be an L^2-martingale. Then

$$\left\|\int_0^t f(s)dX_s\right\|_2^2 \leq \int_0^t \|f(s)\|_\infty^2 d\mu_X(s)$$

where $\mu_X(a,b) = \int_a^b \|\tilde{X}_s\|_2^2 ds$.

<u>Remark</u>. For simple functions, $f = \sum_{k=1}^n h_{k-1}\chi_{[t_{k-1},t_k]}$ the integral is defined just to be the sum

$$\int fdX_s = \sum_k f_{k-1}\Delta X_k$$

where $\Delta X_k = X_{t_k} - X_{t_{k-1}}$.

<u>Proof</u>.

$$\left\|\int_0^t f(s)dX_s\right\|_2^2 = \sum_{k,j} \mathbb{E}(\Delta X_k^* h_{k-1} h_{j-1}\Delta X_j)$$

$$= \sum_k \mathbb{E}(\Delta X_k^* |h_{k-1}|^2 \Delta X_k)$$

using the martingale property

$$= \sum_k (\mathbb{E}|h_{k-1}|^2 \Delta X_k \Delta X_k^*) \leq \sum_k \|h_{k-1}\|_\infty^2 \|\Delta X_k^*\|_2^2$$

$$= \sum_k \|h_{k-1}\|_\infty^2 \|\Delta X_k\|_2^2 = \sum_k \|h_{k-1}\|_\infty^2 \mathbb{E}(|X_{t_k}|^2 - |X_{t_{k-1}}|^2)$$

using the martingale property

$$= \sum_k \|h_{k-1}\|_\infty^2 \int_{t_{k-1}}^{t_k} \|\beta\tilde{X}(s)\|_2^2 ds$$

since the terms involving Z_t go out

$$= \sum_k \|h_{k-1}\|_\infty^2 \int_{t_{k-1}}^{t_k} \|\tilde{X}(s)\|_2^2 ds$$

as β is isometric

$$= \int_0^t \|f(s)\|_\infty^2 d\mu_X(s). \qquad \square$$

This contractive property is then used in the standard fashion to extend the integral to the completion of the simple functions in the norm given by the right hand side. For this class of integrands f, the integral is defined as the limit of the integrals of approximating simple functions; and this definition gives the <u>same</u> answer as that given by the chain rule: $\int f dX = \int f\tilde{X} d\Psi$. This result we call the Radon-Nikodym theorem.

The increasing part of M_t^2, where M_t is a classical (Brownian) martingale, is called the bracket process, $\langle M_t, M_t \rangle$. We can make an analogous definition.

<u>Definition</u>. Let $X_t = X_0 + \int_0^t \tilde{X} d\Psi$ and $Y = Y_0 + \int_0^t \tilde{Y} d\Psi$ be L^2-martingales with respect to the Clifford filtration $(\mathfrak{A}_t)$. The bracket $\langle X_t, Y_t \rangle$ is the $L^1(\mathfrak{A})$-process

$$\langle X_t, Y_t \rangle = \int_0^t \beta(\tilde{Y}(s))^* \beta(\tilde{X}(s)) ds, \qquad t \geq 0.$$

We show that $\langle X,Y \rangle$ has bounded variation [19], and obeys

$$\langle X_t, \int^t f(s) dY_s \rangle = \langle \int^t f(s)^* dX_s, Y_t \rangle$$

$$\int_0^t f d\langle X,Y \rangle = \langle X_t, (\int_0^t f(s) dY_s^*)^* \rangle$$

where $\int \ldots d\langle X,Y \rangle$ is Bartle's bilinear integral [20]. The latter relation has, as corollary, the following characterization of stochastic integral: Let (X_t) be an $L^2(\mathfrak{A})$-martingale, and let $f: \mathbb{R}^+ \to L^\infty(\mathfrak{A})$ be, locally, the almost-everywhere limit of a uniformly bounded L^∞ step-function. Then $\int_0^t f dX$ is the unique centred $L^2(\mathfrak{A})$-martingale, Z_t^* say, such that

$$\int_0^t f d\langle Y, X^* \rangle = \langle Y_t, Z_t \rangle.$$

This characterization might be used to define the stochastic integral relative to a wider class than martingales.

Generalizations in another direction, to factors of type II_∞ and type III, have been made in [21], [22].

§7. Stochastic Differential Equations

In [16] we study differential equations of the form

$$dX_t = F(X_t,t)d\Psi_t + G(X_t,t)dt$$

subject to the initial condition $X_0 = X(0)$, a given element of $L^2(\mathfrak{A})$. Of course as usual, this equation is to mean that the integral form

$$X_t = X(0) + \int_0^t F(X_s,s)d\Psi_s + \int_0^t G(X_s,s)ds$$

holds for each t, where $\int \ldots d\Psi_s$ is the Ito-Clifford integral.

But now we see an extra difficulty in the quantum case: what is to be meant by the functions $F(X_t,t), G(X_t,t)$ of the operators X_t? If X_t is self-adjoint for all t, or even merely normal, then one might expect to be able to define $F(X_t)$ by the functional calculus. So we might consider limiting our discussion to equations whose initial datum X(0), and solution X(t) were self-adjoint. But we would also like to develop the theory to cope with multi-component processes $X_1(t),\ldots,X_n(t)$, and these will not in general commute, even if they are self-adjoint. So we are unable to use the functional calculus and must resort to the concept "F is a function of X_t if F is affiliated to the W*-algebra generated by X_t".

In order to solve the equation, we use Picard's iteration

$$X_0 = X(0)$$

$$X_1(t) = X(0) + \int_0^t F(X(0),s)d\Psi_s + \int_0^t G(X(0),s)ds$$

$$X_{n+1}(t) = X(0) + \int_0^t F(X_n(s),s)d\Psi_s + \int_0^t G(X_n(s),s)ds$$

To proceed thus, we see that F and G must be well-defined for all approximate solutions $X(0), X_1(s),\ldots$ not just for the exact solutions X(t). In practice we take F,G to be given maps defined for all operators $X \in L^2(\mathfrak{A})$, for each $t \geq 0$, and we say that F is <u>adapted</u> if F(X,t) lies in the W*-algebra generated by X. Then if (X_t) is adapted (to $(\mathfrak{A}_t)$ in the earlier sense) then $F(X_t,t)$ is adapted to $(\mathfrak{A}_t)$. Thus, if F is adapted, then the iterated solutions make sense as Ito-Clifford integrals.

We say F, as above, obeys the Lipschitz condition if for $X,Y \in L^2(\mathfrak{A})$:

$$\|F(X,t)-F(Y,t)\|_2 \leq K\|X-Y\|_2$$

independent of t if $0 \leq t \leq T$ say. Then we have the following theorem [16].

Suppose F and G obey Lipschitz conditions; then for each X_0 there is a unique continuous adapted L^2-process $(X(t))_{t\geq 0}$ satisfying

$$dX_t = F(X_t,t)d\Psi_t + G(X_t,t)dt,$$

with $X(0) = X_0$.

We consider $X_n(t)$ of Picard's iteration, and estimate

$$\|X_{n+1}(t)-X_n(t)\|_2 = \|\int_0^t (F(X_n(s),s)-F(X_{n-1}(s),s))d\Psi_s + \int_0^t (G(X_n(s),s)-G(X_{n-1}(s),s))ds\|_2$$

The stochastic integral is dominated by

$$\|\int_0^t (F(X_n(s),s)-F(X_{n-1}(s),s))d\Psi_s\|_2 = (\int_0^t \|F(X_n(s),s)-F(X_{n-1}(s),s)\|_2^2 ds)^{\frac{1}{2}}$$

(by Ito's isometry)

$$\leq K(\int_0^t \|X_n(s)-X_{n-1}(s)\|^2 ds)^{\frac{1}{2}}.$$

The integral involving G gives a similar estimate. Using $(a+b) \leq 2(a^2+b^2)$ then gives

$$\|X_{n+1}(t)-X_n(t)\|_2^2 \leq C\int_0^t \|X_n(s) - X_{n-1}(s)\|_2^2 ds.$$

Iteration then gives

$$\|X_{n+1}(t)-X_n(t)\|_2^2 \leq c^n\int_0^t ds_1\int_0^{s_1} \cdots \int^{s_{n-2}} ds_{n-1}\|X_1(s_{n-1})-X_0(s_{n-1})\|_2^2$$

$$\leq \frac{ct^{n-1}}{(n-1)!}$$

Hence $\|X_{n+1}(t)-X_n(t)\|_2 \leq \frac{ct^{n-1/2}}{\sqrt{(n-1)!}}$.

Hence $X_n = \sum^n (X_j-X_{j-1})$ converges in L_2 as $n \to \infty$. One shows by substitution that the limit obeys the given equation, and the initial condition.

Unfortunately the polynomial $(X_t)^n$ does not obey this operator Lipschitz condition for all $X_t \in L^2$, and so we cannot apply the theorem to equations of motion with polynomial interactions. Indeed, functions obeying the Lipschitz condition in our sense are hard to come by. There are some! [16]

We also show that the equations are stable, in that small changes in X_0 or F or G (in the L^2 sense) lead to small changes in the solution (in $0 \leq t \leq T < \infty$).

Our existence theorem can be extended to stochastic differential equations driven by more general martingales [23]. Of physical importance is the generalization to non-martingales [24]. The idea here is that if we want the noise to be at a finite temperature, rather than the ∞ temperature of the tracial state, we must restrict the spectrum of the noise to the positive-energy subspace, [2]. We replace white noise by coloured noise. Then the noise is no longer a martingale. However, if the noise is regular, in the sense that it is almost a martingale, then we can solve the stochastic differential equations, following McShane's book with suitable modifications for the anti-commuting nature of the variables [25]. Some results can also be got for the quasi-free type III case [22].

Another classical field is the connection between diffusion equations and stochastic processes. In its abstract form this is the dilation problem - when can a diffusion contraction semigroup be dilated to a unitary motion, It is solved by the martingale method [26]. A Fermion version of the martingale problem is presented in [27].

§8. Hitting times, stopping times

An important technique in classical probability is to consider a process X_t and a random variable T, called a stopping time, and to consider the stopped process $X_{t\wedge T}$ where $t \wedge T$ means the minimum of t and T. A stopping time is a random variable $T \geq 0$ such that the event $\{\omega: 0 \leq T(\omega) \leq \lambda\}$ is $\mathfrak{F}_\lambda$-measurable, where $(\mathfrak{F}_\lambda)$ is the filtration. The first result of the theory is then that the stopped process $X_{t\wedge T}$ is a martingale if (X_t) is a martingale.

In the quantum version of this there is no difficulty defining a stopping time: let $(\mathfrak{A}_t)$ be the filtration of W*-algebras; then a 'stopping time' is a positive operator T affiliated to $\mathfrak{A} = \bigvee_{t\geq 0} \mathfrak{A}_t$, such that the spectral resolution of T for $[0,\lambda]$ belongs to $\mathfrak{A}_\lambda$. Thus if $T = \int_0^\infty \lambda dE(\lambda)$, then $E(\lambda) \in \mathfrak{A}_\lambda$, $\lambda \geq 0$.

Barnett and Lyons [28] show how to reformulate a "stopped process" X_T and the "process stopped by T", $X_{t\wedge T}$, for any L^2-martingale (X_t); suppose first, that T has a finite discrete spectrum (λ_j). In the classical case, where

$$T = \sum_{j=1}^{n} \lambda_j (E_j - E_{j-1}), \text{ we have}$$

$$X_T = \sum_{j=1}^{n} (E_j - E_{j-1}) X_{\lambda_j}$$

$$X_{T\wedge t} = \sum_{\lambda_j < t} (E_j - E_{j-1}) X_{\lambda_j} + (1-E_t) X_t$$

(in the commutative case, the order does not matter).

In the case where the spectrum has jumps and a continuous part, these formulae look like integrals with respect to the increasing family of projections (E_λ) (taken, as usual, to be right continuous). To make sense of this concept, Barnett and Lyons first "sum by parts', and define

$$\int_0 X_s dE_s = (X_\infty E_\infty - X_0 E_0) - \int_0^\infty dX_s E_s$$

$$= X_\infty - \int_0^\infty dX_s E_s$$

in our context. This makes sense whenever the left stochastic integral $\int dX_s E_s$ makes sense (it is $(\int E_s dX_s^*)^*$). They go on to show that this is the case whenever X_s is an L^2-martingale or more generally any process obeying the standing hypotheses of [24], in the context of the Clifford filtration $(\mathfrak{A}_t)_{t\geq 0}$. The stopped random variable is then

$$X_T = \int_0^\infty X_s dE_s = X_\infty - \int_0^\infty dX_s E_s$$

and the stopped process is

$$X_{T\wedge t} = \int_0^t X_s dE_s + (1-E_t) X_t$$

$$= X_t E_t - \int_0^t dX_s E_s + X_t - E_t X_t$$

$$= X_t - \int_0^t dX_s E_s .$$

From the final form it is obvious that if (X_t) is an L^2-martingale and T a stopping time, then $X_{T\wedge t}$ is a martingale. More elaborate versions of the optional stopping theorem can be found in [28].

We end this brief review by the remark that stopping times are quite common in a quantum stochastic process. Take for example the quasi-free representations of the CCR (the boson theory) described in [22]. As emphasized by Hudson, the non-abelian algebras of the boson theory contain many abelian subalgebras, each a copy of a Gaussian process with independent increments. Such an abelian subalgebra defines a classical process which has many stopping times. A fortiori, such a classical stopping time is a stopping time for the full quantum theory.

It would be interesting to stop the momentum Brownian motion $P_t = \frac{A^*(t)-A(t)}{i}$ with respect to a stopping time for the position

Brownian motion $Q_t = A^*+A(t)$.

Concrete examples of stopping times for the Clifford filtration $(\mathfrak{A}_t)$ remain to be found.

I thank L. Accardi, W. von Waldenfels and G. Gorini for the hsopitality of the 2nd Workshop on Quantum Probability, Heidelberg 1984, and Kendal Anderson for typing it.

References

[1] C. Barnett, R.F. Streater and I.F. Wilde. The Ito-Clifford Integral. Jour. Funct. Anal. 48, 172-212 (1982).

[2] R.F. Streater. The damped oscillator with quantum noise. J. Phys. A15, 1477-1486 (1982).

[3] I.R. Senitzky, Phys. Rev. 119, 670 (1960); ibid A3 421 (1970).

[4] M. Lax. Phys. Rev. 145, 111-129 (1965).

[5] E. Nelson, Dynamical Theories of Brownian Motion. Princeton Lecture Notes, Princeton University Press.

[6] F. Guerra and P. Ruggiero: New interpretation of the Euclidean Markov field in the framework of physical space-time. Phys. Rev. Lett. 31, 1022 (1972).

[7] G. Parisi and Y. Wu. Scientia Sinica 24, 483 (1981).

[8] R.F. Streater. Current Commutation Relations and Continuous Tensor Products. Nuovo Cimento 53, 487 (1968).
R.F. Streater, Current Commutation Relations, Continuous Tensor Products and Infinitely Divisible Group Representations; in Local Quantum Theory (R. Jost, Ed.), Academic Press 1969.

[9] T. Hida. Brownian Motion. Springer Verlag, N.Y., Berlin, Heidelberg 1980.
R.L. Hudson and R.F. Streater. Examples of quantum martingales. Phys. Lett. 85A, 64-66 (1981).
R.L. Hudson and R.F. Streater. Ito's rule is the chain rule with Wick ordering. Phys. Lett. 86A, 277-279 (1981).
R.L. Hudson and R.F. Streater, Non-commutative martingales and stochastic integrals in Fock space; in Lecture Notes in Physics 1973, Springer 1981.

[10] D. Mathon and R.F. Streater. Infinitely Divisible Representations of Clifford Algebras. Z. für Wahr. verw. Geb. 20, 308-316 (1971).

[11] H. Umegaki, Conditional Expectation in an Operator Algebra. II. Tohuku Math. J. 8, 86-100 (1956).

[12] D.E. Evans and J.T. Lewis. Commun. Dublin Institute for Advanced Studies A, 24 (1977).

[13] I.E. Segal. A non-commutative extension of abstract integration. Ann. of Math. 57, 401-457 (1953). Ibid, 58, 595-596 (1953). E. Nelson. Notes on non-commutative integration. Jour. Functl. Anal. 15, 103-116 (1974).

[14] J. Dixmier. Formes linéares sur un anneau d'opérateurs. Bull. Soc. Math. France 81, 9-39 (1953).

[15] C. Barnett, R.F. Streater and I.F. Wilde. The Ito-Clifford Integral III: Markov property of solutions to stochastic differential equations. Commun. Math. Phys. 89, 13-17 (1983).

16] C. Barnett, R.F. Streater and I.F. Wilde. The Ito-Clifford Integral II, Stochastic differential equations. J. Lond. Math. Soc. (2), 27, 373-384 (1983).

17] R.F. Streater. Quantum Stochastic Integrals, Acta Physica Austriaca, Suppl.XXVI, 53-74 (1984).

18] C. Barnett. Supermartingales on semi-finite von Neumann algebras. J. Lond. Math. Soc. (2), 24, 175-181 (1981).

19] C. Barnett, R.F. Streater and I.F. Wilde. The Ito-Clifford Integral IV: A Radon-Nikodym Theorem and Bracket Processes. J. Operator Theory, 11, 255-271 (1984).

20] R.G. Bartle. A general bilinear vector integral. Studia Math. 15, 337-352 (1956).

21] C. Barnett, R.F. Streater and I.F. Wilde. Stochastic integrals in an arbitrary probability gauge space. Math. Proc. Camb. Phil. Soc. 94, 541 (1983).

22] C. Barnett, R.F. Streater and I.F. Wilde. Quasi-free quantum stochastic integrals for the CAR and CCR. J. Functl. Anal. 52, 17-47 (1983).

23] R.F. Streater. Quantum Stochastic Processsss. Rome II Conference on Quantum Probability (L. Accardi, A. Frigerio and V. Gorini, Eds.).

24] C. Barnett, R.F. Streater and I.F. Wilde. Quantum Stochastic Integrals under Standing Hypotheses; submitted to J. Lond. Math. Soc.

25] E.J. McShane. Stochastic calculus and stochastic models. Academic Press, N.Y. 1974.

26] D. Stroock and V.S. Varadhan. Multidimensional Diffusion Processes (Springer), 1979.

27] H. Hasegawa and R.F. Streater. Stochastic Schrödinger and Heisenberg equations: a martingale problem in quantum stochastic processes. J. Phys. A. L697-L703 (1983).

28] C. Barnett and T. Lyons. "Stopping non-commutative processes", Imperial College preprint, 1984.

DETAILED BALANCE AND EQUILIBRIUM

A. Verbeure

Instituut voor Theoretische Fysica
Katholieke Universiteit Leuven
B-3030 Leuven , Belgium

ABSTRACT

In these lecture we aim at a didactically clean presentation of the notion of detailed belance which is understandable both for mathematicians as well as for physicists. We concentrate mainly on its relation with equilibrium states for quantum systems and discuss in particular the relaxation to equilibrium.

I. INTRODUCTION

The characterization of thermal equilibrium states for infinite systems started in mathematical physics more or less twenty years ago. For classical lattice systems the equilibrium conditions are given by the so-called DLR-equations [1,2] . For quantum systems it turned out that the KMS-conditions are the equilibrium conditions [3] . Also the equilibrium conditions for continuous classical systems were established [4] .

The physical motivation for these conditions began some ten years ago by the study of their stability properties. Without being exhaustive we mention the proof of the dynamical stability of the equilibrium states, i.e. the stability with respect to conservative perturbations [5,6] ; next we mention the proof of the thermodynamic stability of the equilibrium states, i.e. their stability with respect to dissipative perturbations both local and global ones [7,8] . All these characterizations are within the frame work of equilibrium statistical mechanics.

The next task is to give a dynamical characterization of equilibrium states. In particular, one has in mind the approach to equilibrium; what is the mechanism of any state evolving to an equilibrium state. In the physics literature about lattice spin systems the Glauber model [9] was the prototype of a dynamics governing the phenomenon of return to equilibrium.

As far as rigorous results are concerned, probably the most advanced result, is the proof that any stationary state under the Glauber dynamics is a Gibbs state [10] (proved in one and two dimensions).

The Glauber dynamics is a special case of an evolution satisfying the detailed balance condition. For classical systems this notion is known for a long time in the physics literature [11,12] . It appears mainly in connection with Markovian systems described by the Fokker-Planck equation. One can visualize this concept as follows. A system moves in phase space from one point to another by continuous or discrete jumps. One computes the probability that at a certain time the system being in a point x_0, it makes a jump to another point x_1. The system is in detailed balance if the transition probability for the inverse process is exactly the same. One also speaks about microreversibility or local equilibrium. Not only one gets a stationary system, but the stationarity is established by balancing in detail every micro transition with its inverse.

It is clear that this notion is typically a classical one. It is not immediately clear how to generalize this notion to quantum systems. In the next section we write a version of detailed balance which is general enough to be valid also for quantum systems. Furthermore, we discuss our results obtained in collaboration with J. Quaegebeur and G. Stragier [13-15].

II. THE NOTION OF DETAILED BALANCE

In order to introduce the notion of detailed balance in a mathematically rigorous way, we start with the simplest classical system one can imagine. Consider a system with a finite number of events (finite phase space) labeled by $\Sigma = \{1,\ldots,n\}$ in a state ω, described by the probabilities $p = (p_1,\ldots,p_n)$, $0 \leq p_i \leq 1$, $\Sigma_i\, p_i = 1$.

The observables are the complex functions f on Σ $(f : i \in \Sigma \to f_i \in C)$. The expectation value of the observable f in the state ω is given by $\omega(f) = \sum_{i=1}^{n} p_i f_i$.

This system can have a Hamiltonian, generally described by a real function H on Σ , describing the interactions in the system. At a fixed temperature such a system has an equilibrium state, given by the Gibbs procedure. By no means this Hamiltonian yields a time evolution because the phase space is too small to carry a non commutative structure like a non-vanishing commutator or a Poisson bracket. This system has no natural dynamics.

Therefore consider the following Markov process defined on the state space of the system i.e. in the Schrödinger picture.

$$\frac{dp_i(t)}{dt} = \sum_{k=1}^{n} D_{ki}p_k(t) - D_{ik}\, p_i(t) \tag{1}$$

where the D_{ik} $(i,k = 1,\ldots,n)$ are positive constants suitably normalized, interpreted as the transition probability from the event i to the event k.

In the Heisenberg picture, equation (1) reads: for any observable f

$$\frac{d}{dt} f_i(t) = \sum_{k=1}^{n} D_{ik}\,(f_k(t) - f_i(t)) \tag{2}$$

The Markov process (1) or (2) satisfies the <u>detailed balance condition with respect to the state ω or p</u> if for all $i,k \in (1,\ldots,n)$ one has

$$D_{ik}\, p_i = D_{ki}\, p_k \tag{3}$$

or in words there is a detailed balance between transitions from the phase point i to the point k and its inverse, taking into account the system is in the state ω .

An immediate consequence of this definition yields that

$$\frac{dp_i}{dt} = 0$$

and for all f:

$$\frac{d}{dt}\,\omega_t(f) = \sum_{i=1}^{n} \frac{dp_i}{dt}\, f_i = 0$$

which means that the state is invariant for the Markov process. It is clear that more is needed in order to have the converse statement, which would be that if a state is stationary under a Markov process then the process is detailed balance with respect to that state.

Definition (3) of detailed balance is clearly a classical one. In order to find a generalization also valid for quantum systems one proceeds as follows to a more abstract form of it.

Let $<.,.>$ be the scalar product

$$\langle f,g\rangle = \sum_{i=1}^{n} p_i(t=0)\,\bar{f}_i g_i = \omega(\bar{f}g)$$

for the functions f,g on Σ . Let L be the generator of the Markov process:

$$(Lf)_i = \sum_{k=1}^{n} D_{ik}(f_k - f_i)$$

then definition (3) implies for all f and g:

$$\langle f,Lg\rangle = \langle Lf,g\rangle \tag{4,a}$$

or

$$\omega(fLg) = \omega(L(f)g) \tag{4,b}$$

Hence (3) implies (4).
Conversely, take in (4) :

$$f_\ell = \delta_{\ell i} \text{ and } g_\ell = \delta_{\ell k} \quad ; \ \ell,i,k \in (1,\ldots,n)$$

then (4) yields (3).

Hence we have that the condition of detailed balance with respect to a state ω is equivalent to the selfadjointness of the generator L of the Markov semigroup acting on the GNS-representation space induced by the state ω .

It is clear that the more abstract definition (4) can be used for quantum systems.

Let us first write it down for finite quantum systems, i.e. let $\mathcal{H}$ be a Hilbert space and consider $A = B(\mathcal{H})$, all bounded operators on $\mathcal{H}$ as the algebra of observables of the system. Take ρ a density matrix on $\mathcal{H}$ and denote by ω its corresponding state

$$\omega(x) = \mathrm{tr}\,\rho\, x \ , \ x \in A$$

This state defines a scalar product on A :

$$\forall x,y \in A \rightarrow \langle x,y\rangle_\rho = \mathrm{Tr}\,\rho\, x^* y = \omega(x^* y)$$

Suppose that one has given a semigroup time evolution in the Heisenberg picture with generator L:

$$\frac{dx(t)}{dt} = L\,x(t) \ , \ x \in A$$

with x(0) = x. In accordance with (4), this evolution is said to satisfy the detailed balance property with respect to the state ω if for all $x,y \in A$, holds:

$$\langle x,Ly\rangle_\rho = \langle Lx,y\rangle_\rho \tag{5}$$

or $\omega_\rho(x^* Ly) = \omega_\rho((Lx)^* y)$

Remark also that if we have a detailed balance evolution then the state must be stationary, indeed take in (5) $x = 1$, then one gets $\mathrm{Tr}\rho\ (L(y)) = 0,\ y \in A$.

Note also that if L is the generator of a Hamiltonian evolution i.e. $L = i\ [H,.]$ where H is a selfadjoint operator on $\mathcal{H}$, then

$$\mathrm{Tr}\ (L(y)) = 0 \text{ for } y \in A$$

implies $[H,\rho] = 0$, but then

$$\langle x, Ly\rangle_\rho = - \langle Lx, y\rangle_\rho$$

This means that a Hamiltonian evolution is excluded from the definition (5).

In order to include also the Hamiltonian evolution some authors [16,17] adapt the definition (5) as follows. A generator L of a semigroup satisfies the condition of detailed balance with respect to the state ω if L can be decomposed in a selfadjoint L_s and an antiselfadjoint part L_a such that

$$L = L_a + L_s \tag{6}$$

with $\omega_\rho(x\ L_a y) = - \omega_\rho(L_a(x)y)$

$$\omega_\rho(x\ L_s y) = \omega_\rho(L_s(x)y)$$

For the sake of completeness we mention the definition of detailed balance using the time inversion operator [18,19]

Let σ be an antilinear map of $\mathcal{B}(\mathcal{H})$ such that with ρ a density matrix: $\sigma(\rho) = \rho$ and $\omega_\rho(\sigma(x)\sigma(y)) = \overline{\omega_\rho(xy)}$, $x,y \in \mathcal{B}(\mathcal{H})$ and $L = L_a + L_s$ as in (6) and such that

$$\sigma L_a \sigma = - L_a$$

$$\sigma L_s \sigma = L_s$$

then

$$\begin{aligned}\omega_\rho(xLy) &= - \omega_\rho(L_a(x)y) + \omega_\rho(L_s(x)y)\\ &= \omega_\rho((\sigma L_a \sigma)(x)y) + \omega_\rho((\sigma L_s \sigma)(x)y)\\ &= \omega_\rho((\sigma L \sigma)(x)y)\\ &= \omega_\rho(\sigma(y^*)L\ \sigma(x^*))\end{aligned}$$

or

$$\omega_\rho(x\ L\ y) = \omega_\rho(\sigma(y^*)L\sigma(x^*)) \tag{7}$$

Clearly σ is a generalization of the time reversal operation and several authors use (7) as the definition of detailed balance.

In this lecture we are specially interested in the detailed balance evolutions with respect to a thermodynamic equilibrium state and we will concentrate ourselves on quantum systems.

Now we proceed to the rigorous formulation of our framework.

As usual a reversible C^*-dynamical system is given by a pair $(A, (\tau_t)_{r\in\mathbb{R}})$, where A is the C^*-algebra of observables and $(\tau_t)_{t\in\mathbb{R}}$ a one-parameter strongly continuous group of $*$-automorphisms representing the quantum dynamical time evolution.

Such a system may have one or more equilibrium or KMS-states ω_β at a given inverse temperature β i.e. for $x,y \in A_\tau$ (the set of analytic elements in A for τ) one has

$$\omega_\beta(x\ \tau_{i\beta}\ y) = \omega_\beta(yx) \tag{8}$$

For such systems one defines a detailed balance evolution $(\gamma_t^\beta)_{t\in\mathbb{R}}$ with respect to the equilibrium state(s) ω_β as follows: it is a one-parametergroup of strongly continuous, unity preserving positive maps of A such that for all ω_β and all $x,y \in A$ holds:

$$\omega_\beta(x\ \gamma_t^\beta(y)) = \omega_\beta(\gamma_t^\beta\ (x)y)\ ,\quad t \in \mathbb{R}^+ \tag{9}$$

In fact we will not work with the semigroup γ_t^β but with its generator denoted by L ($\gamma_t^\beta = \exp t L$). The properties of the semigroup translated to the generator are then: L is a linear map from a dense $*$-algebra $\mathcal{D}(L)$ into A satisfying

$$L(1) = 0 \quad \text{(unity preserving)} \tag{10}$$

$$L(x^*) = L(x)^* \tag{11}$$

$$L(x^*x) - L(x)^*x - x^*L(x) \geq 0,\ x \in \mathcal{D}(L) \tag{12}$$

(dissipativity condition)

$$\omega_\beta(L(x)y) = \omega_\beta(xL(y)) \text{ for all } \beta\text{ - KMS-states } \omega_\beta \tag{13}$$

(detailed balance condition)

Mutatis mutandis one can formulate the detailed balance evolution in the W^*-dynamical system approach.

Which are the questions to be asked? First of all one can ask for the solutions γ_t^β satisfying (9) or the solutions L satisfying (10)-(13). There are surely trivial solutions, but we are interested in those solutions describing the phenomenon of return to equilibrium i.e. if ω is any state, does the limit t tending to infinity of $\omega.\gamma_t^\beta$ exist and eventually is the limit a β-KMS-state?

The next question to be answered is about the rate of approach to equilibrium i.e. for any observable $x \in A$, determine the function

$$g(t,\beta,x) = \omega(\gamma_t^\beta(x)) - \omega_\beta(x)$$

When do we have an exponential decay, a power law decay? For systems showing a phase transition, do we have critical slowing down?

We do not have a solution to all these questions, but we do have a class of non trivial solutions (see section III). We do have a characterization of the equilibrium states in terms of the solutions (see section V), and we have analyzed the approach to equilibrium (long time behaviour) for the ideal Bose gas (section VI).

III. DETAILED BALANCE EVOLUTIONS

Here we construct a class of solutions of the equations (10-13) in the case that the dynamics (τ_t) of the system is L^1-asymptotically abelian, this means that there exists a τ-invariant norm dense $\star$-subalgebra A_o of A such that for all $x,y \in A_o$ holds

$$\int_{-\infty}^{\infty} ds \, \|[\tau_s x, y]\| < \infty$$

If the dynamical system (A,τ_t) does not satisfy this condition, it does not mean that we are unable to give solutions. For an example see the next section.

The construction we make is a generalization of the semigroups obtained in [20] for finite systems coupled to free heat baths.

Denote by F_β the set of complex functions f on C satisfying the following conditions:

a) analytic in the set $D_\beta = \{z \in C \mid 0 < \mathrm{Im} z < \beta\}$
b) bounded and continuous on the closure $\overline{D}_\beta$
c) for all $t \in \mathbb{R}$: $f(-t+i\beta) = f(t)$
d) the map $t \in \mathbb{R} \to f(t)$ is of positive type
e) $f \in L^1(\mathbb{R}, dt)$.

Denote by $\mathcal{L}(\tau,\beta)$ the strong closure ($L_\alpha \to L$ if for all $x \in A_o$, $\|L_\alpha x - Lx\| \underset{\alpha}{\to} 0$) of the convex hull generated by the maps

$$L_x^f(y) = \int dt\, ds\, f(t) \{\tau_s(x) \quad [y,\tau_{s+t}x] + [\tau_s x,y]\, \tau_{s+t}\, x\}$$

for $f \in F_\beta$, $x = x^* \in A_o$.

Remark that L_x^f is well defined , the s-integral exists because of the L^1-asymptotic condition and the t-integral because of f being L^1.

Suppose that the evolution τ has a discrete spectrum then the maps L_x^f can be defined by replacing the s-integral by the mean over s.

The properties (10)-(13) are checked by inspection and therefore we proved that all elements $\mathcal{L}(\tau,\beta)$ are dissipative maps satisfying the detailed balance condition with respect to the equilibrium state.

IV. DETAILED BALANCE AND EQUILIBRIUM

Now we discuss the problem whether the class $\mathcal{L}(\tau,\beta)$ of maps describes the evolution to equilibrium. This is a very difficult problem. There is first of all the exponentiability of the maps L_x^f to $\gamma_t = \exp t\, L_x^f$ and afterwards there is the problem of the limit t tending to infinity. We have no results on that. However, we have some results about a weaker statement. In particular, one can ask whether the following conjecture holds: let ω be any state invariant for all maps L_x^f i.e. $\omega.\, L_x^f = 0$ for all $f \in F_\beta$, $x = x^* \in A_o$, then the state ω must be a β - KMS-state or an equilibrium state at inverse temperature β . This conjecture is proved for the following systems:

1. finite systems [13]
2. ideal Bose gas (see section IV)
3. ideal Fermi gas [13]
4. for all systems under the following conditions on the state (see [21])

   i) ω is L^1-clustering

   ii) $\exists$ $x,y \in A_o$ such that

$$\int ds \quad \omega(x^* \quad [\alpha_s(y),x]) \neq 0$$

As far as these notes are concerned we limit ourselves to the proof of the conjecture for finite systems.

Let $A = M_n$ the set of n x n matrices, $\tau_t = \exp it \operatorname{ad} H$ with $H = H^* \in A$; and $\{\varepsilon_1,\ldots,\varepsilon_n\}$ is the spectrum of H.

Let $\{E_{k\ell} | k,\ell=1,\ldots,n\}$ the set of matrix units of A for a basis diagonalizing the Hamiltonian, then

$$\tau_t E_{k\ell} = E_{k\ell} \exp i(\epsilon_k - \epsilon_\ell)t$$

Consider the detailed balance generators L_x^f with $f \in F_\beta$, $x = x^* \in M_n$

$$L_x^f(y) = \int dt \; M_s \; f(t) \; \{\tau_s x \, [y, \tau_{s+t}(x)] + [\tau_s x, y] \; \tau_{s+t} x \}$$

$y \in A$

where

$$M_s \; y(s) = \lim_{R\to\infty} \frac{1}{2R} \int_{-R}^{R} ds$$

Proposition

If ω is a state of A, such that $\omega . L_x^f = 0$ for all $f \in F_\beta$, $x = x^* \in A$ then ω is the equilibrium state.

Proof:

Take first $x = E_{kk}$

$y = E_{k\ell}$ with $k \neq \ell$

$f \in F_\beta$ with $\hat{f}(0) \neq 0$

one computes

$$L_x^f(y) = - (2\pi)^{1/2} \hat{f}(0) E_{k\ell}$$

hence

$$0 = \omega(L_x^f(y))$$

yields $\omega(E_{k\ell}) = 0$

Take now

$x = E_{k\ell} + E_{\ell k}$

$y = E_{kk}$

$f \in F_\beta$ with $\hat{f}(\epsilon_k - \epsilon_\ell) \neq 0$

then compute again

$$L_x^f(y) = (2\pi)^{1/2} \, 2 \, \{- \hat{f}(\epsilon_k - \epsilon_\ell) E_{kk} + \hat{f}(\epsilon_\ell - \epsilon_k) E_{\ell\ell}\}$$

As $\hat{f}(p) = e^{\beta p} \hat{f}(-p)$

$0 = \omega(L_x^f(y))$

implies

$$\frac{\omega(E_{kk})}{\omega(E_{\ell\ell})} = e^{-\beta(\epsilon_k - \epsilon_\ell)}$$

for all $k,\ell = 1,\dots,n$

or $\omega(E_{kk})\, e^{\beta\varepsilon_k} = \omega(E_{\ell\ell})e^{\beta\varepsilon_\ell} = \lambda$

As

$$1 = \omega(1) = \sum_k \omega(E_{kk})$$

$$\lambda = \frac{1}{\mathrm{Tr}\, e^{-\beta H}}$$

and $\omega(E_{kk}) = e^{-\beta\varepsilon_k}/\mathrm{Tr}\, e^{-\beta H}$

therefore

$$\omega(x) = \frac{\mathrm{tr}\, e^{-\beta H} x}{\mathrm{Tr}\, e^{-\beta H}}$$

and ω is the equilibrium state.

V. RELAXATION OF THE IDEAL BOSE GAS

Here we study the function

$$\omega(\gamma_t^\beta(z)) - \omega_\beta\,(z) = \phi(\omega,t,\beta,z)$$

for large t, and its dependence on the values of β . We have not yet general results but we have studied the ideal Bose gas, which is particularly interesting because of the presence of a phase transition.

Consider the equilibrium states of the ideal Bose gas at inverse temperature $\beta = 1/kT$ and at density $\bar\rho$ [22] . We distinguish the following cases.

(a) $\omega_\beta = \omega_n$ the normal (n) state at $T > T_c$, the critical temperature. The creation and annihilation operators for this state are denoted by $a_n^\pm(g)$, $g \in S$(Schwartz functions).

(b) $\omega_\beta = \omega_c$ the condensate (c) state at $T < T_c$ and for $\bar\rho > \rho_c$, the critical density. Here we have the creation and annihilation operators $a_c^\pm\,(g)$, $g \in S$ for the excited bosons, and $\alpha_c^\pm$ the creation and annihilation operators of a condensate boson [23] .

The linear dissipative detailed balance maps in the W^* dynamical system set up are now

$$L^f_{n,c,g} = \int dt\, ds\, f(t)\, \{\tau_s^{n,c}(x_{n,c})\, [\cdot,\tau_{s+t}^{n,c}\,(x_{n,c})]$$

$$+ [\tau_s^{n,c}(x_{n,c}),\cdot\,]\ \tau_{s+t}^{n,c}(x_{n,c})\}$$

where

$\tau^{n,c}$ is the time evolution in $\omega_{n,c}$

$x_n = a^+(g) + a_n(g)$

$x_c = \alpha_c a_c^+(g) + \alpha_c^+ a_c(g)$

$f \in F_\beta$

The $L^f_{n,c;g}$ are exponentiable, yielding one parameter semigroups of unity preserving, completely positive maps of the CCR-algebra.

At $T > T_c$, if ω is a locally perturbed state of ω_n then for all $f \in F_\beta$, as one has a soluble model one computes that

$$\omega(e^{tL^f_{n;g}}(a_n^+(h)a_n(h)) - \omega_n(a_n^+(h)a_n(h)) \approx e^{-C(\beta)t} \text{ for large } t$$

with $C(\beta) > 0$, this means exponential decay for the two-point function.

Furthermore, as $C(\beta) \approx (T-T_c)^2$ if $T \to T_c$ one gets the effect of critical slowing down with a relaxation time of the order $(T-T_c)^{-2}$ around T_c.
This phenomenon was recently proved in [24] by considering the Bose gas in interaction with a heat bath. Their strategy was to derive a non linear evolution equation. The non linearity is a consequence of the mean field type coupling with the heat bath and is, as we showed irrelevant for the description of the phenomenon of critical slowing down.

For $T < T_c$ one checks that if ω is a locally perturbed state of ω_c then there exists an element $f \in F_\beta$ and a function $g \in S$ such that

$$\omega(e^{tL^f_{c;g}}(a_c^+(h)a_c(h)a_c(h)) - \omega_c(a_c^+(h)a_c(h) \approx \frac{1}{t}$$

this means, polynomial behaviour for the approach to equilibrium below the critical temperature.

REFERENCES

[1] R.L. Dobrushin; Theory Prob. Appl.13,197 (1968).
[2] O.E. Lanford, D. Ruelle; Comm. Math.Phys.13, 194 (1969).
[3] R. Haag, N.M. Hugenholtz, M. Winnink; Comm. Math.Phys.5,215 (1967).
[4] G. Gallavotti, E. Verboven; Il Nuovo Cimento 28,274 (1975).
[5] R. Haag, D. Kastler, E.B. Trych-Pohlmeyer; Commun.Math. Phys.38, 173 (1974).
[6] M. Aizenman, G. Gallavotti, S. Goldstein, J. Lebowitz; Commun. Math. Phys. 48,1 (1976).
[7] G. Sewell; Commun. Math. Phys. 55, 53 (1977).
[8] M. Fannes, A. Verbeure; J. Math. Phys. 19, 558 (1978).
[9] R.J. Glauber; J. Math. Phys. 4, 294 (1963).
[10] R.A. Holley, D.W. Stroock; Commun. Math. Phys. 55,37 (1977).
[11] R.M. Fowler; Statistical Mechanics, 2d edition; Cambridge University Press 1955.
[12] S.R. De Groot, P. Mazur; Non-equilibrium thermodynamics; North-Holland 1962.
[13] J. Quaegebeur, G. Stragier, A. Verbeure; Ann. Inst. H. Poincaré 41, 25 (1984).
[14] J. Quaegebeur, A. Verbeure; Relaxation of the Bose gas; preprint KUL-TF-84/9, to appear in Lett. Math. Phys.
[15] J. Quaegebeur, G. Stragier; Detailed Balance for Continuous Classical Systems; preprint-KUL-TF-83/23.
[16] A. Kossakowski, A. Frigerio, V. Gorini, M. Verri; Commun. Math.Phys. 57, 97 (1977).
[17] R. Alicki; Reports Math. Phys. 10, 249 (1976).
[18] G.S. Agarwal; Z. Physik 258, 409 (1973).
[19] W.A. Majewski; J. Math. Phys. 25, 614 (1984).
[20] E.B. Davies; Commun. Math.Phys. 39, 91 (1974).
[21] G. Stragier; Ph. D. Thesis KUL 1984.
[22] O. Bratteli, D.W. Robinson; Operator Algebras and Quantum Statistical Mechanics II; Springer-Verlag New-York, Heidelberg, Berlin, 1979.
[23] M. Fannes, J.V. Pulé, A. Verbeure; Helv.Phys. Acta 55,391 (1982).
[24] E. Buffet, Ph. de Smedt, J.V. Pulé; On the dynamics of the open Bose gas; Ann.of Physics, to appear.

SPONTANEOUS LIGHT EMISSION DESCRIBED BY A QUANTUM STOCHASTIC DIFFERENTIAL EQUATION

Wilhelm von Waldenfels
Universität Heidelberg
Institut für Angewandte Mathematik
Im Neuenheimer Feld 294
6900 Heidelberg 1
Federal Republic of Germany

Abstract

We study spontaneous emission from a two level atom in Wigner-Weisskopf approximation . Considering the radiation field as a heatbath of temperature 0, the Schrödinger equation of the atom may be considered as a quantum stochastic differential equation. The equation can be solved in different ways, in the first place by the Stratonovich method identical to the usual solution in the second place by the Ito integral introduced by Hudson and Parthasarathy [4] and in the third place by the multiplicative Ito integral used already in [8] in the positive temperature case. We restrict ourselves to the Stratonovich and multiplicative Ito solutions. The multiplication Ito solution enables a very intuitive description not only in the one-photon case but in the multi-photon case too. We obtain the same solutions as Maassen [6], who used his theory of kernels. Most results of this paper are already contained in a preprint [7].

§ 1. Introduction

We start in section 2 by recalling Wigner and Weisskopfs method [10] for calculating the natural shape of a spectral line. A two level atom is coupled to a continuous set of oscillators labelled by frequency forming a model for the radiation field. If the atom is at time 0 in the upper state, the amplitude a(t) of the upper state in an interaction representation is given by the equation

$$\frac{d}{dt}\, a(t) = -\int_0^t k(t-s)a(s)ds$$

In the Wigner-Weisskopf approximation $a(t)$ is approximated by $\exp(-\kappa t)$, hence $k(t)$ by a multiple of a δ-function. This implies the Lorentzian line profile (Cauchy distribution).

In section 3 we calculate the time evolution of the system and perform then the Wigner-Weisskopf approximation. It is useful to label photons not any more by their frequency but by a formal time parameter τ. In the interaction representation the Hamiltonian becomes

(1.1) $$H(t) = \sigma_+\int g(t-\tau)B_\tau d\tau+\sigma_-\int \bar{g}(t-\tau)B^*_\tau d\tau$$

Denote the two levels of the atom by $|+\rangle$ and $|-\rangle$. Then $\sigma_+=|+\rangle\langle-|$ and $\sigma_-=|-\rangle\langle+|$. The operators B_τ and B^*_τ are the annihilation and creation operators of a photon with formal time parameter τ, the commutator is $[B_\tau,B^*_{\tau'}]=\delta(\tau-\tau')$. The time evolution is given by

(1.2) $$\frac{d}{dt} U(t,s) = -i H(t)U(t,s), \; U(s,s) = I.$$

The Hilbert space of the atom is $\mathcal{H}_{at}=\mathbb{C}^2$ spanned by $|+\rangle$ and $|-\rangle$, the Hilbert space of the radiation field is $\mathcal{H}_{rad}=\Gamma(L^2(\mathbb{R}))$, the Fock space associated to $L^2(\mathbb{R})$. The Hilbert space of the total system is $\Gamma(L^2(\mathbb{R}))\otimes\mathbb{C}^2$. An essential property of the Hamiltonian is that it leaves invariant the subspaces $\mathcal{V}_0,\mathcal{V}_1,\ldots$ of $\Gamma(L^2(\mathbb{R}))\otimes\mathbb{C}^2$, where $\mathcal{V}_0=\Gamma_0\otimes|-\rangle$, and $\mathcal{V}_1=\Gamma_0\otimes|+\rangle\oplus|\Gamma_1\otimes|-\rangle,\ldots,\mathcal{V}_n=\Gamma_{n-1}\otimes|+\rangle\oplus\Gamma_n\otimes|-\rangle$. The physically most interesting case is the atom in the upper level and no photon or the atom in the lower level and one photon. We calculate the restriction of $U(t,s)$ to $\mathcal{V}_1$ and perform then the limit $g(t)\to\alpha\delta(t)$. Then

(1.3) $$\begin{aligned} U(t,s) &= L(t-s)\,|\emptyset,+\rangle\langle\emptyset,+| \\ &-i\alpha\int_s^t d\tau\; L(t-\tau)\,|\emptyset,+\rangle\langle\tau,-| - i\bar{\alpha}\int_s^t d\tau\; L(\tau-s)\,|\tau,-\rangle\langle\emptyset,+| \\ &-|\alpha|^2\int_s^t d\tau_1\int_{\tau_1}^t d\tau_2 L(\tau_2-\tau_1)\,|\tau_2,-\rangle\langle\tau_1,-| + \int_{-\infty}^{+\infty} d\tau\,|\tau,-\rangle\langle\tau,-| \end{aligned}$$

with

$$L(t) = \begin{cases} \exp -\kappa t & \text{for } t \geq 0 \\ \exp +\bar{\kappa} t & \text{for } t \leq 0. \end{cases}$$

The state $|\emptyset,+\rangle$ is the photon vacuum and the atom in the upper level, the vector $|\tau,-\rangle$ signifies one photon with parameter τ and the atom in the lower level, and $|\alpha|^2=2\,\mathrm{Re}\,\kappa$. For $\alpha=i$ and $\kappa=1/2$ this is exactly H. Maassen's result.

We interpretate now the situation in a quantum stochastic way: We consider the radiation field as a heathbath with temperature 0, the field B_t,B^*_t may be regarded as white quantum noise and $F(t)=\int g(t-\tau)B_\tau d\tau$ as coloured quantum noise. Equation (1.2) together with (1.1) gets then a quantum stochastic differential equation. As long as the function g is square integrable the equation may be integrated straight forward and no ambiguities arise. If however, we perform Wigner-Weisskopf approximation and replace $g(t)$ by $\alpha\delta(t)$, we run into problems familiar from stochastic differential equations. Write at first formally

$$H(t) = \alpha\sigma_+ B_t + \bar{\alpha}\sigma_- B_t^* \tag{1.4}$$

Then equation (1.2) does not make sense any more. For e.g.

$$\langle \emptyset,+| \iint_{s\leq\tau_1\leq\tau_2\leq t} H(\tau_2)H(\tau_1)d\tau_1 d\tau_2 |\emptyset,+\rangle = |\alpha|^2 \iint_{s\leq\tau_1\leq\tau_2\leq t} d\tau_1 d\tau_2 \delta(\tau_1-\tau_2).$$

The integral, however, is not defined, because the domain of integration is a triangle and the singularity of the δ-function is situated on one side of the triangle.

There are different ways out of the dilemma. The first one is to calculate the solution with a non degenerate $g(t)$ and replace at the end $g(t)$ by $\alpha\delta(t)$. As this method yields the Stratonovich solution [2] in the classical one-dimensional case, we call it Stratonovich solution in our case too. It is given by (1.3) and coincides with the usual solutions. Remark that into (1.3) enters not only the constant α, approximately given by $\int g(t)dt$, but $\kappa=(1/2)|\alpha|^2+i\omega_s$ approximately given by $\int_0^\infty dt\int d\tau g(t-\tau)\bar{g}(-\tau)$. The imagniary ω_s part of κ represents the line shift. As the other approaches start from the limiting equation (1.4) they do not contain the line shift. If one wants to include it, one has to add a drift term into the stochastic differential equation cf. [1] We shall not deal with this problem in this paper.

It is possible to simplify equation (1.4). The phase factor of α may be absorbed into the definition of $|+\rangle$. The modulus $|\alpha|$ can be made to 1 by a changement of time scale so (1.4) may be replaced by

$$H(t) = \sigma_+ B_t + \sigma_- B_t^* . \tag{1.5}$$

In the Stratonovich solution B_t was smeared out by $\int g(t-\tau)B_\tau d\tau$. Another way of smoothing B_t is averaging over short time intervals. This leads to the multiplicative Ito solution (section 4). Fix two times $t<s$ and a partition $z=\{t_o=s<t_1<\ldots<t_n=t\}$. Define $H(\Delta_k)=\int_{t_{k-1}}^{t_k} H(\tau)d\tau = \sigma_+ B(\Delta_k)+\sigma_- B(\Delta_k)^*$ with $\Delta_k=[t_{k-1},t_k]$ and $B(\Delta_k)=\int_{\Delta_k} B(\tau)d\tau$. The operators $(B(\Delta_k),B(\Delta_k)^*)$ are independent pairs of creation and annihilation operators. If B_t were complex white noise the quantities $(B(\Delta_k),B(\Delta_k)^*)$ were independent pairs of gaussian variables and the $\exp(-iH(\Delta_k))$ were stochastic variables taking values in the unitary group $U(2)$. Hence $\exp(-iH(\Delta_k))\ldots\exp(-iH(\Delta_1))$ were the candidate for the solution of (1.2) with $H(t)$ defined by (1.5). In the quantum case we proceed in an analogue way. Define $U_z(t,s)=\exp(-iH(\Delta_k)\ldots\exp(-iH(\Delta_1))$. In §4 we consider at

first the one photon case described by the subspace ϑ_1 of $\Gamma(L^2(\mathbb{R}))\otimes\mathbb{C}^2$ It can be easily shown that $U_z(t,s)$ converges in the strong operator topology to $U(t,s)$ given by (1.3) with $\alpha=1$ and $\kappa=1/2$. We discuss then the multi-photon case using graphs and obtain the same results as Maassen [6]. The second part of §4 is heuristic but to my opinion it can be made rigorous without big effort.

The Ito-integral introduced by Hudson and Pathasarathy [4] is still nearer to the classical Ito integral. Introduce $A_t=\int_0^t B_\tau d\tau$ and $G(t)=-i\sigma_+A_t \ -i\sigma_-A_t^*-\frac{1}{2}\pi_+t$ with $\pi_+=|+\rangle\langle+|$. Let $z=\{t_o=s<\ldots<t_n=t\}$ be a partition of the interval $s<t$. Call $\Delta G_k=G(t_k)-G(t_{k-1})$. The assertion that $U(t,s)$ is the solution of the Ito equation

$$(1.6) \qquad d_tU(t,s) = dG(t)U(t,s)$$

with $U(s,s)=I$ is equivalent to the assertion, that $I+\sum_{k=1}^{n}\Delta G_kU(t_{k-1},s)$ converges to zero if the mesh of z goes to 0. In fact using $U(t,s)$ as given by (1.3) with $\alpha=1$ and $\kappa=1/2$ this can be easily shown, even that convergence takes place in operator norm. The drift term $-1/2\,\pi_+dt$ is necessary to insure that the solution is unitary. As the result is straight forward and to a big extend contained in Hudson's and Pathasarathy's paper [4] we shall not deal with it.

§ 2. The Wigner-Weisskopf theory of natural line shape

We establish at first equation (1.1) of the introduction in frequency representation. We consider the photon states $|\emptyset\rangle, |\omega\rangle, |\omega_1,\omega_2\rangle,\ldots$ where $|\emptyset\rangle$ denotes the photon vacuum and $|\omega\rangle$ denotes the presence of one photon of frequency $\omega+\omega_o$, where ω_o is the energy difference between two atomic levels. These states form a (generalized) basis of the Fock space $\Gamma(L^2(\mathbb{R}))=\mathfrak{H}_{rad}$. The Hilbert space for the atom is $\mathfrak{H}_{at}=\mathbb{C}^2$. It is spanned by the states $|+\rangle$ and $|-\rangle$. Call

$$(2.1) \qquad \sigma_+ = |+\rangle\langle-|;\quad \sigma_-=|-\rangle\langle+|;\quad \pi_+=|+\rangle\langle+|$$

The Hamiltonian of the atom is

$$(2.2) \qquad H_{at} = \omega_o\pi_+$$

the Hamiltonian of the radiation field is

$$(2.3) \qquad H_{rad} = \int(\omega_o+\omega)B_\omega^*B_\omega d\omega,$$

where B_ω, B_ω^* are the usual annihilation and creation operators with the commutation relation $[B_\omega, B_{\omega'}^*] = \delta(\omega-\omega')$. The Hamiltonian of the interaction is in rotating wave approximation (cf. [3]):

(2.4) $$H_{int} = \int g(\omega) B_\omega \sigma_+ d\omega + \int \bar{g}(\omega) B_\omega^* \sigma_- d\omega \ ,$$

where the function $g(\omega)$ is supposed to be square integrable. It is a consequence of the rotating wave approximation that the subspaces ϑ_o, $\vartheta_1, \ldots$ of the total Hilbert space $\mathcal{H} = \mathcal{H}_{rad} \otimes \mathcal{H}_{at} = \Gamma(L^2(\mathbb{R})) \otimes \mathbb{C}^2$ stay invariant under $H_{tot} = H_{at} + H_{rad} + H_{int}$. Here ϑ_o is the space spanned by $|\emptyset,->$ and ϑ_k is $\Gamma_{k-1} \otimes |+> \oplus \Gamma_k \otimes |->$, i.e. (k-1) photons are present and the atom is in the upper state of k photons are present and the atom is in the lower state. The total Hamiltonian H_{tot} splits into $H_o + H_1$, with

$$H_o = \omega_o (\pi_+ + \int d\omega B_\omega^* B_\omega) \, ,$$

(this is the number of photons plus the number of atoms in the upper state multiplied by ω_o) and

$$H_1 = \int \omega B_\omega^* B_\omega d\omega + H_{int} .$$

As H_o and H_1 commute we disregard the time evolution due to H_o and consider only the non-trivial time evolution generated by H_1. Call $H_{10} = \int \omega B_\omega^* B_\omega d\omega$ and use the interaction representation with respect to H_{10}. In this representation the time evolution is given by

(2.5) $$\frac{d}{dt} U(t,s) = -iH(t)U(t,s), \quad U(s,s) = 1$$

with

(2.6) $$H(t) = (\int g(\omega) e^{-i\omega t} B_\omega) \sigma_+ + (\int \bar{g}(\omega) e^{+i\omega t} d\omega) \sigma_-$$

This is equation (1.1) in frequency representation.

Assume that at time 0 the system is in state $|\emptyset,+>$. Then it will stay all time in ϑ_1. The state vector at time t is

$$|\psi(t)> = a(t)|\emptyset,+> + \int d\omega c_\omega(t) |\omega,->.$$

Schrödinger's equation

$$\frac{d}{dt} |\psi(t)> = -i \, H(t) |\psi(t)>$$

yields

(2.7) $$\frac{d}{dt}\, a(t) = -i\int g(\omega)e^{-i\omega t}c_{\omega}(t)d\omega$$

(2.8) $$\frac{d}{dt}\, c_{\omega}(t) = -i\ a(t)\bar{g}(\omega)e^{+i\omega t}$$

Integrating (2.8) and inserting into (2.7) one obtains

(2.9) $$\frac{d}{dt}\, a(t) = -\int_0^t k(t-s)a(s)ds$$

with

(2.10) $$k(t) = \int |g(\omega)|^2 e^{-i\omega t}dt.$$

The essential assumption of Wigner and Weisskopf is the Ansatz $a(t)=\exp(-\kappa t)$, or $\dot{a}(t)=-\kappa\ a(t)$. This implies that $k(t)$ is a multiple of a δ-function, and that $|g(\omega)|^2$ is a constant. Introduce a parameter $\varepsilon>0$, replace $g(\omega)$ by $g_{\varepsilon}(\omega)=g(\varepsilon\omega)$ and consequently $k(t)$ by $k_{\varepsilon}(t)=(1/\varepsilon)k(\frac{t}{\varepsilon})$ and go $\varepsilon\downarrow 0$. As it will be shown in the next section the amplitude $a(t)$ converges to

(2.11) $$a(t) = \exp(-\kappa t)$$

with

(2.12) $$\kappa = \int_0^{\infty} k(t)dt = \frac{\gamma}{2} + i\omega_s$$

(2.13) $$\gamma = \int_{-\infty}^{+\infty} k(t)dt = 2\,\mathrm{Re}\,\kappa = 2\pi|g(0)|^2$$

(2.14) $$\omega_s = \mathrm{Im}\,\kappa = -\int |g(\omega)|^2\, \frac{P}{\omega}\, d\omega$$

where P/ω denotes the Schwartz distribution called "principle value $1/\omega$" We shall not discuss the regularity assumptions which enable the convergence $\varepsilon\downarrow 0$.

For $\varepsilon\downarrow 0$ equation (2.8) becomes

$$\frac{d}{dt}\, c_{\omega}(t) = -i\bar{g}(0)e^{(+i\omega-\kappa)t}\ .$$

hence

$$c_{\omega}(\infty) = \frac{i\bar{g}(0)}{\kappa-i\omega}$$

The line shape is

$$(2.17) \qquad |c_\omega(\infty)|^2 = \frac{\gamma}{2\pi}\frac{1}{(\omega-\omega_s)^2+(\gamma/2)^2}$$

using (2.13). This is the well-known Lorentz-profile with half-width $\gamma/2$ and line shift ω_s. The formula is due to Wigner and Weisskopf. Remark that ω is not the frequency but the frequency difference to ω_o. Remark too that negative frequencies $\omega+\omega_o$ come in showing that the transition $\varepsilon\downarrow 0$ leads to physical impossible results and can therefore be correct only approximatively.

§ 3. Calculation of the time evolution

We want to calculate the restriction of $U(t,s)$ given on the space $\vartheta_1=\Gamma_o\otimes|+\rangle\oplus\Gamma_1\otimes|-\rangle$ by (2.5) resp. (1.2) with $H(t)$ given by (2.6) and go then to the limit in the same way as we did for the calculation of the natural line shape.

We want to get rid of Fourier transform and switch from the frequency representation to a formal time representation, which is more suitable as well for mathematical treatment as for physical interpretation. Introduce

$$|\tau\rangle = \frac{1}{\sqrt{2\pi}}\int e^{-i\omega\tau}|\omega\rangle d\omega$$

$$B_\tau = \frac{1}{\sqrt{2\pi}}\int e^{-i\omega\tau}B_\omega d\omega$$

$$g(t) = \frac{1}{\sqrt{2\pi}}\int g(\omega)e^{-i\omega t}d\omega.$$

Then (2.6) is replaced by

$$(3.1) \qquad H(t) = \sigma_+\int g(t-\tau)B_\tau d\tau+\sigma_-\int\bar{g}(t-\tau)B_\tau^* d\tau.$$

This is eq. (1.1).

An arbitrary vector of ϑ_1 has the form

$$|\psi\rangle = a|\emptyset,+\rangle+\int c_\tau|\tau,-\rangle d\tau.$$

As

$$H(t)|\psi\rangle = a\int\bar{g}(t-\tau)|\tau,-\rangle d\tau + \int g(t-\tau)c_\tau d\tau|\emptyset,+\rangle$$

one has

$$\|H(t)|\psi\rangle\|^2 \leq (|a|^2+\int|c_\tau|^2 d\tau)\int|g(t)|^2 dt$$

Hence the operator norm of H(t) on ϑ_1 satisfies the inequality

(3.2) $$\|H(t)\| \leq \|g\|_{L^2} .$$

A similar reasoning yields

(3.3) $$\|H(t)-H(t')\|^2 \leq \int|g(t-\tau)-g(t'-\tau)|^2 d\tau.$$

So $t \to H(t)$ is continuous in operator norm and the equation

(3.4) $$d/dt\ U(t,s) = -i\ H(t)U(t,s), \quad U(s,s)=1$$

can be solved on ϑ_1 without difficulty yielding a unitary evolution continuous in operator norm.

We want to calculate U(t,s) more explicitely. Choose

$$|\psi(s)\rangle = a(s)|\emptyset,+\rangle+\int c_\tau(s)|\tau,-\rangle d\tau$$

and define

$$|\psi(t)\rangle = U(t,s)|\psi(s)\rangle = a(t)|\emptyset,+\rangle+\int c_\tau(t)|\tau,-\rangle d\tau.$$

Then

(3.5) $$\dot{a}(t) = -i\int g(t-\tau)c_\tau(t)d\tau$$

(3.6) $$\frac{d}{dt}c_\tau(t) = -i\bar{g}(t-\tau)a(t).$$

Integrate (3.6)

(3.7) $$c_\tau(t) = c_\tau(s)-i\int_s^t \bar{g}(u-\tau)a(u)du$$

and insert into (3.5). Then

(3.8) $$\dot{a}(t) = -i\int g(t-\tau)c_\tau(s)d\tau-\int_s^t k(t-u)a(u)du$$

with

(3.9) $$k(t) = \int g(t-\tau)\bar{g}(-\tau)d\tau = \int|g(\omega)|^2 e^{-i\omega t}dt,$$

cf. (2.10). Denote by K(t) the solution of the linear integro-differential equation

$$(3.10) \qquad \frac{d}{dt} K(t) = -\int_0^t k(t-\tau)K(\tau)d\tau, \; K(0) = 1.$$

Then (3.8) has the solution

$$(3.11) \qquad a(t) = a(s)K(t-s)-i\int_0^t du\; K(t-u)\int d\tau\; g(u-\tau)c_\tau(s)$$

Equation (3.7) and (3.11) reduce the problem of solving (3.4) to simple quadratures, if K(t) is known.

We want now to calculate U(t,s) in the Wigner-Weisskopf limit. We formulate the limiting procedure a bit more general than in §2. Consider a sequence $g_\nu \in L^2 \wedge L^1(\mathbb{R})$ such that

$$(3.12) \qquad \int g_\nu(t)f(t)dt \to \alpha f(0)$$

or any $f \in C_b(\mathbb{R})$, the space of all continuous bounded functions on $\mathbb{R}$. Define $k_\nu(t)$ as in (3.9) and assume that

$$(3.13) \qquad \int_0^\infty k_\nu(t)f(t)dt \to \kappa f(0)$$

$$(3.14) \qquad \int_0^\infty |k_\nu(t)|f(t)dt \to \tilde{\kappa} f(0)$$

for any $f \in C_b(\mathbb{R})$.

<u>Lemma 3.1.</u> Call K_ν the solution of (3.12) with k(t) replaced by $k_\nu(t)$. The functions $K_\nu(t)$ are bounded by 1 and converge to

$$(3.15) \qquad L(t) = \begin{cases} \exp(-\kappa t) & t \geq 0 \\ \exp \bar{\kappa} t & t \leq 0 \end{cases}$$

uniformly on any compact subset of $\mathbb{R}$.

<u>Proof.</u> Equation (3.11) yields that $K_\nu(t-s) = \langle \emptyset,+|U_\nu(t,s)|\emptyset,+\rangle$. Hence $|K_\nu'| \leq 1$. Consider at first $t \geq 0$. Equation (3.10) can be written

$$\theta(t)K_\nu(t) = \theta(t)-\int_0^t K_\nu(t-\tau)\ell_\nu(\tau)d\tau = \theta(t)-((K_\nu\theta)*\ell_\nu)(\tau)$$

where $\theta(t)=\begin{cases} 1 & t\geq 0 \\ 0 & t<0 \end{cases}$ and $\ell_\nu(t)=\theta(t)\int_0^t k_\nu(\tau)d\tau$ and the $*$ denotes convolution. So

$$\theta K_\nu = \theta-\theta*\ell_\nu+\theta*\ell_\nu*\ell_\nu-\theta*\ell_\nu^{*3}+\dots$$

Call $\kappa_\nu=\int_0^\infty k_\nu(\tau)d\tau$. This integral exists because $g_\nu \in L^1$. Then

$$\theta(t)\exp(-\kappa_\nu t) = \theta(t)-\kappa_\nu\theta^{*2}(t)+\kappa_\nu^2\theta^{*3}(t)-+\ldots$$

Subtraction yields

$$(\theta K_\nu)(t)-\theta(t)\exp(-\kappa_\nu t) =$$
$$(\ell_\nu-\kappa_\nu\theta)*(-\theta+\theta*(\ell_\nu+\kappa_\nu\theta)-\theta*(\ell_\nu^{*2}+\kappa_\nu\theta*\ell_\nu + \kappa_\nu^2\theta^{*2})+\ldots)$$

By (3.14) there exists a constant $0<C<\infty$ such that $|\ell_\nu(t)|\leq C\theta(t)$ and $|\kappa_\nu|\leq C$ for all t and ν. Hence

$$|(\theta K_\nu)(t)-\theta(t)\exp(-\kappa_\nu t)|\leq|\ell_\nu-\kappa_\nu\theta|*(\theta+2C\theta^{*2}+3C^2\theta^{*3}+\ldots)$$
$$\leq|\ell_\nu-\kappa_\nu\theta|*e^{2Ct},$$

as $\theta^{*n}=t^{n-1}/(n-1)!$. The last expression is

$$\leq\int_0^t d\tau e^{2C(t-\tau)}\int_\tau^\infty d\sigma|k_\nu(\sigma)|$$

$$\leq e^{2Ct}\int_0^\infty d\tau\ e^{-2C\tau}\int_\tau^\infty d\sigma|k_\nu(\sigma)|$$

$$=e^{2Ct}\int_0^\infty d\sigma|k_\nu(\sigma)|\int_0^\sigma d\tau\ e^{-2C\tau}$$

$$=e^{2Ct}\int_0^\infty d\sigma|k_\nu(\sigma)|(1-e^{-2C\sigma})/2C$$

This converges to zero by (3.14) uniformly on every compact subset of $\mathbb{R}_+$. As $k_\nu(-t)=\overline{k_\nu(t)}$ for $\mathbb{R}_-$ everything goes the same way.

<u>Theorem 3.1.</u> For $\nu\to\infty$ the unitary evolutions $U_\nu(t,s)$ restricted to ϑ_1 converge to the unitary evolution $U(t,s)$ in the operator norm on ϑ_1 uniformly on every finite time interval in t and s with

$$U(t,s) = L(t-s)|\emptyset,+\rangle\langle\emptyset,+| \tag{3.16}$$
$$-i\alpha\int_s^t d\tau L(t-\tau)|\emptyset,+\rangle\langle\tau,-|-i\bar{\alpha}\int_s^t d\tau\ L(\tau-s)|\tau,-\rangle\langle\emptyset,+|$$
$$+\int d\tau|\tau,-\rangle\langle\tau,-| - \gamma\int_s^t d\tau_2\int_s^{\tau_2} d\tau_1 L(\tau_2-\tau_1)|\tau_2,-\rangle\langle\tau_1,-|$$

where $\gamma=|\alpha|^2=2\,\mathrm{Re}\,\kappa$ and $L(t)$ is given by (3.15).

<u>Proof.</u> We start from (3.11) and insert the index ν

(3.17) $$a_\nu(t) = a(s)K_\nu(t-s)-i\int_0^t du K_\nu(t-u)\int d\tau g_\nu(u-\tau)c_\tau(s)$$

and claim that for $\nu\to\infty$ the quantity $a_\nu(t)$ converges to

(3.18) $$a(t) = a(s)L(t-s)-i\alpha\int_s^t d\tau L(t-\tau)c_\tau(s)$$

uniformly on compact sets in t and s and on the unit ball in $L^2(\mathbb{R})$ with respect to the function $\tau\to c_\tau(s)$. The first term poses no problems. For the second term we have to consider the difference

$$\int_s^t du K_\nu(t-u)g_\nu(u-\tau)-\alpha \mathbb{1}_s^t(\tau)L(t-\tau),$$

where $\mathbb{1}_s^t(\tau) = \begin{cases} 1 & s<\tau<t \\ -1 & t<\tau<s \\ 0 & \text{else} \end{cases}$. Call

$h_\nu:h_\nu(u)=\mathbb{1}_s^t(u)K_\nu(t-u)$ and $h:h(u)=\mathbb{1}_s^t(u)L(t-u)$ then the difference may be written

$$h_\nu*g_\nu-\alpha h = (h_\nu-h)*g_\nu+(h*g_\nu-\alpha h)$$

The L^2-norm of the first term may be estimated by

$$\sup\{|K_\nu(t-u)-L(t-u)|:u\in[s,t]\}\,\|\,|\mathbb{1}_s^t|*|g_\nu|\,\|_{L^2}\to 0$$

as $\|g_\nu\|_{L^1}$ stays bounded by (3.12) and the uniform boundedness theorem and as $\|\,|\mathbb{1}_s^t|*|g_\nu|\,\|_{L^2}\le\|\mathbb{1}_s^t\|_{L^2}\,\|g_\nu\|_{L^1}$. The L^2-norm of the second term goes to zero because of (3.12) So finally the difference converges to 0 in L^2-norm. So $a_\nu(t)$ converges to $a(t)$ uniformly on the unitball of $L^2(\mathbb{R})$. It is easy to see that the convergence is uniform on bounded t and s sets as well.

Go to (3.7) and insert the index ν.

(3.19) $$c_{\nu,\tau}(t) = c_\tau(s)-i\int_s^t \bar{g}_\nu(u-\tau)a(u)du$$

By a same kind of reasoning it is clear that this quantity converges to

(3.20) $$c_\tau(t) = c_\tau(s)-i\bar{\alpha}a(\tau)\mathbb{1}_s^t(\tau)$$

uniformly on the same sets. Equations (3.18) and (3.20) are equivalent to (3.16).

As the $U_\nu(t,s)$ are unitary evolutions, the limit $U(t,s)$ in operator norm is a unitary evolution too.

Theorem 3.2. The function $t \to U(t,s)$ with $U(t,s)$ given by theorem 3.1 is Hölder continuous of order 1/2 in the operator norm.

Proof. As

$$U(t+\Delta t,s)-U(t,s) = U(t,t+\Delta t)-I)U(t,s)$$

it is enough to consider the first factor on the right side. Now for small Δt

$$\begin{aligned} U(t,t+\Delta t) &= |\emptyset,+\rangle\langle\emptyset,+| \\ &-i\alpha\int_t^{t+\Delta t} d\tau\,|\emptyset,+\rangle\langle\tau,-| - i\bar{\alpha}\int_t^{t+\Delta t}|\tau,-\rangle\langle\emptyset,+| \\ &+\int d\tau\,|\tau,-\rangle\langle\tau,-| + O(\Delta t) \\ &= I - i\alpha|\emptyset,+\rangle\langle\Delta,-| - i\bar{\alpha}|\Delta,-\rangle\langle\emptyset,+| + O(\Delta t) \end{aligned}$$

with $|\Delta\rangle = \int_t^{t+\Delta t}|\tau\rangle d\tau$. As the norm of $|\Delta\rangle$ equals $\sqrt{\Delta t}$, the norm of the operators $|\emptyset,+\rangle\langle\Delta,+|$ and $|\Delta,-\rangle\langle\emptyset,+|$ equal $\sqrt{\Delta t}$ too.

§ 4. The multiplicative Ito solution

We start by some auxiliary considerations. Let b,b^* be a pair of boson cration and annihilation operators acting on the states $|0\rangle,|1\rangle,|2\rangle$, by $b^*|n\rangle = \sqrt{n+1}|n+1\rangle$ and $b|n\rangle=\sqrt{n}|n-1\rangle$. Let $\mathfrak{h}$ be the Hilbert space spanned by $|0\rangle,|1\rangle,|2\rangle,\ldots$ Let $\mathbb{C}^2$ be the Hilbert space of the atom spanned by $|+\rangle$ and $|-\rangle$. Call $h=b\sigma_+ + b^*\sigma_-$. The subspaces $d_0, d_1,\ldots$ of $\mathfrak{h}\otimes\mathbb{C}^2$ are invariant under h, where d_0 is spanned by $|0,-\rangle$, and d_1 is spanned by $|0,+\rangle$ and $|1,-\rangle$ and generally d_k is spanned by $|k-1,+\rangle$ and $|k,-\rangle$. One has $h|0,-\rangle=0$ and

$$h\begin{pmatrix}|k-1,+\rangle \\ |k,-\rangle\end{pmatrix} = \sqrt{k}\begin{pmatrix}0 & 1 \\ 1 & 0\end{pmatrix}\begin{matrix}|k-1,+\rangle \\ |k,-\rangle\end{matrix} \tag{4.1}$$

Hence

$$e^{-ith}\begin{pmatrix}|k-1,+\rangle \\ |k,-\rangle\end{pmatrix} = \begin{pmatrix}\cos\sqrt{k}t & -i\sin\sqrt{k}t \\ -i\sin\sqrt{k}t & \cos\sqrt{k}t\end{pmatrix}\begin{pmatrix}|k-1,+\rangle \\ |k,-\rangle\end{pmatrix}. \tag{4.2}$$

for $k\geq 1$ and $e^{-ith}|0,-\rangle=|0,-\rangle$.

Let $\Delta=[t',t''[$ be an interval of $\mathbb{R}$, call $\delta=t''-t'$ and consider the Fock-space $\Gamma(L^2(\mathbb{R}))$. Denote by $|k\Delta\rangle$ the element $B(\Delta)^{*k}/\sqrt{k!}|\emptyset\rangle$ with $B(\Delta)=\int_{t'}^{t''}B_\tau d\tau$. The calculation above can be applied with $b=\delta^{-1/2}B(\Delta)$ and $|k\rangle=\delta^{-k/2}|k\Delta\rangle$. Then $H(\Delta)=\sigma_+B(\Delta)+\sigma_-B(\Delta)^*$ gets $\sqrt{\delta}h$. So

$$e^{-iH(\Delta)}|\emptyset,-\rangle = |\emptyset,-\rangle \tag{4.3}$$

(4.4) $$e^{-iH(\Delta)}|\emptyset,+\rangle = \cos\sqrt{\delta}|\emptyset,+\rangle - \frac{i}{\sqrt{\delta}}\sin\sqrt{\delta}|\Delta,-\rangle$$

(4.5) $$e^{-iH(\Delta)}|\Delta,-\rangle = -i(\sin\sqrt{\delta})\sqrt{\delta}|\emptyset,+\rangle+\cos\sqrt{\delta}|\Delta,-\rangle$$

(4.6) $$e^{-iH(\Delta)}|\Delta,+\rangle = \cos\sqrt{2\delta}|\Delta,+\rangle - \frac{i}{\sqrt{\delta}}\sin\sqrt{2\delta}|2\delta,-\rangle$$

Let $s<t$ and $z=\{s=t_o<t_1<\ldots<t_n=t\}$. Define $\Delta_k=[t_{k-1},t_k[$ and $\delta_k=t_k-t_{n-1}$ and put

(4.7) $$U_z(t,s) = e^{-iH(\Delta_n)} \ldots e^{-iH(\Delta_1)}$$

Now

(4.8) $$U_z(t,s)|\emptyset,-\rangle = |\emptyset,-\rangle$$

and

$$U_z(t,s)|\emptyset,+\rangle = \cos\sqrt{\delta_1}\ldots\cos\sqrt{\delta_n}|\emptyset,+\rangle$$
$$-i\sum_{k=1}^{n}\cos\sqrt{\delta_1}\ldots\cos\sqrt{\delta_{k-1}}\frac{\sin\sqrt{\delta_k}}{\sqrt{\delta_k}}|\Delta_k,-\rangle$$

as $$e^{-iH(\Delta_\ell)}|\Delta_k,-\rangle = |\Delta_k,-\rangle \quad \text{for } \ell\neq k.$$

This follows from above, as $|\Delta_k,-\rangle$ can be written $|1\Delta_k\rangle\otimes|0\Delta_\ell\rangle\otimes|-\rangle$.
By the same way

$$U_z(t,s)|\Delta_k,-\rangle = \cos\sqrt{\delta_k}|\Delta_k,-\rangle$$

$$-i\sqrt{\delta_k}\sin\sqrt{\delta_k}\, e^{-iH(\Delta_n)}\ldots e^{-iH(\Delta_{k+1})}|\emptyset,+\rangle$$

$$=\cos\sqrt{\delta_k}|\Delta_k,-\rangle-i\sqrt{\delta_k}\sin\sqrt{\delta_k}\cos\sqrt{\delta_{k+1}}\ldots\cos\sqrt{\delta_n}|\emptyset,+\rangle$$

$$-\sum_{\ell=k+1}^{n}\sqrt{\delta_k}\sin\sqrt{\delta_k}\cos\sqrt{\delta_{k+1}}\ldots\cos\sqrt{\delta_{\ell-1}}\sin\sqrt{\delta_\ell}/\sqrt{\delta_\ell}|\Delta_\ell,-\rangle.$$

Let f be a stepfunction adapted to z, $f=\sum c_k 1\!\!1_{\Delta_k}$. Then $|f,-\rangle=\sum c_k|\Delta_k,-\rangle$ and

$$U_z(t,s)|f,-\rangle = \sum_{k=1}^{n}\cos\sqrt{\delta_k}c_k|\Delta_k,-\rangle$$
$$-i\sum_{k=1}^{n}c_k\sqrt{\delta_k}\sin\sqrt{\delta_k}\cos\sqrt{\delta_{k+1}}\ldots\cos\sqrt{\delta_n}|\emptyset,+\rangle$$

$$-\sum_{1\le k<\ell\le n}c_k\sqrt{\delta_k}\sin\sqrt{\delta_k}\cos\sqrt{\delta_{k+1}}\ldots\cos\sqrt{\delta_{\ell-1}}\sin\sqrt{\delta_\ell}/\sqrt{\delta_\ell}|\Delta_\ell,-\rangle.$$

It is easy to guess the limit of this expression, if the mesh of z goes to 0.

$$U(t,s)|\emptyset,+\rangle = e^{-(t-s)/2}|\emptyset,+\rangle - i\int_s^t e^{-(\tau-s)/2}|\tau,-\rangle d\tau$$

$$U(t,s)|f,-\rangle = |f,-\rangle - i\int_s^t e^{-(t-\tau)/2} f(\tau)d\tau|\emptyset,+\rangle$$

$$- \iint_{s\leq\tau_1\leq\tau_2\leq t} e^{-1/2(\tau_2-\tau_1)} f(\tau_1)|\tau_2,-\rangle d\tau_1 d\tau_2.$$

In fact, it can be easily shown that for a stepfunction f adapted to z the following inequalities hold: there exists a function g(t) such that

$$\|(U_z(t,s)-U(t,s))|\emptyset,+\rangle\| \leq g(t-s)\|z\|$$

$$\|(U_z(t,s)-U(t,s))|f,-\rangle\| \leq g(t-s)\|z\| \|f\|$$

with $\|z\| = \max_k (t_k - t_{k-1})$. Let f be arbitrary L^2-function on $[s,t]$. Define $f_z = \sum \mathbb{1}_{\Delta_k} \frac{1}{\delta_k}\int_{\Delta_k} f(\tau)d\tau$. Then $\|f_z - f\|_{L^2} \to 0$ for $\|z\| \downarrow 0$. From there one concludes

<u>Proposition 4.1.</u> The restriction of $U_z(t,s)$ on the subspace ϑ_1 of $\Gamma(L^2(\mathbb{R}))\otimes\mathbb{C}^2$ spanned by $|\emptyset,+\rangle$ and $|\tau,-\rangle$, $\tau\in\mathbb{R}$ converges to $U(t,s)$ in the strong operator topology, where $U(t,s)$ is given by

$$(4.9)\qquad U(t,s) = e^{-(t-s)/2}|\emptyset,+\rangle\langle\emptyset,+| - i\int_s^t e^{-(t-\tau)/2}|\emptyset,+\rangle\langle\tau,-|d\tau$$
$$-i\int_s^t e^{-(\tau-s)/2}|\tau,-\rangle\langle\emptyset,+|d\tau$$
$$- \iint_{s\leq\tau_1\leq\tau_2\leq t} e^{-(\tau_2-\tau_1)/2}|\tau_2,-\rangle\langle\tau_1,-|d\tau_1 d\tau_2$$
$$+\int_{-\infty}^{+\infty} d\tau|\tau,-\rangle\langle\tau,-|.$$

This is the result indicated in the introduction.

We want now to discuss states with possibly more than one photon. Choose again a partition z of $[s,t]$ as above. We have to calculate expressions of the form

$$\langle q_1\Delta_1,\ldots,q_n\Delta_n,i|U_z(t,s)|p_1\Delta_1,\ldots,p_n\Delta_n,j\rangle$$

with $i,j=\pm$ and where

$$|p_1\Delta_1,\ldots,p_n\Delta_n\rangle = \frac{1}{\sqrt{p_1!}}B(\Delta_\Gamma)^{*p_1}\ldots\frac{1}{\sqrt{p_n!}}B(\Delta_n)^{*p_n}|\emptyset\rangle.$$

The matrix element is equal to

$$<\emptyset,i|\overleftarrow{\prod_k} \frac{1}{\sqrt{q_k}!} B(\Delta_k)^{p_k} \exp(-iH(\Delta_k)) \frac{1}{\sqrt{p_k}!} B(\Delta_k)^{*p_k}|\emptyset,j>$$

$$= (\overleftarrow{\Pi} V_{q_k,p_k}(\Delta_k))_{i,j}$$

where the 2×2-matrices V are given by

$$(V_{q_k,p_k}(\Delta_k))_{i,j} = <\emptyset,i|\frac{1}{\sqrt{q_k}!} B(\Delta_k)^{q_k} \exp(-iH(\Delta_k)) \frac{1}{\sqrt{p_k}!} B(\Delta_k)^{*p_k}|\emptyset,j>.$$

So finally

$$(4.10) \qquad <q_1\Delta_1,\ldots,q_n\Delta_n,i|U_z(t,s)|p_1\Delta_1,\ldots,p_n\Delta_n,j> = (\overleftarrow{\Pi} V_{q_k,p_k}(\Delta_k))_{i,j}$$

In order to calculate the total matrix elements we have only to calculate the matrices V, which are depending only on the interval Δ_k and on the numbers p_k and q_k of incoming and outgoing photons with time parameters in Δ_k.

Our aim is to calculate a matrix element of U(t,s) of the form

$$(4.11) \qquad <\sigma_1,\ldots,\sigma_v,i|U(t,s)|\tau_1,\ldots,\tau_u,j>$$

where $|\sigma_1,\ldots,\sigma_v>$ is a state with v photons, $\sigma_1<\ldots<\sigma_v$ and $|\tau_1,\ldots,\tau_u>$ a state with u photons, $\tau_1<\ldots<\tau_u$. As these states are not ordinary Hilbert space vectors we will encounter difficulties. On the other side these states are very intuitive.

We want to illustrate the matrix element (4.11) by a graph

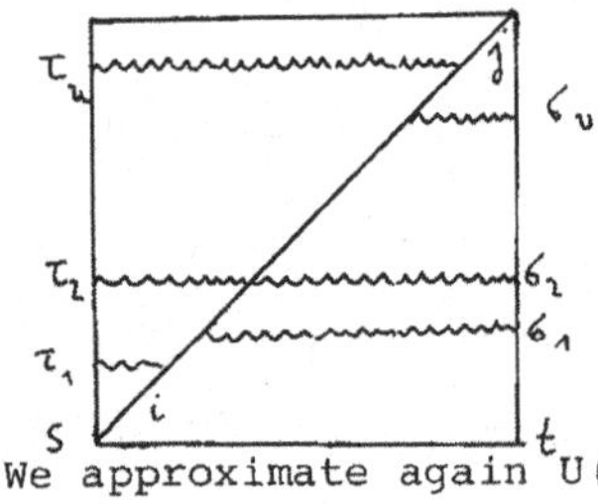

The horizontal line is the time axis. On the vertical line the formal time parameters of the photons are indicated. The diagonal signifies the atom. Photons with parameter τ interact only with the atom at the time τ.

We approximate again U(t,s) by $U_z(t,s)$ and choose z fine enough that the τ_i and σ_i are separated by many empty subintervals. We approximate $|\tau_1,\ldots,\tau_u>$ by $(\delta_{k_1}\ldots\delta_{k_u})^{-1}|\Delta_{k_1},\ldots,\Delta_{k_u}> = (\delta_{k_1}\ldots\delta_{k_u})^{-1}B^*(\Delta_{k_1})\ldots$ $\ldots B^*(\Delta_{k_u})|\emptyset> = (\delta_1^{p_1}\ldots\delta_n^{p_n})^{-1}|p_1\cdot\Delta_1,\ldots p_n\cdot\Delta_n>$ with $\tau_i\in\Delta_{k_i}$. The p_j are 1 or 0 if p_j is one of the k_i or not, resp. Similarly $|\sigma_1,\ldots,\sigma_n>$ is approximately by $(\delta_1^{q_1}\ldots\delta_n^{q_n})^{-1}|q_1\cdot\Delta_1,\ldots q_n\cdot\Delta_n>$.

Finally our matrix element (4.11) is approximated by

$$(4.12) \qquad (\Pi_{k=1}^n \delta_k^{-(p_k+q_k)} V_{q_k,p_k}(\Delta_k))_{i,j}$$

using (4.10).

A knot of the graph is a point of the diagonal, where a photon line is coming in, going out or crossing. We discuss different pieces of the graph

1) Segments of the diagonal without any knots. Here the matrices of the form

$$V_{00}(\Delta) = \begin{pmatrix} \cos\sqrt{\delta} & 0 \\ 0 & 1 \end{pmatrix}$$

have to be used. This shows at first, that in such a segment no change of the atomic level may occur. So we have only two possibilities

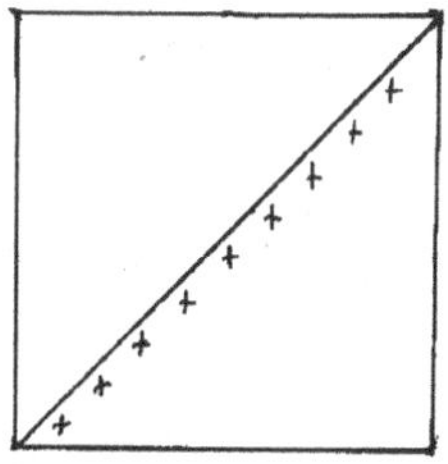

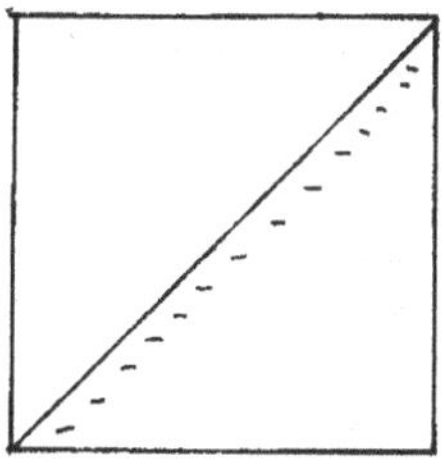

Considering that there are many intervals Δ_k contained in $[t',t'']$ the contribution of the first graph is $e^{-(t''-t')/2}$, the contribution of the second one is 1.

2) Knots.

a) One incoming photon. As

$$\frac{1}{\delta} V_{01}(\Delta) = \frac{1}{\delta} \begin{pmatrix} 0 & -i\sqrt{\delta}\,\sin\sqrt{\delta} \\ 0 & 0 \end{pmatrix} \approx \begin{pmatrix} 0 & -i \\ 0 & 0 \end{pmatrix}$$

only the following situation can occur: the photon is absorbed and the atom goes from the lower to the upper level. Contribution: factor (-i)

b) One outgoing photon. As

$$\frac{1}{\delta} V_{10}(\Delta) = \frac{1}{\delta} \begin{pmatrix} 0 & 0 \\ -i\sqrt{\delta}\,\sin\sqrt{\delta} & 0 \end{pmatrix} \approx \begin{pmatrix} 0 & 0 \\ -i & 0 \end{pmatrix}$$

only the following situation occurs: the photon is emitted and the atom jumps from the upper to the lower level. Contribution: factor (-i)

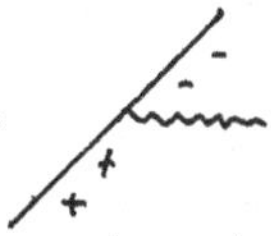

c) One crossing photon

We have to consider

$$\frac{1}{\delta^2}\, V_{11}(\Delta) = \frac{1}{\delta^2}\begin{pmatrix} \delta\cos\sqrt{2\delta} & 0 \\ 0 & \delta\cos\sqrt{\delta} \end{pmatrix} \approx \frac{1}{\delta}\begin{pmatrix} 1 & 0 \\ 0 & 1 \end{pmatrix}$$

The photon passes without changing the state of the atom. There are two possibilities

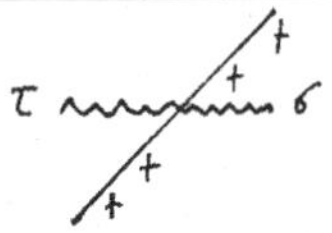

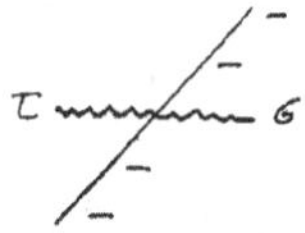

The contribution is $\delta(\tau-\sigma)$.

We are now in the position to discuss concrete examples.

(G0) No photon, atom in lower state

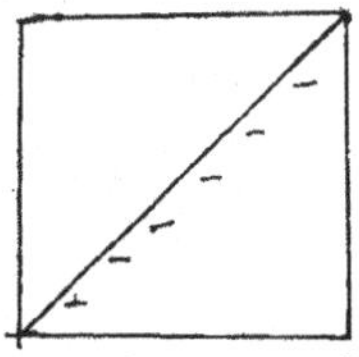

$U(t,s) = 1.$

(G1) No photon and atom in upper state or one photon and the atom in lower state. We have 5 possible graphs corresponding to the five terms of (4.9).

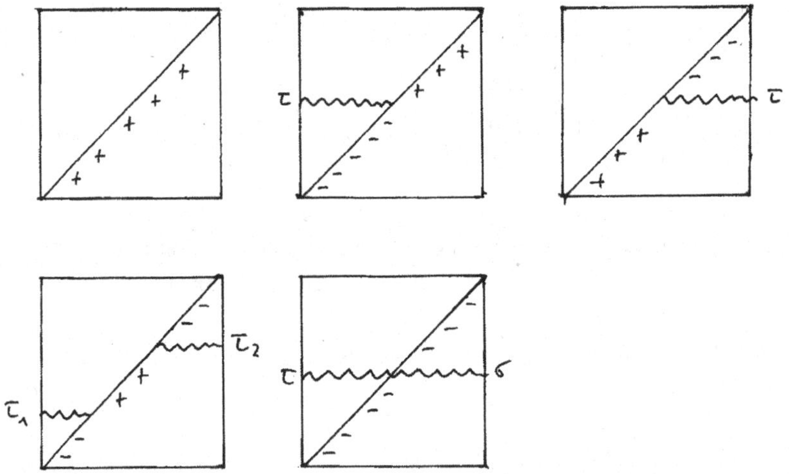

Remark that the last graph of (G1) coincides with the graph of (G0) if we omit the photon line.

(G2) One photon and the atom in the upper state or two photons and the atom in the lower state.

We get at first 5 graphs by simply adding a photon line crossing the diagonal to the graphs of (G1). We obtain furthermore the following 4 diagrams.

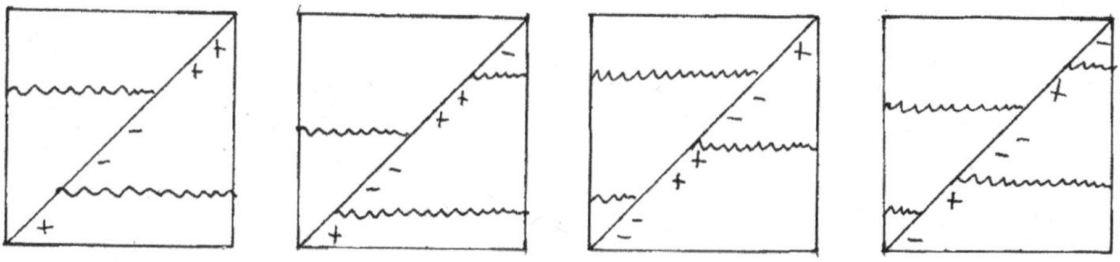

They correspond to the following four terms

$$<\sigma,+|U(t,s)|\tau,+> \;=\; (-i)^2 \exp(-1/2)(t-\tau+\sigma-s)$$

$$<\sigma_1,\sigma_2,-|U(t,s)|\tau,+> \;=\; (-i)^3 \exp(-1/2)(\sigma_2-\tau+\sigma_1-s)$$

$$<\sigma,+|U(t.s)|\tau_1,\tau_2,-> \;=\; (-i)^3 \exp(-1/2)(t-\tau_2+\sigma-\tau_1)$$

$$<\sigma_1,\sigma_2,-|U(t,s)|\tau_1,\tau_2,-> \;=\; (-i)^4 \exp(-1/2)(\sigma_2-\tau_2+\sigma_1-\tau_1)$$

(Gn) (n-1) photons and the atom in the upper state or n photons and the atom in the lower state. We consider at first the diagrams where no photon crosses the diagonal. There are again four of them corresponding to the four matrix elements

a)

$$<\sigma_1,\ldots,\sigma_{n-1},+|U(t,s)|\tau_1,\ldots,\tau_{n-1},+>$$

$$= (-i)^{2n-2} \exp(-1/2)((t-\tau_{n-1})+(\sigma_{n-1}-\tau_{n-2})+\ldots+(\sigma_2-\tau_1)+(\sigma_1-s))$$

where $s<\sigma_1<\tau_1<\ldots<\sigma_{n-1}<\tau_{n-1}<t$.

b)

$$<\sigma_1,\ldots,\sigma_n,-|U(t,s)|\tau_1,\ldots,\tau_{n-1},+>$$

$$= (-i)^{2n-1} \exp(-1/2)((\sigma_n-\tau_{n-1})+\ldots+(\sigma_2-\tau_1)+(\sigma_1-s))$$

with $s<\sigma_1<\tau_1<\sigma_2<\ldots<\tau_{n-1}<\tau_n<t$

c)

$$<\sigma_1,\ldots,\sigma_{n-1},+|U(t,s)|\tau_1,\ldots,\tau_n,->$$

$$= (-i)^{2n-1} \exp(-1/2)((t-\tau_n)+(\sigma_{n-1}-\tau_{n-1})+\ldots+(\sigma_2-\tau_1))$$

with $s<\tau_1<\sigma_1<\tau_2<\sigma_2<\ldots<\sigma_{n-1}<\tau_n<t$

d)

$$<\sigma_1,\ldots,\sigma_n,-|U(t,s)|\tau_1,\ldots,\tau_n,->$$

$$= (-i)^{2n} \exp(-1/2)((\sigma_n-\tau_n)+\ldots+(\sigma_1-\tau_1))$$

with $\tau_1<\sigma_1<\tau_2<\sigma_2<\ldots<\tau_n<\sigma_n$.

Denote by $\tilde{U}(t,s)$ the part of $U(t,s)$ described by graphs without any photon lines crossing the diagram. Let $\Omega_n(s,t)\subset\mathbb{R}^n$ be the subset of all n-tuples τ with $s<\tau_1<\ldots<\tau_n<t$. Let $\sigma,\tau\in\Omega_n(s,t)$. Then

$$\langle\sigma,-|U(t,s)|\tau,-\rangle = \sum_{k=0}^{n} \sum_{\substack{\alpha\subset\sigma \\ \beta\subset\tau \\ \#\alpha=\#\beta=k}} \langle\alpha|\beta\rangle\langle\sigma\setminus\alpha,-|\tilde{U}(t,s)|\tau\setminus\beta,-\rangle$$

where $\alpha\subset\sigma$ and $\beta\subset\tau$ correspond to the photon lines crossing the diagonal and $\langle\alpha|\beta\rangle=\delta(\alpha_1-\beta_1)\ldots\delta(\alpha_n-\beta_n)$. Similar relations hold for the other matrix elements. The last formula is equivalent to the proposition, that $\tilde{U}(t,s)$ is the kernel of $U(t,s)$ in the sense of Maassen [6] . The kernel $\tilde{U}(t,s)$ and Maassen's kernel $u(t,s)$ coincide except a trivial phase factor, which can be absorbed into the definition of $|+\rangle$.

Literature

[1] Accardi, L.: Quantum Stochastic Processes. Talk given at the Koszeg conference on "Random fields and rigorous results in statistical mechanics", August 1984. To appear.

[2] Arnold, L.: Stochastic differential equations, New York, Wiley (1974)

[3] Haken, H.: Laser Theory, Handbuch für Physik, Vol. XXV/2C. Springer-Verlag, Berlin, Heidelberg, New York, 1970.

[4] Hudson, R.L. and Parthasarathy, K.R.: Quantum Ito's Formula and Stochastic Evolutions, Commun.Math.Phys. 93, 301-323 (1984).

[5] Louisell, W.H.: Quantum statistical properties of radiation, New York, Wiley, 1973.

[6] Maassen, H.: Quantum Markov processes on Fock space described by integral kernels. This volume p.

[7] von Waldenfels, W.: Light emission and absorption as a quantum stochastic process, Preprint Nr. 176, Sonderforschungsbereich 123, Institut für Angewandte Mathematik der Universität Heidelberg, 1982.

[8] von Waldenfels, W.: Ito solution of the linear quantum stochastic differential equation describing light emission and absorption, Proceedings of "Quantum Probability and applications to the Quantum Theory of Irreversible Processes". Villa Mondragone 1982 Springer Lecture Notes in Mathematics 1055.

[9] von Waldenfels, W.: Stratonovich solution of a quantum stochastic differential equation describing light emission and absorption, "Stochastic aspects of Classical and Quantum Systems". Proceedings, Marseille 1983, p. 155. Lecture Notes in Mathematics 1109.

[10] Weißkopf, V., Wigner, E.: Berechnung der natürlichen Linienbreite aufgrund der Dirac'schen Lichttheorie. Z.Phys. 63, 54 (1930).